THE
EARTH CARE MANUAL

A Permaculture Handbook
For Britain
&
Other Temperate Climates

Patrick Whitefield

Permanent Publications

Published by:

THE QUEEN'S AWARDS
FOR ENTERPRISE:
SUSTAINABLE DEVELOPMENT
2008

Permanent Publications
Hyden House Limited
The Sustainability Centre
East Meon
Hampshire
GU32 1HR
UK
Tel: 01730 823 311 or 0845 458 4150 (local rate UK only)
Fax: 01730 823 322
Overseas: (international code + 44 - 1730)
Email: info@permaculture.co.uk
Web: www.permaculture.co.uk

Published in association with
Permaculture Association (Britain), BCM Permaculture Association, London WC1N 3XX
Tel: 0845 458 1805 Email: office@permaculture.org.uk Web: www.permaculture.org.uk

First published 2004
© 2004 Patrick Whitefield
Reprinted 2006, 2007 and 2009

The right of Patrick Whitefield to be identified as author of this work has been asserted by
him in accordance with the Copyright, Designs and Patents Act 1988

Edited by
Maddy Harland

Designed and typeset by
Tim Harland

Cover design by
John Adams

Cover image
Planetary Vision Limited

Line drawings by
Pascale Féat

Case study design drawings in Chapter 13 by
Paul Wagland

FSC
Mixed Sources
Product group from well-managed
forests, controlled sources and
recycled wood or fiber
Cert no. SGS-COC-2953
www.fsc.org
© 1996 Forest Stewardship Council

Printed in the UK by CPI Antony Rowe, Chippenham, Wiltshire

Printed on 75% recycled FSC Certified paper

The Forest Stewardship Council (FSC) is a non-profit international organisation established to
promote the responsible management of the world's forests. Products carrying the FSC label are
independently certified to assure consumers that they come from forests that are managed to
meet the social, economic and ecological needs of present and future generations.

British Library Cataloguing-in-Publication Data
A catalogue record for this book is available from the British Library

ISBN 978 1 85623 021 6

This book is much more about solutions than about problems, more about what we can do in the present situation than about how we came to be in it in the first place. Yet there's no escaping the fact that the Earth is in a dire state, and getting worse. In the twenty-three years I've been actively involved in the ecological movement almost every aspect of planetary health has got worse.

This raises the question: Is it all worth it? If we do our best to heal the Earth and make our place in her a sustainable one, is there a good chance that we will succeed? Or is it a forlorn hope? It's a big question, and one which can lead to depression if we look at the facts honestly and dispassionately. But to my mind it's the wrong question. Even if we could answer it – and we can never know anything about the future for certain – it would beg the question, How do I want to live my life?.

Here I find the teaching of Mahatma Gandhi very useful. One of his precepts was that of non-attachment to the fruits of our labour. All we can do in life is to make sure that we play our own part in it the best way we can. Much as we would like to, we can never do more than that. Everything we do is so complex, and relies for its ultimate completion on so many different people and natural forces, that we can never take responsibility for the final outcome of our actions. We can only take responsibility for our actions themselves.

So my answer to the question, How do I want to live my life? is that I want to be a part of the solution rather than a part of the problem.

Contents

Acknowledgements   viii
About This Book   ix
How To Use This Book   x

PART ONE:
BASICS

Chapter 1
WHAT IS PERMACULTURE?   3
    A brief history   4
    A sense of ethics   5
    Quit while you're ahead   9
    Permaculture and the green movement 10

Chapter 2
THE PRINCIPLES OF PERMACULTURE   13
Wild Soil   14
    No-till   14
    Covered soil   14
    Perennials   14
Diversity   16
    Species diversity   18
    Genetic diversity   20
    Ecological diversity   21
    Cultural diversity   22
Multi-Dimensional Design   22
    Stacking   22
    Succession   24
    Edge   24
Relative Location   26
    The Wider Picture   26
Key Planning Tools   27
    Zoning   27
    Networking   28
    Sectoring   28
    Elevation planning   28
    Integrating the four   30
Small-Scale   30
    ...and diversity   30
    ...and yield   31
Input-Output   32
    Linking   32
    Multiple outputs   32
Energy Flows   33
    Energy in use and embodied energy 33
    Biological resources   34
Wholes   35
    Designing in wholes   36
Can Permaculture Feed the World?   37

Chapter 3
SOIL   39
Soil Fertility   39
    Soil texture   41
    Organic matter   42
    Soil life   46
    Plant nutrients   48
To Till or Not to Till?   50
    Reasons for tilling   50
    Reasons for not tilling   52
Assessing Soil   58
    Indicator plants   58
    The soil profile   60
    Soil types   61
    Texture test   64
    Chemical analysis   64

Chapter 4
CLIMATE & MICROCLIMATE   67
Climate   67
    Continental and maritime climates 67
    Regional climates   68
Global Climate Change   71
    Causes   71
    Consequences   71
    Action   73
Microclimate   74
    Wind   75
    Light   76
    Temperature   76
    Frost   78
    Moisture   79
    Combined effects   81
    Observing microclimates   81
Windbreaks   83
    Principles of windbreak design   84
    Coastal windbreaks   88

Chapter 5
WATER   91
The Water Cycle   91
    Water in the landscape   92
Water in the House & Garden   93
    Using less   94
    Re-using   96
    Collecting rainwater   97
Water on Farms   99
Sewage Treatment   101
Aquaculture   106
    Fish   107

Plants, broadscale 108
Domestic scale 110
Pond siting & construction 111
Broadscale 112
Domestic scale 116

Chapter 6
ENERGY & MATERIALS 119
Energy Fundamentals 119
Entropy 119
A new energy environment 121
Conservation 121
Asking the Right Questions 122
Energy Use 126
In transport 126
In the home 130
Use of Materials 132
Industrial ecology 133
Timber 133
Guest Essay:
Ecodesign: there's more to it than it seems 135
by Peter Harper

PART TWO:
APPLICATIONS

Chapter 7
BUILDINGS 141
Settlement Design 141
Density 142
Layout 145
Low Energy Design 147
Passive solar design 148
Heating with wood 152
Biotecture 153
Windbreaks 153
Green buildings 154
Some Ecological Styles 157
Choosing Materials 159
Renovation 162
Case Study:
An Eco-renovation Project 166
by Tim and Maddy Harland

Chapter 8
GARDENS 169
Why Grow Food? 169
Zoning 170
Indoors 172
Window boxes and balconies 173
Conservatories 174
Patios and courtyards 175
Gardens and allotments 176
Larger gardens 181
Community gardens 183
Sector & Elevation 184
Sector 184
Elevation 190

Original Permaculture 191
No-dig 191
Mulching 194
Perennial Vegetables 197
Polyculture 201
Stacking 203
Forest garden 204
Golden Rules 206
Case Studies:
A Small Productive Garden 209
Community Gardening in the City 212

Chapter 9
FRUIT, NUTS & POULTRY 215
Nuts 216
Designing a nuttery 217
Nuts one by one 222
Fruit 225
Choosing what to grow 226
Table of fruit trees and shrubs 228
Where and how? 231
Microclimate 237
Plant diversity 240
Animals in the orchard 244
Poultry 248
Chicken forage systems 251
Guest Essay:
Own-Root Fruit Trees 254
by Phil Corbett

Chapter 10
FARMS & FOOD LINKS 257
Organic farming 257
Animals 258
Omnivorous or vegan? 258
Choosing animals 261
Grazing animals 263
Arable & Vegetable Crops 264
Diversity in the field 264
Grain 267
Bicropping vegetables 273
Agroforestry 275
Principles 275
Grassland with trees 278
Alley cropping 280
Hedgerows 281
Shelterbelts 282
Food Links 285
Food co-op 286
Farm shop 287
Local directory 288
Markets 288
Box scheme 289
Subscription farming 291
Community supported agriculture 292
Case Study:
A Permaculture Farm 294

Chapter 11
WOODLAND   299
Outputs of Trees & Woodland   299
    Ecological functions   299
    Multiple outputs   300
Existing Woodland   309
    Assessment of woods   309
    Working with existing woods:
        A. Coppice   313
        B. High forest   317
New Woodlands   321
    Site   321
    Species   323
    Establishment   327
    Grants   328
Urban & Community Woodland   328
    Urban trees and woods   328
    Community action   334
Case Study:
    Living in the Woods   336
Table of Trees for Timber & Wood   338

Chapter 12
BIODIVERSITY   341
Basics   341
    Why?   341
    How?   343
Ecosystems in Britain   345
    Rural ecosystems   345
    Urban ecosystems   348
    Assessment   352
Working with Rural Ecosystems   356
Working with Urban Ecosystems   362
    Community action   362
    Wildlife gardening   364
Native or exotic?   368

PART THREE:
DESIGNING

Chapter 13
THE DESIGN PROCESS   375
    Before we start   376
Base Map   377
Site Survey   378
Design Questionnaire   382
Evaluation   385
Design Proposals   386
    Aims   387
    Concepts   387
    Details   387
Re-evaluation   388
Case Study:
    Design Process, an example   389

Chapter 14
DESIGN SKILLS & METHODS   401
Surveying Skills   401
    Distance   401
    Angles   402
    Elevation   404
Mapping   408
    Basics   408
    Ordnance Survey maps   410
    Drawing skills   411
Listening Skills   414
    The land   414
    The people   418
Design Methods   419
    Options and decisions   419
    Design tools   420
    Planning for Real®   422

NOTES   425

APPENDICES

    A. Further Reading   439
    B. Glossary   445
    C. Scientific Names   451
    D. Organisations   455
    E. Suppliers of Permaculture Requisites   461
    F. Metric–Imperial Conversions   463

    INDEX   465

The Author

Patrick Whitefield is a permaculture teacher, writer, designer and consulting editor for *Permaculture Magazine*. He is the author of the mini-classic, *Permaculture in a Nutshell*, which has been translated into four languages, and *How to Make a Forest Garden*.

He grew up on a smallholding in rural Somerset and qualified in agriculture at Shuttleworth College, Bedfordshire. He has wide experience of farming in Britain, the Middle East and Africa.

He has experience in many diverse areas. These include organic gardening, nature conservation, country crafts and green politics. Patrick has found that his mixed experiences have led him to the logical conclusion of permaculture and are directly relevant to his present work. He inspires respect, affection and a good measure of action wherever he imparts his considerable knowledge.

Acknowledgements

First of all I would like to acknowledge my debt to Bill Mollison and David Holmgren, who first devised the concept of permaculture, and to the many others who have originated the material which I have collected together in this book. If it was based only on my own personal experience it would be a much shorter book!

The following people have read and commented on various parts of the book, pointed out mistakes and made suggestions for improvements: Alison Malcom, Ben Law, Caroline Haynes, Cristina Crossingham, Claire Roper, Jimmie Hepburn, Mark Moodie, Martin Wolfe, Matt Dunwell, Michael Littlewood, Pauline Pears, Peter Harper, Phil Corbett, Phil Hinton, Rebecca Laughton, Rob Hopkins, Robin Roper and Tony Currivan. Sarah Lagden has read the entire text and made many useful suggestions for changes. I'm grateful to them all for their time and expertise, but I take full responsibility for any errors which may remain.

My special thanks go to my publishers, Maddy and Tim Harland and their team. Working with them is a truly co-operative venture, full of mutual support and genuine friendship. I hope and trust that the quality of our relationship is reflected in the all-round quality of this book. Equally special thanks to Cathy, my wife, who has put up with a lot while I've been writing it.

About This Book

The aim of this book is to provide practical information on permaculture for people living in Britain and other temperate countries. It came to be written for two reasons.

Firstly, I wrote it to support my main work, which is teaching permaculture to adults. The standard course, the Permaculture Design Course, is only 72 hours long, and you can't pack a great deal of information into that amount of time. The course gives people the experience of permaculture design and, I hope, inspires them, but it has always left me with the feeling that the students could use a lot more information. This book contains that information.

Secondly, it exists to supplement the existing books on permaculture, almost all of which are by Australian writers. Australia is mainly a tropical country – its latitude makes it equivalent to a stretch of land from Spain to Nigeria – and there has long been a need for a full-length permaculture book for the temperate world. The principles travel well enough but the details of what you can do in a tropical climate are not relevant here.

I trust this book will be of use to people in all temperate countries but I have written it primarily from a British point of view. One of the chief characteristics of sustainable solutions to world problems is that they are local, tailored to the unique conditions of each place. I, immersed as I am in British ecology and culture, can only write with confidence about Britain – and even here I must allow for wide differences between different parts of the country and even between neighbouring localities. You, whether you live in Britain or another country, are best able to judge what is applicable to your home and what is not.

Even if you live just round the corner from me, I'm not here to tell you what to do. My aim is to give you the wherewithal to make your own decisions. This includes information, but perhaps more importantly it includes concepts and perspectives. The most important thing to know in any situation is which questions to ask. Then at least we have a chance of coming up with the right answer. If we ask the wrong questions we have no chance.

In fact permaculture covers such a wide field of human activity that it would be impossible to give all the detailed information you may need in one book. But most of that information is readily available elsewhere, and I have concentrated here on the particular contributions which permaculture can make. For example, in the chapter on gardening you will not find detailed information on how to grow carrots and peas, but you will find details on vegetable polycultures, perennial vegetables, no-dig gardening, the application of permaculture design principles to the garden and so on.

You may already have the relevant skills, as a gardener, farmer, forester, architect or whatever, or you may be a complete beginner. This book should work for you either way. Adding the permaculture perspective to existing skills gives a powerful combination. Equally it's a good perspective to start from, and an understanding of permaculture can help you recognise the wheat and the chaff in the sometimes bewildering mass of knowledge on offer these days.

One area of activity which is usually included in permaculture but which I have not covered in this book is economics. This is partly because I have no personal knowledge or skill in this area and partly because an excellent book already exists on the subject of green economics, Richard Douthwaite's *Short Circuit*. Nothing I could write on the subject could match it, and if I tried to I would only end up plagiarising it. I warmly recommend it to anyone who's interested in sustainability and community development. It's a vibrant, readable book, and requires no previous knowledge of the subject. (See Appendix A, Further Reading.)

My aim throughout this book is to give information which is practical and reliable. With an innovative subject like permaculture this can present a problem. So many of the ideas in permaculture are new, or depart so radically from the mainstream of our culture, that they have not been tried and tested as thoroughly as the conventional ideas which they seek to replace. This by no means applies to everything in the book. For example, raised bed gardening has been practised for hundreds if not thousands of years and its advantages are well known, whereas growing crops through a permanent understorey of clover is new and experimental. I have tried to make plain in the text whether I'm talking about something which is well-established and known or something which is on the cutting edge of our knowledge.

Nevertheless there has to be an element of experiment in something as radical as permaculture. There is absolutely no doubt that it represents the direction we must go in if we're going to survive on this planet. But, when virtually all scientific and commercial development has been in the opposite direction, it's inevitable that there will be doubts about some of the details. If you want to play safe you can go a fair way along the path to permaculture, but if you want to go the whole way I have to say, "Welcome to the experiment."

How To Use This Book

If you choose to start at page one and read this book right through you'll find that it has a logical sequence and each chapter builds on what has gone before. But it can also be dipped into.

The breadth of subjects covered by permaculture, and the fact that all these subjects are interconnected, means that there are many topics which could easily fit into a number of different chapters. Rather than repeat them unnecessarily I have had to choose where to place them. This means there may be something to interest you in chapters which you would not at first think were relevant to you.

For example, if your main interest is in gardening and you never envisage growing an orchard or farming, you may still find ideas and information in the chapters on orchards and farms which are relevant to you. One such is the technique of growing crops, including vegetables, through a permanent sward of clover, which is described in Chapter 10, Farms and Food Links. Similarly, if your interest is in smallholding you may find much of interest in Chapter 8, though its main focus is on domestic gardening. In either case you will find relevant information in Chapters 3 and 4, which deal with soil and microclimate.

So I invite you to dip around: you never know what you might find. However I do suggest you read the first two chapters, What is Permaculture? and The Principles of Permaculture, first. The rest of the book is based on the ideas presented in these two chapters, and can be understood more fully in the light of them.

The Glossary, Appendix B, is designed to help readers who want to dip into the book. It gives definitions of words and phrases which may be unfamiliar to the general reader. Many of them will have already been defined in the text, but looking them up in the glossary is quicker and easier than using the index to find the original entry.

The fact that so many topics could have fitted into a number of different places has led to a fair number of cross-references in the text. You should not need to use these in order to understand any passage you're reading, but you might like to follow them up if you want to get a fuller picture of the subject in hand.

Appendix A gives suggestions for Further Reading, with a general section and a section for each chapter. All the books listed are practical how-to books which give the detailed information you need to put permaculture into action on the ground. (The one exception to this is the book *Permaculture One*, which is included more for historical interest.)

Sources of information on more specialised subjects are to be found in the Notes on pp425-437, referred to by superscript numbers in the text. To some extent the notes refer to supporting evidence for statements made in the text. But their main purpose is to point the way to information which may only be of interest to a minority of readers. There are also a few notes which are comments of my own which I have left out of the main text so as not to break the flow.

The list of Scientific Names, Appendix C, gives the botanical or zoological name of every plant and invertebrate animal mentioned in the text. This may be particularly useful to readers who don't have English as their first language, and to American readers who may know some plants by different names.

I have used metric measurements throughout. If you're more familiar with the old imperial measurements don't feel intimidated. It's really quite easy once you get acquainted with a few basic equivalents, and I've given these in Appendix F.

FEEDBACK

If you, the reader, should find any mistakes or misinformation in this book, or have any suggestions for improvements, please contact me, by letter or email, via the publishers. This will help to improve the quality of any future editions.

Patrick Whitefield
Glastonbury, Somerset
May 2002

PART ONE

BASICS

Chapter 1

WHAT IS PERMACULTURE?

Permaculture means different things to different people, but at root it means taking natural ecosystems as the model for our own human habitats. Natural ecosystems are, almost by definition, sustainable, and if we can understand the way they work we can use that understanding to make our own lives more sustainable. An insight into how natural ecosystems work can be had by comparing a wild woodland, which is the natural vegetation of this country, with a wheat field.

The wildwood, containing trees, shrubs, herbaceous plants and climbers, has a much greater biomass than the wheat field, which is only about half a metre high and contains just one type of plant. Not only that but the amount of new biomass produced each year is much greater.[1]

The wildwood needs no external resources other than sunshine, rain and the rock from which it makes its own soil. By contrast the wheat field needs an annual round of ploughing, harrowing, sowing, manuring, weeding, pest control and disease control, and far from accumulating soil it suffers a net loss through soil erosion.

There's a catch, of course. About half of the biomass in the wheat field is edible grain, and even the straw is useful to us, whereas the wildwood contains a very low proportion of human food, and we can only make use of a certain amount of timber. How happy we would be if we could create systems which combine the yield and self-reliance of the wildwood with the highly edible nature of the wheat field!

This is the inspiration for permaculture, but the practice of it is more than simply making direct imitations of natural ecosystems. We can do that: a forest garden is a direct imitation of a natural woodland with the wild plants replaced by edible fruits and vegetables, and there is a place for forest gardens in permaculture. But there is a great deal more to permaculture than that. We go beyond the surface appearance of ecosystems and look at the principles on which they operate. If we can understand these principles and apply them more widely, we can get the benefits of natural ecosystems in buildings, gardens, farms and whole settlements which don't outwardly resemble natural ecosystems at all.

So what makes natural ecosystems work? One of their best-known attributes is diversity. This does not simply mean a high number of species. You could collect together a thousand randomly assorted species of plants and animals and put them down in one place but you wouldn't get an ecosystem. You'd get a mess. What makes an ecosystem work is the diversity of beneficial relationships between its components, the plants, animals, microbes and non-living components which make it up.

For example, one species of plant provides shelter for another, protecting it from wind and maintaining the humidity it requires; another plant, by co-operating with certain bacteria, provides nitrogen for itself and its neighbours; fungi extract minerals from the soil and exchange them with green plants for sugars; insects pollinate plants and receive nectar in exchange; closely related species of animals consume slightly different diets and thus avoid competition with each other; water-loving plants and animals colonise marshy ground, making it productive without any need for drainage.

Every ecosystem is composed of a vast web of beneficial relationships of this kind. This web of relationships is the fundamental principle which enables natural ecosystems to be highly productive without the level of inputs needed by our wheat field. It's also the central idea of permaculture, and it can be applied both to food production and to a wide range of other human activities.

To illustrate this central idea, let's look at an example: the choice between having a free-standing greenhouse in a domestic garden or placing a greenhouse against the south-facing wall of the house, as a conservatory.

If the greenhouse is free-standing, at night it will radiate heat out in all directions and cool down rapidly, but if it's alongside the wall of the house much of the day's heat will be absorbed by the wall and re-radiated at night to keep the greenhouse warm. This can keep it frost-free all winter without the need to burn fuel and will keep plants growing during the cooler seasons of the year.

The house itself can also gain some of the heat collected by the greenhouse during the day, again reducing the need to burn fuel.

One of the greatest benefits of the house-conservatory combination is in the relationship between the people in the house and the plants in the greenhouse. This is especially true when the conservatory is used to bring on young plants in the spring. It's much easier to give them the care and attention they need when all you have to do is to pass from one room in your house to another rather than make a trip outside, especially in inclement weather. The plants receive all the attention they need, and getting them off to a good start when they're young is essential to successful crops.

The people also benefit directly from having a pleasant, semi-outdoor place to be in when the weather is too cold or wet to enjoy being outside.

John and Donna Butterworth produce organic fruit trees in their nursery in Ayrshire, specialising in old Scottish varieties. As they are in a rather frost-prone area the conservatory which runs along the front of their house is important for their business. It also reduces the heating season in their house by a full two months.

This example shows how a network of beneficial relationships can go to make up a system which minimises the inputs, both of fossil fuel and of human effort, and maximises the outputs. In practice careful design will be needed to make sure the system works as intended, and of course a conservatory will not be appropriate for every household. But it illustrates the central idea of permaculture.

Since our aim is to set up a network of beneficial relationships, where we place things in relation to each other is very much of the essence. In the greenhouse example the benefits depend on the two structures actually being joined together. This means that permaculture is basically a matter of design. The aim is to put the maximum effort into the initial design of a system so as to save unnecessary effort once the system is in operation. Permaculture can be described as: careful thought followed by minimum action, rather than hasty action followed by long-term regrets.

Beneficial Relationships

↓

Relative Location

↓

Design

A Brief History

Permaculture, or something very close to it, has been practised for thousands of years in various parts of the world, and still is here and there, by people who have never heard the word permaculture. For example, the Chagga people of northern Tanzania and the inhabitants of the Kandy area of Sri Lanka both cultivate gardens which are modified versions of the natural forest vegetation. Productive trees, vines, shrubs, herbs and vegetables grow together, each providing the conditions the other needs to grow in. These gardens suffer no soil erosion, require no heavy energy input, and production is very high. They provide the people with all or much of their food, most of their medicines and fibres, and some cash crop, all on a very small area of land.

But the conscious idea of permaculture and the word itself were formulated in the 1970s by two Australians, Bill Mollison and David Holmgren. They set out their ideas in a book, *Permaculture One, a perennial agriculture for human settlements*. The opening paragraphs of the book state that:

> Permaculture is a word we have coined for an integrated, evolving system of perennial or self-perpetuating plant and animal species useful to man [sic]. It is, in essence, a complete agricultural ecosystem. . . It would suit any climatic region, and is designed to fit in urban situations.
>
> We jointly evolved the system in the first place as an attempt to improve extant agricultural practices, both those of Western agribusiness, and the peasant grain culture of the third world. The former system is energy-expensive, mechanistic and destructive of soil structure and quality. The latter makes drudges of men [sic], and combined with itinerant herding, deserts of what once were forests. Perhaps we seek the Garden of Eden, and why not? We believe that a low-energy, high-yielding agriculture is a possible aim for the whole world, and that it needs only human energy and intellect to achieve this.[2]

Permaculture has developed over time, and continues to change and develop. In its original form, as laid out in *Permaculture One*, the emphasis was on a fairly direct imitation of natural ecosystems. I call this 'original permaculture'. A garden or farm designed along the lines of original permaculture actually looks like a natural ecosystem. The soil is not ploughed or dug and is always kept covered, while the majority of crop plants are perennials, especially trees. The classic example is a forest garden.

But, as we have seen, what makes a natural ecosystem work is not its components but the network of beneficial relationships which knits them together, and this principle can be used to make any human-made systems more efficient and sustainable whether they actually look like a natural ecosystem or not. The originators of permaculture discovered that they had invented a design system which could be applied much more widely, and permaculture moved on from its original conception.

In *Permaculture, A Designer's Manual*, Bill Mollison wrote that a typical small farm or village may already contain all the components necessary for a sustainable, self-reliant system. They only need to be re-arranged so as to create the harmonious relationships between them in order to make this potential a reality.[3] I call this approach 'design permaculture'.

Permaculture can include plantings which resemble natural ecosystems, like this forest garden.

But a traditional organic vegetable garden like this can equally be part of a permaculture design if it meets the needs of both land and people.

By the early 1990s, when Bill Mollison and Reny Slay wrote the *Introduction to Permaculture*, permaculture had been redefined as:

> ...a design system for creating sustainable human environments. The word itself is a contraction not only of **perma**nent agri**culture** but also of permanent culture, as cultures cannot survive for long without a sustainable agricultural base and landuse ethic. On one level, permaculture deals with plants, animals, buildings, and infrastructures (water, energy, communications).

However, permaculture is not about these elements themselves, but rather about the relationships we can create between them by the way we place them in the landscape.[4]

Permaculture is still evolving, and evolving in different ways in different parts of the world. Here in Britain some people see it as a mainly urban activity, reflecting the reality that most of us live in cities.

There is just as much scope for applying permaculture design in cities as there is in the country. The quantity of food which can be grown in cities is clearly less than in the country, but the value of that food is much higher because it's grown where the people who eat it live. This is a central theme of permaculture. But permaculture as a design system can be applied to much else besides food growing, including buildings, settlement design and community development. Many British permaculturists have made a deliberate decision to remain in the city and to take part in the greening of it. (See for example the box, Living In The City, p143.)

On the other hand a large proportion of the keen permaculturists I meet are aiming to get hold of a piece of land in the country or to join a land-based community. Many are already there. Permaculture offers a way to make this a reality for more and more people. (See for example the case study, A Permaculture Farm, p294.) The present methods of mechanised and chemicalised farming and forestry are leading to a countryside which is either depopulated or inhabited mainly by commuters. Permaculture offers a way of repopulating the countryside with people making their living in a way that is both productive and in harmony with nature.

Whether it's practised in the city or in the country, permaculture is at heart a matter of living more sustainably. It's not just about the way we feed and house ourselves and make our living but about whole lifestyles. It addresses transport, energy use, water resources, waste reduction and aesthetics. The central aim of permaculture is to reduce our ecological impact. Or more precisely it's to turn our negative ecological impact into a positive one.

Everyone has their own definition of permaculture. I've even heard the word used to characterise the purist end of the 'green' spectrum, which makes all others look pale by comparison. A few years ago you could certainly say that permaculture was the whacky end of the movement, which made organics look sane by comparison. These days, when so many of the ideas which we were talking about ten years ago have become mainstream, this description hardly fits any more.

A Sense of Ethics

At the heart of permaculture is a fundamental desire to do what we believe to be right, to be part of the solution rather than part of the problem, in other words a sense of ethics. The ethics of permaculture can be summed up as:

Earthcare, Peoplecare and Fair Shares

Earthcare

Care of the Earth is the fundamental motive for permaculture, and there are two ways of looking at it. One could be called environmentalism, the other ecology – or deep ecology, to distinguish it from the academic science of ecology. These two do not represent different groups of people but different ways of thinking and feeling. Both are valid, and one person may think or feel either way from time to time.

Environmentalism sees humans as essentially separate from the rest of nature. The very word environment means something surrounding us, not ourselves. Environmentalism speaks of plants, animals, water bodies and soil as resources. In other words it sees them primarily in terms of their usefulness to humans. Wise stewardship is the theme, with humans very much in control and at the centre of things.

Ecology sees humans as part of nature, one species among many. It holds that all beings have rights, not just humans. Although we are very powerful and have a great capacity to change and destroy, ultimately we are not in control.

It's possible to trace a progression through human ethics from the most basic to the most advanced (so far):

- The egotistical is the least developed: good is what benefits me.

- The tribal divides humanity into an in-group and an out-group: good is what benefits my in-group, my family or tribe. Racism, slavery, war, adversarial politics and competitive sports are based to varying degrees on this ethic.

- The humanitarian gives equal worth to all humans: good is what benefits the most people most of the time. This is probably the dominant ethic in the world at the moment, at least in theory.

- The ecological embraces humans, other sentient beings and the health of the planet as a whole: good is what benefits the whole Earth. To embrace this ethic is the next necessary step in the evolution of human culture. Without this ethic we will surely perish.

Each of these ethical statements includes the one before it. All of us probably act according to all four of them at different times. Environmentalism is essentially part of the humanitarian ethic. It sees caring for the Earth as a matter of human self-interest. Indeed caring for the Earth is in our interest, not only for our ultimate survival but for our quality of life in the short term, and as a human I find it only natural to consider my own kind before all others. Yet at a deeper level I know that it is fundamentally right to care for the Earth, whether it benefits us or not.

REDUCTIO AD ABSURDUM

Stating the ecological ethic can lead to rather silly debates around subjects like 'Does the smallpox virus have a right to exist?'

Certainly there is a conflict between our natural desire to defend our species against others which can harm us and a belief in the right of all species to exist. It's much the same as the conflict between my desire to do the best I can for myself and my family and my belief in equal rights for all human beings. Sometimes these two will be in conflict and I will have to try to make a wise decision.

The need to make wise decisions in complex situations is part of what it is to be human. The idea of a simple set of rules which we can obey without having to think is attractive but unrealistic.

Humility is an essential part of the ecological ethic. We are just one species among the web of species which makes up the Earth, albeit an astonishingly powerful one. When we alter natural systems to meet our own ends the change all too often means a loss of diversity, stability and health in the system.

The implications of this are:

- We need to leave as much of the Earth's surface as possible undisturbed by human activity so as to preserve the health of the planet as a whole. Since permaculture can enable us to meet our needs on a small area of land it can help us towards this aim.

- The systems we construct to meet our needs should depart as little as possible from natural ecosystems – which takes us right back to the original inspiration of permaculture.

We also need to acknowledge that we can never fully understand natural ecosystems, at least not by the scientific method of examining the parts individually and thereby hoping to understand the whole. The parts are too many, their interactions too complex, and the whole too vast for this. Reductionist methods are useful, and should not be totally discarded on the grounds that they are incomplete. But sometimes we must respect and rely on our intuition, and at other times simply acknowledge that we do not know.

Peoplecare

It is possible to care for the Earth without caring for the people. In China, for example, the land has grown cereals for four thousand years without depleting its fertility. This is certainly a sustainable agriculture, but it has been achieved at the cost of unremitting drudgery for the vast majority of the people.

There is a common assumption that there are only two ways to produce food, or any other material need: hard human toil, or the level of fossil fuel use that is characteristic of our present economy. Permaculture design offers another way, a way which is kind to both planet and people, a way based on intelligent design.

In any permaculture design the needs of the land and the needs of the people are given equal weight. A design which is perfect ecologically but does not meet the real needs of the people involved will not succeed. At least it won't in a free society, where people have the choice whether to implement it or not.

Permaculture can be fun. Local schoolchildren put down mulch at the Drift Permaculture Project, Newcastle-upon-Tyne.

In fact it's becoming increasingly clear that the technical solutions to problems are very much easier to come by than the interpersonal ones. On the whole we know what we need to do on a physical level to care for the Earth. We may not have the ultimate answers but we know of enough positive changes we can make to keep us busy for a generation or two. The things that get in the way of implementing them are human emotions, such as fear, greed and rivalry.

At heart the ecological problem is not a technical one at all but an emotional and spiritual one. This book contains a great deal of technical information which can help us live more sustainable lives, but I know it won't help us if we do not address our emotional and spiritual lives too.

Many people consider that the problem lies with political and economic institutions and their policies, rather than the actions and feelings of individuals. But ultimately the two are the same, as society, including the big power bases within it, grows from the billions of day-to-day actions of ordinary people. We get the institutions we deserve.

Be that as it may, permaculture is much more about taking matters into our own hands than about persuading the powers that be to change things. It is about making changes in our own lifestyles rather than demanding that others do it for us. This doesn't mean that political action is a waste of time. There are many things which can only be done at the political level, and this will probably be so for the foreseeable future. It does mean that our first reaction to any problem or challenge is not 'Something must be done!' but 'What can we do about it?'

Fair Shares
This third permaculture ethic is the result of combining the first two. It asks: just how much of the Earth's bounty can each of us use in order to both maintain the planet in a sustainable state and allow all her people the opportunity for a reasonable material standard of living?

Our ecological impact is a product of three factors: the size of our population, our level of consumption, and the efficiency of the technology we use to achieve that level of consumption.

population x consumption x technology = ecological impact

POPULATION, CULTURE AND THE ECONOMY

What causes population to grow is a very complex and controversial subject. But some light is thrown on the subject by the changes which have taken place over the past two decades in Ladakh, in northern India.[5] For many centuries the people of this isolated Himalayan region depended for their livelihood on the local ecosystem, and their population remained steady, in balance with the ecosystem. Now that the Indian government has introduced a cash economy, many Ladakhis depend on resources brought from far off, and their ability to obtain them depends only on their access to money. Thus there is no longer a direct relationship between the local ecosystem and local population, and thus no immediate need to keep population within bounds. So it is rising.

All conventional 'development', in both North and South, is aimed at increasing people's involvement in the global cash economy, replacing local production for local needs with long-distance trade. Here in the north it's called economic growth, and it increasingly separates us from the resources on which we depend for survival. Only by reconnecting ourselves with local resources can we move towards a sustainable society.

When people in industrialised countries discuss sustainability the population problem is usually central to the discussion. This is quite an easy subject for us in the industrialised North to discuss because most population growth is taking place in the South. We feel it's their problem rather than ours and, though it is a very serious one, it's not something which requires us to take any action or to question our lifestyles.

Yet on average each one of us in the North consumes many times as much as one person in the South. Per capita energy consumption gives a rough guide to overall consumption, and the average northerner uses between 10 and 35 times as much as the average southerner,[6] while one US citizen uses 100 times more fuel per year than a Bangladeshi.[7] To put it another way, the industrialised countries, with 20% of the Earth's population, already have an 'ecological footprint' greater than the capacity of the entire planet.[8] (See box, Our Ecological Footprint.)

Leaner, more efficient technology is often held to be the answer. The technology already exists to reduce the resources used for each unit of material consumption to a quarter of what we now use, and some scientists believe that a reduction to one tenth is possible.[9]

Yet increased efficiency only saves resources if we don't use the savings to enable ourselves to consume still more. For example, a commuter who switches from driving to work to cycling will save both energy and money. If they invest the money saved in something which will be of positive gain to the planet, like buying organic vegetables instead of conventional ones, then that person's total ecological impact will be reduced.

But if they spend it on consumer goods the likelihood is that their net ecological impact will be unchanged or even increased. Studies have shown that improved energy efficiency in industry has actually generated more economic growth, and thus led to more use of energy and materials rather than less.[12]

On its own, efficiency appears to be counter-productive. It's an essential ingredient of sustainability, but there's no substitute for grasping the nettle of our profligate level of consumption. All the permaculture technologies described in this book are valueless if each of us does not take responsibility for reducing our personal consumption of material goods.

In addition to the direct impact of our own consumption there is an indirect effect which is perhaps even more important. Although the North presently consumes something like twice as much overall as the South, that is set to change. Given the population growth which is inevitable in southern countries in the medium term, even at a moderate level of *per capita* consumption, the ecological impact of the South will be vastly greater than that of the North before long. By one calculation their impact could be ten times ours towards the end of this century. This assumes a level of consumption in the South that only gives the minimum of a good diet, clean water, decent housing and education for all, not a car and air-travel culture like ours.[13] We can't possibly expect the South to accept such modest material aims while we in the North continue as we are now.

What all this means in terms of actual levels of consumption here in Britain has been calculated by Friends of the Earth.[14] In order to consume at a level which is both within the Earth's sustainable limits and allows all the people to have their fair share, we need to cut our consumption of all major resources. By the year 2050 we need to have cut the 1990 level by the following percentages:

	%
Energy	-88
Land (including overseas land from which we import food)	-27
Timber and timber products (including paper)	-73
Water	-15
Steel	-83
Cement	-72
Construction aggregates	-50
PEOPLE	?

Given the timescale and the available technology these are not impossible aims. Shifting taxes from labour to resource use has an important part to play in achieving them, but so do our personal lifestyles.

OUR ECOLOGICAL FOOTPRINT[10]

Ecological footprint analysis is a simple tool which enables us to put ball-park figures on our level of sustainability. It can be calculated for a nation, region, city or even a single household. It does this by measuring all ecological impacts in terms of land area.

First the built-up area is measured. To this is added the area of land needed to grow the food, timber and other fibres consumed. Energy use can be converted to land by calculating the area of growing vegetation needed to absorb the CO_2 produced. Land degraded by mining, quarrying and waste disposal is included, as is a share of the world's stock of undisturbed wilderness.

London, for example, has been calculated to have a footprint of nearly 20 million hectares, some 125 times its surface area, and equivalent to almost 95% of Britain's productive land.[11] Of course much of the land used by London is outside Britain. Over half of it is made up by the area needed to absorb the CO_2, and this is left for future generations to find.

I feel it's important not to feel guilty about our lifestyles. The time to make a specific change is when the positive desire to do so grows to the point where it's greater than the discomfort of giving up an old habit. If we take the trouble to learn about the ecological impact of our daily lives the process of lifestyle change happens naturally. Then it's not a self-imposed penance but a process of liberation.

Nevertheless the concept of fair shares is a challenging one to us who are so used to the assumptions of the affluent society. It's all about limits, about acknowledging the fact that we live on a finite planet and accepting that that imposes limits on our physical appetites. To our culture the concept of placing a limit on something, at least for those who can pay, is anathema. But an acceptance of limits can also be a liberating concept. Once we have accepted our physical limits we can relax and concentrate on the things which really make life worth living: its spiritual, emotional, artistic and social aspects.

Quit While You're Ahead[15]

There's a whole range of things we can do to make our lifestyles greener, but if we try to achieve a hundred percent ecological purity our lives could become extremely tedious. Recycling everything down to the last little yoghurt pot would hardly leave time for anything else in life.

Peter Harper of the Centre for Alternative Technology has come up with the 'quit while you're ahead' principle to deal with this situation. What it means is: concentrate on lifestyle changes which have a big ecological impact or which are easy to make, and don't bother too much about the ones which give a poor return on the effort invested.

So which lifestyle changes do give the best return on effort? First we need to look at which areas of our lives – food, energy and so on – have the most impact, and then at the specific things we can do to reduce our impact in each of these areas.

Several attempts have been made to quantify our ecological impact and assign percentages to the different areas of life. Those I have seen fall into one of two camps. The book *Our Ecological Footprint*[16] gives the

breakdown shown in the first pie chart. This general pattern is supported by other studies, including the authoritative Danish study *Environmental Impacts of Household Activities*, which actually gives the food system a higher share at 36%.

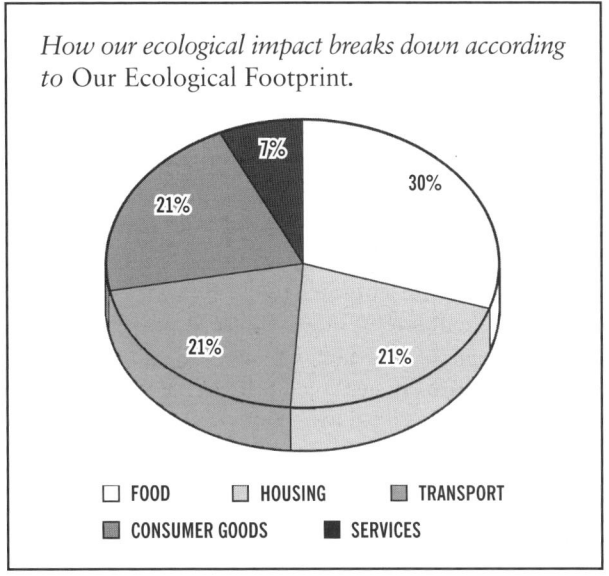

How our ecological impact breaks down according to Our Ecological Footprint.

The Dutch Enviro-Meter questionnaire comes up with a very different picture, as shown in the second pie chart. Here food has been knocked down to third place and transport takes number one. This running order is echoed by at least one other study.

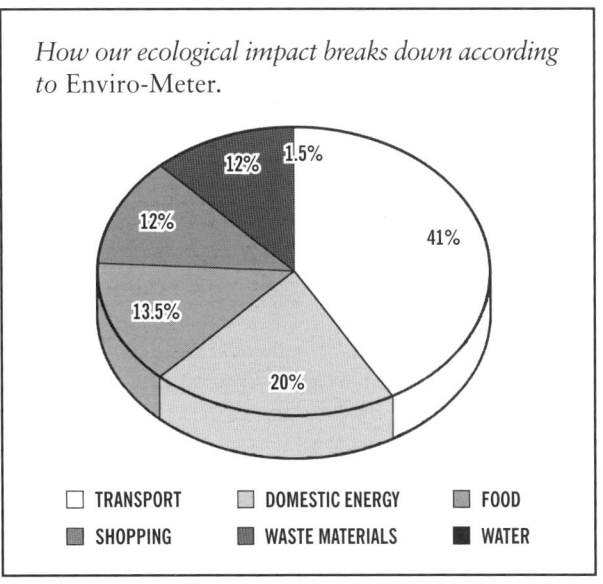

How our ecological impact breaks down according to Enviro-Meter.

The problem with comparing sets of figures like these is that they don't use exactly the same categories, and I suspect that much of the difference in their conclusions comes from how the categories are defined. Most of the ecological impact of food is generated in transport, processing and retailing, not on the farm. If you include flying green beans in from Kenya under transport, then food will get off very lightly.

There's no doubt that the total impact of the food system, as opposed to farming alone, is very high. I've even heard it suggested that walking can have a higher ecological cost per mile than driving, if the extra exercise causes us to eat more food. I haven't seen any figures to support this claim but it's not implausible – if we eat food from the mainstream food system. It has been calculated that by the time food reaches our plate, for each calorie of energy in the food, ten calories of fossil fuel energy have been expended.[17] I don't mean to suggest that we should drive rather than walk, but that we should grow our own food or buy locally-grown organic food.

There is also a qualitative difference between food and the other categories: food is one thing that we simply cannot do without. Those kinds of ecological degradation which specifically impair our future ability to feed ourselves, such as soil erosion, must be taken especially seriously.

With the other categories there is more agreement. All the figures I have seen ascribe a very high proportion of our impact to energy consumption, both in transport and in the home, and low ones to waste and water. Waste and water don't even get categories of their own in some of the studies, and where they do they average around ten and five percent respectively. Of course these are generalisations. If you live in a low rainfall area water will be much more important. But it puts the recycling of yoghurt pots into perspective.

As to specific things we can do, the trade-off between effectiveness and the investment required is a very personal matter. The investment we make may be in terms of time, such as that spent sorting waste for recycling; of money, such as the extra cost of organic food; or of giving up something we like, such as holidays in the sun.

On the other hand it may not be a sacrifice at all but something we really enjoy, like eating locally-grown, organic food in season, or best of all growing some of our own food. The reward for greening is often an immediate improvement in lifestyle, as well as the knowledge that we are helping the Earth.

I make no attempt to classify lifestyle changes according to the amount of effort they require; that's something each person can only decide for themself.

In the table, Possible Lifestyle Changes, I have listed some of the more effective things we can do, with the most effective ones at the top of each list and the less effective at the bottom. The order is very approximate. There's plenty of room for debate, for example over whether it's more important to eat organic food or locally-produced food. It also depends on degree, for example switching to a very fuel-efficient car could have more effect than a slight reduction in car use.

Most of these things are discussed later in the book.

Permaculture and the Green Movement

People sometimes ask "What's the difference between permaculture and organic growing?" As the previous section should illustrate, permaculture is about a lot more than growing food. But in as much as it is about growing food, the short answer is that permaculture is more about design and organics more about method. Although there is some overlap, the two are complementary.

However, if you take either to its logical conclusion it inevitably includes the other. If you take organic growing to its logical conclusion, sooner or later you have to address things like energy consumption and no-till methods, which are central to permaculture but have not been given much attention by organic growers. Similarly, you don't have to go far down the permaculture road before you realise that using synthetic chemicals on the land is self-defeating.

The same goes for appropriate technology, nature conservation and environmental campaigning. Any one of these becomes futile if practised on its own without reference to the whole. Each inevitably includes the others. This doesn't mean that each

POSSIBLE LIFESTYLE CHANGES			
Food etc	Transport	Domestic Energy	Others
Grow some of your own	Don't fly	Live in a small or terraced house	Buy second hand
Eat local food	Become car-free	More efficient heating system	Recycle aluminium
Less meat	Don't live in the country	Lower the thermostat	Compost organic wastes
Eat organic	Join a car-share scheme	Increase insulation	Use ecological paints etc
No tobacco	Reduce car use	Service the boiler	Use less water
Less alcohol	Fuel-efficient car	Draught-strip	Recycle paper
Eat seasonally	Keep the car tuned	Buy green electricity	

of us must personally practise every one of them. We would spread ourselves too thin. But the nature conservationist who thinks organics is for cranks and the eco-activist who lives in an energy-wasting home are absurdities. Although people like this were comparatively common even five years ago I'm happy to say they are now endangered species, if not entirely extinct.

This doesn't mean there are no differences of emphasis between these various branches of the green movement. The 'food miles' debate, about the advantages of eating locally grown food, is an example. Although the Soil Association has been at the forefront of the local food links campaign and has done more than any other organisation to set up actual box schemes on the ground, organics as such remains detached from the food miles issue.

It's still possible to import fruit and vegetables from thousands of miles away by air and quite legally call it 'organic'. Given the enormous ecological damage caused by air transport this is clearly absurd. The amount of harm caused by the transport is likely to be much more than that caused by growing the food locally with conventional pesticides, to say nothing of the social and economic harm done to communities by globalisation of this kind.

Organics, like appropriate technology, nature conservation and environmental campaigning, looks at one part of the picture, in this case the method of growing food. Permaculture takes a wider view, looking at all the inputs to a system, all the outputs flowing from it, both desirable and undesirable, and its relationship to the whole. This is not to say that many practitioners of organics, appropriate technology and so on don't also take a wide view. The difference is that it's intrinsic to permaculture.

Thus, as well as being a design methodology, permaculture can be seen as an overall framework which brings together many diverse green ideas in a coherent pattern. Since it's more about the relationships between things than about the things themselves it's well suited to this role. Indeed, the vast majority of the specific ideas and practices which come under the heading of permaculture are not unique to it and were not originated by people who call themselves permaculturists.

Some of these are traditional practices, such as coppicing, intercropping fruit trees with vegetables, and raised bed gardening. Others are the result of recent work in organic growing and ecological design, including many intensive vegetable techniques developed by people like Joy Larkcom and the HDRA, passive solar house design, vegetable box schemes and community supported agriculture.

By including these ideas and many others in this book I lay no claim to them in the name of permaculture. On the contrary, I give thanks to those who have gone before us who have made it possible for us to add our own small contribution to the sum of human knowledge and, dare I say it, wisdom.

Chapter 2

THE PRINCIPLES OF PERMACULTURE

The principles described in this chapter are the foundation on which permaculture is built. All the specific practices which are described in the rest of this book are examples of one or more of these principles in action.

The Earth is enormously and wonderfully varied. Physical, biological and cultural conditions are different from one place to another, and what is appropriate in one country or region is not necessarily appropriate in another. The principles of permaculture are broad principles, not detailed prescriptions. They are intended to be used in combination with local knowledge, and the results will look very different from place to place.

The essence of permaculture is to work with what is already there: firstly to preserve what is best, secondly to enhance what is there, and lastly to introduce new things. This is a low-energy approach, making minimum changes for maximum effect, working in co-operation with both natural forces and human communities. Not only will solutions be different from region to region but from one locality to the next and even from one household to the next. Subtle differences of microclimate, soil and vegetation are taken into account, and so are the differences between the needs, preferences and lifestyles of different people.

Even the principles themselves are not carved in tablets of stone. Ask any two permaculturists what the principles are and you'll get two slightly different lists, although all are based on the work of Bill Mollison and his collaborators. My list is given in the mind map below.

Although the principles are presented here as separate concepts, in reality they are all parts of a whole, and the boundary between one concept and the next is often indistinct. Many of the examples given to illustrate particular principles in this chapter could equally be used as examples of other principles. For instance, growing a mixture of tall and short plants together is an example both of Stacking and of Species Diversity. Reality often does not conform to the classifications we divide it up into, but a framework helps us to understand the myriad relationships which make up the real world. Although imperfect, it's a useful tool.

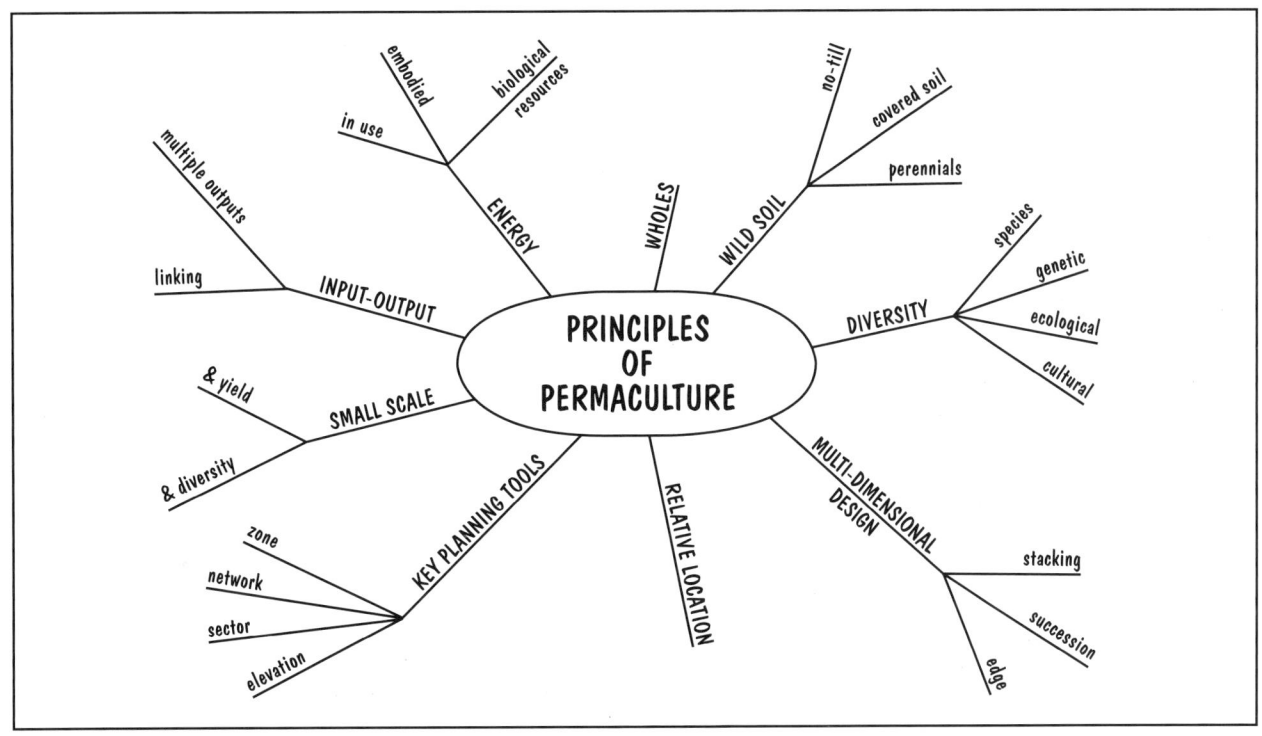

WILD SOIL

French-speaking permaculturists have coined the term *sol sauvage*, which has no real English equivalent but which I have translated as wild soil. As well as meaning a soil which has not been affected by human activity, it stands for an approach to growing crops which maintains the soil in a condition as close to its natural state as possible. It's a matter of working with nature rather than against.

No-till

Tilling means cultivating the soil, whether by ploughing a field or digging a garden. In nature this kind of soil disturbance is rare and usually localised, as when a wild pig digs for roots, or a tree is blown over. Although such disturbances are quite normal, they never account for more than a tiny proportion of the land at any one time. Yet we take tilling for granted as the normal way to prepare the ground for growing plants. The advantages and disadvantages of tilling are dealt with in detail in Chapter 3, but in general there are two good reasons for growing crops without it whenever possible.

The bed system of gardening, seen here at the Camphill Community at Oaklands Park, Gloucestershire, allows annual vegetables to be grown year after year without digging the soil. (See p192.)

Firstly, it takes a great deal of energy to move all that soil – over 7,000 tonnes of it per hectare – whether it's done by human or animal power or with fossil fuels.

Secondly, disturbing the soil, and especially inverting it, destroys its natural fertility. It destroys soil organic matter and structure, lays the soil open to erosion and disrupts soil life, killing off many of the vital micro-organisms and reducing earthworm populations. By choosing to till we are declining the gifts of the soil and committing ourselves to supplying that fertility ourselves.

This is not a dogma. In fact no permaculture principle should be taken as a dogma. No one is saying,

'Thou shalt not till'. What we are saying is that wherever there's a choice of doing something by tilling or by a no-till method, there has to be a compelling reason to choose the tilling option.

Covered Soil

Keeping the soil covered goes along with avoiding tilling. Much of the damage which can come from tilling the soil, especially soil erosion, is caused by the exposure of bare soil to the elements. This is also discussed in Chapter 3. Even if we choose to till we do well to cover the soil as quickly as possible, either with living ground cover or a non-living mulch.

Perennials

The logical extension of not tilling is to grow perennial plants instead of annuals. There are well-established ways of growing annuals without tilling both in the garden and on the farm. But growing perennials reduces the energy required even more by removing the need to sow and plant every year. (See p123-124.) Weeds are also less of a problem where the soil is constantly occupied by mature plants.

Growing perennials rather than annuals is also a direct imitation of nature. Annuals are rare in natural ecosystems. They normally need bare soil to get established, and are quick-repair specialists, able to move in rapidly and heal the wound. But they are soon succeeded by perennials. All natural ecosystems but the very youngest are dominated by perennial plants.

Perennial plants can be either woody – that is trees and shrubs – or herbaceous. Examples of herbaceous perennials are the grasses and herbs which make up natural prairies and traditional pastures. Here in Britain, and over much of the Earth's surface, the natural climax vegetation is woodland: that is what the land would eventually develop into if we left it entirely to its own devices. This suggests that growing tree crops is the ultimate in working with nature.

The reasons why humanity as a whole has, since the invention of agriculture, chosen to rely on herbaceous annuals rather than nut-bearing trees for the bulk of our food are lost in the mists of time. But the root of it probably has to do with the length of a tree's life cycle.

When people first started to grow food it's likely there was some degree of urgency about the matter, at least enough urgency to make them want to choose plants which gave a yield in the first year rather than ones which take several years to begin yielding. The speed with which annuals can be improved by selective breeding must have reinforced their dominance. A crop which has a new generation every year is obviously quicker to improve than one which doesn't complete its life cycle for a century.

CAN PERENNIALS YIELD AS MUCH AS ANNUALS?

As far as tree crops go the answer is an emphatic yes. Walnuts can yield around 7.5 tonnes per hectare, which is comparable with the typical yield of cereals in our climate. Considering that nut crops have received only a fraction of the plant breeding effort that annual grains have, this suggests that they may actually be able to yield more than cereals.

The total lifetime yield of tree crops is reduced compared to annuals because they don't reach full production for a number of years after planting. But this is counterbalanced by the fact that herbaceous crops can be grown between the trees while they are still young and only occupying part of the space they will eventually take up.

What about herbaceous perennial grains and perennial vegetables? Edible herbaceous perennials, other than a few luxury crops like strawberries and asparagus, have hardly been touched by plant breeders, so any direct comparison is sure to show annuals outyielding perennials. But we can make some predictions based on observing how the plants grow and on the few studies of the subject that have been made.

The conventional theory is that annuals can yield more because they don't have to put a proportion of the year's food supply aside to keep the plant body alive over winter. An annual plant puts as much of its energy as possible into the seed, which represents its entire future, leaving as little as possible in the dried-up remains of its vegetative body at the end of the season. In grain crops like wheat and barley, the proportion of the plant's energy in the seed has been increased as much as possible by selective breeding.

With annual vegetables the situation is slightly different. Both by plant breeding and by the way we grow them we have induced them to produce much more vegetative growth than their wild ancestors would. We harvest this vegetative growth before they start turning the energy stored in it into seed. Thus we harvest the leaves of a lettuce before it bolts, and capture all of the energy it has produced before it gets a chance to use it for reproduction.

Biennials, such as carrots, grow vegetatively in their first year and only set seed in the second. By harvesting them before the beginning of their second year we likewise take all the food they have produced.

On the other hand, a perennial grain plant must divide its energies each year between seed and a food storage organ, often a rootstock, which will enable it to survive over winter and grow again next year. Some biologists believe that this makes it impossible for a perennial grain to yield as much as an annual, given the same conditions of soil, climate and so on.

Similarly a perennial vegetable must divide its year's energy production between the leaves we harvest and a food store to carry it over the winter. In addition, many perennial vegetables set seed each year, so there can be a three-way split of energy.

However, although we can't harvest all the food produced by a perennial crop, it's in a position to produce much more in total than an annual. This is because it starts out at the beginning of each growing season as a full-sized, mature plant, with the energy resources to start growing at maximum rate as soon as the temperature and other conditions are right. Annual and biennial crops start the season as a seed, with tiny reserves of energy by comparison. They must build up their body size and rate of production bit by bit, from a very low starting point. All the unused ground between young plants early in the year represents unused growing potential.

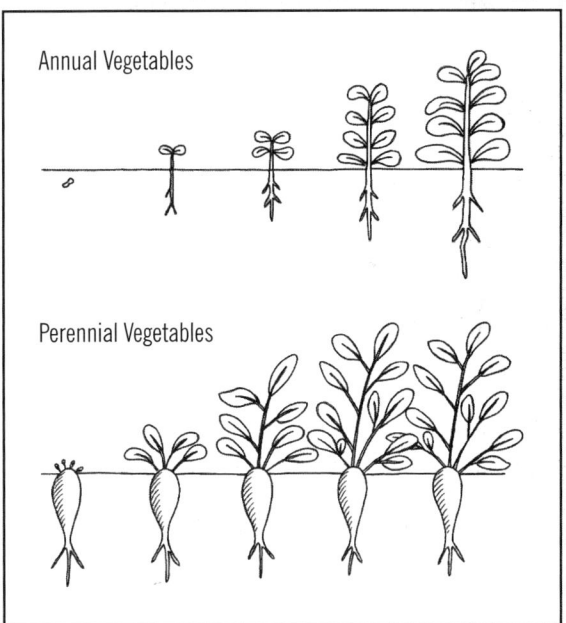

Growth patterns of annual and perennial vegetables during spring and summer.

So in theory there's both a yield advantage and a disadvantage in being perennial. This is not something which has been much studied by scientists, so there's not a great deal of empirical evidence to say whether the advantage outweighs the disadvantage. But the few studies which have been done on wild perennials suggest that they do not in fact sacrifice yield of seed in order to provide energy for overwintering.[1] There's no sound reason why we should not expect perennials to yield as well as annuals.

Once started, it's easy to see how the habit of grain culture became self-reinforcing: we like eating what we're used to eating; we get good at growing what we're used to growing; and we improve the crop plants by constant selection. Thus the more we grow and eat cereals, the more attractive they become.

Another reason which is sometimes suggested is the ease of harvesting cereals compared to nuts. But considering the amount of backbreaking work that goes into growing a cereal crop, unless you have the benefit of fossil-fuel technology, this seems unlikely.

The reasons why our ancestors chose not to use herbaceous perennials are less easy to guess. The question of whether they are inherently lower yielding is discussed in the box on the previous page, Can Perennials Yield as Much as Annuals?

The fact that we now feed the world largely with annual cereals does not mean that this is necessarily the best way to do it forever. From an ecological point of view it certainly is not. We now have a unique opportunity to develop perennial systems. For once in human history we don't have to worry where the next meal is coming from. We have an unprecedented level of food security, founded on a one-off expenditure of fossil fuel. We have the leeway to set up for future generations a tree-based agriculture which can feed them but which doesn't need to feed us in the short term.

DIVERSITY

Diversity lies at the very heart of permaculture. We see it in natural ecosystems and we know intuitively that our own systems will be more healthy, productive and sustainable if they are more diverse. There's also plenty of empirical evidence to support this intuitive knowledge.

Four aspects of diversity are important in permaculture. For the purposes of this discussion they are defined as:

- **Species diversity** – the number of different species of plants and/or animals which are grown and kept in a cultivated ecosystem; a mixture of species grown together is known as a polyculture, the opposite to monoculture

- **Genetic diversity** – the number of different varieties of plants and/or animals which are grown and kept.

- **Ecological diversity** – the total number of ecosystems and wild plant and animal species on Earth.

- **Cultural diversity** – which includes all kinds of human activity, both food growing and others.

Before looking at these individually, let's see what we can learn about diversity from observing nature.

SPECIES AND VARIETIES[2]

Although these two words are often used interchangably, they mean two quite different things:

- A species is a group of plants or animals which can interbreed among themselves and produce fertile offspring.

- A variety is a group of individuals of the same species which show enough common characteristics, different from those of other members of the species, to be regarded as a distinct group. To some extent varieties represent the genetic diversity within the species. In animal species the term race or breed is used instead of variety.

For example:

Species	Varieties/Races/Breeds
Human	African, European
Pig	Large White, Tamworth
Carrot	Early Nantes, Autumn King

Diversity in an ecosystem is a measure of the number of different **niches** which are filled. The niche of a plant or animal species is its function in an ecosystem in relation to other species. It's the sum total of that species' relationships with other organisms and with the soil and climate. It includes how the species gets its food, what other species eat it or parasitise it, what diseases it suffers from, what range of temperature, humidity and acidity it can tolerate, how it avoids competition with other species and so on – in other words the way in which it is specialised. To make an analogy, if an organism's habitat is its address, its niche is its occupation, the way it makes its living.[3]

The niche concept can be illustrated by comparing a few characteristics of two organisms which occupy very different niches in the same woodland habitat, the ash tree, and the wild garlic or ramsons.

	Ash	**Wild Garlic**
Height	25-30m	30cm
Protection from grazing/browsing	Height	Unpalatable to wild animals
Pollination	Wind	Insect
Season of leafing	Late May to October	March to early June
Protection from frost	Late leafing	Frost-tolerant leaves

Wild garlic flourishes in woodland. It leafs early in spring and flowers in May before dying back to its bulb as the trees come into full leaf.

It's easy to see how these characteristics interact. For example, the high-growing ash has better access to wind for pollination, whereas the low-growing garlic occupies a sheltered niche where insects can forage undisturbed by the wind. Of special interest to permaculturists is the way in which two plants avoid competition for light by coming into leaf at different times of year. They thus share the resource of sunlight rather than competing for it.

There's a great contrast between the niches of these two plants, but what about species whose niches are very similar?

Observations of ecosystems all over the world have led to the conclusion that no two species can occupy exactly the same niche in a natural ecosystem. An example of this is the displacement of the native red squirrel by the introduced grey squirrel over much of Britain. There's no direct aggression involved but their niches are so nearly the same that when the two live in the same habitat the stronger of the two, the grey, out-competes the weaker, the red, which dies out.[4]

The three species of British woodpecker, by contrast, have sufficiently different niches that they can co-exist in the same habitat. They each find their food at different levels above ground and have somewhat different preferences for nesting sites. Each species is using a different part of the resource base, although there is some overlap.[5]

This is the key to diversity: species avoid competition with each other by using different parts of the total resource base.

The advantages of copying this arrangement in cultivated ecosystems is clear. For example, sheep eat grass when it's short and cattle when it's longer, and what one leaves the other eats. Farmers reckon that if you have a herd of cows you can also keep the same number of sheep without any increase in acreage. Wheat and beans occupy slightly different niches, too, and they reckon that if you plant 50 acres of mixed wheat and beans you get 60 acres worth.[6] The same gains can be

had from combining orchard and garden crops with different niches.

It would be a mistake to think that maximum productivity and maximum diversity always go hand in hand in nature. Reedbeds are typically the most productive ecosystem in our climate, yet they usually contain only one plant species. Nature conservationists are only too aware of the fact that the quickest way to reduce biodiversity in a grassland is to apply fertiliser. Productivity increases, but a few vigorous grasses and herbs respond much more readily to the increased fertility, while the others are out-competed and die out. Both reedbeds and artificially fertilised grasslands have much higher levels of plant nutrients in them than most natural ecosystems, and it's generally true that an above average level of nutrients is correlated with low diversity.

The effect of high nutrients levels on diversity is revealed in this downland field. The flat grassland has had fertiliser applied and is devoid of wildflowers, but the bank is too steep for tractors, so no fertiliser has been spread there and an abundant population of cowslip flowers in the spring.

At the other extreme, very infertile land with correspondingly low productivity often has a low level of diversity, because not many species can survive the conditions. Upland moors are an example.

At moderate levels of fertility there is evidence that diversity and productivity do go hand in hand. A series of experiments with North American prairie species on a nitrogen-limited soil showed that more diverse mixtures of species were more productive than simpler mixtures. The more diverse mixtures were found to retain and use nitrogen more efficiently and this is probably the key to their greater productivity. More diversity of rooting depth means that there are active roots in a greater depth of soil, and more diversity of annual growth cycles means there are active roots in the soil for a greater part of the year. In other words more niches are filled. Soil nitrogen is soon lost from an ecosystem if it's not taken up by plants. It's frequently the limiting factor to production in organic growing.[7]

These results are what we would expect from niche theory: diversity leading to better overall utilisation of the resource, leading in turn to greater productivity. But it doesn't happen where the effect is masked by an over-abundance of nutrients, as in the reedbed or on artificially fertilised farmland. In other words diversity is the alternative to the massive applications of plant nutrients which is characteristic of conventional, chemical-based agriculture.

Maximum yield is not everything. The stability of a system, that is its ability to withstand stress, is equally important to those who rely on it for survival. Another study of prairie vegetation compared the stability of more and less diverse communities. After the most severe drought in 50 years the more diverse plant communities recovered their pre-drought biomass more quickly than the simpler ones. Less diverse systems are dominated by a few plants which grow very well in typical conditions, whereas more diverse ones also contain species which do better in untypical conditions, such as drought. These are the species which take over and keep productivity up when conditions depart from normal.[8]

DIVERSITY-STABILITY vs. SPECIES REDUNDANCY

Ecologists take differing views as to whether diversity really does make ecosystems more stable or not.

- In one corner is the diversity-stability hypothesis, which states that there is a positive correlation between the two.

- In the other is the species redundancy hypothesis, which states that only a few species are vital to the functioning of an ecosystem; that it can lose most of its component species and still be able to recover from traumas like drought, fire, unseasonable frosts or defoliation by insects.

This is an important debate. If species redundancy is right, and can be applied to the Earth as a whole as well as to individual ecosystems, it means we can destroy most of the Earth's biodiversity and still survive. One of the main reasons for preserving global biodiversity is gone.

The second study of North American prairie described above is among the growing body of evidence which supports diversity-stability. But the debate is by no means over. In a way the debate is meaningless, because the only way we can find out whether the planet needs its diversity of life or not is to drastically simplify it and see what happens. It would be a one-off experiment, because if diversity-stability turns out to be right we will all die. Unfortunately that experiment is just what we are doing with the Earth now.

It would be equally wrong to suppose that undisturbed plant and animal communities are always more diverse than those which are subject to some stress. It depends on the nature of the stress. Industrial pollution usually reduces diversity, as relatively few species can cope with the introduced chemicals. But grazing tends to increase the diversity of herbaceous plants. Without grazing the most vigorous plant species grow unchecked and crowd out the smaller, less vigorous ones. But herbivores will munch the vigorous plants down as fast, or faster, than the smaller ones. With all the plants reduced to much the same size, they all have much the same chance to grow. Predation has a similar effect on communities of herbivorous animals.

The one factor which is consistent in its effect on the diversity of ecosystems is the warmth of the climate. As a general rule the most diverse ecosystems are found in the tropical lowlands, and diversity decreases towards the poles and with increasing altitude.

Species Diversity

Estimates of the number of edible plants on Earth vary, but the total is probably somewhere between 35,000 and 70,000. Some 7,000 are known to have been grown as crops during recorded history. Yet today 90% of our food comes from only 20 plants worldwide, and 60% of it from just three: rice, maize and wheat. These crops can have very high yields and support large numbers of people. But relying on such a small number of plants makes us vulnerable to major new crop diseases, and to climate change. If any one of these crops was to run into serious trouble it would make a hole in our food supplies which could not be filled in the short term.

Paradoxically, the potential for diversity is greater now than it was in previous ages because crops have been introduced from one part of the world to another. We can grow much more varied polycultures than our ancestors ever could, and this is one of the important differences between pre-industrial peasant farming and post-industrial permaculture. Many of the non-European plants we can now grow are familiar ones such as potatoes or runner beans, but many more are unfamiliar. Plants For A Future (see Appendix D, Organisations) have created a database of 6,000 plants of potential use in Britain, and 4,000 of these are edible. The scope for increasing species diversity is enormous.

However, the fact that something will grow here doesn't necessarily mean that it will thrive and produce a useful yield. The work of assessing these plants, and breeding varieties suitable for growing here, has only just started. In the meantime whether to grow them or not depends on your objectives. Where the main priority is to produce a secure supply of food the well-known crop plants have a double advantage. Firstly, they are likely to be more reliable in themselves, and secondly, we know more about how to grow them successfully here. If experimentation is a priority then more unusual kinds can be included in a design. Many people may want to do both, especially those who have more than enough land to meet their basic needs.

The benefits of crop diversity are summarised in the mind map below. Diversity here can mean either intimate polycultures or a diversity of monocrops grown on rotation. Some of the benefits apply to both, others to one only.

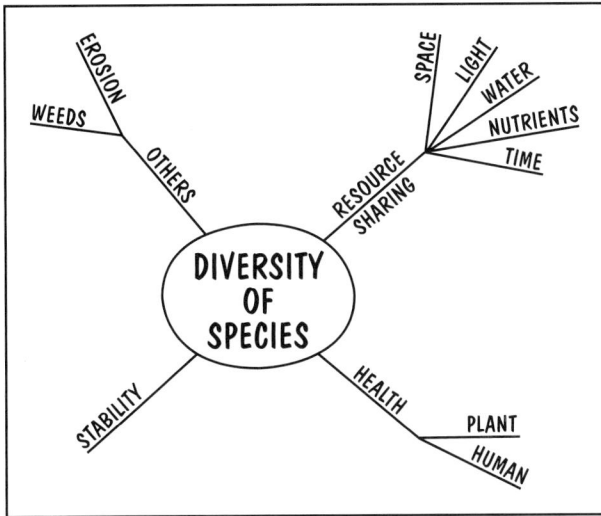

Resource Sharing. The aim of polyculture design is to put together plants which use a different part of the resource base to meet their needs. Tall thin plants and short spreading ones use different parts of the space available and do not compete as much for light as plants which are the same shape.

Different rooting depths make for more complete use of the soil, water and mineral nutrients. Each plant species requires a slightly different mix of nutrients, so a mixture of plants can make better use of the range of nutrients present in the soil. In addition some plants, mainly legumes, actually benefit their neighbours by fixing atmospheric nitrogen. (See pp48-49.) A mixture of early and late season plants will make more even use of all these resources through the year.

An example is a mixed crop of garlic and lettuces, which makes complementary use of the resources of space and time. Garlic is tall and thin, lettuce short and bushy, so neither casts significant shade on the other. Garlic can be sown in the autumn and harvested the following midsummer. Lettuce can be planted between the garlic plants in spring, and the garlic will have been harvested before the lettuces grow to full size and need all the space in the bed.

Health. Many pests and diseases of plants are host-specific, that is they can only affect one species of plant. An insect pest or fungal spore alighting in a monoculture of its host can spread rapidly in every direction and cause severe damage. If it lands in a polyculture its progress is barred on every hand by plants which it cannot infect and serious damage is unlikely.

There are also some plants which put out chemicals which boost the growth of their neighbours. For example, corncockle growing amongst wheat can increase the yield of the crop by up to five percent. This is called an allelopathic effect. Unfortunately, for every positive allelopathic effect like this there are many more which are detrimental to neighbouring plants. One of the skills of designing a polyculture is to minimise negative allelopathic effects by careful choice of plants. (See p218.)

Stability. Every growing season is different. One year the weather pattern will favour one crop, another year it will favour another. Some years will be very difficult for all crops and only the toughest will survive. If we could predict what each year was going to be like we could maximise our yield by planting a monoculture of the most suitable plant. But we can't. A diversity of crops may not hit the jackpot but it will most likely be more stable from year to year, and the average yield over the years will be more than that of a monoculture. Including a few really tough plants which can yield under almost any conditions, such as Jerusalem artichokes, is a good insurance against disaster.

In our culture we place a very high value on getting maximum yield, but subsistence cultures tend to be at least as concerned about reliability of yield. Whether our present economy is about to collapse and become a subsistence one, as has largely happened in Russia over the past decade, is impossible to say. But the one clear message which stands out in all predictions on global climate change due to the greenhouse effect is that we will get more extremes of weather. Preparing for seasons in which old reliables become unreliable is a sensible precaution.

Others. Better weed suppression is another advantage of polyculture. Since polycultures make more complete use of the available sunlight resource there is less light left for the weeds. By the same token less of the soil is exposed to the action of the rain, and growing a polyculture can reduce or eliminate soil erosion.

So much for the theory. What about the practice? Do polycultures on average yield more than monocultures? In his book, *Agroecology*, Miguel Altieri has made a survey of experimental work on a range of polycultures in different parts of the world. This showed spectacular advantages for polycultures, with land equivalent ratios ranging from 1.26 to 2.51. (See box overleaf, Land Equivalent Ratio.) What's more, the gain was greater with more components in the mix. Mixtures of two crops had LERs of between one and two, while mixtures of three crops had LERs of between two and three.[9] Details of some polyculture experiments done in Britain are given in Chapter 10. (See p266.)

LAND EQUIVALENT RATIO

The yield advantage of a polyculture over a monoculture is usually expressed as the land equivalent ratio (LER). This is the land area of monoculture required to produce the same amount as one hectare of polyculture. If the LER is greater than one the polyculture outyields, if it's less than one it underyields.

For example, take two adjacent 10ha fields with identical soils and microclimate. One is sown to a mixture of wheat and beans, the other to two separate monocultures, one of wheat and the other of beans. Suppose the biculture yields 60 tonnes of mixed grain and the monocultures a total of 50 tonnes. The amount of land needed to grow 60 tonnes of grain in monoculture would be 12ha. This is 1.2 times the size of the field in which the biculture was grown. Therefore we say that the biculture has an LER of 1.2.

Although polyculture can raise yields and help to control pests and diseases neither outcome is a foregone conclusion, and there are plenty of examples of polyculture trials which had negative results.[10] In order to outyield a monoculture, a polyculture must be carefully designed so that the positive interactions between the component species outweigh any negative ones which may occur.

From a farmer's point of view the down side of polyculture is that the work of growing the crop is more intricate than that for a monoculture. This is much less of a disadvantage in small-scale production than in large-scale mechanised farming, and the ideal place to practise polyculture is in a home garden.

Genetic Diversity[11]

Many of the advantages of growing polycultures of species can be had from growing mixtures of varieties. For example, mixtures of cereal varieties can greatly reduce the incidence of fungus diseases, while a mixture of varieties can help ensure yield stability both on the farm and in the garden. But our main concern with genetic diversity must be with the rate of extinction of varieties we are now witnessing worldwide.

In terms of human survival, the loss of genetic diversity is as serious as the loss of the tropical rain forests. All over the world traditional varieties of crops, which have evolved over hundreds or thousands of years to fit the conditions of a particular locality, are being discarded and replaced with a very few modern varieties.

The new varieties can yield more but they are dependent on fossil-fuel based inputs – chemical fertilisers, pesticides and so on. Where they have been introduced, as part of the so-called Green Revolution, food security has in many cases increased, but so has population, and that population is now dependent on the continued supply of these inputs. In addition, the cost of the inputs required are often beyond the means of smaller farmers, and many have been driven off the land and ended up as urban slum-dwellers.

The old varieties are essential if we are ever to develop permaculture, or any other ecological way of growing food, as they are the only ones which are not dependent on external inputs. What's more they are the only reserve of genetic material for future plant breeding. If, say, a new strain of disease arises which the current varieties are not resistant to, the only place plant breeders can go to find resistance is the old varieties – or wild ones, which are disappearing fast as more and more wild land is taken over for human use. We are becoming very vulnerable.

According to the UN Food and Agriculture Organisation three quarters of the cultivated varieties which have been bred over the past 10,000 years have been lost during the twentieth century.[12] In the Netherlands, one potato variety now accounts for nearly 80% of the crop. The real level of diversity is often less than the number of varieties still in cultivation would suggest, as many of them share common parents and have similar genetic makeup in many respects, including resistance to disease. This is particularly true of the handful of wheat varieties currently grown in Europe.

The erosion of diversity is part of a general cultural and economic trend towards uniformity in all aspects of life, but it is often hurried along by the policies of governments. Within the European Union any crop variety must be registered on a national list before its seed can be legally sold. In Britain it costs something like £7,000 to register a variety and keep it on the list for ten years, and in France it costs four times that much. You have to sell an awful lot of seed packets to get that money back on a vegetable variety, so only the few most popular varieties get on the list. In practice this means the ones best suited to commercial growing. Varieties that don't get on the list become extinct if no-one takes the effort to save them. As time goes by more and more of them are struck off.

Genetic engineering can be seen as an extension of the trend towards the domination of agriculture by industry. It puts yet more power into the hands of the huge multinational companies which produce both seed and chemicals, who are the only people who can afford to do it. But it's more than that. All previous plant breeding, including that of the modern Green Revolution varieties, has been a matter of selecting the best strains from the naturally occurring diversity of existing species. Genetic engineering does something which can never happen in nature: genes are taken from one species and artificially implanted into another, which may be totally unrelated to it.

No-one knows the long-term ecological consequences, and it's irreversible. If a genetically modified organism can survive and reproduce without human help it can never be recalled. Results so far suggest that the effects can be very damaging ecologically. For example, potatoes which were bred to resist aphids also decreased the population of ladybirds, the natural predators of aphids. So genetic engineering doesn't just replace sustainable farming and growing, it makes them less possible. It can be seen as an act which crosses a major boundary which we know intuitively we must not cross.

Just a few of the enormous range of tomatoes bred by gardeners and farmers over thousands of years.

Solutions

Governments have recognised that the loss of genetic diversity is a problem, and have reacted by setting up gene banks, which are basically cold stores containing seeds. There are various problems associated with gene banks. One is that the viability of the seeds depends on constant monitoring by technicians, and an absence of power cuts. There are doubts about how many of the seeds are still alive in many gene banks.

Another is that when varieties are not in everyday use on farms and in gardens people forget what they're like. Plant breeders looking for new material may find nothing but a name or number to identify a variety held in store. They would have to grow out all the varieties held there over a number of years to find the qualities they were looking for. If varieties are in regular use, their characteristics are known. Simply preserving the genes without the knowledge of how to use them is not much use.

Another drawback of gene banks is that the genetic material is 'fossilised'. When plants are regularly grown they are constantly evolving, adapting to gradual changes in the climate and in the genetic makeup of pests and diseases. The changes are slow and small. The variety will still be the same variety, but imperceptibly changed to be more fitted to its environment. Only by growing in the soil can plants maintain that contact with the rest of the Earth which is essential to healthy life.

Gene banks are better than nothing, but there is no doubt that the best place for plant varieties is in the fields and gardens of people who grow them to provide food and other needs. This is just what is being done by various small bands of seed-savers scattered around the world. Often desperately overworked, non-funded rather than under-funded, working in a political and economic environment that is fundamentally hostile to their values, these groups are busy collecting old varieties, growing them, exchanging seed and keeping alive a thread that has run deep in our culture since the invention of agriculture thousands of years ago.

The principal seed-saving group in Britain is the Heritage Seed Library of the HDRA. (See Appendix D, Organisations.) For a small fee you can become a member of their seed club and then receive free seeds of rare varieties to the value of your fee. This is a way of getting round the ban on selling non-listed seeds. Joining the club is a small but valuable contribution any ordinary gardener can make to this priceless work.

In return you get the chance to grow vegetables which were bred for taste and hardiness rather than for commercial profitability under a chemical regime. They are often more suitable for home gardeners in other ways too. For example a commercial grower wants all their peas to ripen at the same time so they can all be harvested on one day, whereas a home gardener wants them to ripen over a long period to give continuity of supply.

There's a great potential for co-operation between gene banks and the grass-roots groups, with the former able to provide scientific facilities and the latter growing out the seeds that lie dormant in the banks.[13] It's vitally important work. Of all the things on Earth, biodiversity is the most fragile. It can only survive as long as the individual living beings which carry it survive. When they die it's gone forever. We can never recreate it, and we need it for our own survival.

Solutions to the problem of genetic engineering must be more political in nature. Since profit is the motive for it we can completely undermine it by refusing to buy genetically engineered products. Consumer boycotts have already had a significant effect on slowing down the development of genetically engineered crops. Direct action, including destroying the crops in the field, can also be effective. More information can be had from the Genetic Engineering Network. (See Appendix D, Organisations.)

Ecological Diversity

So far we have looked at the diversity of cultivated plants and domestic animals within our own productive systems. But equally important – in fact more important to the future of the Earth – are the myriad of wild species making up the natural and semi-natural ecosystems, and co-existing with us in our gardens, farms, plantations and towns. This is the subject of Chapter 12 so no detailed discussion is needed here. Suffice to say that by far the greatest part of our interaction with wild species and ecosystems is not direct but indirect, in the form of the products we consume. (See p343.)

Cultural Diversity

The principle of diversity holds true at every scale from individual plants and animals upwards. At the largest scale diversity means preserving and working with the unique character of different localities. The hallmark of modern industrial culture is uniformity. The same glass and steel skyscraper is built everywhere from the frozen north to the sands of Arabia, and vast quantities of fossil fuel are thrown at it to keep it warm in the one place and cool in the other.

Ecological building or agriculture are things which cannot be taught and learned centrally. The principles can be taught, but the practice must be in tune with the unique characteristics of an individual place, and of the people who live there.

This is partly a practical matter of getting the physical details right, but it also has an emotional side. Most of us experience our connection with the Earth largely in terms of our relationship with the place where we live. An important part of that relationship is founded on the distinctiveness of the place. As more and more places all over the country and all over the world start to look and feel very similar, that relationship becomes eroded. We start to care less for our home locality, and less for the Earth as a whole.

The principle of diversity can also be applied to the way we make our living. Relying on a single source of supply for any of our needs is a risky way to live. A network of multiple sources gives greater security. Just as crop diversity can give greater food security and reduce reliance on external inputs, having a number of different part-time jobs and enterprises rather than one full-time one can give economic security and greater independence.

Firstly, in a job market where whole swathes of jobs cease to exist as industries die or become automated, it can buffer against the stress of periodic unemployment and retraining. People with polycultural incomes only lose part of their living if one of the things they do folds up under them.

Secondly, having more than one job gives you the opportunity of doing something which is only marginally profitable under present economic conditions, though in real terms it may be more sound than most well-paid work. An example is small-scale farming, combined with some kind of teleworking. While it becomes increasingly difficult to make a living on even a medium-sized farm let alone a small one, it is becoming increasingly easy to work hundreds of miles away from employers or clients with a computer and an internet connection.[14]

It is possible to have too much diversity. A farm or smallholding can benefit from having a diversity of enterprises much as an ecosystem can benefit from having a diversity of species. But you can only divide your attention so many ways at once. If there are too many things demanding your attention some of them won't get enough, and something which is badly done is more likely to make a loss than a positive contribution to the farm economy. The same is true of a crop polyculture. A combination of two crops which is well known and understood may in practice be more productive than a mix of five crops which is just too complex to manage effectively.

The thing to aim for is not the maximum level of diversity but the optimum. But this is likely to be much greater than is normal in this day and age.

MULTI-DIMENSIONAL DESIGN

This concept, like those of Wild Soil and Diversity, is based on direct imitation of ecosystems. Most agriculture is virtually two-dimensional, consisting of low-growing field crops. Stacking introduces the third dimension, the vertical, succession works with the fourth dimension, time, while edge is about boundaries between different parts of a system.

Stacking

Stacking is multi-layer growing. At its most basic it's an imitation of a natural woodland, which usually has at least three productive layers: trees, shrubs and herbaceous plants. A stacked system is potentially much more productive than a single-layer crop such as a field of wheat.

PRODUCTIVITY OF SOME ECOSYSTEM TYPES, I[15]		
Ecosystem Type	Net Primary Productivity (g dry organic matter/m²/yr)	
	Average	Range
Tropical rainforest	2,200	1,000-3,500
Temperate deciduous forest	1,200	600-2,500
Scattered trees or shrubs (all climates)	700	250-1,200
Temperate grassland	600	200-1,500
Cultivated land (all climates)	650	100-3,500

The figures in the table reveal how the productivity of land ecosystems is broadly related to the tree cover. Tropical rain forests, with an abundance of both rain and heat, are naturally more productive than any temperate ecosystem, and are included here for comparison. The cultivated land includes some highly productive crops, such as irrigated rice and sugar cane, which can only be grown in tropical climates. No temperate crops would fall in the higher end of the range given. The three ecosystem types in the centre of the table show an increase in average productivity with increasing density of tree cover.

A stack of poplars and grassland can produce more than either could alone. Even when the trees are nearly ready for felling, as here, there's still a useful yield of grass.

The higher productivity of more wooded land is associated with the higher rainfall of those areas. Trees will grow wherever there's enough water for them, while grass can survive in much drier climates. The natural vegetation of Britain is woodland, yet our agriculture is based on herbaceous crops whose ancestral home is the dry and relatively unproductive steppes of western Asia. It would appear that we are forgoing the potentially higher yield of stacked systems.

A major reason for the high productivity of woodland is that more ecological niches are available than in a grassland. This not only allows for greater diversity but for greater overall productivity, as different organisms use different parts of the total resource base available. We have already seen how ash trees and wild garlic avoid competition for sunlight by coming into leaf at different times of year. (See pp16-17.) This is part of the general pattern of leafing in temperate deciduous woodlands. In general the herbaceous plants, the lowest layer, come into leaf first, the shrubs next and the trees last of all.

We can make use of this pattern in designing our own stacked systems. For example, pasture grasses carry out 60% of their annual photosynthesis before ash trees come into leaf, so the two can be combined in a stack which is more productive overall than either planted alone. Combinations of tree and field crops like this are known as agroforestry.

It's the overall yield which is important in stacked systems, not necessarily the yield of a single component. Thus the yield of either timber or grass may be greater when each is grown in a monoculture, but the total yield of the biculture will be greater than that of either monoculture.

A similar combination on a garden scale is the interplanting of a tall, thin vegetable with a short bushy one. (See Colour Section, 1.) Neither casts significant shade on the other, and together they make more complete use of the space than either could alone. The total three-dimensional space is used more effectively than it could be in a monoculture of either. We have already seen the example of garlic and lettuces under the principle of Species Diversity. (See p19.)

These are simple two-plant combinations, but stacked systems can be far more complex. The most complex is a woodland or forest garden, which is a direct copy of a natural woodland with a wide variety of fruit trees and shrubs and perennial vegetables and herbs.

Using vertical space. A pear tree is trained up the south-facing wall of this house.

Training fruit trees or other productive plants up walls is another way of working in three dimensions and thus another form of stacking. This is particularly useful in urban areas or other situations where space is short. People often say there's not enough room to grow a serious amount of food in towns, but this is only true when you think in two dimensions. There is space on walls, balconies, flat roofs and even gently sloping roofs.

It would be wrong to exaggerate the quantity of food which can be grown in these situations. But it's very high value food, both because it's grown just where the people who eat it live, and because favourable microclimates can be used to grow tender and out-of season crops.

Succession

A stacked system is not static. One consisting of perennial plants changes through the years as the larger but slower-growing trees gradually take over the leading productive role from the herbaceous plants which dominate at first. A typical natural succession in our climate may broadly follow the sequence: bare ground, annual plants, herbaceous perennials, shrubs and pioneer trees, mature woodland. To get the full productivity from stacked systems it's necessary to work consciously with succession.

For example, young trees take up very little in the way of space and resources. If young standard fruit trees are planted out into a market garden it will be many years before they compete significantly with the vegetable crops. During that time they will benefit from the weed-free conditions of the vegetable garden and from the attention they receive. The land is being doubly productive, producing vegetables and also bringing on a young orchard.

This simple two-stage succession could have more stages added in. The annual vegetables could be replaced with perennials as the roots of the young trees grow out and occupy more ground. As the shade level increases soft fruit, which is relatively shade tolerant, can be planted beneath the trees, and edible woodland herbs, such as wild garlic, can be planted underneath. When the trees are fully grown the shade and root competition may be too much for the soft fruit to yield well, but by now they will have reached the end of their productive life anyway, and they can be removed. (See illustration, A Cultivated Succession.)

Stacked systems consisting of annuals can display a similar change through time, though of course it will be on a shorter scale. An example is interplanting relatively slow-growing cabbage with the faster-growing lettuce which is harvested before the cabbages take up all the space.[16] (See p201.)

Edge

The most productive part of a natural ecosystem is often on the edge, where one kind of ecosystem meets another, a phenomenon known as the edge effect.

In our dull climate this is especially true of stacked systems, where the main limitation on productivity of the lower layers is light. As we have just seen, the conflict between the layers can be minimised by working with the annual cycles of the different plants and with year-

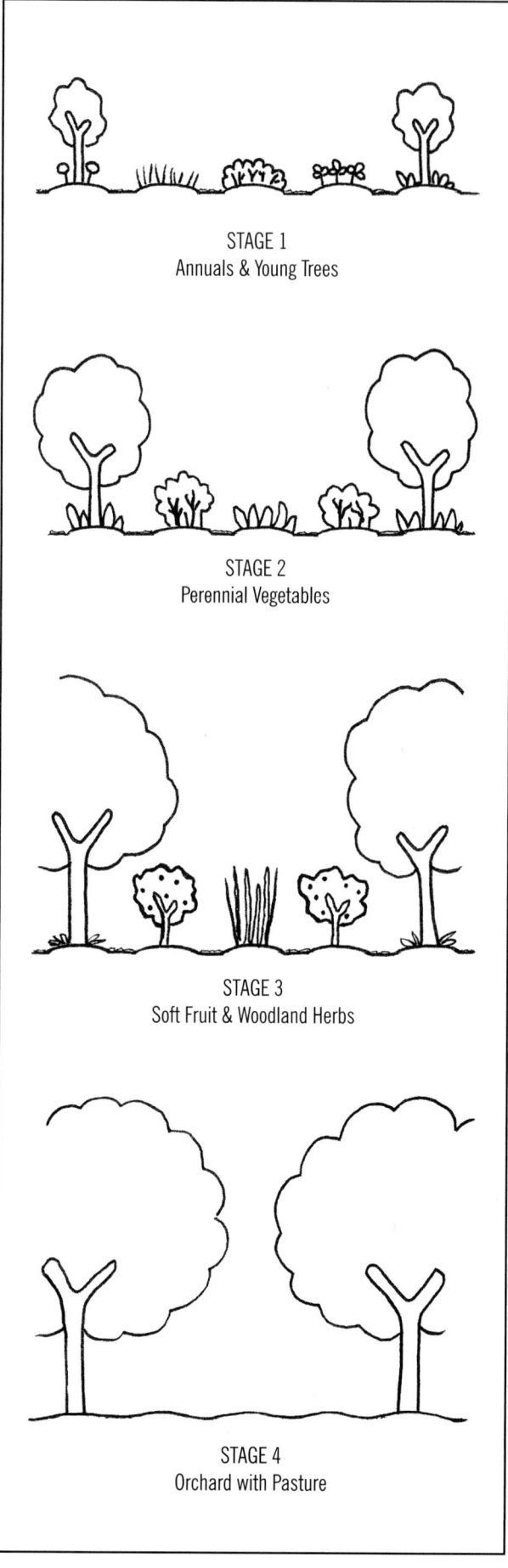

STAGE 1
Annuals & Young Trees

STAGE 2
Perennial Vegetables

STAGE 3
Soft Fruit & Woodland Herbs

STAGE 4
Orchard with Pasture

A Cultivated Succession

on-year succession. Designing in plenty of edge can minimise it further.

The edge between woodland and grassland has some of the advantages of both ecosystems. It has the three-dimensional structure of the woodland, and the lack of shading enjoyed by the grassland. The edge is where woodland shrubs thrive and produce most of their fruit. Animals such as birds are able to forage from the wide variety of food sources offered by the two environments, and make use of the nesting cover offered by the wood. Edges themselves are often very diverse. Many of the species from both ecosystems live there, together with some that only live on the edge, giving more species diversity than the interior of either ecosystem.

In our own cultivated systems we can make use of the edge effect when designing productive landscapes. Instead of arranging farmland and woodland in large separate blocks we can establish an intimate mixture of woods and fields, which can be much more productive than either alone. The edge effect can also operate on the interface between individual crops as well as between two ecosystems. Crops grown in alternate strips can show an increase in yield over crops grown in large single-species plots.

The woodland edge can also be reproduced to great effect on a garden scale. A woodland garden will be most productive if it is given the maximum edge so that sunlight can penetrate to the lower layers. It can also enhance the microclimate for vegetables and herbs growing in the sheltered area along its edge.

Gradual edges are more productive than abrupt ones (see illustration below left). This is as true of the edge between water and land as of that between woodland and field crops, and shallow water is an edge habitat par excellence.

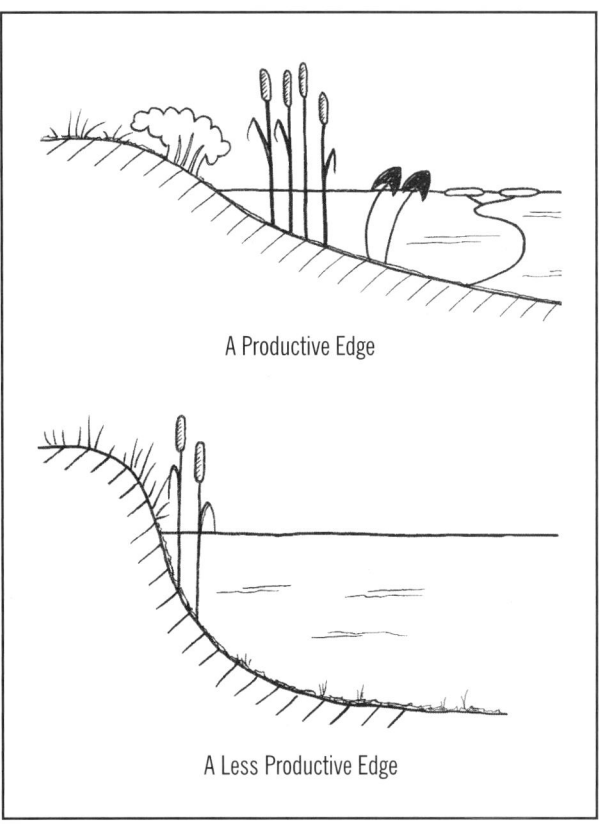

A Productive Edge

A Less Productive Edge

Plants growing in shallow water have optimum access to the three mediums of water, light and soil. Dryland plants in almost any climate will experience some loss of production from time to time due to lack of water, but plants growing in a permanent marsh or body of water will not. Deeper water is less productive because the resources of air and soil are separated by a greater distance. Plant nutrients sink to the bottom but oxygen and light are concentrated near the surface. Plants either live on the bottom and suffer reduced levels of oxygen and light, or float at the surface and starve for nutrients.

As the table overleaf shows, on a global scale ecosystems of shallow water tend to be the most productive. Seaweed beds and reefs can be even more productive than tropical rain forests, while swamps and estuaries can match their yield in the same climatic zone. Temperate forests, the most productive land ecosystem of cooler regions, cannot match the productivity of semi-aquatic ecosystems growing in the same climate. As a general rule, semi-aquatic ecosystems are the most productive ones in any climate. In tropical coastal areas this means mangrove swamps. In temperate inland areas it means reedbeds.

Estuaries are on the edge between sea and freshwater, and receive a constant input of nutrients eroded and

A Productive Edge

A Less Productive Edge

PRODUCTIVITY OF SOME ECOSYSTEM TYPES, II[17]		
Ecosystem Type	Net Primary Productivity (g dry organic matter/m²/yr)	
	Average	Range
Tropical rainforest	2,200	1,000-3,500
Temperate deciduous forest	1,200	600-2,500
Seaweed beds and reefs	2,500	500-4,000
Swamp and marsh	2,000	800-3,500
Estuaries	1,500	200-3,500
Open ocean	125	2-400
Cultivated land (all climates)	650	100-3,500

leached from the watersheds of their rivers. The tides constantly replenish them with oxygen and both plant and animal food. Some British estuaries are as productive as a tropical rain forest. The nutrient-starved plankton of the open oceans produce very little per square metre. But in total they account for almost a quarter of the Earth's production of organic matter, because they cover two thirds of the planet's surface.

Nothing illustrates more dramatically than these figures the enormous importance of the edge effect.

An intimate mixture of water and dry land in the landscape can be extremely productive. Fish, for example, get a great deal of their food from marginal plants which overhang the water's edge, and a pond which has a high ratio of edge to area will be more productive than one with less edge.

Finally it has to be said that maximum edge is not always desirable. For example, where a garden is situated in the middle of an area of creeping weeds such as couch and creeping buttercup, the less edge the better! The same applies to a vegetable plot bordered by dense vegetation which can harbour slugs. Houses are another example. A terrace house, which has little in the way of outside edge and shares walls with two other

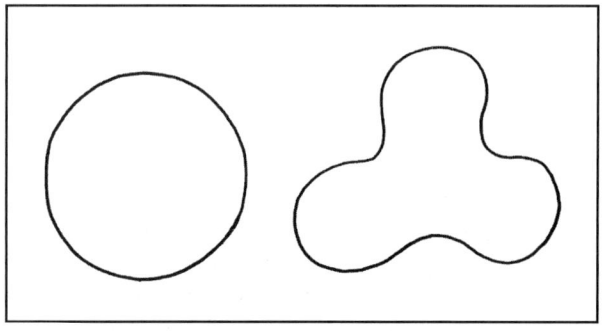

These two ponds have the same surface area. Other things being equal, the one on the right will be more productive.

houses, loses much less heat than a detached bungalow, which has a great deal of edge compared to its volume. The terrace house is thus much more energy efficient.

RELATIVE LOCATION

In a way this principle is the mother of all the others. If our aim is to create a network of beneficial relationships between the components of our systems, where we put things is very much of the essence. (See pp3-4.)

Nowhere is this principle more clearly illustrated than in a combined chicken-greenhouse. This is a chicken house with a greenhouse attached to its south side. The heat of the chickens' bodies is enough to keep the greenhouse free of frost on the coldest of nights, while the greenhouse helps to warm the chickens up on a cold winter's morning, and it may be possible to use the carbon dioxide breathed out by the chickens to enhance plant growth in the greenhouse.

Thus the waste body heat from the chickens saves fossil fuel which would otherwise be used to heat the greenhouse. In a large-scale battery or broiler house the heat and the CO_2 become pollutants which must be removed by a forced ventilation system, powered by electricity generated from more fossil fuels.

The beneficial relationships between the chickens and the plants in the chicken-greenhouse are only possible because of their relative location, together in the same building. If they were each housed in separate buildings, however close together, these links could not be made.

The placement of wall-trained fruit trees is another example. A south-facing wall is an ideal place to grow tender fruit such as peaches and figs. The wall gives them shelter, stored heat and reflected light. They repay being grown in such a favoured position, in fact in most parts of Britain they are hardly worth growing in the open or against a less sunny wall.

A north-facing wall, on the other hand, denies most of the Sun's light to trees trained against it. But there are some fruits, such as early-season cooking apples and morello cherries, which can do well enough there, and will not benefit greatly by being planted in a sunnier position. Walls facing other directions suit trees which fall between these two extremes.

Careful placing of fruit trees relative to the available walls in the garden will not only allow the widest range of fruits to be grown, but will give the greatest potential yield from a mixture of trees.

The Wider Picture

One of the most important applications of the principle of relative location is the relationship between where people live and where their food is grown. There are many ecological advantages in people growing their own food in their own back gardens, compared to buying supermarket food which may have been grown hundreds or thousands of miles away:[18]

- The energy use and pollution involved in transport are eliminated.

- Energy-intensive mechanised farming is replaced by attention-intensive gardening.

- The need for processing and additives is reduced or eliminated.

- The need for packaging is eliminated.

- The wastage of food which occurs in the distribution system is eliminated.

- Food is eaten fresh, enhancing its nutritional value.

- People have more control over how their food is grown.

- Less land is needed for food production. (See pp31-32.)

- Gardening is a healthy occupation for people with sedentary jobs, and helps put us back in touch with nature.

- Optimum use can be made of the organic matter and plant nutrients which accumulate in cities, eg by home composting.

It has been calculated that the average UK family diet leads to the production of 9 tonnes of CO_2 per year, which is twice as much as the annual output of a typical new house.[19] Growing food at home does away with virtually all its energy cost, so the potential energy savings from growing our own are considerable, and energy use is often a fair indicator of all-round ecological impact.

Enabling more people in towns to grow at least some of their own food is one of the main themes of permaculture. Buying local food direct from the producer is the next best thing to growing your own, and this is also central to permaculture. It's a matter of making beneficial relationships between people, and these relationships are just as important as those between plants and animals in the internal design of a farm or garden.

Much the same applies to our other material needs apart from food. As a general rule the closer to home they come from the less the ecological cost.

KEY PLANNING TOOLS

This is a design tool-kit containing four tools which can be used together to find the best placement for elements in a design. The four tools are: zone, network, sector and elevation. In this context the four words have the following specific meanings:

- The **zone** in which a piece of land falls is a measure of how much human attention it receives – in plain terms, how close it is to the back door.

- **Network** analysis looks at the relationships on a site where there is more than one centre of human attention.

- A **sector** is an area affected by an influence coming onto the site from outside, such as wind or sunshine.

- The **elevation** characteristics of a site include the degree of slope, the direction in which the slope faces (its aspect), its height above sea level and its height relative to the surrounding land.

The key planning tools can be used in three ways: to analyse the existing situation, to develop a new design, or as a check on a new design proposal. It's convenient to describe the four factors separately but in practice they must be used together, because it's the interactions between them which give the full benefits.

Zoning

The principle of zoning is that whatever needs the most human attention should be placed nearest to the centre of human activity. Thus an intensive vegetable garden, which needs daily attention, is best placed in view of the kitchen window, whereas a timber plantation, which may not need visiting from one year to the next, can be far away. To place the plantation near the house would be a waste of a high-value site from a zoning point of view.

Of course this is just common sense, especially when the examples chosen are as different from each other as these two. But it's remarkable how uncommon common sense can be. It's not unknown for a market gardener to live miles from their garden and commute there each day by truck. This is not only wasteful of energy, it's stressful, and makes it more difficult for the gardener to give the plants the attention they need.

There are six conceptual zones into which land can be divided. They are based on the kinds of land which may be found on a farm but they are applicable, with some modification, to all human settlements.

Zone 0
The house itself.

Zone I
The home garden, including: intensive vegetable beds, salads and herbs, wall-trained and other intensively-grown fruit. Tender and exotic plants are most likely to be found in this zone. Productivity is very high and zone I is usually a net importer of soil fertility from the rest of the landscape. Human influence on the landscape is very high.

Zone II

Orchards, poultry runs, housing for other animals, workshops and maincrop vegetables which require more space than is available in zone I. In urban situations allotments are the typical zone II.

Urban Zone II. Andy Waterman in his productive allotment.

Zone III

Farmland, including field-scale crops and pasture, ideally integrated with productive water and small, intensively managed belts of woodland. Much of the produce from this zone is for sale rather than for home consumption.

Zone IV

Rough grazing and woodland. Human influence on the landscape is much reduced in this zone, and the majority of the plants are native. The value of yields for human use is relatively low.

Zone V

Wilderness. In Britain there is no land which can truly be described as wilderness, completely untouched by human hand. Here we think of Zone V as land where the interests of wild plants and animals take top priority, and yields of produce for human use are only taken when to do so benefits the wild species, as when a flower meadow is mown for hay. (See Chapter 12.) Every design, however small, should have its zone V, if only a bird feeding station.

We all have relationships with land in each of these zones, whether direct or indirect. For most of us the relationship with zones II to V is an indirect one, but a crucial one. We influence what kind of farming is practised by what kind of food we choose to buy, we influence the world's forests by the wood and paper products we use, and many products we can buy involve the destruction of wilderness.

The discussion of growing food in cities under Relative Location can be seen as an application of the principle of zoning. (See pp26-27.)

Networking

The concept of zoning assumes a situation in which there is one centre of human attention and activity. On some sites there is more than one, and though one may be more important than the others it's necessary to consider the flows between them all.

An example is given by David Holmgren from the design of his own smallholding in Australia.[20] In addition to the house there's a group of buildings including shed, barn and poultry house, and a site for a possible second house in the future. In the design process these three centres of attention and activity were seen in relation to the three main areas of productive land: the gardens, the orchard and two ponds. The resulting analysis guided the design of access ways and fencing.

Sectoring

While zone and network analysis are about the relationships between the land and on-site human energy, sector analysis is about its relationships with influences coming from outside. These include: wind, sunshine, flows of water, pollution, neighbours and views. The principle of sectoring is to place things so that they have the best possible relationships with these influences.

Taking the example of the timber plantation again, we have already decided it should go far away from the centre of human attention, in zone IV. Considering sector, we might decide to place it so that it gives shelter from the prevailing wind or a particularly damaging one. In most of Britain these come from the south west and north east respectively. Say there was a particularly good view to the south west on our site, that might be enough to clinch the decision in favour of a site to the north east.

Sectoring is to a great extent working with micro-climates. A microclimate is the climate of a small area, anything from a parish to a couple of square metres. Examples of microclimate factors are: wind, sunshine, humidity and frost risk. Many microclimate factors are influenced by the elevation of the land. Indeed there is a great deal of interaction between sector and elevation.

Elevation Planning

This is a matter of placing things in relation to the landform. It can best be illustrated by looking at the profile of the typical landform in humid climates such as ours, an S shaped curve, or concavo-convex curve. In reality a slope is usually composed of a complex of S shaped curves of different sizes, but here a single curve is shown for the sake of clarity. (See illustration on facing page.)

The highest land is exposed, and may be suitable for wind generation of electricity. This land is often suitable for grazing, and maybe for arable crops. It's an ideal place to store water, as water can be directed from here to any part of the landscape by gravity. Water stored here can also be used to generate electricity. On the other hand there's less potential to collect large volumes of water here than there is lower in the landscape.

As we go down the slope it becomes steeper. There's a greater potential for soil erosion on steep slopes, and this must have a major influence on what we do with sloping land. Arable farming should be confined to the most gentle slopes, but grazing can cause soil erosion too, largely by the wearing away of herbage by animals' feet, and the steepest slopes are best used for continuous-cover woodland.

Just below the point where the convex slope turns into a concave one the ground starts to level out. This is sometimes called the key point, or key area, and a number of things can happen here.

It's a particularly good place to collect and store water for three reasons:

- Water storage is always more efficient on relatively flat land, as you store more water for a given height of dam.

- There's a much greater potential for collecting water at this point than there is higher up.

- There's still a considerable part of the landscape which can be watered by gravity from this point, and this may include the land with the deepest soils.

These three factors make key points particularly suitable for storing water within a gravity-fed irrigation system. (See pp112-114.)

It's also a good place to site a house, below the winds of the exposed hilltops, and above the frost pockets and flooding risk of the valley floor. It's usually the warmest part of the landscape, especially on a south-facing slope. If the steeper land immediately above the house site is wooded, the trees will have a modifying effect on the microclimate of the house, keeping it warmer in winter and cooler in summer, as well as possibly sheltering it from wind.

For much the same reasons this is the ideal part of the landscape for orchards, which are sensitive to both wind and frost.

The area immediately below this is often ideal for arable cropping. The slopes are gentle, and drainage is likely to be better than on the very flat land at the bottom. The soil will be relatively deep, with the accumulated fertility which has been carried down the slope by the natural processes of erosion, usually increased by human activity in the past.

The bottom land is often poorly drained and subject to flooding, and is usually more prone to late frosts. A prime use for this land is meadow, that is grassland used for hay and silage. Permanent grass can stand poor drainage much better than arable crops, and it benefits from the plant nutrients and organic matter brought down in flood water and left behind when the floods subside. High-yield willow and poplar plantations are another good use of bottom land.

Each part of the landscape offers both opportunities and constraints.

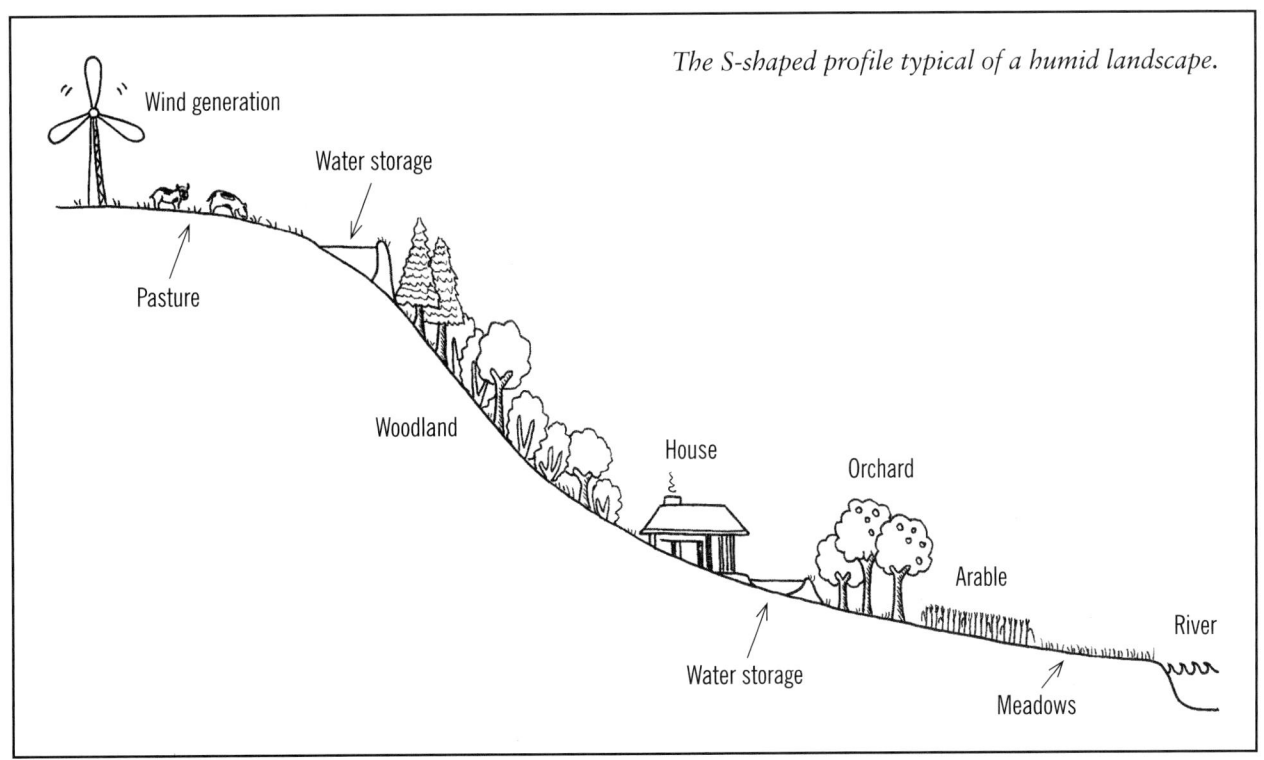

The S-shaped profile typical of a humid landscape.

Wind generation

Water storage

Pasture

Woodland

House

Orchard

Arable

River

Water storage

Meadows

A GARDEN EXAMPLE

The key planning tools are presented here mainly in the context of farm scale design, but they can be equally useful on the scale of a domestic garden. As an example let's look at siting a compost bin.

Zone. There will be frequent trips from the house with kitchen waste, so the bin should not be too far from the back door. (Let's assume we are competent compost makers so it won't be smelly.)

Network. Inputs from the garden may be less frequent but probably more bulky. What part of the garden will most compost material come from? Is there a comfrey patch?

Sector. Plants need light to grow, and shade is often a limiting factor in gardens. A compost bin is one element in the garden that doesn't need sunlight, so it can go in a shady place.

Elevation. Finished compost has about ten percent of the weight and bulk of the fresh ingredients. If the garden is on a slope I would place the bin at the bottom rather than the top, so the fresh garden waste gets carried downhill and the finished product uphill. On the other hand, an experienced gardener I know says he would always place it at the top, because filling the bin is a casual job done in dribs and drabs whereas emptying it is a major task which you do all at once and it involves a lot of heavy barrowing.

Which just goes to show there are no hard and fast rules. Much depends on the individual preferences of the gardener. The important point is not to come up with a set of standard answers, but to make sure that the relevant questions are asked. In some gardens, getting the bin in the right place could save a great deal of work.

Integrating the Four

Looking again at the example of siting a timber plantation, network and elevation factors, such as extraction routes for the timber and the degree of slope of the land, may have to be put into the balance along with considerations of zone and sector. Protecting a steep hillside from erosion may be a high priority and this may take precedence over other considerations. The answer will be unique to the individual landscape, and influenced by the aims and preferences of the people living in it.

There is in fact a fifth factor which influences the siting of elements in the landscape, and that is soil type. To some extent this has been down-played in permaculture because it's always possible to improve soil, whereas bad placement is there for ever – unless you're prepared to demolish things and start again. But siting things on the soil that suits them can save a lot of work and expense. For example, most fruit trees do badly on a shallow soil over chalk, and a pond needs a liner if it's not sited on a clay soil. Fruit on chalk and ponds on light soils can be made to work, but why not choose something more suitable to begin with?

David Holmgren has found experience has led him to pay more attention to the soil:

> ...my strong interest in physical design aspects of the project have led to less than adequate attention to [soil] matters. It is very clear that having established a design framework for sustainable production, the key to achieving the potential will involve more focus on the traditional concerns of biological soil husbandry. This emphasises the complementary nature of permaculture to organic and biodynamic methods.[21]

In previous ages landscapes were designed according to the principles of the key planning tools, not consciously but because people had no choice. Today, with an abundance of under-priced fossil fuels, we can ignore this and treat the land as though it were homogenous, but only by throwing energy at every problem. If we work with the land, using every piece of ground for what it's best suited to, we can reduce the level of inputs needed to maintain a highly productive landscape. In planning for a sustainable low-energy future we need to use these principles consciously. (See Colour Section, 2.)

SMALL SCALE

...and Diversity

The scale at which we do things is crucial to permaculture. Many of the principles are difficult or impossible to implement on a large scale.

For example, elevation planning suggests that when crops are grown on a slope they should be grown on the contour to minimise erosion. This is fairly easy to achieve in a garden where the work is done by hand, but as soon as machinery is used it becomes very attractive to work up and down the slope. Using machinery across the slope is difficult, especially with row crops like vegetables, as the tractor and machinery tends to slip downhill. Whole rows of cabbages can be accidentally hoed out when the intention was to weed them.

The principle which is most affected by small scale is that of diversity. Some applications of the diversity principle can be used on any scale. For example, a mixture of cereal varieties can be grown in a field of any size. But in general diversity becomes easier the smaller the scale.

The ideal vegetable garden from a biological point of view contains an intimate mixture of different plants, growing in small drifts or mixtures, making maximum use of the advantages of diversity. A home gardener may not mind very much if it takes five minutes instead of two to pick the vegetables for the evening meal. In fact it may be a pleasant diversion after a day spent on other kinds of work.

Commercial growing is quite another thing. Profit margins are paper thin and the main cost is labour, whether the gardener's own or paid. The biggest job is harvesting, and if you get an order for a dozen crates of this and a dozen nets of that you can't afford to wander about gathering a leaf here and a leaf there. You have to get your head down and pick.

Some compromises are possible. Where vegetables are grown on a bed system, each bed may be restricted to a single crop, but with no two adjacent beds growing the same crop. This kind of market garden will reap some of the plant health benefits of diversity, and there may be a slight yield advantage due to the edge effect

Mandy Pullen's organic market garden is worked by hand and the beds are laid out on the contour, with terracing here at the top of the garden where the slope steepens.

This market garden is also organic but somewhat larger, and is worked with a tractor. This makes working up and down the slope the most attractive option.

between beds. But it will take more work to run than a garden where all the beds containing a single crop are placed together.

It will also be less easy to mechanise. Overhead irrigation, for example, will water a block of several beds at a time. If each crop is planted together in a block the irrigation can be applied to the crop that needs it most at any particular time. Where beds are mixed it can only be applied to a mix of crops, some of which will need watering and some of which will not. Even where each crop is planted in a block it's still possible to get some of the plant health benefits of diversity by growing different varieties in adjacent beds (see pp264-265) but this is a far cry from the permacultural ideal.

...and Yield

Small scale growing does not just offer the possibility of greater diversity, it's also inherently more productive.

This may sound surprising in light of the oft-repeated claims that large-scale European and North American agriculture is the most efficient in the world. But that efficiency is very narrowly defined. Large-scale agriculture is efficient in terms of production per person employed on the land – though it doesn't come out so well if you include those employed in the factories which make the machinery, fertilisers and so on, and those employed in food processing and distribution.

However it's extremely inefficient in terms of production per unit of energy input. In terms of energy used on the farm – including fuel, machinery, fertilisers and so on – something like half a calorie of food energy is produced for every calorie of energy used. This compares, for example, with more than forty calories of output per calorie of input by Chinese peasant farmers in the 1930s.[22] (Most crops for direct human consumption yield more energy than they consume, but the average is brought down by the inherent inefficiency of animal production. See pp119-120.)

But what happens on the farm is only part of the story. The distribution system associated with large-scale production is even more energy-inefficient. By the time the food gets to our plates energy has been spent on processing, transport, retailing, driving to the super-market to buy it and cooking it. The net yield is only one-tenth of a calorie of food energy per calorie of energy invested in it.[23]

Large-scale agriculture is also less productive than small-scale growing in terms of yield per square metre. The 1980 World Census on Agriculture carried out by the Food and Agricultural Organisation of the UN found that small farms outyielded larger ones. Typically, farms of between a half and six hectares in extent were around four times as productive as farms of over 15 hectares. In some cases they were 12 times as productive.[24]

The yield advantage applies as much to home gardens as it does to small farms. This was shown in a study of British farms, market gardens and home gardens made in 1956.[25] It found that value of food output per cultivated acre from different kinds of land was:

- good farmland   £ 45
- market gardens   £110
- home gardens   £300

Monetary values were used rather than weight of food so that foods of different quality could be compared. Farm-gate prices were taken for the farm and market garden outputs, and retail value – about twice farm-gate value at the time of the survey – for the home garden output. This reflects the saving in cost, both financial and ecological, of home-grown produce, as well as the lack of waste and the extra value of eating fresh food. Even if the same prices were used, the home gardens would out-produce both farms and market gardens on a per-cultivated-acre basis.

Vegetables and fruit were grown on 14% of the average house plot, the remainder being house, lawn, flowers etc. This gave an output of £42 per house plot acre, which was similar to that of farmland, allowing for uncultivated land on farms. Since house plots covered half the area of new urban development food was grown on 7% of the total area, giving an output of £21 per acre of new townscape.

These figures cannot be used uncritically today. Since the 1950s productivity has certainly gone up faster on farms than it has in home gardens, although this increase has been due to using methods which are completely unsustainable. Houses are now built at much higher densities, and many suburban dwellers would be hard put to devote as much land to food as was possible in 1956. It's interesting to note that 66% of the area of the average house plot in the study was potentially cultivable, though only 14% was down to fruit and vegetables, so the potential for urban food production was hardly being tapped even then.

The general point is inescapable: small areas of ground cultivated by amateur home gardeners produced more food than larger areas cultivated by professional farmers and growers. How can such a thing be? The answer lies in the greater amount of attention which can be given to smaller areas. In a home garden the scale of operation is the individual plant, while on a farm it's the acre.

If gardening is so much more productive than farming, do we need farms at all? Of course we can't transform our economic and social system overnight, so we will certainly need them well into the future. But in the long term the answer is no, we don't need farms. Indeed the present high-energy, low-yield food system can only be kept going by the subsidy of massively under-priced fossil fuels and cannot last indefinitely.

A vision of a sustainable future would include a far greater proportion of people growing at least some of their own food. Many of them would grow all of it and work part time for their other needs, some would grow a surplus for sale and others would grow nothing at all. There is no one model that fits everybody. The landscape would be one of gardens and smallholdings, with larger areas returned to wild vegetation or productive woodland.

There are situations in which small is not beautiful. An example is nut growing in areas with a grey squirrel population. One or two trees on their own will be stripped of nuts by the squirrels, but a larger plantation will only lose a proportion of the crop to them. (See pp222-224.) We could rename this principle Appropriate Scale, but in the vast majority of cases the appropriate scale is much smaller than what is common today.

INPUT-OUTPUT

Linking

As we saw in Chapter 1, the essence of what makes an ecosystem work is the network of beneficial relationships between its components. In order to allow these relationships to happen it's necessary to place things so that the output of one can easily become the input to another. This is linking.

The first stage of the process is a matter of including elements in the design which can make beneficial links. The discussion of poultry in Chapter 9 gives many examples of the links which can be made between chickens and other elements in a farm, smallholding or garden. (See pp248-250.) They can help cultivate the vegetable beds, control pests in the orchard, make use of household food scraps and waste grain, or even heat the greenhouse. If any of these things are included in a design along with chickens then useful links can be made which will help to meet the chickens' needs and make use of the range of outputs they offer. The more these links are made the less the system needs external inputs and the more productive it becomes.

The second stage is to put the elements together in such a way that the relationships happen without too much effort. If this is not done, making the link may be more trouble than it's worth, and the advantages of diversity will be lost. For example, if the chicken run is too far from the orchard it may not be worth the bother of taking them there in winter to pick overwintering pests out of the soil.

An example of linking in an urban industrial context is given in Chapter 6. (See p133.)

Multiple Outputs

This principle states that every plant, animal and structure in a permaculture design should serve as many functions as possible, in other words it should have a multiple output. There are two ways of achieving this:

• Seeing the full potential of plants, animals and structures which we already have or are familiar with.

• Choosing new ones which intrinsically have more than one function.

As we have just seen, chickens are an example of an animal which has many more outputs than they are usually credited with. For every extra output that is realised the value of keeping chickens goes up, and there is little or no loss of the principal yield or increase in the level of inputs. This is largely because the extra yields come from things the birds do naturally whether we make use of them or not. All we need to do is make the link.

A roof can always be seen as a source of fresh water as well as something to keep the house dry, a wall as a place to grow food or paint a mural as well as the structure which holds up the house. How many functions can you think of for a hedge? I came up with eleven, though not all of them would apply to the same hedge at the same time.

An example of a plant which intrinsically has multiple outputs is the false acacia or black locust tree. (See Colour Section, 4.) Originally from North America, it is often planted as an ornamental in Britain. It's the acacia after which so many Acacia Avenues have been named. In addition to being visually attractive it provides: fodder for bees, edible seed for chickens, edible foliage for grazing animals, timber which is durable without chemical preservatives, and nitrogen fixation. (See pp48-49.)

If you want to plant a tree to provide any one of these outputs, including visual beauty, you can choose false acacia and reap one or more of the other benefits as well. The more outputs we take from this one tree the fewer trees we need to plant in order to meet our needs.

Some of the yields may not be wholly compatible. For instance, for optimum timber production trees are felled when still in their youth, when they have many decades of fodder production still ahead of them. (See p220.) In practice the trees will be grown for one primary product and the others will be regarded as a bonus.

In permaculture we're not looking for the maximum yield of a single product but for the maximum yield of the whole system. Battery hens may produce more eggs per bird than hens kept in a permaculture style, but that's all they produce. False acacia trees are not the highest yielding of all the durable timber trees we can grow, but they produce much else besides.

Multiple output can only work in a diverse system, because most of the extra outputs are indirect ones. The chicken's pest control activities yield extra fruit in the orchard, and the false acacia's flowers yield honey. These benefits can only be realised by harvesting another part of the system.

It's a way of getting the most out of what we put in. If we can get two or more outputs from the level of input which would have given us only one, we can get a higher yield off the same area of land, or meet our needs off a smaller area.

There are some things which are worth growing for a single output. Potatoes are an example. Although it is possible to compost the haulm and they can have a beneficial effect on soil structure and weed populations, when it comes down it we grow spuds almost entirely to get a bulk yield of carbohydrate. They are so useful for this that they're worth growing despite their lack of other outputs. There are exceptions to every rule.

ENERGY FLOWS

To some extent energy is seen as the currency of sustainability. There's a fairly close correlation between the level of energy use and overall ecological impact. The vast majority of our energy presently comes from fossil fuels, and whenever fossil fuels are used there is a corresponding amount of pollution produced. In addition, the use and degradation of other resources go hand in hand with energy use. Much of permaculture is centred around reducing energy use to a sustainable level.

Energy In Use and Embodied Energy

In practice there are two forms of energy: energy in use and embodied energy.

• Energy in use is what we normally think of as energy, the wherewithal to move things or raise temperatures. Examples are a lump of coal or the water in a hydro-electric reservoir.

• Embodied energy is the energy which has been used to produce materials. An example is the energy used to make a car, as opposed to the fuel used to run it. Another is the concentration of plant nutrients in a fertile soil. Either human or natural energy, or a combination of the two, has been expended to make that soil more fertile than the average.

Pure energy cannot be cycled. It can only pass through a system, whether this is a natural ecosystem or a human-made agricultural or industrial system. Once it has been used it can never be reconstituted. Our first priority, if we are to reduce our ecological impact, is to use less of it, and the most attractive way of doing this is to use it more efficiently.

When energy is changed from one form to another there are two energy outputs, the main intended output of high quality energy, and an output of less useful energy, usually in the form of low-temperature heat.

For example, when a chicken eats her food the

high-quality energy in the food is converted partly to high-quality energy in the form of body weight, movement and eggs, and partly into low temperature heat. The latter is the body heat of the chicken, and as we have seen that can be used to heat a greenhouse. Another example is the low-temperature heat which goes up the cooling towers when electricity is generated from fossil fuels. This is normally wasted but it can be used for heating buildings. (See pp123 and 125.)

Eventually all the energy in the original chicken food or fossil fuel is lost to the system as very low temperature heat. But in a well designed system we will have got as much work out of it as we can before it leaves.

Embodied energy, on the other hand, can be cycled. In a natural ecosystem almost all of it is. While the energy of the Sun passes straight through an ecosystem, and is eventually radiated back to the cosmos as very low temperature heat, mineral nutrients are largely cycled within the system. A small fraction of the minerals flow out, mainly in rivers, and are replaced either from weathering of rocks or in dust deposited by the wind, but the majority is cycled locally.

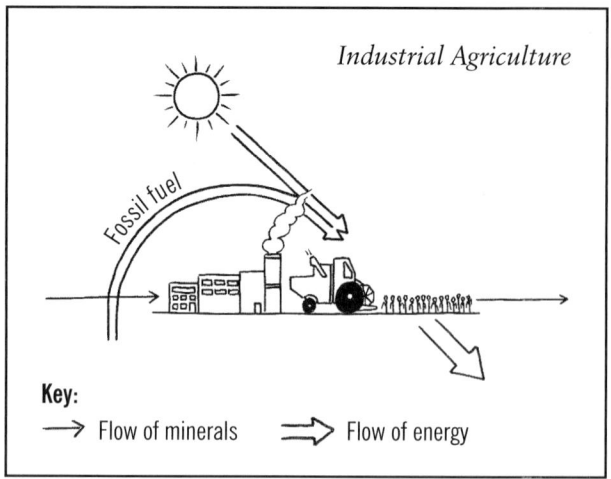

Industrial Agriculture

Key:
⟶ Flow of minerals   ⟹ Flow of energy

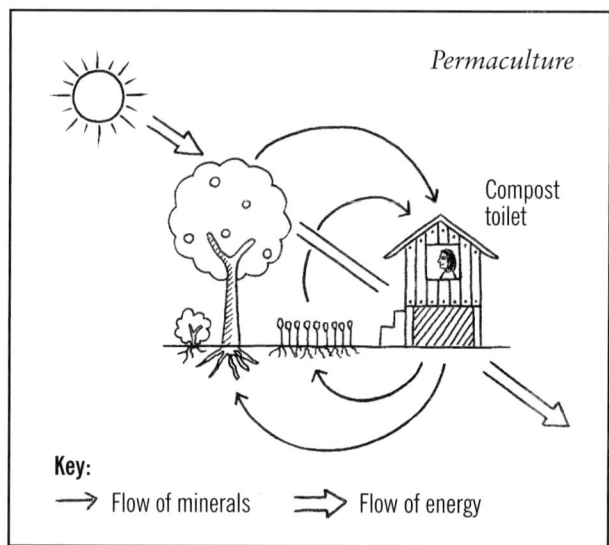

Permaculture

Compost toilet

Key:
⟶ Flow of minerals   ⟹ Flow of energy

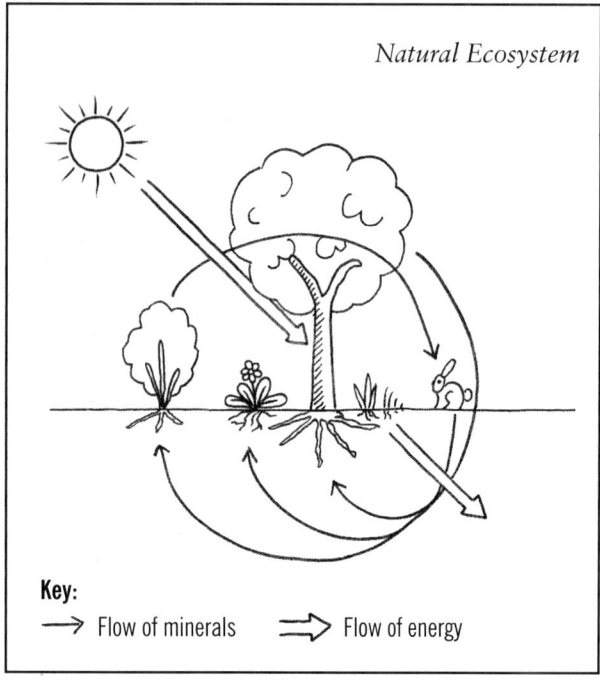

Natural Ecosystem

Key:
⟶ Flow of minerals   ⟹ Flow of energy

In industrial agriculture, not only is the energy flow greatly increased but the nutrient cycle is turned into a straight line flow. The minerals are mined from the ground, or extracted from the air at a great cost in fossil fuel, made into fertilisers, applied to the soil, taken up by plants, eaten by people, flushed down the toilet and disposed of as cheaply as possible. These extra outputs of the system become pollutants. (See box, Entrophication of Water Bodies, p102.)

In a permaculture system the nutrients in human urine and faeces are ideally kept within the system, using hygienic composting technology, and the cyclic nature of the system is restored.

Biological Resources

The notion that energy cannot be recycled can seem a bit daunting, considering the degree to which we rely on finite fossil fuels. Indeed there is no such thing as an infinite supply of energy. But the Sun will go on burning for billions of years more, and on any human time scale we can regard its energy as infinite.

Plants get their energy directly from the Sun, and animals get theirs from eating plants. Any time we use a plant or an animal to perform a task which we would otherwise do by chemical or mechanical means we're using solar energy. Such plants or animals are known as biological resources.

An example is the use of pest-predator attractant plants for pest control in the garden. The adults of the predatory insect species feed on the flowers of certain plants, many of which are culinary herbs, while the larvae eat pests. (See pp242-245.) Frogs eat slugs, and so do ducks. Either one can be introduced into a garden to great advantage where slugs are a problem. (See pp110-111.)

Another example of an animal resource is the 'chicken tractor'. This is not a hundred or so chickens tied up to the front of a plough with someone cracking a whip behind! It's simply chickens doing what comes naturally, pecking and scratching. If they're confined to a specific piece of ground for a short time they will remove the weeds, plus some of the pests, and give it a light dressing of manure. Then the ground only needs a light surface cultivation and it's ready for planting. (See pp248-249.)

In most of these examples the plants and animals have other uses. In fact the uses mentioned here may be subsidiary to the main one of food production. We are simply taking the behaviour of the plant or animal concerned and arranging for it to take place where it's of benefit to ourselves. Chickens will always peck and scratch given the chance, so they might as well do it where it will save us some work.

These examples also involve living plants and animals, rather than dead plant and animal resources. There are advantages to this. For example, growing a hedge for a windbreak rather than erecting a plastic mesh saves the energy and pollution cost of producing the plastic. But the hedge also has a lower ecological cost than a wooden paling, brought from afar at a cost in energy and treated with chemical preservatives. What's more, the hedge replaces itself free of charge, which makes it cheaper in the long run, though it may be more expensive and slower to establish in the first place.

It's early days on the Springfield Community Garden in Bradford, and the plastic mesh windbreak round these beehives gives instant protection from wind on an exposed site. It's a worthwhile stopgap till a living windbreak can be established.

Even so there can be great advantages to using non-living biological resources in place of chemical or mineral resources. Timber, for example, can be 30 times stronger than steel on the basis of strength per unit of embodied energy. Almost all the energy content of the timber was provided free by the Sun. Only a little sawing and transport are needed to make it usable on a building site, compared to the enormous energy input of steel production. Of course much depends on whether the timber is locally grown or imported.

In other words, when we choose a biological resource we are harvesting sunshine. But this is only true when we consume that resource at a rate no greater than it's being produced. If we live off the capital rather than the income, we are not harvesting a resource but mining it. Under these circumstances timber, or any other potentially renewable resource, becomes a non-renewable resource, just like a fossil fuel or mineral. We need to plant at the same rate as we consume. (See box, The Story of the College Roof.)

THE STORY OF THE COLLEGE ROOF

Some years ago, the story goes, the roof of the great hall of one of the colleges at Oxford University needed repair. The college administrators called in a building contractor who specialised in restoring ancient buildings and asked his opinion. He sucked his teeth, shook his head and told them they needed some pretty special oak to make a proper job of it, and he wouldn't know where to get hold of such timber these days.

Now this college had kept much of the landed estates it was originally endowed with in the middle ages, and it still employed a forester to look after the woodland on the estate. The forester was the only person the administrators knew who had any knowledge of timber, so they called him into the office to ask if he knew where such timber might be found.

"Well" said the forester, "When this hall was built, some hundreds of years ago, the builders knew that the roof would need repairing about now. They also knew that the timber you need for this kind of roof takes about the same length of time to grow. So they planted some oaks on the estate. They're ready now."

I don't know whether this story is true in its details or not, and it really doesn't matter. It illustrates a way of working which we can learn much from.

WHOLES

Perhaps the greatest single lesson that the study of ecology has had for humankind is that things work in wholes.

The characteristics of an ecosystem cannot be predicted from adding up the characteristics of its parts. It's an integrated system in which all the parts interact to form a complex whole. In fact it has become clear that the Earth herself works in a very similar fashion, with all her different components interacting in a way that maintains conditions suitable for life. This is the Gaia Theory. (See box, Gaia, overleaf.)

GAIA

Some twenty years ago the scientist and inventor James Lovelock was employed by NASA to design a means of detecting whether life exists on Mars. They envisaged some equipment which they could put on their Mars lander to report back to Earth by radio. Lovelock thought about the problem and realised that you could tell there is no life on Mars by simply looking at the proportion of gasses in its atmosphere. For example Mars has 99% carbon dioxide, compared to 0.03% on Earth.

The Martian atmosphere is at the equilibrium you would expect from the working out of purely chemical processes. The Earth's atmosphere is equally stable, but its makeup, in terms of the proportions of different gasses present, could never persist without the presence of life. It's kept in that state by the biological processes of the living organisms here. That mix of gasses is also one of the conditions which allows life to exist and thrive here. In other words it's the interactions of the living organisms on Earth which make life itself possible.

This applies not only to the atmosphere but to the temperature and all the other conditions which living things require. Lovelock realised that the Earth is something very similar to a single living organism, with the different life forms on the planet interacting in a way similar to the organs in a plant or animal body. He gave the name of the Greek mother goddess, Gaia, to the idea – the being – which he had discovered.

NASA were not pleased, because his discovery took away one of the main reasons for sending a probe to Mars. But Gaia theory had been born.[26]

The fact that everything is connected leads to some fundamental realisations about our relationships with the rest of the system. One of these is:

You can never do one thing

In other words whatever we do has many consequences in addition to the one we intended. A simple action like turning on a tap to fill the bath has consequences all the way from the ecology of the rivers in the place where the water was extracted to the extra load on the sewage system, including the use of energy and chemicals in the water system. On the other hand, planting a tree may have a whole web of beneficial consequences apart from providing us with timber – it may also have some consequences which we consider undesirable, such as casting shade or extracting too much water from the soil.

Another one is:

Everything has multiple causes

We like to ascribe single causes to things which happen. It's much easier on our minds to keep things that simple. But in practice, both in the realm of ecology and in human affairs, events and phenomena are always the result of a complex of causes.

Soil erosion is an example. Climate, weather, soil texture and landform all affect the potential for erosion, while a whole range of agricultural practices can effect the extent to which that potential is realised. Whether individual farmers adopt practices which increase or decrease the likelihood of erosion depends on an equally wide range of political, economic, financial, social, emotional and technical factors. There is no single answer to a problem of this kind, least of all a single technical fix.

Sometimes, in the interests of brevity, we speak of a single important cause of a problem as though it's the only one, and as though all the other causes will remain unchanged if we do something to change this one. At such times it's necessary to have a little voice at the back of our minds reminding us that this is only an approximation of the truth, and possibly a dangerous one.

A third realisation, which comes up when we think of throwing away our rubbish, is:

There is no 'away'

Whatever we do with the pollutants we create they end up somewhere, affecting some part of the huge, interdependent system of which we are part. One day the boomerang will come back and affect us.

Designing In Wholes

The practical implications of this for permaculture design are two:

• Look at the whole system first and last.

• What we think of as the whole system is really part of a bigger system.

I have heard people say things like "Ours is quite a conventional garden, but we've got a little permaculture area down at the end." They are mistaken. The essence of permaculture lies not in the use of individual techniques, but in looking at the system as a whole and seeing how its components can be best designed to enable it to work harmoniously.

To the casual observer a permaculture garden may or may not look different from any other garden. What makes it a permaculture garden is that all its components

– whether they be lawn and flower beds or forest garden and vegetable beds – are placed in such a way that they make the best possible use of the resources present and interact in a harmonious way.

Even where the task in hand is to make a detailed design for one part of the garden only, the way to start is by looking at the garden as a whole, noting the relationships between that part of the garden and the rest of it. These relationships will inform the detailed design work. At the end of the design process it's important to check how the proposed changes will effect the garden as a whole.

But the garden itself is not the whole system; it's part of a household. The household is a useful unit of design. There's a high degree of integration within it and relatively clear boundaries between it and the outside world. It includes the house, the garden and the people within it. We can also include the purchasing patterns and transport choices of the people, as well as their work and other activities.

All of these things are relevant to the design of the garden. The house has resources to offer, including rainwater and grey water for irrigation, organic matter in the form of waste paper, and plant nutrients in the form of urine. These can all become pollution outputs from the system if they're not used within it. The level of human attention received by a home garden means that yields are usually high compared to more extensive agriculture. Fruit and vegetables which are grown in the garden will replace food which would otherwise be bought, and the ecological impact of home-grown food is almost always less than that of purchased food. If the garden is attractive as a recreational space it may reduce the number of car trips made for pleasure.

Of course the household is not the whole system either. Everything we do interacts with the Earth as a whole, and these interactions are constantly in our minds as we practice permaculture design.

CAN PERMACULTURE FEED THE WORLD?

In looking at the principles of permaculture I have often mentioned the levels of yield which can be expected from applying them. There's little doubt that well-designed permaculture systems can yield at least as much as conventional high-input agriculture.

Yet yield per hectare is not necessarily the most important thing. For us as individuals, families or communities the yield per unit of input can be important.

We may have a set amount of land, and we probably don't want to work much harder than we have to – and it usually takes as much work to grow a low-yielding crop as a high-yielding one. So yield per unit area may be important to us personally. But on a world scale we have massive overproduction. There is more than enough food produced in the world now for all its inhabitants. According to the United Nations' World Food Programme, we are already producing one and a half times the amount of food needed to give everyone an adequate diet. People are certainly going hungry, but this is because of unequal distribution, not an absolute lack of food.

Food is exported from poor countries to the rich, often to service debt payments. Much of this food is used to feed farm animals. When we eat the meat, eggs or milk produced from these animals we get the benefit of around ten percent of the nutritional value in the original plant food – such is the inefficiency of an animal-based diet. (See pp119-120.) Many of us suffer from obesity.

We need not be obsessed with yield in the short term. Social justice would do far more for the world's hungry than any increase in agricultural yields, which would only be swallowed up by the rich. But in the medium term the world population is set to rise by a factor which will dwarf the present food surplus. In the future every square metre of cultivated land will have to produce to the full.

The evidence we have suggests that permaculture certainly can meet the forecast needs. A switch to smaller-scale production, which is central to permaculture, would on it's own increase productivity by a massive leap. But to ask whether permaculture can feed the world is really asking the wrong question. Rather we should ask whether the present industrial form of agriculture can feed the world in the future.

The answer is almost certainly no. Firstly, it is dependent on a one-way flow of non-renewable resources. Secondly, it destroys land by soil erosion, desertification and salination of irrigated land. In the next few decades one third of the world's agricultural land may well be degraded to the point where it can no longer produce food.[27]

We have a window of opportunity. We have a chance to change our way of feeding, clothing and housing ourselves to one which is both high-yielding and sustainable. The future can be permanently abundant.

Chapter 3

SOIL

This chapter has two aims. Firstly, it's intended to give the basic information on soil which a practising permaculturist needs to know, or at least have access to. This is equally relevant to gardeners, farmers and those working with the land in any other way. Secondly, for those who already know about the soil, it presents a permaculture perspective, which is very different from the conventional approach and has a somewhat different emphasis from the Organic.

There is nothing on Earth more precious than the soil. It's the mother of all plants, and through them the animals, ourselves and our civilisation. Without soil there would be nothing but the wind blowing over bare rock. With soil all richness is possible. We need to look after it.

The bio-dynamic people describe soil as the meeting place of heavenly and earthly forces, and this is true in a physical sense as well as in a spiritual one. Soil is not a purely mineral substance but a mixture of earth and air: in the soil Heaven and Earth literally interpenetrate each other. Something like a quarter of the volume of a healthy soil may be air and mineral matter may account for less than half.

Water and organic matter are the other constituents. The soil water merges seamlessly into the water of nearby streams and lakes, while the organic matter is part of the web of life which encompasses soil, atmosphere and open water in an ever-cycling flow of energy. Soil water may be around a quarter of the total volume, while organic matter is usually less than ten percent – but its importance far outweighs its modest volume.

These are typical proportions in a fertile soil. A compacted soil will be more than half solid mineral matter; in a waterlogged soil all the pore space is filled with water, excluding air; and a chemically-farmed arable soil may have as little as 1% organic matter in it.

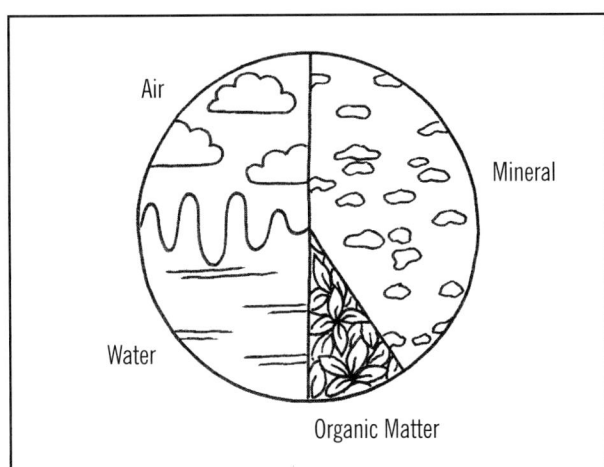

The composition of a fertile soil.

Soil is the meeting place of Earth, Air, Water and Life. But it's more than just a mixture, it's a complex ecosystem in which each component influences and is influenced by every other. Breaking soil down into its constituent parts and looking at each individually is necessary because our minds aren't big enough to grasp the whole thing at once. But it's also necessary to remember that soil is a whole, not just the sum of its parts, and we can never completely understand it by looking at its parts.

Before reading the rest of this chapter, I suggest you go out and make physical contact with some soil. Do one or more of the following:

* Rest your hands on the soil for a minute or two. Feel it, look at it, then bend down and smell it. Dig up a handful of soil. How does it feel?

* Dig a hole, about 30cm deep, or more if the digging is easy. Take care to leave one face of the soil unsmeared by your spade. Does the soil smell different at a lower level? Does it feel different at different levels? What plant or animal life can you see? Are there differences in the colour of the soil?

* Find two or more places with different vegetation (eg garden, woodland, grassland, arable field), preferably near each other. Take a trowelful or more of soil from the top 15cm of each, including any surface litter and plants. How do they differ? Bring them back to where you are reading and have them beside you as you read the chapter.

SOIL FERTILITY

It is not the purpose of people on earth to reduce all soils to perfectly balanced, well-drained, irrigated and mulched market gardens, although this is achievable and necessary on the 4% of the earth we need for our food production... Without poorly drained, naturally deficient, leached, acidic or alkaline sites, many of the plant species on earth would disappear. They have evolved in response to just such difficult conditions, and have specialised to occupy less than perfect soil sites.

Bill Mollison[1]

Very often the quickest way to get an increase in crop yield from a piece of land is to increase the level of soluble plant nutrients in the soil, which can be done most easily by applying chemical fertiliser. This has led many people to equate soil fertility with the level of nutrients in the soil. Nothing could be further from the truth.

Soil fertility is the ability of the soil to support abundant and healthy plant growth. It's influenced by many factors, which are all to some extent interdependent. The main factors of soil fertility are given in the mind map below.

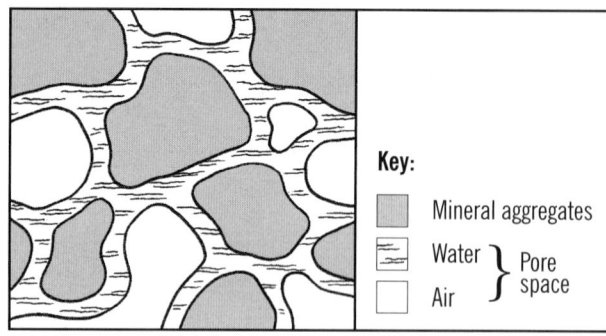

Soil structure.

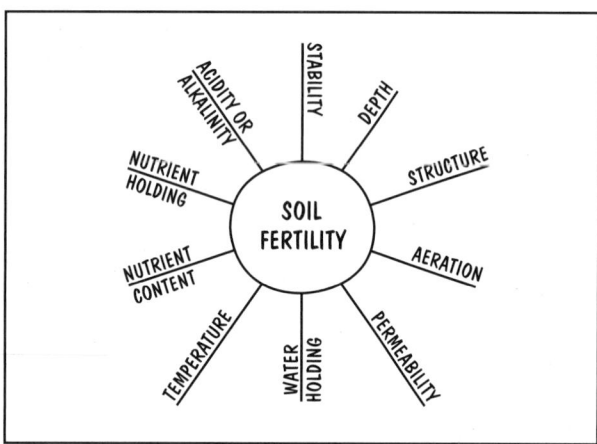

Depth. The deeper the soil the more of it there is for plants to grow in. In a shallow soil they will run out of nutrients and water more quickly than in a deep soil, and large plants such as trees may find insufficient anchorage. A high water table often limits the depth of usable soil more effectively than the presence of bedrock near the surface. Some roots can penetrate the more crumbly kinds of rock but they can't cope with the complete lack of oxygen in a waterlogged layer. However, a high water table can often be amended by drainage.

Structure. The structure of a soil is the way in which the individual soil particles are held together in crumbs, blocks or other aggregates. The spaces between the aggregates are known as pores. In a well structured soil the pores form a continuous network in the soil through which plant roots, air and water can penetrate. (See illustration, Soil Structure, top right.)

Compaction destroys good structure and makes the soil impermeable. It can be caused by working or treading on the soil when it's wet.

Aeration is important because the soil life needs to breathe. This includes plant roots, animals such as earthworms, and the micro-organisms which play

such a vital part in the life of the soil. In addition to compaction, two things which can interfere with aeration are capping and waterlogging. Capping is the formation of a hard crust on the soil surface when rain is followed by dry weather. Waterlogging is when the soil pores are full of water and air is excluded. Excessive aeration is also possible, causing increased loss of organic matter from the soil by oxidation.

Permeablity to water goes hand in hand with aeration. A fertile soil has good structure at the surface to allow water to enter easily, and good structure below to allow it to drain away. A capped soil is less permeable to water, and a poorly drained one soon becomes waterlogged. As well as excluding air, waterlogging leads to a loss of nitrogen from the soil, and inadequate drainage is a very common cause of poor plant growth.

The majority of the land of Britain is potentially subject to bad drainage, and although structural measures such as ditches and pipe drains have a part to play, a well-structured soil is essential to good drainage. Often it's enough on its own.

Water Holding Capacity. A fertile soil is able to hold water like a sponge and make it slowly available to plants over a long time. If water is lost too quickly the plants will soon suffer in a drought. Plants need water not only as a nutrient but also to replace the constant stream of water vapor lost from the leaves to the atmosphere. If the supply from the roots can't keep up the plant suffers moisture stress and growth is reduced, even before obvious signs of wilting are visible.

Temperature. The temperature of the soil is affected by its capacity to hold water. A wet soil takes longer to warm up in spring because more energy is needed to raise the temperature of water than dry mineral matter.

Nutrient Content. The nutrients needed by plants fall into two groups. Carbon, hydrogen and oxygen, are needed in large quantities and come from carbon dioxide in the air and from the soil water. Mineral nutrients are needed in much smaller quantities.

They include nitrogen, phosphorous, potassium and 15 other elements, some of which are only needed in traces (micro-nutrients or trace elements). When we speak of plant nutrients in the soil we mean the mineral nutrients. They are taken up by plants in solution with the soil water, and may be stored in the soil in less soluble forms.

Nutrient Holding Capacity. The ability of a soil to hold on to soluble nutrients is as important as the amount of nutrients it contains. Soluble nutrients may be washed downwards and out of reach of plant roots by water passing through the soil, a process called leaching. A fertile soil resists leaching.

Acidity and Alkalinity are measured on the pH scale:
pH0 = extremely acid
pH7 = neutral
pH14 = extremely alkaline
A pH of between 6 and 7.2, ideally around 6.5, is appropriate in most British soils. Outside of this range plant nutrients can be locked up in unavailable forms and soil organisms become inactive or die out.

Stability. A stable soil is able to resist erosion. Good soil structure enables water to infiltrate into the soil rather than flow along the surface, taking soil with it. Many other factors influence erosion, but soil structure is absolutely fundamental.

Soil Texture

Texture is a measure of the size of the individual mineral particles in the soil, as distinct from structure, which is about how the individual particles are joined together to form aggregates. Leaving aside stones, sand are the biggest particles, clay the smallest and silt are intermediate.

Each of these has its distinctive strengths and weaknesses relative to the different factors of soil fertility. The characteristics of each are given in the table, Soil Texture Characteristics. All soils are a mixture of all three, and the characteristics of the soil will be strongly influenced by which predominates.

The clay fraction is important in all soils, because clay particles have a unique chemical nature which is not shared by sand and silt particles. They have an electrical charge which enables them to hold onto nutrients, and, in the presence of calcium, to each other. This ability to stick together, or flocculate, leads to the formation of soil aggregates, and thus to soil structure. The supply of calcium to aid structure is one reason for liming the soil.

Clay soils are difficult but rewarding. Raw clays are hard to work with, often changing straight from an unworkable goo to concrete-like hardness in the spring. But with their high nutrient content they can be the most fertile of soils if a good structure can be developed and maintained. The key to this, in addition

SOIL TEXTURE CHARACTERISTICS			
	Sand	Silt	Clay
Structure	Structure is less important in sandier soils. Not prone to compaction	Needs structure, but silt particles cannot form aggregates. Prone to compaction	Needs structure. Clay particles are able to form aggregates. Prone to compaction
Aeration	Very good, can be excessive	Very poor without good structure	Poor without good structure
Permeability	Very permeable, drainage usually good	Very poor without good structure. Especially prone to capping	Poor without good structure. Drainage is slow
Water Holding	Usually poor, prone to drought	Excessive. Waterlogging a problem	Good
Temperature	Warms quickly in spring	Slow to warm in spring	Slow to warm in spring
Nutrient Content	Low	Low	Can be high, depending on the kind of clay minerals
Nutrient Holding	None. Leaching a problem	None	Clay can hold nutrients against leaching
pH	Usually acid	Variable	Usually alkaline
Stability	Fine sand erodible, coarse sand less so	Prone to erosion	Resistant to erosion only if well structured

to liming, is adding as much organic matter as possible over a number of years.

Sandy soils are often preferred for gardening because they warm up quickly and can be cultivated almost whenever you like without fear of compaction. But nutrients and water holding capacity must be built up, again by adding organic matter. They are often called light soils, because they are easy to cultivate, in contrast to silty and clay soils which are known as heavy soils.

Silty soils may be fertile but unless they have a sufficient proportion of clay in them they are difficult to work and very prone to problems of poor structure.

A loam is a soil with a good mixture of sand silt and clay, any two or all three. Loams combine many of the advantages of the different textures, and minimise the disadvantages, particularly in their water relations. They hold more water than sands, and so are less vulnerable to drought. Although they hold less total water than clays they hold it at lower pressures, so a higher proportion of it is available to plants. Thus a loam may hold as much or even more useful water than a clay, but have a lower total water content, making it a warmer soil, and reducing drainage and compaction problems. Most of the soils in Britain are loams of one kind or another.

Organic Matter

Soil organic matter is the dead remains of plants, and to a lesser extent of animals and their manure. It includes a whole range of materials, from freshly fallen litter through partially decomposed material to humus. Humus is a relatively stable, black, jelly-like substance, which remains in the soil for a number of years before finally oxidising to carbon dioxide and water. The average turnover time for organic matter in British soils is around 18 years, though humus thousands of years old has been detected. A healthy soil needs organic matter in all its stages of decomposition.

The remarkable thing about organic matter is that it can help correct almost all of the imperfections of both heavy and light soils.

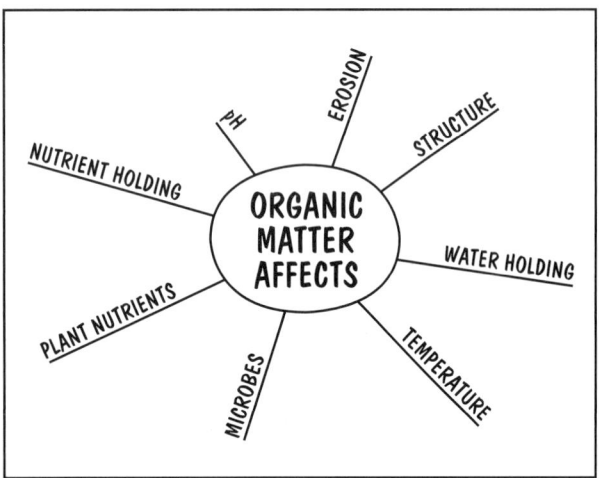

Structure. Organic matter is absolutely crucial to soil structure, and thus to aeration and permeability. Humus shares some of the chemical properties of clay and greatly increases the formation of aggregates. Less-decomposed organic material is food for soil life, particularly earthworms. Heavy soils are less prone to compaction when they contain higher levels of organic matter.

Water Holding Capacity of soils is increased both by the improvement in soil structure and by the sponge-like nature of humus. Humus can also increase the availability of water in heavy soils by increasing the proportion of larger pores, which hold water at lower pressures than smaller ones.

Temperature. As organic matter is dark it absorbs light energy and can help to warm the soil in the spring. But the improved drainage due to better structure probably has more effect.

Microbes. Organic matter provides food for the microbes and other soil life which are a vital part of the soil ecosystem.

Plant Nutrients. It contains mineral nutrients, though nutrient content varies according to the source of the material. In fact most soils contain enough mineral nutrients to grow crops for hundreds of years. The rate at which they become available to plants is the limiting factor. Microbial activity is what makes them available, and our main means of encouraging this is by feeding the microbes and ensuring good soil structure, both of which can be done by adding organic matter.

Nutrient Holding Capacity. Humus has between twice and ten times the capacity to hold on to plant nutrients that clays have. So even a relatively small increase in organic matter can make a big difference to the soil's nutrient holding capacity.

pH is the one factor which organic matter tends to worsen, because the products of decomposition are slightly acid. There is at present no biological alternative to liming for raising pH. Limestone, chalk, shell sand and calcified seaweed are the sources of the calcium carbonate used for liming. The first two are non-renewable resources, and so are the latter two at the rate they are currently being used. Under natural conditions falling pH is countered by earthworms and deep rooted plants bringing up calcium from deeper layers.

Erosion. By improving structure and permeability, a high organic matter content enables even a bare soil to resist erosion.

I don't believe in panaceas, but if ever anything approached being a true panacea it's organic matter in soil.

The proportion of organic matter in different soils varies widely. A soil covered with long-term perennial vegetation such as permanent pasture usually has around 5-10%, while in some arable soils it can be as low as 1-2%. Temporary perennial vegetation, such as a grass and clover ley grown as part of an arable rotation, is intermediate. A level of around 3.5% is required for stable soil structure in most soils. The organic matter content of garden soils is extremely variable, depending on how they have been treated.

The two main influences on organic matter content are the rate at which it's added and the rate at which it's removed. In rotational farming the greatest source of addition is from the constant turnover of grass and clover roots during the ley phase of the rotation. As the roots die they rot in situ and add organic matter evenly throughout the rooting depth. The addition of manure never amounts to as much.

Some organic market gardeners include a ley in their rotation. It takes some of the land out of production for a while, but if the land can be spared it's worth it for the gain in soil fertility. Short-term green manures, which are cultivated in when they are green and sappy, can play a part, but they do not add to the long-term organic matter level of the soil.

With home gardens the situation is different. The area is small enough that it's usually possible to get hold of enough manure, compost and other organic materials to significantly raise the organic matter content over time, and there's rarely enough space to spare for a ley course in the vegetable rotation. (Lawns don't often accumulate much organic matter in the soil. The top growth is kept so short that the grass roots never grow big enough to contribute a significant amount.)

Home gardeners should not be shy about importing any organic matter they can lay their hands on into their gardens. Although local self-reliance is an important aim, zone I has been a net importer of fertility throughout history and there's nothing inherently unsustainable about this. In towns and cities there's usually the opportunity to take in unused outputs from other parts of the urban system, such as leaves and lawn mowings from the local council. In many areas it's possible to buy soil conditioner made from composted municipal waste quite cheaply, and in some areas it's free.

However, in the long run, the organic matter level in the soil is more affected by the rate of loss than by the rate of addition. In other words it's easier to lose it than to get it back again. It's lost mainly by the process of oxidation, and anything which slows down or speeds up the rate of oxidation has an effect. Cultivation speeds up the oxidation of organic matter by exposing the soil to excessive air.

The best illustration of this is the much higher level of organic matter in the soil under permanent pasture than in a well-farmed arable soil. Even a long-established organic arable farm, using rotations and returning manure to the soil, typically has an organic matter of around 3.5% in its soil,[2] whereas a permanent pasture,

as we have just seen, may average twice that. It's the oxidation of organic matter whenever the soil is ploughed, rather than the amount of organic matter which is added, which puts a limit on the long-term level of humus in the soil.

The same is true in a garden. However much organic matter is added it will be lost if the garden is frequently dug. Digging can be the best way to add organic matter to a raw soil. But once the soil is in good heart it's much better not to disturb it.

Composting

Used pallets make a good compost bin. The Drift Permaculture Project, Newcastle-upon-Tyne.

An important question about organic matter is, 'Does it need to be composted before being added to the soil, or could the same results be had by using the available organic matter in an uncomposted state?' The reasons for composting are summarised in mind map below.

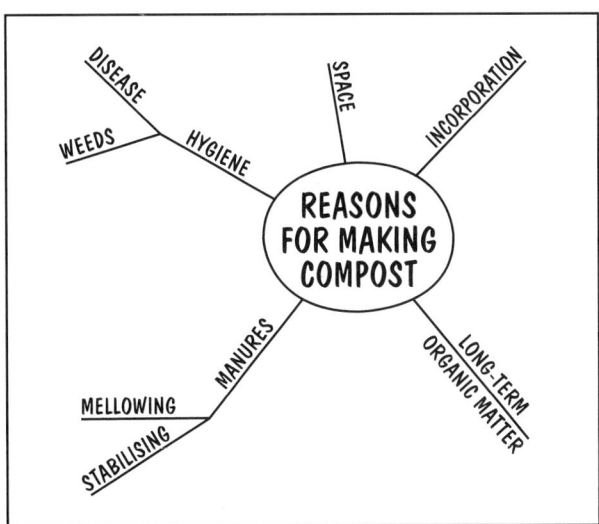

Incorporation. If undecomposed organic matter is dug into the soil the soil bacteria immediately start breaking it down, and in order to do this they need nitrogen. A young, sappy green manure crop has enough nitrogen in it. But to decompose more mature organic matter, such as straw, the bacteria must draw on the soil's supply of that nutrient. They are more efficient at extracting nitrogen from the soil than plants are, so the plants suffer a temporary deficiency of nitrogen. Eventually, when the decomposition is complete, the bacteria die back to their previous population and the nitrogen becomes available to plants. But for a season or more you get stunted little plants with yellow leaves.

If the organic matter is simply placed on the surface of the soil, it's taken down bit by bit by earthworms. There's never so much raw organic matter in the soil at one time that the nitrogen gets seriously depleted.

This surface mulching of uncomposted material is an alternative to making compost. Apart from saving a great deal of work, both in making compost and incorporating it into the soil, there are some positive reasons for choosing this method of recycling organic material to the soil.

Most of the soil microbes live very near the soil surface, and they need feeding. The process by which they break their food down is central to all the processes in the soil. In particular it appears to be associated with the process by which nutrients stored in the mineral fraction of the soil become more available to plants. It mainly takes place here, at the edge between soil and air.

Another reason for placing uncomposted material on the surface is that earthworms love it. The deep burrowing earthworms get most of their food from the surface and drag it down into their burrows. Look at a woodland floor in autumn or early winter. Where worms are active you will see fallen leaves in the process of disappearing down their holes. On farmland, applying manure to the surface appears to promote more earthworm activity than incorporating it into the soil.[3]

Long-Term Organic Matter. There is some evidence to show that composted organic matter has a greater effect on increasing the long-term humus level in the soil than uncomposted.[4]

Manures. The mineral nutrients in uncomposted manures are more soluble and thus more available to plants in the short term. Those in composted manures are less available in the year of application but are held against leaching and are available over a longer period of time. The decision whether to compost manure or not can depend on whether you want a quick boost for the current crop or a long-term increase in nutrient status and organic matter.

The soluble nutrients in raw animal manure are caustic and will 'burn' plants, so manures are hard to use in a garden if they're not composted first.

In composting, faeces and urine are combined with high-carbon material such as straw, which both stabilises them against loss and mellows them so they will not harm the plants.

Human urine is a valuable source of nutrients for the garden. It's very caustic, and can only be used either as a very dilute liquid manure (mixed with water at 20:1)

or when composted with high-carbon material. As a liquid manure it acts in much the same way as a chemical fertiliser, bypassing and to some extent disrupting the biological processes of the soil. So it only has a limited application in organic growing in this form.

In order to compost urine large amounts of high-carbon material such as straw or cardboard are needed. (See box, The Carbon:Nitrogen Balance.) In most gardens the supply of high-carbon materials will be the limiting factor to the amount of urine which can be used. An efficient method of composting urine is described in Chapter 5. (See pp105-106.)

THE CARBON:NITROGEN BALANCE

Humus in the soil, or in mature compost, contains carbon and nitrogen in the ratio 12:1. This ratio is remarkably constant, regardless of what materials the humus was originally made from.

The ideal carbon to nitrogen ratio in the fresh ingredients of a compost heap is 25:1, as much of the carbon is lost through respiration during the composting process. If there's less carbon than this the surplus nitrogen will be lost as ammonia gas. If there's less nitrogen the composting process will slow down or stop.

This means it's difficult to make use of materials which are high in either carbon or nitrogen without combining them with something else. Adding urine to sawdust is a way of making useful humus out of two materials which are hard to use on their own.

Even where less extreme materials are used for composting it's necessary to have a balance of high-carbon and high-nitrogen materials in the heap. Farmyard manure, which is a mixture of straw, faeces and urine, has a good balance. Most domestic compost heaps, composed largely of kitchen scraps, lawn mowings and young weeds, are very short of carbon. Paper and card can be added to redress the balance, but on acid soils too much paper could possibly lead to problems as it contains aluminium and this can cause lockup of phosphorous in the soil. (See box, Is Paper Mulch Safe?, p197.)

Hygiene. A really hot compost heap can kill weed seeds, though not the roots of tough perennial weeds. Disease organisms may also be killed, but this can't be relied on, and diseased material is best disposed of elsewhere or burnt.

Space. In an intensive garden where all the space is taken up with crops, there may be no space to spread large quantities of unrotted organic matter on the soil surface as a mulch. It can be placed between the plants, but then it can become a haven for slugs. A mulch of well-made compost does not harbour slugs.

Choosing a Composting System

Broadly speaking there are four ways of handling the organic matter input to a garden:

- hot compost
- cold compost
- worm compost
- uncomposted mulch

Hot compost is the deluxe method. It's the only way to kill weed seeds, it produces a first class product, and it's quick – it can take as little as two months in summer. But making successful hot compost is an exacting task. The requirements for success are summarised in the mind map below.

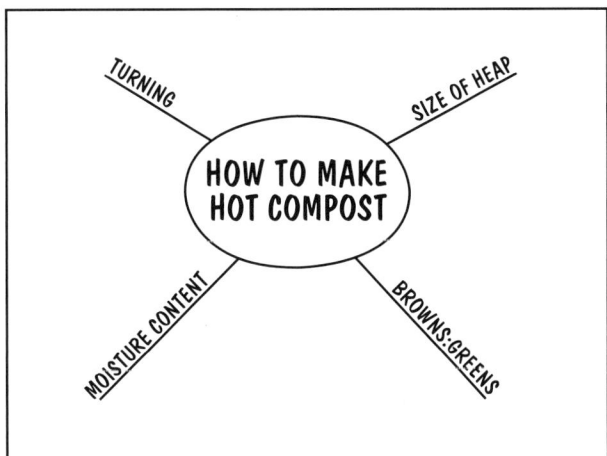

A small heap has a high ratio of edge to volume so it loses heat more quickly than a larger heap and never reaches a high temperature. One cubic metre is the minimum volume for a hot heap, and all the material needs to be fresh at the time the heap is made. If you can't get this much material together at one time don't even think of making hot compost.

Browns are fibrous materials, like paper and mature plant material. Greens are soft materials, like kitchen scraps and young plant material. It's essential to get a mixture of browns and greens for two reasons. Firstly, browns give a structure to the mix which allows air in and keeps the heap aerobic. Without them it becomes a slimy and smelly mess. Secondly, browns are high in carbon and greens are high in nitrogen, and a balance between the two gives a good carbon to nitrogen ratio. (See box, Common Compost Ingredients.)

If the heap is too wet air will be excluded; if it's too dry the micro-organisms won't have enough water. Moist but not wet is the key.

A good heap will heat up in a few days. When it cools down again it needs turning. There are two reasons for this. Firstly, the action of turning reintroduces air and gets the aerobic bacteria going again. Secondly, the edges of the heap, which didn't heat up first time round, must be turned into the centre of the new heap.

Cold compost is much less demanding. Material can be added as it becomes available and the heap need not be turned. But it is still essential to get a good mix of greens and browns, and the moisture content needs to be right. An advantage of cold composting is that the losses of nitrogen which inevitably happen during the composting process tend to be less than in hot composting.

Worm composting requires care and attention. But it's the one kind of composting which can be done with greens alone. Only a small quantity of material can be added to the bin each day, or the worms get overwhelmed. The bin can be very small and fits easily into a crowded urban environment.

The advantages and disadvantages of mulching uncomposted material have been discussed above, at the beginning of this section.

To sum up:

- Hot composting is ideal on farms and market gardens, but very few domestic gardens can accumulate a cubic metre of material all ready to go at one time. Browns can be stored in advance but greens will start to rot anaerobically if stored.

- Cold composting makes good use of the materials available in a typical household. The biggest source of carbon in most households is used paper and card. The biggest source of nitrogen and other nutrients is urine. (Comparatively little of our nutrient output is in faeces.) The rest is mostly greens: kitchen scraps, weeds and lawn mowings.

COMMON COMPOST INGREDIENTS

Greens	Browns
Animal faeces	Autumn leaves
Comfrey	Bracken[5]
Feathers	Cardboard
Green, sappy prunings	Mature grass*
Kitchen scraps	Mature weeds*
Lawn mowings	Mature nettles*
Urine	Paper
Seaweed	Sawdust**
Young nettles	Straw
Young weeds	Woody prunings, shredded**

* Hot heap needed to kill seeds
** Very slow to decompose, best composted separately with only urine added

- Worm composting is ideal for a household with only a balcony or a small patio for gardening. It takes up little space and is ideal for making use of a constant small input of kitchen scraps and not much else.

- Mulching uncomposted material is the most laid back method. It requires the least work of all and is suited to a garden of perennial vegetables and fruit, where there are few young plants which would be vulnerable to slugs. It can't make use of urine, but this can be handled separately. (See Manures on p44.)

Soil Life

The soil is teeming with life. In a teaspoon full of *healthy* soil there are more living beings than there

---

MY COMPOST HEAP

I make cold compost in an old coal bunker which was made redundant long ago when the house was converted to gas.

Almost all the organic waste materials from the household go into it: kitchen and garden waste, both of which are mostly greens, paper and cardboard, which make up the majority of the browns, and urine. The only things excluded are perennial weeds, annual weeds which have seeded, and anything woody. Wood is very slow indeed to decompose and locks up nitrogen in the process.

I fill it with alternate layers of greens and browns, so as to keep the heap reasonably well aerated. A layer of browns goes in – the paper scrunched up, the cardboard shredded – and is soaked with urine. Greens are added till the browns are more or less covered, then it's time for another layer of browns. The supply of urine is always more than the heap can take, and the easiest mistake is to use too much,

leading to a wet, anaerobic heap, especially in winter. Good drainage at the bottom is essential.

It takes two years to fill the bunker. Then I take off the unrotted top of the heap, which goes back in to start the next batch. The top half of what remains is rough compost, suitable for surface application. The bottom half is of excellent quality, though I say it myself, suitable for digging into the soil or adding to a potting mixture.

The soil was low in organic matter when we moved in, and I have no compunction about bringing in compost from outside to supplement what I make. This includes donkey manure from a neighbour and municipal compost from the County Council recycling centre. But once the soil is in good heart it should be able to keep going on home-made compost. Of course even this is not all home-grown, as all the paper and much of our food comes in from outside.

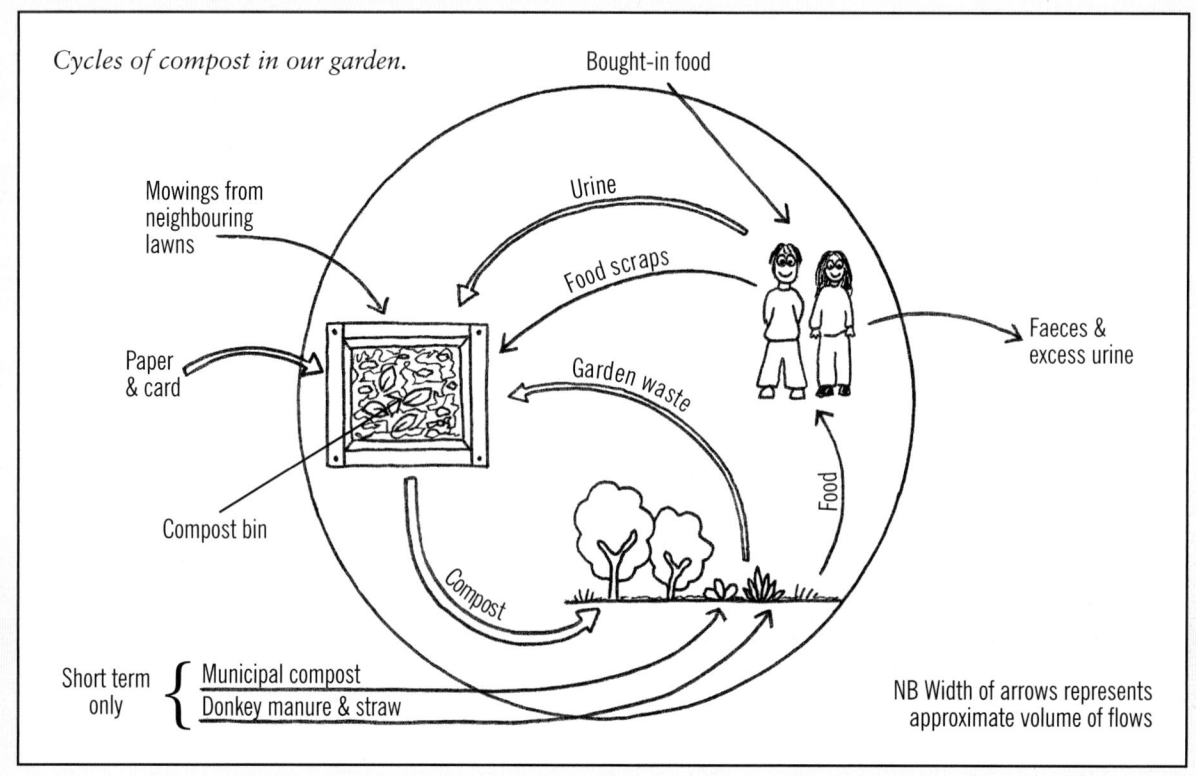

Cycles of compost in our garden.

Bought-in food

Mowings from neighbouring lawns

Urine

Food scraps

Paper & card

Garden waste

Faeces & excess urine

Food

Compost bin

Compost

Short term only { Municipal compost / Donkey manure & straw

NB Width of arrows represents approximate volume of flows

are humans on Earth. Most of these are microbes, including fungi, bacteria, algae and protozoa. They are absolutely central to the processes of the soil. They:

- Decompose raw organic matter into humus.

- Make mineral nutrients available to plants.

- Contribute to soil structure by producing gums which stick soil particles together.

- Fix nitrogen, carbon and sulphur from the air.

- When present in a natural balance, promote plant health by preventing excessive growth of pathogens.

Microbes are mainly active in the rhizosphere. This is the edge between plant roots and the soil solution, usually extending some three millimetres from the roots. The roots exude organic substances into the rhizosphere which encourage and feed the microbes. Roots of perennial plants are more effective in this respect than those of annuals.[6] This is one of the reasons why perennial plants such as grass leys have such an improving effect on soil structure.

The larger creatures include a whole range of invertebrates including nematodes, springtails, worms and bumble bees in their nests, and even mammals such as moles. All of these play a part in the vital soil processes, cycling nutrients to plants and maintaining structure. But pride of place goes to the earthworm.

Earthworms

I have made much of the importance of organic matter in the soil, but in order to be effective it needs to be intimately mixed with the mineral soil particles. In a temperate soil the vast majority of this mixing is done by earthworms, and it's hard to exaggerate their importance. They take organic matter into their guts along with mineral particles, and what comes out at the back end is an intimate mixture of clay, lime and digested organic matter, the perfect soil aggregate. In a single year as much as 40 tonnes of earth (dry weight) can pass through the collective guts of a healthy worm population on one hectare. This is equivalent to 0.5cm of soil depth. They also enrich the topsoil with nutrients from the subsoil, and their burrows enhance drainage, aeration and root penetration.

The main thing which encourages a high worm population is a good supply of organic matter, including some fresh material on the soil surface. They also need a pH of at least 4.5, preferably over 5. Some chemical fertilisers and pesticides can kill them, and worm numbers are higher in organically farmed or gardened soil than where chemicals are used. But tilling the soil kills far more worms than chemicals ever do. The effect of tilling on earthworm populations is discussed later in this chapter. (See p57.)

The New Zealand flatworm is a predator of earthworms. It has become established in Britain over the past few decades, and is slowly spreading across the country. At first it appeared to exterminate earthworms wherever it occurs, but now it seems that at least some earthworms have survived in some of the places where flatworms have been longest established. The degree of wipeout seems to be worse in wetter areas.

Nevertheless finding some biological control for the flatworm would appear to be crucial to the future of sustainable food production in this country. Ground and rove beetles kill them, and these may be responsible for the survival of earthworms in infected areas.[7]

The presence of moles is an indicator of a high earthworm population, as their main food is earthworms.

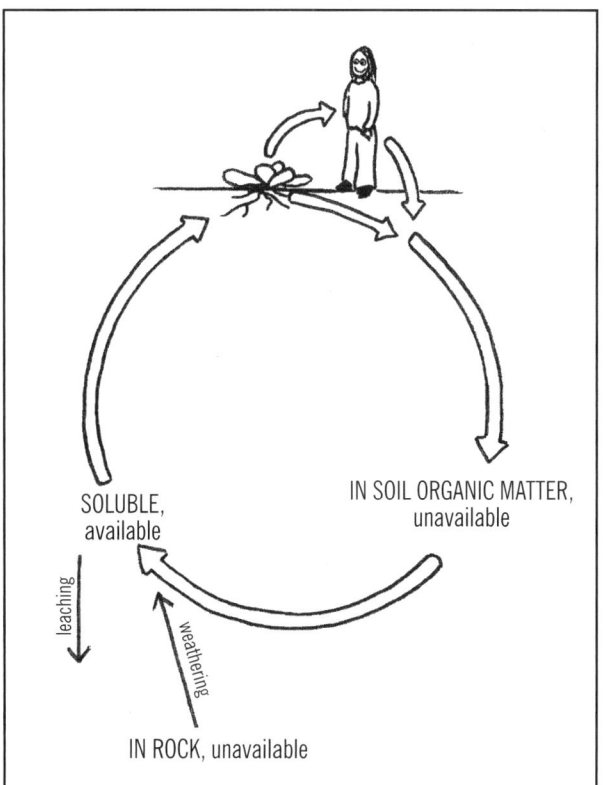

The general plant nutrient cycle, simplified.

Plant Nutrients

Microbes and earthworms are largely responsible for nutrient cycling as well as soil structure. By breaking down organic matter into simpler substances they make the nutrients in it available to plants. There is also a store of nutrients in the mineral fraction of the soil, and the organic acids released in the decomposition process play a part in making these nutrients more available.

Nitrogen is different from the other nutrients in that it's not stored in rocks or in the mineral particles of the soil, but mainly in the air, which is 78% nitrogen. What nitrogen is stored in the soil is in the organic matter. Losses in the system are made good by certain microbes which can 'fix' nitrogen from the air. Some of these are free-living in the soil and others live in a symbiotic relationship with certain plants. Nitrogen can be lost from the soil by leaching and by denitrification. The latter is a process of conversion back to gaseous forms, and can be a serious problem if the soil is waterlogged during the growing season.

Runner beans are sensitive to drought and appreciate plenty of organic matter in the soil beneath them. As they are nitrogen fixers they can handle large amounts of high-carbon organic matter, and pure newspaper can be dug into a trench before sowing or planting out.

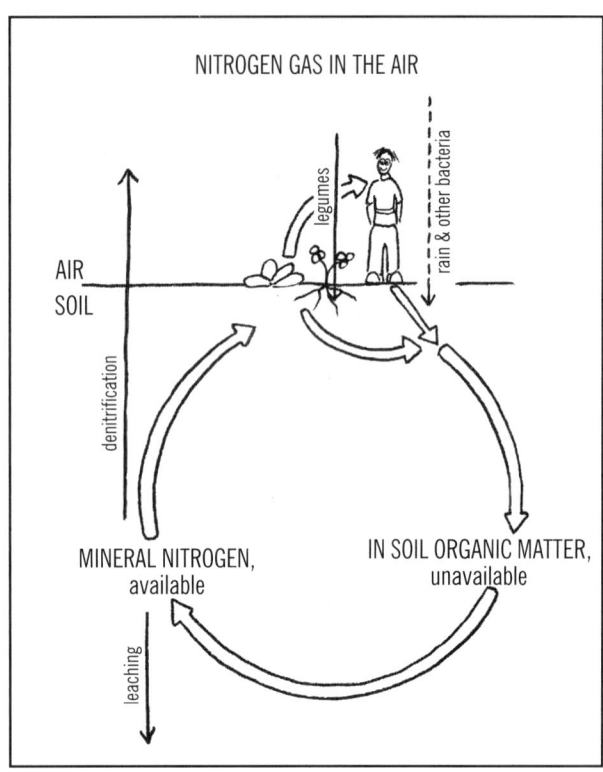

The nitrogen cycle, simplified.

Nitrogen Fixers[8]

Nitrogen-fixing bacteria of the genus *Rhizobium* live in nodules on the roots of plants of the legume family. The legumes include clovers, peas, beans, some shrubs such as gorse and broom, and trees, although no trees native to Britain. Another group of plants, known as actinorhizal, have a similar relationship with microbes of the *Frankia* genus. These include alders, bog myrtle and *Eleagnus* species.

The *Rhizobium* bacteria are somewhat specific in

their choice of host. The legumes can be divided into groups on the basis of which strains of *Rhizobium* they can combine with. A number of different strains can form nodules on the roots of any member of a group, though some of these strains are much better at fixing nitrogen than others. Where there are none of the right strains present it's necessary to inoculate the seed with a compatible strain. This is most likely to be necessary when no member of the relevant group has been grown on the site in question before. It's usually considered necessary when lucerne (alfalfa) is grown in this country, but never, for example, for clovers. Both native and non-native plants can belong to the same group, so some exotic plants may be able to find the right bacteria without inoculation.[9]

There are also some legumes which never form nodules and thus do not fix nitrogen. The honey locust tree is an example.

The actinorhizal plants are less specific in the strains of *Frankia* which they will combine with, and inoculation should never be necessary for these plants.

In nature many nitrogen fixing plants grow mainly in soils which are deficient in nitrogen. These can be infertile, sandy soils, for example most woody legumes and *Eleagnus*, or very wet soils, for example common alder and bog myrtle. They are often pioneering plants, colonising areas which have lost their plant cover in the recent past and are thus low in nitrogen because they are low in organic matter.

In cultivation nitrogen fixers can be used in similar situations, such as reclaiming coal tips. But they are more often used to add nitrogen to productive cultivated areas in gardens, farms and, occasionally, tree plantations.

Other plants can benefit from the nitrogen fixed by these plants in three ways.

Firstly, they can absorb nitrogen directly from the nodules in the soil when some of the legume's roots die. This happens on a large scale when the plant suffers some trauma, such as defoliation by a grazing animal or an increase in shading. It happens on a smaller scale due to the regular turnover of the roots of perennial nitrogen fixers. But most of the nitrogen fixed by annual legumes, such as beans, ends up in the seed of the legumes. Typically some 90% is in the seed and only 10% remains for the following crop.

Secondly, they can benefit from the extra nitrogen in the legume's body when it dies or sheds its leaves and these are decomposed in the soil. The neighbouring plants stand to gain most if the legume is prevented from setting seed.

Thirdly, nitrogen and other nutrients can be passed from one plant to another by means of mycorrhizas. These are combined structures made up of the roots of plants and the bodies of fungi. (See pp56-57.)

Thus a legume which lives out its life without a setback and produces a good crop of seed will only provide a little nitrogen for other plants. It will benefit them more if its growth is checked in some way. This is what happens when sheep or cattle graze a pasture of mixed grass and clover, when a gardener digs in a green manure crop, or when an alder tree is coppiced. This fact must be borne in mind when designing any cultivated system in which these plants are included as a source of nitrogen for other plants.

The free-living nitrogen-fixing microbes, those which don't have a symbiotic relationship with plants, cannot fix anything like the amount of nitrogen which Rhizobium and Frankia can, but they are constantly present in a healthy soil. They can be encouraged by all the practices which lead to a generally healthy soil.

Dynamic Accumulators[11]

The other nutrients, which are stored in the ground in mineral form, can be made more available by growing dynamic accumulators. These are plants which have a particular ability to take up certain nutrients from the soil. These nutrients become available to other plants when the dynamic accumulators die and decompose. They can't perform miracles on soils which are very deficient in a particular nutrient, but they have an important part to play in increasing the general availability of nutrients in most soils.

SOME USEFUL NITROGEN FIXERS					
Plant	Form	Soil	Climate	Shade Tolerance	Other Uses
Common alder	tree	wet to moist	moist if soil is not wet	moderate	low-quality wood
Italian alder	tree	moist to dry	tolerant	moderate	low-quality wood
Grey alder	tree	wet to dry	very cold tolerant	moderate	low-quality wood
False acacia	tree, suckering	poor, sandy	warm	no	durable timber; bee, chicken and ruminant forage
Siberian pea & bladder senna	small tree	poor, sandy	ideally, cold dry winter & hot summer	no	chicken forage
Eleagnus, various species	trees & shrubs	poor, dry	various	moderate to very	fruit of moderate quality
Gorse	shrub	poor, dry	tolerant	no	low windbreak
Broom	shrub	dry, acid	tolerant	no	decorative
Lucerne	p. tall herb.	fertile, well drained	tolerates drought	no	very high fodder production
White clover	p. low herb.	most	tolerant	no	fodder, ground cover
Alsike clover	p. herb.	wet or acid	wet &/or cold	no	fodder
Field beans	a. herb.	most	tolerant	no	grain

p = perennial   a = annual   herb. = herbaceous

Notes under 'Soil' indicate preferred conditions, all will tolerate a wider range of conditions.

A wide range of green manures for garden use, many of which are nitrogen-fixing, are described in good seed catalogues.[10]

Including them in grassland is the natural way of providing grazing animals with the balance of micro-nutrients which they need.

Foremost among them is comfrey, which is famous as an accumulator of potassium. It also accumulates calcium, iron and a little phosphorous and produces a high yield of organic matter with very little care. It does need a high level of nitrogen to give of its best, and this nitrogen is also accumulated and passed on to whatever plants are fertilised with its leaves. It's almost impossible to remove once established, though, so a comfrey patch must be sited with care. Russian comfrey should always be used in preference to the native common comfrey, as it's more productive and doesn't spread by seeding as the native species does. Neither spreads by its roots.

It's normally grown in gardens, but grazing animals will eat it if brought up on a diet containing it, and there is certainly a potential for using it on farms and smallholdings. The recommended garden variety is Bocking 14, and for animal fodder, Bocking 4. Chicory is another accumulator of potassium.

Phosphorous is accumulated by many perennial herbaceous legumes, including clovers, vetches and lucerne, and also by chicory. Mustards and buckwheat are two annual green manures which accumulate phosphate, and buckwheat also accumulates calcium.

Minor and trace elements are accumulated by a whole range of deep-rooted perennial herbs. Chicory, ribwort plantain, yarrow, salad burnet, sheep's parsley, catsear and caraway are commonly included by organic farmers in leys to increase the mineral content of the herbage – and thus ultimately the soil via the animals' dung. They can all be used in the garden and added to the compost heap or mulch.[12]

In general any big, leafy, richly green plant will be extracting plenty of nutrients and can be cut for compost and mulch. Some of these are not strictly dynamic accumulators, as they grow where nutrients are abundant rather than being specially efficient at extracting nutrients from an average soil. Nettles, which only grow where there is an unusually high concentration of phosphate in the soil, are an example.

Families who are largely self-sufficient in food from their own holding need to be aware of the possibility of trace element (micro-nutrient) deficiency. Every soil has its own mix of nutrient minerals, and the balance in any particular soil may not be the same as the balance required by humans. Those of us who eat a diet brought in from many sources are unlikely to suffer any deficiency.

TO TILL OR NOT TO TILL?[13]

Tilling means cultivating the soil, in the sense of moving it with a plough, spade or other implement. We have already had a look at the permaculture attitude to tilling in Chapter 2. (See p14.) Here we look in more detail at the reasons why we may or may not choose to till the soil. Some specific no-till methods for garden and farm are described in Chapters 8 and 10.

There's a whole range of different ways of tilling the soil, from digging or ploughing at one end of the spectrum to so-called minimal tillage methods at the other. In fact some disturbance of the soil, however minimal, is always necessary for growing crops, if only enough to place the seed or transplant in the soil. There are a whole range of different options between the two extremes, and the question is really more one of degree than a straight 'Do we or don't we?'.

However there is a major difference between those methods of tilling which invert the soil, broadly speaking ploughing and digging, and those which don't. Turning the whole soil profile upside down is one of the most violent and disruptive things we can do to it. So to a great extent the discussion concentrates on the difference between digging and ploughing on the one hand and other means of tillage on the other.

Reasons for Tilling

There are some good reasons for tilling, both the inverting and non-inverting kinds, otherwise people wouldn't do it. However, yield comparisons which have been made suggest that no-till or reduced tillage methods can produce similar or greater yields than ploughing on the farm or digging in the garden.[14]

The reasons for tilling are summarised in the mind map below.

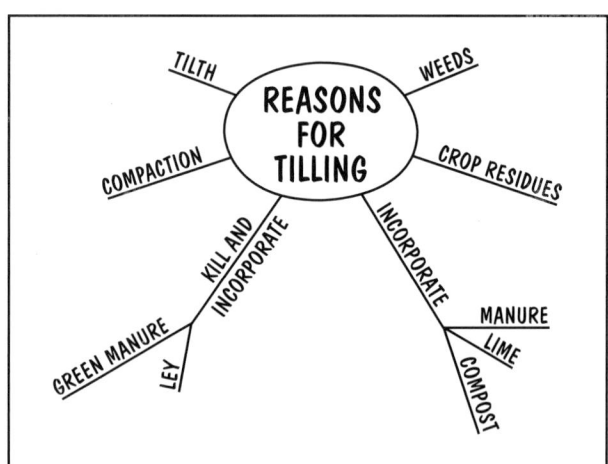

Weeds. Annual weeds and some perennials may be killed by burying them. But in the process dormant weed seeds are brought to the surface, where they germinate, so inverting the soil can cause as much of a weed problem as it solves. Mulching and hoeing are alternatives for weed control in the garden, shallow tined cultivation on the farm, and living cover crops have a part to play on both scales. (See for example the bicrop method of growing cereals and vegetables, pp268-270 and 273-275.)

Killing of annual weeds is probably the biggest single reason for ploughing, especially on organic farms where herbicides are not used. Another is to kill 'volunteers', self-sown crop seedlings which are in effect weeds and can also carry over disease from one crop to the next.

Crop Residues. In a garden, crop residues can be left on the surface as a mulch or removed to the compost heap. On a farm they may interfere with operations such as drilling (sowing) if left on the surface, and ploughing is the only effective way to bury them in situ. But there are seed drills which are specially designed to sow into a surface layer of crop residues. Residues may also carry disease over from one crop to the next, and harbour slugs.

DEEP TINED CULTIVATION

The aim of this kind of cultivation is to break up under-surface compaction and enhance the natural structure of the soil without churning the soil up, let alone inverting it. Thus it's a way of achieving one of the main aims of tillage whilst avoiding many of its disadvantages.

In a garden this can be done with a fork, simply by thrusting it as deep as it will go into the ground and pulling it back till the soil opens a little. A steel handle is preferable to a wooden one, which will sooner or later break under this treatment. A tarmac fork, available from builders' merchants, is just like a garden fork but with a steel handle.

For covering somewhat larger areas, but still working by hand, a broadfork is a useful tool. As far as I know these are not available in Britain, but they could easily be made by a blacksmith.[15] It's possible to cover quite large areas in a short time with a broadfork, and it's an invaluable tool on a small-scale market garden where mechanisation is not used. Eliot Coleman reckons he can cultivate an acre (0.4ha) a year without too much strain, and of course it should not be necessary to cultivate the whole garden each year.[16]

On a field scale there are a number of implements based on the traditional subsoiler, a heavy tool with a single, deep acting tine. The Howard Paraplow, McConnel Shakaerator and Wallace Soil Reconditioning Unit are implements with three to five tines, of lighter weight than the subsoiler's one, working at a shallower depth but still below the surface. As the tines pass through the soil they lift it and slightly shatter it, enhancing the natural fissures. Drainage, aeration and root penetration are all enhanced by this treatment.[17]

It is worth pointing out that this kind of implement is not the same as a mole plough or a chisel plough. A mole plough has a bullet-shaped foot instead of a tine, which makes a circular tunnel through the soil well below cultivation depth, specifically for drainage. A chisel plough is a multi-tined surface-working implement designed to thoroughly stir up the topsoil.[18]

All tined cultivation, whether deep or shallow, tends to leave plant residues on the soil surface.

Mild cases of compaction can be relieved by biological means. Where time is not of the essence, perhaps between the trees of a developing orchard, monocarpic plants can be used. These are plants which, like biennials, only flower once in their lives but which may live for more than two years before they flower and die. Burdock, hogweed, angelica and alexanders are deep rooted monocarpics. They have the advantage over annuals and biennials of being able to grow well on soils of low fertility.

In a garden with a heavy clay soil broad beans and the bigger brassicas can help. The broad beans must be sown in situ to get the benefit of their powerful taproot.

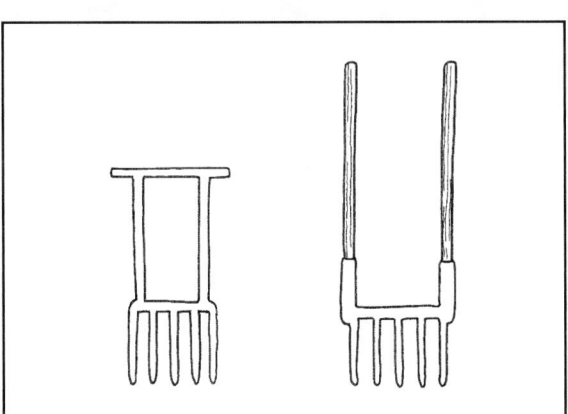

Short- and long-handled broadforks.

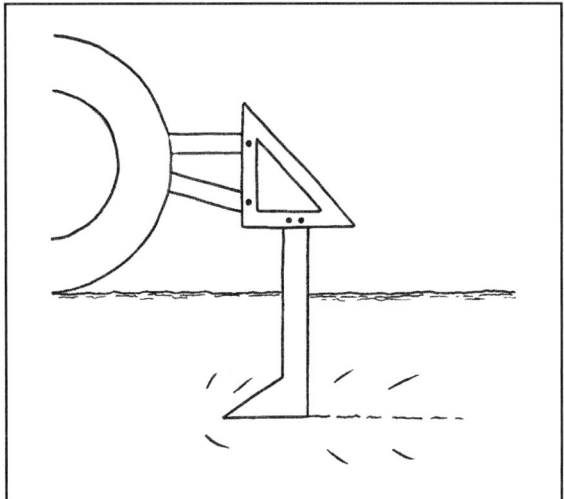

The traditional subsoiler.

Incorporating soil amendments may be worthwhile on heavy land, where there is less downward movement of material through the soil. On lighter land they will be carried down into the soil from the surface. Their effect may be felt more slowly when applied to the surface but it will last longer. Simply burying them at the bottom of a plough furrow or digging trench does little good. Vigorous mixing cultivations are needed to incorporate them evenly throughout the topsoil.

Digging to incorporate organic matter is something which is often well worthwhile in a new garden with a raw soil low in organic matter. But once the soil is in good heart it should not be necessary, even on heavier soils. (See p191.)

To *Kill and Incorporate* a ley (temporary grassland) or a long-term green manure crop like lucerne, there is really no alternative to the plough, the spade or the rotavator. Short term green manures can be hoed off and left on the surface.

Compaction. Although tilling the soil can relieve compaction, some kinds of cultivation can actually cause it. The bottom surface of a plough body can smear the soil as it moves along, leaving a compacted layer of soil which is more or less impenetrable by plant roots, water and air. This is known as a plough pan. Rotavator blades, as they spin through the soil, are always travelling horizontally at the bottom of their stroke. This can also cause a compacted layer, known as a rotavator pan. These pans are more likely to form in a clay soil, and especially when the soil is wet. Not cultivating when the soil is too wet is essential to avoid all forms of compaction, including the formation of pans.

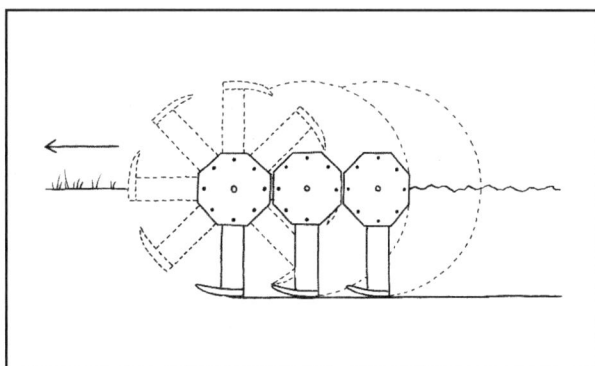

How a rotavator pan can form.

Deep tined implements are much more effective at relieving compaction and they need not disturb the surface soil. (See box, Deep Tined Cultivation.) On heavy soils which are prone to compaction some sort of cultivation may be needed every few years. But the less pressure that's put on the soil surface, either by wheels on farmland or by feet in gardens, the less compaction is caused. The main advantage of the bed system of gardening is that it keeps the gardener's feet off the soil and thus avoids the need for digging. (See p192.)

Well cared-for soils with a loamy texture and a good organic matter content need never suffer from compaction. Cultivation is often needed to get a badly-treated soil into good condition but should rarely be necessary after that.

Tilth. It's normally necessary to create a tilth in order to get even and reliable germination of seeds. This can be done simply by lightly cultivating the top two to five centimetres of soil. In some cases this constant cultivation of the same layer of soil can lead to loss of structure at the surface, and it may be necessary to plough or dig up fresh soil from below every few years. But this should not be a problem on soils with stable structure, that is those with a high clay or organic matter content. Getting enough organic matter in the soil to allow continuous surface cultivation should never be a problem in a garden, though it may be on a broad scale.

A traditional practice on heavy soils is to dig or plough in autumn so as to allow the frosts of winter to form a tilth for spring sowing. This does work but the tilth formed by the biological activity under a winter mulch can be as good.

Some crops really don't do well in an undug or unploughed soil. Carrots are the one which most readily spring to mind. But this is something which does vary from soil to soil and from season to season.

Reasons for Not Tilling

The reasons for not tilling are summarised in the mind map below.

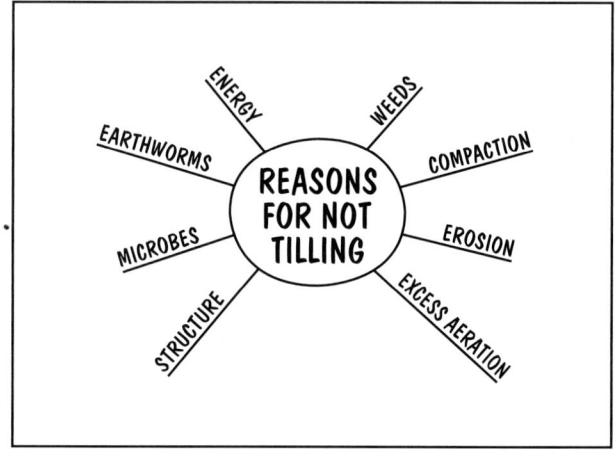

We have already seen how the problems of weeds and soil compaction may be increased by cultivations.

Soil Erosion[19]

At the beginning of this chapter I said that soil is the mother of all plants but it's equally true to say that plants are the mother of all soil. When life emerged from the oceans there was only raw rock to grow on. Plants have played the key role in creating all the soil there is, and it's their presence which protects the soil from destruction. This reciprocal relationship is a part of the web of relationships which make up Gaia.

Soil can be washed away by rain or blown away by wind. Wind erosion, though it can be locally severe, is confined to some five percent of the farmland in Britain, mainly in eastern England and Scotland. Most of this land is fertile in other respects, and the best use for it is surely agroforestry, combining field crops with hedges, windbreaks and belts of trees. Keeping the soil covered as much as possible and maintaining good soil structure are also important in preventing wind erosion.

Water erosion, though for long denied, is much more widespread, and 44% of British farmland is either experiencing water erosion or at risk. The rate of soil formation is typically 0.2 tonnes per hectare per year, and a similar rate of erosion is normal under natural conditions. Let's compare that with some of the rates of erosion now going on.

In the USA until recently the average rate of soil loss from arable land was 40 t/ha/ann. Now all farm subsidies are linked to soil conservation measures, including minimum tillage methods, and the rate has reduced to some 15 t/ha/ann.[20] This can only be described as a change from catastrophic to very severe. One third of the soil which was on that land when Christopher Columbus arrived is now at the bottom of the sea. Plough agriculture can only be a temporary phenomenon at that rate of loss.

Here in Western Europe, where rain is much more gentle and more evenly distributed through the year,

SOIL EROSION, A SUMMARY	
Natural Causes	**Solutions**
Climate characterised by intense rainstorms	• change to perennial vegetation* or minimum tillage / no-till methods
Vulnerable soil texture, ie high in silt or fine sand, low in clay	• change to perennial vegetation* • mild cases – increase organic matter content of soil
Excessive slope. Length of slope, steepness and area of water catchment all affect the speed and volume of water moving over the surface	• severe slopes – change to perennial vegetation* • broadscale, moderate slopes – install hedgerows or swales (see p100) on contour • garden, severe slopes – install terraces • garden, moderate slopes – beds on contour
* eg woodland, pasture, clover bicrop, orchard, forest garden, perennial vegetables or ornamentals	
Human Causes	**Solutions**
Poor soil structure, causing slow infiltration of rainwater	• increase organic matter, encourage earthworms • reduce or eliminate cultivations, especially digging or ploughing
Bare soil – raindrops destroy surface structure	• cover with dense vegetation or mulch • mild cases – improving soil structure may be enough
Compaction caused by heavy machinery	• small, mixed farms use smaller machinery than large, monocultural ones
Working, wheeling or treading soil when wet	• avoid overcommitment • reduce cultivations • in garden, adopt bed system
Working up and down the slope	• work on contour (may be difficult and dangerous with farm-scale machinery)
Removal of hedgerows	• replace them
Path or track which channels water	• place paths and tracks along ridges, avoid placing them in side valleys and other low places which can collect water

This maize field, seen after a heavy spring rainstorm, is big, collects water from the slope above, and has much bare soil in the springtime. Bicropping would help in this situation.

Even small areas of bare soil can be vulnerable. This vegetable garden was hit by a spring rainstorm just before sowing time and received the runoff from a large pasture field uphill of it. Since this picture was taken a diversion drain has been dug at the top of the garden to intercept runoff from the field and protect the garden from erosion.

There's no erosion in Ella and Andy Portman's mulched, raised bed garden, even in spring when there are few plants to cover the soil.

plough agriculture has gone on for thousands of years, so you would have thought that soil loss must be at a sustainable level. It may have been in the past, but the changes in farming methods over the past few decades have been massive, and, by one estimate, present farming methods will cause a loss of fertility equivalent to one third of present yields in a hundred years' time.

The worst recorded cases in Britain involve losses of up to 200 tonnes of soil per hectare per year. These are exceptional, but losses ranging between one and 40t/ha/ann occur on between 5 and 15 percent of UK arable land.[21] Some soil scientists assert that a loss rate of 2 t/ha/ann is 'tolerable'. But with the rate of soil formation averaging a tenth of that, to whom is such a rate tolerable?

A problem with soil erosion is that most of the losses are not felt by the person who causes them. In the short term, a little extra fertiliser can easily compensate for the yield loss due to erosion. The main losers are future generations of farmers, foresters and gardeners, and those who receive the run-off. The costs of the off-farm effects of soil erosion are actually greater than the on-farm costs. These include: flooding of houses and roads with muddy water; silting of rivers; damage to aquatic ecosystems from increased turbidity, pesticides, and eutrophication; and extra purification costs for the water supply industry. One estimate from the USA put off-farm costs at 75 times the on-farm costs.

This is a good example of a valuable resource becoming a pollutant when it's in the wrong place. It's also an example of how the true costs of our food are externalised. In other words no one in the food system, from farmer to final consumer, actually pays the price for soil erosion, although others must, either now or in the future.

The fundamental cause of water erosion is water flowing along the surface of the soil rather than infiltrating. The main reason for water not infiltrating into the soil is loss of structure at the surface, and the commonest cause of that is raindrops falling on bare soil, especially if that soil has weak structure. The greater the speed and volume of the water, the more soil it carries with it, so the slope of the land and the size of fields both affect the rate of erosion.

In forestry, erosion is increased by the practice of clear-felling large areas at a time, and the use of heavy machinery which compacts the soil surface. On upland grazing land the main cause is overgrazing, which both compacts the soil and removes vegetation.

The main causes for soil erosion on farms and in gardens are summarised in the table, Soil Erosion, A Summary, on the previous page, together with pointers to the solutions. More detail is given in the relevant chapters.

Organic growing eliminates a great deal of erosion, especially by increasing the organic matter in the soil, but not all of it. In one study an organic farm was compared with a neighbouring conventional farm on the same soil. Between 1948 and 1985 the conventional farm lost 21cm depth of topsoil, a third of what had been there at the beginning of the period. Meanwhile the organic farm lost 5cm, or just under a twelfth of what had been there. This is a great improvement, but is it enough of one?

One of the prime aims of permaculture is to reduce soil erosion to the rate of replacement or less, as it is in a stable ecosystem. Maintaining structure and organic matter in the soil is extremely important, but keeping the soil covered is even more important. If the soil is habitually left bare when heavy or continuous rain falls, a net loss of soil is more or less inevitable in most parts of the country.

Excessive Aeration: The Ethylene Cycle[22]

The amount of air in a freshly dug or ploughed soil is far more than could ever occur in a natural undisturbed soil. This amount of air disrupts a natural rhythm of great importance in the soil, the ethylene cycle.

The cycle takes place in the air-filled pores of the rhizosphere, that thin layer where plant roots and soil meet. At the beginning of the cycle the air in these pores has a high concentration of oxygen in it. The microbes which break down organic matter, and those which cause plant diseases, are mainly aerobic, that is they breathe oxygen. They reproduce rapidly in the

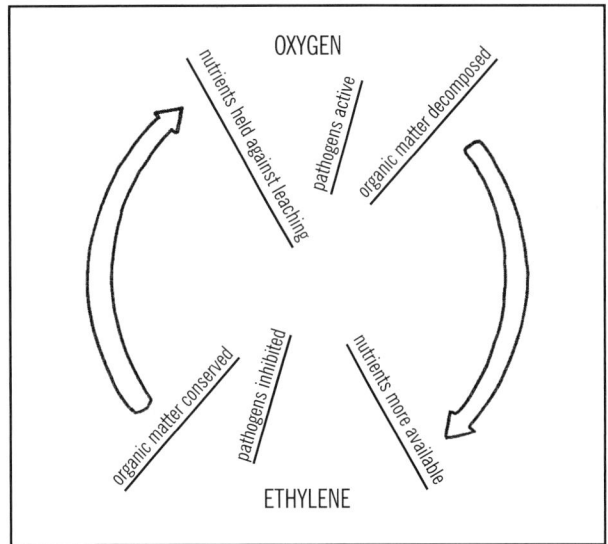

The ethylene cycle.

rhizosphere, encouraged both by the level of oxygen and by the exudates from plant roots.

Sooner or later microsites occur in which the oxygen is all used up. In the absence of oxygen, anaerobic microbes – those which are active only in the absence

TREES AND SOIL EROSION

Contrary to popular expectation, soil does not usually form more quickly under forest than under other kinds of vegetation. In fact soil formation is typically two to three times slower under forest than under farmland.[23] But trees can have an important role in protecting the existing soil from erosion.

All vegetation slows down falling raindrops and thus protects the soil, and trees, with their multiple layers of leaves, can do this very effectively. But the height of the canopy and what vegetation grows below it are both important. Raindrops may coalesce on the canopy surface to form larger, more damaging drops which can regain much of their original velocity by the time they hit the ground. In a fall of seven metres they can regain 90% of their velocity.[24] Thus the effectiveness of trees in preventing erosion may be more dependent on the shrubs, herbaceous plants and leaf litter below them than on the trees themselves.

Unfenced clumps of trees in pasture can actually lead to increased erosion, especially on slopes. The grazing animals will spend a lot of their time in the shelter of the trees and graze, browse or trample away all the under-vegetation leaving bare soil.

Conifer plantations, which suppress all growth beneath them, suffer considerable erosion when they're clear felled. Mature beech trees also suppress everything beneath them and should never be clear felled on slopes, though their thick leaf litter gives

more protection than does a layer of conifer needles. Regardless of the species of trees grown, systems of forestry which avoid clear felling are an important means of preventing erosion. (See pp318-320.)

As well as protecting the soil from the impact of raindrops, trees can slow down or stop the movement of water along the surface. This is partly a function of the leaf litter and the strongly humified upper layer of topsoil, which act as a great sponge, absorbing water and slowly releasing it, and partly of the roots, which help to keep the soil open and allow water to infiltrate. Hedges and narrow bands of trees along the contour can be as effective as continuous woodland if they're close enough together to prevent water building up speed and volume between one and the next. Control of erosion can be one of the benefits of agroforestry.

Landslips are a specific, and dramatic, kind of erosion. Waterlogging of the soil is usually one of the causes. Trees are the best defence against landslips: the roots anchor the soil in place, and the soil will be drier under trees because they use more water than smaller plants. On treeless slopes landslips can be caused by rainstorms one tenth the size that would cause a slip on a similar forested slope. In Britain landslips happen mostly in upland areas.

The effect on the soil of leaf litter from different species is described in Chapter 11. (See p326.)

of oxygen – take over. They produce ethylene gas, which inactivates but does not kill the aerobes, including both decomposers and pathogens.

In a healthy soil there is a constant alternation between aerobic and anaerobic conditions at a host of microsites adjacent to the plant roots. Thus the ethylene cycle regulates the rate of decomposition of organic matter and checks the vigour of plant pathogens.

In the aerobic phase mineral nutrients are less available to plants and less susceptible to leaching, while in the anaerobic phase they are more available and more easily leached. So the cycle also prevents excessive loss of nutrients while making them sufficiently available to plants.

In an excessively aerated soil conditions remain aerobic and the ethylene cycle cannot operate. All freshly tilled soils are excessively aerated, and the most important consequence of this is that most of the organic matter is lost to the soil by the process of oxidation, which converts it to carbon dioxide. Since organic matter is the key to natural soil fertility this process has a very harmful effect on the soil. It's also a major source of the carbon dioxide which is causing global climate change. (See p72.) In addition, excessive aeration allows plant pathogens to go unchecked and reduces the availability of plant nutrients.

Soil Structure

Anything which reduces the organic matter content of soil, or the earthworm population (see Earthworms on next page), is bad for soil structure. But the mechanical action of cultivation itself can destroy soil structure too. This is especially so if the soil is cultivated when too wet, but also when it's too dry or when violent implements such as rotavators are used. Tined implements which enhance soil structure by opening up natural lines of cleavage do less harm than ones which chop or slice the soil.

Capping rarely happens when the soil surface is well covered with plant material, even on susceptible soils. Keeping a continuous cover of perennial vegetation, mulching, or using a cultivation method which leaves plant residues on the surface are ways of preventing capping. Good soil structure helps,

---

SOWING SEEDS ON A CAPPING SOIL

Anyone who grows annual vegetables will find it hard to avoid sowing seeds into a bare soil. On susceptible soils a cap can be hard enough to prevent the emergence of seeds. This can be avoided by the following method:

- make a seed drill in the soil
- water it gently but thoroughly
- sow the seeds immediately
- cover them with fine damp compost or sand to the appropriate depth for their type

---

but this can be difficult to achieve on the silty soils which are the most prone to capping. At the very least bare soil should be avoided on these soils when heavy or continuous rain is likely, and bare soil should not be watered or irrigated.

Soil Life

Some 80% of the soil microbes live in the top 5cm of soil. If they are buried by digging or ploughing most of them die, and the population has to be rebuilt over time. These microbes are the powerhouse of soil fertility. By continually knocking them back we are weakening the soil's ability to generate its own fertility.

Where soil inversion is necessary, for example when turning in a ley, the shallower this can be done the better. Although any inversion of the soil is disruptive to soil life, ploughing to 12cm is less disruptive than ploughing to 24cm. In the Netherlands a shallow-working plough, the Ecoplow, has been developed to do just this.[25] Bringing up subsoil must be avoided at all costs. Double digging, when done properly, does not involve inverting the soil at all.

Mycorrhizas.[26] One life form which is particularly sensitive to cultivation is the mycorrhizas. A mycorrhiza is a symbiotic association of plant roots and fungi, together making what is in effect a joint organism.

The plant provides the fungus with organic food while the fungus supplies the plant with nutrients, especially phosphorous. Phosphorous is a mineral which is present in the soil largely in insoluble forms which are not readily available to plants. Mycorrhizal fungi are better at taking up the less soluble forms of phosphorous than are plants. They can also regulate the plant's nutrient uptake, actually decreasing that of some elements in order to prevent toxicity and keep a healthy balance. In addition they can increase the plants' ability to take up water and make them more resistant to disease.

Much of the efficiency of the fungus is due to the small diameter of its hyphae, the thread-like organs which penetrate the soil, compared to plant roots. They can be one hundredth the diameter of the smallest root, so they can cover a much greater volume of soil for the same expenditure of energy. In some cases a mycorrhizal association can in effect increase a plant's root surface area by up to 100,000 times. This is especially important for phosphate, which doesn't move around in the soil much and must be sought out.

Mycorrhizal interactions are complex and involve many soil micro-organisms as well as the two principal actors. Disease control in particular involves a web of interacting organisms of which the plant and fungus are only two. These relationships are also associated with the formation of nitrogen-fixing nodules on legumes.[27]

Where there is a high level of soluble phosphate in the soil, as when chemical fertilisers are added, the activity of the fungus decreases. In fact, adding chemical fertilisers can sometimes actually reduce the plants' uptake of phosphorous. Under these conditions the fungus is of no benefit to the plant and becomes parasitic. Where the level of soluble phosphate is low, as it is under natural conditions or where insoluble rock phosphate has been applied, the increased uptake of phosphorous by the plant is well worth the expenditure of organic food. Where there is virtually no phosphate in the soil the fungus again becomes a net loss to the plant, but this condition is rare.

Most plant families can form mycorrhizas, the crucifer or cabbage family being a notable exception. Many fungi are involved, and many of them can associate with a wide variety of plants. This means that phosphorous and other nutrients can be shared between plants of different species via a mycorrhizal bridge. Nitrogen fixed by legumes can be shared with other plants in this way.

Tilling does not necessarily kill mycorrhizas outright. When perennial vegetation is broken up and cultivated the mycorrhizal population tends to decline steadily over a number of years. There may even be sparse populations of mycorrhizas in soils which are cultivated annually over many years, mainly arising from short-lived recolonisation by airborne spores. When tilling ceases it may take a number of years for the mycorrhizal population to rebuild.

Plants can be inoculated with mycorrhizas, and if the soil has no native population of mycorrhizas this can lead to increased germination rates, vigour and yield. Where native strains exist they may be more effective than introduced ones, and then inoculation can result in a decrease in plant vigour. Probably the best way to reintroduce mycorrhizas to disturbed soils is to bring in small quantities of soil from an undisturbed site nearby. This is particularly valuable when planting trees, as they are often very dependent on mycorrhizas, especially the conifers. The ideal is not to disrupt mycorrhizas or other soil microbes in the first place.

Present day agriculture is dependent on chemical phosphate fertilisers manufactured from rock phosphate which is mined in North Africa and Russia. Conventional agronomists have no alternative to the policy of depending on this non-renewable resource. Encouraging mycorrhiza would seem to be the only practical way of maintaining yields without the input of this non-renewable resource.

Earthworms. As we have seen, earthworms are absolutely central to soil fertility in temperate regions, and tilling has more of an effect in reducing their population than any other factor. (See p47.)

A healthy population in a permanent pasture can be as high as two million worms per hectare and weigh as much as 1500kg, while chemically treated arable land can support as little as 20kg per hectare. These are the extremes, but there are always more earthworms in permanent grassland than in arable land, usually at least ten times as many.

The main reason why tilling does away with earthworms is the reduction in organic matter which it causes, but the physical action of tilling has an effect too. Rotavation and other violent forms of tilling kill earthworms outright. Ploughing and digging expose them to desiccation and predators, especially birds. Any form of cultivation disrupts their burrow system and destroys the natural soil structure which they have evolved to live in.

Surface-only or deep tined cultivations cause less of a reduction in earthworm numbers than ploughing or digging, mainly because they cause less excessive aeration and thus less loss of organic matter. One comparison between direct drilling and deep ploughing found there were one and a half to four times as many worms in the direct drilled land.[28] Direct drilling not only means that cultivation is minimised, but it leaves the organic matter where the worms like it – on the surface. The difference between a dug garden and one run on the no-dig method would be similar.

What's true for worms is true for all the other soil creatures, most of which are beneficial on balance. Even moles, which can be an infernal nuisance, play a part in drainage and aeration and are welcomed by some farmers and gardeners.

Work and Energy

In the garden, saving the labour of digging is often the main reason for adopting no-till methods. Although some people enjoy the exercise of digging, it is very hard work – hard enough to put many more people off gardening and to make it impossible for those with less than robust health. It can make all the difference between gardening being an accessible, attractive occupation and one that's really out of the question.

Since permaculture aims towards gardening becoming a mass activity, the general adoption of no-dig methods must be central to permaculture if for that reason only.

On the farm, however, the amount of energy saved simply by substituting direct drilling for ploughing, within a chemical-intensive, mechanised agricultural system should not be exaggerated.

In grain production, the embodied energy in fertiliser can amount to half the total energy expended on the farm. The amount of machinery on the farm, and thus the embodied energy in it, may be much the same with either method. In one study the total energy saving was only 11.3%, although the saving in tractor fuel was 88.5%.[29]

In any case, the energy used on farms only amounts to a small proportion of the total energy used in the food production and supply system. (See pp129-130.) Only a reappraisal of the whole system will give any meaningful reduction in energy use, though reducing tillage can play a part.

Conclusion

To sum up the tilling debate, there are reasons both for and against using the most drastic forms of tillage, digging and ploughing. Most of the reasons against are connected with preserving the natural fertility of the soil and reducing energy expenditure, while those in favour are mainly about achieving a specific aim in the immediate term, such as burying crop residues and preparing a seedbed. It may often be seen as a choice between long-term sustainability and short-term expediency. This is not to say that we must never till, but that we should be aware of the tradeoffs we are making when we decide to go one way or the other.

The response of conventional arable farming to this dilemma is to ignore it. Soil fertility, in its fullest sense, is disregarded and the soil is treated as an inert medium to which chemicals are added. If continual cultivations destroy soil structure then larger machines and more powerful tractors are brought in, creating a vicious cycle of ever-more-compacted soils and ever-bigger machines.

The traditional answer to the dilemma, and that taken by organic farmers, is crop rotation. Annual grain crops are alternated with perennial grasses, an exploitative phase with a restorative one – in effect a see-saw of soil fertility. In permaculture we aim to remove the need for ploughing by using minimal cultivation techniques, and to reduce the need for rotation by combining grain crops with fertility-building crops in the same field at the same time. (See Chapter 10, pp268-270 and 273-275.)

On a garden scale there is very little need for digging at all, except in the case of a raw clay or silt soil which needs organic matter incorporated into it. This is a once-and-for-all operation. Though it may take a number of years to get the soil into good heart, there is no further need for digging once this has been achieved. (See Chapter 8, p191.)

ASSESSING SOIL

One of the first steps in any permaculture design is to find out what the soil is like. Every soil has its strong points and weak points, assets and potential problems. These can be optimised by choosing crops and other land uses which are best suited to that particular soil.

In some places the soil is uniform over large areas, and in others it can vary greatly over short distances, even from one end of a garden to the other. The texture may be completely different, or where it is the same, factors like depth and drainage may be greatly influenced by topography. The way the soil has been previously treated will also affect it, and this may vary from one part of a site to another. It's never safe to assume anything about the soil from a superficial look. A close examination is always worthwhile.

Indicator Plants

The first thing to do in assessing a soil is to look at the plants which are already growing there, not the ones which have been planted but the ones which choose to grow there of their own accord. This means different things in different kinds of vegetation:

- In gardens and arable land it means the weeds. (Though how well the crop plants are doing may also provide some clues. For example, a patch of stunted, yellow plants may indicate an area of poor drainage.)

- In semi-natural vegetation, such as moorland, 'unimproved' grassland, semi-natural woodland and ancient hedgerows the majority of the plants will be self-selected, though human action, such as grazing pressure or coppicing, will have influenced which ones have survived.

- On abandoned or 'waste' land all the plants are self-selected, but human influence may be strong in that we have determined what seed parents are available. In inner city areas where there are few seed parents nearby, the ability to colonise over long distances may have more influence on what is growing there than the nature of the soil.

It's never safe to extrapolate. An ancient woodland lying beside a field may have a very different soil, indeed this may be the very reason why the one has been cultivated and the other left as woodland. Ancient hedges can give a better clue, but as hedges often follow ditches they may run through a particularly wet part of the land.

Also, the soil may not be the dominant influence on which wild plants grow in a specific place. In gardens the dominant influence is weeding, and the plants which survive there are ones which are difficult to eradicate. Plants which would indicate quite different soil conditions if found in a wilder context can be seen growing side by side in a garden.

Most plants can be found over a wide range of conditions, and these plants tell us little or nothing about the soil. Only certain ones can be used as indicators. Never rely on an individual plant to tell you anything about the soil. It may have got there by chance. Look for strong populations of indicator species, obviously thriving. Only rely on a single species when it's a very strong indicator, such as rushes. In most cases it's better to look for characteristic communities of plants. Plants growing in untypical places, such as paths, should not be counted. Long-lived perennial plants are more reliable indicators of the fundamental nature of the soil than annuals or short-lived perennials.

Reading indicator plants is a skill which takes some time to develop. It's also something which requires local knowledge, because plants can behave differently in different localities. The table, Soil Indicator Plants is offered as a starting point, not as a definitive list of the best indicators in your area. Published lists of indicator plants tend to contradict each other, which suggests they are less than reliable. I've based this selection on a combination of my own observations and any consensus I can find in other published lists.

The most reliable thing which plants can indicate is whether the soil is wet or dry. This is obviously useful during a dry time, when all the land looks dry. Some

These rushes indicate a wet patch in the field. Note that it's not the lowest part of the field, which we might otherwise assume to be the wettest.

These rushes indicate compaction in the ruts left by the tractor which hauled the wood out of this recently-felled woodland. They're young, indicating the compaction is recent, but it can persist for decades in a situation like this.

SOME SOIL INDICATOR PLANTS[30]			
Trees & Shrubs			
Wet	*Well Drained*	*Acid*	*Alkaline*
Common alder*	Bramble	Rhododendron*	Wild clematis*
Willows†	Wild privet	Heather*	Wayfaring tree
Guelder rose	Wayfaring tree	Bilberry, blaeberry*	Yew
Bog myrtle	Juniper	Broom	Whitebeam
		Scots pine	
Dry			*High Nutrients*
Gorse			Elder*
Broom			Wild cherry, gean
Herbaceous Plants			
Wet	*Well Drained*	*Acid*	*Alkaline*
Rushes*	Bracken*	Sheeps sorrel*	Charlock
Horsetail	Campions	Common sorrel	Rockrose
Silverweed	Sheepsbit	Lesser stitchwort	Salad burnet
Creeping buttercup	Agrimony	Tormentil	Common poppy
Coltsfoot	Salad burnet		Wild basil
Moist	*Severely Compacted*	*High Nutrients*	*Fertile*
Mosses*	Silverweed	Common orache	Stinging nettle*
Chickweed	Great plantain	Charlock	Fat hen*
Great willowherb		Cleavers	Chickweed*
Meadowsweet			Annual mercury
Common fleabane		*Low nutrients*	Annual nettle
Ragged robin		Wild legumes	Black nightshade
		Spurreys	
* Most reliable indicators. † Not goat willow.			
Where a plant is listed under more than one heading, it may indicate either condition, not necessarily both.			

caution is needed though: most herbaceous plants which grow in wet soil may also grow in a soil which is temporarily compacted but usually well drained. pH is less reliable, so a larger number of plants and of species should be looked for. It is also not unknown for an acid topsoil to overlie an alkaline subsoil. Then the deep rooted plants will tell you something different from the shallow rooted ones.

Some animals can also give clues about the soil. The main animals to look out for are earthworms. They indicate that drainage, pH and organic matter content are all reasonable. Worm casts on the surface show they are present. In grassland, a layer of undecomposed leaf litter on the surface indicates that they are not. Moles live mainly on earthworms, so their presence also indicates a healthy population of worms.

The Soil Profile

Virtually all soils show differences between the surface layers and deeper layers, giving a distinct profile as you go down. The best way to examine the soil profile is by means of the Two-Spade sampling method.

Choose a representative place to take your sample, not a gateway or former compost heap site for example. To make sampling representative it may be necessary to take more than one sample. Note changes in plant species and topography, as they may indicate parts of the sampling area with different soil conditions.

If the soil is not stony or too dry, the amount of force needed to push the spade in can be a clue to any compaction problems.

- Push the first spade into the ground until its top is flush with the soil surface, using a side to side motion – a backwards and forwards motion will affect the sample. Ideally this spade should be flat and rectangular, but a normal garden spade will do.

- With the second spade, dig a pit in front of the first spade, as shown in the illustration below. Note that the width of the pit is greater than that of the first spade.

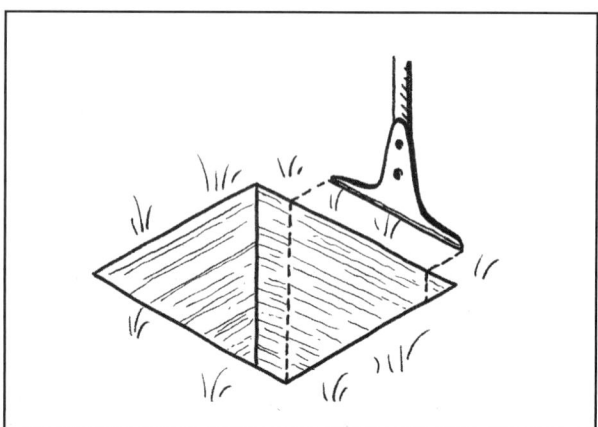

The Two-Spade sampling method.

- Carefully cut the sides of the sample. It should measure about 30cm x 20cm x 10cm.

- Lift the sample out of the ground on the first spade, holding it in place with the second spade to prevent disintegration.

If you only have one spade, dig the pit first and support the exposed face as you push the spade in to cut the sample.

Examine the soil for the following clues about its fertility:

Structure. The structure of a fertile soil changes as you go down (see illustration opposite):

- Near the surface lies the most fertile soil, ideally with a crumb structure of rounded aggregates of 2-5mm size. Crumb aggregates are very permeable to air, water and roots and hold a high proportion of available water. Granular structure is similar, but the aggregates are smaller and it's less fertile.

- Lower down, the structure may change to blocky. These are larger aggregates, 5-50mm in size. The smaller and more rounded blocky aggregates are more fertile than larger or more angular ones.

- In deep clay soils there may be a prismatic layer below this. These aggregates are taller than they are wide and are associated with drainage. Very little in the way of roots, water or air penetrates them.

- A massive structure consists of large blocks of soil with no fissures. It is impenetrable and extremely infertile. Sometimes a mildly compacted soil may appear to be massive when it's not. Hold a largish piece of the soil in both hands and gently pull it apart. If it comes apart along natural lines of cleavage it's not massive.

- A platy structure is almost always the result of compaction caused by cultivating the soil when it's too wet. The aggregates are horizontal plates, which may run together to form a continuous pan, and they are a barrier to root penetration and drainage.

Colour. The colour of the soil can be a useful indicator. A soil with an overall red colour has a high iron content. Iron is only needed in minute quantities as a plant nutrient, but it is essential to the operation of the ethylene cycle. Red soils tend to be fertile.

Humus is black or dark brown, so broadly speaking the darker the soil the higher the humus content. In many garden soils some of the characteristic black colour must be put down to ash and soot being put on them over the years. But even then it's a sign of intensive manuring so the organic content may be high too.

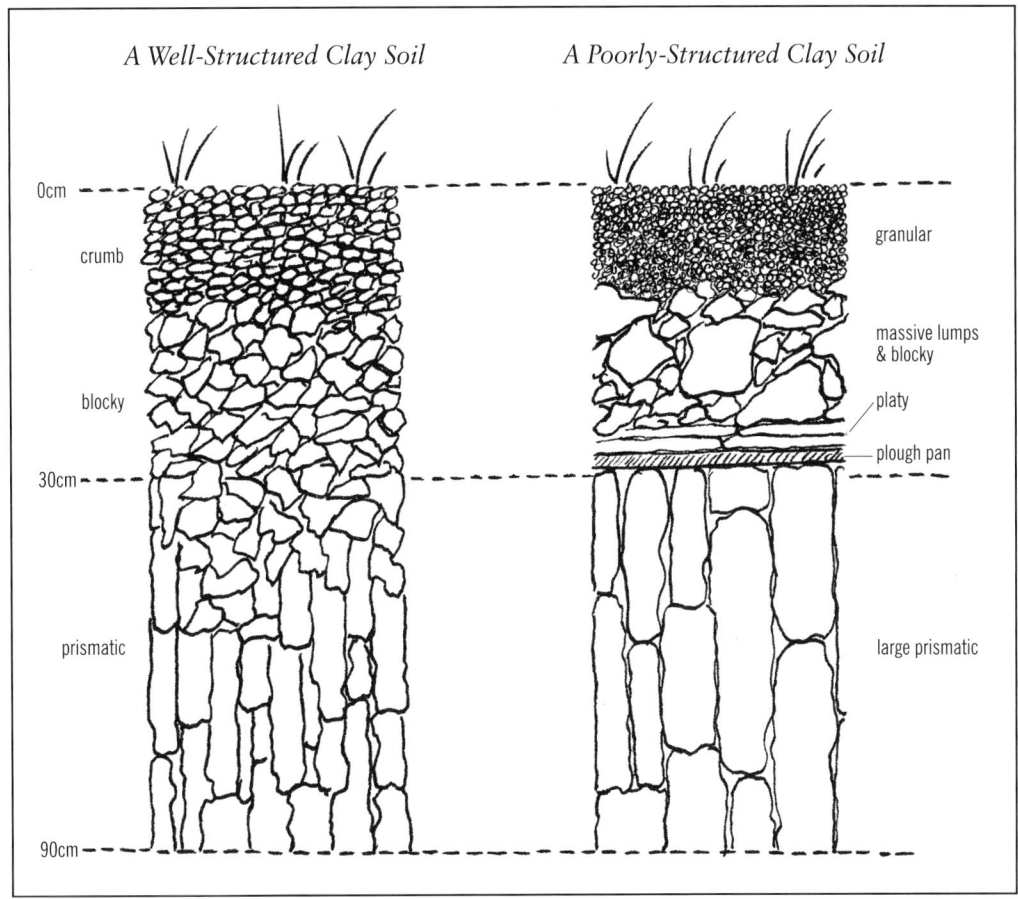

A Well-Structured Clay Soil   A Poorly-Structured Clay Soil

0cm

crumb

blocky

granular

massive lumps
& blocky

platy

plough pan

30cm

prismatic

large prismatic

90cm

Waterlogging excludes air from the soil. In the absence of oxygen the red coloured iron compounds are replaced by blue or blue-grey compounds. A layer of blue clay in the soil profile indicates that this level is normally waterlogged. A layer of grey which is mottled with rusty orange, indicates a layer that is alternately waterlogged and drained.

A smell of putrefaction indicates anaerobic conditions caused by waterlogging.

Soil Life. The way roots grow can tell much about the soil. Short, stubby roots, especially ones which grow downwards then turn a right angle and grow horizontally, are obviously unable to penetrate very far. This may be due to compaction or periodic waterlogging. Deep, finely ramified roots which are able to penetrate the soil aggregates indicate a fertile soil. Of course it's important to compare like with like. A carrot and a wheat plant will have very different roots in the same soil!

Any legumes can be checked for nodulation. Nodules are little balls about 1mm in diameter attached to the root. If they are white they are inactive, if pink they are active. Activity usually starts in late spring as the temperature rises.

The number of worm channels and their depth are worth noting. This is more worthwhile than looking out for actual worms as the number you see may be more a matter of chance than anything else.

The presence of undecayed organic matter indicates a low level of biological activity in the soil, unless it has only been added recently.

Soil Types

Natural soils have developed their individual characteristics over long periods of time due to a range of influences. One of these is the parent material, the rock from which they are formed. For example a sandstone gives a sandy, usually acid, soil, while limestone and chalk give a more alkaline soil. The vegetation also has a big influence, for example coniferous trees and heather both tend to acidify the soil through the products of decomposition of their leaf litter.

Often the biggest influence on the soil is its relationship with water. A gley is the name given to a soil which is strongly influenced by waterlogging, often with blue or grey anaerobic layers lower down in its profile. A brown earth is a well-drained, fertile soil, slightly leached but not excessively so, usually formed under deciduous woodland. A podsol is an excessively leached soil, usually strongly acid, formed in areas of high rainfall, sandy soils, acidifying vegetation, or all three together.

These soil types can be found in uncultivated areas, such as ancient woodland and rough grazing land. But most of the soils we are likely to find and work with on farms and in gardens have been so modified by human activity that their natural origins are largely obscured. Our ancestors have drained, limed, manured and cultivated over the years with the aim of bringing all soils as close as possible to the ideal condition for crop growth. Indeed any natural soils we do encounter are much best left alone, as they are a vital part of the dwindling biodiversity of the planet.

In urban areas a completely different range of soil types is found. Many of us will be working with these from time to time and they warrant a closer look.

Urban Soils[31]

An urban soil can be defined as one which has been profoundly changed by human activity, bearing little or no resemblance to the kinds of soil found in wild or agricultural areas. They are often quite uninfluenced by the underlying geology.

These soils can sometimes be found in the country, just as rural soils can be found in towns, in places such as parks and cemeteries which have never been built on. But the latter are never quite the same as country soils. In particular they are likely to be severely compacted by human feet. In fact compaction is a major characteristic of most urban soils. Some silty soils can even be compacted by the vibration of the traffic on a nearby road.

The three main types of soil found in urban areas are:

- Raw soils.
- Topsoiled sites.
- Deep humus soils.

Raw soils include:

- brick rubble
- other aggregates such as cinder, ash and railway ballast
- subsoil, either in situ or transported
- chemical wastes

Ecologically speaking they are young soils. That is they lie right at the beginning of the succession process which starts with bare rock and ends, in this country, with mature woodland. At first they are undifferentiated vertically (unless by random layering of different raw materials) and totally lacking in organic matter. The profile and the organic matter level both develop as plants colonise them.

The degree of compaction is important. Drainage and aeration are normally excellent in brick rubble, but if it's combined with clay it can be compacted into a substance as impermeable as concrete. Subsoil, whether moved or not, may have been compacted by heavy machinery, especially if worked when wet. Deep ripping with a heavy crawler tractor can be beneficial on any severely compacted soil, but where this is not feasible it's possible to grow crops in raised beds on top of a solid substrate. Soil and compost must be brought in, and excess water must be able to drain away across the top of the compacted layer since it can't go through it. This kind of growing is often very successful. (See case study, A Small Productive Garden, p209.)

The chemical nature of the material is also important. Brick rubble, which is perhaps the commonest of the raw soil types, is highly alkaline, from the mortar. It has high levels of phosphorous, potassium and other nutrients, from the clay the bricks are made of. The other aggregates are usually much lower in nutrients and more acid. The chemical nature of raw subsoil varies, but nutrients are likely to be relatively unavailable. Industrial wastes may contain toxic chemicals which make it impossible to grow food on them.

Nitrogen is very low indeed in all raw soils, and if it's not deliberately introduced it only builds up slowly as plants colonise and organic matter builds up. It's often the main limiting factor to plant growth. Thus a mixture of grass and clover can often be established directly on raw brick waste where pure grass cannot.

Eventually these soils can become quite fertile if left to themselves, though it can take decades rather than years if earthworms are not present. In practice earthworms are often introduced by accident in the root ball of planted trees, but there is no reason why they should not be deliberately introduced. But adding a layer of organic matter to the surface of a raw substrate can work wonders. (See box, Transforming Raw Slate.)

Although pH is strongly influenced by the nature of the parent material, older urban soils are likely to be acid due to acid rain, even those developed on brick rubble.

Topsoiled sites have supposedly fertile topsoil added to a regulation depth: 15cm where grass is to be sown, and 30-35cm for shrubs. Of dubious origin, and often severely compacted during the laying process, this topsoil sometimes doesn't help at all. In one study of 44 topsoils, half were no better than brick waste, 15 no better than raw colliery shale, and one prevented plant growth altogether.[33] In most of these samples low nitrogen levels were largely responsible for the infertility of the topsoil. Low nitrogen can be detected on grassed topsoil sites by a yellow sward with bright green patches where dogs have peed. (See illustration on page 64.)

Deep humus soils are characteristic of gardens and allotments. They have an abnormally thick topsoil, dark with good crumb structure, due to incorporation of organic matter and ash over the years, often accompanied by double digging. Deep humus can also form where organic rubbish such as night soil or tannery waste has been dumped over many years. These soils may overlie the remains of buildings, which can effect both the drainage and pH. On the whole they are very fertile.

TRANSFORMING RAW SLATE

The Centre for Alternative Technology, near Machynlleth in Powys, is built on the waste tip of a former slate quarry. In their organic gardens they have a demonstration consisting of four small plots, each growing the same kinds of vegetables on different kinds of soil.

Two of the plots have 22cm of slate dust for their 'soil', the other two have 30cm of good fertile topsoil. One plot from each pair is mulched with 5cm of compost each year, and one from each is not. This gives four different treatments for comparison.

Plant growth is pretty meagre on the slate dust without compost, but hardly any better on the topsoil without compost. The vegetables on the plot with both topsoil and compost are as healthy and productive as any in the whole garden, but the ones growing on slate dust and compost are very nearly as good. The results show that while adding topsoil makes a difference it's a small one compared to adding organic matter.

A large proportion of the compost oxidises away, and unlike the topsoil it has to be constantly topped up. But in the process a stable topsoil has been created out of sterile slate dust. The message of this experiment is: if you have no real soil, it's more worthwhile to bring in organic matter than to bring in topsoil. It has wide relevance for many urban growing situations.

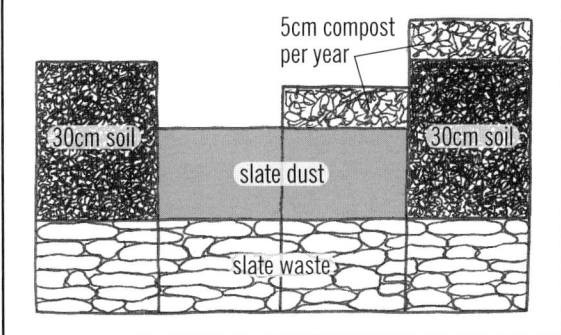

The four treatments.

Slate dust.

Topsoil.

Slate dust and compost.

Topsoil and compost.

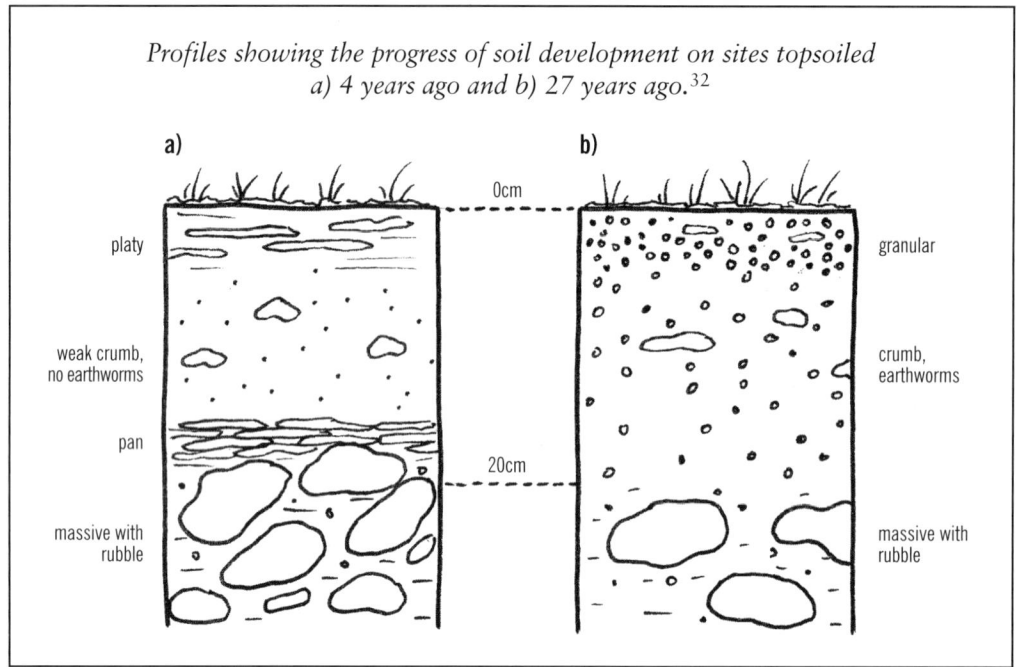

Profiles showing the progress of soil development on sites topsoiled a) 4 years ago and b) 27 years ago.[32]

Texture Test

The nature of a soil is largely determined by a combination of its profile, as discussed in the previous two sections, and its texture. All soils contain a mixture of sand, silt and clay, and what we mean in practice by texture is the relative proportions of the three. Only the most extreme soils with very little of any other particle size are known simply as a sand, a silt or a clay. These are fairly rare, and most soils are loams of one kind or another.

Note that the last word in the description of a soil's texture carries the most weight and the first one the least. Thus a loamy sand has more sand in it than a sandy loam, while a sandy clay loam is richer in clay than in sand.

It's easy to get a good idea of the soil texture by means of a simple finger test. (See box, How To Test Texture.) With practice this can become almost as accurate as laboratory tests.

A high organic matter content in a mineral soil can effect the feel of it, making a clay feel less sticky and a sand more silty. By and large soils of all textures will feel more loamy if they have a high organic matter content. A high lime content can have the same effect.

Chemical Analysis

Various kinds of soil testing kits are available from garden centres. Whether they are accurate or not is doubtful. Probably the pH tests are more reliable than the nutrient level tests.

ADAS, the national agricultural advisory service will test samples sent into them for pH and the major nutrients. A much more comprehensive, and slightly more expensive, service is provided by Elm Farm

Research Centre. (See Appendix D, Organisations.) In addition to pH and major nutrients they also test for soil texture, organic matter, nutrient holding capacity, availability of phosphorus, and levels of some key minor nutrients. They also provide an explanation of the figures and recommendations for soil amendments.

This kind of analysis can be well worth while, especially if you have just moved to a new place or are thinking of changing land use. Land that grows good grass, for example, may not be suitable for an orchard if it is strongly alkaline.

The method for taking a sample for chemical analysis is as follows:

- Check whether the area you want to sample has more than one soil type or texture within it. If it has, take separate samples from the areas covered by each soil type.

- Take at least ten samples from each area of up to 10ha. This applies even to a small garden, where different kinds of soil amendment may have been applied to comparatively small areas.

- Avoid untypical areas such as gateways and paths.

- Take the samples from random spots scattered all over the area.

- Take each sample evenly from the top 20cm of soil, ie the sample should contain equal amounts of soil from every depth from 0 to 20cm.

- Thoroughly mix the samples on a clean flat surface, removing any stones and fresh organic matter. The coning and quartering method is a good way to do this. (See box, Coning and Quartering on p66.)

HOW TO TEST TEXTURE

Take a good teaspoonful of soil from a representative part of the garden or field. Remove any stones, roots or other pieces of unrotted organic matter from it. Moisten it and knead it till it reaches its maximum cohesion. In other words, if it will form a ball, get it to the moisture content and consistency at which it will form the strongest ball possible.

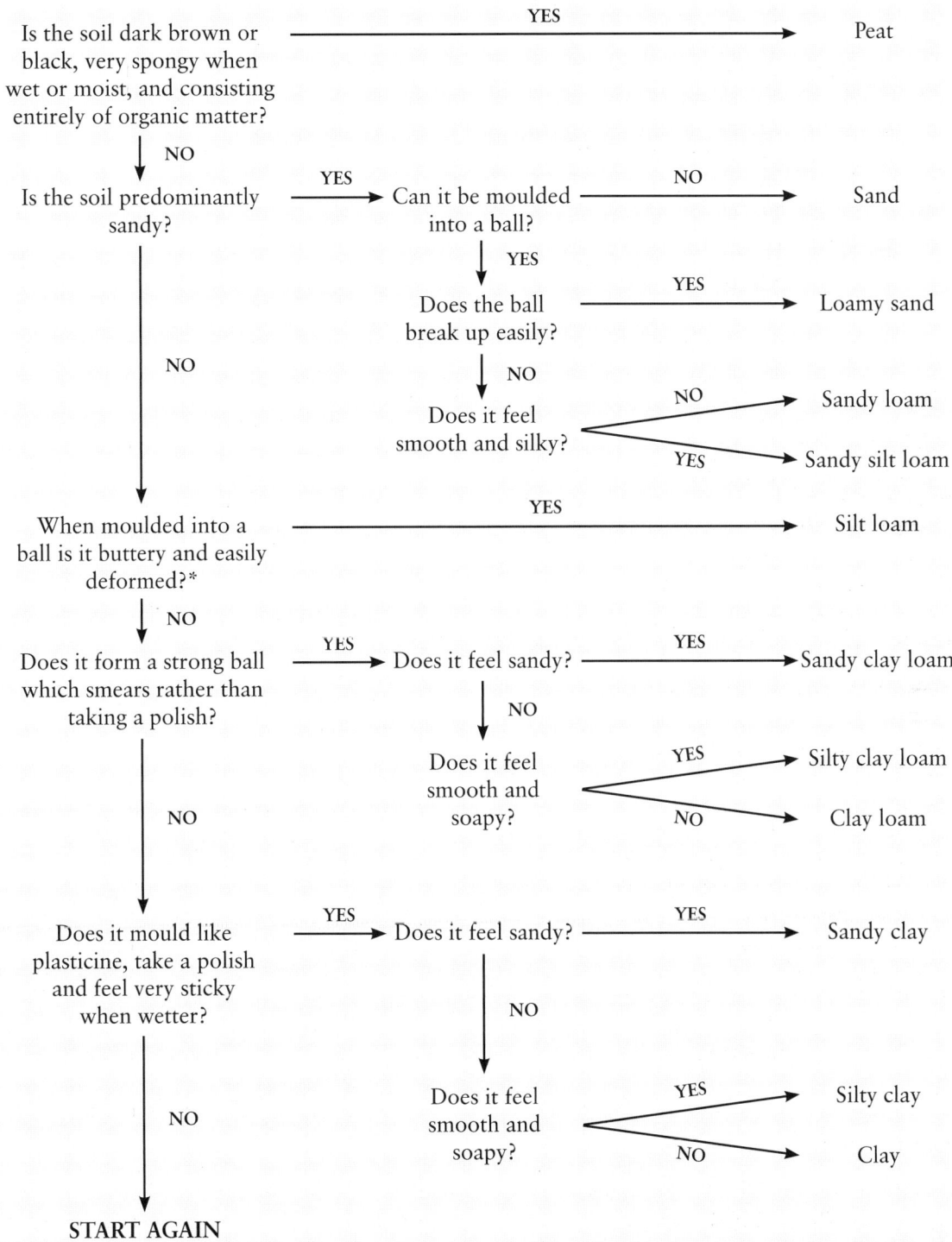

Is the soil dark brown or black, very spongy when wet or moist, and consisting entirely of organic matter? — YES → Peat

↓ NO

Is the soil predominantly sandy? — YES → Can it be moulded into a ball? — NO → Sand

↓ YES

Does the ball break up easily? — YES → Loamy sand

↓ NO

Does it feel smooth and silky? — NO → Sandy loam / YES → Sandy silt loam

↓ NO

When moulded into a ball is it buttery and easily deformed?* — YES → Silt loam

↓ NO

Does it form a strong ball which smears rather than taking a polish? — YES → Does it feel sandy? — YES → Sandy clay loam

↓ NO

Does it feel smooth and soapy? — YES → Silty clay loam / NO → Clay loam

↓ NO

Does it mould like plasticine, take a polish and feel very sticky when wetter? — YES → Does it feel sandy? — YES → Sandy clay

↓ NO

Does it feel smooth and soapy? — YES → Silty clay / NO → Clay

↓ NO

START AGAIN

*   Any ball of moist soil can be deformed and inexperienced soil testers are often inclined to answer yes to this question for all samples. The question is relative. When you have tried a few samples you will get an idea of what constitutes 'easily deformed' and 'a strong ball'.

- Take the amount of soil required for analysis from this mixture. (The testing organisation will tell you how much they require.)

CONING AND QUARTERING

1. Put all the samples together on a clean flat surface, remove any stones and pieces of fresh organic matter, and crumble any large soil aggregates down to give a uniform size.

2. Put all the soil on one side of the work area, and progressively move it to the other side in double handfuls, placing each on top of the previous one, so that the soil forms a cone-shaped pile.

3. Divide the cone into four quarters, and remove two opposite quarters.

4. Put the two remaining quarters together, and repeat stages 2 and 3 until only the amount of soil required for the chemical analysis remains.

Chapter 4

CLIMATE & MICROCLIMATE

Our effect on the climate, via the greenhouse effect, is probably the biggest problem facing the Earth today. At the same time, working sensitively with the local climate and microclimates is one of the main ways in which we can make our human habitats more sustainable and productive.

Working in harmony with the local climate and microclimates is one of the main themes of permaculture. It's especially important in food production, but also has a significant role in the design of housing and public open spaces. It is also something which has been largely ignored in the past century. Why bother to plant a windbreak when you can get a quicker and cheaper response by putting on an extra few bags of fertiliser? Why place housing in a naturally warm spot when you can turn the thermostat up a notch or two? These have been, and still are, the attitudes of the fossil fuel age. We now know that a different approach is needed, and this chapter sets out the basic information required for a more sustainable approach to landscape design from the point of view of both climate and microclimate.

CLIMATE

Nobody knows exactly how many edible or otherwise useful plants can be grown successfully in our climate, but there's little doubt that we can grow far more than we currently do. Plants for a Future is a charitable organisation researching plants which may be useful for permaculture in Britain. (See Appendix D, Organisations.) In addition to growing many of them on their experimental site in Cornwall they have a database of some 6,000 species which are potentially useful in Britain. This is a very high total compared to the number which is cultivated now, yet by world standards it's pretty low.

It's an observed fact of ecology that a far greater diversity of species can live in hot climates than in cooler ones.[1] To quote a famous example, the ecologist Edward O Wilson once counted 43 species of ant in a single tree in the Peruvian rainforest, which is approximately the same number as are found in the entire British Isles.[2] This diversity gradient, decreasing as you go away from the equator, applies equally to wild plants and animals and to useful domestic species.

The comparative lack of diversity which is possible in our climate leads some people to question whether permaculture, which originated in Australia, is feasible here. After all, the Plants For A Future database casts its net very wide, and some of those species will be of fairly marginal use. The number of really useful species which can be included in a tropical permaculture must be much greater than it is here.

This misses the point. True, our relative lack of potential diversity is a limitation, but it's a limitation on any system of gardening, farming or forestry, not just permaculture. It is not the total number of components which is important in permaculture, but the number of beneficial relationships which can be set up between them. This can be done with a comparatively small number of components, as long as they are well understood, well chosen and well placed.

In fact permaculture designs are more likely to be successful if they're based on reliable and well-known plants. There is room for experiment. Indeed for some people experimenting with new species is a main motive for growing plants. But those of us who want to feed ourselves or make a living from what we grow are well advised to make tried and tested plants our mainstay. This does not exclude having a few experimental plants as well, but it's important to be clear about the distinction between a plant or crop which is to be relied upon and one which is experimental.

Continental and Maritime Climates

It takes more energy to heat up and cool down water than it does dry land. This means that bodies of water change temperature more slowly in response to changes in solar radiation than does land. Large bodies of water, especially seas and oceans, have a big influence on nearby land: the temperature of islands and coastal areas fluctuates much less between summer and winter than does that of inland areas.

Here in Britain we have a maritime climate, with a difference between summer and winter temperatures of around 12°C. This is very little compared to most temperate parts of the world, though the west coast of Ireland is even more maritime, with a difference as little as 5°C. By contrast the plains of Russia have an annual range of as much as 30°C. Maritime climates are also more cloudy than continental ones, with a higher rainfall, and the rain tends to be softer and more evenly distributed through the year.

Our maritime climate affects many aspects of permaculture, including building design, renewable energy generation and soil erosion potential. But the effect on plant choice deserves special mention, especially when it comes to trying unusual species of fruits and nuts.

Plants are frequently rated for winter hardiness according to the United States Department of Agriculture (USDA) hardiness zone system. This indicates the minimum winter temperature which a given species or variety can survive. It's a useful system in the continental United States, where winter cold is the main climatic limitation on the range of perennial plants. But it has less meaning in a maritime climate, where the climatic limitations are less straightforward.

Many trees and shrubs which can stand winters much colder than anything they are likely to find in Britain may not get enough summer heat and sunshine to fruit properly here. In fact some species actually need the chilling of a really cold winter in order to fruit the next summer, and we don't get that every year. Many of these plants also need a reasonably hot summer in order to ripen the wood of their new growth. If this does not happen the new twigs can be damaged even by a relatively mild winter. Some of them are more harmed by winter wet than by winter cold, especially when wet weather and cold weather alternate.

Thus a plant which is rated by the USDA system to survive much colder winters than we get may fail to produce any fruit here, or suffer damage in a relatively mild winter.

The lack of contrast between the summer and winter temperatures can cause problems at flowering time for many fruit and nut species. Where winters are very cold and summers very hot, spring is a time of rapid and constant increase in temperature. But in a British spring an early mild spell is often followed by a late frost, and plants may be induced to flower early, only to have their flowers destroyed by the frost.

The maritime climate has the advantage over the continental when it comes to vegetables. Some perennial vegetables can survive a mild winter but not a really cold one, and others which die down in a cold winter stay green in a mild one. This not only lengthens the picking season but enables them to make better use of early spring sunshine, so they can be grown more successfully under deciduous trees and shrubs. (See pp16-17.)

Mild winters also make it easier to grow self-seeding biennial vegetables. Swiss chard and leaf beet, for example, produce leaves in their first year and seed in their second. Where the plants can survive the winter they will self-seed, but in regions with cold winters they may be killed off and so they must be hand-sown each year.

Annuals and biennials will have a longer picking season in more maritime climates, and some kinds will provide fresh leaves throughout the year without the protection of cloches or a greenhouse. This applies both to the familiar winter and spring greens, such as leeks and sprouting broccoli, and to self-seeding salads, like land cress and lamb's lettuce.

In addition, any plant which is actively growing during the winter can make use of plant nutrients in the soil which would otherwise be leached out by winter rains, thus conserving the fertility of the soil.

Each climate has its own advantages and disadvantages. While a bit of wild experimentation is no bad thing in the vegetable garden, when it comes to planting long-lived plants like fruit trees it pays to know what the pros and cons of your local climate are before planting. It may be a decade or more before a mistake is revealed, and by then you will have invested a great deal of time, energy, money and land in it.

Regional Climates

Within Britain there are marked regional differences in climate. These refine further the influence of our national climate on building styles, energy generation and plant selection. The fact that a plant grows well in Aberdeenshire does not necessarily mean it will succeed in Pembrokeshire. The Siberian pea shrubs, for example, are plants which need a cold, relatively dry winter and are much easier to grow successfully in the east of Britain than in the west.

Differences between regional climates are caused by a number of factors which interact with each other. The main ones are summarized in the mind map below.

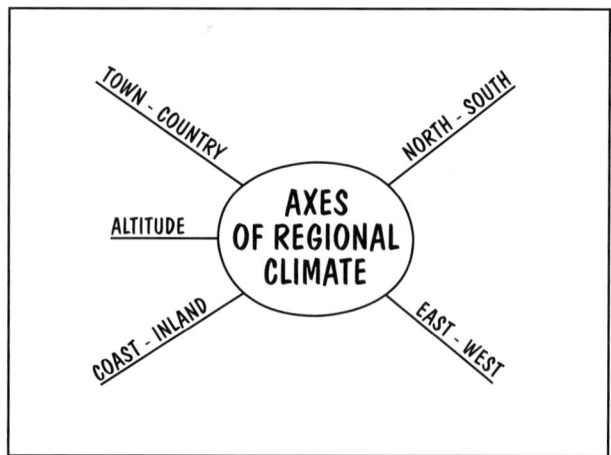

See also the climate data maps on p70. These may be used as a rough guide to the suitability of your area for energy generation, rainfall collection and cropping. More detailed information can be had from the Met Office. (See Appendix D, Organisations.)[3]

North-South. This axis has a dominant effect on the summer temperature. The further north you go the cooler the summer, and the number of days on which buildings need heating rises.

East-West. By contrast, winter temperatures are affected more by how far you are along the east-west axis than the north-south. Winter temperatures on the Isle of Lewis are similar to those on the Isle of Wight, while East Anglia is colder than both.

There's also a gradient from west to east in the average difference between summer and winter temperatures. At Aberystwyth it's 11°C, while at Reading it's 13°C. This may not sound like much of a difference, but it's enough to have a significant effect on the plants which can be grown successfully.

The reason for the east-west difference is not just that the Atlantic Ocean is bigger than the North Sea, and thus has a bigger moderating effect. It's also caused by the warming effect of the North Atlantic Drift, the warm water current, an extension of the Gulf Stream, which washes over the shores of these islands. This warming effect is highly significant in the winter, but in summer the effect of direct radiation overrides it.

Rainfall is much higher in the west, partly because the prevailing westerly winds bring moist air in from the ocean, but also because that's where most of the high ground is. (See Altitude, below.) Because these are the prevailing winds the average windspeed throughout the year is higher in the west, making wind power a better proposition there.

The growing season, the number of days in the year with a temperature of 5°C or more, is affected by both north-south and east-west axes.

Coast-Inland. All coastal areas tend to be warmer in winter and cooler in summer than inland areas. (This effect accounts for some of the difference between Aberystwyth and Reading in annual temperature difference.) Coastal areas also have more total sunshine and stronger winds.

Very close to the coast there are few frosts, or in some cases none at all, and plants which require an almost total absence of frost, such as figs, can be grown with confidence. But the winds are even stronger and, most importantly from a plant growing point of view, contain salt. This makes growing trees more difficult, and at the same time more important in order to provide shelter for other plants.

Altitude. This factor is probably the most significant of all in British conditions. It's quite possible to climb a local hill in springtime and find the season a fortnight behind the valley below, then to drive a hundred miles or more to the north and see very little difference between two places at the same altitude.

The difference in temperature is around 1°C for each 200m of altitude, but this difference is magnified by the other factors which change as you go up. Cloudiness increases, reducing both heat and the light which plants

The snow line is a visual reminder of the importance of altitude. It can either mark the height at which rain turned to snow on a wet day, or the height at which precipitation started on a cold day.

need for photosynthesis. Rainfall increases, leading to a further drop in temperature, an increase in fungus diseases, and often an excessively wet soil. The wind gets stronger, multiplying the temperature drop by the wind chill factor.

Because there's a more varied topography in hilly areas there are more variations in microclimate. For example, some high places may be sheltered from wind, while others may have it intensified by the funnelling effect of a narrow valley. On an open plain the wind will blow with much the same force everywhere.

Areas to leeward of high hills or mountains may have less rainfall than is typical for the region. For example the rainfall is higher on the plain of Lancashire than on the plain of Cheshire, which lies to leeward of the Welsh mountains.

Town-Country.[4] The biggest difference between urban and rural areas is in temperature. The centre of a large city like London may have an average temperature 0.5-1.5°C greater than the nearest rural areas, rising to a maximum of 10°C warmer in narrow streets and courtyards. City centres are usually warmer than the countryside on two days out of three and on four nights out of five. The growing season can be two to three weeks longer, with an average of ten more frost-free weeks in the year.

Rainfall can be 5-10% more. Most of this falls as intense summer storms which tend to run off quickly, so it makes little difference to plants growing in the soil, but it is available for collection as a general water supply. But air pollution is greater in cities, often making rainwater unsafe to drink. Air pollution also reduces sunlight levels by 15-20% compared to the country. This can limit the range of plants which can be grown, especially as many urban growing sites are shaded by buildings for part of the day. Wind speed is on average 10-20% less than in the country, but it can be increased locally by funnelling between buildings or by turbulence at the foot of high-rise buildings.

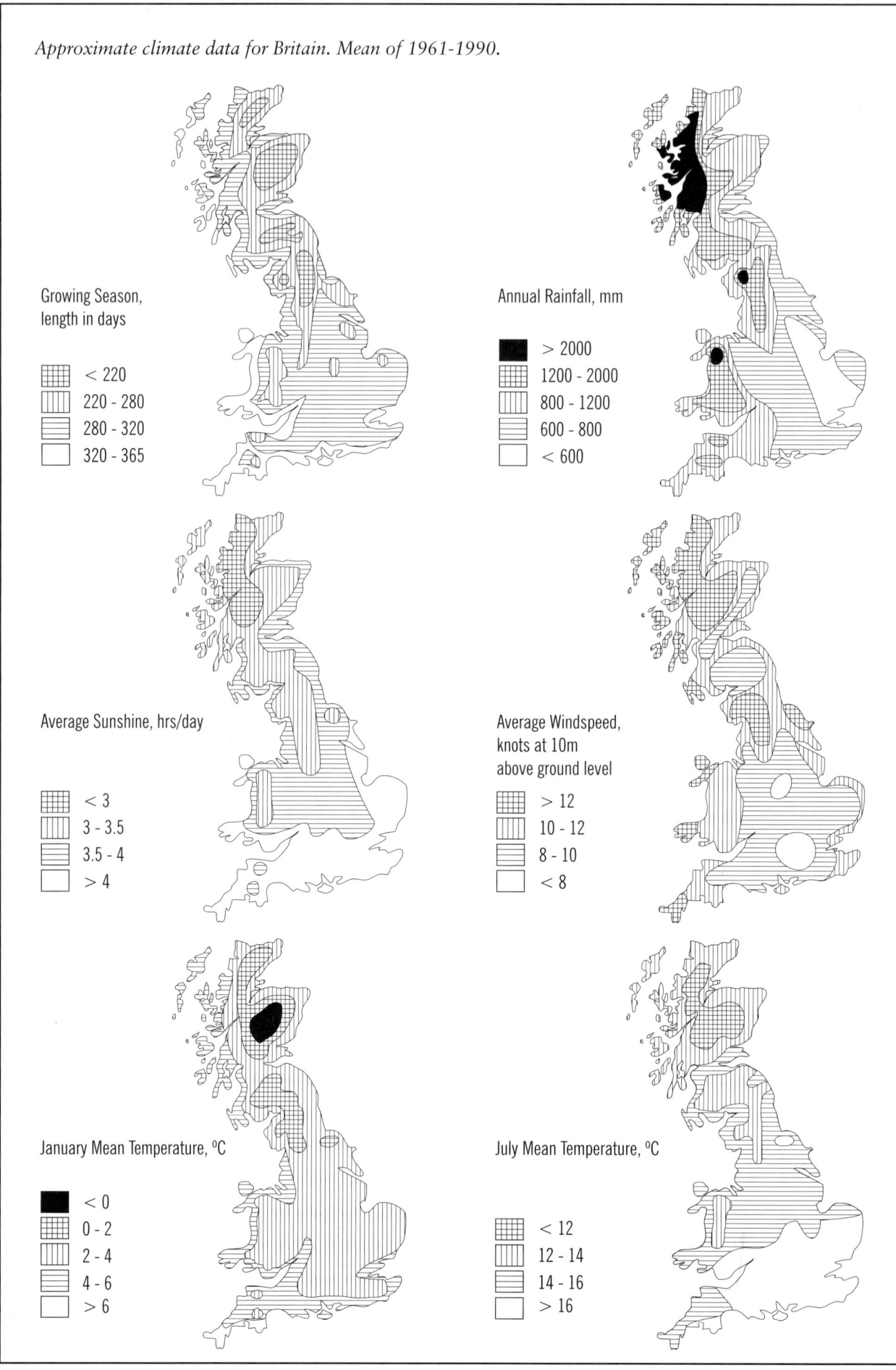

Approximate climate data for Britain. Mean of 1961-1990.

Growing Season,
length in days

< 220
220 - 280
280 - 320
320 - 365

Annual Rainfall, mm

> 2000
1200 - 2000
800 - 1200
600 - 800
< 600

Average Sunshine, hrs/day

< 3
3 - 3.5
3.5 - 4
> 4

Average Windspeed,
knots at 10m
above ground level

> 12
10 - 12
8 - 10
< 8

January Mean Temperature, °C

< 0
0 - 2
2 - 4
4 - 6
> 6

July Mean Temperature, °C

< 12
12 - 14
14 - 16
> 16

GLOBAL CLIMATE CHANGE[5]

Anything which may be said about climate must be qualified by the fact that the climate is changing in response to the enhanced greenhouse effect.

The greenhouse effect is a natural phenomenon. Energy comes to us from the Sun mainly in the form of light, which has a short wavelength and can pass through the atmosphere easily. The Earth re-radiates this energy mainly in the form of heat, which has a long wavelength. Some of this heat passes out into space but some of it is absorbed by certain gases in the atmosphere. (The glass of a greenhouse works in the same way as these gases, letting in light but trapping some of the heat.) Without the greenhouse effect the Earth would be so cold that hardly any living things could survive.

The enhanced greenhouse effect is the correct term for that portion of the greenhouse effect which is due to human activity, mainly the release of various gases into the atmosphere. But in common speech we normally drop the word enhanced, and the greenhouse effect is taken to mean the bit we are responsible for.

There can no longer be any doubt that the Earth's atmosphere is warming fast in response to our activities. It's impossible to know for certain how fast, but the scientific consensus is that, over the next 100 years, the mean temperature of the Earth will rise by between 1.4°C and 5.8°C.[6] This compares with a difference of 5°C between the coldest point of the last ice age and the present. It's a huge change, and an extremely rapid one by biological standards.

This is the nub of the problem. It's not that the climate is changing but that it's changing so fast. Climate is never static but a change as rapid as the one which is now underway is very hard to adapt to in time. Many species of wild plants and animals will not survive the change. We humans will probably survive as a species but many millions of us will die or become impoverished.

Causes

The principal gases contributing to the enhanced greenhouse effect are:

Carbon dioxide. This causes 50-60% of the enhanced greenhouse effect, two thirds coming from burning fossil fuels, one third from deforestation and loss of soil organic matter. It can persist in the atmosphere for hundreds of years. (See box, The Carbon Cycle, overleaf.)

Methane. This causes around 15% of the effect. It comes from a wide variety of sources, wherever organic matter is broken down in anaerobic conditions. A significant source is the digestive processes of cattle and other ruminants. (See p259.) It only persists in the atmosphere for about a decade, but each molecule is many times more potent than one of CO_2. Thus the net greenhouse effect of each molecule of methane over 100 years has been calculated as 11 times that of CO_2.

CFCs and HCFCs. These cause some 15% of the effect. Formerly used in fridges, air conditioning and aerosols, they are now banned in developed countries because of their effect on the ozone layer. But they are persistent and extremely potent, and the net effect of one molecule over 100 years is thousands of times that of CO_2. Their main replacement in fridges, HFCs, are also highly potent greenhouse gases, though harmless to the ozone layer. Fridges using hydrocarbons, which do no harm to the ozone layer and cause much less global warming, are now available.

Nitrous oxide. This causes some 5% of the effect, mainly from agricultural sources. It is persistent and potent, with an effect per molecule over 100 years some hundreds of times that of CO_2. It also depletes the ozone layer.

The overall process of climate change is very complex. There are many effects, such as the melting of ice caps and increased take-up of CO_2 by plants, which may speed up the process or slow it down. No-one can predict with any confidence what the net effect of these numerous interactions will be. Climate change is a phenomenon so complex that traditional logic is overwhelmed in the attempt to understand it and we must have faith in our intuition.

Consequences

Although there is considerable certainty about the global trend, it's much less easy to be certain about how the greenhouse effect will change the climate in different parts of the world. While the global average temperature rises, some regions may actually get cooler, and changes in rainfall patterns may be as significant as changes in temperature. Climate change is a much better phrase than global warming to describe the changes that are taking place in response to the greenhouse effect.

Here in Britain there are at least two possible futures. In one of them the British climate will heat up and become about as warm as that of southern France, with wetter winters and drier summers than now.

The other possibility is that our climate will actually get colder. The overall rise in global temperature may cause ocean currents to change their courses. At present the climate of Western Europe is kept much warmer than other places on the same latitude by the North Atlantic Drift, an extension of the Gulf Stream. If it changes course it will leave Britain and Ireland with a climate similar to that enjoyed by Iceland today.

Whatever the overall changes in climate may be it's clear there will be a general increase in extreme events: droughts, floods and violent storms. In fact these extremes of weather have already started. Here in Britain records for the wettest, driest and warmest months and years since records began are frequently being broken, while hurricanes, typhoons and floods of unprecedented ferocity are becoming frequent in the tropics. Worldwide there was more damage caused by extreme weather events during the year 1998 than in the whole decade of the 1980s,[7] and it has been estimated that in two generations damage to property will exceed Gross World Product.[8]

Looking at global food supply, the production potential of the Earth may not change greatly in the long run, as some areas become more productive and others less so. But in the short and medium term there are likely to be disastrous crashes in food production. Suppose, for example, that rainfall in the Sahel – the semi-arid region to the south of the Sahara desert – decreased to

the point that the whole region became unfarmable. It would be little comfort to the inhabitants to know that the northern limit of grain growing in Russia had extended some distance towards the pole.

Many major grain exporting regions are vulnerable to climate change. Up to 30% of the cropland in the western USA could go out of production by 2030 by one estimate. Again there will be little comfort in the knowledge that as one area becomes unsuitable for cultivation another becomes more suitable. The infrastructure necessary to support the kind of agriculture which can export large amounts of grain is considerable. There would be a long time lag before the infrastructure could be created to bring new land into production in new areas. Many of these new areas, such as Siberia and northern Canada, have infertile soils,

THE CARBON CYCLE

Carbon dioxide makes up about 0.03% of the atmosphere by volume. It's absorbed by green plants in the process of photosynthesis. In photosynthesis plants combine CO_2 with water, using the energy of the Sun, to form the organic carbon compounds which make up their bodies, and thus also the bodies of all animals and microbes. Oxygen is released to the atmosphere. All of these living organisms (except for some microbes) get their energy by respiration. In respiration oxygen is absorbed from the atmosphere and the carbon compounds are split into CO_2 and water, releasing energy for use by the organisms. The CO_2 is returned to the atmosphere.

Thus living things represent a store, or 'sink', of carbon, keeping it out of the atmosphere. When they die and their bodies decompose carbon is returned to the atmosphere, mainly as CO_2 from the respiration of the decomposing microbes. Trees, being larger and longer lived than other living things remove more CO_2 from the atmosphere than other kinds of vegetation and keep it out of the atmosphere for longer.

Part of this organic matter decomposes very slowly and is stored in the soil. Thus soil is another carbon sink. The total amount of carbon stored in the soil organic matter is greater than that stored in living plants in temperate ecosystems. In temperate forests it's about twice as much and in temperate grassland some 30 times as much.[9]

Tilling the soil exposes the soil organic matter to oxygen and greatly increases the rate of oxidation, that is the rate at which the decomposing microbes turn organic matter back to CO_2 and water as they respire. When virgin soils are ploughed CO_2 which has been fixed for thousands of years is released to the atmosphere. Peat soils are almost pure organic matter. When they are ploughed or drained oxygen

floods in and they rapidly oxidise. This happens when moorlands are planted with conifers, and the amount of CO_2 released by ploughing greatly exceeds that fixed by the growing trees. In addition some methane is released. Repeated tilling of agricultural soils is a constant source of atmospheric CO_2.

Worldwide, soil carbon is estimated to amount to three times as much as that stored in the vegetation in ecosystems on land. Changing to minimum tillage farming systems would have a significant effect on the enhanced greenhouse effect.[10]

Fossil fuels are made from organic matter which has been dead for a very long time. In fact peat can be seen either as a highly organic soil or as the youngest of the fossil fuels. Burning is fundamentally the same process as respiration: oxygen combining with the carbon in organic matter to release CO_2, water and energy. By burning fossil fuels at the rate we do now we are converting carbon which has been stored over a very long time back to CO_2 in a very short time.

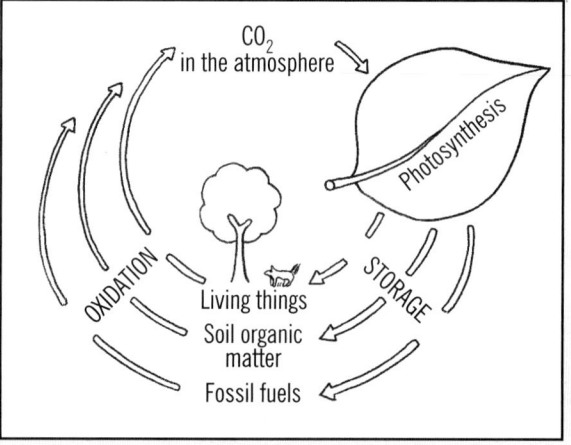

The carbon cycle.

and there would be another time lag before fertility is built up.

Timber and other tree crops will be particularly badly effected because of their long life span. Trees planted today may mature in a climate to which they are quite unsuited.

As sea levels rise, more due to the expansion of water as it warms than to the melting of ice caps, parts of many low-lying countries will be flooded. This includes heavily populated areas like Bangladesh and the Netherlands, and cities like London. The Maldives are one of six nations which will disappear altogether. The number of displaced people will make all previous refugee problems look small by comparison.

Drastic as the effect will be on humans, it's not likely to lead to the destruction of the whole species. The same cannot be said for the wild species of plants and animals with whom we share the planet. (See box, Global Climate Change and Biodiversity.)

Action

It has been calculated that an average US citizen uses 100 times more fuel per year than a Bangladeshi.[11]

While rising population is a factor in global warming, and everyone on Earth has a part to play in reducing greenhouse emissions, this simple statistic shows clearly where most of the pollution is coming from, and where action most needs to be taken. The first priority must be to reduce our output of greenhouse gases here in the industrialised North. The scientific consensus is that we need to reduce our output of greenhouse gases by 60-80%.

Many of the things which need to be done can only be done on a political level, and it behoves all of us who care about the future to make our voices heard in the political sphere. But real significant change will only come about if we also take action in our own lives. There are a number of ways in which we can alter our patterns of consumption and production to reduce climate change:

- Reduce our consumption of material goods, especially those involving high levels of energy use or with a high embodied energy content. Reducing car and air travel is almost certainly the first thing to look at here. (See Chapter 6.)

- Use the cleanest and most efficient sources of energy to meet our needs. (See Chapter 6.)

- Stop consuming goods whose production leads to deforestation. (See Consuming Forests, overleaf.)

- Change to no-till and minimum tillage systems of farming and gardening, so as to reduce the rate of oxidation of soil organic matter.

- Avoid anaerobic decomposition of organic matter, unless the resulting methane is collected and burnt

---

GLOBAL CLIMATE CHANGE & BIODIVERSITY[12]

As the Earth warms, the climatic zones are moving bit by bit towards the poles – northward here in the northern hemisphere. Many species of plants and animals can only survive by migrating polewards. Most trees of temperate areas can migrate at a rate of 20-40km per century, but the migration rate needed to keep up with climate change is around 300km per century. Many herbaceous plants and animals, especially invertebrates, are equally slow to move.

Others can migrate faster, but these are mainly the opportunistic species which are already widespread and tend to do well in the disturbed conditions created by humans. The slow migrators include many of the less common species, which are already closer to extinction than the versatile pioneers.

Ecosystems will not move as a whole but new ones will form. These will be less diverse than their predecessors, firstly because many species which can't migrate will perish, and secondly because some of the successful migrants will be too successful and become invasive, as so often happens when species become established in alien environments. (See p369.)

The problem is compounded by the fact that so much of the Earth's surface is now occupied by our farms, plantations, towns and roads. In many areas migration of plants and sedentary animal species is not possible at all. Mobile animals, such as birds and some mammals, may be able to cross these barriers, but many of these will become extinct anyway as the habitat they need for survival disappears.

One such animal is the Sunderbans tiger, which will become extinct as its mangrove habitat in Bangladesh is swallowed up by the rising sea. All coastal habitats are threatened in this way, especially where people have constructed sea defences which prevent coastal ecosystems migrating inland with the rise in sea level. There is evidence to suggest that virgin or old growth forests may be more resistant to climate change than more disturbed habitats. They are also better carbon stores than plantations and managed woodlands because of the dead wood they contain and the high humus content in their soil. This is yet another reason for leaving undisturbed what has so far escaped human attention.

as a fuel. Aerobic decomposition and burning give off CO_2, which is a much less potent greenhouse gas. Anything put in a dustbin ends up in landfill, where any decomposition which takes place is anaerobic. Mulching or aerobic composting are the best uses for unwanted organic materials.

- Reduce consumption of animal products. Ruminants – cattle, sheep and goats – give off methane as a product of their digestive systems. Non-ruminants – pigs and chicken – may be indirect consumers of forests. (See below.)

- Avoid any product containing CFCs or HCFCs, or made with them, including some expanded plastic foams.

- Plant trees. (See below.)

Consuming Forests

No-one who cares for the Earth would buy mahogany furniture any more, or consume more paper than they need to. But important as it is in some areas, logging is not the major cause of deforestation world-wide. The biggest single cause is clearance for agriculture and ranching.[13]

Much of the hamburger meat consumed in North America, for example, comes from land recently cleared from rainforest in South and Central America. Other forests in South America are being cleared to grow soya beans. A tiny proportion of the soya beans we import in Europe are eaten by vegetarians. The vast majority are fed to animals, especially pigs and chickens, and we end up eating about ten percent of the food value in them, such is the inefficiency of eating high up the food chain. (See pp119-120.)

Consuming products which come from the intact virgin forest, like Brazil nuts or wild African honey, is in principle a good thing as it gives the inhabitants a motive for keeping the forest intact. Unfortunately, the Brazil nut trade is notorious for the abuse and exploitation of the workers involved; and the commercialisation of honey production in Zambia has led to the trees which are used to make hives being overexploited.

The best way to ensure that we don't cause ecological damage, whether global warming or any other kind, is to eat from as close to home as possible.

Tree Planting and Global Warming

The fact that they take CO_2 out of the atmosphere is one good reason among many for planting trees. But the rate at which they can compensate for our profligate lifestyle must be put in perspective.

Our present forest stock absorbs only one percent of our CO_2 output.[14] It has been calculated that even if every single person on Earth was to plant and tend 1,000 trees a year till all available land was reforested it would only absorb three years' worth of the CO_2 we produce by burning fossil fuels.[15]

This applies only when new trees are planted where none grew before. It assumes that existing forests will be kept intact, with a neutral CO_2 balance. Nor, as we have already seen, does it apply when new plantations are made on carbon-rich soils, which is where the majority of current afforestation in Britain is taking pace. (See box, The Carbon Cycle.)

On the other hand, by one account if we turned our entire agriculture over to tree crops it would absorb enough CO_2 to absorb all our present output.[16] But a wholesale change in our agricultural economy and food culture is almost certainly not possible in the timescale available to us.

However, where we have a choice between meeting our needs by planting trees or by other means, we can take the CO_2 balance into consideration when making the decision. Products we could get from trees include carbohydrate from chestnuts instead of from cereals, protein from walnuts instead of from beef or soya, structural materials from timber instead of from steel, to name but three.

As well as actually absorbing carbon, a tree-based economy would also reduce the use of fossil fuels. There is 30 times less embodied fossil fuel energy in timber than in steel on a strength for strength basis,[17] and far less energy is used in growing a tree crop than an annual crop. The latter point is illustrated by the relative energy costs of producing biofuels from trees and annual crops. (See p124.)

The longer the body of the tree remains intact after it has finished growing, the longer the carbon is kept out of the atmosphere. Thus a tree which lives on in the timbers of a house will be many times more effective than one which is pulped for paper the day it's felled.

Conifer monocultures are not part of the solution to global climate change, but developing an integrated woodland economy certainly is.

The good news about global warming is that virtually all the changes we need to make in order to deal with it are worth making for other reasons too. Cast your eye up the list above and see if you can find a single one which would not be worth doing for at least one other good reason. Some of them, such as increasing energy efficiency, are profitable even in the limited terms of conventional economics.

MICROCLIMATE

A microclimate is the climate of a small area. This could be on the scale of many hectares, such as the microclimate of the north side of a hill compared with that of its south side, down to a few metres or even centimetres, such as that on top of a raised bed compared with that on the path beside it. Differences can be significant on these scales and all others in between.

There are three ways of working with microclimates:

- make the best use of existing ones
- enhance existing ones
- create new ones

They are listed in order of priority. The more we change things the more energy we need to expend, the less we change things the more we are working with the existing landscape. Placing things in the microclimate which best suits them is the first step in harmonious landscape design. (See Colour Section, 2, 3, 5, 6, 11 and 12.)

Enhancing a microclimate: adding a cold frame to this south-west-facing wall means that even earlier crops can be grown here.

Working with microclimates in all three ways enables us to:

- Increase the range of plants and animals which can be grown or kept on a particular site.

- Increase the productivity of plants and animals, by both increasing growth rate and decreasing disease and damage.

- Make fuller use of the landscape by bringing inhospitable areas into use.

- Improve living conditions for people.

- Reduce energy requirements, especially for space heating.

- Reduce the need for other inputs, eg cut down the need for irrigation by sheltering crops from drying winds.

- Reduce pollution effects, eg noise.

- Increase the diversity of wild plants and animals which can live on the site.

Details on working with microclimates in specific situations are given in the relevant chapters. Here the

general principles are given. The main microclimate factors are given in the mind map below.

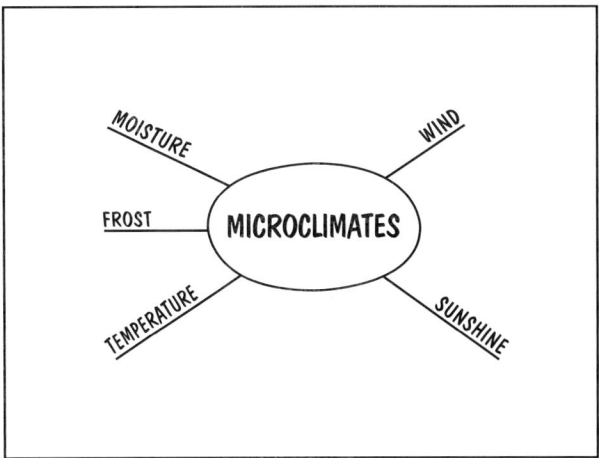

Wind

Strong winds can be put to good use, pumping water and generating electricity, but in most cases shelter from winds is more beneficial than exposure to them. Buildings, public open spaces, gardens, orchards, farm crops and animals all benefit from reduced wind speed in a number of ways:

- Reduced wind chill.

- Less moisture loss from plants and soil, reducing the need for irrigation

- Increased temperature at ground level due to:
  - less evaporation;
  - less mixing of warm low-level air with cooler high-level air.

- Less damage to plants, buildings etc.

- Shelter from driving rain for animals and buildings.

- More comfort for humans and animals.

The first priority in relation to wind, as with any microclimate factor, is to minimise its effect by careful placement of buildings and plantings, placing those things which need most shelter in the most sheltered positions. Alignment of structures can also reduce the harmful effects of wind, and make the most of any positive effects. For example, aligning a house end-on to the prevailing wind reduces chilling. Doing the same with a polytunnel facilitates ventilation, which is the main means of controlling fungus diseases in an organic system. Working with the microclimate in these ways means we get the maximum effect for the least effort.

These measures are not always enough to get the degree of shelter needed, and planting windbreaks in many cases yields an excellent return on the energy invested, especially when the windbreak has multiple outputs. Windbreaks are such an important aspect of permaculture design that they are described separately in the final part of this chapter.

The local landform has a big modifying effect on wind direction. In a steep sided valley the wind will blow either up or down the valley, regardless of what's going on high above. Thus in a valley which runs northwest to southeast the wind at ground level will almost always come from one of these directions, even though the prevailing wind above is southwest. The occasional very strong wind may burst over the valley wall and cause a whirlwind in the bottom, but this is rare. If the valley is aligned the same way as the wind it will intensify that wind, and it may actually be windier in the valley than on nearby hilltops.

Urban streets, especially ones of terraced houses, can funnel winds in both of these ways.

In a moderately steep valley the wind may blow from any direction, but winds aligned with the valley will blow more strongly than those which blow across it. In undulating land windward slopes and hilltops are more windy than sheltered lee slopes, though there may be turbulent eddies on the lee side, similar to those in the lee of a solid windbreak. (See p84.)

At night time, if there is little overall wind, a down-hill wind is often generated by cold air sinking. This is known as a katabatic wind. Even if it's very gentle it may have quite an impact because it's cold. A house which is set broadside on to a katabatic wind will dam up the cold air and be cooled down by it, whereas a house aligned with the wind will allow it to slip by.

Exposed places are obviously windy but the worst wind problems can often be in places which at first sight seem quite sheltered. This is because wherever wind has to pass through a narrow constriction it speeds up, just as a river speeds up where its channel narrows. The gap between two buildings is just such a constriction, and the most intense winds can often be experienced in built-up areas.

Any gap like this must be carefully investigated to see if in fact it does speed up the wind, and by how much. If so it can be ameliorated with a windbreak. When any buildings or tree plantings are being planned the aim must be to avoid creating this kind of gap. (See pp88 and 186.)

Light

Sunlight is essential to plant growth. In a temperate maritime climate it is often in short supply and can be a limiting factor to plant growth. This is especially so in stacked systems and in urban areas. Ways of working with light and shade in housing, gardens, orchards, agroforestry and woodland are given in the respective chapters.[18]

Shade cast by hills or mountains lying to the south of a site can have an important effect on human wellbeing.

Painting walls white can increase the light in a shady urban environment, as here at Riverside Community Gardens in Cardiff.

I know a village in the Highlands of Scotland where the Sun never rises for several weeks in the winter. But the effect on plant growth is much less, as the light which is lost is mainly in the winter, when plants are hardly growing, or in the early morning and late evening during the rest of the year. Plants can make some use of indirect light and this is still there when the Sun is not visible.

The aspect of a slope affects the amount of sunlight falling on it. (See box, Aspect, on p80.)

Temperature

Daytime temperatures are mainly determined by the amount of sunlight falling on the place in question, whereas night-time temperatures are determined more by the extent to which daytime heat is stored in the landscape and slowly released at night. Daytime temperatures are increased by concentrating solar energy, night-time temperatures by increasing the heat storage capacity in the area. A microclimate which is warmer than surrounding areas during the day will not necessarily be warmer at night, and vice versa.

When light energy is reflected it remains in the form of light, but when it's absorbed by an object it's converted to heat. Any object which receives energy in the form of sunlight will reflect a proportion of it and absorb the rest. The colour and texture of the object's surface have a big effect on how much is reflected and how much absorbed. Smooth, light-coloured surfaces reflect more than rough, dark ones. The degree to which a surface reflects light or absorbs it is known as its albedo.

Daytime temperatures can be increased by reflecting sunlight from a wider area onto a smaller one. This is in part how a suntrap works, although in practice the decrease in wind chill due to the shelter they give usually has more influence on the effective temperature. Suntraps can be made from buildings and walls, or trees and shrubs. The albedo of these objects is crucial to

SOME ALBEDOS[19]	
	Reflected %
'The perfect reflector'	100
Smooth white paint	96
Clean fresh snow	75-95
White gravel	50-93
Calm water, Sun at low angle	50-80
Sand dunes	30-40
Sandy soil	15-40
Dry hay	20-40
Wood edges	5-40
Rough water, Sun at high angle	8-15
Young oaks	18
Young pines	14
Dark soil	7-10
Fir forest	10
'The perfect black body'	0

A patio is a good place for heat-loving plants like figs and aubergines. These aubergines even appreciate the extra heat of a home-made frame.

places for growing summer crops which don't like too much heat, such as leafy salads, but unsuitable for winter crops, such as leeks and kale. When observing microclimates it's important to remember that a place which gets plenty of sunlight in summer may receive none at all in winter, as the angle of the Sun is much lower.

the effectiveness of the suntrap, the more reflective the surfaces the more effective they are. Sunlight can be reflected onto plants from the ground as well as from surrounding vertical surfaces, so the paving in a courtyard or patio has an effect as well as the walls.

Much the same happens when plants are grown against a south-facing wall, the wall reflects light onto the body of the plant. Here again, the low windspeed near the surface of the wall plays a large part too.

Night-time temperatures can be increased by absorbing the Sun's energy during the day and re-radiating it as heat during the night. The mass of an object influences its capacity to store heat. Thus, masonry, concrete, rocks and water all make good heat stores, especially if they or their containers are dark in colour.

Unless the heat store is something vast, like a sea or major lake, the effect is usually very local. In the open air it only extends for centimetres rather than metres from, say, a massive wall. So it has application in gardening rather than farming. It means that wall-trained fruit trees may stay frost-free on nights with a light frost, as well as enjoying increased daytime temperatures. In an enclosed space the heat is less quickly dissipated, so the effect can be used to great advantage in heating buildings, and this is the principle on which passive solar heating works. (See p148.)

Shady places obviously tend to be colder than those which receive more direct sunlight. They may be good

TREES & TEMPERATURE

Trees can have a moderating effect on the temperature of the microclimate.

Like all plants, they take up water with their roots and lose it to the atmosphere through their leaves, a process known as transpiration. Because they are such large plants they transpire much more water than others, especially when the weather is hot. When water evaporates it cools, and the evaporation of water from the leaves of trees can have a significant cooling effect on people, animals and buildings.

Condensation, by contrast, causes the temperature to rise. At night-time there is often a surplus of water vapour in the air, but condensation only takes place when there is a surface for the water to condense onto. The leaves of trees provide an enormous surface area for condensation, and in winter even the bare twigs provide a significant area.

Trees also even out the extremes of temperature by virtue of the water content of their leaves. As it takes more energy to change the temperature of water than most dry substances, trees tend to keep a more even temperature than either soil or buildings. This evening-out of temperature extremes is just one of the reasons why trees are so beneficial in a built environment and to grazing animals on farms.

Frost

Frost is a particularly important temperature effect. Many annual plants, like runner beans and members of the pumpkin family, can be killed by a single frost, so the number of frost-free days in the year can determine the growing season for these, and this may limit the range of plants which can be grown. Fruit and nut trees can have a whole year's production wiped out by a frost at the wrong time of year. Other perennials, including most of our native broadleaf timber trees, are vulnerable to hard frosts when young and may be hard to establish in frosty places.

Although there are more frosts, and more severe ones, in regions with generally colder climates, microclimate is more important than regional variations in determining whether frost will be a problem in any particular place. This is because it's not the overall winter frosts which cause the problem but the patchy spring frosts.

On very cold nights in the middle of winter every part of the landscape is frosted. These frosts do no real harm, as the plants and animals are adapted to frost at this time of year. In fact these frosts are beneficial to the health of plants, animals and humans alike. It's the frosts which happen in late spring which do the damage, even if it's just half a degree of frost, because neither we nor the plants are expecting it then. Fruit trees are in blossom, and gardeners have planted out frost-tender annuals. These late frosts tend to be light ones, and they affect some parts of the landscape while leaving other parts unscathed.

The combination of landform, vegetation and buildings influences where frosts will occur by affecting:

- how fast the temperature falls at night
- the downhill flow of cold air
- the rate at which cold air is mixed with warmer air

Temperature Fall. Heat loss at night is caused by the radiation of stored heat from the ground into the sky. The rate of heat loss is in proportion to the amount of sky visible from the point in question. Thus any position near a wall, a paling fence or a hedge will cool down marginally slower than a nearby spot in the open, and may stay frost free when the open position freezes, though the effect may be small, extending centimetres from the fence rather than metres.

A position in the angle of two walls will cool down yet more slowly, and the frost-free area will be correspondingly larger. A confined area between buildings may remain totally frost free on all but the coldest nights, whether the buildings are heated or not. The canopies of trees, including deciduous ones which are not in leaf, can greatly reduce upward radiation, and the ground beneath them may avoid all spring frosts, allowing early growth of any plants growing beneath them. This is particularly beneficial to the perennial vegetables in a forest garden.

Temperature fall is also affected by the presence of significant heat stores. Thus a solid brick wall will be more effective than a paling at keeping frost off any trees trained up it. As noted above, the effect of heat stores is also to be measured in centimetres, but the combination of these two effects may be significant on a garden scale, especially for wall-trained fruit.

The temperature falls fastest nearest the soil surface, and the air at ground level will reach freezing point first. When this happens it's known as a ground frost. If cooling continues freezing point will be reached at higher levels. When the air at 1.2m above ground falls to 0°C it's called an air frost.

Downhill Flow. Cold air sinks. It flows downhill, perhaps more like treacle than water, and settles in any low part of the landscape which it cannot easily flow out of. On a still night this can result in frost forming in the low places while the higher parts, especially those on a slope, remain frost free. The places where frost forms are called frost pockets. They can occur on almost any scale, from whole valleys to little patches a couple of metres wide and only a few centimetres lower than their surroundings. They are especially common at the bottom of north-facing slopes.

Areas with a generally flat topography may not have distinct frost pockets, but they are likely to be generally more frosty than sloping sites because there is nowhere for the cold air to flow away to.

Although most damage is done by late spring frosts, it's worth remembering that where there's a frost pocket in spring the winter frosts will also be more intense. Any plants which may be damaged by severe winter frosts, such as herbaceous perennials which do not die down in winter, should be kept away from frost pockets.

Frost pockets tend to be very reliable. That is they always occur in the same places and usually with the same boundaries. Knowing where they are is a vital prerequisite to permaculture design. To some extent they can be predicted, but the only sure way of knowing where they are is by observation.

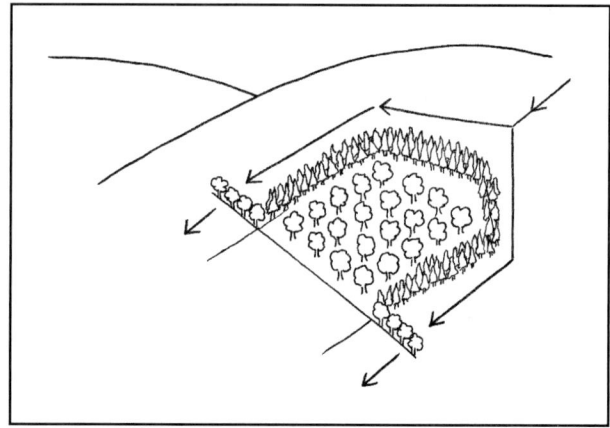

Frost protection can be given by deflecting cold air away from an orchard by means of tree rows.

The ideal solution is not to place anything which would suffer from frost in a frost pocket. But this is not always possible. For example, all the land you have may be in a frost pocket. Or there may be other good reasons for choosing that place to grow the crop in question, for example low lying places often have the deepest, most fertile soil and may be warm suntraps in summer. (See, for example, the case study, A Permaculture Farm, p297.) In these cases it may be possible to alleviate the frost pocket.

This can be done by stopping the downward flow of cold air from above. A single wall, hedge or tree belt across the slope may be enough to do this if it is tall and impermeable enough. (An impermeable barrier is bad from a windbreaking point of view, but a hedge which is impermeable at the base and permeable above is alright as a windbreak.) In many cases a number of cross-slope barriers is more likely to be effective than a single one. An alternative is to deflect the cold air away from the area to be protected with a barrier or barriers at an angle to the slope.

Where an obstruction to airflow like a wall or thick evergreen hedge runs across the slope below the site it can dam up the cold air and actually cause a frost pocket to form. It can often be enough to make a gap in the barrier, or if it's a hedge simply to thin out the lower branches, to allow the frost to drain away. Of course this remedy only works if the land below the obstruction is frost free. If the whole area is one big frost pocket making a hole in the hedge won't make any difference.

If the frost pocket is relatively shallow and you want to grow fruit trees there, it may be possible to avoid the problem by growing standard trees with tall trunks rather than bush trees. On a sloping site it may be possible to grow standard trees lower down, bush trees in the middle of the slope and soft fruit higher up.

Frost pockets may not be the places with the coldest average temperature in the winter. These are often places which are shaded from the Sun during the daytime, which is not the same thing as a frost pocket. If anything, these areas can be good sites for plants which are sensitive to spring frosts, as a slow thaw is less harmful than a quick one. But they are the worst places for winter crops as they stay cold all day. (See Observing Microclimates, pp81-83.)

Air Mixing. In our mild climate the temperature rarely falls to freezing when there is any wind about. On a cold night the air nearest the ground is colder than that up above. Wind mixes the two, preventing the lower air reaching freezing point, unless it's so cold that the air at all levels is below freezing, which is rare. This means that sheltered areas are at more risk of frost than more exposed areas.

The relationship between wind and frost is something to be borne in mind both when observing the landscape and when designing new elements in it. A spot which looks like a beautiful suntrap on a warm summer's day can often be a real frost pocket in the spring. A windbreak which stops the wind completely can increase the risk of frost, and may even do more harm than good to an orchard it's protecting.

Moisture

It's important to appreciate the difference between soil moisture and air humidity. The two don't necessarily go together. For example, large trees and woodland tend to dry out the soil because they transpire so much water through their leaves, yet the air in their shelter is usually more moist than that in the open.

A part of a garden shaded by a wall is likely to have more moisture in both soil and air than an unshaded one. This can be a disadvantage for a gardener waiting for the soil to dry out sufficiently to make a seedbed in spring, but it can be an advantage for leafy crops in the summer. (See Colour Section, 5 and 6.)

If the shade is provided by large trees the soil is likely to be drier than in surrounding areas, but smaller trees, including most fruit trees, use much less water and may not have a great drying effect on the soil beneath them.

Directly at the base of a wall both air and soil are usually dry. There are two reasons for this. Firstly, the wall acts like a wick and draws water out of the soil, even if it has a damp proof course, and quickly loses it to the air by evaporation. Secondly, the wall or the building it forms part of can shield the soil from the rain.

Wind increases the speed of both evaporation from the soil and transpiration from plants' leaves, so sheltered places remain moist for longer than more exposed ones.

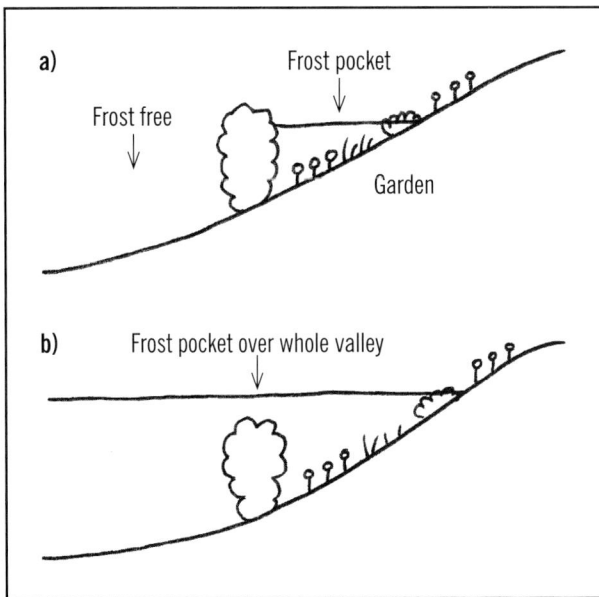

In a) the frost pocket can be cured by thinning or making a gap in the hedge at the bottom of the garden. In b) this would not help at all.

Condensation of mist and fog can sometimes contribute a significant amount of water to the soil. On the Mendip Hills in Somerset farmers try to leave the grass on at least some of their fields fairly long as summer comes on. There may be a shortage of rain in the summer, but the Mendips are famous for their mists, and these usually provide enough moisture to keep the grass growing, as long as there are leaves for it to condense onto.

The potential yield from condensation was brought home to me one foggy winter's day in Devon. A large beech tree stood beside a country lane, and the fog

ASPECT

The aspect of a hillside, bank, wall or woodland edge is the direction it faces in relation to north. Each aspect has its typical microclimate characteristics.

The fact that the Sun's rays strike a north-facing slope at a lower angle than they do a south-facing slope means that a northerly slope is cooler, slower to warm up both in spring and in the morning, shadier and moister. Much the same amount of rain falls on both sides of the hill, but the slower evaporation on the northerly side means it stays wetter. A hillside facing the prevailing wind, which is usually south-west in Britain, will dry out more quickly than a sheltered one, as both evaporation from the soil and transpiration from plants will be increased. Woodland, which needs water but is not as sensitive as other crops to temperature, is well placed on a north facing slope.

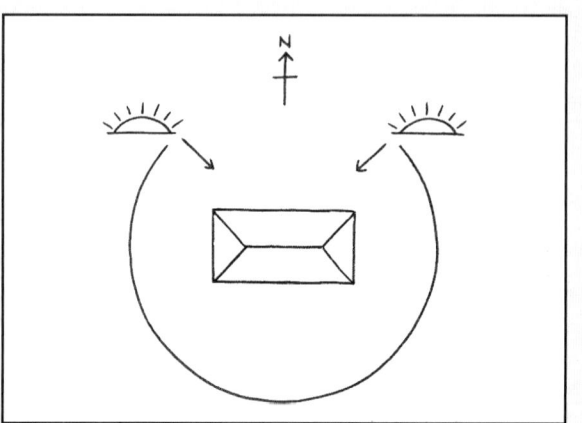

Sunrise and sunset at midsummer.

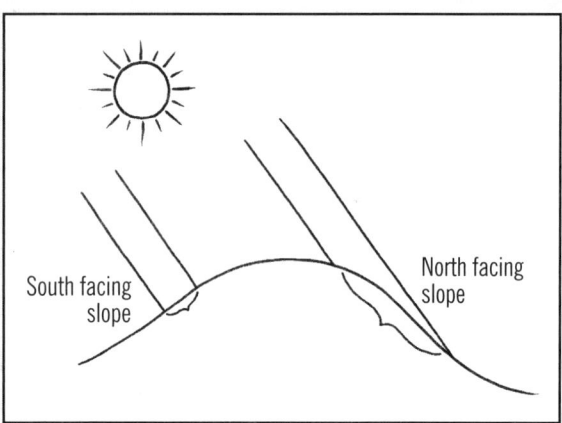

A north facing wall is shady, but it does receive light in the early morning and evening during spring and summer, because between 21st March and 21st September the Sun rises and sets to the north of due east and west. It also receives indirect light at all times of the day. Some shade tolerant plants will do very well there but only as long as there are no nearby obstructions to the indirect light. Since the amount of light is minimal anyway, any extra interference in it will have a major effect.

The dryness which always occurs at the base of a wall is, on balance, much the same at different aspects. The base of a north or east facing wall is shielded from most of the rain, which comes from the south west, so it receives less water. But south and west facing walls are hotter and so lose more water by evaporation.

There is an important difference between east to

south-east facing aspects and west to south-west ones. The easterly aspects warm up more quickly in the morning and this can be a disadvantage for fruits and other frost-sensitive plants. Much of the damage done by frost happens when the ice crystals thaw rather than when they are formed. The quicker this happens the more damage occurs, so damage can be significantly more on an easterly wall or slope than a westerly one.

In addition, the westerly aspect is warmer than the easterly, because the Sun is hotter in the afternoon than in the morning, so more heat-demanding plants are better placed on this aspect.

In northern areas the growing season is short, temperatures are cooler and the Sun shines at a lower angle. For all these reasons a south or south-west facing slope can be a valuable growing area. On a garden scale it can be worth modifying the profile of the land to give it a more southerly aspect. It has been calculated that a 5% increase in south-facing slope is equivalent to being 50km further south.[20]

In more southerly areas it's rarely worthwhile to do this to a whole garden. But constructing a south-facing bank in one part of the garden to provide a special microclimate for early spring vegetables can give a good return on the effort invested. The improvement in microclimate won't be as much as could be had with a greenhouse or polytunnel, but it's a permanent feature, which once installed will never need maintenance or replacement.

The importance of aspect to orchard design is discussed in Chapter 4. (See p238.)

was condensing on it so fast that water ran down the trunk and made an appreciable flow along the tarmac of the road. That was what condensed on the twigs alone. If the tree had been in leaf no doubt the flow would have been even greater.

However the yield of water from condensation should not be exaggerated. The amount condensed on the leaves of grass is never enough to cause overland flow. The 'dew ponds' on the chalk downs of southern England were actually filled by rainfall, not dew.

Combined Effects

It's always important to consider how the different microclimate factors interact. They may act to reinforce each other or to compensate for each other.

For example, herbaceous plants which are somewhat frost tender are more likely to succumb if frosty conditions alternate with wet. Thus the microclimate of a south-facing wall will help such plants survive the winter both by reducing the incidence and severity of frosts, and by virtue of the drier soil. Conversely an area which is both poorly drained and a frost pocket will be worse for them than an area which is either one or the other.

A steep south-west facing slope may be a difficult place to establish trees for several mutually reinforcing reasons. Firstly, the steepness of the slope will mean the soil tends to dry out quickly. Secondly, the prevailing south-westerly winds will dry the soil and increase transpiration from the trees. Thirdly, the slope is facing towards the hottest sunshine of the day, which will also both dry the soil and increase the trees' water consumption.

These microclimate factors interact with other factors, such as soil. In this case, on a steep slope, it's likely to be thin and thus have a limited capacity to store water. Trees planted on a site like this without extra special care are unlikely to succeed.

Good all-round conditions can sometimes make up for one adverse factor which it's impossible or impractical to improve. For example, there's nothing we can do to lessen excessive rainfall or increase sunshine in a wet and cloudy area. But we can give shelter from wind – making sure there's still enough air movement to reduce the chance of fungus diseases – and pay particular attention to soil fertility, especially to drainage. The plants will be more vigorous and thus more able to put up with the wet and lack of sunshine.

Observing Microclimates

There's a useful rule of thumb in permaculture design, which says "don't do anything for the first twelve months," sometimes known as the Twelve Month Rule. It doesn't mean do absolutely nothing, but nothing permanent which commits you to a particular course of action. So, for example, if you move into a new house, go ahead and grow all the annuals you want, but don't plant any trees, build a shed or put down paths. After you've been there for a whole year, seen every season, and had time to think about it, you will know the place well enough to have a good idea of what permanent changes will really work well.

If you make any permanent changes prematurely and then realise they were mistaken, you're faced with a hard choice. Either you discount the energy you've already invested in the place and put in more energy to dismantle and rearrange it, or you live for the rest of your life with a design which you know isn't as good as it could be.

The chance to observe microclimates is one of the main reasons for the Twelve Month Rule. In some cases you can tell at a glance what sort of microclimate to expect: a strong wind in the gap between two houses, for example, or a frost pocket in a deep hollow. But to predict the frequency, intensity and boundaries of even these obvious ones is unreliable, and the less obvious will be missed altogether. What's more, it's notoriously difficult to imagine what a place you see in summer is like in winter and vice versa. There's no substitute for observing and experiencing the microclimates as the year goes by.

Your own body is a sensitive instrument for measuring microclimates. Conventional instruments can be used to measure temperatures, winds and so on, but they are expensive and time-consuming, and do not give the all-round synergistic picture which your body can. It's not so good at quantifying temperature, windspeed and so on, but it is very sensitive to its own comfort, and most plants and animals like much the same conditions as we do.

The best way to learn about microclimates is to walk around your farm or garden and stop in particular places for long enough to really experience what it feels like there. The comparisons between sheltered and exposed places, sunny and shady and so on are much greater than we normally think as we buzz about our daily business, ignoring all the little signals our bodies are giving us. Of course there is a great deal of difference between different days with varied weather conditions. You need to visit each place under a variety of conditions, sunny and dull, wet and dry, with the wind in all different directions and at different strengths.

Just to get the feel of it try leaning against a south-facing wall on a sunny day for ten minutes. Then stand in the open a few paces from it for the same time. Do the same on a windy day. Still, dull days don't tell us as much about the microclimate.

Clues from Plants. The direction and strength of the prevailing wind is revealed by the way it affects the shape of trees, a phenomenon known as wind-flagging. In very windy places the effect is obvious but it's worth looking out for it where the winds are more moderate. A moderate prevailing wind can have a significant effect

It's perfectly obvious which direction the wind comes from here, and how strongly.

The wind blows more gently here, but careful observation reveals that this tree too is shaped by the wind.

on plants, animals and buildings, and in hilly country it may not come from the direction you expect. You may need to look at a tree from every angle before you see signs of flagging, because when the effect is slight it's only visible from a position at right angles to the prevailing wind.

Useful though wind-flagging can be as a clue to the direction of the prevailing wind, the most damaging wind is not necessarily the most frequent one. In Britain it rarely is. One fierce north-easterly can do more damage to plants than a whole year of mild south-westerlies. Windflagging is just one clue to the whole wind-picture of an area.

Looking at frost-sensitive plants just after the first frost of the autumn can reveal where the frost pockets are and where plants are protected from frost. You can only get a really accurate picture if the same frost-tender plants are growing all over the area. It can be worth deliberately planting the same plant in all parts of a garden to see how they fare. Nasturtiums are useful for this.

On the morning of the first frost of autumn this nasturtium is unaffected in its niche by the house, though its leaves are curled and dying just a metre away in the open garden. This gives a clue to the relative microclimates.

Different plants die at different temperatures. Nasturtiums usually die at the first frost, the above-ground parts of fennel die back not long after, and lemon balm comes next. It's not possible to put an exact figure on the temperature at which each succumbs, as much depends on other factors, such as the condition of the plant and the soil. But they can give a comparative scale.

In the spring, the blooming of the first flowers may reveal the warmest microclimate, if the same kind are growing over a wide area but only flowering in one or two places. The amount of growth put on over winter by autumn-sown broad beans can give a picture of the winter warmth in different parts of the garden. (See Colour Section, 11 and 12.)

Frost, Fog, Snow and Fallen Leaves. When you go to check where the frost pockets are after a frost, make sure you go early in the morning. If you go after an hour of sunshine what you will see are the places where the frost thaws out more slowly. There might have been an overall frost rather than one that effects the frost pockets only, and the places which are still frosty in mid-morning may not be in frost pockets at all.

This is a very easy mistake to make. I have often been told "The frost pocket is over there," by someone indicating an area that could not possibly be a frost pocket but is shaded from the low winter Sun.

Mist and fog often collect in the bottom of valleys, leaving the hills clear. There may be a fairly close match between the area covered by mist and the frost pockets, and in a winter with no frosts observing this may be a good second best to observing the frost pockets themselves. Obviously this is only useful for large-scale frost pockets. The small ones just a few metres across are very hard to detect by any means but direct experience.

If snow falls in windless conditions so that it forms an even layer all over, watching how it melts can give clues about the relative temperature of different parts of the landscape. It's hard to tell much if the snow has drifted because it will lie longest in the places where it's thickest rather than those which warm up fastest. These winter observations may not apply directly to the summer, as the angle of the Sun makes a big difference to the patterns of sunshine and shade.

Drifting snow can tell a great deal about the windy and sheltered places, and so can autumn leaves. Both get deposited where the wind is less intense.

Observing microclimates is part of the skill of reading the landscape. It's a skill that can only be acquired by doing it and the ideas given here are a starting point rather than a complete kit. It's great fun, and once you start looking for it you will see more information about microclimates in the landscape than you probably ever thought existed.

WINDBREAKS

Windbreak is a general term for all structures, living or non-living, which are used to reduce windspeed. A shelterbelt is a windbreak consisting of more than one row of trees, and the term windbreak is often reserved for a single line of trees. A windbreak or shelterbelt consisting mainly of shrubs is better called a hedge.

Details of windbreaks for special situations are given elsewhere in the book. (For houses and urban areas on p153, for gardens on p186, for orchards on p238 and for farms on p282.) This section covers the principles of windbreak design.

Windbreaks have great possibilities for multiple outputs in addition to their main function of reducing windspeed. Full lists of the functions and uses of trees and woodland can be found in Chapter 11. (See pp299-309.) The mind map below is a check list of outputs for windbreak designers.

There are negative effects from windbreaks as well as positive ones, and these are given in the mind map below on the right.

The roots of windbreak plants compete with crop plants for both water and plant nutrients. The roots of most trees extend at least one and a half times as far as the branches, and in sandy soils as much as three times.[21] Productive plants can be kept back from the windbreak and the edge strip used as a road, path or wildlife area. Where space allows roots can be pruned back by regular underground cultivations. In gardens the competition effect can be greatly reduced by avoiding the most competitive trees and shrubs, though some of these may be useful on a broad scale. They include willows, poplars, eucalyptus, leyland cypress and garden privet.

Root competition is much more of a problem where space is at a premium. On broadscale farmland the loss near the windbreak is more than compensated for by the overall yield and it's not necessary to treat the edge of the field any differently from the rest of it. In a garden it's much more important to avoid competitive trees and to site a path near the hedge.

Shading varies according to the time of year and time of day, and to the orientation of the windbreak. A north-south orientation generally gives the least reduction in yield due to shading. Broadly speaking the area significantly effected extends to the same distance from the windbreak as the height of the windbreak itself.

Frost risk can be increased by windbreaks in two ways: firstly by reducing overall air movement, and secondly by damming up the downward flow of cold air and creating a frost pocket. (See section on Frost, p78.)

Nothing dries things out like the wind. Stopping it from doing so can be a help when crops are suffering from moisture stress, and generally reduce the need for watering in gardens. But in other circumstances it can increase fungus diseases and slow down the drying of hay or grain crops at harvest time.

While windbreaks can provide habitat for useful wild species, such as pest predator insects, they can also be home to creatures which may be less welcome, including foxes, crows, rabbits and slugs. Some trees and shrubs harbour diseases of crops. But most diseases are host-specific, so as long as certain trees are avoided there should be no problem. (See box, Bad Neighbours, overleaf.)

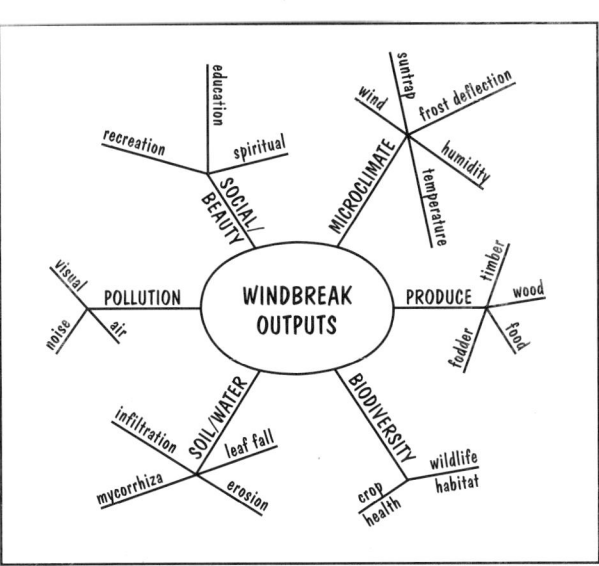

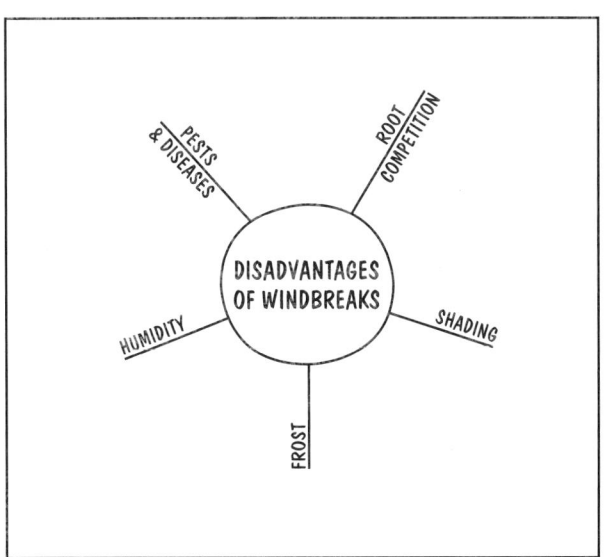

BAD NEIGHBOURS

A number of trees and shrubs can harbour pests or diseases which also affect crop plants. Some which are specific to orchards are given in Chapter 9. (See p240.) The main ones which might affect general garden and farm crops are given here.

Tree or Shrub	Disease or Pest	Affected Crops
Hazels and seedling conifers	*Botriytis cinera*	cereals fruit, vegetables, etc
Prunus serotina	aphid	beets
Privet, *Rubus*, roses	*Verticillium* wilt	pasture, potato, plum, vegetables
Spindle	aphid	bean
Hawthorn	fireblight	all fruits, especially pear

Cross infection is not inevitable, and if these species are present it's not essential to remove them. But they are best avoided in new plantings if the crops in question are to be grown.

Principles of Windbreak Design

The principles of windbreak design are summarised in the mind map below.

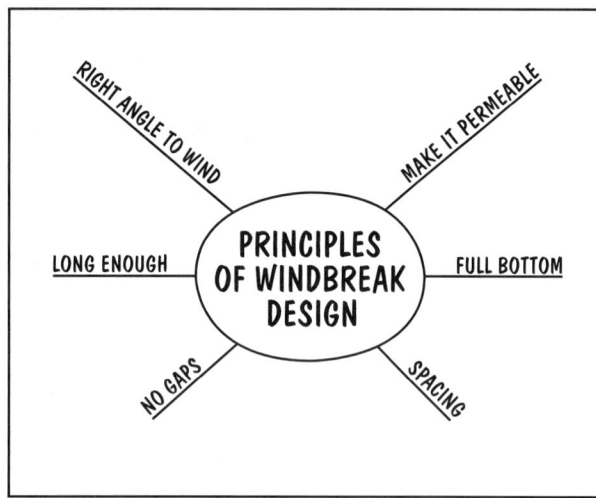

Make It Permeable
A barrier which is completely impermeable gives very good shelter immediately in its lee but can actually intensify wind damage only a short distance downwind. All the air is deflected up over the obstruction, causing it to speed up. On the downwind side a partial vacuum is formed and the speeded-up wind rushes in, causing eddies which can be more destructive than the original wind was. On the upwind side wind rebounds from the barrier and blows strongly away from it at ground level, often causing damage to plants growing there.

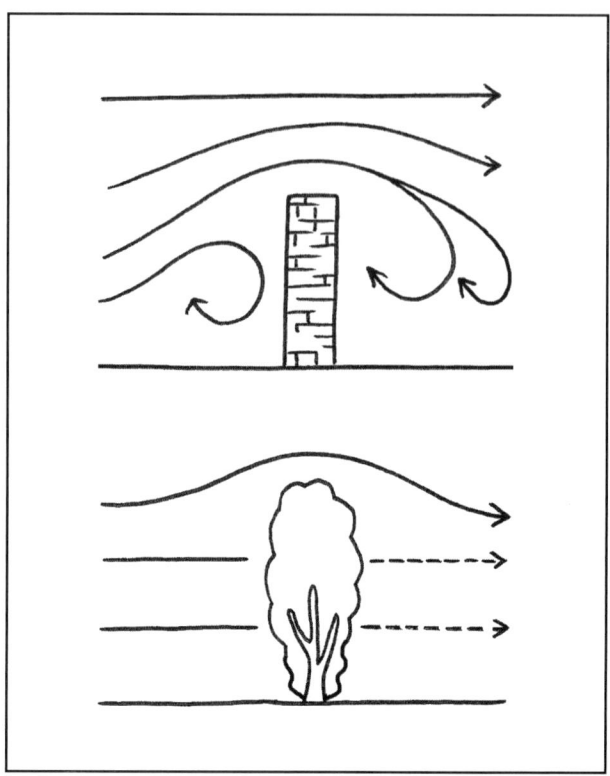

Dense and permeable windbreaks.

A permeable windbreak allows some of the air to pass through it. The reduction in windspeed immediately in the lee of the windbreak is not so great, but there are no gusty eddies and the effect is felt much further downwind than it is with a dense or solid barrier.

The ideal windbreak for buildings and crops is 20 to 50% permeable. That is, if you stand in front of the mature windbreak the mesh of leaves, stems and twigs should cover 50 to 80% of the view. If you're planting a living windbreak it's not possible to be this accurate, but the figures give something to aim for. (This does not necessarily apply to shelter for animals. See p284.)

Thus brick walls and hedges or belts of dense conifers make poor windbreaks, while openwork palings and mixed, twiggy trees or shrubs make good ones. Where a continuous belt of conifers exists, removing every other one can improve it as a windbreak.

It's more important that the upper part of the windbreak is permeable than the lower part. An existing brick wall can be improved by growing trees beside it. In effect the trees become the upper part of the windbreak, and as long as this is permeable problems with destructive eddies will be minimised. Victorian walled gardens were never without their sheltering trees.

Purpose-made windbreaks of perforated plastic can

be useful in the short term while a living windbreak is established in their shelter.

A windbreak with a triangular profile is aerodynamic, and most of the wind goes straight over the top of it even if it has a permeable texture. A vertical sided windbreak is more effective. This is unfortunate, as the extra edge in the triangular profile gives plenty of opportunity for fruit and nut production in the windbreak itself. A good compromise is often a vertical edge to windward and a sloping edge on the lee side.

A wide block of woodland is effectively a dense windbreak, and gives the same kind of shelter.

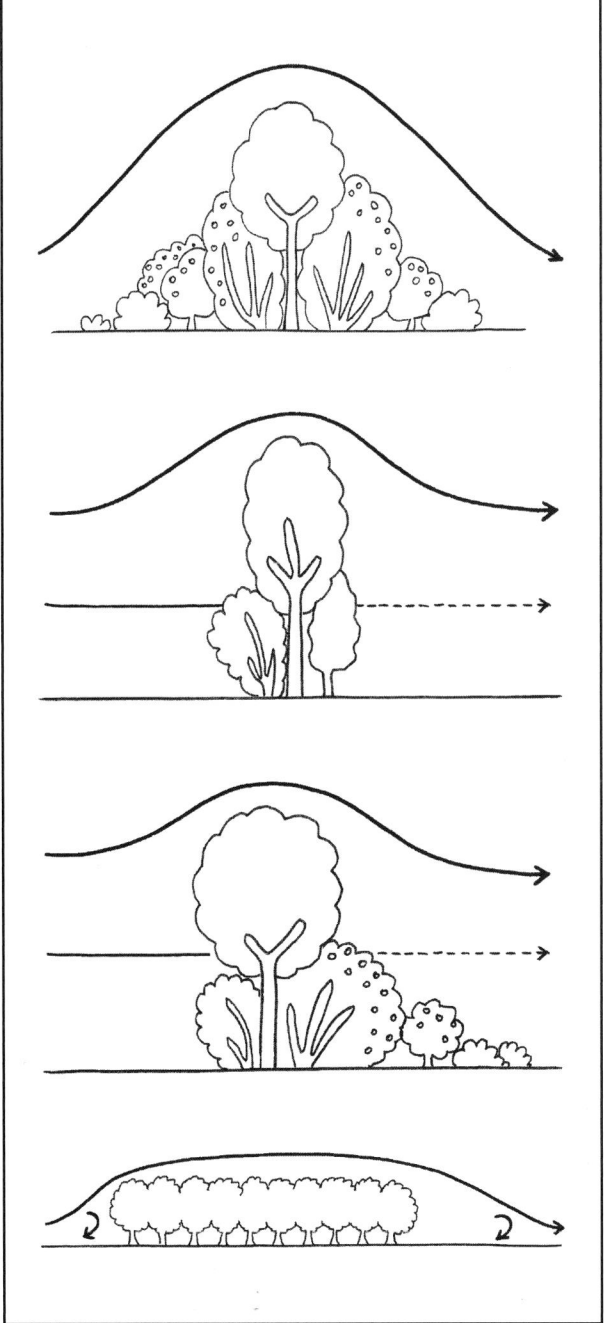

The profile of a windbreak, as much as its composition, determines whether it will slow the wind or deflect it upwards.

Full At The Bottom

As well as being too dense, a windbreak can be too open. If the lower part of it is nothing more than a line of tree trunks the wind is funnelled through the gap between the trees' crowns and the ground and speeds up. Although it may give a little shelter at some distance downwind, immediately to the lee of it the windspeed is greater than in the open.

This often happens as a windbreak matures and it can be corrected by planting shade-tolerant shrubs, such as *Eleagnus* species, beneath the trees. It can be avoided by incorporating shrubs as well as trees into the windbreak from the start. In a multiple-row shelterbelt these must be shade tolerant species, though in a single line windbreak this is less important. It can also be avoided by choosing trees which characteristically retain their lower branches as they grow and thus keep a bushy bottom. Some poplar cultivars have this form.

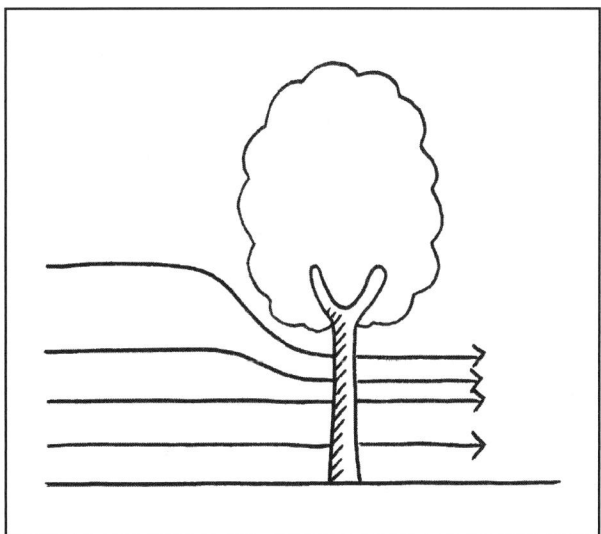

A windbreak with no 'bottom' actually speeds up the wind.

The principle here is always to aim for the bottom to be denser than the top.

Spacing

In order to decide how far apart to place windbreaks it's necessary to know how far the shelter it gives will extend.

Of course the protected area tails off gradually, so the answer depends to some extent on the degree of protection required. Since the wind's energy varies roughly according to the cube of its velocity, a small decrease in wind speed gives a significant improvement in conditions. Thus, while a 50% reduction in windspeed is usually considered ideal, 20% is still useful.

The height of the object which is sheltered is relevant, because the height of the shelter given, as well as the reduction in windspeed, tails off with increasing distance from the windbreak. This needs to be taken into account when designing a windbreak to shelter buildings.

The distance from the windbreak which is effectively sheltered is normally expressed as so many times the height of the windbreak. For a permeable windbreak as described above, the following rules of thumb can be used:[22]

- For tall structures such as buildings, the area within which windspeed is reduced by 50% or more extends to 6 times the height downwind, and a 20% reduction extends to 12 times the height.

- At ground level, ie for crops, farm animals and recreational areas, a 50% reduction extends to 10 times the height downwind, and a 20% reduction to 20 times.

- A daytime temperature increase of up to 1.5ºC may be experienced within 10 times the height downwind, due to reduced evaporation and to reduced mixing of warm air with colder air from higher levels.

- Upwind, ie in the direction the wind is coming from, effective protection extends to some 2-4 times the height.

- Complete control of wind erosion on susceptible soils is only achieved when the distance between windbreaks is 10 times the height or less.

For a dense windbreak the rule of thumb is:

- A 20% reduction in windspeed at ground level extends to 10 times its height downwind.

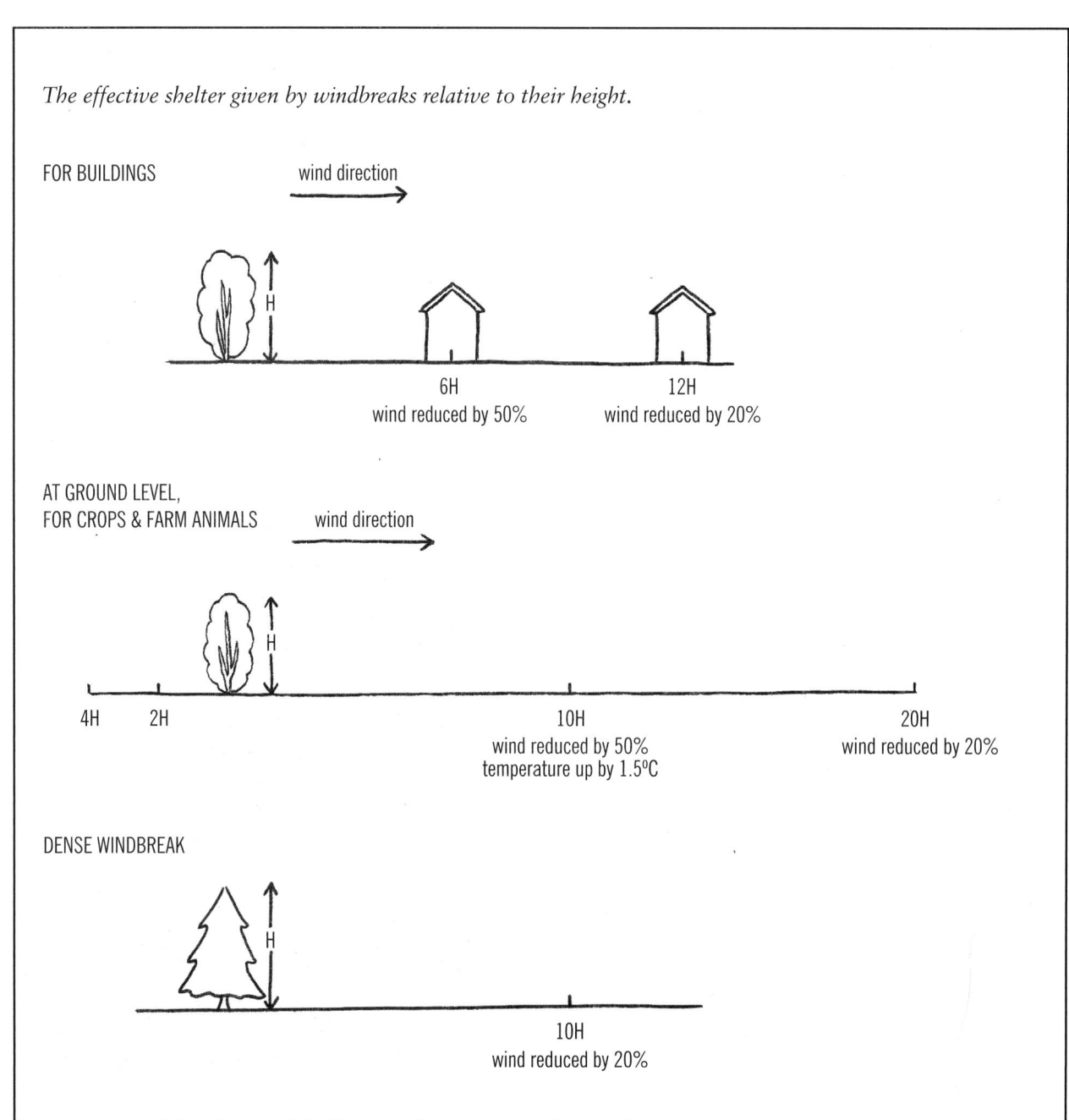

The effective shelter given by windbreaks relative to their height.

FOR BUILDINGS

wind direction

6H
wind reduced by 50%

12H
wind reduced by 20%

AT GROUND LEVEL,
FOR CROPS & FARM ANIMALS

wind direction

4H   2H

10H
wind reduced by 50%
temperature up by 1.5ºC

20H
wind reduced by 20%

DENSE WINDBREAK

10H
wind reduced by 20%

HOW HIGH WILL IT GROW?

Predicting the height of trees is not an exact science. Their final height and growth rate will depend on the soil and microclimate, and especially on wind itself. The more you need a windbreak the more difficult it is to grow one! Under adverse conditions trees may not reach half their potential height. Control of grass competition is essential to fast growth.

As a general rule the quicker a tree grows the shorter its total lifespan. Thus a mixture of fast and slow growing species may give the best overall value.

Mature height of tree species

30m or more	Sweet chestnut, beech, limes, oaks, poplars (black, grey and hybrid) sycamore, most conifers.
25-30m	Ash, wild cherry, holm oak, hornbeam, willows (crack and white), white poplars, black locust.
15-25m	Alders, aspen, birches, Norway maple, yew.
to 15m	Bird cherry, hawthorn, holly, field maple, rowan, wild service, whitebeam.
to 10m	Cherry plum, damson, crab apple, elder, hazel, juniper (erect form), willows (goat and grey).
to 5m	Blackthorn, buckthorns, dogwood, *Eleagnus*, guelder rose, wayfaring tree, Siberian pea, spindle.
to 2.5m	Sea buckthorn, gorse, juniper (prostrate form), wild privet, wild roses, ramanas rose, purple willow.

Growth per year in first 20 years

200cm	Some poplar cultivars.
100cm	Poplars, willows.
60-100cm	Alders, ash, birches, Douglas fir, Sitka spruce, hybrid larch.
50-60cm	Field maple, Norway maple, sycamore, hornbeam, sweet chestnut, wild cherry, European larch, western red cedar.
50cm or less	Oak, beech, limes, pines, Norway spruce, yew, holly.

These figures are for favourable conditions in a timber plantation, and vertical growth may be less in a windbreak, where trees can put more growth into branches.

Within this 20 year period some trees have distinctive growth patterns. Alders and western red cedar put on very fast growth in the first few years and then slow down, while oak and beech start very slowly and then speed up.

This ash wood is some 3km from the sea, growing on a steep valley side, with the upper edge towards the sea. The slope and soil conditions are similar at the top of the wood and the bottom, but the size of the trees is affected dramatically by their exposure to the wind. The photograph on the left shows the top edge of the wood and photograph on the right the bottom edge.

No Gaps

Where there is a gap in a windbreak the wind speeds up to get through it, and the area down-wind of the gap will be windier than it would if there was no windbreak. But sometimes it's necessary to have gaps in windbreaks for access, and there are two ways of avoiding these becoming wind tunnels. The first is to plant a baffle across the gap. The second, where a series of windbreaks is being installed across the landscape, is to make sure that no two gaps are in a direct line with each other. In very windy areas it may be worthwhile to combine both these methods.

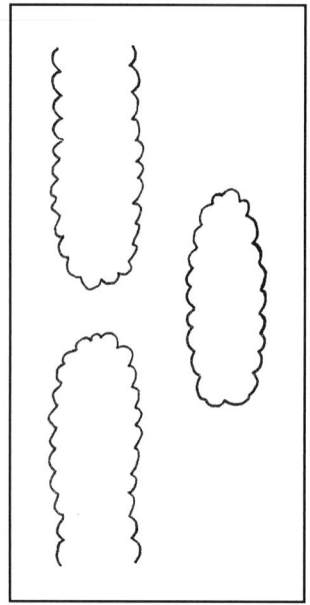

Closing a gap with a baffle.

Long Enough

Where wind flows round the end of a windbreak it speeds up, just as it does when it flows over the top of an impermeable windbreak. The overall length of a windbreak should not be less than ten times its height, ideally not less than 25 times. One authority suggests that a windbreak should be 15m longer than the area to be protected, to avoid the eddies at either end.[23] In an ideal world the issue of length would not arise, because windbreaks should not stand isolated in the landscape but be part of a network without sudden ends and beginnings.

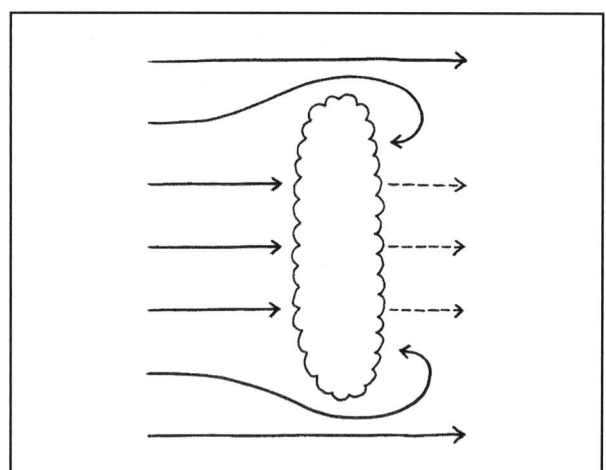

Eddies round the end of a windbreak.

Right Angle To Prevailing Wind

A windbreak at an angle to the wind can actually cause the wind to speed up along the windward face. In most parts of Britain the prevailing wind is south-westerly

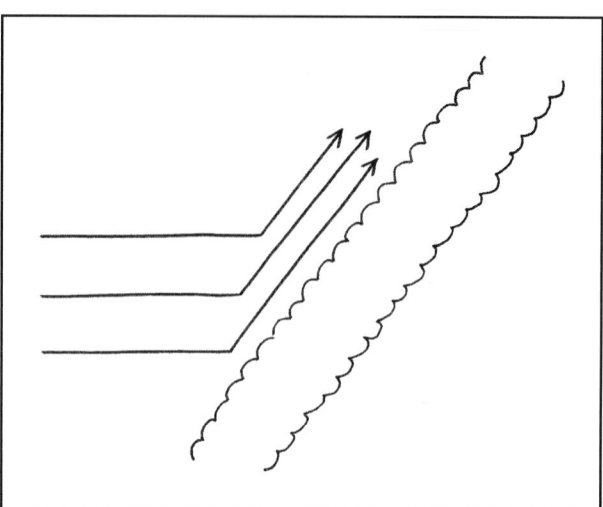

The effect of a windbreak at an angle to the wind.

and the most damaging one the north-easterly, so a windbreak aligned south-east to north-west is well placed to intercept both.

Where a whole network of windbreaks is being installed the angle to the prevailing wind should not be a problem, because there is nowhere a long enough uninterrupted stretch of windbreak for this kind of intensified wind to build up. In fact where the existing field pattern is on some other alignment there are good reasons for working with that rather than changing it. Firstly, it's usually more economical to work with what's already there than to remove it and start again. Secondly, there is a value in maintaining continuity with the historical grain of the landscape.

Coastal Windbreaks

Salt winds are particularly damaging to plants. As well as the mechanical effects of the wind itself – and coastal winds tend to be stronger than inland ones, having blown unimpeded over miles of sea – there are the chemical effects of salt. The worst damage occurs within 5-6.5km (3-4miles) of the sea, depending on the local topography. But occasional storms can bring salt winds further inland.[24]

Protection from salt-laden winds is both more important than protection from inland winds and more difficult, as there are a limited number of windbreak plants which can stand salt winds themselves.

The trees and shrubs listed in the box are more resistant to salt than most. However, if you want to plant a windbreak in a coastal area local knowledge is invaluable. This can be had either from asking experienced local gardeners and foresters, or simply from observing what is growing. For example, on Orkney a mixture of local willows and salmonberry was observed to grow well and subsequently proved effective in a seaside shelterbelt.

Regardless of species, strains of trees and shrubs

SOME PLANTS FOR COASTAL WINDBREAKS[26]

Trees		Shrubs	
Sycamore	the toughest and most coast-tolerant of the broadleafs	*Eleagnus*	most kinds have edible berries; all can fix nitrogen
Ash	less hardy, and only comparatively tolerant of salt	Siberian pea	prefers cold, dry winter, edible seed
		Elder	tough, easy to propagate
Holm oak	broadleaved evergreen, for warmer areas, can be very invasive, do not plant near semi-natural vegetation	Gorse	as first barrier, but can increase fire risk
		Ramanas rose	tolerates severe wind and poor soils, edible fruit
Lodgepole pine	tolerates cold climate and poor soils, including peats	Beach plum	tolerates poor, sandy soil, edible fruit
Corsican, Monterey and maritime pine	for warmer areas	Tamarisks	can take salt spray, mild areas only
Austrian pine	for chalky soils	Salt bush	for warmer areas, edible leaves
Sitka spruce		Blackthorn	tough but invasive, use with care
Cupressus macrocarpa	for warm areas		
Eleagnus (see Shrubs)	some kinds grow to tree size	Sea buckthorn	very invasive, not recommended

Note: Warmer areas include the south and west coasts of Britain.

which originate from coastal areas are likely to be better adapted to salty winds than those from inland. Indeed some foresters believe the origin of the planting material is more important than species choice.[25]

The key to the success of a coastal windbreak lies in the row of trees or shrubs closest to the sea. This must be composed of only the hardiest and most salt tolerant species. Somewhat more marginal kinds can go in subsequent rows. It is recommended that the seaward row should always be of shrubs and should be included even where only a single row of trees is grown.

Planting material for coastal windbreaks should be small. Little plants of 30-45cm suffer less because they are closer to the surface where frictional drag reduces windspeed. Some temporary shelter on the windward side also helps, but it should be fairly open so the plants are conditioned to their windy environment from the start.

Any seaside shelterbelt becomes streamlined, which means that the filtering effect is minimised and they are rarely effective to more than 12 times the height downwind. This can be improved by combining vigorous and less vigorous species to give a less even surface, or even erecting sections of trellis in the shelterbelt so some trees can grow taller than others. It also means that height can only be achieved by having as many rows of trees as possible, with each one benefiting from the shelter of the one to seaward.

Chapter 5

WATER

Although we often take it for granted, water is as necessary to our survival as food, and there is an ecological cost both in supplying it and in purifying it after we have used it. Most of us have little direct control over the supply of our water or the disposal of our sewage. But we can all reduce our impact by using less water and there are many tried and tested ways of doing this. Some of us can further reduce our impact by adopting equally tried and tested alternatives to the mains supply and sewage services.

Sustainable aquaculture is less well-known territory and this part of the chapter is more of an 'invitation to the experiment'. But it's a very important subject. Commercial sea fishing will probably only last a few more decades at the present rate of over-exploitation, and conventional fish farming is both unsustainable and inhumane. Yet water can produce far more food per square metre than land, and the potential of aquaculture, of both fish and food plants, is enormous.

THE WATER CYCLE

Clean water is an increasingly scarce resource. A litre of water can already cost as much as a litre of petrol: a bottle of mineral water in a supermarket is as much per litre as the petrol in the pump outside. This is not quite a fair comparison because the one is sold packaged in small quantities while the other is sold in bulk. But it makes you wonder which will be the most precious commodity a generation from now.

Already oil tankers have been used to transport Scottish water to Mallorca and Norwegian water to Los Angeles. Ten percent of the load is bottled and sold to pay for the trip and the rest mixed with the local supply to bring it up to a quality fit for human consumption.

There is more water than dry land on Earth, but 97% of it is salt and only 3% fresh, and only 0.003% of the total is available for consumption. The rest of the fresh water is either tied up in the ice caps of Greenland and Antarctica or in aquifers too deep to tap economically, or is too polluted to use. Even this tiny fraction adds up to 10 million litres per person.[1] The problem is that it is unevenly distributed. A quarter of it, for example, is in Lake Baikal in Siberia (which itself is now suffering from pollution).

Water is a cyclic resource. It's constantly evaporated from the seas and other water bodies, falls as rain, and drains back to the seas. But we only interact with part of the cycle. It flows into our system, whether house-and-garden, farm, village or town, and out again. This means it's effectively a linear resource, like energy-in-use, rather than a cyclic one like plant nutrients. (See p34.) So our aim must be to use it economically, and reuse it if possible before it leaves the system.

It is in theory possible to recycle water on a single site. But pumping is needed, and it's almost always more energy efficient to make use of the great natural pump in the sky. The other requirement for cycling water is of course purification, and this should be done in any case, whether the water is recycled on site or not. It's a good principle to pass water on to your down-stream neighbour in at least as good a condition as your received it from upstream or from the sky.

There are, broadly speaking, two kinds of water resource available to us: surface flow, and ground water. Surface flow includes rain falling on roofs, runoff from the land and hard surfaces such as roads, and the water in streams, rivers and lakes. Ground water includes water stored in aquifers at various depths beneath the surface, and underground streams.

Surface flows are renewed on cycles that can be measured in days or years, for aquifers the time-scale is years to centuries. There are also aquifers which are not being renewed at all, known as fossil aquifers. There is a large one beneath Arabia. It was filled by rain which fell on the Sahara in an age of higher rainfall and flowed along the gently sloping strata till it came to rest. Clearly this is not being renewed, but it's being pumped out like there was no tomorrow.

This kind of groundwater mining also occurs in aquifers which are being recharged but at a rate slower than the extraction rate. The rate of extraction from the extensive Ogallala aquifer in the western USA averages eight times the recharge rate, and in some areas it's 100 times as great.[2] This area is too dry for rain-fed farming, but it produces much of North America's exportable surplus of food.

The aquifers in south east England are also pumped at a rate which has little respect for their rate of recharge.

In recent years some rivers have run dry as a consequence of the lowered water table. At the time of writing they are mostly flowing again, but this is due to a succession of wet winters rather than a moderation of demand. If demand continues to rise against a background of fluctuating supply, what appears to be an intermittent shortage caused by the weather will soon be revealed as a permanent one.

If we are to live off our water income rather than off capital, surface flows must be our first choice of water supply. But extracting large quantities of water from rivers also has its ecological cost, particularly to the wildlife that lives in the rivers. The sources of supply with the least negative impact are by and large the small, local ones. Harvesting rainwater from our roofs is probably the most ecologically benign source of water available to us.

Water in the Landscape

The way we design the landscape as a whole has a big effect on the nature of surface flows and the recharging of aquifers. A natural landscape, especially one containing wetlands and extensive woodland, acts like a sponge. Heavy rainfall is absorbed, avoiding floods, and is then released slowly over a relatively long time, avoiding severe fluctuations in the level of rivers and lakes. The soil rarely suffers from drought and the water which flows in the streams is crystal clear.

A bare landscape of hedgeless arable fields, hard surfaces and buildings is the other extreme. Water rushes straight off, causing flood in one season followed by drought in the next. It takes soil with it, not only depleting the fertility of the land but silting up rivers and other waterways.

Undeveloped river flood plains are important sponges in the landscape but these days they are often taken for housing and industrial development. At first sight this makes economic sense, as many existing towns and cities have grown up on rivers. But covering the flood plain with hard surfaces and installing storm water drains means that the water which was absorbed and gently released is now sent downstream as quickly as possible, where it becomes a problem for the next downstream settlement. Installing land drains so that arable crops can be grown on the fertile soils of flood plains has a similar effect.

Catastrophic floods in riverside settlements are not natural disasters, they are human-made ones. They are caused by building in places which are likely to flood, by replacing marshes and meanders with hard surfaces, arable fields and straightened drain-like rivers, and by increasing the intensity of rainfall by the enhanced greenhouse effect.

We can avoid these problems by good landscape design, including appropriate tree planting. Trees slow down water in the landscape. Infiltration is increased by the roots of trees opening up the soil, and water is absorbed and stored by the leaf litter layer and the humus-enriched woodland soil.

Large areas of trees are not necessary, though they may be desirable for other reasons. Indeed, since woodland transpires 15% more water than grassland, large areas of plantation can reduce the flow to reservoirs and aquifers by that amount. Carefully placed belts of trees can have an effect out of all proportion to the area of land they occupy.

Some of the key places to plant trees for this purpose, or to allow natural regeneration, are:

- in strips along the contour
- along the banks of rivers, streams and lakes
- in the bottoms of gullies and small valleys
- on steep slopes
- on soils prone to erosion

The last two need not always be covered by continuous woodland. On medium slopes and moderately vulnerable soils a series of tree belts or even hedgerows may have a very beneficial effect. The trees may have multiple purposes – food, fodder, timber, coppice etc – but harvesting should not, of course, ever involve clear felling. Belts of trees along the banks of water bodies will also intercept and utilise plant nutrient run-off from arable land, preventing eutrophication (over-enrichment) of the water. (See p102.)

The preservation of wetland and marsh is equally important. Natural wetlands not only regulate the flow of water but purify it. Whether upstream or downstream from settlements they confer great benefits on us if we choose to leave them alone. They are also, of course, important habitats for wild animals and plants. Using a wetland or marsh for direct production of aquatic crops may reduce its ability to absorb and store water, and this must be put in the balance if we are considering converting them to aquaculture.

Artificial land drainage should not be ruled out. Poorly drained soil can only produce a limited range of produce. Meadow and summer pasture were two traditional uses for flood plains; growing willows, poplars and alders for fuel is another alternative. Poorly drained areas can be used for crops like these without reducing their water holding capacity.

But there are very extensive areas of land in the temperate world where drainage is imperfect but the land is not so wet as to have a high value for regulating water in the landscape. There are few human food crops, either herbaceous or trees, which grow well on poorly drained soil – though many wetland plants have an unused potential as food crops. (See p108-109.) So draining these soils, which are often potentially fertile, is an option to be considered. Indeed most of them have already been drained, and the choice is more often between maintaining the current drainage system or allowing it to deteriorate than one of draining hitherto undrained land.

As for building in flood plains, this is usually a matter

of adding to existing cities and towns. In the long term, a more sustainable economy will in any case be based on a more dispersed settlement pattern, with most new development taking place away from existing centres. (See pp141-145.) Improving water relations is just one good reason among many for decentralising the economy.

Existing settlements can be made to absorb more water by collecting rainfall off roofs and storing it for use. The water storage tanks act like the soil in an unmodified landscape, absorbing rain as it falls and releasing it steadily over time. If the water is used for watering gardens or released to the soil after other uses it helps to recharge aquifers and maintain the flow of rivers in the summer. Collecting rain water also benefits the wider landscape by reducing the demand on distant water sources, whether rivers or aquifers.

In large cities the level of air pollution may make rainwater unsuitable for drinking. But in any case the density of population in urban areas is too great to allow roof water to meet all our needs, so rainwater can be used for non-drinking uses and drinking water can still be piped in from outside. In many cities this will be enough to bring the demand for external water supply down to a sustainable level.

Turf roofs can also play a small part in slowing down the flow of water through built-up areas. Although the thin layer of soil on a turf roof can't hold as much water as a deep natural soil, it is certainly better than a tile roof leading straight to a drain.

WATER IN THE HOUSE & GARDEN

The first question to ask about any resource, water included, is not 'How can we get more?' but 'How can we reduce our level of consumption?'

Water, like energy, can be consumed both directly and as embodied water in the products we buy. So reducing water consumption involves reducing our consumption of material goods overall, not just of what comes out of the tap. 70% of all water used worldwide is used for irrigation of agricultural crops.[3] This suggests that a high proportion of the embodied water we consume is in the agricultural produce we import.

Just how much embodied water we import overall is hard to calculate. But to take one example, the embodied water in our cotton imports is put at 1.5 billion cubic metres per year, a staggering total.[4] (See box, Cotton, A Natural Fibre?) With the decline of manufacturing in our own economy an increasing proportion of the embodied water we consume is in the form of imports. Many of these come from parts of the world where the water situation is more critical than it is here.

Within Britain, households account for some 40% of all water used and 65% of all treated water.[5] The remaining 60% is split between industry and the service sector, with electricity generation by far the biggest single use. Since irrigation is not common in this country, agriculture uses only one percent of the total.

Domestic consumption in Britain probably averages around 160l per person per day.[6] A breakdown of specific uses within the household is given in the pie chart overleaf.[7] This shows that the largest single use of potable water is to flush toilets, and this is obviously the place to start when we look at how to use less water at home.

There are three ways or reducing our demand on the mains water supply:

• Using less water by increasing efficiency within the household.

• Collecting rainwater from the roof, or, if we should be so lucky, making use of a well or spring on our land.

COTTON, A NATURAL FIBRE?

Most of the water used in cotton production is used in irrigating the crop, and it is very thirsty. On average it takes 10,500 litres of water to produce a single kilo of cotton.[8]

In the days of the Soviet Union a cotton monoculture was decreed over vast areas of Central Asia. Water was taken from the rivers which flow into the Aral Sea, and so much has been extracted from them that the sea has now shrunk to a third of its former size. Areas which once provided the local people with fish are now a sandy desert.

Cotton also has more pesticides used on it than any other major crop, except perhaps tobacco. So high is the burden of pesticides in the soil and water of the cotton area of Central Asia that the rate of birth deformities there is among the highest in the world.

Organic cotton is available. It may still be irrigated but, since organic crops can't be grown in a monoculture, growing it does not place such an intolerable burden on the water supply of any one area, and of course pesticides are not used. It's more expensive, but like so many ecological products most of the extra price is due to the fact that it's traded in relatively small quantities rather than the extra cost of production. The more we buy it the cheaper it will get.

The most sustainable alternative is to look to fibres we can grow here in Britain, like hemp and linen. But before that we need to look at how much clothing we consume. Most people in our culture consume far more than necessary due to the concept of fashion, which leads to clothes being thrown away long before they wear out.

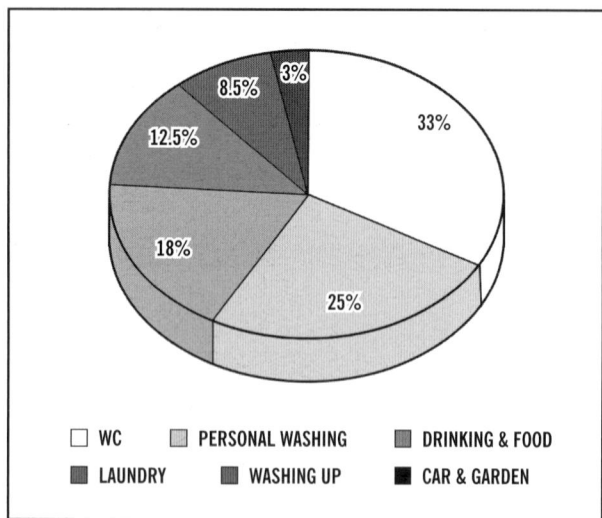

□ WC	■ PERSONAL WASHING	■ DRINKING & FOOD
■ LAUNDRY	■ WASHING UP	■ CAR & GARDEN

How we use water in the home.

- Reusing grey water – ie water which has already been used, other than that used for flushing the toilet.

Using less is the top priority. The yield of rainwater from the roof is only likely to be enough to meet all our needs when we have reduced our consumption way below the national average.

Some of the things we can do cost money, and in the short term these will only pay for themselves in households where water is metered. In the long run they will become a necessary part of a sustainable lifestyle, and many people who can afford to are starting to install them now. It's one of the best investments we can make in our future. But in any case it makes sense to do the simplest and cheapest things first.

Using Less

In the House
Simply adopting an attitude of water awareness will lead to a reduction in use. Where water meters have been trialed in Britain consumption has gone down by some ten percent without any other changes. But there are a number of specific things we can do. The most practical way to look at these is to group them according to how much they may cost.[9]

Almost Free. Water companies are castigated for the high proportion of water lost from leaking mains, but not every householder is immune from the same critic- ism. A dripping tap may seem a trivial thing, but over days or weeks of dripping 24 hours a day it can waste a serious amount of water.

Most toilets can flush efficiently with less than a full cistern of water, and something can be placed in the cistern to reduce its effective volume. This can be done with a Hippo, a plastic bag which can be filled with water and put in the cistern. Many water companies provide them on request. A plastic bottle filled with water, with

the lid screwed on, will work as well, though it's not as adjustable as a Hippo.

A shower normally uses less water than a bath, though a shallow bath will use less than luxuriating under the shower for half an hour.

Not buying a dishwasher actually saves you money. As well as using much more water than washing up by hand, dishwashers use much more energy, as they use higher temperatures to make up for the lack of physical scrubbing.

Make sure the clothes washing machine always has a full load. Recently, plastic balls filled with a substance called zeolite have been marketed as an ecological alternative to detergent. People have done trials with them, comparing them with detergent and with nothing at all, and it may turn out that the most useful thing about these balls is that they've revealed that with lightly soiled clothes you can use water on its own in a modern machine. That means you can cut out the rinse cycle and save a lot of water. It also makes the grey water more usable in the garden.

Tens of Pounds. A dead-leg is the length of pipe between the boiler and the hot water tap. Water is wasted by running off cold before the hot comes through. Lagging a dead-leg keeps the water in it hot for a considerable time. If the dead leg is particularly long it could be worth re-plumbing. It may be possible to reduce its length by redesigning its route, if not, fitting a narrower gauge pipe reduces the volume of water held in it. (This job may come into the next cost band, depending on the nature of your plumbing.)

Spray taps use less water for hand-washing and low-flow heads can be fitted to showers.

Howard Johns

The Ifö low-flush toilet is no different to use from any other dual-flush toilet.

Kits are available from hardware shops to convert a standard toilet to a dual flush one. But the load on the toilet can be reduced more by saving urine to use in the garden instead of peeing in the toilet. (See pp105-106.) A simple covered bucket in the loo will not smell if it's emptied frequently. Alternatively you can keep it outside and nip out for a pee whenever the weather's not too wet or cold. It's not necessary to use it every time you go in order to get a worthwhile reduction in water use – and a worthwhile yield of urine.

Hundreds of Pounds. There are two alternatives to a conventional toilet, a compost toilet or a low-flush model. Compost toilets are the ideal, as they don't use any water at all and retain your personal output of soil fertility for productive use. But they're not easy to install in most houses, and require a certain cultural shift which not everyone is prepared to make.

Conventional toilets have a flush of around 9l, and most new ones of 7.5l. But the most efficient flush toilet currently available has a dual flush of 4 and 2l. (The first time I used one I found the experience of so much being flushed away by so little quite surreal.) There are a number of different models on the market and some of them are little more expensive than conventional toilets. From an ecological point of view the savings are so great that it may be worth installing one even if your old toilet doesn't need replacing, despite the embodied energy in the old one.

Given that they're much easier to install and more socially acceptable than compost loos, low-flush toilets could have much more impact on overall water use in the medium term. 100% of the population saving two thirds of the water they use to flush the toilet would amount to much more than the possibly 10% who would be prepared to use a compost toilet saving all of it.

If you need a new washing machine, shop around for the most efficient one. There's great deal of difference in both water and energy consumption between the best and the worst, and the more efficient ones don't necessarily cost any more. Even if your water is not metered you'll save money by heating up a smaller volume of water.

In the Garden

Although the average family uses less than five percent of their household water for the garden and car washing combined, most people reading this book probably use considerably more in the garden. Productive vegetable gardening can be very water-intensive. The main ways of reducing water use in the garden are summarised in the mind map below.

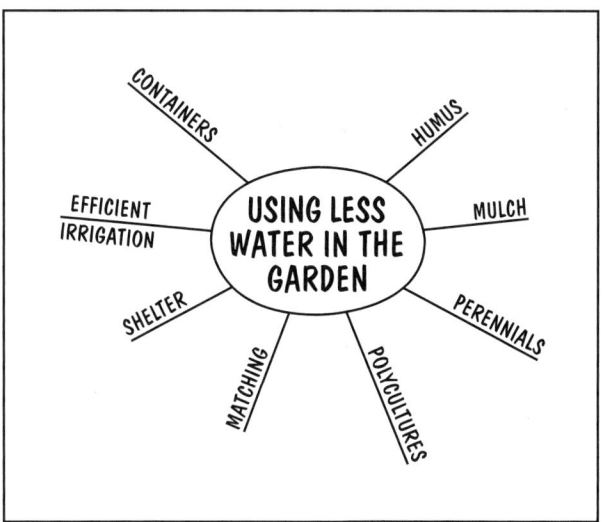

Humus greatly increases the soil's ability to hold water. A soil which is high in humus will hold onto much more of the water it receives than one which contains little. In a rich, dark humic soil less water is lost to drainage and plants can go for much longer without watering. Because humus improves soil structure it also ensures that more water enters the soil rather than running off on the surface. Adding plenty of organic matter to the soil is absolutely fundamental to reducing the need for watering.

Mulch applied to the soil between the plants will reduce evaporation from the soil, and thus ensure that most of the water in the soil is used for plant growth. This can cut the need for summer watering in half. Mulching techniques are described in Chapter 8. (See p194-197.)

Perennial Plants can often make better use of the existing water in the soil as they have an extensive root system which fills the soil throughout the year. Young annuals need watering when the surface soil is dry even if there is plenty of water deep down. It's difficult to water the soil thoroughly but only to a very shallow depth,

Gail Cousens with her Clivus Multrum compost toilet. Although this is a deluxe model it does require a little more attention than a flush model.

Howard Johns

thus much of the water is wasted as it percolates below the depth of their roots. Much is also wasted by falling on unoccupied soil between the small plants.

Polycultures, either of annuals or perennials, tend to make better use of the water in the soil than mono-cultures. A variety of root shapes and depths makes more complete use of the water, and a variety of annual growth patterns makes better use of it through time.

Matching plants to the wetness or dryness of different parts of the garden can reduce watering need. On the whole vegetables need more water than ornamentals, and some vegetables need more than others. Broadly speaking leafy vegetables need most water, fruiting vegetables such as beans and squashes need water mainly at flowering and fruit set time, while roots and onions need least water.[10]

Of course the need to grow annual vegetables on rotation means it may not be possible to give them ideal conditions every year, but where there is a big difference between one part of the garden and another it may be possible to set up two separate rotations. There's much more scope with perennials, whether fruit, vegetables or ornamentals.

In areas with low rainfall and/or very sandy soils, drought tolerant ornamentals can be grown, leaving more water for the vegetables and fruit.

Shelter from the wind reduces both evaporation from the soil and transpiration from plants' leaves. This is one of the main reasons for sheltering plants from wind.

Efficient Watering Systems, including drip irrigation and seep hose, can dramatically reduce the amount of water needed by targeting it more accurately than hosepipes or sprinklers. Swales may be appropriate in some gardens on steep slopes. (See p100.)

Containers. Plants grown in containers, either pots or tubs, take much more water than plants grown in the open ground. (See p175.)

Re-using

Grey water includes the output from the bath, shower, kitchen sink, washing machine and hand basin, in other words any used water which is not mixed with faeces or urine. The idea of re-using it for a purpose which doesn't need clean water is very attractive to permaculturists, but in practice it turns out to be rather difficult to use.

Purifying water to drinkable standard at great expense and then using it to flush the toilet is one of the great absurdities of modern life, but there are two problems with using grey water to flush toilets. One is that it becomes smelly and dangerous to health if it's not used within 24 hours. The other is that it's not often produced

upstairs from the toilet so that it can be used by gravity. The only really efficient way of dealing with these problems is a sealed system containing filters and a pump.

Such systems have been developed, and one of them was trialed on the Isle of Wight, one of the areas in Britain where metering of domestic water has been introduced experimentally. Feedback from householders was said to be good, and the claimed payback period is ten years. But there's little doubt that the money invested would be more effective in reducing the use of mains water if it was spent on water economy measures such as low flush toilets and on rainwater collection.

There's more scope for using grey water for garden irrigation, as much simpler systems are possible. In addition the garden plants benefit from the nutrients in the dirty water, while the soil micro-organisms purify the water.

The key is to keep it simple. If you run a hotel with its own laundry and extensive gardens in a climate with a summer drought, such as southern California, a high-tech sealed, pumped system may pay handsomely both financially and ecologically. If you have a medium sized domestic garden in rainy Britain and the main thing you water is vegetables, only the very simplest and cheapest ways of using grey water will be worthwhile.

The simplest thing is just to take the washing-up bowl out and pour its contents at the base of a fruit tree. Siphoning water out of the bath with a garden hose is more worthwhile, as the volume is greater. Of course the bath must be uphill from the garden, but as most of us have upstairs bathrooms this is usually the case. Anything beyond this requires modifications of the plumbing, and when you get to this point you should consider whether this really is the best thing you can do with your time and money to reduce your ecological footprint.

If you're seriously interested in using grey water in your garden I suggest you consult the books listed in the Further Reading section for this chapter.[11] Meanwhile the main points to note are:

- I can't recommend you use it on vegetables, but if you do so, avoid getting it on the parts of plants which are eaten and avoid salads altogether.

- Fruit trees are a good place to use it as their edible parts are up above splash level.

- If you store it in a tank don't keep if for more than 24 hours.

- Any system involving perforated pipes will eventually clog up. Delivery to a mulched surface with a hose is the most practical method for plumbed systems.

- Avoid bleach, softening and whitening agents, 'biological' detergents and substances containing boron; soap is better than detergent and soft soap better than hard – this is to minimise sodium, which harms soil structure.

- If you need to water regularly through the summer don't use grey water as the sole source of water to any one patch of soil, or you risk a buildup of undesirable chemicals such as sodium.

- Grey water will raise the pH, so avoid acid-loving plants.

The best use for grey water in some gardens may be to grow reeds for mulch material. Any gardener who regularly uses mulch will find that demand usually outstrips supply, and we have already seen how the most productive ecosystems in temperate climates are reedbeds. (See pp25-26.) The common reed can thrive on a diet of high-sodium grey water, and because the plants are not for human consumption there is no problem about contamination.

This kind of reedbed is very different in aim from one intended primarily to purify the grey water (see pp103-105), and therefore different in design. It may well be bigger, as its size will be determined by the amount of mulch required or the amount of space available, whereas a reedbed primarily for purification can be quite small. In a production reedbed the water will be consumed rather than cleaned, and the reeds may be harvested more frequently. A horizontal flow design is adequate, and gravel is not needed. Reeds are invasive, especially in wet soil, so the water should be well contained by a low bank or wall.

This kind of informal reedbed could be ideal for turning grey water into useful mulch.

Collecting Rainwater

Collecting rainwater for garden use is something which can be done in most households. It's always worthwhile to get the largest storage tank you can afford and have room for. The little rainwater butts sold in garden centres only hold enough for a day or two's watering in hot dry weather and are very little use. If possible the tank should be sited in the highest part of the garden so it can be used by gravity with a hose pipe rather than being bucketed. Whether this will be possible will

The rainwater collection system at Tim and Maddy Harland's house. When the tank is full rainwater is automatically diverted to the downpipe by the filter device.

depend on the height of the delivery end of the gutter, but in general the higher the tank is placed the less work you will have in using the water.

The next stage up from garden-only use is a dual water system for the house. This can either be one where the mains are kept for potable uses and rainwater used for everything else, or one where the mains come in when the rainwater runs out. Neither of these is likely to be financially worthwhile unless the water is metered as they involve some duplication of equipment and costs.

Completely replacing the mains with rainwater can be a realistic prospect in areas where air pollution is not a problem. Whether the yield from the roof will be enough to completely replace the mains depends on:

- the amount of water the family uses
- the annual rainfall
- the roof area of the house

Unless you have a metered water supply it can be difficult to calculate your actual water consumption. The national average is around 160l per person per day, of which a quarter is used for drinking, cooking and washing up. (See pie chart on p94.) The potential for reducing consumption by adopting some of the measures suggested in the previous sections is considerable. (See box, A Self-Sufficient Family, overleaf.)

To calculate the annual yield of the roof you need to know:

- The area of the roof. This can be taken as the same as the floor area. The side of a double pitched roof facing the prevailing rainy wind will yield more than the lee side, but the two will balance each other out.

- The runoff coefficient. This allows for the proportion of the rainfall which evaporates, leaks or overflows before it reaches the tank. For pitched roofs in Britain it may taken as 0.7, and for flat roofs, which lose more by evaporation, as 0.5.

- The efficiency of the filter. This is fitted in the downpipe to keep coarse dirt out of the tank, and it may waste a proportion of the water. The supplier of the filter should be able to give the expected efficiency coefficient. It is typically 0.8 or 0.9.

- The annual rainfall. Rainfall figures should not be taken too literally for a number of reasons. Firstly, the nearest recording station may have significantly different rainfall to your site, especially in hilly country. Secondly, the amount of rain which falls in any one year varies widely above and below the average. Thirdly, climate change is a big unknown. But they give something to go on, and you can make separate calculations for worst case or best case scenarios as well as calculating the average.

  The rainfall map in Chapter 4 will give you a rough idea of the figure for your area. (See p70.) More detailed local rainfall figures can be had from the Met Office. (See Appendix D, Organisations.)

The formula is:

roof area in m² x runoff coefficient x filter efficiency x rainfall in mm = annual yield in litres

The resulting figure can be divided by 1,000 to give the annual yield in cubic metres. For example:

70m² x 0.7 x 0.9 x 750mm = 33,075l or 33m³

Comparing the yield with the demand will show whether it's feasible to aim for self-sufficiency or not.

As well as calculating the annual yield, the distribution through the year also needs to be taken into account. Most of our rain falls in the winter, and if self-sufficiency is the aim there must be enough storage to last through the longest drought that can be expected in summertime. This is a simple enough calculation:

daily requirement in litres x longest expected period without significant rain, in days = size of tank in litres

The tricky part is deciding what figure to put on the longest expected drought. Monthly figures can be obtained, but the same reservations apply as with the annual total. Global climate change is expected to have more effect on the seasonal pattern of rainfall than on the annual totals, with more in the winter and less in the summer. If this prediction turns out to be true storage capacity based on present figures will be inadequate.

The quality of roof water is obviously a concern. A primary filter is necessary to keep dirt, leaves and bird droppings out of the tank. If any organic matter is present it will decompose and the water will become foul. People often ask if water in the tank will go 'stagnant', because it's not moving. The answer is that it will if there's organic matter in it, but not if there isn't. The condition we call stagnant water is caused by the decomposition of organic matter in the absence of oxygen. Without organic matter this can't happen.

Air pollution in urban areas may mean that the rainwater is unfit to drink. In the USA drinking rainwater within 30 miles (48km) of urban centres is not recommended. In practice much depends on the prevailing wind. There's a great difference between being upwind and downwind of the pollution source. If in doubt professional advice should be sought.

Although I personally have no qualms about drinking rainwater off my roof, I cannot recommend it for drinking without secondary treatment. Professional help should be sought before designing a rainwater system for potable supply.[12]

A SELF-SUFFICIENT FAMILY[13]

Until recently, Brenda and Robert Vale and their two children lived in an eco-house they designed themselves in Southwell, Nottinghamshire. Their water consumption was 35l per person per day. This is only 22% of the national average but, with a waterless toilet and a range of conservation measures it still gave them a shower each per day and 'very occasional car washing'. This made their total consumption 140l per day, or 51m³ per year.

The average annual rainfall at Southwell is 600mm, which is lower than in most parts of the country. But their house, including conservatory, was big for a family home at 140m².

Using the formula for rainfall yield, their annual total is:

140 x 0.7 x 0.9 x 600 ÷ 1000 = 84m³

Their storage capacity turned out to be overdesigned. At the height of the 1995 drought they had only used 9.5m³ of stored water, which was a third of their total storage capacity.

If the Vales lived in a more average sized house of say 70m² their water supply would fall to 42m³ and would not meet their needs. On the other hand, if they lived in a wetter part of the country the smaller roof would comfortably meet their needs. The average rainfall in central Wales, for example, is more than double what they get at Southwell.

Water On Farms

Water may be used on farms for irrigation, animal drinking, wildlife habitats, aquaculture, electricity generation and domestic consumption.

Irrigation can be a big user of water. If farm-scale crops, as opposed to a garden, are irrigated the volume used for irrigation will be much more than all other on-farm uses put together. Although irrigation accounts for most of the water used by people worldwide, here in well-watered Britain all agricultural uses together account for only around one percent of water consumption.[14] But the irrigation which does take place is mostly in the driest parts of the country at the driest time of year, so its impact on the regional water supply can be considerable.

Irrigation can be extremely worthwhile, especially on high value crops, and where surface flow rather than ground water is used. But it's not without its problems, and the first question to ask about a piece of land where the soil water is limited is not 'How can we irrigate it?' but 'What will thrive here without irrigation?'

Similarly, there is much land which can only grow conventional crops if some form of artificial drainage is installed. Again the first question is not 'How can I drain it?' but 'What will thrive here without drainage?'

When all things have been considered it may be best to go ahead and irrigate or drain. But the first option to consider is always the one of going with what the land has to offer rather than putting in energy to change it.

The Keyline System [15]

Keylining is the name given by an Australian farmer, PA Yeomans, to the comprehensive system he worked out for 'drought-proofing' farms. It uses a combination of undersurface cultivation and a network of ponds to distribute water more evenly throughout the soil of the farm, and to store it for irrigation or any other use.

A deep tined cultivator (see p51) is used to create underground channels in the soil through which water can move. Where a soil water deficit is expected the channels are made so that they fall from the wetter parts of the landscape to the drier. Since the wetter parts are usually lower lying than the drier this may seem impossible. But in many landscapes it is possible to go downhill from relatively high up a minor valley to relatively low down on a minor ridge. This is particularly so in gently undulating country where field sizes are large.

It's not necessary to measure the fall. Simply driving the tractor slightly downhill by eye is enough. The implement should be raised out of the soil before reaching the point of the ridge, because, if two sets of channels approaching each other from opposite sides of a ridge should meet, the water may come up to the surface at that point and cause an erosion gully.

In country where wetness is more of a problem than dryness, the fall on the lines of cultivation can be reversed, so as to lead the water towards the valleys, where it can be led away by a ditch or drain.

Ponds in this system are not sited so as to intercept a perennial stream, they are filled by diversion drains which intercept runoff from the land. A fall of 1:600 to 1:1,500 is recommended for diversion drains, the more gentle fall on more erodible soils.[16] Dam walls are of earth construction. Ponds are only practical where the soil is sufficiently heavy to hold water without the use of a plastic liner. (See p113.)

The keyline system got its name from the fact that the best pond sites are often at the keypoint in a side valley (see p29), and that a series of keypoints, roughly parallel to a main valley, often form a gently descending line through the landscape. In such a case the overflow channel or spillway of one pond can become the beginning of the diversion drain which feeds the next. This line of ponds is the keyline. The land lying below it, which usually includes the deepest, most fertile soils, can be irrigated from the ponds by gravity.

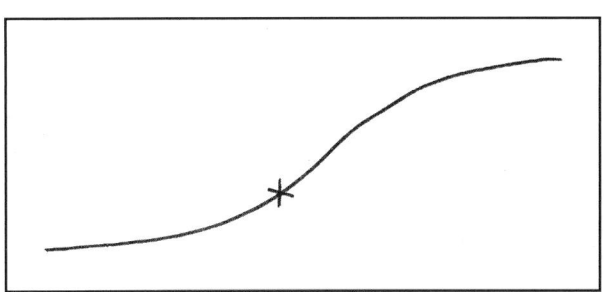

Keypoint – just below the point where the convex slope turns into a concave slope.

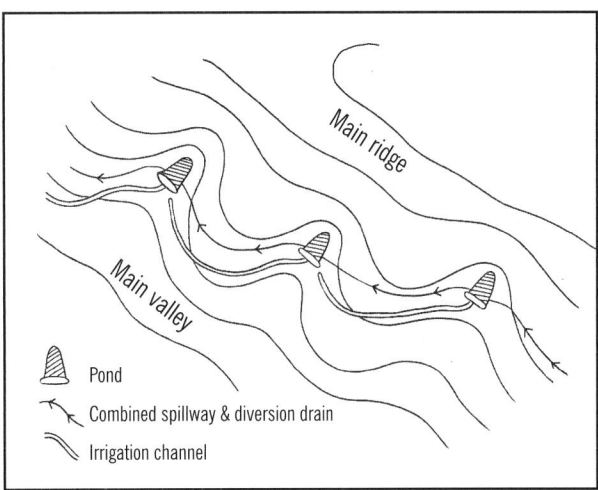

The Keyline System. All runoff from the land uphill from the diversion drains can be collected, while all land downhill from the irrigation channels can be irrigated, and this is likely to be where the most fertile soils are. The entire system works by gravity.

Whether it's worthwhile installing a full keyline system depends on a variety of factors, including soil type, climate and the kind of outputs wanted from the land. In many areas of Britain it may be worth using deep tined cultivations but not installing a series of ponds.

If climate change due to the greenhouse effect leads to a decrease in summer rain the ponds may become worthwhile too.

Swales

The overall aim of the keyline system is to slow down runoff and to store as much water as possible in the landscape. A somewhat different approach to achieving the same end is the use of swales. A swale is a broad, shallow furrow, running exactly along the contour. It is designed to intercept moving surface water and allow it to infiltrate into the ground, in other words to stop it moving. It is the exact opposite to a ditch, whose function is to move water, either to drain the soil or to supply water for irrigation. A ditch has a fall on it, while a swale does not.

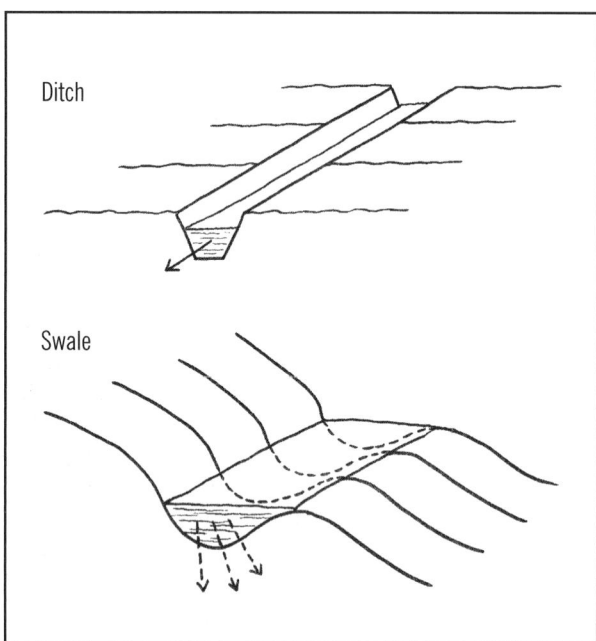

The effect of swales is to take runoff water, which would otherwise be lost to the land, and store it in the soil. There it can be used by plants and move slowly through the ground, evening out drought and flood and keeping springs flowing for longer. Cutting runoff also cuts soil erosion. Swales are an immediate measure which can halt erosion without a drastic change to the farming system.

Swales are rarely installed singly but usually as a series down the slope. The swale system needs to have enough capacity to absorb the maximum expected runoff without overflowing. If a swale overflows a gully can start at that point and the swale can end up causing more erosion than it cures. The capacity of the system is affected by both the size of the swales and the distance between them. On a farm scale the width of a swale, including the bank, is rarely less than two metres overall, though on a garden scale they can be much smaller.

To increase infiltration the bottom of a swale may be ripped, planted with deep rooting plants such as

umbellifers, or, in urban areas, covered with stones or gravel. In dry areas the bottom of a swale may be a good place to grow summer vegetables or other moisture demanding crops. The bank may become an access road or path, and in hilly country a swale system may be the first step towards establishing terraces.

Swales are of use in regions where during the growing season long dry spells alternate with intense, heavy rain. A dry soil resists wetting, so infiltration is low under these conditions and much of the rainfall goes straight off the land. This weather pattern is more characteristic of continental climates than of maritime ones such as that of western Europe, but the greenhouse effect may make it much more common here in the near future. (See p71-72.)

Swales are best seen as a temporary feature in the landscape. The banks should be planted up with trees and shrubs so they can become hedgerows or tree belts. These permanent biological structures will take over the function which the mechanically-built swales have served in the short term.

Swales can be combined with some of the features of the keyline system in a design tailored to fit the specific needs of an individual landscape. Both swales and keylining must be seen as part of the overall approach to water in the landscape described at the beginning of this chapter.

Ben Law with the bunyip water level he used to mark out this swale. (See p406.)

The Hydraulic Ram[17]

This beautifully simple device is a water-powered water pump. It uses the energy in a flowing river or stream to raise a small proportion of the water in that stream to a point much higher in the landscape. It only has two moving parts and once installed can keep pumping for a century or more.

Water is taken from a point upstream from the ram, ideally from a supply head of 1-2m, though it can work with a supply head of half a metre or less. With the waste valve open, water flows through the valve box. This flow closes the waste valve, causing pressure in the valve box to rise. The rising pressure opens the delivery

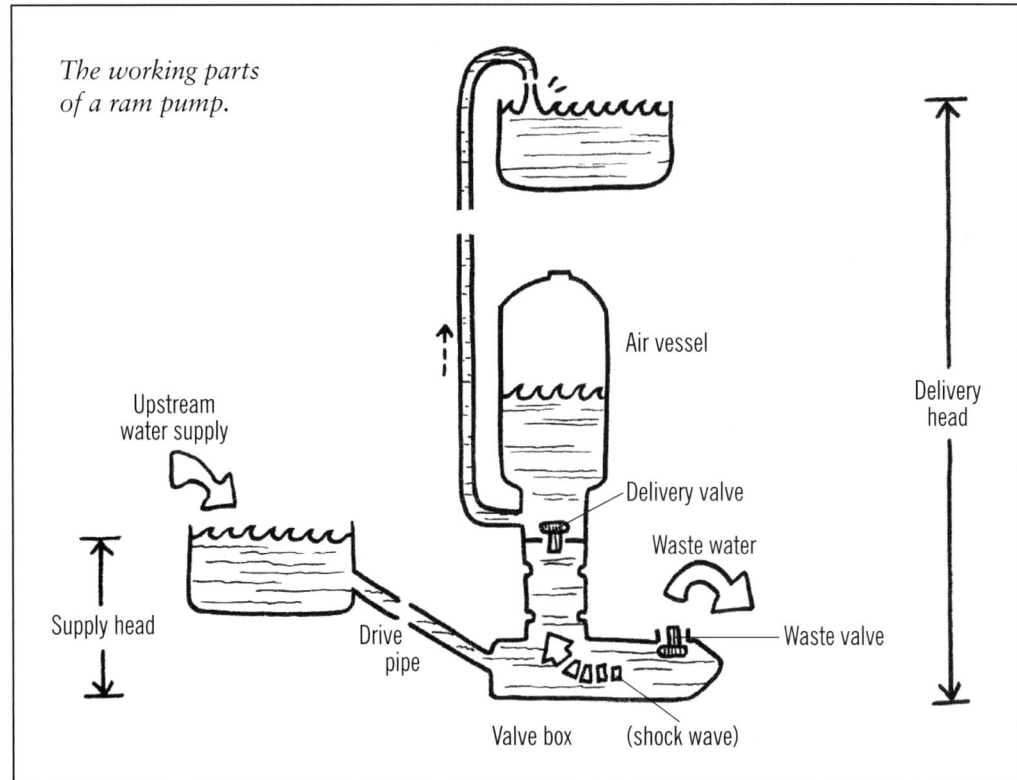

The working parts of a ram pump.

Upstream water supply

Supply head

Drive pipe

Valve box

(shock wave)

Air vessel

Delivery valve

Waste water

Waste valve

Delivery head

valve and water enters the air vessel. This compresses the air in the upper part of the vessel, and, since the delivery valve is a non-return valve, the compressed air forces water up the delivery pipe. Once the flow in the valve box has ceased the waste valve opens, water flows through it again and the cycle can start again.

A hydraulic ram supplied from a head of 2m can deliver water to head of 60m. Of course the higher the water is pumped, the slower the rate of pumping. But since the ram goes on pumping 24 hours a day, 365 days a year, as long as the stream is running, the total amount of water raised can be considerable. Ram pumps, as they are also known, are usually used for domestic water supply or to provide drinking water for farm animals.

SEWAGE TREATMENT

In former times, alchemists tried to turn ordinary materials into gold. Modern society is inclined to laugh at the very idea, but is itself based on something foolish; to take pristine natural resources and turn them into pollution. It is a kind of alchemy in reverse.

Peter Harper & Dave Thorpe[18]

The resources referred to in this quotation are clean drinking water and the human manure itself. Both are useful on their own but mixed together they become a problem. From an ecological perspective it's absurd that we should go to such lengths to collect and purify water to drinking standards and then mix a large proportion of it with our excreta. This not only contaminates the hard-won water, but multiplies the volume of potentially toxic product to be dealt with.

The permacultural approach to sewage treatment is, firstly, to reduce or eliminate the use of water and, secondly, to recover as many of the useful outputs as possible. As we have already noted, there is probably more mileage in reducing water use by means of low-flush toilets than in eliminating it altogether by adopting compost toilets, at least in the short term. (See p95.) But compost toilets are probably the ideal to aim for, as they also enable us to make full use of the plant nutrients in human excreta.

The main aim of all sewage treatment is to decompose the complex organic molecules of which excreta are composed into simple molecules and a residue of harmless organic matter. In this process the pathogens are destroyed. There are two kinds of chemical process by which this can be done, aerobic and anaerobic. (See table overleaf.)

In practice very few decomposing processes are totally aerobic or anaerobic, but they can be predominantly one or the other. Anaerobic decomposition has the advantage that it can produce methane, a useful fuel. But if the methane is not collected it escapes to the atmosphere, where it is a greenhouse gas, many times more potent than CO_2. (See p71.) This means that where we have a choice aerobic processes have a big plus in their favour unless the methane is reliably collected and used in place of fossil fuels.

The main questions to ask about a sewage system are:

- Is it suitable for a single household or for a whole community?

AEROBIC	ANAEROBIC
• Takes place in the presence of air.	• Takes place in the absence of air.
• Carried out by aerobic micro-organisms, which use oxygen in respiration.	• Carried out by anaerobic micro-organisms, which get their energy by means other than aerobic respiration.
• Heat is produced, though it is negligible in most human sewage systems.	• Heat is not produced.
• Carbon combines with oxygen to form CO_2.	• Carbon combines with hydrogen to form methane.
• Sweet smelling.	• Foul smelling.

• What are the inputs, in addition to human excreta?

• What are the useful outputs?

• What are the polluting outputs?

Note that a potentially useful output which is unused usually becomes a pollutant. Methane is the obvious example.

Five kinds of sewage treatment system are considered below. All except the compost toilet are based on water and involve flush toilets.

Mains

This is the conventional type of sewage treatment plant which serves virtually all cities, towns and large villages in the developed world. By definition the mains serve a large number of households in one plant.

Most older systems take rainwater off roofs and roads along with the sewage. This leads to a widely fluctuating rate of flow, and when heavy rain falls there comes a point at which the system can't handle the volume, and raw sewage is discharged into waterways. Modern developments have a separate storm water drainage system. Rainwater is still treated as a problem to be got rid of rather than a resource, but without overloading the sewers.

Some sewers also take industrial effluent. Most organic chemicals (compounds of carbon) can be broken down by the decomposition process, but heavy metals cannot. If heavy metals are present in the soil at high concentrations they can directly contaminate human food. At lower concentrations they can be toxic to the soil microbes which are so vital to the soil ecosystem, such as the *Rhizobia* which fix nitrogen in combination with legumes.[19] In recent years industries have cleaned up their act, and most heavy metals now come from domestic sources. These include zinc from vitamin pills and other pharmaceuticals and copper from plumbing systems, and from roads, in the form of cadmium from tyre wear.[20]

At the sewage works the sewage is divided into liquid and sludge. The liquid is treated aerobically. Various mechanical devices are used to get a thin film of liquid into contact with the air so that the aerobic micro-organisms can get to work. An example is the rotating arm device, often visible from the road as you pass a sewage works, which sprinkles the liquid over a thick bed of clinker. As the liquid percolates down it forms a film over the pieces of clinker. Once it's purified the water may still contain nitrogen and phosphorous, and thus may cause eutrophication in water bodies which the purified effluent flows into. (See box.)

The sludge is digested anaerobically, and in some cases methane is collected and used to generate the electricity used at the sewage works. The residue is sometimes used

---

EUTROPHICATION OF WATER BODIES

Eutrophication is the enrichment of a body of water (or an area of land) with plant nutrients. As a natural process it goes on at a relatively slow rate and tends to lead to increased productivity of plants and animals in the water. Human activity can greatly speed up the process, with dramatic consequences.

Large quantities of nitrogen and phosphorous can be added to bodies of water both from industrial and sewage effluent and from over-fertilised farmland. As sewage and industrial outputs are cleaned up, the agricultural sources are becoming the most important. Nitrogen, which is highly soluble, is leached from the soil in the drainage water, whereas phosphorous, which is largely insoluble, enters the water via soil erosion.

The increase in nutrients can cause a sudden increase in the growth of algae. In extreme cases they grow to such an extent that they use up all the oxygen in the water, whereupon they, and all other aerobic life in the water, die. When all the rotting organic matter is finally decomposed oxygen levels can rebuild and aerobic organisms can recolonise the water, but biodiversity may be seriously reduced for a long time.

Strictly speaking such human-caused events are known as cultural eutrophication, but in common speech the 'cultural' is usually dropped.

as an agricultural manure and sometimes sent to landfill. Unfortunately, the possibility of heavy metal contamination means that using sewage sludge as a manure is prohibited under organic certification rules.

Septic Tank

This is the conventional option for houses which can't easily be connected to mains sewage, usually in isolated rural situations. It's not the same as a cess pit, which is just a holding tank for raw sewage; it's a carefully designed system for the anaerobic decomposition of sewage.

In the tank some solids are digested and the remainder either sink to the bottom as a sludge or float on top as a crust. The liquid flows out to a leachfield, a network of underground perforated pipes, where the soil organisms complete the purification process. The leachfield must be kept well away from any pond or stream, and if the house is near such a water body it may be necessary to site the leachfield uphill from the septic tank to give enough distance, which will of course require some pumping.

A reedbed (see below) can take the place of the leachfield. It may take up less space but it will usually need more head, which on some sites will mean pumping. It's also not as maintenance-free as a leachfield. Reedbeds are sometimes installed as a cure for a septic tank system that isn't working properly.

The solids need to be removed periodically – from once in six months to once in six years – and taken by tanker to a sewage works. With the older kind of tank, rectangular and made from masonry, the crust can be removed separately and composted with straw or bracken then used in a similar way to compost from a compost toilet.[21] Reclaiming some of the nutrients from the leachfield may be possible by growing crops such as comfrey and nettles on it and using them for mulch and compost.

Overall a septic tank is really designed as a disposal system and recovering nutrients from it is not easy. As a disposal system it doesn't have a high ecological cost, and where one is working well it's best left alone. Reducing the volume of water throughput increases the efficiency of a septic tank, so low-flush toilets and other indoor water saving measures can help it.

Methane Digestor

This is an anaerobic system. Since it has an additional output compared to other systems, in the form of energy, it's attractive at first sight. But the inputs are correspondingly high, and the system is not without its limitations.

In tropical countries small-scale, low-tech digestors can work very well on the output of a nuclear family, one cow, and the plant waste from a smallholding. But a certain ambient temperature is needed for the process to operate efficiently. In temperate countries this often means that some of the gas produced must be used to heat the process, and it's only viable with large volumes of sewage, because a small volume has a high ratio of edge to volume, so heat is lost more quickly. Thus a digestor is never an option for a single household here in Europe.

Some digestors work well processing the manure from large pig or dairy units, but these are not the kind of farming favoured by permaculture principles. For domestic sewage in our climate they can only be considered on a municipal, or at least a neighbourhood, scale. The capital cost of the equipment is high, and the operation needs careful monitoring.

The solid residue after digestion forms a high quality compost and can be marketed as such. Indeed some digestors currently in operation regard this as their main output and methane as a by-product. As with all wet sewage processes, there is always some output of nitrogen and phosphorous in the effluent water.

Reedbed

Like the clinker beds used in the mains, this is an aerobic system. Water is trickled over a bed of pebbles so that it forms a thin film in which aerobic bacteria can get to work. Reeds are grown in the bed of pebbles,

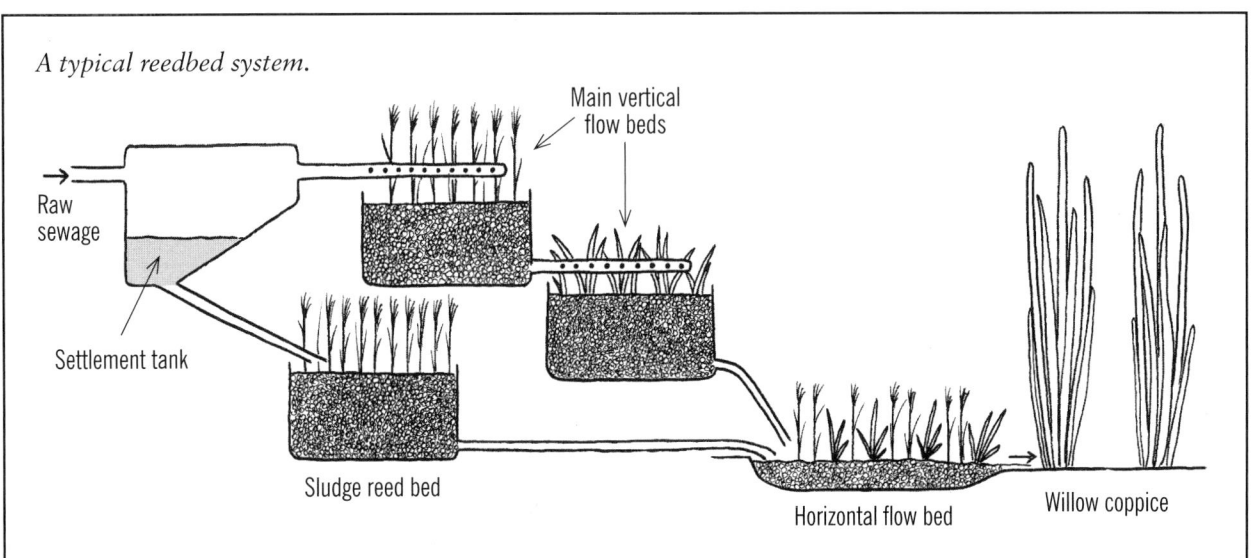

A typical reedbed system.

Raw sewage

Settlement tank

Main vertical flow beds

Sludge reed bed

Horizontal flow bed

Willow coppice

but they are not actually essential to the functioning of the system. They help in a number of ways, perhaps most importantly by creating those habitats which are so rich in microbial life in the rhizosphere zone close to their roots.

The common reed is the plant most often used in reedbeds; yellow flag iris and reedmace are used occasionally. These are very competitive plants which normally form pure stands in the wild. If more than one species was planted together in the same bed it would only be a matter of time before the one most perfectly fitted to the environment would outcompete the others. So diversity in reedbeds is normally achieved by having a series of beds of different species.

Reedbeds can be made to any scale, from a single household to municipal size. In practice they have often been shown to perform as well as or better than state-of-the-art mechanical systems.[22] They're also used to treat industrial effluent of all kinds, including the waste from some of ICI's chemical plants. Horizontal flow reedbeds (see illustration on previous page) are also often added to the end of conventional sewage works systems as a final 'polishing' stage.

A typical reedbed system for treating domestic sewage is shown in the illustration on the previous page. Although a professional design is needed to ensure the system works, the installation itself can be cheap. No unusually expensive materials are required, and a handy person with some bricklaying skills could do the whole job. A reedbed system is often suitable for a small community, such as a hamlet, hotel, or visitor centre, especially where the land slopes away from the settlement to allow the whole system to run by gravity.

There is a small output of reeds from a reedbed sewage works. These can be used for mulch, but they can't be used for thatching as the high nitrogen content makes them decompose quickly. Virtually all the organic matter in the sewage is oxidised.

The reeds will take up some of the nitrogen and phosphorous in the sewage, but there will still be some output. A good way to absorb this is to use the effluent water to irrigate a crop, preferably a non-food crop such as willow coppice. This will of course only work during that part of the year when the willows are actively growing, but this has been found to be sufficient to prevent eutrophication in practice.

WET SYSTEMS

WET (Wetland Ecosystem Treatment) systems are biological water treatment systems which differ from reedbeds in a number of ways.

They are usually more extensive, covering a larger area of ground and constructed with earthworks rather than brickwork. No materials other than plants are brought onto the site so embodied energy is low. A wide diversity of wetland plants is used. In order to prevent one competitive species taking over and crowding out everything else, the system is constructed with the widest possible diversity of habitats, including ditches, ponds, boggy areas and less boggy areas. In addition, the more competitive plants are harvested regularly, which keeps them in check and allows the less competitive ones to thrive.

Harvesting of biomass is an integral part of WET systems. Both woody and herbaceous plants are harvested for fuel, craft materials and mulch. WET is thus an output-oriented system, turning effluent into useful produce. This means that the area taken up by a system, though larger than that needed for a more conventional reedbed, is productive land and may be seen as an addition to the productive area of a holding, not a loss of it. It can also provide good wildlife habitat, particularly for birds, and the flow of water in the landscape is slowed down by the construction of a new wetland. (See p92.)

WET systems are the brainchild of Jay Abrahams, a permaculturist based in Herefordshire. He has installed them in commercial, light industrial and residential situations, large and small. The installation cost can be as little as ten percent of that for a high-tech treatment system. Even if the produce of the system is not harvested the saving in interest on the capital cost will provide a greater return per hectare than any agricultural land use. They have been working well for over ten years now. One of the earliest systems treats the sewage from a busy pub, which has changed hands three times since the system was installed. The present owners were unaware of its existence, but it was still working perfectly despite receiving no maintenance.

This WET system treats the effluent from Westons Cider factory in Herefordshire.

Where compost toilets are installed reedbeds can be used to clean the grey water. A single small horizontal flow reedbed should be adequate, though a larger one is an option if the reeds are seen as a useful output. (See p97.)

Reedbeds need a certain amount of attention to ensure they operate efficiently. It's probably a job more suited to a gardener than an engineer. (See Colour Section 7 and 9.)

Compost Toilet

This is the ultimate ecological sewage solution. No water is used and the return of plant nutrients and humus to the soil is maximised. What's more, because it works on a household scale, the nutrients are cycled locally and people are able to take responsibility for their own shit. But it's not without its disadvantages.

Firstly there's the cultural problem of getting large numbers of people to accept a toilet which doesn't use water. (See p95.)

Secondly, it's hard to make one which is compact enough to fit easily inside a typical bathroom. There are some dry toilets which simply dehydrate the faeces, but this is not the same thing. A true compost toilet needs enough space beneath the seat for the chamber which holds the composting material. An innovative compact model has been designed by Andy Warren, and one has been installed in the AtEIC building at the Centre for Alternative Technology.[23] But even this requires some space beneath the floor for the composting chamber. It's easy enough to build into new houses, and this is now beginning to happen.

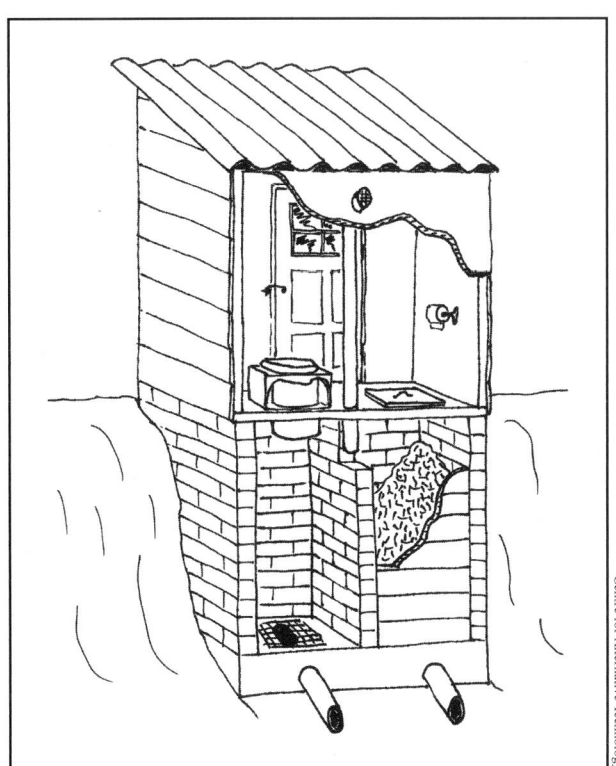

Twin vault compost toilet.

Centre for Alternative Technology

But it's hard to retrofit into existing houses, which is where most of us live.

Thirdly, a compost toilet is part of a complete sewage system, not all of it. Most of them take faeces only, or at most a small amount of urine, and all of them leave the grey water to be handled separately. But if urine can be handled, either by the compost toilet or separately, it does mean that the grey water becomes a relatively benign and harmless substance instead of the toxic sewage produced when a flush toilet is used.

The classic compost toilet design is the twin-vault. (See illustration below left.) The pedestal is movable and can be placed above either vault in turn, so one pile is maturing while the other is accumulating. This avoids both the need to handle uncomposted faeces and the possibility of mature material being contaminated by fresh. It's easy to see why this design is impractical inside a house, though it can be built on as an extension or put in an outhouse.

The key to successful, smell-free, aerobic composting is to keep the material dry. This is achieved partly by adding 'soak' each time the toilet is used. This is a dry, fibrous material such as sawdust, straw or shredded paper, which not only keeps the compost dry and open but also adds carbon to balance the high level of nitrogen in the manure. (See p44.) Keeping the compost dry is also the reason why little or no urine can be allowed into a compost toilet. With the model illustrated here you have to pee elsewhere, but the Warren toilet separates the urine automatically and this is one of it's great advantages.

In either case it's really worthwhile using the urine, as it contains the lion's share of the nutrients in our total output, especially nitrogen. The typical output of nutrients per day for an adult are:[24]

	Faeces	Urine
Nitrogen	3g	8g
Phosphorous	2g	2g
Potassium	1g	2g
Calcium	2g	2g

Urine is free of pathogens when fresh, but it's more difficult to collect than faeces because it needs an enormous amount of soak to balance out its high water and nitrogen contents. A small amount may be used fresh as a liquid manure, but it can only be stored by composting it, and the amount of soak available will in most cases be the limiting factor to the amount which can be saved.

Cardboard is a good source of carbon. The easiest way to use it is to take a plastic box and fill it with sheets of corrugated cardboard with the flutes arranged vertically. Pee on it, or empty a pee pot onto it, either whenever convenient or when it dries out. Sawdust or a bale of straw, placed with the cut ends uppermost, will do as well. It's much best to use fresh urine. Stale urine is smelly, only works if it's diluted, and soon collects a load of pathogens. To start the process off a little fresh plant material such as lawn mowings or comfrey leaves is a great help, as the decomposer organisms need fully formed protein to get started.

An alternative approach is to add both high carbon material and urine to the garden compost heap. (See box, My Compost Heap, p46.) But this is unlikely to use up more than a small proportion of the family's urine.

Compost made with urine is perfectly safe and can be used as any other garden compost on edible crops. Fully composted faeces should be no more toxic than normal garden soil. But to be totally on the safe side it's as well to use them only where they can't come into contact with food, that is on ornamental plants or on fruit trees and bushes.

AQUACULTURE

Aquaculture is the growing of human food, both plant and animal, in water. Thus it includes fish farming, but the kind of fish farming which may find a place in permaculture is very different from that practised today.

Current fish farming has much in common with battery egg or broiler chicken production: monoculture of one species, food brought in from afar at a high energy and ecological cost, animals kept in very crowded conditions with no opportunity to practice normal behaviour, routine use of unpleasant chemicals to control parasites, and often a high pollution output. The salmon farms in the sea lochs of Scotland cause a great deal of ecological damage, largely due to masses of fish manure, uneaten food and highly toxic pesticides accumulating on the sea bed beneath the cages.

The fish which are kept, salmon in salt water and trout in fresh, are carnivores. They are fed pelleted food made from fish such as sand eels which have been taken from the wider sea. Thus they are right at the top of the trophic pyramid, or in other words high in the food chain. This means that in general we would get ten times the amount of food from an equivalent resource base by eating herbivorous fish, or at least omnivores which get most of their energy from plant material. (See pp119-120.) At the same time, over-fishing of sand eels has been associated with drastic falls in sea bird populations.

In terms of both negative ecological impact and energy inefficiency aquaculture of this kind is totally unsustainable. But water is inherently more productive than dry land, and we can take advantage of this greater productivity in a sustainable way by:

- Eating lower down the food chain – plants and herbivorous or omnivorous fish.

- Creating beneficial relationships between aquaculture and other elements in the landscape.

- Using polyculture to fill as many niches as possible.

According to Bill Mollison[25] water can be anything from 4 to 20 times as productive as land given the same inputs of energy or nutrients. The main reasons for this are summarised in the mind map below.

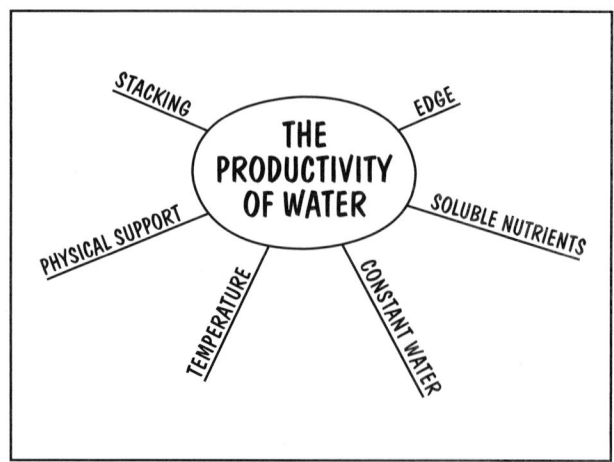

Edge. Shallow water is an edge environment par excellence, being the meeting place of water, soil and air. Its remarkable productivity has been discussed in Chapter 2. (See pp25-26.) This edge effect applies to both plant and animal life.

Plants growing beside the water play an important role in the productivity of freshwater ecosystems. Various studies have shown that, even where there is plenty of plant growth in the water, material falling in from outside is the source of the majority of the plant food in the water.[26] As trees are so much bigger than any other plants, leaf fall from waterside trees can play an important part in sustainable aquaculture. Carnivores may also benefit. For example much of the food for wild trout in the autumn comes from caterpillars falling from the leaves of alder trees.

However, excessive leaf fall, shading and excessive shelter are all bad for the pond ecosystem, so the relationship between trees and water must be carefully designed. (See pp114-115.)

Soluble Nutrients. Plant growth in water is enhanced by the fact that many plant nutrients are in a much more soluble form than are nutrients in soil. This makes them more readily available to the plants and more easily circulated to different parts of the pond.

Constant Water. Only in the most extreme circumstances, when a pond or river dries up completely, do water plants suffer any check to their growth through lack of water. Land plants, on the other hand, are likely to suffer from moisture stress quite frequently during the growing season. Production is lowered as soon as water in the soil falls below the optimum level, not only when there is an obvious drought.

Temperature Buffering. As water warms up and cools down more slowly than air, plants and animals in water enjoy a more even temperature than those on land. In addition to this buffering effect, fish have the advantage of being cold blooded. This means that when the temperature does fall they simply slow down their metabolism, unlike warm blooded mammals and birds, which use up much of their food energy just to keep warm.

Physical Support. In comparison to land animals fish spend little energy in supporting their bodies because they are held up by the water. This means they can devote a much higher proportion of their food to putting on edible flesh than land animals can.

Stacking. Fish can use the three-dimensional space of water in a way that land animals cannot do in the air. This allows them to make a more thorough use of the resources there, as each part of the water is associated with a different food source. In other words there are more potential niches to be filled in a body of water than on a piece of land of equivalent size.

Fish

Sustainable fish farming is more suited to the scale of a farm or smallholding than a domestic garden. Not only is it difficult to raise fish in the limited environment of a small pond, but to be sustainable aquaculture needs to be integrated with other activities. It's the network of useful links between inputs and outputs which can replace the importation of high-energy-cost food, and these are usually easier to achieve on a relatively large scale.

The beneficial relationship between trees and fish has already been mentioned. Fish can also make use of the outputs from land animals, namely their dung. In some parts of the world pig and poultry houses are built out over fish ponds so that the dung falls in the water. This may be eaten directly by some kinds of fish, but its main contribution is as a manure which feeds the plankton on which the fish feed.

The bed of the pond accumulates nutrients and humus. It's possible to make use of this for crop production by dredging, but a more energy-efficient option in many cases may be to grow crops in situ. This may be in the form of a polyculture combining fish and water plants, or by draining the pond periodically and growing dryland crops on the bed for a few years. The latter was known in medieval England and has been done in tropical countries in more recent times. Continuity of fish production can be maintained by having several ponds which are drained and cropped on rotation.

The outfall water from a fish pond will from time to time be high in plant nutrients. Wherever possible this water should be used to irrigate crops, which will make productive use of the nutrients, rather than be returned to a natural watercourse where it may cause eutrophication.

Fish polyculture can potentially make use of a wide range of niches. A possible species mix based on the carp family and making complementary use of the range of natural foods in a pond is:

Common carp	mainly zooplankton (microscopic animals)
Silver carp	phytoplankton (microscopic algae)
Bighead carp	larger algae and some zooplankton
Grass carp	pond vegetation

Except for the common carp these species need a warmer environment than ours to reproduce, perhaps a small pond covered with a polytunnel, but all of them live and grow well here. It's not enough to put them all in a pond and let them get on with it. They need to be closely watched, and numbers of each species carefully matched to the available food supply by harvesting some or introducing more as appropriate. If uneaten food accumulates in the pond it's broken down by aerobic bacteria, and these can so deplete the oxygen in the water that in extreme cases the fish die.

The reason why freshwater aquaculture in western Europe is at present based on trout rather than carp is that we western Europeans prefer their taste, though carp is a great delicacy in eastern Europe. Carp have a reputation for tasting of mud, and indeed they do taste muddy when first caught from a pond or slow moving river. There are two ways of getting rid of the muddy taste. One is to marinate the fish in salt water for 12 hours. The other is to keep them in a tank of clean water before killing them; but opinions vary as to how long this needs to be done for, from a few days to months.

The main reason for eschewing trout is their carnivorous diet, which has been discussed above. It may be possible to create a sustainable aquaculture for trout where all their food is produced on site, though the stocking rate would be low compared to today's intensive systems.[27] But there is another reason for preferring carp and that is water supply.

Trout need a cool water temperature and plenty of oxygen, and these can only be provided by a copious supply of fresh water. A pond soon warms up in summer, and the oxygen is depleted both by use and by the fact that cold water can hold more of it than warmer water. So trout farming is only possible on sites where there is a vigorous stream or river, whereas a carp pond can be sited on almost any farm or smallholding.

For those who want to start profitable but sustainable fish farming in the short term, or to provide their family with a reliable supply of fish, concentrating on the common carp is a more reliable approach than a complex polyculture. This is not as monocultural as it sounds, because the pond will also be inhabited by a wide range of other species, both plant and animal. A carp pond could be compared to a traditional species-rich pasture grazed by beef cattle – except that it's much more productive.

The great virtue of common carp is that they are omnivorous scavengers. They can take small, widely dispersed packages of protein which would otherwise be beyond our reach and turn them into a large edible package without any input from us. They don't need to be fed at all by the farmer, but if they are it's usually just with grain, because they can handle the protein side of their diet very well without help from us. In effect this is turning carbohydrate, a food of medium value to humans, into protein, a high value food for us. What a contrast to mainstream fish farming!

Three ponds, each 5x20m, can give an output of 50kg of fish per year without any external input of fish food. They can be harvested in rotation to give a steady yield.[28]

Harvesting can be either by netting or by draining the pond. For efficient netting the pond really needs to be rectangular. The more irregular the shape the easier it is for the fish to double back round the ends of the net. It's very easy to end up with nothing at all in the net when netting a typical farm pond. A rectangle is of course the least productive shape for a pond, but there's little point in growing the fish if you can't catch them.

Harvesting by draining is only an option where there is enough fall on the land, and a 'monk' needs to be designed into it from the start. The pond doesn't need to be completely drained to harvest the fish. The water level just needs to come down low enough so that you can see what fish are there and take out what you need with reasonable ease.

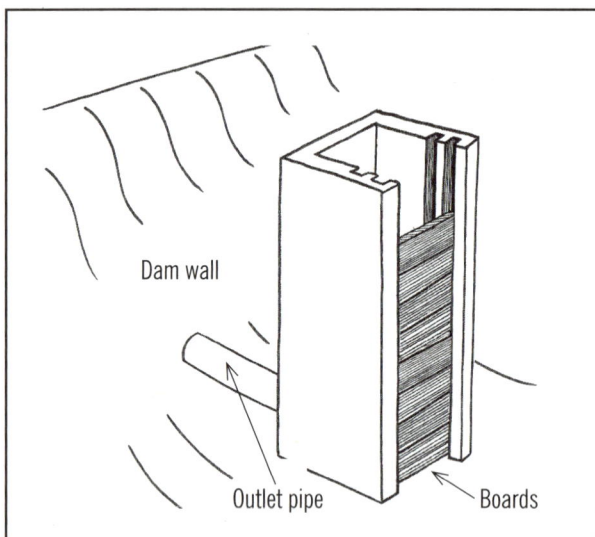

A 'monk' water outlet.

A quick draining followed by immediate refilling is usually less disruptive to the pond ecosystem, and thus to productivity, than netting, which can do much damage to the bottom. This, together with the possibility of designing edge into the pond, makes draining the favoured choice where the topography allows it.

At present the market for carp as food is limited. In fact fish for restocking angling ponds fetch twice the price of table fish, and this is the main commercial outlet at the moment. But a large proportion of the catch could go for home consumption. Perhaps the most important yield of a system like this would be experience of keeping fish, which must be an essential requirement for anyone who wants to design a more polycultural aquaculture for the future.

Before introducing fish into a pond a permit must be obtained from the local office of the Environment Agency. This applies to all water bodies larger than a domestic garden pond. It's a wise law, as unthinking introductions of exotics can cause the extinction of native species (see p369), and even bringing natives from another water body may run the risk of spreading disease.

AN INTEGRATED FISH FARM[29]

An example of an integrated aquaculture system, incorporating cattle, geese, carp, reeds and watercress is the Aquafarm, established by Roy Watkins on a 5.5ha site in Cheshire.

A vital link in the chain is the tiny daphnia or water flea, which is the main food for carp, and is found most commonly in ponds where cattle have access. The system starts with liquid manure from a dairy farm, which is stored in a pond for two months at a temperature of not less than 10°C in order to allow the daphnia to breed. Carp are then introduced, and grown as a commercial crop. The nutrient-rich effluent water from the carp is passed into a reedbed, and the water output from the reedbed is sufficiently purified to be used to irrigate watercress for human consumption.

The reeds can be harvested, though when grown in a nutrient-rich situation like this they are not suitable for thatching as the carbon: nitrogen ratio is just right for decomposition. (See p44.) The purification of the dairy effluent is another useful output, and geese are also kept on the carp pond. They graze the pond margin and their faeces fertilise the pond and contribute to the food chain which supports the carp.

Plants, Broadscale

The potential for farm-scale production of edible water plants is considerable. Many water plants have roots, tubers or rhizomes which are high in starch or sugars. Reedmace (see box) is one example, but there are many others, occupying a variety of habitats, including boggy ground, the water's edge, shallow water and deeper water.

Some of the most productive are the emergent plants, those which grow in shallow water and tend to be tall. These include reedmace, common reed and yellow flag iris. They also tend to be very competitive and both reedmace and reed are usually found in pure stands in the wild. The key to designing a polyculture of aquatic plants is to provide a diversity of different habitats. If there is a variety of depths of water, together with marshy soil of varying degrees of wetness, different plants will dominate different areas.

As the example of reedmace shows, yields can be high, and are potentially even higher if the skills of plant breeders can be brought to bear on aquatic edibles. The big problem must surely be in the harvesting. Getting at the rhizomes under water would be difficult and energy-intensive, either by hand or by machine. The solution may be to grow them in shallow beds which can be drained for harvesting.

Experimental work needs to be done to establish whether the energy ratio of input to output would in the end be better than that of cereals. But there would be other advantages to growing aquatic plants, including providing wildlife habitat and regulating the flow of water in the landscape. (See p92.) Still water and low-lying marshy ground are nutrient sinks, places where nutrients which are washed out of the rest of the landscape end up. Using them for food production is a good way of making use of those nutrients before they are lost to the landscape altogether, and of reducing the need for inputs of fertiliser from outside the system.

One solution to the harvesting problem would be to use the plants as pig forage and let the pigs do the harvesting themselves. Pigs love a wallow, and this could work well. This would still mean draining the pond or bed, but not so thoroughly, as pigs could probably handle mud better than human or mechanical harvesters.

Another solution is to keep production small-scale, so that roots and tubers can be harvested from the bank with hand tools without any need to drain the pond. This could work well in a chinampa system. (See p115.) Ultimately small scale production is the permaculture ideal in any case.

REEDMACE

Reedmace is one of the most promising broadscale aquaculture plants. It's a vigorous plant up to 3m tall, and grows in wet soil or water not more than 15cm deep. It's very invasive, but it can be confined at the edge of a pond by deeper water on one side and dry land on the other.

The main yield is the rhizome, which has a starchy core surrounded by a layer of spongy tissue. The outer layer can be easily peeled off while the rhizome is still wet. The core can then be dried and ground into a flour which has a similar nutritional makeup to other sources of starch. It's said to be very palatable in puddings and when used 50/50 with wheat in biscuits. One American study recorded a yield of 6t/ha of finished flour.[30] This came from a wild reedmace bed and it's not clear whether it represents several years' accumulated production or what could be grown in a single year. Either way it's an impressive yield for a wild plant which has not been selected for high yield over thousands of years, as the cultivated cereals and tuber crops have.

The rhizome cores can also be cooked like potatoes, or macerated and boiled to yield a sweet syrup. Various parts of the plant can be eaten as a vegetable, including the young shoots in the spring, the peeled bases of mature stems, and the immature flowering stems, which are said to taste like sweet corn. The pollen can be added to flours as a protein supplement.

The leaves and stems can be used for thatching, mats and chair seats. Like most emergent water plants, it produces a great volume of these and any surplus makes useful mulch or compost material. The cigar-like seed heads are perhaps the most remarkable part of the plant. Pull a mature one apart and see how the volume of the material increases ten- or a hundred-fold as it's released from the tight confines of the spike. It has great powers of absorption, insulation and buoyancy, and is useful for pillows, wound dressings, babies' nappies and building insulation. Altogether a truly remarkable plant.[31]

Domestic Scale

A garden pond has many uses. Wildlife, frogs for slug control and beauty are reasons enough for installing one, but many of the plants which may be grown in a garden pond are edible and plant aquaculture can be combined with other outputs.

In rural areas it's important that the water supply for plant aquaculture does not come from anywhere where sheep graze, or there is a chance that people eating the water plants may contract liver fluke, a parasite of both humans and sheep. Cooking the food kills the parasite but I would not advise anyone to rely on this.

Slug Predators

Frogs and ducks are both useful for slug control, but you can't keep both because ducks also eat frogs. The two can co-exist in a large farm pond but not in the size of pond which is possible in most gardens. It's one of those choices between a high input / high output approach and low input / low output.

Ducks, if properly managed, can almost eliminate slugs in a garden and produce eggs into the bargain, but they need looking after and feeding. Frogs need no attention at all once you've established them, but they only reduce the slug problem, they don't get rid of it.

Nevertheless a healthy frog population in the garden can make the difference between hardly being able to grow a thing, and just having to take some sensible precautions with the more susceptible plants. They're especially valuable in a garden with lots of perennials, which provide just the right cover for slugs and snails.

Frogs need a pond for breeding. It can be tiny, as little as 1m by 50cm and 30cm deep, but of course it doesn't matter if it's much bigger. The important thing is that there should be some water not more than 60cm deep in it, as frogs like to lay their spawn in shallow water, and a gently sloping side where the frogs can crawl out. If an old sink or other container is used to form the pond a plank or a pile of stones can provide access.

The best water for the pond is from an existing pond, as this will contain the microscopic spores and eggs of the plants and animals which tadpoles eat. Above all make sure the water is not polluted, as frogs are very sensitive to pollution. Some water plants should be added too. These will both feed the animals on which the older tadpoles prey, and give cover for the tadpoles against predators which may want to eat them.

At this point it may be enough to sit back and wait for frogs to arrive and start laying. In fact, if the algae that the tadpoles feed on are present in the water, the adult frogs can smell it from a considerable distance away and will seek out the pond to lay in.

SOME EDIBLE WATER PLANTS FOR THE GARDEN[32]			
Plant	Size *(height)*	Habitat*	Main Edible Parts
Arrowhead	to 1m	water 30-60cm deep/ wet soil	tuber, cooked
Bulrush	to 2.5m	wet soil, shallow water/ deep water	starchy root, raw or cooked; buds at end of rhizomes, raw
Floating sweetgrass	45cm	wet soil, shallow water	seed, sweet taste, a bit small and fiddly to harvest
Flowering rush	1m	wet soil, water to 30cm deep	tuber, cooked
Fringed water lily	small, floating	pond edges; can be invasive	leaves and stems, cooked
Golden saxifrage	low	wet soil, deep shade	leaves, raw or cooked[†]
Watercress	small	shallow, slow-moving water or ponds/wet soil	leaves, raw or cooked
Water lily, yellow	floating	deep water, to 2.5m, slow moving or still	starchy root, cooked; bitter taste, improves with cooking

* Preferred habitat / will also grow in.

† Not terribly tasty, but virtually the only food plant for this habitat.

All the plants listed here are perennial and native to Britain. There are exotic species of arrowheads (*Sagittaria* spp) but they have smaller tubers.

If you want to be sure of getting frogs in the first year, though, it's best to find some spawn and put it in. Most spawn is laid during March and April, though in mild years, and especially in mild areas such as Cornwall, it may be found as early as February. The best source is probably from a friend or neighbour who has frogs in their own garden pond, or from the county Wildlife Trust. Collecting from the wild should be done with caution, because frog numbers have fallen catastrophically in recent years. But spawn can be taken from places where there is an obvious excess, or from puddles which will obviously dry up before the tadpoles hatch and mature.

The tadpoles become frogs some time during June, July or early August. Then they leave the water and it doesn't matter if the pond dries up. They need plenty of dense, shady, low-level vegetation to hide them from predators and keep their skin moist. In fact they like just the sort of habitat that slugs do! The following spring there should be a noticeable drop in the slug population.

Unfortunately cats love to kill frogs, and if cats frequent your garden they may make it impossible to build up a decent frog population. Toads, contrary to popular opinion, rarely eat slugs. In fact they eat beetles, which do eat slugs. But they do eat snails, so they may be worthwhile in a garden with more snails than slugs. They lay in water over 60cm deep and their spawn is laid in long thin strips, which makes it quite easy to distinguish from the amorphous mass of frog spawn.

Ducks, as well as eating frogs, will also eliminate all plant life in a small pond. So if you want to keep ducks and grow water plants you need two ponds, or a pond with a secure fence down the middle of it. In fact the ducks need a pen of their own where they can live all the time they're not on slug patrol, because given unlimited access to the garden they will eat the vegetables.

Indian runner ducks.

Their pond should be in their permanent pen. They can manage with a small tank, such as an old sink, but they do like to swim around on a proper pond if at all possible. They must also have access to clean water to prevent eye disease, so if their pond is small and muddy they will need a bucket of clean water each day.

Most breeds of ducks are more interested in plant food than animal and will eat the vegetables in a garden in preference to the slugs and snails. Khaki Campbells have the most carnivorous tastes and are the best breed for slug and snail control. Indian Runners come second. The best time to let them into the garden is when you are there gardening. Keep an eye on them, and when they finish the slug course and start on your greens pen them up again.

Some plants will need netting to keep the ducks off them. But their favourite vegetables are young brassicas, and these have not yet been planted out in the spring, when the slugs are worst. In the winter, if there's an area of garden with no vegetables in it the ducks can be given free range over it and they will search out the slugs' eggs with a zeal that's a joy to behold. This gives the garden a clean sheet at the beginning of the growing season, with the only slugs being immigrants from outside.

Free ranging during the winter can cause compaction at the soil surface but this is easily corrected with a light forking, and the ducks do love to range. If you're keeping ducks anyway and don't have broad acres for them to roam on, setting them free to forage in the garden now and then is only being fair.

POND SITING & CONSTRUCTION

A love of water is part of human nature, and almost everyone wants a pond in their permaculture design. I understand the feeling. I've been lucky enough to live in a spot where I could wake up with the sound of running water in my ears, and see the moon reflected in still water before I went to sleep. But not every place is so endowed and in many places trying to create what isn't there means working against the landscape, not with it.

In places where the soil, water supply and topography are favourable, only some small changes are necessary to create a pond, and it will soon become a harmonious part of the landscape. But on a well-drained soil, or where there is no water supply other than runoff from roofs, roads or land, or on a sloping site with no natural hollows, a great deal of energy will be needed to create and maintain a pond. It will always look like what it is, an imposition on the landscape.

This applies more to large or medium-sized ponds than to small garden ponds. Gardens are highly artificial, humanised places anyway, and the amount of energy required to make a little pond of a couple of square metres is small compared to the value it can give. But large ponds, even large garden ponds, are better designed with the natural flow of the land rather than against it.

Broadscale

Pond design, like all permaculture design, is a dialogue between the designer and the land. Whether to have a pond, and if so of what kind, depends both on the reasons for wanting one and on the characteristics of the landscape.

So in deciding whether to install a pond the two basic questions are: 'Why do we need or want a pond?' and 'Is there a suitable site?'

The functions of a pond can include:

* aquaculture, of both fish and plants
* irrigation
* wildlife
* water for domestic animals
* fire fighting supply
* electricity generation
* beauty and spiritual qualities
* swimming and boating
* microclimate modification

Not all of these will apply to every pond, but the more yields that can be had from a pond the more worthwhile it is.

Siting

The questions to ask in looking for a suitable pond site, or sites, are given in the mind map. It's always a good idea to identify all the good sites on a piece of land. Two small ponds can be more versatile than one larger one. Even if only one pond is contemplated in the short term more may be wanted in the future, and it's a good idea to keep potential sites free of buildings, roads or trees in the meantime.

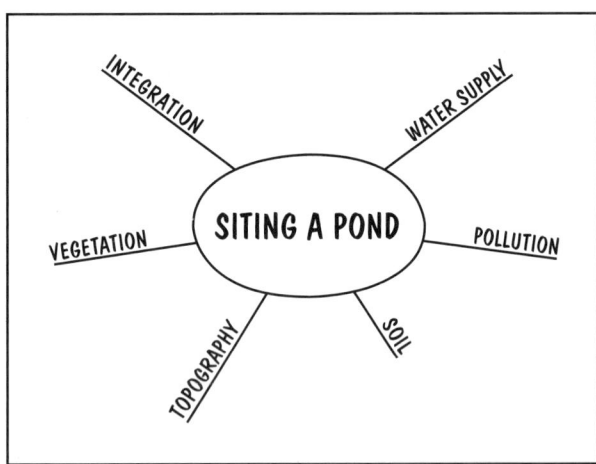

Water Supply. There are basically two kinds of pond: one which is filled from an external source and keeps the water in because its bottom and sides are watertight, and one which is formed from a hole dug into the water table. I call the first a sealed pond and the second a water table pond.

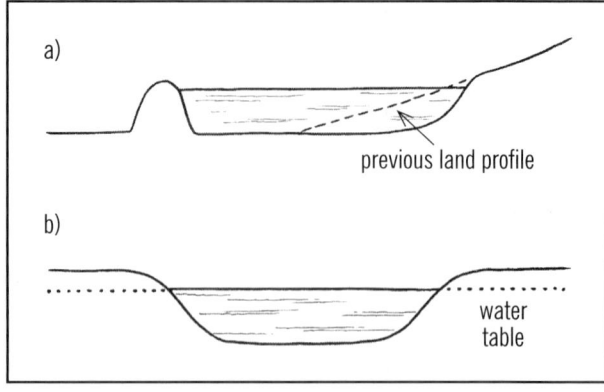

a) A sealed pond holds the water against drainage and is usually formed by making a dam on the lower side out of soil excavated from the upper side of the pond. b) The water table is the level beneath which the soil is waterlogged. A hole dug into the water table will fill with water.

Most ponds are of the first kind, because it gives the pond-maker much more choice about where to put the pond and more control over the water level. There are not that many places where the water table is consistently high, and most of these marshy places are low-lying, which means that the prospects of using the water by gravity are limited.

Water tables fluctuate with the seasons and a water table pond may only contain water for part of the year. A seasonal pond can be quite valuable for wildlife (see p362) and for some useful semi-aquatic plants, such as reedmace. But for some other functions, such as irrigation or swimming, the whole purpose of a pond is to store water from the winter for use in the summer. Others, such as fish-keeping and fire-fighting, need a constant supply through the year.

A water table pond which is fed by seepage from a nearby perennial stream or river is more likely to stay constant through the year than one which is dug into generally marshy ground.

The best way to tell what actually happens to the water table through the seasons is to dig a small hole to the proposed depth of the pond and see how it fills and empties through the year. It's important to remember that there can be great variations from year to year. The hole must be dug to the full depth of the proposed pond. A shallow hole may be in a 'perched' water table, where the surface soil is kept waterlogged by a thin compacted layer which lies over a freely-draining soil below. Summer is the best time to dig the hole, unless you like digging under water!

Sealed ponds collect water from the wider landscape and hold it against gravity in a watertight bowl. For large ponds the source of supply may be:

* surface runoff
* a spring or springs
* a stream or river

It's easy to overestimate the potential yield from runoff. Most rainfall soaks into the ground unless it falls on hard surfaces such as roofs and roads – and water from a road is very likely to be polluted. In most rural areas there is only significant runoff on a few days in the year, when the soil is saturated and heavy rain still falls. Any pond which relies entirely on runoff is likely to be small and seasonal, but runoff can be a useful supplement to other sources. If global climate change gives us a more extreme climate, with longer dry spells and heavier rainstorms, collecting runoff may become much more significant.

Often a large puddle or area of standing water seen in winter will suggest that spot as a pond site. If the only source of water is runoff – and it usually is – it's important to remember that the volume of water will be less at other times of the year, and it will almost certainly dry up in summer.

A good spring is an ideal source. Before relying on it to feed a pond its flow should be checked through at least a whole year, and preferably measured. If any kind of pollution is suspected it must be tested. The temperature of the water is also important if the pond is intended for carp growing, as they grow more slowly in cooler water.

Any pond which takes water from a river, stream, spring or borehole at a rate of 20m^3 a day or more needs a licence from the Environment Agency (EA). This applies whether the water is used or returned straight to the stream from the outflow of the pond. Stream-fed ponds can be either on-stream or off-stream, and a water table pond which is fed by seepage from a river counts as the latter from a licensing point of view. The EA should be contacted well in advance as they take a minimum of six months to issue a licence. They can also give advice on pond design and what grants may be available.

Of course any water source must be higher than the proposed water level of the pond when full. If there's the slightest doubt about the relative levels they must be checked with levelling equipment. Our eyes make adjustments to the general trend of the landscape, and even surveyors of many years' experience will tell you that you can't judge levels by eye. In most cases a 'bunyip' water level is ideal for this job. (See p406.)

Pollution. Aquatic ecosystems are much more sensitive to pollution than dryland ones.

Non-point sources of pollution, such as runoff from roads and chemically farmed land, can usually be absorbed by a belt of perennial vegetation on the uphill side of the pond. Uncut grass, shrubs or trees are suitable, and a width of 5m is usually adequate. This is often a good place for a willow coppice. Point sources, such as the leachfield from a septic tank, may make the site unsuitable for a pond altogether unless they can be removed.

Soil. Only heavy clay soils can hold water in a pond and it's as well to get advice from a professional pond installer about whether a particular soil is suitable. But for a preliminary survey the clay ball test (see box below) gives a good rule of thumb.

Where the topsoil is loamy there may be clay in the subsoil and this is always worth investigating. A deposit of suitable clay a couple of metres down can be excavated and used to line a pond. Bringing raw clay in from outside is rarely worthwhile because of the bulk and weight. But since most of the weight and much of the bulk is in the water it holds, it can be worthwhile bringing in dried bentonite clay for small, high-value projects. In general a thin layer of clay over a well-drained substrate like chalk or a sandy soil is not a good

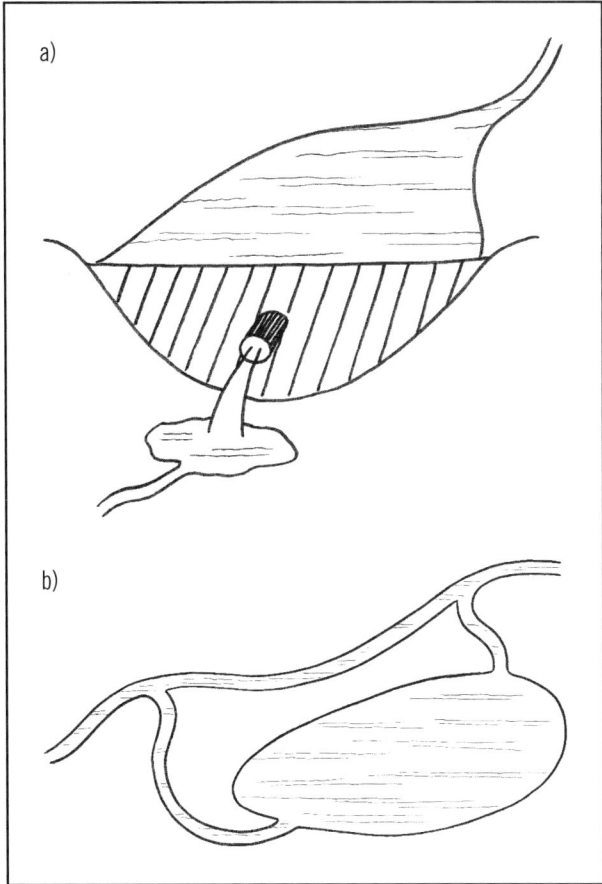

a) An on-stream pond. b) An off-stream pond. On-stream ponds disrupt the movement of aquatic wildlife and tend to silt up, therefore they are not recommended.

THE CLAY BALL TEST

With a little practice you can get a good idea of the soil texture by using the test sequence on p65. But a short cut is the clay ball test:

Take about a desert spoonful of the soil, knead it till it loses all structure, moistening it if necessary, and mould it into a ball. Throw it up into the air. If it doesn't break when it hits the ground it's probably near enough to pure clay to make a pond.

idea, because if the water level in the pond falls and the clay dries out it will crack and will leak when the water supply is resumed.

On all non-clay soils an artificial liner is needed for a sealed pond. As soon as we start bringing in artificial liners we know we're beginning to depart from the natural flow of the landscape. It's expensive in both energy and cash terms. It's also only temporary, as the most durable liners only last some 50 years. This is not to say that liners are never appropriate, but that it's only worthwhile to use them on relatively small ponds with a relatively high value, such as an irrigation supply for high value crops or high-output aquaculture.

A water table pond can of course be made in soil of any texture.

Topography. The first requirement is a relatively level site.

For a sealed pond with a dam, the more level the site the more water it can store for a given amount of earth moving.

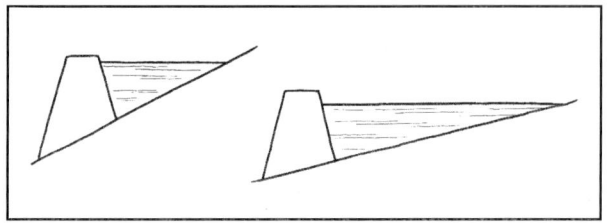

As a rule of thumb slopes of more than 5% are likely to be uneconomical. A water table pond needs an almost completely level site, because building up one side into a dam wall will have no effect on the water level and storage capacity. But land with a high water table is usually flat anyway.

The second requirement is that the pond site must be low enough in the landscape relative to its water supply. If it's spring-fed the position of the spring or springs gives the highest possible level at which it can be sited. If runoff is to provide all or much of the water supply, the lower it is in the landscape the larger the catchment area is likely to be. But the catchment area can be greatly increased by constructing diversion drains. (See Keylining, p99.)

These two requirements are most likely to be found in a low lying place. But if one of the functions of the pond is to provide water for use outside the pond, say for irrigation, a third requirement is to place it above the area where the water is to be used. Then it can be used by gravity, saving the energy cost of pumping.

How effective this can be does depend to some extent on the pressure of water needed. Some irrigation equipment requires a relatively high pressure, and if the head – the vertical distance between the surface of the pond and the point at which the water is used – is limited this may restrict the kind of irrigation equipment which can be used. High-output sprinkler irrigation needs a water pressure which may be hard to get from gravity

Mandy Pullen uses this low-pressure sprayline irrigation equipment in her market garden. It works by gravity with a three metre head from a farm pond. It has a relatively low output, but because the garden is small it's sufficient and no pumping is necessary.

alone unless the head is very high. Where the head is more modest there are two options, either to use the more low-output sprayline equipment or to supplement gravity with a little pumping. Even if supplementary pumping is necessary, the size of the pump and the amount of fuel needed will be less than if the water had to pumped up from below.

The conventional approach to siting an irrigation pond is to choose a place low down in the landscape and then pump the water to wherever it's needed. The permaculture approach is to find a site which is both low enough in the landscape to collect enough water, and high enough that the water can be drawn off by gravity. The gross output may be less, but the net output from an energy point of view will be greater.

Where there's a choice of more than one site, it's worth remembering that the energy cost of pumping is something which will have to be borne for the entire life of the pond, whereas earth-moving is a one-off cost. From a permacultural perspective this may tip the balance in favour of a site which is less level but high enough to give the required pressure at the point of use.

A pond whose main function is aesthetic looks best in a relatively low-lying place, where a pond might naturally occur. But there must be somewhere for the water to drain to when the pond is full and it goes on raining. As for a fish pond, it's a great advantage if it can be drained, both for harvesting and to make use of the rich deposit of silt on the bottom. This means the land must fall away downstream of the pond at least as far as its maximum depth.

Vegetation. Waterside trees can supply food for fish and nutrients for aquatic plants. (See p106). They can also provide useful shelter. Shelter not only keeps the pond warm but on a large pond with a long fetch it can prevent the bank being eroded by waves. On the other hand there are a number of disadvantages to siting a pond too close to trees or planting too many around it.

Firstly, if the leaf fall is excessive it can cause a rapid depletion of oxygen as the aerobic bacteria decompose the fresh organic matter – an effect similar to that of cultural eutrophication. Secondly, most pond life needs light to thrive, so shading should be avoided. Thirdly, oxygenation of still water is greatly enhanced by the action of wind on the surface, so too much shelter is not wanted. Fourthly, existing trees can be damaged or killed if their roots are damaged in the digging of the pond. Equally, the lining of the pond, whether puddled clay or plastic, can be punctured by the roots of trees.

It's a matter of proportion. The larger the pond the more leaves it can cope with, and indeed profit from. Thus smaller ponds should generally be sited well away from trees if possible, and certainly not overhung by them, whereas a large well-stocked aquaculture pond should benefit from trees along its edge. Placing the pond to the south of existing trees, or planting trees to the north of the pond, avoids the problem of shading. As most of our wind comes from the south-west, this arrangement doesn't give too much shelter, and many of the leaves from the trees will be blown away from the pond.

Although ponds make good wildlife habitats, it's important not to site them in places which are already high in biodiversity. (See p352.)

Integration. Where there is space for more than one pond it can be an advantage to construct them in series, so the outflow of one feeds the inflow of the other. This makes it possible to control the water level of the lower pond, especially to keep it at a constant level through the summer. This can be particularly useful if there are chinampas on the edge of the pond. (See next column.)

However, where fish are an important output it's much better to arrange ponds in parallel. This prevents the spread of disease from one pond to another, allows the water supply to each pond to be regulated independently, and allows ponds to be drained and emptied independently.

Ponds intended for farm animals to drink from are best sited where they can be accessed from a number of fields so that maximum use is made of a single pond.

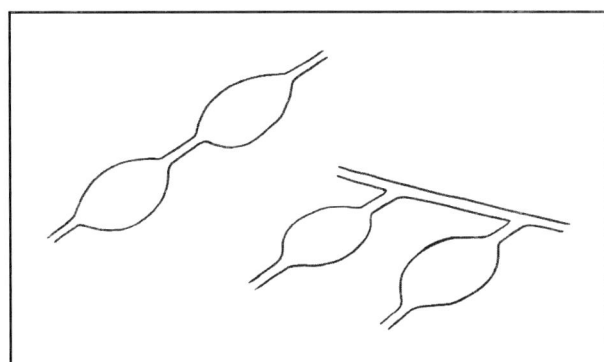

Ponds in series and parallel.

But unrestricted access by animals means a lot of trampling, which reduces or eliminates their value for wildlife, aquaculture and even irrigation – the latter because muddy water will block sprinkler nozzles. If the pond is to have other functions it's best to restrict animal access to part of the pond only or to provide a different source for their drinking water.

In rural areas where the mains water supply is not adequate for fighting fires, closeness to buildings may be a factor to consider. A clear run for laying out hoses between the pond and the place where fire appliances would stop is necessary.

A pond full of valuable fish is best sited near the house, or at least where it gets plenty of attention. This is not only good zoning in a general way but will also be a deterrent against poaching.

Last but not least, a pond is best sited where its beauty can be appreciated by people. This is perhaps more obvious in the context of garden ponds than large-scale ones. But all water has a special magic, and even the most utilitarian ponds can develop great beauty quite soon after construction if they are well designed. Any pond which can be seen easily from the house or a frequently used path is more valuable than one which can't. Beauty, after all, is a valuable yield.

Design and Construction

Building a large pond, especially one which requires a dam wall, is a specialist task and it's well worth getting advice from an experienced pond-maker. As ever when dealing with a specialist, it's worth having a good grasp of the principles before consulting them, and Bill Mollison covers the subject comprehensively in *Permaculture – A Designers' Manual.*[33]

The highly productive edge between water and land can be increased by making the shoreline curvy. A row of promontories, like the fingers of a hand, giving alternate strips of land and water, gives the maximum edge. These are sometimes referred to as chinampas, after a traditional Mexican system where wetlands were turned into strips of alternating open water and dry land. This can be a highly productive system, but there is really no scope for using it on a large scale in Britain, as our remaining wetlands are a precious resource, both for biodiversity and for the regulation of water in the landscape.

However a row of small chinampas on the edge of a pond can be useful. If the water level can be kept constant through the growing season the plants on the land need no other irrigation, and if they do need it the water is near at hand. Plant debris from the land can enrich the water strips and dredgings from the water can fertilise the land.

An island will both increase edge and provide a fox-proof nesting site for ducks. But anything which increases edge will make a pond more difficult to net – chinampas and islands impossibly so. A pond intended for serious aquaculture really needs to be rectangular unless it can be drained for harvesting.

These chinampas at Ragmans Lane Farm were originally intended to grow vegetables which would not need hand watering, but the water level falls too much in summer for this. The willows which have replaced the vegetables yield at a rate equivalent to 25t/ha/ann, and are used both to fuel a masonry stove and to provide planting stock for willow sculpture and hedges.

A gently shelving shore, at least in one part of the pond, allows amphibious animals such as frogs, to crawl in and out.

Having different depths of water in different parts of the pond gives a variety of habitats and will support the greatest diversity of plants and animals. If one part is made as deep as possible it will be more likely to stay unfrozen in winter and be less likely to dry out in a drought. 2m is ideal. Shallow areas are particularly productive and maximum productivity in our climate is found in water of around 75cm. However fish are more vulnerable to predators such as herons at this depth, and deeper water may give a higher net yield in a fish pond.

Although productivity is an advantage if aquaculture is the main aim, it's a disadvantage if clean water for irrigation is the top priority. In the latter case a uniformly deep pond is the best design, and a few grass carp will help to keep the water clear of bits which might block the nozzles.

All ponds need some means of dealing with excess water. Sealed ponds fed by a constant supply, such as a spring or stream, will need a regular overflow, usually a pipe passing through the dam wall. This will determine the normal maximum water level. In addition, all ponds need a spillway for times of flood. This must be broad and shallow, well grassed over, and cut into undisturbed soil, not into a dam wall or other made-up ground, or it will quickly erode when water starts to spill.

Constructing a water-table pond is mainly a matter of digging a hole in the ground, preferably in summer or autumn, when the water table is as low as possible. But this kind of pond is likely to have much more spoil to be got rid of, as there is no dam wall to make use of it. Some thought needs to go into how it can be used creatively. Possibilities include cob building, making a

bank to give shelter, and turning level land into a gentle south-facing slope for early cropping. The topsoil should always be kept and relaid in the finished pond, and care should be taken never to simply dump or spread the subsoil on top of the topsoil outside the pond.

There is often more scope for choice in designing a sealed pond, though in many cases the level at which the water supply is delivered will set an upper limit to the water level of the pond when full. Design ideas can be tested for feasibility with a 'bunyip' water level. It's impossible to tell just where the water will come to by eye alone.

There are three possible ways of lining a pond: clay, concrete and plastic liners. Clay is only suitable for larger ponds. It inevitably leaks a bit and only the greater ratio of volume to lining of a large pond can handle this constant loss. A clay-lined pond needs puddling and this is usually done with the excavating machinery. Concrete is really too expensive, and very difficult to repair if it develops a crack.

If a plastic liner is used, three kinds are available:

- Butyl, the most expensive, toughest and longest lasting, usually guaranteed for 50 years.

- Polythene, the cheaper alternative, more difficult to install than butyl and less long-lasting.

- PVC, which is a harmful substance and not recommended.

Plants will arrive of their own accord, often as seeds on the feet of water birds. Where plant aquaculture is not one of the aims, simply allowing this to happen can be a good plan. Where native plants are used it's best to get them from a nearby pond, especially if nature conservation is an aim. (See pp370-371.) But plants should only be taken from sites where they are abundant enough not to miss a few individuals. April and May are the best planting months for aquatic plants.

In a medium to large farm pond a wide variety of animals and plants can live together. For example, ducks may find all their food needs for much of the year from the pond vegetation and small creatures such as tadpoles and young fish, without seriously depleting either.

Domestic Scale

A garden pond is not likely to have as many functions as a larger farm pond. The most usual ones are:

- beauty and spiritual qualities
- plant aquaculture
- slug control – frogs or ducks
- wildlife

In a small pond ducks are not really compatible with any of the other functions. (See p111.)

This garden pond, two metres by one, is home to a variety of wildlife, including newts, and helps greatly in slug control.

Siting

The siting of a garden pond can usually be pretty flexible. They're rarely large enough for the location of the water supply to affect siting, and they're almost always lined, so the soil type is not critical. The main considerations are:

- aesthetics
- overhanging vegetation
- sunlight

From the aesthetic point of view a pond is best sited where it can be easily seen, either from the house or from the main leisure area of the garden. It's also best placed where a pond looks natural, that's to say in a low lying part of the garden. A possible problem with a lined pond in a low place is that if the water table rises it can push the plastic sheet up from below and make a mess of the pond. It may be necessary to choose between a water table pond in the most natural-looking position or a lined pond somewhere else. (See p112.)

If the leaves of trees or shrubs fall into the pond they can cause a quick loss of oxygen as the aerobic bacteria go to work on them. This is much the same effect as cultural eutrophication. (See p102.) It's possible to remove the leaves by hand, or keep them out with a net which is rigged in autumn for the purpose. But it's easier to site the pond away from trees if that's possible. A sunny site should always be chosen for preference, as most pond creatures, plant and animal, prefer a sunny environment.

Some large gardens are blessed with a stream or spring, but in most cases the two possible sources of water are the mains and rain water. The advantages of the mains are that it's a constant supply, and because it's under pressure it can easily be taken to any part of the garden with a hosepipe. The disadvantages are that the water is chlorinated, and in many areas it has a high level of plant nutrients in it which may cause cultural eutrophication in a small pond. The chlorine will outgas if the water is left to stand in a container for a few hours before adding it to the pond. But it's not possible to get rid of the plant nutrients, and unfortunately eutrophication is encouraged by the warm weather of summer, which is when topping up is most needed.

Rainwater is pure by comparison, but the supply is irregular and comes mainly when it's least needed. If a year-round pond is required the rainwater must be stored in a tank to provide a supply in dry spells during the summer, unless the pond is unusually large and you don't mind it getting very low in summer. But a seasonal pond is quite adequate for frogs and many other kinds of wildlife. Delivery from the downpipe to the pond must be through a pipe or in a lined channel. In an unlined channel the water is most likely to seep away before it reaches the pond.

Design and Construction

The small size of garden ponds does somewhat restrict the possibilities for islands, patches of deep water and wavy edges. But it's worthwhile to design in as much edge and as much variation in depth as possible so as to get maximum diversity of habitat. Even a tiny pond of a metre or less in diameter can provide a valuable addition to the range of habitats in the garden as a whole, and is quite suitable for frogs.

A useful aid to designing the outline of a pond is garden hose or a length of rope laid out along the proposed shoreline. This can be easily moved as different design ideas are tried out. If there is any slope at all each idea must be tested for level. If there is any difference in the level of one side of the pond with another, either the bank must be built up on the lower side or the water level will be only come part way up the bank on the other. For larger ponds a 'bunyip' water level is a good tool for checking this. (See p406.) For smaller ponds a spirit level tied to a straight piece of wood or garden cane, long enough to reach from one side of the pond to the other, is ideal.

If the pond is dug into an impervious soil and doesn't need a liner it will still need puddling. The best way to do this is: dig the pond, invite plenty of energetic people to a summertime party, fill the pond with water, then get the people in the pond tramping around as vigorously as possible till the clay loses all structure. Once puddled, the pond must never be allowed to dry out altogether or the clay will dry and crack. Some vigorous plants, like reedmace and the common reed may make holes in a clay seal, and they're best avoided in smaller ponds where the subsoil is well drained.

Concrete is sometimes used to line garden ponds,

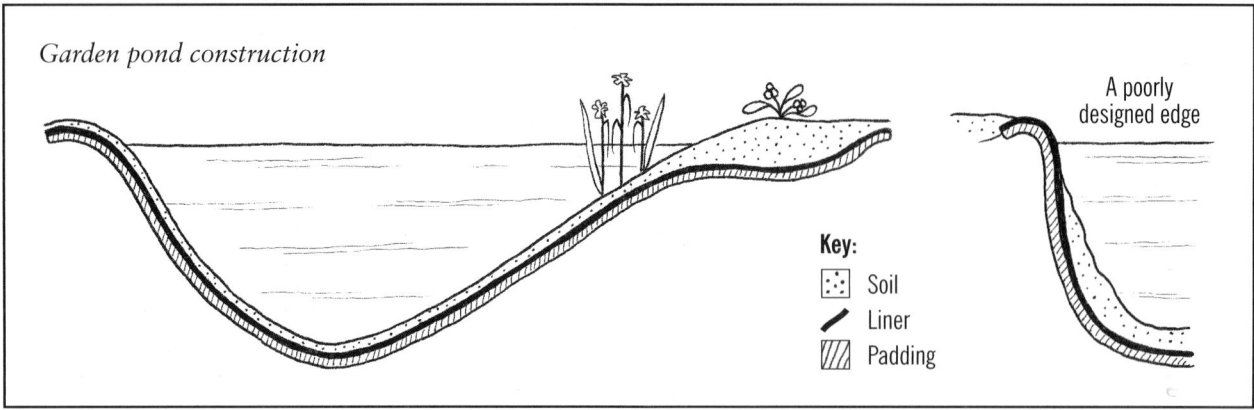

Garden pond construction

A poorly designed edge

Key:
[:.:] Soil
/ Liner
/// Padding

but it cracks easily and is best avoided. Pre-formed fibreglass pond liners are also available, but they give little choice of pond size or design and look horrendous. A flexible liner is much the best solution for soils which are not clayey enough to be puddled.

The procedure for fitting a plastic liner is as follows:

- Dig the hole at least 15cm deeper than the pond will finally be.

- Remove any stones and other sharp objects from the soil in the hole which could puncture the liner.

- Place a layer of padding, such as old carpet, wetted cardboard, or pure sand over the soil to protect the lining.

- Put in the lining.

- Cover it with 15cm of stone-free soil; in small to medium ponds, which have a low volume to surface area ratio, use subsoil to avoid eutrophication, in large ponds use topsoil for extra fertility; as well as supporting plants, the soil protects polythene from ultra-violet light.

- If filling the pond by hand or hosepipe, place a flat stone or tile in the bottom and gently pour water onto this, to avoid disturbing the soil.

The liner can be extended under patches of adjacent soil to make marshy areas. Small chinampas along the shore of a lined pond are best constructed over the top of the liner rather than trying to line each channel individually. In fact if each channel was lined the soil of the chinampas would not benefit from automatic irrigation, which is one of the main advantages of chinampas.

At least some of the water for the first filling should come from an existing pond. This will contain seeds, eggs and spores of plants and animals which can colonise the new pond. Most species of fish predate on frogs and other amphibians, so they are best kept out if amphibians are wanted.

It's usually worthwhile to introduce some plants. They are less likely to colonise a small garden pond by natural processes than a larger one in open country.

Chapter 6

ENERGY & MATERIALS

In a world where our use of energy is largely responsible for the biggest threat to our existence, global climate change, it's surprising how little the average educated person knows about energy. On second thoughts perhaps it's not surprising. We might not be in the situation we are in if we were energy-literate. The aim of this chapter is to give the basic knowledge about energy needed to run a household or a planet. It starts with the fundamentals, the ABC of energy, then it looks at the questions which need to be asked when making any energy decision, before coming to the practical measures we can take to improve our energy performance.

There is usually a close correlation between energy use and the consumption of other materials, indeed energy consumption is often a fair indicator of our overall ecological impact. But reducing the one does not inevitably reduce the others, and the last part of the chapter outlines some ideas on use of materials.

ENERGY FUNDAMENTALS[1]

Entropy

There are two fundamental facts about energy, known as the first and second laws of energy.[2] The first law states that:

Energy is neither created nor destroyed. It can only be transformed from one form to another

The second law states that:

Whenever energy is transformed it is always changed from a more usable state into a less usable state

Imagine a lump of coal. It represents highly concentrated energy. We can use that energy for a variety of purposes, such as heating up a room. When the coal has been burnt and the room heated, the energy has not disappeared. It's all still there, but instead of being concentrated in a lump of coal it's dispersed in the form of increased temperature in the air of the room.

The energy in the heated air is still useful – it keeps the people in the room warm – but it is less concentrated, less storable and less versatile than the energy in the original lump of coal. It is less useful. As the room gradually cools the energy seeps out of the building and imperceptibly warms the air outside. The energy is soon dissipated and goes to add to the temperature of the Earth by an immeasurably small amount. Eventually it is radiated out from the Earth and adds to the background temperature of the universe by an even tinier degree.

Nothing is lost in the process, except usability.

The starting point – lump of coal and unheated room – can be described as an ordered situation with regard to energy, compared to the end point – no lump of coal and an imperceptibly heated universe overall. Order here is seen as a pattern, a pattern of contrast between areas of high and low energy concentration. Disorder is seen as a formless, random situation in which the level of energy is the same everywhere.

Energy transactions, that is to say physical events of any kind, can only happen when you start with an ordered situation. Because of the second law, energy can only flow from high concentrations to low. When there is no energy gradient between one part of a system and another nothing can happen.

The word entropy is used to describe the unusable energy that arises from an energy transaction. Thus an ordered situation is described as being a low-entropy situation and a disordered one a high-entropy situation.

When coal is burned to generate electricity as little as 30% of the energy in the coal is converted to electricity. The other 70% is converted to waste heat and dissipated into the atmosphere or the sea. This waste heat represents the entropy generated by the transaction.

When an animal eats plant food, around 10% of the energy in the food goes towards growth and repair of the animal's body, some 10% goes to fuel the animal's activity and as much as 80% is lost as waste heat.[3] If another animal, such as a human being, then eats that animal it only consumes 10% of the food value that was present in the original plant food. The loss of usable food energy represents the entropy in this transaction.

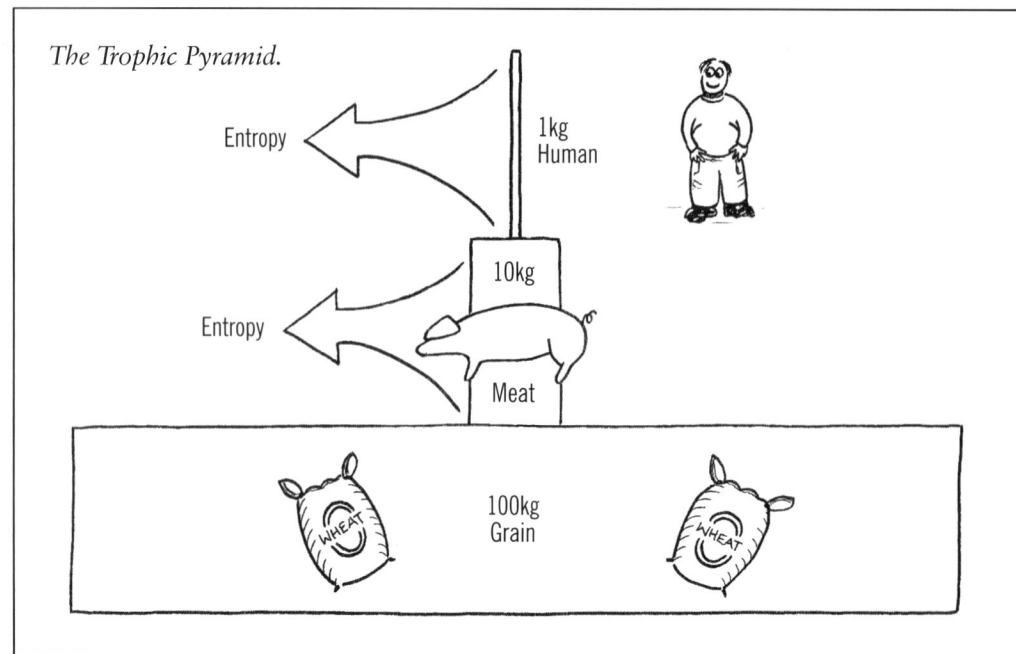

The Trophic Pyramid.

Entropy

1kg
Human

10kg
Meat

Entropy

100kg
Grain

In both these cases energy is being transformed from one kind to another. Purely from an energy point of view, it would be more efficient to use the coal and eat the plants than to use the electricity and eat the animal, though there may be other considerations.

Entropy is the result of the second law of energy in action. Indeed some people refer to the second law as the entropy law. The entropy law applies to the embodied energy in material things as much as it does to pure energy.

Imagine a knife blade. It represents order: steel has been concentrated into one place and fashioned into a useful tool. It took a lot of energy to do that, including the geological energy that concentrated the iron ore in one place, and the human and fossil fuel energy that went into mining, refining, manufacturing and marketing the knife. Whenever the knife is sharpened tiny filings of that steel are broken off the knife and become so widely dispersed in the environment that they're unusable. Eventually the knife will be worn away to a point where it's no longer usable. All the atoms of the steel in the original knife still exist but they are now in a state of entropy.

The steel left in the remains of the blade can be recycled as scrap iron but the filings from sharpening cannot. These filings, plus the energy expended in the recycling process represent the entropy in the recycling transaction.

If the entropy law is seen as a constant tendency to disorder, living things seem at first glance to break it. Their energy transactions – photosynthesis in the case of plants or the eating of food in the case of animals – certainly appear to result in an increase in order: the growth and reproduction of organisms.

However, that is to look at the organism and its immediate surroundings only. Plants get their energy from the Sun. They absorb a small fraction of the solar energy which falls on their leaves for photosynthesis, and the rest is lost as waste heat. Their increase in order is only achieved as a result of a greater decrease in order in the Sun. Animals acquire all their energy from plants. As we have just seen, most of the plant energy is lost as entropy during the transaction, so their increase

SPIRITUAL ENERGY

The two laws of energy apply to every physical thing in the universe. This means that the total amount of physical energy in the universe is limited in quantity and it is continually deteriorating in quality. Some people find this depressing. It certainly flies in the face of the world view held by our culture, which sees history as an inevitable increase in material wealth, which we call progress.

There have even been attempts to prove that the entropy law does not always apply in the physical world, but they have not been successful.

However, the two laws only apply to physical energy. Spiritual energy works in quite a different way. In fact it works in the opposite way: the more it's used, the more it grows. The more we love, the more love grows in the universe.

We need not feel depressed by the law of entropy. Human happiness is not the same thing as ever-increasing material wealth. Once our basic needs are satisfied it's not material things which bring real happiness into our lives. It's the emotional and spiritual aspects of life, and there is no limit to these.

in order is also achieved only as a result of greater disorder elsewhere.

Having said that, the energy of the Sun is inexhaustible on any human time scale, so in effect we can regard photosynthesis as an endless source of increasing order on Earth. But it's endless in the sense of being endlessly repeatable. It doesn't mean that we can have a limitless supply of energy at any one time.

A New Energy Environment

Thirty years ago it began to dawn on us that one day we would have to start living on renewable energy rather than the limited supply of fossil fuels. Now it's clear that the limit on our use of fossil fuels is not so much the prospect of running out of them but the increase in entropy caused by using them. Global climate change and many other kinds of pollution are simply part of the disorder resulting from burning fossil fuels. In fact all pollution can be regarded as a form of entropy.

If we continue to use fossil fuels at the present rate we will probably render the Earth uninhabitable, or very nearly so, long before we've mined the last lump of coal or pumped the last drop of oil.

Renewable energy almost exclusively means solar energy, which we can harvest either directly or indirectly.[4] In any one year the total amount of solar radiation reaching the Earth's surface is greater than our entire reserves of fossil fuels. About a third of it is reflected straight back into space. The remaining two-thirds warms up the atmosphere and the land, evaporates water from the seas which can then fall as rain, generates wind, or is used by green plants for photosynthesis. We can capture solar energy in all these forms, as direct solar, water power, wind power and biofuels respectively.

However, although solar energy is much more abundant than fossil energy it's also much more diffuse. It's spread thinly over the Earth and must be concentrated before it can be used. Fossil fuels, though limited in supply, are highly concentrated, a job done for us by natural forces over geological time. Concentrating solar energy is an energy transaction and is accompanied by an inevitable increase in entropy. In other words it costs a lot of energy to concentrate solar energy to the point where we can use it.

This fact has two major implications. The first is that solar technology is most efficient when it's decentralised. It's usually more efficient to collect solar energy at a household or neighbourhood level and use it there than to concentrate it in large power stations and then distribute it. The present large-scale centralised industries and huge crowded cities are unsuited to an energy supply which is spread relatively thinly over the whole globe. They will probably not survive the transition to a fully solar age.

The second is that solar energy cannot hope to provide us with the highly concentrated energy we are used to in anything like the quantities we are used to having it. The entropy involved in concentrating it is just too great.

There is a close parallel between our energy history and the development of an ecosystem. In the early stages of succession the plants and animals which succeed are those which maximise their throughput of energy. This is a phase of colonisation, a 'frontier economy' in which resources are abundant relative to the organisms available to use them. As the ecosystem matures and more and more niches are filled, the species which succeed best are those which use energy more economically.

We are still practising a frontier economy on a planet which is already 'full'. Changing our technology is a priority, but an even higher priority is to change our expectations.

Solar technology is constantly improving, and undoubtedly we will be able to concentrate solar energy more effectively in the future than we can now. But technology cannot create energy. It can only change it from one form to another, and the entropy loss is inevitable.

Energy can't be considered on its own. Suppose the laws of energy turn out not to be true after all, and we discover a supply of energy which is totally non-polluting, highly concentrated, available in any quantity we want, and available indefinitely. It would be a disaster. Whenever we use energy we use other resources, both mineral and biological. Not only would we very soon run out of many critical mineral resources but the output of pollution from using them would shoot up. As to the damage we could do to the Earth's ecosystems, it would be proportional to our supply of energy – infinite.

Conservation

All of this points clearly to the importance of using less energy. The key to our energy problems is not to produce more renewables in the vain hope of reproducing the level of energy use we now enjoy. It's to reduce consumption to the level which the Earth's life support systems can handle.

There are two aspects to energy conservation:

- increasing energy efficiency
- reducing consumption

The potential for increasing energy efficiency is enormous, and saving energy almost always gives better value for money than producing more. (See box overleaf.)

There is some consensus that a good third of the energy used in Britain could be saved quickly and easily, using existing technology and without the slightest change in lifestyle. Rather than imposing a cost on industry the required changes would actually increase profits. Another third could be saved at slight financial cost but with no decline in performance.[5]

DON'T SWITCH IT ON

An impressive example of the power of conservation comes from the American energy expert Amory Lovins. He asks the question, 'What's the most profitable thing to do with a brand new power station?' and comes up with the answer: Don't switch it on, and use the money that would have been spent on the running costs on energy conservation. The cost of saving a kilowatt hour of energy in the USA by carrying out insulation, draught-proofing and other conservation measures is half a cent. The running costs alone of a nuclear power station come to four cents a kilowatt-hour, never mind the capital cost of building it.[6]

The fact that this is not happening has nothing to do with technology and everything to do with emotions. People are fascinated by power. We use the word power to describe both dominance in human relationships and the use of physical energy. Power excites ambitious people. A new nuclear power station has a glamour that ten thousand insulated lofts can never have. Add to this a general ignorance of the basic facts of energy even among the better educated part of the population, and it's no wonder we don't behave in a rational way when it comes to energy.

It also has to do with the disproportionate political power of the oil companies, especially in the USA.

Any discussion of energy must start with the premise that the most cost-effective source of energy is conservation. Only when all opportunities for reducing our energy consumption have been explored is there any justification for looking at increasing supply.

Although the potential for energy saving by increasing efficiency is huge, in practice the savings get swallowed up by increased consumption, and even lead to increased consumption themselves. (See p8.) There is no alternative to restraining our level of consumption.

Just what level of consumption will be possible in the long term is impossible to say at the moment. In the short term there is little incentive to reduce our energy expenditure as energy is so grossly under-priced. "Imagine what it would be like if all future generations for the next 100,000 years could somehow bid for the oil our generation is using up," writes Jeremy Rifkin.[7] The true price of oil can only be imagined. In these circumstances it's up to each of us to decide how far we will go in simplifying our lifestyles.

ASKING THE RIGHT QUESTIONS

When we're faced with a decision about energy supply or use it's usually fairly easy to acquire a good deal of detailed information on the systems which are available. But making sense of all this detail can be more difficult. To help the process there are a set of fundamental questions which can be asked about any energy source or system. Each energy option has its strengths and weaknesses, and some are more suited to one situation than another. The questions set out below should help to sort out which is the most ecologically benign solution in any particular case.

They are equally relevant on both a personal and a political level, both when considering what kind of domestic energy systems to go for, and in order to play an informed part in public debate on energy matters. Public policy is important, but when it comes down to it the future will be determined by millions of decisions made by ordinary people, decisions like whether to buy a gas cooker or an electric one.

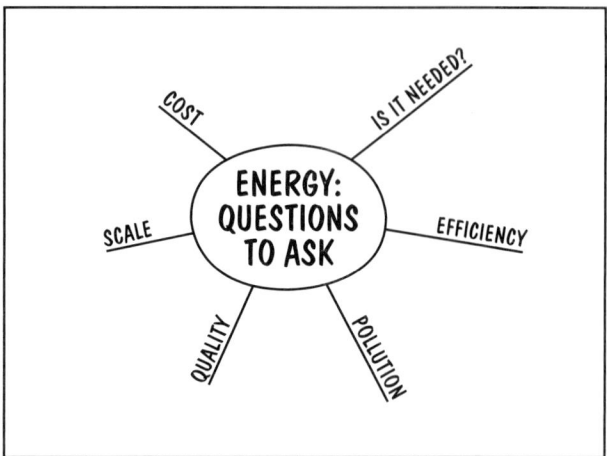

Is It Needed? In view of what I have just written about conservation, the first question must be, 'Do we need this product or service at all?' For example, when considering travelling to work the first thing to consider is not which means of transport is most efficient for commuters, but whether we can arrange our lives so we can work at home or within walking distance.

If we're sure we do need something, we need to ask how much we need. For example is it necessary to heat a room to a constant minimum temperature of 20°C, or would we be just as happy at a lower temperature? The answer probably depends on the age and health of the occupants. But it comes before the question of what the best heating system may be. In fact knowing what temperature we require may be crucial to choosing the appropriate system.

Net Energy Efficiency. All energy systems have a net output of usable energy which is less than the total energy input. This is partly because of the work which must be done to turn the raw energy into a usable form: coal must be mined, oil must be pumped and refined, wind generators must be manufactured, transported, maintained and finally disposed of. It's also because of the loss of usable energy when energy is transformed, as when coal is transformed to electricity. The difference between the total input of energy and the usable output represents the entropy of the system.

Net energy efficiency can be expressed as the percentage of the total input which is delivered as usable energy. The bar chart below shows some examples of the efficiency of different domestic heating systems.[8]

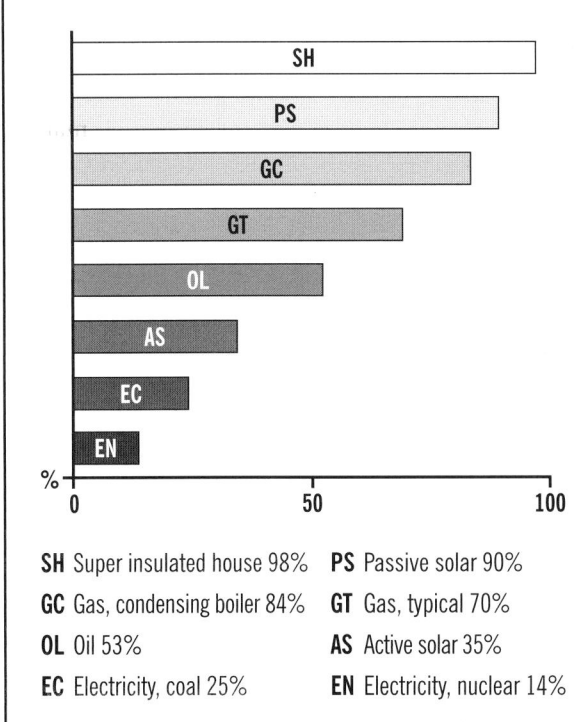

SH Super insulated house 98%   **PS** Passive solar 90%

GC Gas, condensing boiler 84%   **GT** Gas, typical 70%

OL Oil 53%   **AS** Active solar 35%

EC Electricity, coal 25%   **EN** Electricity, nuclear 14%

The figures in the bar chart are revealing. Firstly, active solar heating is inefficient compared to some fossil fuel systems, so renewable energy sources do not necessarily have a high efficiency rating.

It's important to remember that the energy input to any renewable system is, in the present, inevitably fossil fuel. So whenever we consume energy from a 'renewable' system with a low efficiency rate we are primarily consuming fossil fuel. The renewable element is only as great as the efficiency percentage.

Of course in a solar economy all the energy input would be renewable. But, as we have seen, there is a limit to the amount of concentrated energy which can be produced from purely renewable sources. Whether it could ever be enough to produce some of the high-tech renewable energy systems being developed at present remains to be seen.

The same principle applies to nuclear electricity. Its advocates claim it doesn't cause global warming, but they are forgetting all the fossil fuel energy it takes to get the power station up and running and to decommission it. As the chart shows, this input energy is almost as much as the output.

In the future fossil fuels will become progressively less efficient. As we use up the more accessible fossil fuel reserves the efficiency rating will gradually fall, until one day the energy cost of getting a barrel of oil will be a barrel of oil.

The second point illustrated by the table is a tendency for efficiency to fall with increasing sophistication: passive solar heating is much more efficient than active.

Passive solar is heating which arises from the design of the house – south-facing windows, massive heat-storing walls and so on. (See p148.) The only energy input lies in the increased energy cost of building a house this way compared to a standard house. Active solar heating, by contrast, involves a gadget which is added to the building. It's distinguished from passive by the presence of moving parts. Again, the only energy input is in constructing and disposing of the equipment, but it's extra to the structure of the house, and needs to be high-tech in order to perform well.

A third point is the relative inefficiency of electricity when it's generated from fuels which could themselves be used for heating. It's always more efficient to burn fuel directly than to transform it to electricity and then use this as a source of heat. As we have seen, the majority of the energy in the original fuel is converted to low temperature heat, which is usually wasted. More energy is lost in the transmission of the electricity.

Combined heat and power (CHP) generation is a way of greatly increasing the efficiency of thermal electricity generation. Here the waste heat is used for industrial or domestic heating. Without CHP the absolute maximum efficiency of electricity generation from coal is around 47%, with it efficiency can exceed 80%.[9]

When it comes to biofuels – fossil fuel substitutes manufactured from plants – those made from trees have a much higher efficiency percentage than those made from annual crops. The bar chart comparing alternative fuels for cars, overleaf, suggests that the energy cost of producing biofuels from maize is around five times as much as that of producing them from wood. Energy efficiency is just one of the advantages of using perennial plants rather than annuals.

Pollution Output. Carbon dioxide is the big one. The amount of CO_2 produced by different systems depends firstly on the chemical composition of the fuel, and secondly on how efficiently it's used. Coal is virtually all carbon, oil contains both carbon and hydrogen, and natural gas contains both but with a higher proportion of hydrogen than oil. This is reflected in their CO_2 outputs when used for domestic space heating, as shown in the bar chart overleaf.[10]

These figures assume a perfectly working system. It has been suggested that in practice the reduced CO_2 output as a result of switching from coal to gas is more than offset by the leaks in the pipelines. Natural gas is methane, which is more than ten times as potent a greenhouse gas as CO_2, so a little leakage means a lot of global warming.[11]

Anything which increases the efficiency of the system, such as CHP generation, reduces the amount of CO_2 released per unit of useful output.

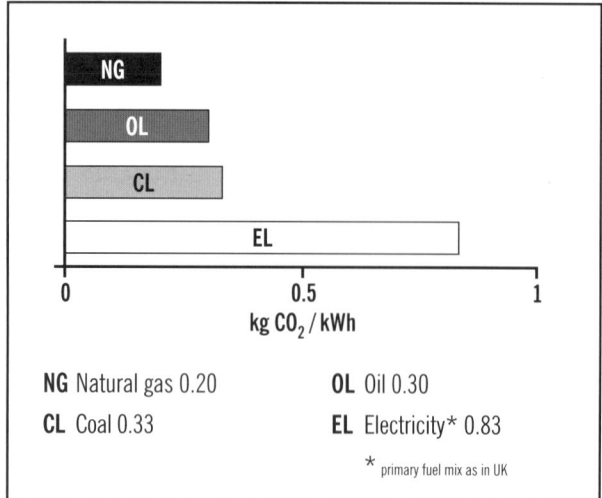

NG Natural gas 0.20   **OL** Oil 0.30

CL Coal 0.33   **EL** Electricity* 0.83

* primary fuel mix as in UK

Biofuels are often hailed as being carbon-dioxide neutral, because for each kilo of carbon released on burning another kilo is fixed by the growing crop. But this is to omit the energy used in growing and processing the crop. As the bar chart below shows, the energy cost of growing an annual crop eats away almost all the advantages of using a biofuel rather than petrol.

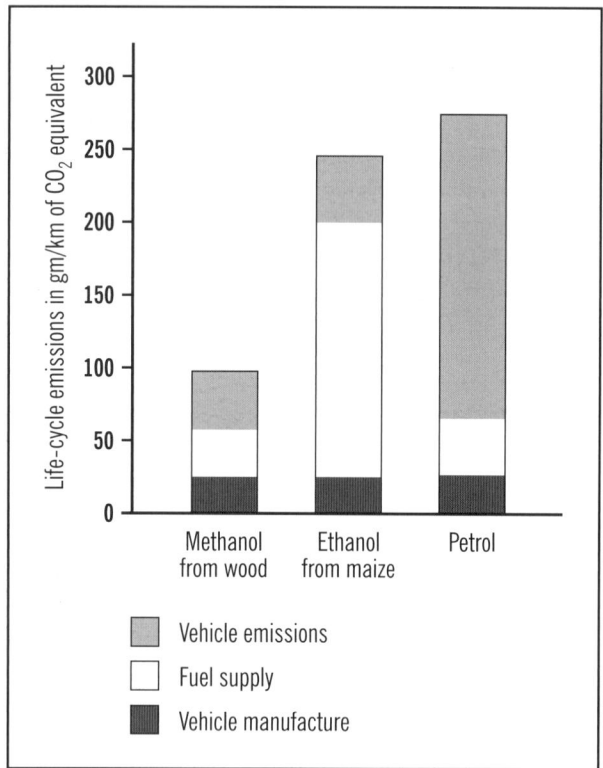

Life-cycle greenhouse gas emissions for alternative fuel cars.[12]

Other pollutants include sulphur dioxide and the oxides of nitrogen, which cause acid rain and come from the burning of both fossil fuels and biofuels.[13] One pollutant which really is in a class on its own is nuclear waste. This continues to be dangerous for tens if not hundreds of thousands of years and must

be looked after by skilled people during this time. The use of nuclear power is based on the assumption that our civilisation will remain intact and fully effective continuously for that length of time. The sketchiest knowledge of human history shows how remotely unlikely this is. Even if there was nothing else whatsoever wrong with nuclear power the waste problem alone would be sufficient reason to rule it out as an energy source.

Insulating buildings is an excellent way of saving energy. But some forms of insulation material have their pollution cost, and careful choice of materials is important. (See p160 and 162.)

Energy Quality. Energy is said to be high quality when you can do many things with it.

Electricity is a high quality energy. You can use it for running electronic equipment, motive power, lighting, high or low temperature heat. Passive solar heating is a low quality energy. You can only use it for space heating. Petrol is another high quality energy. Compact, lightweight and concentrated, you can put it in your tank and drive off for hundreds of miles without refuelling. The same applies to ethanol. Wood, by contrast, is bulky, heavy and burns at a lower temperature.

High quality energy almost always has a lower net energy efficiency. This is simply the working of the entropy law. An increase in order (high quality energy) in one part of a system can only be achieved by a greater decrease in order in another part of the system.

There is also a tendency for high quality energy to be more polluting. Thus electric central heating produces more CO_2 than a system which burns fossil fuels on the spot, which in turn produces more than a system which uses a combination of wood and passive solar gain.

Therefore it behoves us to match the fuel to the job, minimising the amount of high quality energy we have to produce. You can't run a sound system on logs, but it's equally silly to try and heat your home with electricity.

As well as being versatile, high quality energy needs to be available when we want it. Nuclear is a low quality form of electricity from this point of view. A nuclear power station generates electricity at a constant rate throughout the 24 hours, while demand goes up and down. Power stations burning fossil fuels, or biofuels, are more flexible and can be turned up or down with demand.

Solar and wind powered electricity are weather-dependent, and at present can only be stored in batteries, or by selling to the grid when in surplus and buying back when in deficit. The latter is not really storage but it avoids the high ecological and cash cost of batteries. It's the best solution for home generators unless they're in a remote location where connecting to the grid would be too expensive.

Where grid connection is impractical a combination of wind and solar generation can minimise the need for storage, as the weather usually provides us with one or the other at any given time. An efficient fossil or biofuel

A combination of wind and solar electricity with wood for heating make good sense in a remote rural location, such as Ben Law's home in the woods. (See case study, Living In The Woods, p336.)

generator for backup may be the best option in some cases, especially a domestic CHP unit. (See the next section, Scale.)

Because of this unevenness of supply it has been calculated that wind power could only be used for up to 20% of our national power supply.[14] Water is more storable, at least in principle. But small-scale hydro-electric setups may have scant storage and may be vulnerable to summer drought.

However, renewable energy can be stored as hydrogen. Electricity can be used to split water into hydrogen and oxygen. The hydrogen is then stored and used as a high-quality energy source to power vehicles and to generate electricity when direct supplies are short. Of course the entropy involved in this double transformation is considerable, and the task of converting our fossil fuel economy into a hydrogen economy will be huge. But this may be the best course for the future.

Scale. Some energy sources work best at a particular scale of operation, while others are more flexible with regard to scale.

Direct solar energy is best collected at household level. This can be done in the form of passive solar heating, hot water collectors, or photovoltaic cells, which turn sunlight directly into electricity. Sunlight is diffused over the landscape, and to concentrate it in a centralised power station and then rediffuse it would result in enormous losses.

About half of Britain's houses have roofs suitable for solar collection. If they were all fitted with photovoltaic cells this would produce as much electricity as we now use in total, though the time pattern of generation would not match the pattern of demand.[15]

Wind generators do not sit easily on the roofs of houses. They, together with hydro power and biofuels may be a domestic option for smallholders, farmers and rural communities, but hardly for people with only a house and garden to work with. The amount of land needed for domestic fuelwood production is discussed in Chapter 11. (See pp303-304.)

Combined heat and power stations need to be somewhat smaller than the typical large-scale power station feeding the mains, because the heat output can't be transmitted over long distances. The heat loss from the pipes would be prohibitive.

CHP is suited to industries which need electricity and low grade heat in the same proportions that they are produced in when electricity is generated. Alternatively two or more firms with complementary needs can co-operate.

It can also be used for district heating schemes, where the space heating of a number of buildings in the vicinity of the power station is provided by its waste heat. It's hard to retrofit this to an existing area, but there is great potential for new developments, especially where there is a mix of users to even out demand for heat over the day – homes, offices and a school, for example. District heating does not have to be part of a CHP scheme. Even if the heat source is a simple boiler it makes a more efficient use of fuel than having a boiler in each house.

A more novel idea is CHP on a domestic scale. Home CHP units are already in use in Germany, "no larger than a refrigerator and [making] less noise than a dishwasher." They are virtually maintenance free and can pay for themselves in four to five years.[16] They are especially efficient where the peak demands for electricity and space heating coincide, as they do in a cool climate such as ours, and could be the ideal solution for remote locations where connection to the grid is expensive.

The efficiency of these units is said to be 90%.[17] It may be that this kind of highly efficient use of fossil fuels or biofuels is a more productive route to go down in the medium term than the wind, water, wave and Sun path which is so often thought of as *the* alternative to excessive use of fossil fuels.

Wind generators need wide open spaces.

Cost. Obviously you have to consider cash cost, and systems with the least ecological cost are not always the cheapest in pounds and pence. To a great extent this is to do with the economies of scale, which are so great in the current economic system. The more we ask for, and buy, ecologically sound equipment the more it will be produced and the more the price will come down.

But this is looking at the wrong end of the equation. As I have emphasised above, the best 'source' of energy is conservation, and it's almost always the cheapest too.

ENERGY USE

Since conservation is the key to reducing the ecological impact of energy use, the highest priority must go to looking at how we use it and how we can use less.

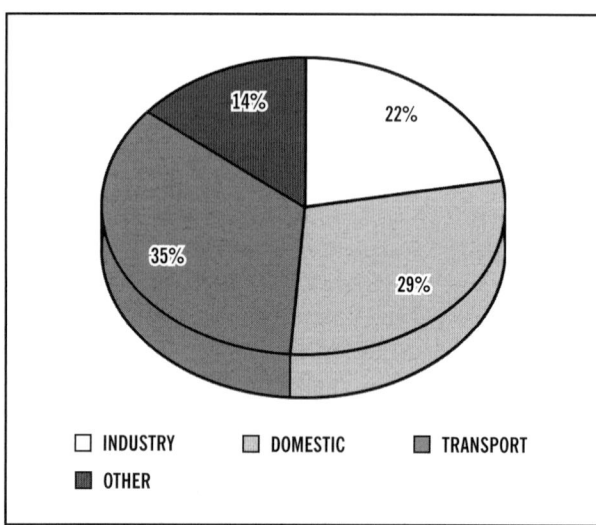

Energy Use by Sector, UK, 1999.[18]

The proportions of energy used by different sectors in Britain are shown in the pie chart above.

These figures possibly underestimate the amount of energy we consume embodied in the manufactured products we buy, as a large proportion of manufactured goods are imported, and the energy used in their manufacture will not show up in the British figures. Nevertheless, even allowing for the decline of British industry, a comparison with figures from the 1980s illustrates the current explosion in energy use for transport. See pie chart in next column.

In fact the figures may disguise the real energy cost of transport. They allocate the energy spent on making cars, planes, ships and trains, on building airports, roads and multi-storey car parks to industry rather than transport. A Spanish study found that more than half the fossil fuel used in Spain was used directly or indirectly for transport, and the same is no doubt true in most developed countries.[20]

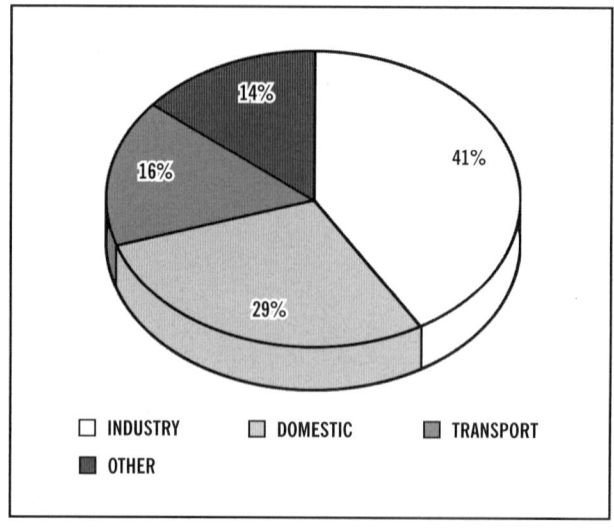

Energy Use by Sector, UK, 1984.[19]

In Transport

Since this is the big one, and growing bigger all the time, it's the first place to look for energy savings.

It's possible to make cars which can run at 180mpg (1.5l/100km) and perform as well as the cars we're driving now. It's a great technological fix, but it wouldn't address the energy and materials used in manufacturing, the land take for roads and car parks, the destruction of communities and all the other effects of road transport.

Switching to public transport, cycling and walking can avoid most of the problems of car travel. The land take for a main line railway, for example, is half that for a motorway, not including car parks; and the number of accidents per passenger mile is much less. A cycle path can carry five times as many people as a road twice as wide, and ten cycles can be parked in the space taken by one car. In purely energy terms, a passenger mile by rail uses half as much as one by car; and there's no reason to suppose that trains could not be improved as much as cars can and stay ahead by the same proportion. But a simple change back to rail is just not on.

This is because cars and lorries have not merely been substituted for railways over the past half century. They have enabled a whole new economy to be created, one that can only be operated by private motor transport. The actual volume of transport, in terms of both people and goods, has vastly increased. But more importantly this growth has been accommodated in a new spatial layout which cannot be serviced by public transport, bike and foot.

We have a major redesign job ahead of us. Fortunately it's a redesign job which most of us would like to see happen for other reasons too. It will involve recreating strong local communities, where people can work, shop and find entertainment close to where they live. In this kind of community, travel, apart from short internal journeys, is an occasional pleasure rather than a daily grind. One of the main aims of good permaculture design must be to make our homes and communities such attractive places that we don't need to travel away from them to make life pleasant and satisfying.

Redesigning our settlement pattern is a long-term solution. In the short to medium term there are many things we can do to reduce the energy cost of transport, and these are summarised in the mind map below.

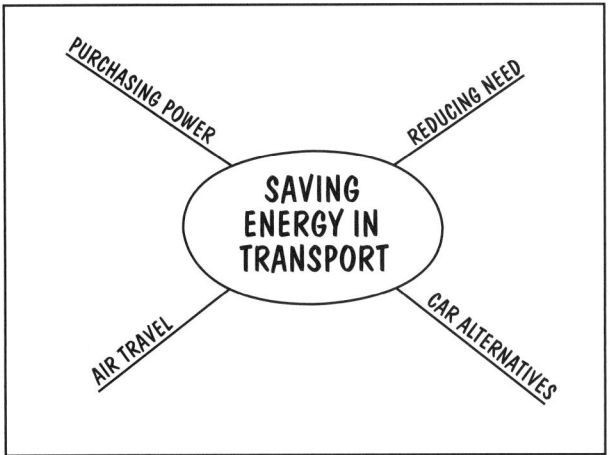

Reducing Need. To a great extent the need to travel can be minimised by choosing to live in a place where work, education, shopping and entertainment are all accessible with little or no travel. This is discussed in Chapter 7. (See pp141-145.) But one change which some of us can make without moving house is to look at the possibility of teleworking.

Many jobs which are now done in offices can be done equally well by people working at home with a computer connection to their employer or clients. Studies have shown that "homeworking requires less than a sixth of the energy used in a 'good practice' office and less than one-eighth of that used in a typical office, when all factors, including travel costs, energy costs in the home and in the office were included." Productivity of homeworkers is higher by 10-20%, and it only takes a 5% increase in productivity to make the initial costs of setting up homeworking worthwhile to an employer.[21]

In the UK there are already a million homeworkers, and the number is increasing by 200,000 a year. The initiative can come from workers as well as from employers, especially if the productivity gains are demonstrated to them. Workers don't need to become completely isolated from one another. Occasional attendance at the office can give face-to-face contact without the need for constant commuting. One firm has a staff to desk ratio of eight to one, an arrangement which saves a great deal of money as well as energy.

Living in a cohousing community can be the ideal complement to homeworking. It provides the day-to-day human contact which many people experience as the most enjoyable thing about going to work. (See p146.)

The weekly drive to the supermarket can be done as well by bike or bus if the shop provides a delivery service.

Indeed orders can be given by phone or internet. Every box of groceries in the delivery van can represent a car journey not made. Again, the initiative can come from customers rather than the company. The more we demand these things the more we'll get them. Supermarkets gain a competitive edge by offering customers the service they want.

The school run has become a vicious spiral: more parents drive their children to school because it's getting more dangerous to walk because there's more traffic on the roads, partly because more parents are driving... Some local authorities in Denmark, and now in Britain, have started to tackle this by getting together with pupils and parents to look at walking routes to schools, identify danger points and make them safer. This can be done by re-routing, traffic calming or giving pedestrians and cycles priority at crucial points.[22]

Once again, the demand for this kind of action can come from parents. We don't have to wait for councils to take the initiative. Sustrans, the organisation which works to set up cycle ways all over the UK, gives support to people who want to organise safe routes to school. (See Appendix D, Organisations.)

Car Alternatives. Those of us who own cars have to use some self-control to leave them at home and use the bus or train. It invariably takes longer, doesn't go door to door, and is less easy for luggage, shopping and small children. Cycling shares some of these disadvantages, but has the advantage of providing enjoyable exercise while making a needed journey.

To a great extent the problem is that the car is already there, and some 80% of the costs are fixed costs which we pay whether we use it or not. The extra cost of a little petrol is usually less than the public transport fare. But taking the plunge of becoming a car-free family can make a big difference to the family finances.

A bit of lateral thinking can put the convenience of car travel in perspective. The speed of it goes down dramatically if you look at all the time you spend earning the money to run the car. If this time is totted up and added proportionately to the time the actual journeys take, the car comes out little faster than a bicycle. If the external costs of driving are included it can even work out faster by bike. (See diagram and box, External Costs, overleaf.)

For those of us who live in the deep country, have poor health or are getting on in years doing without a car may not be a realistic option. But for many of us it may open up the opportunity to take on less work and spend more time doing some of the things we would like to do if only life wasn't so pressured. Hiring a car for the occasional long-distance trip and taking a taxi now and then can help to make the carless option more attractive. Another alternative is car sharing.

Joining a car sharing scheme can give the best of both worlds, combining the independent mobility which you get from owning a car with the financial and ecological advantages of not owning one.

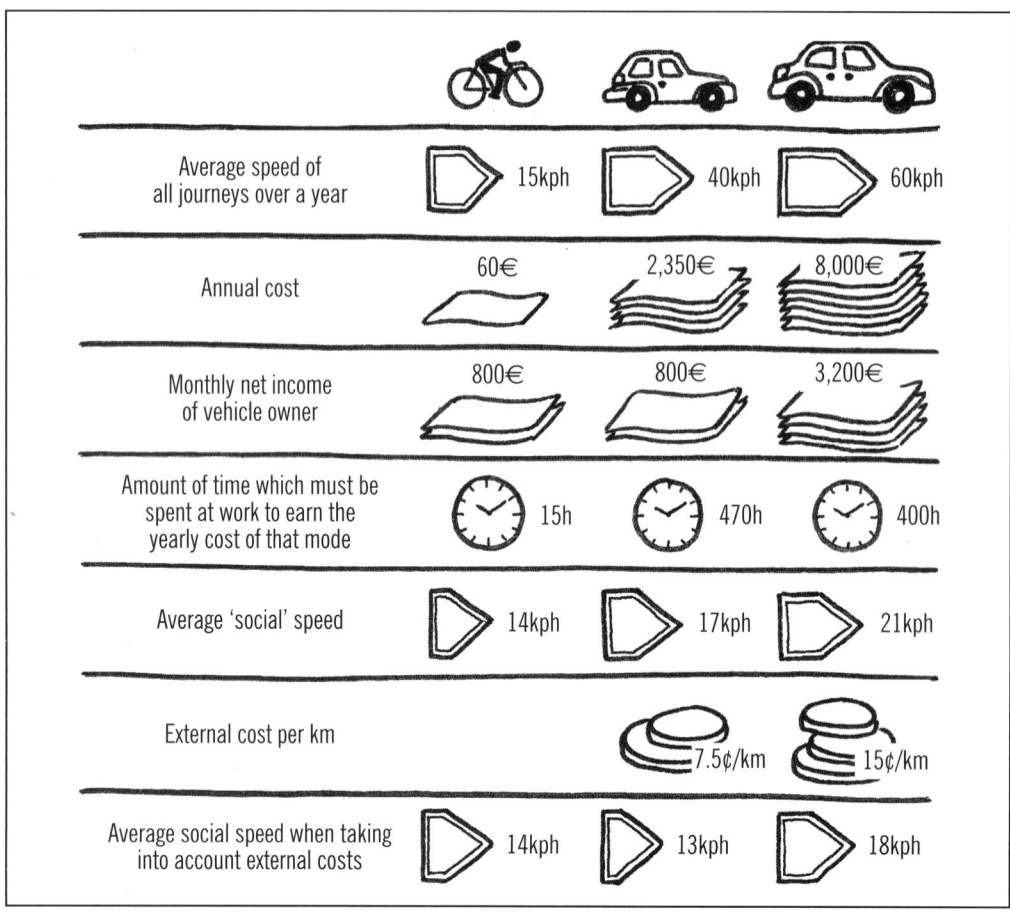

Average 'social speeds' of bicycle and car transport, calculated in Germany.[23]

They can be city-wide organisations, like Edinburgh's City Car Club, which is backed by the City Council and operated by Budget Car Rentals. Or they can be small bottom-up initiatives, like Co-Drive in Leeds, a small group of people who have started up with a single shared car.

The advantage of a large city-wide scheme is that wherever you are you're not far from a car which you can use. But since most journeys start and finish at home it's not a big one, and small schemes can co-operate with neighbouring ones. On the continent

EXTERNAL COSTS

External costs are those costs of producing something which the producer gets away without paying, and which therefore are not passed on to the consumer.

For example, the external costs of transport include: global warming, acid rain, noise pollution, traffic accidents and policing. All of these costs must be paid for, whether by future generations, taxpayers or by people who must simply put up with noise and air pollution. The fact that they are not included in the cost of travelling or of products which have been transported makes transport seem much cheaper than it really is.

The external costs of food include: loss of bio-diversity, soil erosion, chemical pollution, disposal of excess packaging, the external costs of transport and so on.

There are two ways of internalising external costs.

One is taxation. In theory fuel duty and vehicle licences cover at least some of the external costs of transport, such as road construction and land costs. In practice these taxes fall far short of the actual costs, especially with respect to road and air transport.[24] The total external costs of road transport alone in the UK has been estimated at nearly £14 billion a year.[25]

The other way costs can be internalised is for people to choose goods and services which are less ecologically damaging, even though they may cost more. The clearest example of this is organic food. It costs more than conventional food because it internalises some, though by no means all, of the costs which conventional farming leaves for others to pay. Organic food is not expensive: conventional food is unrealistically cheap.

co-operation is international, and a member in Munich, say, can get a train to Amsterdam and use a shared car when they arrive.

Car sharing reverses the cost structure of motoring. 80% of the costs are still fixed costs, but these are shared with other members. With an average of five members per car the fixed cost works out as less than the running cost, instead of four times as much. So instead of having a financial incentive to use the car for every little trip there is a disincentive. If the fixed costs are shared out according to mileage, which they usually are, the extra cost of each trip is the true cost, not just the petrol, so public transport doesn't appear to be unrealistically expensive. All this means that in practice car-sharers use public transport most of the time and only use a car for trips that really warrant it.

Estimates of the cash which can be saved by car sharing vary from £1,000 to £2,000 per year for each member. It has been calculated that each shared car leads to a reduction of 25,000 miles driven per year, which means a reduction of ten tons of CO_2 emissions per year.

Clearly car sharing has more scope in the city than in the country, where public transport is minimal. In fact its success depends on the existence of a decent public transport system, which is not always the case in Britain, even in urban areas. On the other hand, public transport will only improve if we increase the real demand for it by using it. Campaigning for better public transport is worthwhile, but it's a somewhat empty cry if we continue to use our cars at the rate we do now. Information on car sharing can be had from the Community Car Share Network. (See Appendix D, Organisations.)[26]

Boarding an InterCity train at Paddington with a folding bike which weighs around 11.5kg, folds in less than twenty seconds and performs much like an ordinary bicycle. The combined bike/train/bike journey from central London to Somerset takes a little over two hours – considerably quicker than the equivalent time by road and using much less fossil fuel.

David Henshaw, AtoB Magazine

Air Travel. Travelling by air is almost certainly the most ecologically destructive thing we can do.

Aeroplanes produce twice as much CO_2 per passenger mile as cars, and four times as much as trains. The nitrogen and sulphur oxides they emit have some three times the greenhouse effect that they would at ground level. The upper atmosphere is where they have their effect, and as they're short-lived gasses those emitted at ground level have greatly reduced by the time they get up there. Vapour trails can turn into persistent cirrus clouds, which act like a blanket and reflect heat back to Earth. Thus, by one estimate, aviation may be responsible for half of all climate change in the northern hemisphere.[27]

FLYING, A PERSONAL CHOICE

If we're to stabilise the Earth's climate each of us can only produce so much CO_2. It's been calculated that one return trip from Britain to Florida uses up a person's entire lifetime CO_2 allowance.

Air travel is very seductive, and ridiculously cheap – partly because aviation fuel is not taxed.[28] For the cost of a rail ticket from Glasgow to Bristol we can hop onto an aeroplane and be on the shores of the sunny Mediterranean in a couple of hours.

Until recently I didn't travel by air for mere pleasure but would do so to teach permaculture, reasoning that the good I did would outweigh the harm. Now I feel that's a dubious argument, and I've decided not to fly again. I go by train, and I reckon that if I'm not prepared to invest a little more money and the couple of extra days which my trip takes I obviously don't value the journey very highly, and I might as well not go.

It's a personal decision, and I wouldn't try to impose it on anyone else. But I would say it's pretty futile to read the rest of this book and put it into practice, and yet continue to travel by air. The benefit to the Earth from kicking this habit probably outweighs everything else put together.

Purchasing Power. We all consume embodied transport in the material goods we buy. We can reduce this by reducing our overall level of consumption. It's less easy to do so by selective buying, because it's usually impossible to tell what journeys a product and all its components have made in the production process. Production is very fragmented these days, and an item ostensibly made in one place will often have been trucked to and fro between different countries for different stages of its manufacture.

This applies to food, especially processed food,

as much as to any other product. Although the use of energy on farms only accounts for some 1% of our total energy consumption, the transport and packaging of food account for 12%.[29] The growth of transport in the food business is faster than that in any other part of the economy.[30] But food is one product which is easier than others to source from near home. We can grow some ourselves, and there are a number of ways of buying fresh local food straight from the producer. These are discussed in Chapter 10. (See pp285-293.)

In the Home

As we saw, 28% of national energy consumption is used directly in our homes. The pie chart shows what proportions we use for different purposes.

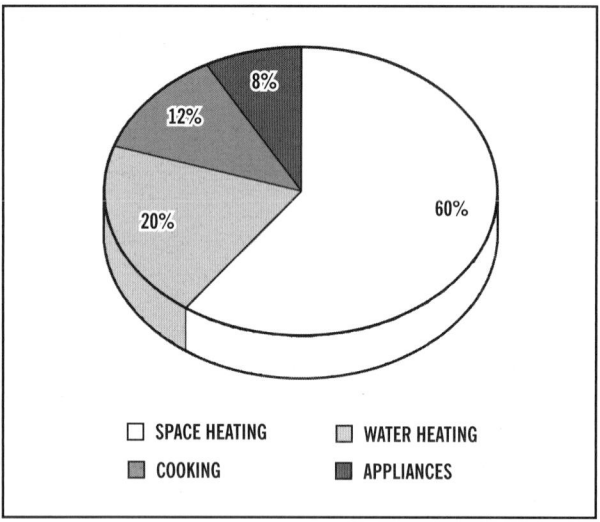

Energy use in the home.[31]

Major work to make the house more energy efficient is discussed under building renovation in Chapter 7, though many of those things, such as improving insulation levels, need not wait for a major renovation project. But energy saving really starts with awareness, and here we look at a few simple things which together

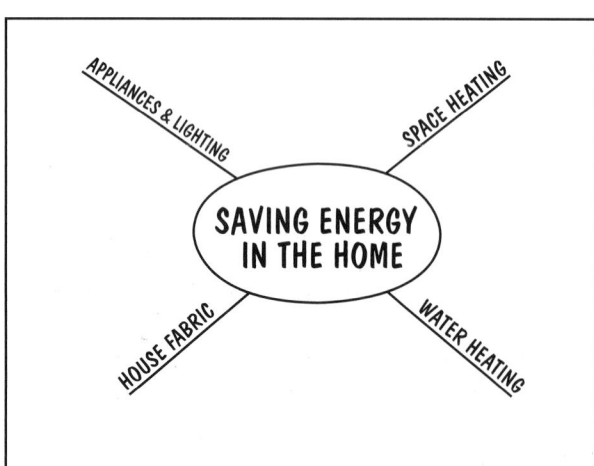

can add up to significant savings. Most of them cost either nothing or a few tens of pounds, and cash payback is typically 1-5 years.[32]

Space Heating. As space heating takes the lion's share of the energy used in the home, the biggest savings can be made here.

The permaculture ideal may be an efficient stove burning home-grown wood. But most of us heat our houses with central heating running on fossil fuels, and have little opportunity to change in the medium term. Some systems are much more efficient than others, but the embodied energy in a central heating system means that it's usually more efficient, both from an energy and a cash point of view, to keep the old system and make the best use of it we can. Broadly speaking, the older the system the more likely that replacement will be worthwhile, but a good energy consultant should be able to do the calculation for any actual system.

The biggest single thing we can do to save energy without investing in expensive new kit is to accept a slightly lower temperature. Every rise in temperature takes more energy than the one before it. Raising the room temperature from 15°C to 20°C takes more energy than raising it from 10°C to 15°C. Thus lowering the thermostat even a few degrees will make a significant difference over the whole heating season. As a rule of thumb, you save 10% of the energy for each 1°C reduction in temperature.

Where the boiler has no control system to keep it at the desired temperature, fitting one is the first step. Timers on the heating control unit are important. Learning to operate them, getting the default settings right, and remembering to re-programme for daily changes of routine will make a big difference by cutting out the waste of heating the house at times when it's not wanted.

Heating rooms which are not in use is equally wasteful. Individual thermostats can be fitted, either in each room or to each radiator, and adjusted according to need. Even where all the rooms need to be heated to the same temperature, room thermostats enable you to make use of passive solar heating. A house with large windows on the west side will get hot on that side in the afternoon while rooms on the east remain quite cool. With individual thermostats it's easy to keep the whole house at a comfortable temperature without wasting any fuel overheating the rooms on the sunny side.

Fitting radiator reflectors is a quick and cheap measure which anyone with central heating can take, even those in short-term rented accommodation. If a radiator backs onto an external wall, putting aluminium foil on the wall behind it will reflect heat back into the building.

For those who don't have central heating, or where a little extra heating is necessary in one room, a paraffin or gas heater is clearly better than an electric one, given the inefficiency of using electricity for space heating.

Water Heating. The easiest way to reduce the amount of energy used to heat water is to heat less of it. Ways of saving water are described in Chapter 5 and those which relate to hot water are relevant here. They include taking showers instead of baths, or smaller baths, and doing something about dead legs. (See p94.)

Both hot water pipes and hot water cylinders can be lagged, that's to say insulated. Lagging the pipes is, as we saw in the previous chapter, one way of dealing with dead legs. Lagging the cylinder is worthwhile even if it already has a standard 40mm insulation jacket or stuck-on foam insulation. Bringing the depth of insulation up to 80mm will pay for itself in a year at most, and still allow enough warmth to escape to heat the airing cupboard.

Tankless, or instant, hot water heaters avoid the problem of heat loss from tanks and pipes altogether. (See p165.)

House Fabric. In many houses draught proofing is the very first thing to look at. If a lot of cold air is leaking into the house there's no point in putting effort into either insulation or heating because the warm air will be whisked away as fast as it accumulates. (This is less so when the heating system works with radiant heat rather than convected heat. The masonry stove is an example. See pp152-153.) Draught proofing is cheap and can pay for itself in weeks in a really draughty house. In a more typical house it may take two or more years to pay back the cost of a very thorough DIY job, and longer if you get it done professionally.

It's important not to make the house totally airtight. People need to breathe, and fuel burning appliances need a constant flow of air, both for combustion and to provide airflow in the flue. Modern gas and oil burners have this built in to them, but older ones need an air source such as a ventilation brick.

The design of curtains can have a big effect on how efficiently heat is used, especially where a radiator is fitted directly under a window. Hot air rises from a radiator in a convection current. If it can pass straight up behind the curtains much of the heat will be lost by conduction through the glass of the window. (See illustration a. below left.) This can be prevented by fixing a shelf above the radiator which the foot of the curtain can rest on, and by fitting a pelmet. Both of these help prevent air passing behind the curtain by stopping the convection current. (See illustration b. below left.)

For people in rented accommodation who may not want to get involved in extensive carpentry, simply having slightly longer curtains and slipping the bottom edge behind the radiator is a help.

Passive solar gain can be considerably increased by regularly cleaning the windows, especially those which receive a lot of light.

Cooking, Appliances and Lighting. Here we are looking at about a fifth of the total energy used in the average house. But as some or all of it is in the form of electricity, which has a low net efficiency and high pollution output, the ecological importance may be greater than that proportion would suggest.

Getting in the habit of switching things off when they're not needed is part of energy awareness. As lights and electronic appliances use quite a small proportion even of the one-fifth we are looking at here, switching off may seem to have more of a symbolic than a real value. But if all the videos, television sets and hi-fis in Britain which are left on standby were switched off it would save the output of two power stations. Every little bit mounts up.

Dishwashers and tumble dryers use a lot of electricity. People who feel they simply have to have them to make life bearable might consider whether it's their over-stressed lifestyle that needs attention. Microwave cookers, on the other hand, use much less energy than other cookers. Many people are put off by the notion of having their food cooked by such means but from an energy point of view they come out well. Most things which are kept in a fridge don't really need refrigeration. A family with a reasonably cool place to store food in the house may not need one at all.

As with central heating systems, it's often more energy efficient to conserve the embodied energy in an appliance than to scrap it and replace it with one which consumes less energy in use. But if you do need to replace something it's worth shopping around. There can be huge differences between the most efficient models and the least efficient, and energy labelling of fridges and washing machines is compulsory.[33] There is also a Europe-wide Ecolabel (see right), which takes account of all ecological impacts, not just energy, but this is voluntary so not all goods have a rating. The more efficient models are not always the most expensive ones.

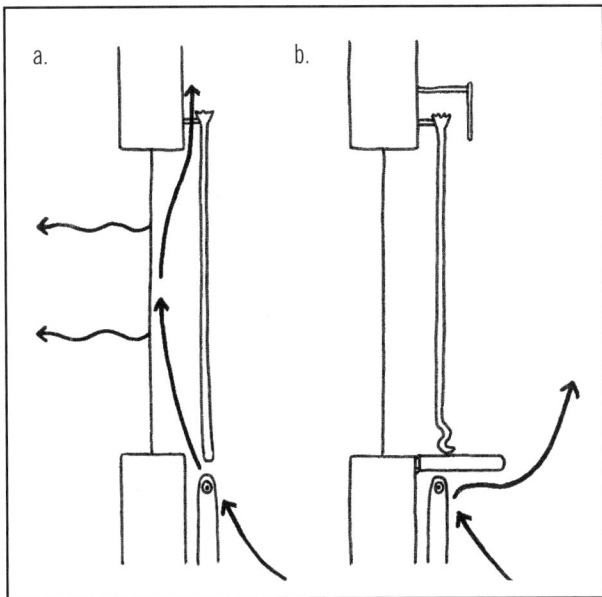

CFCs and HCFCs have been banned from use in fridges as coolants. The main replacements are HFCs, which are harmless to the ozone layer but still very potent greenhouse gasses. There are now some fridges using hydrocarbons and these are much milder greenhouse gasses as well as not harming the ozone layer. At present these hydrocarbons go under the name R600a. Dead fridges should always be disposed of at a local authority tip, where the coolant chemicals will be carefully extracted and recycled so as to keep them out of the atmosphere.

If you should need to replace your cooker, clearly a gas model will be much more efficient than an electric one. Whether it's worth replacing a serviceable electric cooker with a new gas one is another of those questions of embodied energy vs energy in use which can only be answered if you have the relevant facts to hand. These will depend on the make and model of both cookers and a good energy consultant should be able to give you an answer.

Low energy light bulbs save money in the long run and pay for themselves in energy terms in about two years. They do require a bigger cash outlay in the first place than filament bulbs, but even people on low incomes should be able to afford to change over bit by bit as the old bulbs give out. They are also easy to take with you if you move house, though they are fragile and need careful packing.

Much can also be done by careful lighting design. This means things like choosing the minimum wattage for each situation and lighting a specific work area brightly rather than the whole room.

When we've reduced our electricity consumption to a level which is economical but still comfortable it's time to start looking at where we get our electricity from. Those of us who aren't able to generate our own at home can buy renewable energy from the grid.

Most ordinary electricity companies sell renewable electricity, at a slightly higher price, but it may be better not to buy from them. They're obliged to include a proportion of non-fossil-fuel electricity in their output and will do so anyway whether any of us are prepared to pay more for it or not. Thus buying renewables from them may not result in any more renewables being generated. But there are companies which sell renewables only, and these are the best ones to get it from. Details of these companies can be had from the Centre for Alternative Technology. (See Appendix D, Organisations.)

USE OF MATERIALS

If we go back to the model of the natural ecosystem we can see that energy and materials behave quite differently in it. While energy is a one-way flow, materials are largely cycled within it, with only small losses and additions to and from the outside. (See p34.) The result of ignoring this pattern and creating a one-way flow

of materials through our economy is that we turn useful resources into pollution.

The way back to a cyclic pattern is mapped out in the well-known slogan:

Reduce

Repair

Reuse

Recycle

This gives us a set or priorities: each one should be addressed before the one below it.

WyeCycle

Reuse is better than recycling. The energy and pollution costs of recycling plastic and glass containers are so high that it's hardly better than sending them to landfill.

All consumption uses up materials and energy, however much we repair, reuse and recycle. The first and most efficient option is always to ask whether we really need something. Many forms of consumption are inherently resource-greedy. For example, motor transport uses huge quantities of aggregate, cement and tar for road construction. Although much of this is in theory recyclable in practice it is not recycled, as the road network continues to grow and grow.

The choice between repairing things or replacing them is mainly a choice between using human labour or fossil fuel. It's the massive underpricing of energy which makes it cheaper to throw things away than to fix them. Repair keeps almost all of the materials and embodied energy within the system and out of the landfill. It also creates more employment than manufacturing a new one.

The habit of throwing things away is not all due to built-in obsolescence by any means. A survey of UK scrap heaps in 1989 found that a quarter of all washing machines had no obvious fault, and another quarter of them could have been repaired quickly and cheaply. Obviously some people would rather spend money on a brand new one than continue with a perfectly service-able 'old' one. Nevertheless there must be a market for at least some of the perfectly sound goods that get thrown away, with a potential business opportunity in it for permaculturists.

Reusing a glass bottle is, of course, much better than smashing it up and recycling the broken glass. The energy expenditure in recycling, including collection, remanufacturing and redistribution, always makes it the last choice among the four Rs.

Large-scale reusing and recycling can be done much more efficiently in cities, where the materials are not spread so thinly on the ground as they are in smaller towns and the country. Yet the energy cost of collection and redistribution is always a limitation when working on a large scale. This means that reusing and recycling are at their most efficient if they can be done within the household. This may take some inventiveness. It can be fun in a Blue Peterish way, but it's as well to remember the Quit While You're Ahead principle. (See p9.)

Industrial Ecology[34]

In a natural ecosystem nothing is wasted. The output of every plant and animal becomes the input to another. In permaculture systems we aim to imitate this. It's the principle of Linking. (See p32.) In Chapter 4, for example, we saw how the leaf fall from waterside trees and manure from farm animals can, with very little effort in a well designed system, become food for fish. The same principle can be applied to industry, where it's known as industrial symbiosis, or industrial ecology.

A prime example is to be found in the Danish town of Kalundborg, where a network of input-output links has been established between a group of industries and local homes and farms.

There's a power station and many of the links centre around the combined heat and power concept. (See p123.) Heating is provided to local homes and businesses, including a greenhouse and a fish farm, while steam is supplied to a chemical factory and an oil refinery. There's a plasterboard factory which makes use of the gypsum which arises in the power station's sulphur-scrubbing system. The oil refinery supplies the plasterboard factory directly with gas and sends water to the power station. Ash from the power station is used in cement and in local roads, and sludge from the fish farm goes to fertilise local farms.

It took a decade to refine the Kalundborg system and there is no doubt that the most effective way to set up such a system is from scratch. Selecting a group of businesses which are likely to be able to use each others' outputs is more likely to be successful than retrofitting an existing industrial estate.

Nevertheless, a study on an industrial estate in Canada found that the five most frequently used materials were also the five most often wasted: paper, cardboard, wood, oil and electricity. Although many firms recycled paper and cardboard it was rarely used within the same industrial estate. The most basic input required to get industrial ecology going on the estate was information about what materials were available and how they could be linked. Almost all businesses on the site were receptive to the idea.

Timber

One of the recurring themes of permaculture is the idea of using the produce of trees to meet many of the needs we now meet from non-renewable resources. But at present we are consuming timber at an unsustainable rate, so the first step must be to see how we can reduce it.

In Britain, with 1% of the world's population we consume 3% of the world's timber production. It has been calculated that we need to reduce our consumption by 73% in order not to exceed our share of the world's sustainable production.[35]

Almost half of what we consume is in the form of paper and board. Recycling is usually better than sending it to landfill but the energy expended in recycling has its ecological cost.

Some people reckon that burning used paper and making use of the heat comes out better on ecological balance than recycling, although this is not a majority view.[36] It can be done on a municipal scale or at home with a hand press that turns soaked newspapers into briquettes. Composting is another option and is the best use for cardboard, which is hard to recycle. This can also be done either municipally or at home. (For home composting see pp45-46 and pp105-106.)

Both burning and composting can run into the same energy costs as recycling if done at the municipal level. Much greater efficiency can be had by doing either at home for home needs.

Substituting annual fibre crops for wood as the raw material for paper can be useful when the fibre is a by-product from a food crop, such as cereal straw. When fibre is the sole output of the crop, as with hemp, it is in direct competition with food crops. It's hard to see how this represents a gain in ecological terms over using trees, a perennial crop. We have already seen how badly annual energy crops compare with trees as a source of biofuels in terms of energy efficiency and pollution output. (See p124.) No doubt annual fibre crops come out equally badly.[37]

All these things can help reduce the impact of our paper consumption. But when it comes down to it there is no real substitute for using less paper in the first place.

There is unused potential for reusing sawn timber. Furniture refurbishment and reuse, for example, should become increasingly profitable as the landfill tax rises to a more significant rate. This is the kind of work which can make useful self-employment for permaculturists in urban areas, where the raw materials and markets are both near at hand.

Howard Johns

The Brighton and Hove Wood Recycling Project works locally, which makes the reuse of small lots of timber economic.

Building materials can easily be reused if the design of the building allows for it, and with some difficulty if it doesn't. Designing buildings so that the materials can easily be reused when the life of the building is over can help reduce timber consumption in the future. Salvaging timber from existing buildings can have a more immediate effect but the value of the salvaged material is variable. If it's in good condition, free from hidden nails and screws and of useful dimensions it can be used again for building. If not it's only fit for use as a raw material for chipboard production.

Timber is used as packaging for large items such as furniture and car parts, or as pallets for smaller items, and we consume this indirectly when we buy the products. Looking at our level of consumption of all goods is probably the most significant thing we can do to reduce our consumption of timber.

Sources of Timber

Having reduced our level of consumption, we can look at where the timber we still need comes from. Even in the long term we don't necessarily have to think in terms of total self-sufficiency. This is a small, crowded island, and some trade with sparsely populated, well-wooded countries could be sustainable.

There are broadly three different kinds of forest or woodland:

- virgin or 'old-growth' forests
- semi-natural woodlands or secondary forests
- plantations

Virgin forests are not all tropical. Some fragments still remain in temperate timber-exporting regions like North America and Scandinavia, and large areas in Siberia. The value of virgin forest ecosystems is beyond measure, yet they are still being felled. It behoves us to do everything we can to protect them, including supporting campaigning organisations which are working to save them.

Secondary forests, which have grown up by natural regeneration where virgin forests have been felled, can provide a useful source of income to people living in tropical countries. They rarely have the same ecological value as the virgin forest, and can be managed sustainably for timber production. Semi-natural woodlands, such as the coppice woods of Britain, have developed over hundreds or thousands of years with human felling as a dynamic part of the ecosystem. Continuing this relationship can be beneficial to the biodiversity and ecology of the woodland. (See p356.)

Plantations, when they are composed of species which fit the local ecology and are worked with in a sensitive way, can be sustainable and ecologically benign. Sad to say they are rarely either. Usually they consist of large-scale monocultures, often of species which are not native to the area, and are harvested by clear felling. They often have a lower ecological value than the ecosystems which they replace, as when peat moors are planted up with conifers in the Scottish Highlands. (See p72.)

In the past it has been difficult to know whether a timber product has been produced sustainably or not. But there is now a reputable certification scheme which operates in much the same way as certification schemes for organic food. This is run by the Forest Stewardship Council (FSC), and here in Britain the actual certification of individual woods is done by bodies like the Soil Association Woodmark Programme. You can be sure that produce bearing the FSC logo has been produced responsibly, either in Britain or abroad. The criteria for certification

The logo to look out for.

include both ecological and social considerations. A mere statement such as "sourced from sustainably managed forests" is not to be trusted.

One product which is easy to be sure about is barbecue charcoal. Most of it comes from mangrove swamps in the tropics, valuable and diverse ecosystems which are destroyed in the felling. But it's now possible to buy British charcoal. This is either made from offcuts from local timber production or from coppice, in which case harvesting it may actually be helping the woodland ecosystem to remain healthy and diverse.

GUEST ESSAY

ECODESIGN:

There's More To It Than It Seems

by Peter Harper

Peter Harper has been a leading light in the alternative technology movement since the 1970s, and he is now Head of Biology at the Centre for Alternative Technology in Wales. He loves to challenge accepted wisdom, and brings his enquiring mind to bear on all aspects of the sustainability debate, often coming up with answers which contradict our cherished illusions.

The expression 'alternative technology' suggests an array of designable, buildable, buyable chunks of hardware: magic bullets that will solve environmental problems. This is what a gadget-conscious public hopes for and expects.

Alas, it rarely works out so neatly. There are dozens of unquantifiable and conflicting variables to take into account, and a complete solution usually demands a redesign of other elements which are outside the brief. Very often the best answers are an alternative to technology.

I want to illustrate some of the issues in ecodesign by taking as a case study one of the standard clichés of alternative technology, the solar water-heating collector. This is a classic magic bullet, a discrete and highly visible artefact. The concerned citizen imagines that a simple, cheap device on the roof will provide endless amounts of hot water; and that all it takes is a generous proliferation of such gadgets to save the planet.

Would that it were as simple as this! What follows are eight reasons why a more subtle and fundamental approach is necessary.

It's part of a system

The collector is only the visible part of a water heating system that includes pumps, tanks, sensors, control gear, heat exchangers and a great deal of pipework. Designing an effective collector is quite a small part of the overall problem. It's got to be compatible with the rest, and it has to suit the particular application. The full system is different in every case.

The whole package may require liaison with architects, builders, plumbers, electricians and clients. Ecodesign often requires a team effort.

The system is not just technical

Ecodesign is always surrounded by a web of social and legal forces, and these can make a tremendous difference to whether a system is worth doing. The building regulations might change and say you have to install solar water heating in new buildings; or there might be a legionella scare about bacteria in not-very-hot water systems; or some foreign manufacturer might suddenly flood the market with cheap and dodgy systems. Such happenings are quite outside the control of the designer.

The ecodesigner and the market may have different agendas

Little of what we buy is really essential for life; most is symbolic or cultural, and shopping is very often a recreational activity. In the case of solar water heaters we note that many prospective installers do not care what it costs or whether it really does very much: it simply makes them feel better and they earn brownie points with greenie friends. (Shall we call them greenie points? There is a huge market in greenie points.) Yet others are gadgeteers that love complex toys, switches, dials and flashing lights. A solar designer might imagine the required solution is cheap, simple and discreet, but this would not satisfy the real agendas in these two examples. What would you do?

Does it really work?

Enquirers about solar water heating are often disappointed to find that it doesn't replace the existing system, but is simply an add-on. In sunny weather the sun can certainly provide lashings of very hot water. But most of the time it just preheats the cold water so the conventional system has less work to do, thereby saving fuel. That's all it does really, saves a bit of gas or electricity.

This sense of disappointment crops up all the time in ecodesign. Green cleaners and detergents, green pesticides and green cars just don't have the oomph that we have come to expect from industrial products. This is not surprising. It's indiscriminate oomph that is the cause of the whole mess in the first place.

But consider these responses to the question 'Does it really work?':

- Some green products are directly comparable in cost and performance in the raw marketplace. For example the fridges with the highest and lowest energy efficiency have the same performance and the same price. This will become increasingly common and requires us to be well-informed and discriminating shoppers or specifiers.

- Sometimes we have to make more effort to present the whole picture. High efficiency lamps cost ten times more to buy than conventional ones, but they last eight times longer and save so much electricity that they cut your lighting fuel bills in half.

- Otherwise we have to redefine the context so eco-products are not compared directly with their rivals. They are actually doing something different. Organic produce on a local market stall is not simply expensive vegetables but a chance to help the local economy, get to know the grower, become part of a convivial weekly event, exchange gossip and garden lore, perhaps even find an outlet to distribute your own surpluses.

Does it save money?

Solar water heaters used to be advertised under the bold slogan 'free hot water!', but this is seriously misleading. Ingenious and handy people could probably make their own collectors and install a workable system for a few hundred pounds. A fully automatic system installed commercially will cost thousands, and, depending on all sorts of circumstances, will save you between £50 and £200 a year. Is that 'free'? Is it even rational?

Most renewable energy systems are expensive, or appear so, because the playing field is tilted against them at an impossible angle. Conventional fuels are insanely cheap by historical standards. The muscular energy you might expend in a day's physical labour is represented by about a pennyworth of oil or gas.

This is simply environmental asset-stripping. Living off our capital rather than the interest it generates, we can obviously live very high off the hog – for a while. The programme of ecodesign is essentially to work out how to live comfortably off a modest income rather than a millionaire's hoard. This is not something designers can tackle on their own. We need to work collectively towards changes in the agreed rules of society, with instruments such as: carbon taxes, strict emission standards, subsidies for sound practice, labelling schemes, changes in research priorities, public information campaigns – and vast changes in expectations.

Does it have a positive impact on the environment?

True, for every kilowatt-hour of fuel saved by our solar panel there's another pound or so of CO_2 that doesn't go into the atmosphere. But how much CO_2 went into the panel in the first place? You need to know the entire cradle-to-grave story: raw materials, manufacturing, distribution, advertising, installation, operation, maintenance, decommissioning and disposal.

Compare this with the energy you would have used had you not installed the collector. Are you still ahead? Don't take it for granted.

The picture is still not complete without a full environmental impact assessment. Here you are forced to compare unlike values and make choices. What about the visual impact of the panel, or the risks of accidents from installation or maintenance? How do they weigh against the gains on global warming? There is a lot of room for conflict here. At a public enquiry for Wales' first wind farm both proponents and objectors were claiming to speak on behalf of The Environment. Who are the true environmentalists?

Is it the best way to spend money on environmental improvement?

The question to ask is, if you have £3,000 to spend on helping the environment, is a solar water heating system the best way? If it isn't why do it? If you are building a hotel on the south coast, where a large demand for hot water coincides with maximum sunlight, there would be a strong case for integrating solar water heating in the design. But in Britain retrofitted solar water heating comes fairly low on the pecking-order of household energy-reducing measures, and for most households far simpler measures such as tank and pipe lagging, spray and press taps, low-flow shower heads and better controls will save the same energy more cheaply, leaving money to spend on further environmental improvements.

Paul and Judith Otto's entirely home-made solar water heating system cost them just under £800 for materials at 1994 prices. Self-installing a factory made system can save around half the cost of one which is professionally installed.

It is nearly always more cost effective to improve the efficiency with which energy is targeted to its purpose than it is to develop new sources. This is usually also simpler, quicker and has less environmental impact. And the principle applies in other spheres apart from energy.

So is efficiency the answer to the problem of sustainability?

It is at least half the answer, perhaps more. As it is often achieved by technical improvements and clever design, it does not collide with ordinary behaviour or aspirations.

The result is that efficiency improvements can easily spread throughout the whole of society. But of course in the end this won't help if there is constantly rising demand for more stuff and more energy, no matter how efficiently it is delivered. Sooner or later we will have to bite the bullet of living standards and lifestyles, and then there will be much more cultural resistance.

I think we have to be clever about this. There is no need for a puritanical, indiscriminate reduction of living standards, rather we should concentrate on cutting out those bits which don't matter much anyway. A programme of simpler, slower living will come to seem more and more attractive in the mad whirlwind 21st century.

It could be seen as a rational exchange of standard of living for quality of life. Quality of life is what we all really want, yet we only seem to be able to approach it through the vicarious medium of a material standard of living. It seems to me that the globally sustainable society will only come about through a much more direct approach to the quality of life, whatever it means. In this task designers of all kinds must surely play a central part.

Probably this will lead us into something we might call 'lifestyle design' where we are trying to put together rather complex packages of technical solutions, habits and skills, and most crucially of all, rethought aspirations. Sounds a bit like permaculture to me.

PART TWO

SPECIFICS

Chapter 7

BUILDINGS

Permaculture has a number of distinctive contributions to make on the subject of sustainable buildings and the built environment. The first centres around the importance of growing more food in towns and cities, and empowering people to grow at least some of their own food at home. This has implications for town planning which are quite controversial in sustainability circles. The second is a keen interest in the passive solar approach to building, which fits well with the permaculture emphasis on getting the fundamental shape of things right in the first place rather than adding gadgets afterwards. The third is biotecture, using plants as a functional part of buildings. These are the subjects of the first three parts of this chapter.

There are a number of building styles, such as straw bale and green timber framing, which are often attractive to people with permaculture leanings, and there is a brief appraisal of these. But few of us are likely to have the opportunity to build a house from new, and the rest of the chapter focuses on renovation and the choice of building materials, things which many of us are likely to be involved in from time to time. Most of the ideas described under biotecture can be retrofitted to existing housing, so that part of the chapter should also be of interest to those who are not involved in building from new.

SETTLEMENT DESIGN

The most important thing about any house is where it is.

Just imagine a house in the heart of the country, with enough garden to be self-sufficient in vegetables, south-facing for an easy passive solar conversion, with a stream suitable for micro-hydropower, maybe even a coppice for firewood. Wouldn't it be a perfect place to practice permaculture and start really doing something to save the planet!

Possibly not. If this idyllic place is fifteen miles from the workplace, five miles from the children's school, and a similar distance from the shops and friends' houses, and all these journeys are made by car, it would almost certainly be better for the planet to live in a typical city house within cycling distance of everything.

The way things are now, living without a car is probably the second biggest thing most families can do to reduce their ecological impact. (Not flying is the biggest.) In some cases it may be bigger than everything else put together.

On the other hand, people living in a present-day city are almost entirely dependent on external systems for their food, water, energy and other needs. Given the shape and size of contemporary cities, these supply systems are not sustainable, with long transport distances, high energy costs, high losses in transit and

high pollution costs. So which is the best place to live from an ecological point of view?

The answer depends on whether we're talking about the present or the future. In the present, with our current lifestyle, there is little doubt that on balance living in a city will have less ecological impact. Moving goods around takes less energy than moving people around in private cars.

But in the long term we need to design truly sustainable settlements. This means cities and towns which are small enough to get most of their resources locally. It also means repopulating the countryside with new hamlets and smallholdings. More people will work where they live, and more food and manufactured goods will be consumed near where they're produced.

Thus both town and country can become more productive in real terms and more pleasant to live in. A more localised structure will provide the physical framework within which real communities can grow again and social capital be rebuilt. But a dilemma remains for settlement planning in the short term. Do we design for the present, which means compact, high density urban areas, or do we design for the future, which means dispersed low-density settlements. As permaculturists we need to make our voices heard in the planning debate, but exactly what we should say is less than clear. The following section attempts to cast some light on the matter.

Density

Urban Areas

Low density housing has often been favoured by permaculturists because of the opportunities it gives for home food production.

In suburbs with net densities averaging around 24 dwellings per hectare (dph), which is typical of those which were built from the 1920s to the 1950s, the vision of families growing a significant proportion of their own food at home is a real possibility. Allowing for roads and other public spaces, this density gives some 300m^2 per house plot. A survey of suburban London in 1948 found that on average 66% of each house plot was made up by lawns, flowers, shrubs and food crops.[1] All this area is potentially productive, so if a family chose to go all out for maximum food production they might have some 200m^2 available for it.

John Jeavons calculates that in California it's possible to produce a complete diet for one person, plus the needed soil fertility building crops, on 370m^2, using his highly intensive methods.[2] Julia and Michael Guerra reckon, perhaps a little optimistically, that they could feed two people on 250m^2, using no-dig methods here in the

Michael Guerra outside his front door, in a forest of vegetables.

British climate. (See case study, A Small Productive Garden, p209.) Both of these estimates require more land than the 200m^2 per family provided for by the average low-density suburb, but they suggest that a significant proportion of the family's food could be grown at home. In fact the ecological significance is greater than the figures suggest, because the home-grown portion would include most or all of the fruit and vegetables. These foods are 90% water and thus require much more energy to transport per calorie of food than do grains, at 10% moisture.

As well as being good zoning, home gardening allows for maximum linking between household and garden. It's much easier to use compost made from the food scraps, waste paper and urine from a house if the garden is outside the back door rather than in a distant allotment. Using grey or roof water in the garden is only possible when the house and garden are together.

The advantages of low density apply to more than just food. Passive solar heating works better at low densities, as there is less shading of one house by another. In high rainfall areas rainwater collection can meet all domestic needs if the roof is big enough, and a two-storey suburban house clearly has more roof area than a three-storey town house with the same floor area. (See pp97-98.)

The general lack of space in a high-density terrace house puts a damper on a number of ecologically desirable activities. Separation of household waste, which is central to any really efficient recycling programme, is difficult if you only have space for one bin. Further pressure is put on the waste stream if there's no room for composting. Tumble-drying of clothes is much more attractive than a clothes-line if you only have a little patch of back garden, and that is both shady and sheltered from the wind. There's little space for keeping bicycles in most terrace houses other than in the front hallway. Many keen urban cyclists do keep their bikes in the hall, but it can make getting in and out of the house like an obstacle race, and it's enough to put many people off the idea of cycling.

MEASURING HOUSING DENSITY

The density of housing can be expressed as the number of dwellings per hectare.[3]

This can be given either as a net or gross figure. Net density is the density of the housing estate itself. Gross density is the overall density of a settlement or neighbourhood, including parks, schools, allotments, commerce, main roads and infrastructure.

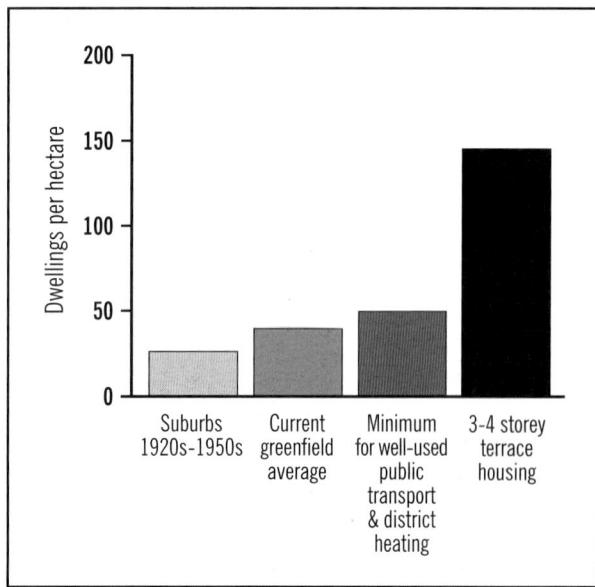

Some housing densities.

Shelterbelts and other urban green spaces can bring multiple benefits to inhabitants of urban areas. (See p83.) But they do take up land, and thus inevitably reduce gross density.

For all these reasons low-density suburbs are often seen as ideal living environments by permaculturists.[4]

Other proponents of sustainable settlement design take the opposite view, focusing on the energy saving benefits of high density. They say that in cities with net densities of 130-160 dph car-free life becomes a real pleasure, with a wide range of work and cultural opportunities within walking distance.[5] The possibility of a truly rich urban existence, in the classical or renaissance sense, is much greater than in widely dispersed suburbs.

This means three to four storey terraced housing on the Georgian or Victorian model. At this density it's possible for about half the dwellings to have a little garden of their own. The scope for food production is insignificant in physical terms, though some people may have the spiritual satisfaction of growing a token bit of their own food.

A less extreme view, still emphasising the advantages of high density, is to aim for an average net density of 40-50 dph with some areas more dense and others less so. This is only a bit more dense than the current average for greenfield sites of 30-35 dph, but it makes

Terraced houses can be pleasant places to live, and they're much more energy efficient than detached houses.

a significant difference. Studies have shown that at 40-50 dph car ownership and car reliance actually do fall, while walking, cycling and public transport use rise.[6]

At this average density some of the housing will be terraced. Terraced houses are more energy efficient than detached or semi-detached because there is less surface from which heat can be lost in relation to volume. They also use less materials in their construction, and so embody less energy. District heating schemes based

LIVING IN THE CITY[7]

Jessica and John Preston will admit to dreaming of a rural retreat now and then, and even to looking eagerly through the small ads in alternative publications. But there's really too much going for them in Leeds to tempt them out.

If they lived in the country it would be hard to manage without two cars. In the city they manage without one at all, except for the occasional hire car for a special trip. Jessica can get to work by a short walk and a train, John by a walk through the park and along the canalside, where his morning is often blessed by the blue flash of a kingfisher.

It took them six years of patient and enjoyable learning to get their allotment fully into production. Now they have put it down to soft fruit and are starting on a second one. Their own back garden was planted up as a wildlife garden by the previous owner, and they are reluctant to disturb it. But they have started cultivating next door's garden, which gives them a handy zone one – and a good place for seed saving, isolated as it is from other vegetable gardens.

The house has a southerly aspect, and all the standard green things have been done, from double glazing to rainwater butts front and rear.

Like most city dwellers, they relish the diversity of shops within easy distance, the cultural diversity and the educational opportunities. Their primary interest is the communities they are part of, including the allotment holders and the wider Leeds environmental networks, which include a nature reserve committee and a demonstration vegetable garden at the conservation centre.

The rural retreat will probably remain a dream forever.

An allotment like this in the city can make the countryside much less alluring.

on combined heat and power (see p125) becomes feasible at these densities. When houses are further apart the increased heat losses in the pipes and the increased need for pumping make them inefficient.[8]

Although there can be a range of garden sizes at this density, there will be few families with space to grow a significant amount of food in their own gardens. Allotments are a compromise. They do have one advantage over home gardens in that people can expand or contract their food-growing as their circumstances change without having to either move house or leave a large garden uncultivated. But they don't have the zoning advantage of home gardens.

Nevertheless, food grown on an allotment has great energy advantages over purchased food, as long as the allotment is within walking distance of home, and ideally within wheelbarrow distance so that organic waste can be recycled on the allotment. As a rule of thumb this means allotments within 200m of every home without a large garden of its own – a far cry from the present situation![9]

Building houses at a higher net density does not necessarily mean that gross densities need to be higher. Proponents of the 40-50 dph approach reckon that all the advantages of high density can be had in a pattern of 'a relatively compact, mixed use built up area surrounded by space for recreation, growing food and fuel, and nature.'[10]

In the short term it's easy to discount the importance of food-growing potential in urban design because so few people do it these days. The main reason for this is the low price we pay for food, because so much of the true cost of food is externalised. When we're paying the full price for what we consume, which we must do if we're to have any future at all, growing food will become attractive to most people. Then there will be a real demand for big gardens.

Buildings last a long time. Whatever we build today is there for future generations as well as our own. Currently buildings are only designed to last some 50-60 years but there is no technical reason why we shouldn't build for ten times that lifespan. There would certainly be a far better return on the embodied energy if we did. On the other hand, a shorter lifetime gives more opportunity to react to changing circumstances by redesigning our towns and cites. Perhaps the best compromise is to build with reusable modular materials. Much of the embodied energy in the materials would be retained, and the portion that would be lost could be a fair trade for the gain in flexibility.

However, even the relatively short life of current buildings will take us well into a time when the economic and ecological situation will be radically different from now. Perhaps we should be designing for the future rather than for the present.

In existing settlements, the practice of infilling is de facto raising gross and net densities, reducing people's opportunity both to grow food and to have contact with nature. What the urban planner sees as a brownfield site the urban resident may see as valuable green space. Large home gardens, both in towns and villages, are all too often sold off as a site for another house. Allotments are being lost to development at such a rate that if the trend continues there will be virtually none left by 2035.[11]

Infilling is imposed by a planning policy which only allows new development within the envelope of existing towns and villages. This is based on the concept that dense settlements surrounded by a largely empty countryside is the ideal. In part this reflects the greater economy of providing mains services to compact built-up areas. But fundamentally it's based on an idea that the countryside should be empty of people.

There's not space here to go into the history of the British landscape. Suffice to say that the ancient pattern of settlement over most of Britain was one of dispersed hamlets and occasional larger settlements. The pattern of nucleated villages and towns surrounded by wide open fields is one which has never existed over most of Britain, and is only about a thousand years old in those parts where it has. Yet it's imposed as the norm.

This is not to say that we shouldn't protect the green belts around cities. It is to say that we should also protect areas of urban 'wilderness', allotment sites and large home gardens.

The place for new developments is in well-designed hamlets, sited to repopulate the deserted farmland, and independent of mains services for energy, water supply and sewage. It would be naive to think that the majority of the population would be keen to move to the country, or suitably skilled to make their living from the land. But this is the direction in which we need to be moving. In Ireland there has been considerable uptake of a scheme which helps unemployed city people to move to the country. It's surely more pleasant to be unemployed in the country than in the bleak environment of some of our low-income housing estates.

We can work politically to change the planning system which favours infilling, and in the meantime public pressure can have an effect on the fate of individual sites. Taking part in, or initiating, campaigns to save allotment sites are among the most important work we can do as permaculturists.[12]

Rural Areas

The key to implementing permaculture in the countryside is repopulation.[13] This includes the breaking up of the present large, mechanised farms into small farms, smallholdings and new hamlets, where energy-intensive production can be replaced with design-intensive and human-attention-intensive production. As well as farming and gardening there can be small scale manufacturing in the countryside, mainly for local needs and/or using local resources, and people teleworking from home. Many people will have polycultural incomes, usually including some food production.[14]

The present cold landscape of wide open, deserted fields and infilled dormitory villages would develop into an intricate landscape, full of diversity, full of people and full of wildlife.

However, a dispersed settlement pattern like this will not be an ecological improvement if we continue with our present lifestyle. These days all of us, town and country dwellers alike, expect what is essentially an urban lifestyle, in terms of shopping, education, entertainment and so on. Those of us who live in the country get it by unrestrained use of our cars.

A sustainable rural lifestyle is one in which staying at home all day every day is the norm, not the exception. The weekly visit to town, perhaps five miles away, the child cycling to school, the evening walk over the fields to visit friends – these are about the limits of travel we can expect.

This kind of lifestyle could be a deeply satisfying one in a thriving, repopulated countryside, but in today's countryside it can be isolated and lonely. We need a change in planning assumptions to make it possible, but equally we need the change in lifestyle to enable planning assumptions to change.

What actually happens in dispersed settlements today does not encourage a change in planning assumptions. New rural housing tends to attract commuters rather than home-workers, and they have longer trips to work than if they lived in town. Even where the new inhabitants are home-workers, in practice they make as many car trips as commuters, though at different times of day. Rural workshops are more often let to town-dwellers who can't find space in town than to local residents. This leads to reverse commuting, and to businesses which are only accessible by car. Not surprisingly some advocates of sustainability come down firmly against new dispersed housing in the countryside.[15]

The way forward is for permaculturists to demonstrate that a sustainable, low-transport life is possible in the countryside now by actually living such a lifestyle. This is the only way to show the planners that not everyone who wants to live in the country is a potential commuter in disguise.

This is starting to happen. In recent years a very small number of planning decisions have been granted in favour of people wanting to build new dwellings in the deep country in order to live a sustainable lifestyle. This often means a long process involving appeals and counter-appeals, and the conditions attached to the permission are usually stringent. But the logic of granting permission to genuine cases is inescapable if sustainability is to be anything more than a meaningless catch-phrase. (See case study, Living in the Woods, p336.)

Community is an essential element of a sustainable lifestyle, whether in town or country. Indeed the growth of our present unsustainable lifestyle has been closely matched by the decline in the network of relationships which exists in true communities. These links can have a direct bearing on sustainability. In rural communities they can include things like running a village minibus with voluntary drivers. This can be an important transitional step while the level of car ownership is still too high to make a commercial service worthwhile.

A hamlet of ten to twenty people can provide a full social life if it's a true community. This can greatly reduce the need to travel outside for social contact and entertainment. The weekly trip to town, whether for shopping or some cultural event, is then quite enough. I speak from personal experience, having lived in a rural community of that size over a number of years.

Layout[16]

The key planning tools provide a useful framework for sustainable settlement design. (See p27.)

Zoning in the permacultural sense means almost the direct opposite of what it means to conventional planners. In conventional terms it means segregating housing, commercial and industrial areas. In permaculture terms it means building integrated neighbourhoods, where people can find work, shops, schools and entertainment within easy reach of their homes. It's the contrast between monoculture and polyculture.

Creating mixed neighbourhoods doesn't only lead to saving energy and reducing pollution. It also makes those neighbourhoods much nicer places to live. There's a greater range of facilities, and a greater sense of community can develop.

The network concept also has a role in settlement design. Placing facilities such as shops, school, health centre and library together in a cluster encourages multi-purpose trips and thus reduces the need to travel. Sharing of resources can also be made easier by placing different elements together. For example, a combined heat and power system can use the heat and electricity generated over 24 hours more evenly if it supplies both daytime users, such as schools and businesses, and night-time users, such as housing.

Open space which is shared by a small group of houses, say five to twenty households, can be a very flexible way of providing space for food growing and other outdoor activities right outside people's back doors. The area can be divided up between whatever uses the current occupants want: allotments, play areas, extra parking and so on. Each house can still have its own private garden, but there is some scope for individual families to increase or decrease the size of their gardens without having to move house or travel to a more distant allotment.[17] This concept of clustered private houses sharing some communal space can lead on to the concept of cohousing. (See below.)

An aspect of the sectoring principle which could, and should, revolutionise the layout of new developments is orientation to the south. This means placing a house so that its longer axis lies east-west and the windows of the main living rooms face south. At present houses are built to face the road, regardless of which way the road is oriented. All houses receive some passive solar heating, whether they're designed with intentional passive solar features or not. Orienting houses towards the south can increase this considerably even without any change in the house design. It also makes the most of natural lighting, which is much more pleasant than electric, and free.

This can be done by aligning roads east-west, especially for terraced housing. Where north-south roads are unavoidable in the layout, detached houses are more able to make use of solar gain. The distance between them can be calculated with reference to the obstruction angle. (See p150.)

Orientation should be as near due south as possible to get maximum solar gain. South-east alignments may intensify the prevailing south-west winds by funnelling them between rows of houses, and south-west facing terraces may suffer from overheating in the afternoon. Shading of terraces to the north by those to the south can be reduced by making the buildings progressively taller as you go north, or leaving gaps in the southerly terrace.

Obviously it's easier to create an integrated solar neighbourhood from scratch than to retrofit an existing one. But the communities we have are largely a result of the piecemeal accumulation of many individual planning decisions. These are guided by local Structure Plans which are drawn up every few years by the local authority after extensive public consultation. All of us can have an input to that process, and bit by bit integrated communities can grow out of the fragmented layout which has been the fashion in the recent past.

Cohousing

Cohousing is a semi-communal way of living which by its very structure encourages the growth of true communities. Typically a cohousing community consists of around 20 households, including both families and single people. Each household has its own private dwelling and there are also shared facilities. Central among these is a common house, with a kitchen and dining room where the whole community can eat together. Other shared facilities may include washing machines, teenagers' and children's rooms, workshops and vegetable gardens.

The first principle of cohousing design is that residents have the choice between the privacy of their own homes or social contact with their neighbours. Although everyone has a kitchen in their own house or flat, in practice most people in cohousing communities choose to eat in the common house most of the time. Cooking for a whole day once a fortnight is so much more pleasant than cooking every day.

Thus social contact is not something which has to be arranged over the telephone and involves a car or bus trip. It happens of its own accord. A key feature is confining cars to the edge of the site, so all the internal area is pedestrianised. This greatly increases the level of day-to-day contact between residents, and of course is ideal for children. Where the dwellings are houses rather than flats each household has its own private garden, so residents always have the choice of public and private open space.

The second principle is that the members participate in the design of the community right from the start. Most are new build projects, as it's much easier to design the appropriate relationships between spaces when starting with a blank sheet. But some successful cohousing communities have been created by recycling redundant industrial buildings, and this does tend to cut both the financial and the ecological cost. Either way, the process of purchase, design and building can take years, and requires a great deal of commitment on the part of the members. But, because individual houses can be smaller, the financial cost per unit tends to be lower than for conventional housing of a similar standard, even allowing for a share in the communal facilities.

The on-going management is also done by the residents. The community structure makes working together and mutual aid easier. Sharing childcare is especially easy, as every child grows up with a familiar relationship with a number of adults. Single parents find they have much more free time than they could get in other living situations. Similarly, cohousing can offer elderly people an integrated lifestyle which combines independence with company and support.

A possible danger of cohousing is that the residents could become isolated from the wider local community, but this need not happen. In some communities the common house has become a valued local resource as a venue for meetings of clubs, adult education classes and the like. One community was deliberately laid out so that the central thoroughfare was also the path which took many local children to and from the village school.

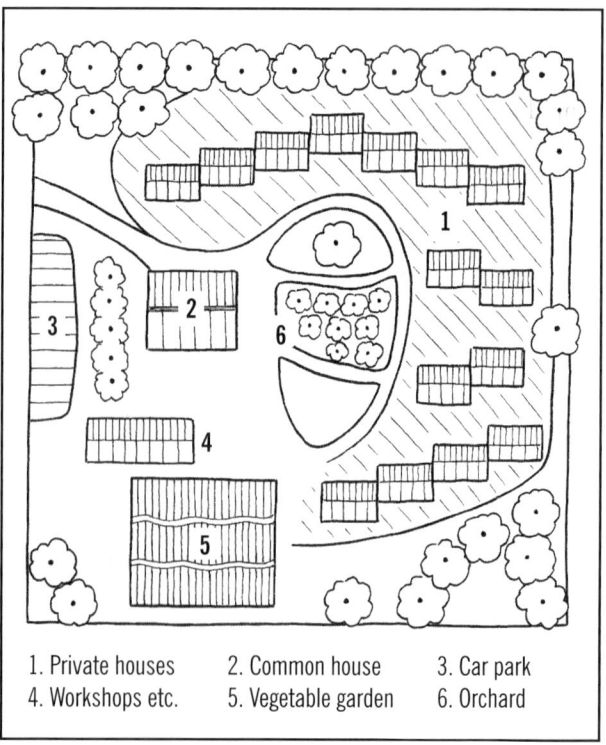

1. Private houses   2. Common house   3. Car park
4. Workshops etc.   5. Vegetable garden   6. Orchard

A possible cohousing layout.

Many cohousing communities have a commitment to making their lives more sustainable. Indeed, what works well socially tends to work well ecologically too. Some renewable energy systems which are impractical for a single household, such as wind generation, may be practical for a cohousing community. District heating can be designed into either new build or retrofitted cohousing projects.

Car sharing is facilitated between members, not just because the car ends up where the next person wants to start from, but because living in a cohousing community engenders the mutual trust which helps sharing arrangements to work. Keeping cars out of the living area means less hard surfaces, which means more infiltration of rainwater. This also gives a financial saving as smaller storm water drains can be installed.

Homeworking is more attractive to cohousing residents than to many other people, because they will not suffer the isolation which puts many office workers off the idea of working at home. (See p127.) There are always plenty of people around to share the coffee break with. One community in the USA has a third of its residents working at home at least part time, even though it's within commuting distance of a big city.

Vegetable gardening, chicken-keeping and other kinds of home food production can be much easier in a community. The mutual support of working together in a group makes gardening a much more attractive proposition to most people. This is, after all, the natural way for human beings to work. The job of shutting up chickens at night and letting them out in the morning, or even milking cows, can be shared out amongst a number of people so it no longer becomes a tie.

Cohousing started in Denmark in the early 1970s and most of the existing communities are there. Although predominantly owner-occupied, some are rented through the Danish equivalent of housing associations. At least one urban housing estate with severe social problems has been converted to cohousing, with positive results. After conversion the estate was fully occupied for the first time in five years, though it has not developed the full community characteristics typical of a cohousing initiated by the residents themselves. Cohousing has spread to a number of other countries in north-west Europe and latterly to the United States. There are no completed projects in Britain at the time of writing but some groups are working towards it.

LOW ENERGY DESIGN

Buildings consume energy in two main ways:

- Embodied energy, including both energy used in the production and transport of materials and that used during construction.

- Energy-in-use, the energy used to run the building during its lifetime, including space heating, water heating etc.

The aim of low energy house design is to minimise both. In most current housing, energy-in-use is by far the greater of the two. Typically it's five to ten times as much as the embodied energy in the first 25 years of a house's life.[18] But in a really efficient house energy-in-use can be brought right down to the point where it's no more than the embodied energy over a similar period.[19] To a small extent this contrast arises because measures which decrease energy-in-use can add a little to the embodied energy. But it's mainly because current housing is built to such low energy-in-use standards.

Thus, when looking at the energy cost of new houses, energy-in-use is much more significant than embodied energy. But when comparing new build with renovation the opposite is true. Renovation saves almost all the embodied energy in the old structure. Although it's possible to get a much better energy-in-use efficiency from new build than from renovation, this will rarely make up for the extra embodied energy. So by and large refurbishing old houses is a much better deal energy-wise than knocking them down and building replacements.

There's also often a social and aesthetic advantage to retaining the physical fabric of existing communities. Destroying the buildings usually means that the community goes too. Keeping them means there's more chance of keeping the delicate web of relationships which link us together, and are such an important part of making life a rich and satisfying experience.

Does It Pay?

Investing in a little extra embodied energy can save a great deal of energy overall. But does it pay as well in cash terms as it does in energy terms? For new buildings the answer is yes.

In one example from the Netherlands, energy efficient houses cost twelve percent more than conventional houses to build but consumed 92% less energy-in-use. The rent for the efficient houses was £11 per month more than for the conventional, while the saving in space heating costs was £24 per month.[20]

In the USA it has been calculated that energy efficient housing typically costs five percent more to build. The payback period for this extra cost is five years, and over 40 years the savings can amount to $50,000 to $100,000.[21]

A study in Scotland compared a low energy house with an otherwise identical house on the basis of Life Cycle Analysis, which looks at all ecological impacts, not just energy. (See p161.) It found that the conventional house had almost four times the impact of the low energy house. The extra capital cost was only 1.1%, and the saving in running costs was 30-35%.[22]

Impressive as these cash gains are, the energy saving is always greater than the cash saving. This is because the cash cost of the extra building materials is only partly energy costs, whereas the cash saving in reduced space heating is all energy costs.

From this we can draw a rule of thumb: if low-energy housing pays at all financially, it will pay very well in energy terms.

So, if energy conservation pays, why isn't everyone building energy-efficient houses? The problem is that the developer pays for the structure and the future occupants pay the fuel bills. It seems that in the present market a slightly lower purchase price is a better selling point than a lifetime of much lower fuel bills. It's up to house buyers to demand energy efficiency, which is certainly in their financial interest.

However renovation, although it's usually more energy efficient than new build, isn't often cheaper. A much higher proportion of the cost of renovation is in labour than in materials. The low price which our economy places on energy and materials tends to make renovation more expensive overall. But, having decided on renovation the same relationship between energy saving and cash saving will apply as with new build.

There are many aspects of building design which affect the overall energy efficiency of the building, some of them are major, others are small details which have a cumulative effect. But broadly speaking there are two approaches to energy-efficient design: one which emphasises passive solar heating and one which emphasises high levels of insulation.

Passive Solar Design

Solar heating can be active or passive. Active solar involves moving parts; typically heat is collected by a solar panel and distributed to where it is needed. Passive solar has no moving parts; the fabric of the building itself collects solar energy, and a passive solar house is one which is intentionally designed to maximise solar gain.

Light energy from the Sun enters the building through the windows and is converted to heat energy when it falls on internal surfaces and is re-radiated

from them. Much of the heat stays within the building because glass is more of a barrier to heat than to light. (The greenhouse effect.) If the surfaces on which the light falls have a good heat storage capacity some of the heat is stored in them and slowly released from them throughout the day and night.

Concrete, stone and brick have a high heat storage capacity, whereas wood has a low one. A house with a high heat storage capacity overall will stay warm with a small input of energy because it loses heat slowly. By the same token it takes more energy to warm it up in the first place as the materials absorb a lot of the available heat. In other words it's a slow response house. (See Colour Section, 10.)

Such a house is ideal for people who occupy it all day. It will stay at a comfortable temperature with a minimal input of energy. But for people who are out all day and come back in the evening it may be less efficient. If the passive solar features are enough to keep it at the desired temperature all evening there's no problem. But if it needs supplementary warming it takes more energy and more time to do so than in a fast response house.

Passive solar heating can be specially effective in commercial buildings because they're occupied mainly during daylight hours, which means heat can be used directly and heat storage is less important, as in this passive solar building on a rural industrial estate at Machynlleth, Powys.

Passive solar has a number of advantages:

• It has a good energy efficiency ratio when compared to other forms of space heating (see p123) largely because it makes use of structures, such as walls and windows, which are needed anyway as part of the house. These may have a greater embodied energy than non-passive-solar equivalents, but the difference is not great and this extra embodied energy is the only energy input to the system other than the Sun's rays.

• Much of the heat is delivered as radiant heat rather than convected heat. (See box, Radiant and Convected Heat.)

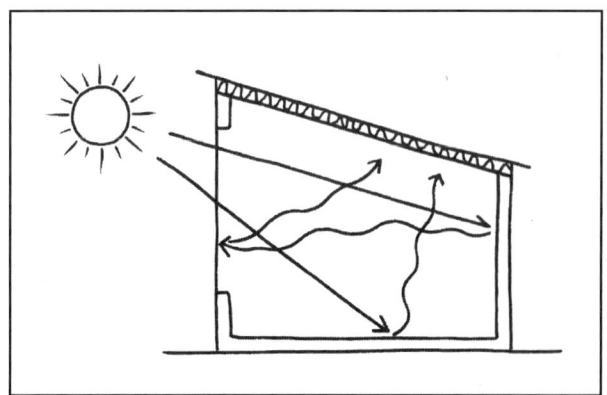

Passive solar heating.

- It combines well with the most efficient form of woodburning stove, the masonry stove, which also delivers mainly radiant heat. (See pp152-153.)

- Being a slow response building, it will stay cooler in summer than a similar house of more light-weight construction.

- Since the temperature in a passive solar house gently fluctuates in tune with natural rhythms, people who live in them develop a comfort tolerance. Provided the temperature changes are gradual, they tend to dress for the season rather than expecting a year-round shirtsleeves environment.[23] This alone saves a great deal of energy.

- Large windows give plenty of natural sunlight, save on electric lighting, and help to reduce the barrier between indoors and out. There's a feel-good factor, not only in the extra sunlight but in a generally more intimate relationship with the weather and the changing seasons.

It also has its limitations:

- It's less useful for high-density housing, which receives more shading than low-density or isolated houses.

- In our cool, cloudy climate it's quite difficult to meet all of a house's heating needs from passive solar alone. In practice the first 30% of the required temperature rise is easy and cheap but additional gains are less so. This is especially so in a super-insulated house, where the heating season is restricted to the coldest months of the year, which are also the cloudiest.

- A slow-response house may or may not be appropriate to the occupants' needs.

- Large windows on the south side of the house can mean less privacy, and the house can become over-heated on a sunny day if the residents are out at work and not on hand to close the blinds.

The main function of insulation is to keep the air inside the building warm, so it works with convected heat rather than radiant heat. One great advantage of insulation over the passive solar approach is that it's usually much easier to retrofit on existing buildings, and as we have seen renovation has both energy and social advantages over new build. High insulation is also the approach to take with timber buildings, as timber has a negligible heat storage capacity, though building with timber has many ecological advantages. (See Materials, pp159-162.) However timber can be combined with passive solar heating if some heat-storing materials are included in the building at those points where sunlight penetrates it.

In fact, although there is a degree of conflict between high levels of insulation and high levels of passive solar gain, in practice the two are used together. All houses experience some solar gain. A few simple measures, like orienting the house to the south and placing larger windows on the south side than the north, will enhance this, even in a lightweight house where the main reliance is placed on insulation.

Similarly, any passive solar design will also include good levels of insulation. But there may be more emphasis on keeping the heat in the structure of the

---

RADIANT AND CONVECTED HEAT

Heat is transmitted in three ways, as conducted, convected or radiant heat.

Conducted heat is passed from one object to another by direct contact, such as when you touch a hot piece of metal with your hand. It plays little part in domestic space heating.

Convected heat moves upwards through a liquid or gaseous body, as when warm air, heated by a radiator, rises to the top of a room.

Radiant heat moves outwards from a relatively hot body to a cooler one and warms it without the intervening air being warmed, as when you feel the warmth of an open fire on your hands.

When we heat a house by convection quite a bit of energy is used just heating up the air, so to that extent it's less efficient than radiated heat. Also, some of the heat is lost as air leaves the house. This is greatest when you deliberately ventilate a room, but in any case there's a constant turnover of air in the room. It can be replaced anything from once to seven times an hour depending on how draughty the room is.

The alternative to losing the heat is to reduce ventilation, giving a stuffy, polluted atmosphere. It is possible to retrieve some of the heat lost in exiting air by means of a heat exchanger, but these have a low net yield of energy unless the house is very air-tight, and they're expensive.

Radiant heat not only avoids these losses, but it makes for a more comfortable environment. People actually feel more comfortable at lower temperatures when heated by radiant heat than by convection. It avoids the rather lifeless feel often associated with central heating.

In practice any heating system will involve a bit of both, but there is a clear advantage to systems in which radiant heat is the predominant form.

building than keeping the air warm. This can be achieved by putting the insulation on the outside of the walls rather than inside, so the heat stored in them radiates into the house rather than out of it.

The difference between the two approaches is one of emphasis.[24]

The opportunities for passive solar heating are clearly much better on a south-facing slope than on a north-facing one. The amount of light striking a vertical surface such as a window is the same regardless of the slope on which it stands, but on a north-facing slope any object to the south of the house will cast a taller shadow. Even quite small objects, such as garden walls and shrubs, will cast significant shade on a northerly slope of as little as 10%.

Whether a particular site will be too shaded for effective passive solar heating can be tested by calculating the obstruction angle. (See illustration, right.) The angle should be no more than 16.5° in the south of Britain and 13.5° in the north. Note that the obstruction angle will be affected by the slope of the land as well as the height of the buildings.

A passive solar house should have a long east-west axis and face as near due south as is practicable.

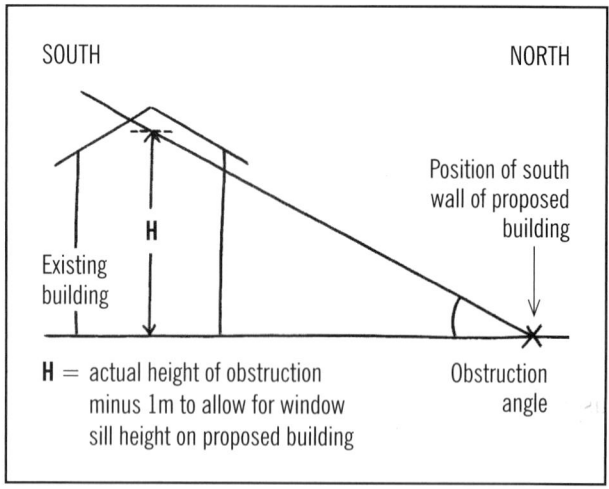

Calculating the obstruction angle.

Within 10° of south is ideal. At 25° around half the potential passive solar gain is lost, and beyond 45° passive solar design is not worthwhile. There may be other microclimate factors which affect orientation, and these need to be balanced with the need for solar gain. (See, for example, katabatic winds, p76.)

A SOLAR HOUSE

An example of a passive solar house is David Holmgren and Su Dennett's home in the mountains of Victoria, Australia. The climate there is much hotter in summer, perhaps a little cooler in winter, and much more sunny than in Britain. But the design principles used could be followed to advantage here.

The house is made from sun-dried bricks, a material with low embodied energy and a good heat storage capacity. It is single storey, with a long east-west axis, and faces due north, the direction of the Sun in the Southern Hemisphere. Over 90% of the north side is glazed, either windows, conservatory or clerestory. A clerestory is a vertical window above the level of the Sun-facing roof which lets light into

the back of the building. (See illustration below left.) During the winter the Sun shines in through the glazing and warms up the walls, the floors, and the soil in the planter in the conservatory. At night these radiate heat into the living space. The conservatory door is left open to allow warm air to flow into the rest of the house. A small wood burner gives supplementary heat on cold nights.

In summer the angle of the Sun is higher, and direct sunlight is largely kept out of the house by overhanging eaves. (See illustration below right.) Warm air rises by convection and is drawn out through the vents as shown. Part of the replacement air is drawn in under the water tank, which has a small cooling effect.

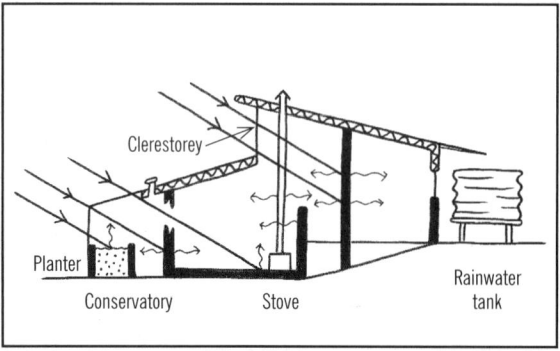

N-S section through the house in winter.

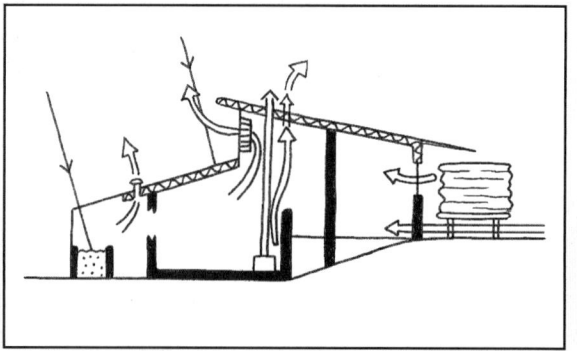

N-S section through the house in summer.

Perhaps surprisingly, passive solar is just as useful in the north of Britain as in the south. Winter days are shorter in the north and the Sun is lower in the sky but the heating season is longer. Thus the useful solar gain through the year is similar in Lerwick and in London.[25]

It's not necessary for a passive solar house to be really massive in construction. Materials can only absorb and release heat at a certain rate. Since the cycle of absorbing and releasing heat is mainly a daily one, heat will only penetrate so far into a heat store, and be given up from the same depth of material. The maximum useful thickness of the best heat storage materials, such as concrete, stone or rammed earth, is 15cm. For brick, which is less effective, it's 10cm.

Conservatories

Adding a conservatory to an existing house is the main way in which passive solar design can be retrofitted to existing houses. A conservatory is also a multi-functional structure, working as an extra room, a source of heat and a place to grow high value food – and thus much loved by permaculturists.

Unfortunately most of the conservatories which are fashionable additions to houses these days are energy sinks rather than contributors. They have no relationship to solar orientation and stick out from the house like a promontory, losing heat in all directions and transmitting very little back into the house. They actually need artificial heating and are a net loss in energy terms. The ideal conservatory is long and relatively shallow, giving a minimum of outside edge and a maximum of edge between it and the house. Alternatively it can be used to fill in a gap or angle between two wings of a house.

When Matt and Jan Dunwell bought Ragman's Lane Farm the farmhouse was U shaped, giving it a very high surface to volume ratio and poor energy efficiency. They filled in part of the U with a solar conservatory. Heat is stored both in the back wall of the conservatory, originally the outside wall of the house, and in the tiled floor. One of the many uses of the conservatory is to ripen squash and pumpkins in the autumn.

The orientation of a conservatory is not as critical as it is for other passive solar options, and the positioning can be designed as much to fit in with the arrangement of rooms as anything else. A south east aspect is best from an energy point of view. It gets the morning Sun, so it warms the house for a longer period of the day, and avoids overheating in the afternoon.

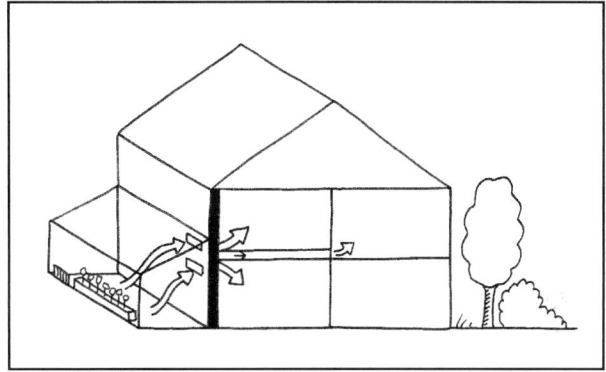

Winter, daytime. The conservatory heats the house by convection.

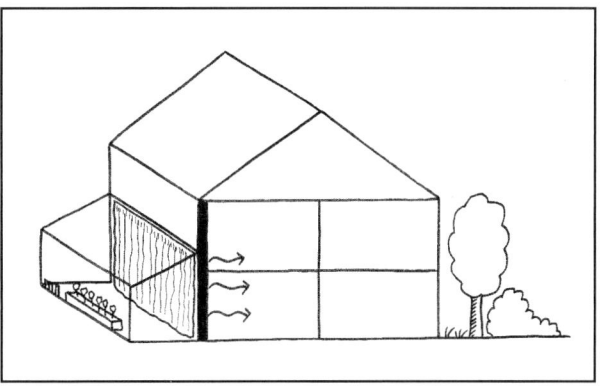

Winter, night-time. The conservatory heats the house by radiant heat.

A conservatory adds heat to the interior of the house mainly by convection. (See top illustration.) During the day air in the conservatory warms, and rises by convection. The convection current draws air in through a vent at the base of the conservatory, and the output at the top is vented into the house. It's easiest to use this air to warm the upper rooms. It can be vented into the top of a downstairs room and circulated within the room by a small fan, though the net usable energy ratio of using electricity to move low-grade heat may be doubtful. It's relatively easy to vent warm air to rooms on the cooler north side of the house.

Alternatively a door or window between conservatory and house can simply be left open.

Radiant heat can be transmitted into the house from the wall which the conservatory backs onto. The usual recommendation is to insulate this wall on the outside to prevent loss of heat from the house to the conservatory. But then the house will receive little heat by radiation because the wall will hardly gain any heat during the day. A compromise is to insulate it with a thick woollen curtain, which is drawn back in the daytime and closed at night. (See bottom illustration on previous page.)

During summer a conservatory can become too hot and heat the house up when more heat is not wanted. Keeping the warm air vents closed during the day will help, and growing tall plants in the conservatory will shade it. But the conservatory can also be used as a solar chimney to actually help cool the house. This can be done by fitting automatic vents at the top of the conservatory which open to the outside when the temperature rises to a certain point. (See illustration below.) These can be operated by a cylinder filled with wax which expands when it warms up, a device often used on garden greenhouses. The convection current draws cool air from the north side of the house, which helps to cool down both house and conservatory. To provide adequate cooling the vents need to be a fairly large proportion of the conservatory roof area.

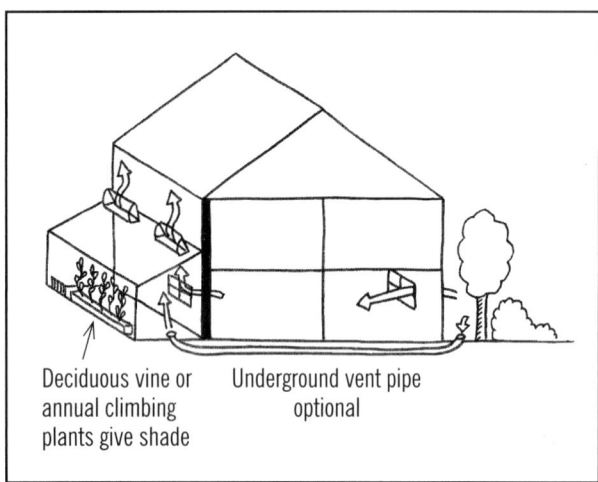

Deciduous vine or annual climbing plants give shade

Underground vent pipe optional

Summer, daytime. The convection current set up by the conservatory is used to draw cool air from the north side of the house.

The comfort tolerance which people acquire in passive solar houses is even greater in conservatories. Because we feel we're almost outside anyway we accept the temperature being a bit closer to what it is outside.

Solar Windowsills

A solar windowsill is in effect a tiny conservatory covering a single window. It's main energy function is to warm the ventilation air, and this can be a valuable contribution to the energy budget of a house or flat. It should be openable, so that you can have direct contact with the fresh air on warm days.

It's probably only worthwhile installing one if it incorporates a window box and thus doubles up as a little greenhouse. Where a windowsill is exposed to wind this may be the only practical way of growing plants there.

On the inside of a window there's rarely enough light to grow anything but shade-tolerant house plants unless it's directly south-facing and/or very large.

Heating with Wood[26]

Heating the house with wood may be impractical for most of us, but where it is practical the most efficient way to do it is with a masonry stove, also known as a ceramic stove because traditionally they were often covered with tiles. As well as being efficient in themselves, these stoves fit very well with passive solar heating, as they work mainly with radiant heat. (See box, p149.)

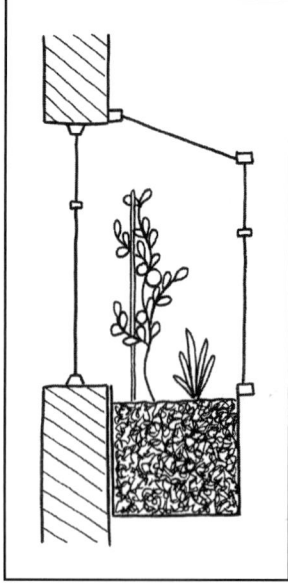

A solar windowsill.

A conventional woodstove is made out of steel or cast iron and burns logs slowly at a relatively low temperature, giving out heat steadily through the day. It heats mainly by convection, warming up the air as it burns. Combustion is incomplete and unburnt volatiles recondense in the chimney as tar or blow away as smoke, causing pollution. Using dry wood helps to minimise these problems, but only up to a point.

The efficiency of a log-burner is typically around 45-60%. Putting a catalytic combuster in it can increase this to 80%, according to the manufacturers, and reduce the unburnt volatiles to the point where it can be used in a smokeless zone. The active ingredient in a catalytic combuster is platinum, a rare and precious metal, and the efficiency and reliability of these gadgets is controversial to say the least.

Masonry stoves achieve 70-90% efficiency and are virtually pollution-free simply by virtue of the way they work. They burn for a short time during the day at a very high temperature. The fabric of the stove, which is of masonry, often covered with tiles, stores the heat of the burn and steadily releases it as radiant heat over the following hours.

The flames and flue gasses are led through a network of flues which bring them into contact with a great volume of heat-storing material, so they give up almost all their heat before leaving the stove. This gives it its high efficiency.

The high temperatures make sure that everything in the wood gets burnt before the flue gasses leave the chimney. If there's no visible smoke 15 minutes after firing up, a wood burner is acceptable for use in a smokeless zone. All masonry stoves should pass this test easily. Stovemaker Reinhardt von Schock reports that his standard model came in at four times better than the minimum standard.

Since most of the heat delivered is in radiant form, people feel comfortable at a slightly lower and more fluctuating temperature with a masonry stove than with a system based on convected heat. In very cold weather two burns a day may be necessary, but in moderately cold weather one in the 24 hours is enough. Each burn lasts for 45 minutes to an hour.

Because a hot burn is needed, thin pieces of wood are more suitable for a masonry stove than logs. This means they can make good use of prunings, brash, slabwood and offcuts which have little economic value otherwise. The high efficiency means that very little wood is needed. The quantity needed is discussed in Chapter 11. (See p304.)

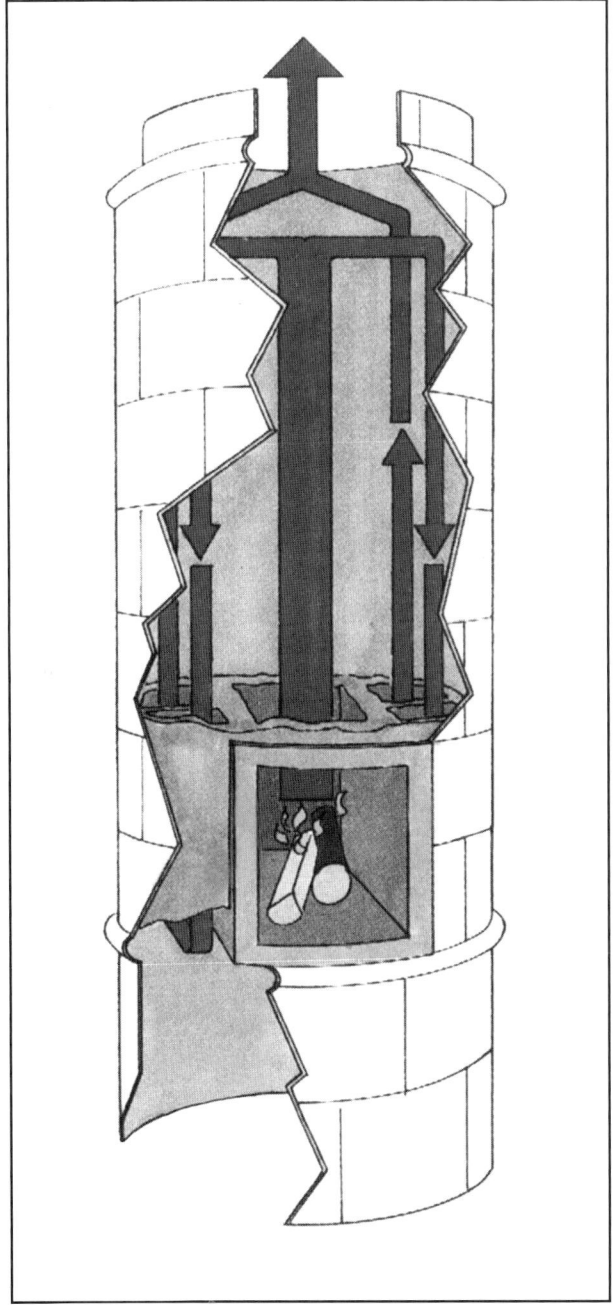

How a masonry stove works.

Masonry stoves can come in all shapes and sizes. They can be fitted into existing houses, but are perhaps most efficient when they are built into a new house. Because radiant heat can only be felt when you can actually see the radiating surface, a masonry stove mainly heats the room it's in. Some of the heat is convected, and this can be let into other rooms through an open door or vent. But the stoves are large, and relatively expensive, so it would not be practical to have one in each room.

But if you build the house around the stove its surface can project into more than one room. In fact one stove could heat all the rooms in a medium-sized house, upstairs and down, if it's built into the centre of the house. This would need a large stove, but the bigger they are the more efficient they are, and the middle of the house is the best place for the heat source as it minimises losses to the outside.

BIOTECTURE

Roger Ulrich, a geographer at the University of Delamore, has produced findings which suggest that trees can reduce costs in hospital care. In a study of patients recovering from a common type of gall bladder operation, he discovered that on average patients with a view of deciduous trees from their beds, spent almost 24 hours less in hospital. They also needed less nursing attention and fewer doses of expensive drugs than patients looking out onto a brick wall.[27]

As well as making buildings more pleasant places to be in aesthetically and emotionally, plants can improve the physical environment and reduce the energy needs of the buildings. Using plants in this way is known as biotecture.

Windbreaks

Planting windbreaks and shelterbelts can give annual energy savings for domestic heating in the range of three to ten percent. The lower end of the range applies to newer houses and the higher to older houses, which tend to be more leaky.[28] Much will also depend on the local microclimate. The savings for commercial buildings, which have a greater volume per surface area, are less. But commercial greenhouses, which lose heat very fast, can save up to 30%.[29]

With such small energy savings in housing, windbreaks are not usually worthwhile on energy saving grounds alone. In fact draught-proofing the house gives much the same energy savings.[30] But of course there are a whole host of other benefits and outputs from planting trees. (See the mind maps on p83 and p303.) In addition windbreaks around houses will give shelter to the gardens, giving a more pleasant microclimate for people and increasing the productivity of fruit and vegetables.

Windbreaks can be used for both rural and urban housing. In both cases the benefits of shelter must be balanced with possible loss of light. A good balance is achieved by placing the windbreak four to five times its own height from the building it shelters.

The illustration below gives more specific recommendations for the different points of the compass. The distance can be a little less to the south, as the Sun is higher at midday. It should be a little further to the south west, so as to make the most of the warm afternoon Sun, which is a little lower. To the north the windbreak can be much closer. The light which comes from the north, in early morning and evening during spring and summer, is at such a low angle that if the windbreak was not to block it it would have to be too far from the building to be of any use.

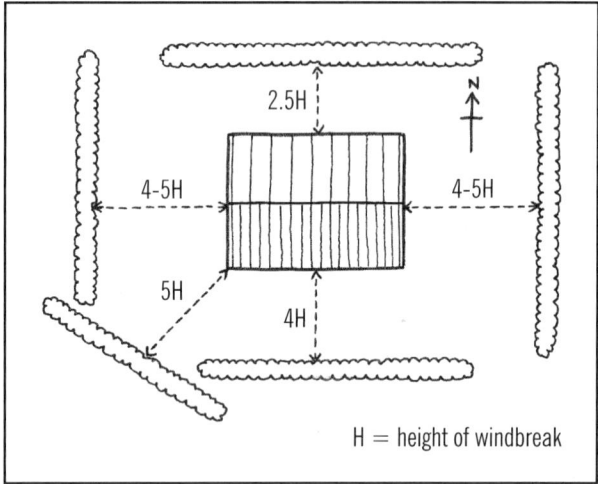

How far to place a windbreak from a house, greenhouse or polytunnel.[31]

The exact placing will depend on the individual circumstances of the site and the priorities of the people living there. Knowledge of the microclimate will reveal where the cold winds come from in the heating season, and views and privacy may be important factors. The slope of the ground also needs to be considered, and it's worth measuring the projected obstruction angle of a proposed windbreak before coming to a decision. (See p150.)

These guidelines are good for greenhouses and polytunnels as well as dwelling houses.

It can actually be beneficial to have a single deciduous tree closer than this on the south side of a building, where it will cast shade in summer but not in winter. This may be useful either in very sunny areas, or if the south side of the house has a high proportion of glazing. In other situations the shade may be excessive; even bare twigs can cast significant shade in the winter. The tree should be placed to give shade at

This house in Norway, with extensive south-facing windows, has been built directly to the north of the old pollard ash tree.

midday but not in the morning or evening. Trees which come into leaf late and lose their leaves early, such as ash and false acacia are the most suitable.

Another possible problem is damage to buildings' foundations from the roots of large trees. The damage is caused by the tree roots extracting water from the subsoil, which then shrinks, causing the foundations to crack. Only clay and peat soils shrink in this way, so there is no problem on sands and sandy loams, and buildings constructed in the past 15 or 20 years should have foundations which can withstand these effects on any soil.

Fruit trees and other small kinds are no problem, but any vigorous species, and especially ash, willow, poplar and eucalyptus, should be kept well away from buildings, by at least a distance equivalent to the height of the mature tree.[32] The heights of different tree species are given in Chapter 4. (See p87.)

Trees can be pollarded or pruned to limit their size, and this will slow down the growth of the roots to some extent. But pollarding should be started early in the tree's life. Pollarding a tree which has grown nearly to maturity as a maiden gives it a shock which, if the tree is suffering stress for any other reason, can kill it. Old trees which have become too big for the space available must be felled rather than pollarded. But this can also damage the foundations as the soil can swell and heave with the unaccustomed water content.

The general principles of windbreak design are given in Chapter 4. (See pp83-88.)

Green Buildings

Green Walls

Vegetation growing up walls can help to cool buildings in summer and keep them warm in winter. It will also provide habitat for birds and other wildlife. Wall-trained fruit trees and cane fruits can produce food at the same time, but where energy saving and increased comfort are the priority true climbers give the best effect.

Summer cooling happens by three means: firstly by shading; secondly by transpiration and evaporation from the leaves (when water evaporates there is a local drop in temperature); and thirdly by insulation, trapping still air between their leaves and the wall. Evergreen climbers can insulate in the winter too, and energy savings of up to 30% have been recorded. But this high figure would only be possible on a building without other insulation and when temperatures are close to freezing.[33]

Contrary to popular belief, a covering of plants protects a wall rather than damaging it, as long as the wall is in reasonable condition in the first place. If plants are grown up a trellis or wires there is no problem in any case. But if plants like ivy are grown directly onto the wall surface they can do damage to a wall with an outer layer which can flake off. Modern bricks can be vulnerable, though not old ones. Bricks used to be fired for a longer time at a lower temperature than they are now. The modern short hot treatment gives a brick that's harder on the outside than inside, and the outside can peel away given the added weight of the plant.[34]

In any case it's better to grow climbers on a support, 10 to 40cm away from the wall, so as to give a good airspace to maximise insulation. Once they reach the eaves they can get under the roof tiles and block gutters, so some trimming is needed every few years.

On south-facing walls deciduous climbers are best, to allow the Sun to heat up the wall in winter. North-facing walls are of course best covered with evergreens. East- and west-facing walls are best treated as south-facing if they catch a lot of Sun and north-facing if they don't. In wet areas evergreen climbers on west or south-west facing walls will help to keep the wall dry, and thus reduce the heat loss of evaporation.

Green Roofs[35]

Green roofs vary in intensity from full roof gardens through turf roofs to a soil-less covering of stonecrop.

Given a strong enough structure, virtually any kind of garden can be grown on a roof. It can be an invaluable resource for people who live or work in urban areas where space at ground level is not available. In most cases the roof will need to be strengthened or built to a stronger specification if new, but on a DIY level a lot can be done with lightweight containers and reclaimed timber to spread the load. It can even be done on a pitched roof. Intensive irrigation is usually necessary over the whole growing season, and the direct effect of wind on plants can be a problem. All in all a roof garden is probably not worthwhile unless you have no other option.[36]

SOME PLANTS FOR GREEN WALLS								
	Decid'/ Everg'	Aspect (underlined = best)	Max Height Metres	Growth Rate	Soil	Edible	Native	Wildlife
Ivy	E	<u>N</u> S E W	30	Slow	Moist, rich	No	Yes	Excellent
Boston ivy*	D	N S <u>E</u> W	15	Fast	Any	No	No	Good
Russian vine	D	N <u>S</u> E W	30	Fast	Any Moist	No	No	Bird nesting only
Hop	D	S <u>E</u> W	6	Very fast	Rich moist	Yes	Yes	Good
Hardy kiwi *Actinidia arguta*	D	<u>S</u> E W	30	Mod	Moist, w. drained	Yes†	No	Good
Hardy kiwi *A. kolomikta*	D	N <u>E</u> W	15	Mod	Well drained	Yes†	No	Good
Grape vine	D	<u>S</u> for fruiting	20	Mod	Loamy, w. drained	Yes	No	Good
Bramble	E/D	N S E W	2‡	Fast	Well drained	Yes	Yes	Excellent
Other fruits	D	Varies	Varies‡	Slow	Loamy, w. drained	**Yes**	Few	Fair

* Virginia creeper is similar.

† Both male and female plants needed. Specify fruiting varieties, especially for *A. kolomikta*.

‡ Not true climbers. Need training.

Andy Goldring

Green roofs needn't be large and expensive. This is a simple one on a garden shed.

Turf roofs are much simpler things altogether. They are lighter, typically with between 5 and 15cm of soil, so roof strengthening may not always be necessary. Irrigation is not usually used, as the plants are selected to be able to stand the tough, droughty conditions. Very drought tolerant plants like stonecrop can manage with minimal soil, while drought tolerant grasses and herbs need a little more. They can be installed on roofs of up to 30% slope.

Whether a turf roof is the best solution in a particular place will depend on a number of factors.

A turf roof will:
• have some insulation value
• provide some wildlife habitat
• look pleasant

But will also:
• need a high-tech water- and root-proof membrane (See table, p160.)

It may:
• reduce pollution
• improve the street microclimate
• slow down and reduce stormwater runoff
• increase the lifespan of a flat roof

But may also:
• need a stronger roof
• be visually inappropriate
• cost more to install

Any vegetation in urban areas can absorb pollution, especially dust particles, and turf roofs are no exception. They can improve street microclimate, particularly in high density urban areas in summer, by relieving temperature inversions. (See illustration right.) This effect can help to disperse concentrations of polluted air, though it does not actually cure the pollution.

Any area of soil and vegetation absorbs and slows down storm water, releasing part of it slowly and returning part to the atmosphere, whereas hard surfaces can create severe problems with runoff. (See p92.) Clearly a thin turf roof doesn't have the water retention capacity of a deep natural soil. But if a proportion of the roofs over a whole city or town were turfed it could make a significant difference to the hydrology of the area.

Although green roofs often cost more than other kinds of roofing, in the case of flat roofs this can be recouped over the years because turf lasts so much longer than conventional flat roofing. The average flat roof has a life expectancy of 10 to 15 years, yet the membrane under the roof garden on one London building was found to be still in excellent condition after 50 years. The weight of a simple turf roof need be no more than the conventional gravel covering.

Much of the protective effect on a flat roof is due to keeping the roof at a more even temperature, reducing damage from both heat and frost, in other words its insulation value. But it wouldn't be worth installing a green roof for its insulation value alone. They usually reduce heat loss by less than ten percent, and there are many more efficient and cheaper ways of achieving that.

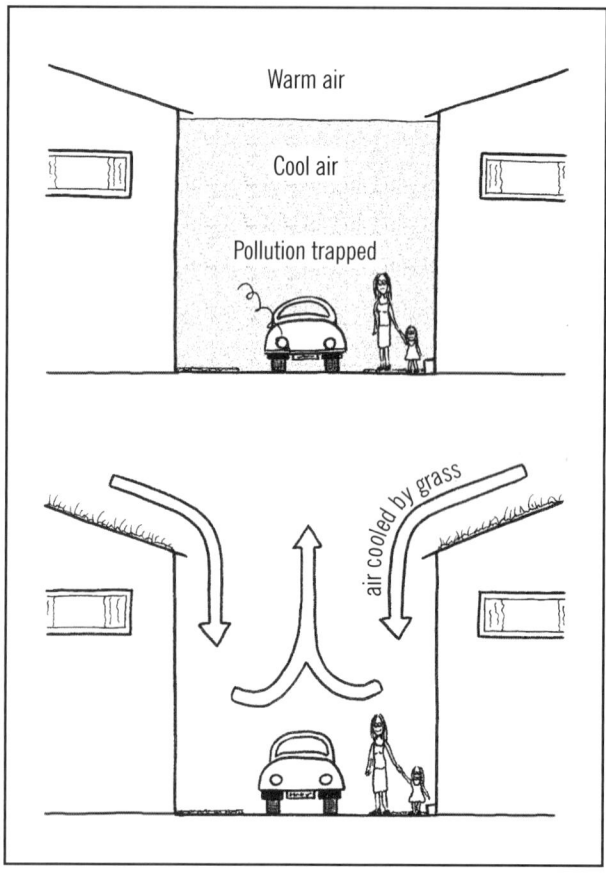

In narrow streets between tall buildings a temperature inversion can form, that is a layer of warm air lying over a layer of cooler air. Since the cool air is heavier than the warm it stays put, the air does not circulate and pollution builds up at street level (top). The grass on a turf roof cools the air immediately above it, mainly by the cooling effect of transpiration and evaporation. This air sinks and restarts a healthy circulation (bottom).

The pleasing appearance of a turf roof can't be accepted as an automatic benefit. Some areas still retain a distinctive local building style. To introduce an alien form, which a turf roof inevitably is in Britain, can dilute that local distinctiveness which is such an important part of quality of life. (See p159.)

TURF ROOFS – HISTORY & FASHION

In Norway in pre-industrial times one of the best roofing materials available was birch bark. It's waterproof, it can be peeled off the tree in relatively large sheets, and lasts a reasonable length of time. All it needed was something to hold it down on the roof. Turf was the best thing available, and the 'turf' roof was born.

Of course it's not really a turf roof at all, it's a birch bark roof, or it was. Today plastic membranes have replaced the birch bark, though in Norway a little frill of bark is sometimes added where it's visible at the edge. The origin of the roof has been forgotten and the material which was once just there to keep the important bit in place is now thought of as the important bit.

The thing about turf roofs is that they look green. Filling the roof with insulation made from recycled newspapers is all very well, but you can't see it. So green roofs have become more of a visual statement than a practical design solution. But are they any worse for that?

SOME ECOLOGICAL STYLES[37]

A number of distinctive ecological styles of building have emerged in recent years. Some are new ideas, others are revivals of traditional styles.

To a great extent they are based on the choice of materials for the main structure. This is the major part of the house in terms of bulk and has a major impact on the energy performance of the completed house. But it's worth noting that all the other components of the house – floors, windows, internal finish, electrics, plumbing etc – may be similar from one style to another. These have their ecological implications and may contain more embodied energy than the more bulky materials. They also form the greatest part of the cost of the structure. So where the main raw material is cheap, as it is in a straw bale house, the finished cost of the house may not be a great deal less than that of an equivalent conventional house.

Green Framing[38]

This is how the vast majority of British houses were built in medieval times. The framework is made from unseasoned timber, usually though not necessarily oak, put together in such a way that it strengthens as

Green framing is most often used for new build, but it can be used for renovations and extensions, such as this conservatory and balcony.

it seasons in situ. The walls can be filled in with any material, from traditional wattle and daub to modern board and insulation. The labour and skill content of green oak framing is high and at present it's often a rich person's fancy.

The energy cost can be very low if the trees are sourced locally. Timber buildings in general can make a very healthy and versatile living environment. Chemical preservatives are not needed in any timber building which is correctly designed and carefully built.

Segal[39]

The Walter Segal system is a simple timber-frame design, mainly used by self-builders. Construction is simple and efficient, requiring less skill and training than other methods. The design is modular and can easily be adapted to individual wants. The houses have a lovely airy, comfortable feel to them and fit well in both urban and rural settings.

Architype Architects

Segal self-build projects usually consist of a row or cluster of houses. The self-builders get together to help each other raise their frames, one a day with the whole group working together. After that there's no work that's too much for one person on their own. Note the foundation pads in the foreground.

The system also has ecological advantages. The building has no extensive footings; each main upright rests on a small pad, which virtually eliminates the need for earth moving. The embodied energy is around half that of a typical British house. It's easy to fit high levels of insulation, and thermal mass can be incorporated if it's wanted. The design is adapted to the standard size of British building materials. This saves wasteful offcuts during construction and makes the materials more easy to reuse at the end of the building's life.

Pole Building[40]

This style uses timber in the round. There are a number of advantages to this. It saves the energy of milling. It preserves the full structural integrity of the wood, giving maximum strength for volume. It makes it easy for anyone with home-grown timber to use it for their own buildings.

When used for an open pole barn the unsawn timber is a major part of the total materials used, but in a building with enclosed walls it becomes less significant, unless you go for a log cabin.

In this traditional Welsh longhouse roundwood has been used in renovation, providing the joists for a new ceiling.

Straw Bale[41]

Settlers on the Great Plains of North America had an abundance of straw but not much timber or stone. It was only common sense to use straw bales as bricks, and a new building technology developed. The houses can be very strong and durable, have an extremely high insulation value, and once the walls are plastered over there is no problem with rot or rodents. They're also very easy and great fun to build, so they're attractive to self-builders.

The embodied energy of straw bales is very low if they are sourced locally, and still pretty good if they have to be brought from the other side of the country, as they must if they are grown in a straw-importing region like the west of Britain. (Even if the straw is actually grown locally, using it will mean more has to be imported to replace it.)

Simon Pratt

Putting the render on the walls of a straw bale building.

Earth Sheltered[42]

Earth sheltered houses are built partly or wholly underground, usually into a south-facing bank. With large south-facing windows, both lighting and passive solar gain can be as good as in an above-ground house. Heat storage is of course excellent. Total running costs are typically 40-70% of an equivalent conventional house. One British example meets all its space heating needs throughout the year by passive solar gain.[43] There is also a saving on exterior maintenance and painting. On the down side a pretty sophisticated membrane is needed to keep out the damp.

Perhaps the greatest asset of an earth-sheltered building is its ability to blend into the landscape. Although a massive structure is needed they are relatively simple to build, and this may be the easiest way to build a house with no need for a heating system other than passive solar.

Earthship

This method, like the straw bale one, uses surplus materials to form the walls – used car tyres and aluminium cans. These are plastered over. The true earthship is a specific design, with the entire south wall a solar window, and the north side dug into the ground, almost an earth-sheltered design. This is suited to the climate of the highlands of New Mexico, sunny with a cold winter, where it was developed.

In Britain a more house-shaped design may go better with both the climate and the landscape. The idea of turning one of the biggest refuse problems in the world into durable housing is applicable anywhere, but this kind of house may prove damp and unhealthy in the British climate.[44]

Cob[45]

Cob is simply making walls with mud. It's a traditional building method in many parts of Britain, notably Devon, where many old cob houses are lived in and come fully up to modern standards of comfort. Subsoil is dug out of the ground nearby, mixed with water, straw and gravel, and built up course by course

Only the tapering wall gives a clue that this cottage is built of cob rather than a more conventional material.

A new cob wall goes up on a foundation of stone.

into walls. The walls are often thicker at the bottom than at the top. After drying out they are plastered, and only their shape reveals that they are not made of stone or brick.

The fossil fuel content of the walls is extremely low, though labour is rather high. The massive walls also have a high heat storage capacity, enough to carry over some of summers' warmth to winter and winters' cool to summer. So cob is an ideal material for a passive solar house.

All of these styles have their ecological merits. But whether a particular style is the ideal one for a particular place and the particular people who are to live in the house is another matter.

Some of the points to consider are:

• Are the materials available locally?

• Does the local built environment have its own character? If so is this style in harmony with it?

• In a rural area, does it blend with the landscape?

• If it's a traditional style, is it traditional in this region?

• Is this style best adapted to meet the physical needs of the building's users? eg is it slow- or fast-response?

• Do they enjoy it aesthetically?

• If self-build, what skills and how much physical energy do the builders have?

• Is it really the most ecological solution in this case?

• How much will it cost?

A number of these questions are about respecting local distinctiveness. Change is natural, and trying to pickle the landscape in some picturesque imitation of the past is both foolish and futile. But a sense of place, of locality, of belonging to somewhere which is not just the same as everywhere else, is important.

As well as adding to our overall quality of life, a sense of place can have a direct impact on sustainability. It helps us to develop a relationship with the place we live, and thus with the land and its myriad inhabitants, and thus ultimately with the Earth as a whole. These relationships are the basis and the origin of the desire to live sustainably. Without them sustainable living is something imposed, at best by our own intellect, at worst from outside, not something which flows from the heart.

CHOOSING MATERIALS

Many of the building styles in the previous section involve the use of unusual and innovative materials. Using something unusual is not an essential part of ecological building, but paying careful attention to the ecological impact of any materials we use most certainly is. It's absolutely central to both new build and renovation.

The most important questions to ask about a material from an ecological point of view are summarised in the mind map below.

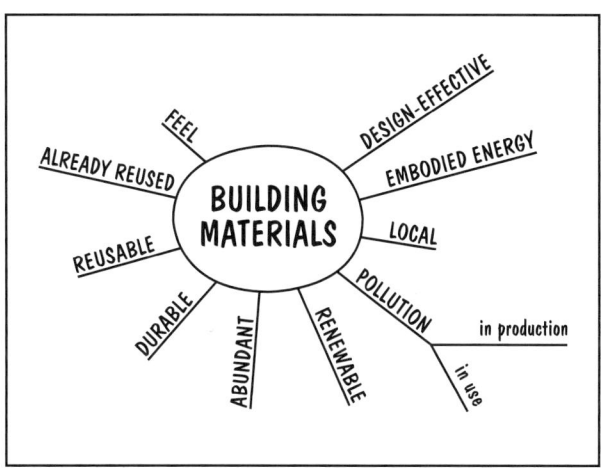

Design-effective. In other words does the material do the job it's asked to do effectively? In general this question is the one to be balanced against all the others put together. There is no perfect material, and the choice will always be a dialogue between its ecological properties and the needs of the individual building.

It can be worth using a material with high embodied energy, or a high overall ecological impact, if it enables the building to consume less energy-in-use. An example is using lots of glass in a passive solar design.

The aesthetic aspect of a material is part of its design effectiveness. If you really love stonework that's something to put in the balance when deciding whether to use it.

Embodied Energy. The embodied energy of a material is made up of energy used in both manufacturing it and transporting it to the site. For example, timber acquires very little embodied energy in manufacturing, but imported timber can have seven times the embodied energy of locally grown air-dried timber. (See table below.)

The amount of material used is also significant. Bricks and tiles have a similar embodied energy per tonne, but twenty times the weight of bricks are used in a house as tiles, so their embodied energy is twenty times as much per house.

It's also worth considering embodied materials, that is materials which were used in the process but are not present in the final building. The large quantity of water used in brick and block laying and concrete work is the most significant of these.

Local. This obviously relates to the previous point, but also to local distinctiveness. (See previous page.)

Pollution, in Production. Since CO_2 is probably the most damaging pollutant on Earth, energy consumption, both embodied and in use, provides a quick check on how polluting a material is. CFCs and HCFCs, both of which damage the ozone layer, have now been phased out for use as foaming agents for expanded polystyrene insulation. But their replacements, HFCs are still extremely potent greenhouse gases.

The destruction of ecosystems can be included under this heading. Uncertified tropical timber and the products of most modern quarries spring to mind.

Pollution, in Use. So many current building materials are toxic that a typical newly-built house should really be classed as toxic waste, and human beings kept out for several years till the bulk of the toxins have finished outgassing. Sick building syndrome is the name given to the cumulative effect of the poisons in modern building materials. (The radiation from computers and bacteria in vent ducts probably play a part too.) Most people are subliminally affected, and may not be aware that they are being poisoned. But sensitive individuals can become acutely ill, for example after the installation of cavity foam insulation.

Natural insulation materials include wool and cork, but the most widely available one at the moment is made from recycled newspapers and marketed under the name Warmcell. Natural insulation materials can be made resistant to rot, insects and fire by treating them with a solution of borax.

Renewable. This does not just mean 'Can it be renewed?' It means 'Is it now being renewed as fast as it's being used?' Tropical hardwoods are certainly not a renewable material by this definition. Nor are monocultures of

EMBODIED ENERGY IN SOME BUILDING MATERIALS[46]		
Material	kWh/m³	kWh/house
Bricks (non-fletton)	1,462	25,600
Lightweight blocks	600	5,600
Plaster & plasterboard	900	5,300
Clay tiles	1,520	1,100
Single layer roof membrane	47,000	18,800
Timber: imported softwood	754	
local airdried	110	
local green oak	220	
Glass	23,000	2,400
Insulation: plastic	1,125	
mineral wool	230	500
cellulose	133	

Sitka spruce, replanted again and again on the same soil, as this is not a practice which can go on for very long. There is now a certification scheme for responsibly grown timber. (See p134.)

Abundant. Although stone is not a renewable resource it's so abundant that we can go on using it indefinitely for building without much chance of seriously depleting it. But the rate at which we're currently using it for road building is another matter. Much of the roadstone for south-east England is quarried from the Mendip hills, and it has been calculated that if they had been quarried at the present rate since Roman times they would no longer exist.[47]

Durable. This affects both the rate of use of a material and the need for maintenance during use.

Reusable. We think of buildings as permanent, but in practice most of them are demolished and replaced before they have worn out. In fact a properly maintained building can last indefinitely, but times change, and the kind of buildings we need changes with them.

Many materials, such as bricks and timber, are reusable but the design of the building can help or hinder reuse. The Segal system, which uses boarding at the standard sizes it is sold in, is an example of a design which makes reuse easy. Recyclable materials, such as glass, are less ideal but better than throw-away ones.

As for rottable materials, the classic example is a cob and thatch house. Once you take the windows out the remaining materials can gently return to the cycles from which they came. The thatch will eventually rot, followed by the unprotected timbers, and the walls will dissolve in the rain. In a few decades only a nettle-covered mound will remain to show that a house was there.

Already Reused. The high cost of labour makes some recycled building materials rather expensive. They tend to fit the image of 'architectural salvage', with a price tag to match. But some real bargains can be had by going round auctions. It's also great fun, and can save a lot of embodied energy as long as you don't have to travel too far.

Feel. This is an intuitive question, and to some extent intuition must play a part in building materials assessment. Even if you are able to get accurate answers to all the other questions, and that is by no means certain, how on Earth do you compare them with each other? How much embodied energy equals how much toxicity or durability? How many square metres of destroyed rainforest equal a tonne of CO_2?

Books are available which attempt to answer these questions using the analysis tool known as Life Cycle Analysis, which quantifies the resource use and pollution associated with different materials, and with the same material from different suppliers.[48] These contain an element of judgement in their assessment, especially

when it comes to giving numerical weightings to very different kinds of ecological impact. They can be no more than a guide. For the individual builder, having consulted the available facts and analyses, the question remains: 'How do I feel about using this stuff? Would I lie down to sleep happily with my head on, beside or surrounded by it?'.

Timber An Example

Let's take timber as an example and see how it performs in answer to these questions. (See Colour Section, 8.)

This ecological house at Ulvik in western Norway is a fine example of what can be done with timber. Architect, Frederica Miller.

Design-effective. You can do almost anything with timber. Buildings can be made almost entirely from it and it's hard to imagine one made entirely without it. It's an insulating material, so it may have more application in buildings based on a high insulation approach than a passive solar one. The skills required to work with wood are relatively easy to acquire. Almost anyone can do simple woodwork.

It's a beautiful material to live with, both in appearance and feel.

Embodied Energy. This is generally low, but there's a big difference between imported and locally grown timber, as the table of embodied energies opposite shows. Kiln-dried timber contains more embodied energy than air-dried, and using it in the round saves the energy of milling.

If timber replaces concrete or masonry, it eliminates the large amount of embodied water that normally goes into a house.

Local. It can be, and whether it is or not makes a big difference.

Pollution, in Production. It's one material which can actually reduce global climate change, but only if it's part of a sustainable production system. Conifers planted on peat moors actually increase greenhouse gasses. (See p72.)

Pollution, in Use. On it's own timber is almost harmless, though some timbers do give out low levels of formaldehyde, and this can be a problem in all-timber houses.

Most timber preservatives are a habit of the chemical age rather than a functional component. If a timber building is constructed properly, so that any part which gets wet can dry naturally, it doesn't need preservative. If it's badly constructed no amount of chemicals will stop it rotting.

Almost all of the composite boards which are made from timber and wood are stuck together with toxic adhesives. (See table, Some Dangerous Building Materials.) The range of non-toxic equivalents is very limited as yet. Planking can be used for some functions, but it's more expensive.

Renewable. It certainly is, though not at the rate we're using it at present.

Abundant. The answer is much the same as the previous one.

Durable. For an organic material timber is remarkably durable, but there's a difference in the durability of different species. The most suitable species for building are noted on the table, Trees for Timber and Wood. (See p338.) As long as the right timber is used for the right function, timber buildings can last for hundreds of years, as many have. The world's oldest wooden building is a pagoda in Japan, some thirteen centuries old.

Reusable. Timber and boards particularly lend themselves to reuse at the end of a building's life.

Already Reused. Maybe.

Feel. I leave you to make your own comment under this heading.

Timber was not chosen at random as an example. It's a building material which has a lot going for it and deserves to be used much more widely in this country. The main reason it hasn't is fashion, habit or inertia, call it what you will.

But there's another point which emerges even more strongly from this brief look at timber, and that's how many maybes there are in it. There's a world of difference between using a locally grown hardwood and one torn from a virgin tropical rain forest, or between planks from a certified source and chipboard from a Sitka spruce monoculture.

Timber is perhaps unusual in coming from such varied sources with such wide differences in ecological impact. But no material can be accepted or written off as a whole. There will be differences between the same material from different sources.

RENOVATION

Simply choosing to renovate an existing house rather than knock it down and build a new one is a step towards sustainability. As we have seen, renovating existing houses makes better use of embodied energy, and usually has a lower overall ecological impact.

The two most important aspects of renovation from an ecological point of view are the choice of materials and designing for low energy-in-use. Choice of materials

SOME DANGEROUS BUILDING MATERIALS		
Material	**Pollutant**	**Alternative**
Cavity foam insulation	Urea formaldehyde	Cellulose fibre, eg Warmcell
Chipboard Medium-density fibreboard (MDF), Sterling board and plywood	Phenol formaldehyde	Hardboard, softboard, mediumboard (not MDF). Low or zero formaldehyde boards are becoming available
Carpets	Formaldehyde	Natural matting, non-treated wool, linoleum
Gloss paints Timber treatments	Volatile solvents	'Natural' paints, eg Nutshell, Auro, Livos
Lead flashing	Lead, if water collected for drinking or garden	Stainless steel
PVC	Mainly in manufacture and disposal	Various, depending on application, eg wooden window frames, other plastics for pipes

has been covered above, so here we concentrate on energy-in-use. Minor ways of saving energy in the home have been given in Chapter 6. (See pp130-132.) Major features which might be part of a renovation project are summarised in the mind map below.

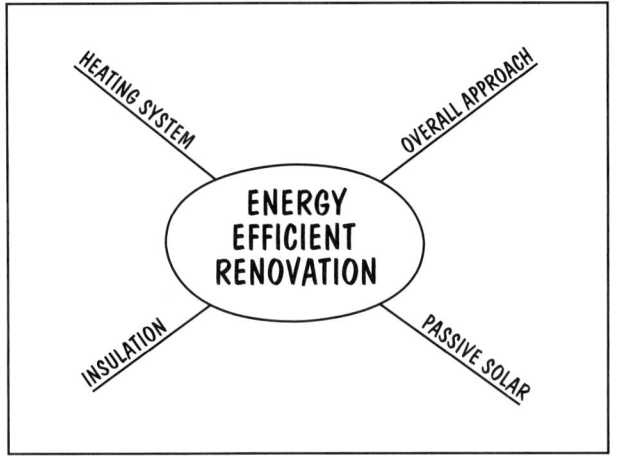

Overall Approach. The first question to consider is whether the emphasis is to be on passive solar heating or high insulation. (See pp148-151.) Factors which will affect the decision include:

• Orientation. Ideally this should be within 25° of south for passive solar.

• Obstruction angle. (See p150.) The aspect of the slope, ie whether north- or south-facing, and the height of any obstructions such as buildings or trees, will both affect this.

• Privacy. A highly-glazed passive solar approach will be more attractive where the south side of a building is not overlooked.

• Existing structure. For example, there may or may not be space for a conservatory, or the house structure may be lightweight with a fast response or heavy with a slow response.

Passive Solar. A conservatory can be added to the outside of a house, or an existing single storey room can have its roof and exterior wall glazed to such an extent that it virtually becomes a conservatory. But there is an important difference between these two options.

A normal conservatory is a halfway house between indoors and outdoors, and is not normally heated except by the Sun. Indeed if it is artificially heated it will lose much of that heat through the glass and may yield a net loss in energy terms. A solar room, as we could call the second option, is still part of the house and will be expected to be at a normal indoor temperature. It will need high performance glazing and a high level of insulation on those surfaces which are not glazed to keep it at that temperature economically.

Where a conservatory or solar room is not appropriate, increasing the size of south-facing windows will increase solar gain. Ideally the light should shine onto surfaces which can store heat. The best materials for heat storage include cob, rammed earth, stone (including slate floors) and concrete. Brick, render and plaster are intermediate, while other materials, including aerated concrete blocks and timber, store little heat. But most of the light entering a room is absorbed by its surfaces whether it's stored for very long or not, so it will at least heat the room while the Sun is shining.

If there's no other reason for altering the wall and replacing the windows other than to increase solar gain, this kind of work is only likely to be worthwhile where the solar access is very good. This means a wall facing within 10° of south with few obstructions. But if the wall needs rebuilding anyway the same rules apply as to new build: within 25° of south is best, but within 45° is still worthwhile.

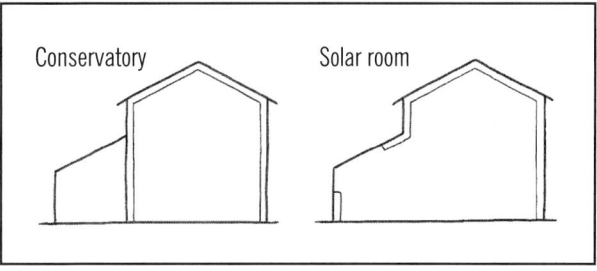

Conservatory and solar room, as in the case study, An Eco-Renovation Project.

Roofs have the advantage as solar collectors of being high up and thus less likely to be overshadowed, but they have the disadvantage of being at a shallow angle. A surface at right angles to the Sun's rays collects more energy than one at any other angle. As the space heating season is from autumn to spring, when the Sun is relatively low in the sky, the best angle for solar space heating at our latitude is 75°, whereas the typical angle of a British roof is around 35°. However in a cloudy climate such as ours much of the solar energy comes from indirect sky light rather than direct sunlight, so a low angle is not such a big disadvantage as it might seem at first.

A vertical surface is closer to the ideal 75° than is a low-pitched roof. So a south-facing dormer window will have the advantage over a simple roof light which is flush with the roof surface. Both these are small and the solar gain from either is not much, so it's not worth choosing a dormer window for that reason alone. But it's something to put in the balance if you're considering installing one for other reasons. A clerestory may be more worthwhile. (See box, A Solar House, p150, and also illustration on next page). It can often throw light back onto a major heat-storing wall, and take both heat and light to the north side of a building.

This recent extension has been given a clerestory window to give full natural lighting to its single large room.

Insulation. Boosting the insulation level of the house will almost always form part of a renovation project, whatever the overall approach. In fact it doesn't need to wait for a major renovation. Improving an inadequately insulated house is a worthwhile project on its own.

The principle of insulation is summed up in the equation: total heat produced in a building = total heat lost to the outside. In other words all the heat ends up outside in the end; when we insulate we're slowing down the flow of heat from inside to outside. In a well insulated house the waste heat from cooking, lighting and people's bodies, together known as free heat gain, become significant sources of heat. In a super-insulated house they can even supply all the heat needed to maintain a comfortable temperature.

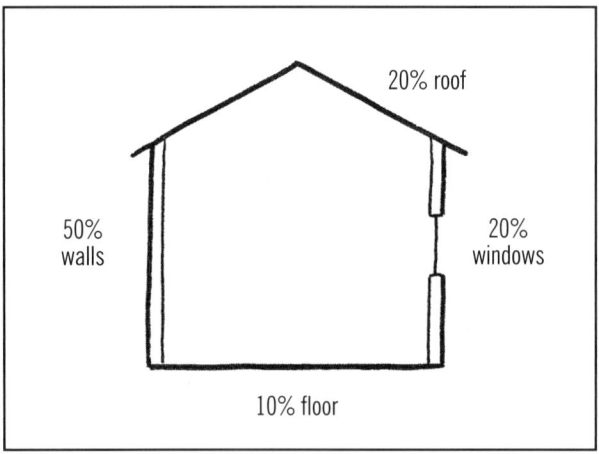

Typical proportions of heat loss from an uninsulated house.

In older houses most heat is lost through excessive ventilation or, in plain language, draughts. Once draughts are sealed, walls are the main source of heat loss, and there are three ways to insulate them: on the inside, on the outside and in the cavity. Houses built since 1930 usually have cavity walls, and filling them with insulation is the easiest and most cost-effective solution. Care must be taken to avoid toxic cavity wall insulation materials. (See p162.)

If walls are insulated on the inside you get a fast response house, and if on the outside a slow response house, with the entire mass of the walls available as a heat store. Fitting internal insulation is a job for a relatively competent DIY-er, though no-one should attempt it without getting sound advice on the techniques and possible pitfalls. External insulation is a professional job and will be more expensive. The insulation material will need weatherproofing with something like render or weatherboarding, and this may change the outside appearance of the house. Nevertheless external insulation can be well worth while as it gives such a big increase in heat storage capacity.

Almost all houses now have loft insulation, usually 100mm or 150mm thick. It's certainly worthwhile to increase this to 300mm, from both a cash and an energy payback point of view. It's important to remember that the more effective the insulation the more important it is to include any water tanks and pipes under the insulated layer, as less heat will be escaping into the loft space to keep them frost-free in winter.

If the loft has potential as living space it's worth considering putting the insulation on the inside of the roof itself rather than on the upper side of the ceiling, but this will be a little more difficult.

CONDENSATION

Stopping condensation is an important benefit of insulation.

Condensation happens when warm air meets a cold surface. Warm air can hold more moisture than cooler air. When the warm air cools down on contact with the cold surface some of the water it contains condenses out onto the surface.

This is quite normal on single glazed windows. When it happens on walls it indicates one of three things:

- There's too much humidity in the indoor air. Kitchen and/or bathroom ventilation is needed.

- The indoor air is at too low a temperature in that part of the house where condensation is occurring, such as in an unused room. Insulating the whole building helps this by evening out the temperature within the building.

- The inside surface of an exterior wall is cold compared to the indoor air. This indicates either inadequate insulation, or cold bridging, where a non-insulating material, such as masonry, protrudes through the insulation layer.

Double glazing is a form of insulation. The air trapped between the sheets of glass acts in just the same way as all the little pockets of air trapped in other insulating materials. As is so often the case, the simple, low-tech versions have a much better payback in both cash and energy terms than the more high-tech options.

In ascending order of sophistication the options are: sticking clingfilm over the windows; fitting rigid plastic sheets; secondary double glazing, ie another sheet of glass fitted over the existing window; factory-made sealed double glazing units; and high-performance units including triple glazing and low-E units which can have a better insulation value than a cavity brick wall. The cash payback rises from around one year to 25 with increasing sophistication, though of course the total energy savings are greater with the more sophisticated kinds. Sealed units are rarely worth installing unless the window needs replacing anyway, and certainly not worth it until the house as a whole has been brought up to a high level of energy efficiency.

As only ten percent of the total heat loss is through the floor, underfloor insulation is another thing which is not worth addressing unless the floor needs attention anyway. Putting extra insulation on top of the floor is easier. Carpet with foam backing, felt or rubber underlay, layers of newspaper, wood blocks and cork tiles can all be used. Filling in any gap between the skirting board and the floor can also give a very good return.

Heating System. In the average house space heating makes up some 60% of the energy in use, and water heating 20%. Solar gain and insulation can reduce the need for space heating but the reduction can never be as great in a retrofit as in a new build.

It's rarely worth while in either energy or cash terms to renew a central heating system when it still has life in it. (See p130.) If it does need replacing the most efficient kind now available is the gas condensing boiler. This extracts every last bit of energy from the fuel by condensing the water vapour in the flue gases and making use of the heat which is produced whenever water condenses.

Maximum efficiency is an amazing 95% and average efficiency year-round is typically 85%. This compares with an average efficiency of around 65% for other modern boilers. The extra cost of choosing a condensing boiler can be paid back in one to ten years.

Underfloor heating can increase the average efficiency of the heating system even more. It works by leading hot water through narrow gauge pipes embedded in concrete below the floor surface. It's not only more efficient but, because heat rises, it gives a more even heat profile to the room, rather than most of the heat ending up near the ceiling. This is especially good for children, who spend most of their time on the floor. But it is expensive to install.

Space heating and hot water are usually supplied by the same system. But if for any reason only hot water is needed a tankless, or instant, water heater is often more efficient than one with a tank. A tankless heater is a wall-mounted unit which switches on in response to the hot tap being turned on. It avoids the heat loss which always occurs when hot water is stored in a tank, however well lagged. A gas one is of course more efficient than an electric one. (See p123.)

Solar water heating is like double glazing in that a simple, unsophisticated system has a much better efficiency ratio than a more sophisticated one. The trouble is that the unsophisticated system has a lower performance, and will not heat the water at all on many days of the year when a high-tech system would give a useful gain. This means the cash payback can be poor. The cost of a full-blown, professionally installed system can be more than halved by making your own, or installing a factory-made unit yourself.[49]

Solar water heating is most useful when a large volume of water needs to be heated, preferably in the summer. This is what it's good at, and it will pay even if the water has to be raised the last few degrees by some other means. A hotel with a mainly summer clientele is the ideal scenario. A single person wanting a hot shower every day of the year could certainly find better ways to help the Earth with the money they could spend on solar water heating. (See guest essay, Ecodesign, on p135.)

CASE STUDY

AN ECO-RENOVATION PROJECT

by Tim and Maddy Harland

We moved into our house on the edge of the South Downs in Hampshire in the mid 1980s, a few years before we discovered permaculture. It was two one-up-one-down 19th century flint cottages knocked together with a hideous 1960s flat-roof extension on the back and side, and a very small garden. It was like living in a split personality: cosy, aesthetic and warm in the front of the house; ugly, cramped, draughty and cold at the back.

By 1995, we had acquired a 'live in' publishing business, which became home to *Permaculture Magazine*, and a third of an acre of adjoining arable land which we designed and planted as a permaculture garden. This has already evolved into a reasonably productive paradise – ecologically diverse with wildflower meadows, 80 fruit and nut trees, native hedgerows, a mulched veggie garden, poultry, and is full of composting systems! But the back of the house was still an eyesore and was literally falling apart...

The roof was leaking, one mouldy ceiling was collapsing, the windows were rotten and threatening to rattle out and the central heating had completely packed up. Our enthusiasm for permaculture had driven us to put everything into setting up Permanent Publications and planting the garden. We had loads of apples, lots of magazines but a freezing house and the kids were getting too many colds in winter. We baulked at the thought of renewing the flat roof which would probably only have a ten year life. Anyway, conventional renovation would cost us dearly and we would be no nearer our ecological goals. What were we to do?

Meeting ecological architect, Andrew Yates of Eco Arc, at Findhorn and a bequest from Maddy's late father was to change everything. We asked Andrew to help us design a retrofit of the house which would integrate front and back, make it energy efficient, incorporate some 'ecological experimentation' and extend a narrow corridor which posed as a kitchen/ diner. We wanted to introduce as much natural light as possible into our computer-bound lives and dissolve the delineation between house and garden. We also wanted to garden indoors in the kitchen all year round making food miles become food inches – a perfect house for cool temperate permaculturists!

With all this in mind, Andrew designed a retrofit of the existing building and a passive solar kitchen extension to the back of the house which was completely in sympathy with our wishes. The following are principles that we included in our design:

The eco-renovation and passive solar kitchen extension and the back of Tim and Maddy Harland's house.

Recycled Materials

As much as possible, we reused on site materials like chalk hardcore, existing bricks and tiles. Many of the bought in materials were recycled.

Resource Locally

Our builder, Paul Lipscombe, is based two miles away. He employs local people and buys as much as possible from local sources. The flints used were gathered by Tim from surrounding fields and only cost us a small payment to the farmer of £10 and a bottle of wine. Flints from a builders merchant would have cost several hundred pounds – a big difference. The timber was also bought from a local, sustainable source. Some additional labour (hard!) was found via the local LETScheme. We also required Andrew to visit only when he was already in the area, visiting Chithurst Monastery, where his design of the Abbot's house was being built.

Energy Efficient Design

The passive solar heating design for the extension works well due to a number of features: the partial glazing in the roof which has a south-westerly aspect; the efficiency of the gas condensing boiler which is only needed in winter; the low 'e' argon filled double glazing used in the solar kitchen and in the upstairs windows; the huge amount of insulation added to the breathing roof space, the walls and the extended floor. We also incorporated further double glazed windows to the same specification throughout the rest of the house, and installed solar hot water tubes on the roof which provide for a considerable amount of our needs for the majority of the year.

Reduce Food Miles

Our new food growing areas also help to feed us all year round, helping to reduce our need for bought in high mileage foods. In summer, the plants have been deliberately chosen to provide shade in the solar kitchen.

Compost Toilet

We also installed a Swedish Septum urine separating compost toilet which replaced a flush toilet upstairs.

The advantages have been: the building regulations officer's official acceptance of the installation; the neighbours' undisguised mirth ("Haven't you heard of Thomas Crapper yet..?"); the fact that the unit sits on the floor and requires no special box beneath it; the solids – mixed with hardwood shavings, a waste product from our local carpenter – do not smell; and the gallons of urine which are carried via a 2.5cm diameter pipe concealed in the wall and collected in jerricans outside the house – we water this down and feed to our fruit and vegetables. This is very good way of introducing nitrogen into a growing system, which requires minimal effort and is free. This additional fertiliser has given us healthy vegetables – sweetcorn crops are particularly enhanced.

The disadvantages have been: having to work from Swedish instructions (no mean feat!); the somewhat crude design – we carefully installed the ventilation pipe and added a fan, making sure it carried any unwanted smells well above the roof, but the design omits a necessary fly screen and we were quickly infested with buzzing Jurassic-type monsters until we put a pair of tights over the fan; difficulty getting the urine to evacuate the collecting bowl due to the angle of the pipe in the original design (Tim adapted it) and sawdust and paper deposits blocking the pipe after a few months (don't asked how we cleared it – Tim deserves a bravery award and required extensive sanitising!); our youngest daughter, when very young, sometimes had difficulty with her aim and soiled the solids bag which resulted in a smelly chemical reaction;

Runner beans provide pleasant green shade inside the kitchen in the heat of the summer.

and our parents' hatred of the contraption – 'your horrible loo'!

Let There Be Light

The final result has had a huge impact on our lives. We are no longer cold in winter and our gas bills are significantly reduced. We have grown all manner of food plants indoors and the patio planters outside have created a microclimate where oriental greens, runner beans, spinach, chard and a selection of unusual edible perennials grow within feet of the kitchen. The house and garden have merged and we no longer feel so cut off from the outside.

Most important of all, we have LIGHT in our main kitchen/dinning/hanging-out space. We can wash-up, garden and eat in an ambient, bright atmosphere which allows us to watch the birds and experience all the seasons and the vagaries of the English climate in comfort. Sunsets, thunderstorms, snowfall and sunny days have never been so appreciated. The acoustics are good too and we have had some great musical parties when even a 100 watt Marshall amp or two at full volume hasn't bothered the neighbours!

Inside the extended passive solar kitchen extension showing planters growing tomatoes and melons.

The solar hot water tubes installed on the roof provide an ample supply for much of the year.

Chapter 8

GARDENS

In Chapter One, I made a distinction between 'original permaculture', and 'design permaculture'. The former imitates natural ecosystems directly by using perennial plants, polycultures, stacking and so on, while the latter is more about the relationships between things than the things themselves. (See pp4-5.) This chapter takes these two separately and looks at how they can be applied in the garden. Of course they can both be applied together, but it's helpful to consider them one at a time because permaculture design can be applied in a conventional type of garden without a hint of original permaculture in it.

The Key Planning Tools, described in Chapter 2 (see p27), are not just a useful design method, they also act as a framework into which most of the design ideas of permaculture fit systematically. So in this chapter I look at design permaculture under the headings 'Zoning', and 'Sector and Elevation'.

The chapter opens with the question why we should bother to grow food in our gardens at all, and it ends with some general guidance about gardening, mainly, though not entirely, for beginners.

WHY GROW FOOD?

In a sustainable society we will grow much more of the food we eat in our own gardens. The ecological advantages of domestic food production have been outlined in Chapter 2. (See p26.) Growing some of our own food is one of the most effective things we can do to reduce our ecological footprint. (See p9.) But there are many other reasons for growing food in our gardens.

It's rarely worthwhile financially. Most of us could earn in an hour or two what we spend on fruit and vegetables in a week, and to be self-sufficient through the whole year would take more time than that, never mind the cost of seeds and the time required to reach the necessary level of skill. Food gardening can be a paying proposition for unemployed people or those on a low income with a large family who eat a lot of fresh food. But for most of us it's not.

So what? We don't do it for the money. But there is a whole range of good reasons for growing food, and they are summarised in the mind map, right.

The food we grow in our own gardens is the freshest we can get anywhere. Even vegetables from a vegetable box scheme (see p289), picked the morning we receive it, will be old by the end of the week. In fact vegetables lose much of their life force within half an hour of picking. What can compare with a salad, gathered after you've prepared the rest of the meal and put straight on the table?

We also know exactly how the food has been grown when we grow it ourselves. Although we can trust the Soil Association organic symbol, or the individual grower we buy our box from, their standards may not be the same as ours. Vegans, for example, may not like the fact that virtually all organic food is part of a mixed farming system involving animals.

Most of us could do with eating more fresh fruit and vegetables, and what could be a greater incentive to do so than having an abundant supply which we have put our own heart and soul into growing? Growing our own can lead us to eating a greater diversity of fresh foods as well as a greater quantity. It's easier to grow a wide variety of edible plants in a home garden than in a commercial setup, and fruits and vegetables which don't travel well are not grown commercially at all.

There are many ways of taking exercise and there's nothing wrong with exercise for its own sake,

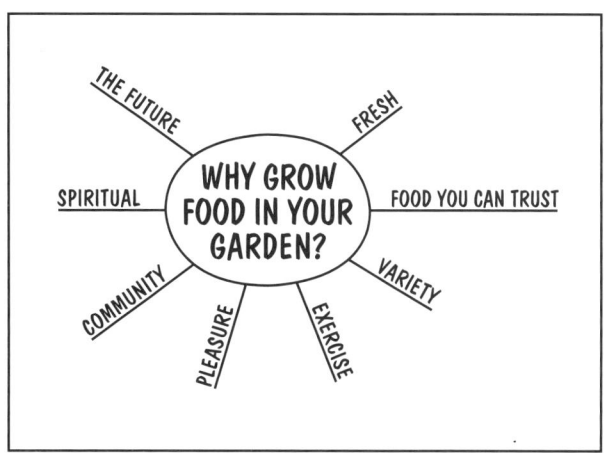

but personally I prefer to take mine by doing something productive. Gardening is a great pleasure. There's the personal satisfaction of doing it and also the social pleasure of becoming part of the community of gardeners. The community aspect is especially strong among allotment growers, but for any gardener it can be a bridge which links us to people we might not make contact with otherwise.

As well as improving our lives as individuals, gardens in communal settings can play a part in community building. Research has shown that "one of the most powerful factors in curbing crime and vandalism on problem housing estates is the presence of a garden. Schools have also reported a decrease in vandalism and an improvement in behaviour when children participate in greening the school grounds."[1]

Gardening may not often be seen as a spiritual activity. But for most of us it's the only direct contact we have with the Earth. We may go for walks in wild or semi-natural places where we can appreciate the beauty and wonder of nature. But we are observers there. In a garden we are participants in the cycle of life. As Karen Christensen has written, "Working in the garden is the only time most of us ever put our hand to the earth, touching the surface of our planet. The environment ceases to be an abstraction when you feel soil crumble between your fingers."[2]

Last but not least, when we're gardening we're doing something positive towards the future of the planet. So many of the things we can do to reduce our ecological impact are a matter of not doing something: not travelling by air, not eating so much meat and so on. Gardening is something we can actually do rather than refrain from doing. There are other active things we can do but gardening is one which is easily available to almost all of us.

By growing some of our own food we're not only making our immediate impact a more positive one, we're also preparing for the future. We're acquiring skills which will have great value for everyone in the future. We're also helping to develop a model of what life must be like if we are to have any future at all.

ZONING

The best fertiliser is the gardener's shadow.
Chinese proverb

Many gardens are long and narrow, in fact this is the most common shape of gardens in both urban and rural areas. The illustration shows a typical layout. There are perhaps a few ornamental shrubs out front and behind the house a flowerbed or two, a largish lawn, maybe a tree, and – down at the end, behind a thick privet hedge – a vegetable patch and greenhouse.

The single thing which would do most to increase the productivity of this garden is to move the vegetables up to where they can be seen from the house.

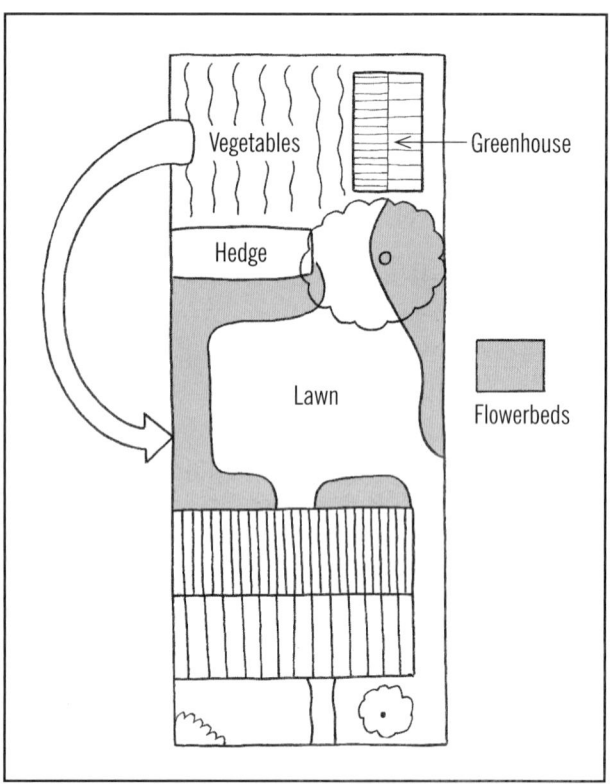

A typical home garden – and the most effective way to make it more productive.

Vegetables which you see every day as part of your normal daily round grow much better. They get the care and attention they need when they need it. Weeds are pulled when they're young, before they have started to compete seriously with the vegetables, and while they can be quickly hoed rather than laboriously pulled. Plants get watered when they need it, not when they have already suffered moisture stress which will inevitably lead to a loss of yield.

You also get to eat more of them. When the vegetables are out of sight you may not know what is ready. Even if you do know, you may be in the thick of preparing a meal, with something boiling on one ring, something frying on another, and a child pulling at your trouser leg; you're just about to go and pick the greens when it starts raining. "Oh damn it," you say and open a tin of beans.

Permaculturists sometimes talk about The Kitchen Window Rule. This says that the most fertile ground is that which can be seen from the kitchen window. Never was a truer word spoken. Unless there's some serious handicap to plant growth in that part of the garden, such as deep shade, more food can be grown and harvested from that part of the garden than any other.

There are some exceptions to this simple rule. If you have a square garden with the house in the middle, the positioning of things may be less crucial than if you have a long one with the house at one end. If your food garden is far from the house and yet you pass through it every day, for example to let your chickens out in the morning and shut them up at night, it is well zoned.

A garden needs to be accessible as well as visible. A friend of mine planted a herb bed in a position that was close to his house but on top of a steep bank. Because it was a climb to get there it was rarely visited and soon became overgrown with weeds. He moved it to a spot a little further from the house but beside a level path and it became much more productive.

Sometimes the soil near the house is less fertile than the rest of the garden. But deficiencies like poor drainage, low organic matter or low levels of plant nutrients can always be improved. Bad layout cannot.

On a small intensive area like a home garden potential yield is high enough that the investment of energy needed to improve the soil will be repaid many times over in increased yields and decreased work.

Some people may object to the idea of growing vegetables close to the house because they want to look out on something attractive, not serried ranks of brussels sprouts and turnips. But edible plants don't have to look ugly. Plants are naturally beautiful things, and edible-ornamental gardening is a style which many people practice with great success. (See box, Edible-Ornamental Gardening.)

EDIBLE ORNAMENTAL GARDENING[3]
(See Colour Section, page III.)

There are many plants which are both edible and ornamental. Some are food plants which are especially attractive to look at, others are ornamentals which also happen to be edible.

The bright red stems of ruby chard, the delicate spirals of Romanesco broccoli and the scarlet flowers of runner beans are some of the more outstanding vegetables. In fact both runner beans and tomatoes were grown purely as ornamentals when they were first introduced from the New World. Most herbs are attractive, and many different varieties have been bred, often with more difference in their appearance than in their taste. The green, purple and variegated sages are examples. Fruit trees are especially attractive when they're in blossom and in many cases the fruit itself is pretty. Some, like the purple cherry plums and purple filberts, have colourful leaves and the bark of the cherry gives visual interest in winter.

Flowers which can be eaten include pot marigold, day lily, pansy and roses but pride of place must go to the nasturtium. The leaves and seed pods are edible as well as the flowers, and once they get going in the summer they grow very vigorously and can be picked quite hard without loss of ground cover or colour.[4]

These plants can be components of an edible-ornamental design, but the essence of the genre is not the components but the composition. Even quite ordinary vegetables can be put together in such a way that the contrast between their shapes, sizes, leaf textures and shades of green make an attractive picture. Imagine, for example, what can be done with parsley, chives, carrots, Swiss chard and sprouting broccoli. An accent of colour added to this, perhaps with a few Lollo Rosso lettuces or by allowing the chives to flower, can make a display which is quite as beautiful as a bank of bedding plants, though somewhat more subtle.

This is the basis of the French potager style of garden. The word literally means vegetable garden but in English it normally implies a vegetable garden designed to please the eye. The overall layout of the garden and the placing of the vegetables themselves form the basis of the design, with herbs, fruit and a few flowers added to set off the natural beauty of the vegetables.

When designing an edible-ornamental garden, or a purely ornamental one, the first things to consider are where the garden will most often be seen from, and the people's personal preferences in beauty. Given this information there are a number of principles which can be used to guide the design:

- Place shorter plants near the front of the view and taller ones at the back.

- Wherever possible combine vertical plants or features with low-growing plants.

- Think of colour and texture combinations. What will look well together?

- Place ornamentals in situations where they will naturally thrive, eg woodland plants in shady places, drought-loving ones in hot, dry places. This will give natural, harmonious combinations, and the plants will grow well with the minimum of care.[5]

- Design for year-round interest. Some plants flower for a relatively short period. Fruits and autumn leaves can give colour later in the year. (See p182 and Colour Section, 25-27.)

- Include some water, if only a tiny pond.

- A view which can't be seen all at once is attractive: a path which curves away from the viewer and disappears can give a feeling of space and mystery to a small garden.

One way to use the principle of zoning is to look at the different kinds of space available for food growing, from indoors to the allotment, and see how each can best be used.

Indoors

It's easy to overestimate the potential for indoor growing. Plants need more light for photosynthesis than we do to see well. A light level that's quite adequate for reading may be pitch dark from a plant's point of view.

The common ornamental houseplants are used for indoor growing partly because they can tolerate low light levels. (See box, Houseplants Clean the Air.) But food plants may do badly where house plants thrive. In general they need to be right up against a window, preferably south-facing. Even plants in a south-facing bow window may lean towards the light if they are not right by the glass, a sign that they would prefer to have more light.

The best use for windowsills is to bring on young plants. They need little space, so they can all be fitted close up against the window, and they will benefit from the warmth, freedom from slugs and extra attention they will get indoors. One way to increase the light for young plants is to grow them in a box, with the side facing the window removed and the other three sides lined with aluminium foil, shiny side towards the plants.

If you have no space to grow food other than a windowsill I wouldn't be discouraged by the limited potential. In fact it can be a rewarding challenge to see just how much you can grow there. The plants recommended overleaf for window boxes are good ones to start with, and the more tender of them will benefit

from being on the inside of the window. If you don't mind losing a little light, taller plants such as tomatoes can make productive use of the vertical space, perhaps placed at either side of a wide window so as not to obstruct too much light.

One food crop which can be grown without any light at all is mushrooms. Being fungi they get their food not from photosynthesis but from the organic substrate on which they live. They're a prime candidate for indoor gardening.

Seed sprouting is another prime candidate. You may think of sprouting as cooking rather than gardening, but by germinating seeds, pulses and grains we actually

Elaine Bruce in her indoor food garden.

HOUSEPLANTS CLEAN THE AIR[6]

Many of the commonly grown house plants originally come from tropical rainforests. The rainforest environment is similar to the inside of a house: light levels are low and the temperature is constantly warm.

The rainforest is also full of the waste products from a mass of fast-growing plants and animals, and almost windless so the waste products are slow to disperse. This has made these plants specialists in handling pollution. Not only can they tolerate it, but they, together with the micro-organisms associated with their roots, can reduce the level of many pollutants in the air by breaking down the polluting substances into smaller, harmless molecules. When we grow them in our houses we benefit directly from this.

Another benefit of indoor plants is the moisture they add to the atmosphere by transpiration – the steady loss of water from the leaf surfaces which is characteristic of all plants. The humidity level in most houses is below the optimum for human health, especially when the heating is on, and plants can correct this. Houseplants can also remove mould spores and other unhealthy microbes from the air.

Different species of houseplants vary in their ability to remove pollutants. Formaldehyde is the commonest indoor air pollutant (see p162), and the Boston fern removes more formaldehyde than any other houseplant, though it's not outstandingly good at removing other pollutants, and needs careful maintenance if it's to be grown successfully.

Other house plants which are almost as effective but easier to keep are: areca palm, lady palm, bamboo palm, rubber plant, dracaena 'Janet Craig', common ivy, dwarf date palm, *Ficus alii* and peace lily.[7]

increase their food value in two important ways. Firstly, the digestibility of young, freshly-germinated plants is much greater than that of dried seeds, so effectively you have more food after sprouting than before. Secondly, the level of vitamins is negligible in dry seeds, but in infant plants it's higher than at any other time in their life. Growing sprouts also saves energy, both in transport and in cooking. Up to 90% of the weight of fresh vegetables is water but seeds are almost all concentrated food. You add the water at home. As for cooking, most sprouts are eaten raw. Ones which are sometimes cooked after sprouting, such as chick peas, take less cooking time than if they had only been soaked.[8]

Window Boxes and Balconies

Of course we can't hope to make a significant contribution to our bulk carbohydrate consumption from a window box. But there are some foods which are worth growing in small containers, whether on a windowsill, balcony, flat roof or any other niche which may be available to urban dwellers. These are foods which either repay being eaten extremely fresh or are expensive to buy.

Fresh herbs qualify on both counts. Salads – eaten seconds after picking, never mind minutes – qualify on freshness.

Even a window ledge can be productive. These three pots of lettuce provided copious salads through the summer.

There are small varieties of lettuce, such as Tom Thumb and Little Gem, which fit quite well in a window box. But the lettuces which make the best use of the limited space are the loose leaf or pick-and-pluck types, such as Salad Bowl and Lollo Rosso, which are also very ornamental. (See Colour Section, 16-17.) With these varieties you pick individual leaves rather than cutting the whole plant at once, and the plant grows more leaves to replace them. This is much more productive than growing a succession of hearted lettuce, as the whole growing area is utilised for a greater proportion of the year. (See illustration in next column.)

Lettuces are not the only salad plants which can be grown on a pick and pluck system. Some of the most useful kinds are listed in the box. (See overleaf, Some Plants for Window Boxes.) All pick and pluck plants should be kept well watered to encourage leafy growth.

An even better way to make use of limited space is the cut-and-come-again method. This is like the mustard and cress some of us used to grow in the airing cupboard as children. The seeds are sown broadcast over a small area of soil, and when the seedlings reach about 5 to 15cm high, depending on the plant, they are cut overall with scissors.

The cut must not be so low as to prevent regrowth. Leaving the plants some 2.5cm high after cutting is usually about right, but the critical thing is not so much the height of cutting as which parts of the plant are left after cutting. On the first cut the seedling leaves must be left. These are the first two leaves produced by the plant, usually oval and different in shape from all the other leaves. On subsequent cuts at least one bud must be left. Buds are formed in the angle between the stem and a leaf, and as long as one of these is still there the plant should regrow. In practice it's a good idea to leave the odd bit of leaf here and there.

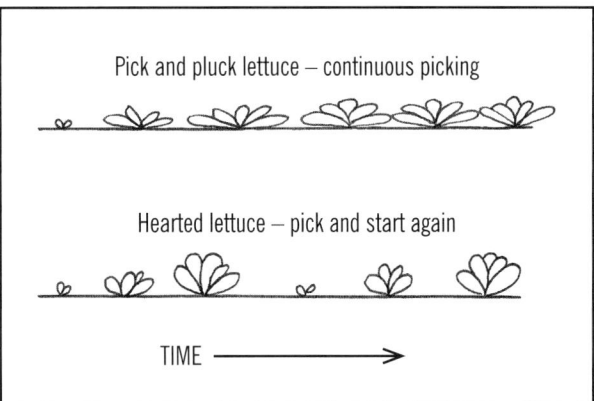

Pick and pluck vegetables make better use of the space available.

You can get anything from three to five cuts, depending on the kinds of plants, time of year and weather conditions. It is generally more successful in spring and autumn than in high summer, when the weather is drier and the plants tend to go to seed. A patch first cut in the autumn can sometimes over-winter and be cut again in spring. Regular watering is always important for cut and come again.

Because the seedlings cover the ground completely almost from the start, production per square inch is high. The nutritional value of the seedlings is also higher than that of older plants; they can have twice the vitamin content. Almost any leafy vegetable can be grown like this. Some kinds which are normally grown for cooking can be eaten as salads when grown as cut-and-come-again because they are much more tender when so young.[9]

Pick-and-pluck and cut-and-come-again are so productive because they utilise all of the growing space for as much of the growing season as possible. A third way to make the maximum use of space is by growing perennials, for example Welsh onions. These onions grow in a bunch and produce new bulbs each autumn by division from the old ones. In spring the full complement of bulbs is there, ready to put on more growth as soon as the temperature warms up, while annual salad onions are only just germinating.

A fourth way to get the most out of a very small area is by growing polycultures. We have seen in Chapter 2 how a polyculture can invariably outyield a monoculture on the same area. (See pp19-20.) On the other hand they do need much more attention, and this is exactly what they will get when grown in containers, which by their nature get a lot of attention. Polycultures keep the ground fully occupied through time by means of a succession of plants, each of which grows to fill the space left by the harvesting of its neighbour.

Garden polycultures are described later in this chapter. (See pp201-202.) These can be modified to fit the conditions of a small container. To some extent size limits the range of plants which can be included. For example, broad beans may be too tall for a window box and cut out too much light, though French beans may fit in.

The importance of using a high-organic-matter growing medium in containers is discussed under Patios and Courtyards, on the next page. On balconies and flat roofs the weight of the containers and the growing medium may be important too. Lightweight aggregates such as vermiculite may be added to the compost to reduce its total weight and increase its water holding capacity.

Wind is a potential problem on roofs, balconies and other high places. Windbreaks may be essential in some cases and will probably help in most others. Trellis is the most practicable windbreak in these situations, where space is almost always at a premium. Hardy plants can be grown up the trellis, but edibles are unlikely to yield well on a trellis in a really windy situation.

An advantage of these high places is that they may avoid the shade problems which are common at ground level in urban situations. Even if a balcony faces north it will get good indirect light as long as it's got a full uninterrupted view of the sky.

Conservatories

A conservatory combines well with a garden in a number of ways. It can extend the range of plants grown to include tender exotics. It can increase the length of the season in both spring and autumn, and produce fresh food right through the winter. But above all it gives the ideal environment for bringing on seedlings at the beginning of the year.

It's not just warmer than a free-standing greenhouse, it gets more attention, especially in the early spring. On a wet and windy March day it's so much more inviting to simply step from one room of your house to another than to put on your raincoat and wellies and go off down the garden path. The difference between a conservatory and a free-standing greenhouse may mean the difference between checking the little plants a couple of times a day or once in two or three days.

That's a big difference in the amount of 'gardeners shadow' the seedlings receive, and the most important thing in the life of any living thing is to get off to a good start. No amount of care in later life can compensate for poor growth in the formative weeks of a plant's life.

SOME PLANTS FOR WINDOW BOXES		
Pick & Pluck Salads	Cut & Come Again	Perennials
Iceplant*	Cress	Herbs
Land cress	Endive	Mitsuba
Lamb's lettuce or corn salad	Leaf chicory	Onions: Welsh
Lettuces, loose leaf	Lettuces, esp. loose leaf	everlasting
Nasturtium	Mustard	tree
Parsley	Oriental greens	Richard the Lionheart**
Rocket	Rocket	Salad burnet
Summer purslane*	Salad rape	Sorrel
Winter purslane or claytonia	Spinach	Turkish rocket
	Summer purslane*	
	Winter purslane	

* Best in warm climate, or protected with a cloche.   ** *Reichardia picroides*.

Patios and Courtyards

A patio is a paved area directly outside a house which has good exposure to the Sun, while a courtyard is more enclosed. For convenience a patio could be defined as receiving direct sunlight for at least half the day throughout the growing season.

A patio has its advantages:

• It gets plenty of attention.

• It often has a particularly warm microclimate.

It also has its disadvantages:

• There is no soil (unless there is some under the paving slabs) so plants must usually be grown in containers

• The heat can be excessive.

• Some patios and courtyards can be subject to destructive winds.

A patio is warm because the paving and the surrounding walls both reflect sunlight onto the plants during the day, and store up some of the daytime heat to release it slowly overnight. Plants which are marginal for outdoor cultivation in your area may do well on a patio. In most of England and Wales these include peaches, figs, kiwi fruit and tomatoes. Many plants which do well in the garden, such as courgettes and squashes, will do even better in the heat of a patio as long as they can be given enough water.

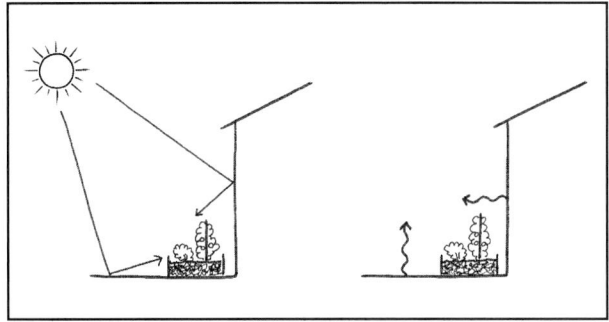

Patios are warmer both by day and by night.

Aromatic herbs do particularly well in a hot, dry environment. Many of them come from the Mediterranean, where summers are hot and dry. They not only grow well in these conditions but produce more of the essential oils which give them their culinary and medicinal value.

Plants grown in containers need much more regular watering than ones grown in the open ground, because the volume of soil available to them is limited. When the weather is hot and dry they can't forage for more water in deeper soil as plants grown in open ground can.

This sunny, south-facing patio is ideal for the tomatoes in pots, but it proved too dry for the runner beans, even with copious watering.

What's more, if the walls of the container are porous water can evaporate from all around, not only from the upper surface. (See Colour Section, 18.)

On the whole larger containers dry out less quickly than smaller ones. Other factors which influence how often they need watering include: the kind of plants, the weather, the microclimate, the size of the plants relative to the containers, and how well the growing medium retains water. In some cases they can need watering daily, or even twice daily in the case of tomatoes. On the whole leafy salad plants such as lettuce need more watering than other kinds, and they may not be suitable for the hot dry conditions of a patio.

The best growing medium for holding water is one that is high in organic matter. Peat does the job beautifully, but using it contributes to the destruction of peat bogs, a valuable and endangered ecosystem, and I would never use it myself. Coir compost is a substitute, but one which is shipped to us from the other side of the world, and so hardly a sustainable alternative.

Home-made garden compost, or commercial composts made from municipal wastes are more local alternatives. But both can become too compacted if used alone. I have had good results with a mixture of 2 parts municipal compost, 2 parts well-structured clay topsoil and 1 part sharp sand.

The soil or growing medium in a container can also become tired in time. Nutrients get used up, structure gradually deteriorates, and there may be a build-up of disease organisms. If you grow the same crop in your containers every year, tomatoes for example, you should change the soil every year. With perennials, or annuals on rotation, you should watch the condition of the crops.

Adding plenty of organic matter or compost can revive a tired container. A thick mulch of comfrey leaves in the autumn is one way to add nutrients and some organic matter without disturbing the soil. Comfrey decomposes quickly, and even in the lower temperatures of winter it will rot down and leave a clear surface by spring.

All containers must have drainage holes in them, and putting a few broken potshards or stones in the bottom of the container will help drainage.

Another point about containers is that frost can penetrate them more easily than it can into open soil, because a much greater surface area is exposed to the cold air. Perennials, even ones we normally think of as frost-hardy, may suffer more from hard frosts in containers than they would in the open soil.

Patio growing is a relatively high-input/high-output form of gardening. The rewards can be great, both in increased yield and in extending the range of plants which can be grown. But the extra watering they need means extra work, and this can be difficult to arrange if you're often away from home.

Courtyards are also places for container growing, but the microclimate is somewhat different. If they're relatively small or narrow compared to the surrounding buildings they tend to warm up and cool down more slowly than surrounding open areas, both on a daily and a seasonal cycle. Thus they're cooler than nearby open areas in the morning and in summer, and warmer in the evening and in the winter. They're usually comparatively dark places, especially if they're relatively narrow. Even if there's an opening to the south they will lose much of the morning and evening sunshine.

This means they're usually best for the more shade-tolerant crops such as salads, other leafy greens and soft fruit, which can often be trained up the walls. (See p185). In fact they may be better than more open patios for these crops as they will dry out less quickly. The relative warmth of winter could be an advantage for winter salads. But light will be even more of a problem at this time of year as the angle of the Sun is lower and it may not get to places which are relatively sunny in summer.

Narrow courtyards may have restricted air movement, leading to a buildup of air pollution, though an opening towards the prevailing wind will alleviate this. Wider ones may be subject to violent eddying winds, as the buildings act much like any barrier which is impervious to wind. (See p84.) As a rule of thumb wind problems are most likely where the width of the courtyard is more than twice the height of the surrounding buildings, which is quite a big courtyard.[10]

Gardens and Allotments

Many people who only have a small garden dismiss it out of hand and think that they must get an allotment if they're going to grow food. But small areas can be extremely productive. (See case study, A Small Productive Garden, p209.)

In terms of yield per square metre and yield per hour of work, a home garden can easily outyield an allotment which is a twenty minute bike ride away. Most allotment holders visit their allotments about once a week. The Kitchen Window Rule applies to the

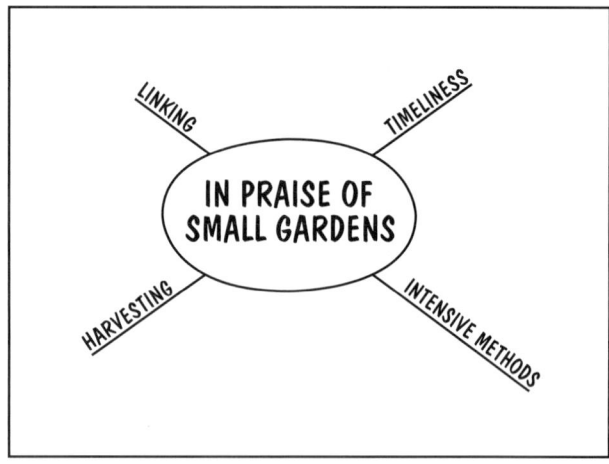

comparison between an allotment and a garden even more strongly than it does when comparing two parts of the same garden. The reasons for this are summarised in the mind map above.

Timeliness. As we have seen, jobs tend to get done when they need to be done rather than days later, when potential yield has already been lost and the job itself may have grown, as when weeds get out of hand.

Intensive Methods. Polycultures, with their intrinsically high yield, are ideally suited to small home gardens. It doesn't take more work to grow a polyculture, but it does take more attention. This can be given more easily when the gardener can keep an eye on what's going on from day to day, and spend the odd few minutes of leisure observing how things are developing.

With the really intensive growing which is possible in a home garden, weeds rarely get a look in. This is partly because the gardener is always on hand to nip them out as soon as they show their heads above ground, and also because of the intensity of production itself. Every piece of ground is occupied by crop plants as soon as it becomes available. Interplanting and other forms of polyculture give much more coverage of the ground than straight monocropping.

More intensive methods of pest control are also possible. Slug traps, for example, need checking daily. In larger home gardens, ducks can be kept for slug control, whereas on an allotment the only option may be a frog pond. Although frogs are worthwhile they are not as effective as ducks or intensive hand methods. (See pp110-111.)

Harvesting. As we have seen, more food is actually harvested when it's grown near the house, and it can be really fresh. I did once hear someone say "I'll just get on my bike and go down to the allotment to pick a lettuce." But he was a retired person and didn't do much else than cultivate his allotment. Most of us don't have the time.

Linking. In addition to the yield advantages, a home garden can benefit much more easily from the outputs of the house. Composting of organic waste from the house is easier, and the use of grey water or rainwater from the roof becomes possible.

Susy Joy grows a mix of vegetables and flowers in the border outside her small town house.

There may be other demands on the space provided by a home garden besides growing food. Space for children and adults to play and relax is one. The area which is needed for this is often overestimated. (See p181.)

If the beauty of the garden is an important consideration one option is the edible/ornamental approach. (See p171.) Another is to place compact ornamental plantings in the most visible parts of the garden.

Something which needs to be put in the balance when deciding whether to grow food in the garden or to take an allotment is the distance to the nearest allotment site where a plot is vacant. If it's only a hundred yards away it will have some of the advantages of a home garden, though not all of them.

Another consideration is the amount of time you would like to spend gardening. An allotment is a commitment, although the amount of time you devote to it can be pretty flexible. Rick Green has been

---

THE LAND NEXT DOOR

It can be possible to effectively double the size of a home garden by taking on a neighbours' garden. Many people don't want to garden or are unable to due to age or poor health, yet don't want to see their garden go wild. They may be more than happy to have the burden taken off their shoulders. The zoning value is almost as great as your own home garden and certainly much better than an allotment.

This kind of arrangement can work well but it does need to be approached with care. The two parties to the agreement may have different ideas about what kind of gardening is going to go on. The garden owners may envisage a neat lawn with a decorative border, rather than intensive food production. A clear agreement from the beginning is important.

---

cultivating his allotment in Birmingham for 76 years, and has kept records of the amount of time he has spent on it. When he was in his working prime some 30 years ago he reckoned that his 10x30m plot "could be run reasonably well in 100 hours a year, while 150 hours produced a good return, and 200 hours a super one."[11]

Mr Green's experience could be used as a rough guide. But no doubt he's an expert gardener, and a beginner may take more time to get the same results. Also, he's talking about an allotment which is up and running. Getting a new allotment or garden to that stage can take time, especially if it has been abandoned for some time before you take it over.

Allotments are valuable for people with no home garden at all, and an excellent resource for those who really want to grow a lot of produce. But I would advise anyone with a garden to be quite sure they are getting the most they can from it before applying for an allotment. This is especially true for people who are new to gardening. It's much easier to learn on a small area than a large one. Even half an allotment can seem very daunting when you get down to working it.

Choosing What to Grow

Obviously it's not possible to grow all your fruit and vegetables in a small garden, and where space is limited it's worth thinking carefully about the selection of crops. The ones which are most likely to be worthwhile are those which are:

- best eaten as fresh as possible
- expensive to buy – or hard to find in the shops
- easy to grow successfully
- low in work requirement
- heavy yielding
- especially liked by the family

Of course not many crops, if any, meet all these criteria, but any one which meets two or more is certainly worth considering. We all have somewhat different priorities. Someone who is new to gardening may put ease of growing first, while someone with a very small garden may put more emphasis on maximum yield, and so on.

Fresh. Salads obviously qualify on the freshness count. There's an enormous variety of salad plants which can be grown in our climate, both annual and perennial, many of them attractive plants to look at.[12] Maincrop potatoes come at the other end of the scale, though some people may think the delight of extra-fresh new potatoes makes them worth the space.

Expensive. Many of the salad plants which can be grown are hard to find in the shops, or expensive if they are available. Rocket and lamb's lettuce are examples, but they are both easy to grow at home. Garlic is expensive too, yet one of the easiest of crops to grow in my experience.

Easy. Ease of growing is something which varies from place to place, from season to season, and from gardener to gardener. Carrots, for example, are not that difficult to grow in a light or medium soil, but a real challenge on a heavy soil. Growing decent cauliflower takes some gardening skill, but in a perfect season, when the weather does just what it's supposed to do, it's not too hard to get a few decent curds. As for different gardeners – personally I've never been able to grow a respectable crop of swedes. The numbers on the table should be taken as only a broad indication of the easier crops, and I wouldn't want to put anyone off trying any vegetable which they really want to grow.

Low work. Some crops are more laborious than others. With some it's the actual growing that's laborious, like carrots, which need painstaking thinning and weeding. But often it's the harvesting and preparation which really take the time. I don't grow brussels sprouts, for example, because I don't like the long fiddly job of preparing them for cooking. Crops which must be planted in succession to get a continuous supply, like headed lettuce and spinach, take much more time to grow than pick-and-pluck crops like loose leaf lettuce and perpetual spinach or Swiss chard which produce for a whole season from one sowing.

But the way the vegetables are grown has at least as much influence on the time taken as does the selection of crops. Doing away with the habit of digging makes a big difference. I've grown potatoes by the digging method and the mulch method and I would never go back to digging now. Weeding can take even more time than digging. Keeping the ground weed-free from the start and never letting annual weeds set seed will save more work than anything else a gardener can do.

The crops which take least work of all to grow are the perennials, including fruits and perennial vegetables. Fruits are discussed in Chapter 9 and perennial vegetables later in this chapter (see pp197-201).

Yield. Yields vary greatly according to soil, microclimate, season and grower. The yields given in the table should be taken as a very rough guide to the sort of quantities you can expect from different crops. Even the relative yields will vary from year to year as each season will be better for some crops than others. But rough as they are, if you want to get the maximum out of a small area, they can suggest which crops to go for.

In larger gardens it is possible to over-produce, and nothing is worse than seeing vegetables you have lovingly grown thrown on the compost heap. Having a rough idea of yields can help to avoid this.

Crops which stay in the ground for a long time effectively have a low yield. Purple sprouting broccoli for example, delicious as it is, is in the ground for the best part of a year. A keen gardener could grow two crops, one summer and one winter, on that ground in the same time. But the yield could be increased by interplanting the broccoli with a fast-growing crop.

(See p201.) Radishes, on the other hand, though they are small, grow quickly. They can be grown as a catch crop, fitted in before or after another crop, so their effective yield can be as high as that from larger roots.

The luxury perennials, asparagus and globe artichoke, not only have a low yield but are in the ground for eighteen months to two years before they start producing. But if taste and expense are your criteria they are certainly worth growing.

Continuity of supply can be as important as the total yield. In a very small garden you are likely to eat everything you grow, and expect to supplement what you grow with bought in vegetables. But with a larger garden it's all too easy to have a glut of vegetables at one time followed by a shortage a few weeks later. In fact achieving perfect continuity of supply is one of the greatest skills of gardening.[13] The problem is greatest with successional crops like radish, spinach and headed lettuce, which gives another reason for growing the pick-and-pluck alternatives.

With crops which are harvested all at once or over a short season, careful storage is the key to continuity. Vegetables which can be stored without freezing or bottling include potatoes, roots, the pumpkin family, white cabbage, onions, garlic and even tomatoes.[14] This does take some care and attention, and a suitable storage space – the mythical 'well-ventilated, cool, frost-free shed' is ideal. If you have doubts about storing vegetables successfully it may be wise to limit the amount of these crops you grow, even in a large garden.

Herbs qualify for inclusion on the freshness count and the perennial ones, which is most of them, are easy to grow and have a low work requirement. But they are one of the easiest things to grow too much of.

Creative recycling: Rudolf Koechli has made a frost-free and rodent-proof root store by sinking an old chest freezer into the ground.

A single sage or rosemary plant can in three or four years grow to a size which would keep a small village supplied. In small gardens where maximum production is an aim these herbs must be kept regularly pruned.

It comes hard to many people to cut and discard something which is potentially edible, and it may be possible to dry the leaves and sell them or give them away. But an unused potential is not a useful yield, and it may be more productive to compost the herb leaves and grow something else in the liberated space. It can be worthwhile to allow the plants to grow big deliberately for their visual effect, but so often they grow big simply because we forget to prune them.

Some herbs, such as lovage and fennel, can't really be kept small by pruning. But many of these attract beneficial insects, and they may earn their space more by virtue of this function than by the produce we harvest from them. (See pp242-244.)

What the Family Likes to Eat. This is to some extent the factor which must be balanced with all the others put together. If there's nothing on Earth you like more than globe artichokes, never mind the low yield and the fact that it's not the easiest plant to grow!

The criteria for choosing which crops to grow on an allotment are quite different. In Key Planning terms we have moved from zone I to zone II. (See pp27-28.) Crops which need attention in large but infrequent doses, and which will keep for a week or more after picking, are best. Potatoes, onions, cabbage and roots are obvious choices, especially if they are maincrops

SOME COMMON VEGETABLES COMPARED[15]									
	Yield	Easy to grow?	Amount of work	Notes		Yield	Easy to grow?	Amount of work	Notes
Asparagus	VL	3	2	60-75 spears/m²	Leeks	H	1	2	
					Lettuce, hearted	L	2	2	
Beans, broad	L	1	1		Lettuce, loose leaf	H	2	2	
Beans, dwarf French	L	2	2		Onions	M	2	2	
Beans, runner	H	2	3		Parsnips	M	2	1	
Beetroot	M	2	1	+ tops	Peas, dwarf	L	3	3	
Broccoli, calabrese & purple sprouting	L	3	2		Potatoes, early	L	1	3	
Brussels sprouts	L	2	3		Potatoes, maincrop	M	1	3	
Cabbage	M	1	2		Pumpkin	L	1	1	
Carrots	H	2	3	difficult on heavy soil	Radishes, summer	L	1	1	
					Radishes, winter	M	1	1	
Cauliflower	L	3	2		Spinach	L	3	3	
Courgettes	H	1	1		Squash	L	1	1	
Garlic	L	1	1		Swede	H	2	1	heaviest yield of all
Globe artichokes	VL	3	2	10-12 heads/m²					
					Sweetcorn	VL	2	3	4 cobs/m²
Jerusalem artichokes	M	1	1	best as perennial	Swiss chard & perpetual spinach	H	1	1	self-seeders
Kale, annual	L	1	2		Tomatoes, outdoor	M	2	3	
Kale, perennial	M	1	1		Turnips	L	2	1	+ tops

Yields: H = heavy, more than 5kg/m²   M = moderate, around 5kg/m²
L = light, around 2.5kg/m²   VL = very light, 1kg/m² or less

Easy?: 1 = easy   2 = moderate   3 = difficult

Work: 1 = light   2 = moderate   3 = heavy

In the small garden of his terraced house Andy Waterman has chickens, a greenhouse and a rich mixture of salads and soft fruit.

On his allotment he grows pumpkins and other large-scale crops, which need much less frequent attention.

for winter storage. If your local allotment rules allow it, and scrumping is not too much of a problem, fruit can also be a good choice.

Obviously a family who only have an allotment will have to grow their whole range of crops there, but it may be a narrower range than they would grow if they had a patch of land at home as well. For those who have both, a garden and an allotment can complement each other well.

Finally, whatever selection of vegetables we would like to grow, we have to fit them into a rotation. If too many of the chosen plants are closely related it may be difficult to create enough of a break between them

to avoid problems of pest and disease buildup in the soil. If that's the case it may be necessary to adjust the selection somewhat. This is what I have had to do in my own garden. (See box.)

Three families of vegetables are especially prone to soil-borne pests and diseases: the brassicas, the onion family, and the potato/tomato family. These should never be grown more than once in four years on the same soil. Although other families are less troubled by soil-borne maladies, all vegetables are best grown in rotation. There are good books on organic gardening which can help you design a healthy rotation. (See Appendix A, Further Reading.)

MY OWN SELECTION

The largest part of our garden is planted with fruit and perennial vegetables, which provide us with all our salads and most of our greens. But we also have a small plot of annuals where we grow:

Broad beans, autumn sown. Very easy to grow, and we love them.

Garlic. Easy to grow, expensive to buy, and we have a variety with very large cloves, which saves a lot of work in the kitchen compared to the garlic you can buy.

Leeks. A reliable and heavy-yielding crop which we eat frequently in winter.

Squash. Not the easiest thing to grow in our slug-infested garden, but we eat a lot of it, and once it gets going it's a good weed-smothering crop.

Swiss chard. Although we have it self-seeding in

the perennial garden, we can never get too much of it, and it gives a fourth course to the rotation; easy to grow and yields well.

Winter radish. Fills a gap in the rotation, and makes a spicy addition to winter salads.

Rotation
The need to rotate the onion family means we don't fit in as many leeks as we would like. But we usually find a spare corner in the perennial garden for a few extra. The other annual vegetable we grow is sweetcorn. It's low-yielding, but we love it and it really does need to be eaten fresh. This fits in here and there in the perennial garden as an upper storey above lower-growing things. It usually does very well like that, and the combined yield of the double-storey crop is quite respectable.

This is a very simple example, but the principles are just the same however wide a range of annual vegetables you want to grow.

Larger Gardens

In the previous section there is an assumption that the home garden is small, but some of us have gardens that are actually too big for our requirements. If after marking out a vegetable patch, some flower beds, a little lawn and perhaps a forest garden there is still some space left, what do we do with it?

The traditional approach is to put it down to lawn, perhaps with a few island beds with shrubs in them. But a lawn, contrary to popular opinion, is a high-maintenance element, and beyond a certain point completely unproductive.

Lawns have had a bad press from permaculturists. Bill Mollison has written a devastating exposé of the amount of fossil fuel, water, pesticides and human labour devoted to them in affluent countries.[16]

He was not objecting to lawns which are actually used for play or relaxation but ones which are only there to look at. He was also writing from the perspective of Australia and North America, where, in most climatic regions, a lawn is a completely unnatural artefact which can only be created and maintained with a great deal of effort.

Here in north-west Europe a lawn is the natural consequence of grazing or mowing almost any piece of land frequently enough to prevent the regeneration of trees and shrubs. The word lawn originally meant an area of grassland within a forest where deer graze. At this level it is an example of semi-natural vegetation, that is an ecosystem which flows naturally from a combination of natural conditions and human activity.

It is possible to spend as much time and resources on a British lawn as on a Californian one. The average British lawn is mown some 30 times a year – approximately the number of Saturdays in the growing season – and many people apply fertilisers and herbicides. This kind of lawn is a very high-maintenance element in a garden, and ecologically dead. But if you don't mind slightly longer grass, a few wild flowers in the lawn, and the possibility of the colour fading a bit during a drought, you can simply mow it four or five times a year and forget about any other kind of treatment.

Grass mowings are a useful output from a lawn when used as a source of fertility and mulch for the rest of the garden. In fact taking them off the lawn will reduce fertility, which reduces grass growth and thus the amount of mowing that's needed. It may also increase the diversity of plants in the lawn, as grasses respond more vigorously to plant nutrients than do wild flowers and tend to crowd them out where the level of nutrients is high.

But the main output of a lawn is recreation. If we look at the garden in the wider context of our total ecological impact, "the 'yield' of a lawn could possibly outrank anything else you get out of the garden if it made the garden so attractive it prevented a few car trips. Lawns are fun. Fun is a yield."[17]

The best approach to lawns is to assess how much you actually need for recreation, and convert the rest either to a more productive or less intensive use such as a meadow. (See below.) Most of us overestimate the space required. A useful exercise is to actually measure the area of lawn used when there is a gathering of people on it, such as an outdoor meal or party. This is the amount of lawn actually needed; any more is purely cosmetic. Without measuring we can only guess how much is needed.

Families with young children may be able to use more lawn space. But few gardens are big enough for bike riding and football, and half the garden space may be as useful as the whole thing for the kind of games they can play there. Again, measuring can be useful. If you have children and a garden which is mostly lawn, it's worth noting down the number of times in the year they actually use it for play, and how much of the lawn is used when they do play on it.

If there is public open space reasonably nearby it may be better to create a different environment for the children in the home garden. (See Shrubberies overleaf and case study, A Small Productive Garden, p209.)

Meadows

One alternative to a lawn is a meadow. At its simplest, turning a lawn into a meadow is just a matter of mowing it once a year instead of every week or two. As the grass grows taller it will become a richer wildlife habitat. Cutting is best done in autumn when the grass has finished growing and both plants and animals have completed their annual cycles. The species composition of the sward will gradually change, and wildflowers can be deliberately introduced to increase biodiversity. (See p366.)

A traditional scythe.

It may be necessary to buy new cutting equipment, because a lawnmower will not cut long grass. Myself I favour the old-fashioned scythe – the two-handed tool you stand up to use, not a little sickle. It's good, satisfying exercise, and if the meadow is large you can spread cutting over a number of days or even weeks. Unlike a strimmer, it's silent and free of vibration.

The mowings can be used for compost, but they will probably contain viable grass seeds, so it needs to be a hot compost if you want to avoid grass seedlings in the vegetable beds when you use the finished product. There should be enough bulk at one time to make a hot heap. But mature grass like this is a 'brown' in composting terms, low in nitrogen and high in carbon, so a source of nitrogen will be needed to balance the carbon. (See p45.)

A close-mown path curving through the middle of the meadow is useful, for two reasons: it makes the point that this is an intentional meadow rather than a case of someone never getting round to mowing the lawn, and it makes it easier to enjoy the meadow.

A mown path like this shows that the long grass is intentional, not just neglected.

A garden meadow can be combined with a small orchard. The grass is short when access is needed to the trees, for picking in autumn and pruning in winter, and mowing for the rest of the year is unnecessary. Grass is a bad companion for young trees or those on very dwarfing rootstocks, but these can be mulched or have a broadleaved ground cover at their base. For mature trees on the more vigorous rootstocks grass can be beneficial. It reduces their vigour, which encourages them to put more energy into fruiting and less into vegetative growth.

Shrubberies

A shrubbery takes a bit more work to install than a meadow, but once it's well established it takes even less work to maintain. The outputs of a shrubbery can include:

- beauty
- wildlife habitat
- play space
- a little food

Although there are many and varied exotic shrubs which can be planted, there is much to be said for native shrubs. They support more wildlife, and, though they may not be as showy as some of the exotics, they have a harmonious beauty, a connectedness to the landscape, especially in rural areas.

By carefully choosing species it's possible to have visual interest throughout the year. A few examples are:[18]

winter
bark and twigs: red dogwood, purple willow, wild cherry, birch (especially *Betula pendula*)
other: wild roses and holly (fruits), hazel (catkins)

spring
blossom: pussy willow, guelder rose, hawthorn, wayfaring tree (a shrub), rowan, crab apple, bird cherry, wild cherry

summer
blossom: elder, wild roses
summer leaves: whitebeam, wild service
fruits: rowan, guelder rose

autumn
leaves and fruits: spindle, guelder rose, field maple, wild cherry, birch, wild service, hawthorn.

With beauties like these who could hanker after exotics? (See Colour Section, 25-27.)

As regards play space, a shrubbery has more to offer than a lawn, except for small children who need to be kept in view. Dens can be made, jungles can be explored, and once the shrubbery has matured a bit, trees can be climbed. When I was a child I certainly preferred this kind of place to a bare lawn.

Some native shrubs bear edible fruit, and some of the wild herbs which can be grown under them are edible too. (See p367.) Although the amount of food will be small, foraging for it is great fun, and food from wild plants tends to have a nutritional quality that cultivated ones lack.

Although I have introduced the shrubbery idea as a solution for over-big gardens, a garden doesn't need to be big to accommodate a shrubbery. The woodland glade concept, with trees and shrubs round the edges grading down to a lawn or other open space in the middle, suits most gardens. It gives a high proportion of edge, and plenty of edge means plenty of blossom and visual interest generally. It's not necessary to surround the garden entirely. Where space is limited shrubs can be planted on one or two sides only. They can be used to form a screen, either for privacy, to block out a busy road or unpleasant view, or as a windbreak.

Care must be taken not to shade areas of the garden which need light, or even the house if the shrubs grow tall.

It's well worth including the occasional tree if there's space. But it's important to check the potential size of the tree first. Wild cherry, although recommended on account of its beauty, can grow as big as an oak. It may need to be coppiced or pollarded to keep it to an appropriate size. This will probably cause it to sucker, which can either be a nuisance or a useful contribution to the shrubbery. An idea of the potential size and growth rate of trees and shrubs can be had from the table in Chapter 4. (See p87.)

Ideally shrubs should be selected to suit the soil and microclimate conditions of the garden: willows and guelder rose in wet soils, wayfaring tree and whitebeam on alkaline soils, rowan in exposed positions and so on. But most shrubs will do well in any reasonably fertile garden soil. Plants should be bought from a native plant specialist. Those from garden centres are likely to be cultivated varieties, or even exotic species of the same genus as a native.

Climbers, such as honeysuckle and ivy, have both visual interest and wildlife value. But they should not be planted till the shrubs and trees have had a few years to get established, otherwise they can smother them.

One of the great advantages of a shrubbery is the low maintenance requirement, far less than even the least intensive lawn. A well-designed shrubbery may not need touching from one year to the next. The table of comparative costs in the box, Public Gardens, is revealing. Since labour makes up the majority of the cost the work requirement is probably quite closely related to the cash cost.

In the establishment phase, before the shrubs have formed a closed canopy, the key to low maintenance is keeping the ground well covered to prevent weeds growing. Initially this is best done with a durable mulch, such as forest bark or chopped reeds over paper. A living ground cover can be established through this mulch. It will need some weeding in the first year or two, but after that it should persist until the shrubs cast too much shade to allow anything much to grow beneath them.

When the shrubs grow big some cutting back may be necessary once in a while. This can be in the form of pruning or coppicing. Sensitive pruning will maintain the visual amenity of the shrubbery better, but coppicing is easier. Some shrubs will need to be pruned or coppiced much sooner than others. Willows, for example, may grow to fill the space available for them in three or four years, while holly may not need cutting in a lifetime.

Community Gardens

Community gardens can provide land for people who have nowhere else to garden, but perhaps more importantly they can play a big part in strengthening local communities, and can even help to create a true community where none exists. The act of getting together to do something as creative as growing plants can be a magic ingredient in community building.

PUBLIC GARDENS

Much of what applies to an over-large private garden applies equally to public gardens and parks.

Lawns can be valuable, especially in dense urban areas. At lunch time on a hot summer's day the office workers pour out and cover every inch of the turf as they soak up the Sun. In other areas they provide much needed open space for both children and adults, but, as in home gardens, a survey of the space actually used may show that the amount of lawn provided is excessive. People who use the park may in fact prefer to see some of it managed in a less formal way too, and it could be worthwhile conducting a public opinion survey before making assumptions about what people want.

Shrubberies and other perennial plantings have ecological advantages over lawns and bedding plants. As well as making better wildlife habitats they receive less input of fossil fuel and chemicals. They also take less cash to maintain, as illustrated in the following table of costs, recorded in the city of Rotterdam.

Landscape Maintenance Costs[19]

Type of Landscape	Comparative Cost (per unit area)
Woodland	17
Shrubs	18
Lawns	33
Hedges	220
Perennials	299
Annuals	619

The layout of shrubberies has some bearing on the maintenance requirement. Although maximising edge has its advantages, both visually and ecologically, it can make mowing the remaining lawn a fiddly job. Intricate edges and small island shrubberies are not as good from this point of view as broad, sweeping edges which can easily accommodate a gang mower.

The main potential for growing food in public places lies in fruit trees, and this is discussed in Chapter 9. (See pp236-237.)

Community gardens are as much about people as about plants. Sheila Lippyat and Madeleine Berry in the Keyford Community Garden, Frome, Somerset.

They can include anything from flowers and vegetables to orchards and wildlife areas. Community orchards are discussed in Chapter 9 (see p236), and community wildlife areas in Chapter 12 (see pp362-364.) What they all have in common is that they are run communally by groups of local residents. In some cases the whole of the garden is worked communally and in others individuals have their own mini-allotment, or there can be a mixture of communal and individual areas.

For a community garden to succeed, two things are usually necessary: a key person or core group who are prepared to put in plenty of effort in the initial stages, and the existence of a real need in the first place. The most obvious need for community gardens is in areas of high density housing where people want to garden but have little or no space of their own. This has been the driving force behind the successful Sheffield Organic Food Initiative. (See case study, Community Gardening in the City, p212.)

In other cases the motive is to tidy up a piece of vacant ground in the neighbourhood or to protect it from other kinds of development. It has to be said that the local residents are sometimes more keen to see something done with the land than to actually do it themselves, especially once the initial enthusiasm has worn off. This kind of garden more often succeeds in high-density areas where there is little other green space, so people are highly motivated to preserve what there is. It's interesting to note that there are few places on Earth with more community gardens than the city of New York.

Much depends on the attitude taken by the local council, who are very often the owners of the land. Some councils actively promote community gardens and have an officer whose duties include helping to set them up. Even councils who are less proactive may be attracted by the fact that community gardens can be an inexpensive option for unused land. In the USA it has been found that the average cost of creating a community garden is just a tenth that of creating a new park of the same size.[20] Support and advice in setting up a garden can be had from the Federation of City Farms and Community Gardens. (See Appendix D, Organisations.)

School gardens are a special kind of community garden, and they can often be the most effective. A school can provide the framework within which a committed person, usually a teacher, can give the continuity of attention which a gardening project needs, and in most cases land can be found in the school grounds.

Gardening can be integrated into the national curriculum, and the organisation Learning Through Landscapes has done much development work on this. (See Appendix D, Organisations.) As well as giving children the opportunity to connect with the living world around them this can be a way of interesting them in improving their diet. In many schools there's a programme called 'Grow it, Cook it, Eat it', and in a Somerset school twelve-year-olds have eaten bread which they have baked from wheat which they have grown themselves.

Reconnecting the next generation with the land in this way is one of the most important things a permaculturist can do. (See Colour Section, 19.)

SECTOR & ELEVATION

Sector

A great deal of sectoring is about working with microclimates. Much of the information in this section builds on that already given in the Microclimate part of Chapter 4. The main sectoring factors relevant in gardens are given in the mind map below.

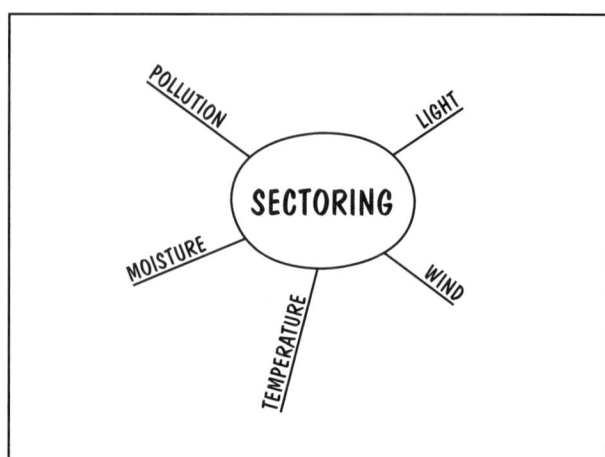

Light and Shade

Very often the main limiting factor to plant growth in domestic gardens is shade, especially in urban areas. To some extent it's possible to predict which will be the sunny and shady areas in a garden simply by looking at the surrounding buildings and trees. But it can also be worthwhile to make a map of where the shade actually falls.

Of course the amount of shade varies, both through the year and through the day, so it's necessary to make observations on a number of occasions. This isn't as big a job as it might seem, for two reasons. Firstly, nothing much grows in the winter, so no observations need to be made then. Secondly, each half of the year is a mirror image of the other, so observations after the summer solstice will only repeat those made before it. Making observations on three days in the year, one each in the months of April, May and June, should give a good picture of the light and shade pattern during the growing season.

To cover the changes in shade through the day, three observations can be made on each of these days, morning, mid-day and afternoon. To get the maximum spread, slightly different times can be chosen on each day. In April, when the day is rather shorter, two observations, one in the morning and one in the afternoon, may be enough.

This makes a total of eight separate observations. They can all be drawn onto the same map of the garden, each one with a slightly different hatching. If this is done carefully, the relative shadiness in each part of the garden can be seen. In the example design in Chapter 13 I simplified my initial shade map to two hatched areas: a) shady for half the day, and b) shady for more than half the day. (See p391.) These are usable categories of shadiness and they match the categories in the table, Some Shade Tolerant Edibles, below.

This kind of map will not be totally accurate. For example some spots shown on the map as relatively shady may get a spell of sunshine in between the times you choose to make your observations. But you will notice these, and the process of making the map will get you focusing on the subject to such an extent that you will know a lot more about light and shady parts of your garden by the time you're finished, even if the map doesn't give a totally accurate picture of them.

There's a difference between the year-round shade cast by buildings and evergreen trees and the seasonal shade of deciduous trees. Perennial vegetables and over-wintering annuals do much of their growing in spring, when the leaves are off the trees, and in some cases these plants can do quite well in partial shade cast by deciduous trees. But much will depend on the growth pattern and light requirements of the individual vegetables and how early the trees in question come into leaf.

Plants can benefit from indirect light as well as direct sunlight. A place which gets half a day's direct sunlight but is open on three sides is a much better growing space than one which gets the same direct sunlight but is surrounded by obstructions. Indirect light can be increased by painting walls white so as to reflect more light onto plants. My own front garden is an example. It receives no direct sunlight before noon but is wide open to indirect light and becomes a real suntrap in the afternoon, with a white wall to the east. I grow annual vegetables there, and light-demanding kinds don't do worse than the more shade tolerant ones.

Shade cast by trees is usually compounded by the competition from their roots, especially for water. (See p190.)

You can use your observations on shade to work out a planting plan. Plants vary greatly in their need for direct sunlight, but there are a few general principles:

SOME SHADE TOLERANT EDIBLES[21]					
6-8 Hours Sun in Summer			4-6 Hours Sun in Summer		
Herbs	*Vegetables*	*Fruit*	*Herbs*	*Vegetables*	*Fruit*

Fennel
Rosemary
Sage
Thyme
	Autumn &
	  winter cabbage
	Beetroot
	Brussels sprouts
	Cauliflower
	Early carrots
	Leeks
	Onion sets
	Parsnips
	Peas
	Potatoes
	Runner beans
	Small tomatoes
	Sprouting
	  broccoli
	Turnips
		Alpine
		  strawberries
		Blackcurrants
		Cooking apples
		Dessert
		  gooseberries
		Kiwi fruit
		Raspberries
		White currants
			Bay
			Chives
			Horseradish
			Mint
			Parsley
				Calabrese
				Chard
				Cress
				Kale
				Kohl rabi
				Lettuce
				Most leafy
				  perennials
				Radish
				Spinach
				Spring cabbage
				Welsh and tree
				  onions
					Cooking
					  gooseberries
					Loganberries
					Morello cherries
					Red currants
					Rhubarb

Note: A full day's sunshine at midsummer is 16 hours					

- Vegetables need more sunlight than most ornamentals.

- Plants with larger leaves are often more shade tolerant than ones with smaller or narrow leaves.

- Soft fruit, and cooking varieties of top fruit, need direct sunlight for around half the day.

- Desert varieties of top fruit need direct sunlight for more than half the day.

Some relatively shade tolerant edibles are listed in the table, and I have written about the light requirements of fruits and perennial vegetables elsewhere.[22]

Placing perennial plants such as fruit in relation to light and shade is straightforward. But annuals must be grown in rotation, so it's not possible to put all the shade tolerant ones in the shady part of the garden permanently. It may be more a matter of dropping some plants from the rotation when their group is in the shady part, or even having two different rotations, one for the sunny part and one for the shady. The latter would be quite difficult to work out, and only worthwhile where shade is a big problem.

The solution most often adopted in practice by gardeners is to put the soft fruit in a relatively shady place and give the annual vegetables the sunniest part.

Wind

Giving shelter to vegetables even from light winds can increase yields by up to 50 per cent.[23] In particular, crops which require insect pollination, such as beans, squashes and fruit, will not do well in an exposed position unless you're lucky enough to get a calm spell at flowering time, as insects can't fly in windy conditions. The general principles of windbreaks have been dealt with in Chapter 4. (See pp83-88.) Here we look at how those principles can be applied in the garden.

The need for a windbreak around an exposed rural garden is obvious. But a garden which is surrounded by buildings may be in just as much need of protection because of the wind tunnel effect, which can be particularly severe where there is a gap between two buildings. An urban garden may be calmer than a rural one when light winds are blowing, but when the wind is stronger it may suffer from occasional gusts which are far stronger and more damaging than the same wind would be in a open situation.

Much the same applies to fruit trees and other plants which are grown against a wall compared to those grown in the open. They are sheltered from winds of all directions as long as the winds are light and steady, but they can be particularly vulnerable to strong or gusty winds.

The first thing to consider is whether it will be sufficient just to place the plants which need most shelter in the most sheltered positions in the garden. If it's not, improving the microclimate with windbreaks, either overall or at critical points, is the next step.

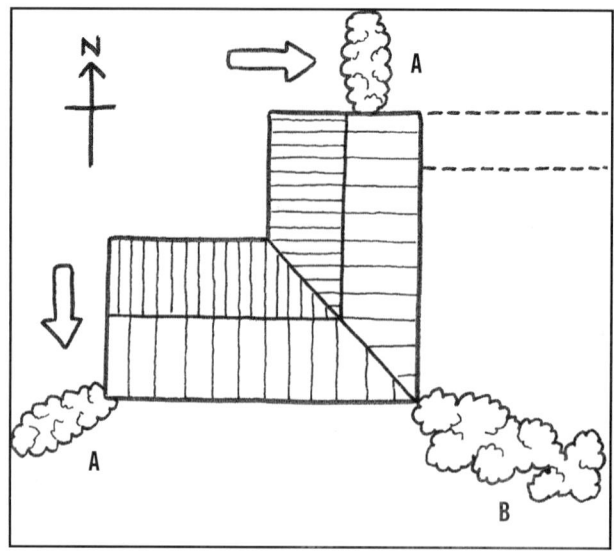

Obvious wind funnels, as at A), should be sealed off. Checking cold northerly winds at B) enhances a southerly aspect.

(See illustration above.) Non-living windbreaks can be useful, either where space is limited or as a temporary shelter while you establish a hedge. Remember that an effective windbreak is a partly permeable one. Trellis can be useful. It gives some shelter when bare and climbers can be grown on it and pruned to give just the right density without taking up much space.

Where there is an existing wall or solid panel fence, adding 30cm or so of open lattice to the top, or growing bushes which will overtop it, can ameliorate the problems of impermeable windbreaks.

A hedge can make a good windbreak but it has two disadvantages: it will cast shade and it will compete with other plants for nutrients and water. These unwanted effects can be minimised by routing a path next to the hedge. However this may not be the best layout from other points of view. For example a path which has a bed on either side of it is twice as useful as one with a bed on one side only. In a small garden you may not want to devote space to a path which doesn't pull its full weight. The priority will vary from garden to garden.

In very small gardens, where the competition from a hedge would effect a relatively large proportion of the growing area a non-living windbreak may be the best option. A trellis with climbing plants grown up it is a compromise, as it takes up less space and most climbers are less competitive than hedge plants.

Root competition from a hedge can be kept to a minimum by avoiding the most competitive hedging plants. These include privet, willow and beech.

Designing a Windbreak. When planning a windbreak, the area of shade which it will cast can be observed at the same time as observing the present shading of the garden, and incorporated into the shade map. (See p185.)

To do this, take a stick the same length as the mature height of the proposed hedge and hold it vertically where the hedge is to be planted. Note where the shadow of the stick falls. Repeat this a couple of times along the line of the hedge and you will have a good idea of how much shade the hedge will cast.

Windbreaks need not be very high. They give good shelter to the area within ten times of their height downwind, so a garden ten metres wide will be well sheltered by hedges just one metre high. If windbreaks are much higher than this the loss due to shading may be greater than the gain due to shelter, though in areas with severe winds heights of around a seventh of the width may give the best trade-off. Of course this applies at ground level. Where fruit trees or other tall crops are the object of the shelter the windbreaks must be correspondingly taller.

Windbreaks for greenhouses, polytunnels and high-value crops which need full Sun can be placed according to the recommendations for houses given

EDIBLE HEDGES

The idea of an edible windbreak or hedge is popular in permaculture, as it embodies the principle of multiple outputs. It means that no land is occupied by plants which are not directly productive, so it can be a valuable feature in a small garden where space is at a premium.

But an edible hedge is not always suitable. Plants in the windbreak will not themselves be sheltered and will yield less than plants which are sheltered. They will also be closer together, with their canopies touching. Since fruit is only produced where the canopy is exposed to the light, this will also restrict the amount of fruit they can produce.

Fruiting trees and shrubs cost much more than ordinary hedging plants, often more than ten times as much, so it would be wasteful to use them in a windbreak where severe winds are expected as they would probably not fruit at all. Fruiting plants are also unlikely to be the best species to resist really severe winds. Don't expect a damson to do the work of a Sitka spruce.

Most fruits are borne on twigs in their second year of growth, so an edible hedge can't be cut every year. This means it may take up more space than a non-edible one would.

Hardy fruits suitable for an edible hedge include:

Damson, especially the variety Farleigh. A smallish tree with dense twiggy growth which gives winter shelter as well as summer.

Cherry plum or myrobalan. A medium-sized tree, its very early blossom brightens up the end of winter; the fruit is sweeter than damsons, but it can't be relied on to fruit every year.

Crab apple. A small tree, useful as a pollinator of domestic apples and a good addition to a mixed jam for its pectin content; wild varieties may be tougher than ornamental ones.

Elders, common, red and American. All species make shrubs or small trees whose flowers and fruits have a multitude of uses; they can inhibit the growth of other plants, particularly brassicas, and occasionally become invasive in a mixed hedge, killing out the other hedging plants; they can be cut back every year without loss of flowers and fruit. The red elder is the hardiest of the three.

Eleagnus spp. A wide variety of evergreen and deciduous trees and shrubs with edible, if not delicious, fruits, some of which are produced in spring when no other fruits are available; they are very wind hardy.[26] (See illustration overleaf.)

Worcesterberry. Like an extra large and extra tough gooseberry; very thorny.

Ramanas rose. A very tough, shrubby rose bearing enormous hips; it suckers, and can be invasive.

Oregon grape. An evergreen with holly-like leaves and yellow flowers in winter; small birds feed from the flowers, and jam can be made from the fruit; very shade tolerant. (See illustration overleaf.)

Hazels. Wild varieties give a more erratic yield of smaller nuts than cultivars, but are cheaper to buy and may be suitable for an edible windbreak; two varieties are needed for pollination; if grey squirrels live nearby they will take all the nuts.

In gardens which are already well sheltered, extra shelter can be given to the vegetables by surrounding fruit trees. These can be espalier or fan trained along wires where space is limited. Cane fruits, such as the hybrid berries, can be used in the same way.

Of the herbaceous plants, Jerusalem artichokes stand out as the best windbreak plant. Since they die down each winter they're not much use as a windbreak for fruit, which needs shelter above all at blossom time, but they're good for summer crops such as tomatoes. They can be invasive and are hard to eradicate once established.

In addition to their other talents, both Eleagnus x ebbingei *(left), and the Oregon grape* (right) *produce edible fruit.*

in Chapter 7. (See p154.) Note that the distances are measured from the windbreak to the edge of the structure or high-value crop, not to the middle. The intervening area can be planted to a lower-value or more shade tolerant crop.[24]

A narrow tapering profile is better for a hedge than a vertical-sided one, as it stops the top of the hedge from shading out the bottom. Pruning of the leading shoots encourages the plants to bush out lower down, and doesn't slow down height growth if it's done lightly.

In addition to providing windbreaks, much can be done by careful design of the internal planting of the garden. Alternating tall and short crops or wind-hardy and sensitive crops, and the placing of occasional shrubs can have a significant effect.[25] But remember that some taller plants, such as runner beans and sweetcorn, are wind-sensitive themselves.

Too much shelter can increase the likelihood of both frost and fungal diseases, so a garden should always be left well-ventilated. Using deciduous trees and shrubs, or a mix of deciduous and evergreens is better from the frost point of view than using evergreens only. Remember that beech, if kept to hedge size, is effectively evergreen because it keeps its brown leaves on over winter. Where winter winds are more of a problem than spring frosts a greater proportion of evergreens is more appropriate.

A hedge consisting of a single line of shrubs makes a better windbreak than a double line, which can get too dense. A spacing of 30cm between plants is suitable for most shrubs and trees, eg hawthorn, hazel, and beech. Bushier or very quick-growing kinds, such as broom, elder and cypresses, should be planted at 50-60cm. Small plants, 40-60cm tall, establish better and grow more quickly than larger ones.

A shrubbery can make a good windbreak as long as it's not so dense as to be impermeable. Mixing shrubs of different heights is a way of ensuring that it will still be permeable once the shrub canopies meet. A few trees can help, but there must not be a gap between the tops of the shrubs and the bottom of the tree canopy as this will funnel winds and increase their speed.

Around larger gardens a single line of trees may be more appropriate than a hedge. A line of trees spaced at 1.5-3m apart according to their vigour, with two or three shrubs between each one is reasonably simple to manage.

Whilst diversity is always a worthwhile aim, it's easier if there's not too much difference in growth habit between the species in a windbreak. More vigorous ones will outgrow the less vigorous and crowd them out. Very vigorous trees like poplar and some willows should be avoided in gardens. Apart from the edible species given in the box, field maple and Italian alder are suitable for milder, more fertile areas, with birch and rowan on more challenging sites. Hawthorn can make a good hedgerow tree in a wide range of conditions.

The intervening shrubs should be shade tolerant, such as the smaller *Eleagnus* species and Oregon grape. Another option for the shrub layer is shade tolerant trees, such as beech, hornbeam or hawthorn, kept trimmed to shrub size. If one of the upper trees dies, one of these can be allowed to grow up and replace it. The disadvantage of this is that beech and hornbeam cast a heavy shade, which could be a problem when they grow bigger. Large species like beech and hornbeam will also need top pruning when they grow big. It's always important to keep the bottom of a windbreak thicker than the top.

Temperature
Shading and wind chill are the main influences on the effective temperature in most gardens, but walls and other hard surfaces usually give rise to the especially hot spots. (See Colour Section, 11-12.)

The heat reflecting and heat-storage effects of patios have already been mentioned. The same effect can be had from any wall, though it will be less intense, and also from a paved path, where it will be less again. Closed-work fences reflect light, though they don't store it as heat. South-facing walls are ideal places for tender fruits and aromatic herbs.

Painting a wall black increases its heat storage capacity, which boosts night-time temperatures and generally keeps the temperature more even. Painting it white increases the amount of solar energy reflected onto the plants growing in front of it, which boosts daytime temperatures. Experiments in Germany with peaches and tomatoes growing in front of both black and white walls found that black walls gave faster growth but white walls gave faster ripening, leading to a higher harvested yield of these crops.[27]

The aspect of a wall, that is the direction in which it faces, and the aspect of the garden as a whole, if it has a slope on it, both affect temperature. This is discussed in Chapter 4. (See box, Aspect, p80.)

Moisture

Good drainage is essential to healthy plant growth and many gardening books contain information on how to drain your garden. The first step, however, is to find out just why a particular patch of soil is wet.

Possible reasons include:

- A leaking underground pipe.

- Runoff from elsewhere adding to the volume of water to be dealt with (see box, A Drainage Problem Solved).

- Drainage impeded by an underground structure, such as a former asphalt path at a lower level (this may cause dry conditions in summer and wet in winter).

- A pan caused by repeated rotavation (see p52).

- A general loss of structure due to poor gardening practice.

- Shading, which may cause one part of the garden to stay wetter than another even where soil conditions are otherwise the same.

- A heavy clay or silt subsoil, especially on a level site.

- A very wet climate.

The remedy will be different in each case. If the cause is a leaking pipe, an underground obstruction or a pan it's a one-off job. If a pan is the problem it may be necessary to double-dig the whole plot. For poor soil structure, adding organic matter and adopting a no-dig, no-tread system of gardening (see pp191-194) is the best remedy. A one-off double digging may be a help to get the ball rolling.

With a really heavy soil or excessive rainfall it may be necessary to install plastic land drains. This is a major job and should only be embarked on if you're sure there's no alternative. In many cases simply laying out the garden in raised beds may give enough improvement in drainage. Improving soil structure can sometimes be enough on its own, but this does take time.

A DRAINAGE PROBLEM SOLVED

When Cindy Engel and her family moved into their smallholding in Suffolk there was a real drainage problem in the zone I area around the house. The soil is a heavy clay, and although the ground slopes away from the house it stood wet for much of the year. This made the effective growing season too short to grow vegetables on that part of the land.

The first thing she did was to put in a couple of raised beds, retained with railway sleepers, where the soil would be unaffected by the poor drainage below. This worked really well for a start, though the area of vegetables was small.

One of the next jobs was to start collecting rainwater from the roofs of the buildings. The house is single-storey and the outbuildings are extensive, so the total area available for water collection was considerable. In fact the buildings take up about half the space in this part of the holding, and this turned out to be the reason for the drainage problem. The soil was having to handle twice the volume of water it would under natural conditions, and being such a heavy soil it couldn't cope. Once the water was flowing into the storage tanks instead of onto the ground the drainage problem disappeared and the garden could expand without the need for specially constructed beds.

It's a classic case of a useful output, in this case rainwater from roofs, becoming a problem when it isn't used.

The raised beds and one of the rainwater tanks: immediate and long term solutions.

Of course the really permacultural approach to a poorly drained soil is not to drain it but to choose plants which actually like poor drainage. There are some edible plants which grow in wet soil or open water. (See Aquaculture on p110.) But virtually all of the vegetables and fruits we normally grow need a well-drained soil. If you have more than enough land and only part of it is poorly drained the best plan may be to grow food on the drier part and reeds or some other bulky plant for mulch on the wet. Even larger wet areas can be planted with willows for firewood, but only the very biggest gardens have space to grow more than a token amount of firewood. (See p304 for areas needed for firewood production.)

The driest soils in a garden are likely to be found at the bases of walls and beneath trees, especially large ones. The roots of a tree commonly spread out one and a half to two times the diameter of the canopy, though the zone of maximum root activity is usually closer to the tree than this.

Mediterranean herbs like a dry summer as well as a hot one, and this is an extra reason for planting them against a south-facing wall. But most plants will need extra mulching and watering at the base of any wall. This can be well worth while if you want to reap the advantages of walls, including the warm microclimate and the vertical growing space they afford. Trees and shrubs are best planted 30-50cm away from a wall and trained back to it so as to avoid the driest soil.

The amount of water transpired by trees depends both on their size and species. Large vigorous kinds, especially willow, poplar and eucalyptus, are very thirsty, whereas fruit trees, especially those on the dwarfing or semi-dwarfing rootstocks which are usually used in gardens, take relatively little water. Trees also cast shade and compete for nutrients, and some trees, such as bay and elder, inhibit the growth of some other plants by chemical means, so the relationship is a complex one. Nor is the tree always the dominant partner in the relationship. Young fruit trees, those on very dwarfing rootstocks or ones which are cordon trained may come off worse in competition with vigorous herbaceous growth.[28]

Ways of using less water for garden irrigation, and the possibilities for using grey water, are discussed in Chapter 5. (See pp95-97.)

Pollution

Many urban gardeners are concerned about the possibility of pollution making their vegetables too toxic to eat. This is a specialist subject and if you have any concerns your first port of call should be your local council, who can give information on where you can have samples tested for specific pollutants. However a few general remarks may be useful.

Produce grown on city rooftops has been shown, by studies in both the USA and Russia, to have one tenth the amount of chemical contamination as produce bought at local markets or grown in suburban gardens.[29]

Personally I have grown vegetables with confidence in a garden separated from a fairly busy road by a thick evergreen hedge some 1.8m high.

The pollutants of most concern are usually the heavy metals, including cadmium from tyre wear on roads, and various ones from industrial sources. Lead from car exhausts has ceased to be a serious pollutant since the introduction of unleaded petrol.

Cadmium could be a problem if a garden is irrigated with water collected from road runoff. Another possible source of heavy metals is where an old garden shed, painted with gloss paint containing lead, has formerly stood on part of a garden or allotment. A significant amount of lead could have leached into the soil, and if the shed was burnt or simply decayed in situ most of it will still be there.

To a certain extent plants have an ability to regulate their uptake of minerals, including heavy metals, from the soil. So at moderate levels of heavy metal pollution it should be safe to eat them, provided they are first washed in a 1% vinegar solution to remove any heavy metals which may be deposited on the surface. Keeping the soil high in organic matter can also help to reduce uptake of heavy metals by increasing the soil's ability to adsorb them. Lead and aluminium are both more soluble at low pH, so keeping the soil neutral to alkaline also helps to reduce toxicity.

However they can be absorbed through plant leaves, so it may be wise to avoid very leafy vegetables such as spinach in areas where airborne heavy metal pollution is likely to occur.

Heavy metals have a more drastic effect on the soil micro-organisms than on plants. Their presence can hinder vital soil processes, such as the breakdown of raw organic matter to humus. This can be a problem in areas which are downwind of heavy industry.[30]

Elevation

In a steep garden the work can be made lighter by working with the slope wherever possible. Elements which need frequent visits are best placed on the same level as each other, and if possible on the same level as the house. This applies especially where things need to be carried. A tool shed is best sited on the middle of the slope, to minimise the uphill climb, either before or after using the tools. The siting of a compost heap in a steep garden has already been discussed. (See p30.)

Protecting Steep Slopes

Steep land is vulnerable to erosion, so it needs to be either kept under perennial vegetation or terraced.

Where a garden is partly steep and partly flat it's usually possible to grow annuals on the flatter areas and perennials on the steeper. But where the whole garden is steep or the flat areas are unsuitable for some other reason, such as shade, you need to consider terracing if you want to grow annuals.

Full-scale terracing is a big job. You have to remove the topsoil and store it on one side, make the terraces, including building retaining walls of some kind, and then return the topsoil. Where the slope is really steep this is the only way to do it, but on moderate slopes you can develop terraces bit by bit.

First mark out raised beds (see p192) across the slope. Place the soil dug out from each path on the uphill side of the path, which is the downhill part of the bed, with a log, a plank or a line of bricks to retain it. Each year as you prepare the beds for sowing or planting rake a little soil downhill from the higher side of the bed to the lower. It's important not to deplete the topsoil on the uphill side, and how much soil you can move each year will depend on the thickness of the topsoil. Plenty of compost and manure should be added to the uphill side of the bed to help the formation of new topsoil there.

Terracing is not always appropriate. In the North of Scotland, for example, where soil warmth is at a premium, some gardens have been deliberately constructed with a south-facing slope to help the soil warm up in spring. In such places it may be better to accept a small annual loss of soil as the price to be paid for a longer growing season.

Frost Pockets

Frost pockets can be small and subtle. They most often occur in relatively flat or low lying gardens and on sloping sites where there is an obstruction to the downhill flow of cold air. How to locate any frost pockets, and what to do about them if one is found, are both covered in Chapter 4. (See pp78-79.)

ORIGINAL PERMACULTURE

As we have seen, permaculture design can be usefully applied to any garden, regardless of the style of gardening. But there are some distinctively permaculture approaches to gardening which go right back to David Holmgren and Bill Mollison's original vision of permaculture, including no-dig gardening, perennial food plants and polyculture. Whether an individual gardener chooses to use one or more of these depends on their situation and personal inclination. But they all deserve consideration where low input, high output and a positive ecological impact are the aims

No-dig

The principles of no-till cultivation are discussed in Chapter 3. (See pp50-58.). Here we look in more detail at how those principles apply in the garden. No-dig gardening is a well established technique. A great fund of information has been built up over the decades by gardeners who have used it, not all of whom by any means would call themselves permaculturists.[31]

On the other hand people wouldn't have been doing it for thousands of years if there was never a need for digging. So what are the specific reasons for digging a garden?

- To relieve compaction.
- To kill weeds.
- To prepare a seedbed.
- To incorporate manures.

Compaction

Sandy soils do not on the whole suffer from compaction problems, thus they are prime candidates for a no-dig approach.

Silty and clay soils may suffer from compaction if they have weak structure. The best way to remedy this is by the addition of organic matter. Mulches applied to the surface will be incorporated into the soil profile by earthworms, but the process is slow in heavy soils and mainly confined to the top few centimetres. Deep-rooted green manures, such as lucerne and grazing rye, will add organic matter throughout the soil profile, but if the structure is poor in the first place their root development may be restricted.

The surest way of adding the required organic matter is to dig in well-rotted organic matter. Compost or manure can be used, but where the soil is already rich in nutrients, which many clay soils are, this could lead to an excess of nutrients by the time enough organic matter has been added. Leaf mould is a low-nutrient alternative. The organic matter must be thoroughly mixed with the soil with a fork. Simply dumping it in the bottom of the digging trench with a spade will do no good. It may be necessary to repeat this for a number of years on a heavy, raw soil.

Where the topsoil is heavy and poorly structured the subsoil will probably be even more so, and double digging is usually worthwhile, at least in the first year. Double digging is a method of loosening the soil to two forks' depth rather than the usual one.[32]

It's important never to dig heavy soils when wet or frosty, as this will do more harm to the structure than good. The best conditions for digging a heavy soil usually occur in early Autumn, just after the dry soil of summer has received its first softening rain. Spring is the ideal time to add nutrient-rich manures so that any soluble nutrients are used by a crop rather than being leached out of the soil by rain. But all too often a raw heavy soil will go straight from goo to concrete as it dries out in the spring.

All this is hard work, but once you've got the soil into good heart you'll never need to dig again – as long as you take care not to undo the good work you have done. The main way in which structure is lost is by compaction, and the main cause of compaction in a garden is treading. Thus an essential part of no-dig gardening is a layout which allows you to tend your plants without ever treading on the soil. This means a bed system. (See box overleaf, The Bed System of Gardening.)

THE BED SYSTEM OF GARDENING[33]

The classic bed system consists of long narrow beds alternating with paths. The topsoil is removed from the paths, where it is not needed, and placed on the beds. On most soils this means that the beds end up being higher than the paths, and hence they are often called raised beds. Sometimes the edges are retained with a plank or a row of bricks, but this is not necessary. In fact it can be a disadvantage because it makes a hiding place for slugs and snails.

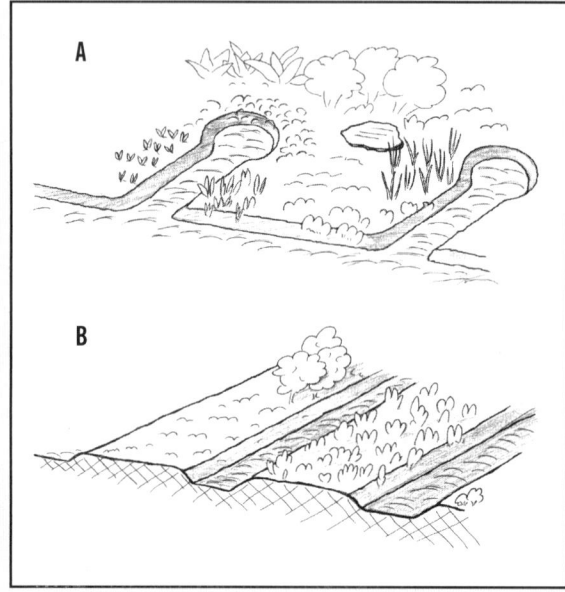

Keyhole raised beds (A) and straight raised beds (B).

The beds should be just the right width for the gardener to be able to reach the centre of each one from the paths on either side without any uncomfortable stretching. Ideally they are tailored to the individual gardener, but the standard width is 1.2m. The paths should be as narrow as possible yet still be comfortable to work from. 50cm is a typical width. Straight beds are not obligatory and the same technique can be used for beds of any shape, including the attractive 'keyhole' shape.

At first sight the bed system seems to devote too high a proportion of the total area to unproductive paths. Yet the overall yield from raised bed gardens, including both beds and paths, is usually a little higher than that from vegetable plots laid out in the traditional rows.

The main reason for this is that the plants on a bed can be grown at equidistant spacing in both directions. Vegetables grown in a garden without beds must be grown in rows far enough apart for the gardeners' feet to pass between them. This means that the vegetables are further apart than their ideal spacing in the between-the-rows dimension, and to make up for it they're closer than their ideal spacing within the rows. Thus, on the one hand they start to compete with each other earlier than need be, and on the other they leave an area of bare, unproductive soil between the rows for longer than need be.

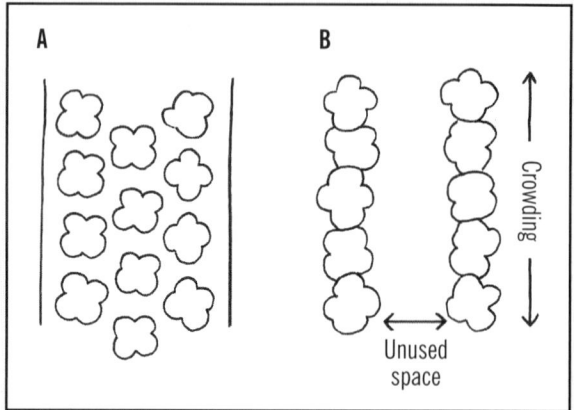

Bed system (A) and rows (B).

Weed control is also easier in a bed, as the plants give complete ground cover earlier in the season.

Another advantage is that rotation planning can be easier. A bed is a permanent and highly visible parcel of land, easy to recognise from year to year and simple to record. The initial layout of a garden can be designed to match the planned rotation, with say four or eight beds of equal size for a four-course rotation.

Beds can be adapted to different conditions. On steep slopes they can become terraces. (See p191.) In northern areas their surface can be tilted towards the south. (See p80.) In dry areas, where the typical raised bed profile may cause the soil to dry out excessively, they can be made flush with the paths. But even so the beds will gradually raise relative to the paths as organic matter is added to them and they develop an open structure, while the paths are compacted by feet. Drying out can be alleviated by giving the beds a concave profile, a design much used in Africa.

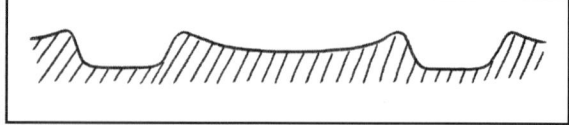

Profile of bed for drier climates.

Although the bed system is essential for no-dig gardening it can also be used in gardens where regular digging is practised. All the other advantages will still apply.

David Holmgren and Su Dennett's intensive raised beds, just outside the kitchen window. This kind of bed gives an even greater depth of enriched topsoil for roots to forage in. The extra height is also useful for people with bad backs. Wheelchair users need them higher again.

I have seen a commercial market garden run on a no-dig raised bed system for a good fifteen years without any loss of yield, and there is no reason why the system should not be continued indefinitely without digging. If some compaction does occur for any reason it can be relieved with a garden fork or broadfork without turning the soil. (See p51.)

Some gardeners have reported declining yields from no-dig crops as the years go by and recommend digging every five years.[34] I suspect this is due to loss of structure because they weren't growing on the bed system. It's unlikely to be due to exhaustion of nutrients in the lower layers of the soil on account of no manure having been dug in. Virtually all the biological activity takes place in the top 5cm of soil, which is constantly recharged with organic matter by earthworms drawing down whatever mulches the gardener applies to the surface. Organic matter is added to the lower layers of soil by the death of plant roots. This can be enhanced by the occasional deep-rooted green manure crop.

It's not always necessary to dig in a lot of organic matter before starting a no-dig garden. If the soil is only slightly compacted and the organic matter content is OK forking without digging may be enough. Where the soil structure is good no preparation is needed at all. It's worth spending some time carefully examining the soil before starting a new garden. Digging in a lot of organic material, and especially double digging, is hard work. It's not something to be done if you don't really need to – nor something to be neglected if you do.

Weeds

You can remove perennial weeds by digging, but it's incredibly slow work, picking every little bit of root and rhizome out of the soil. And you always miss a few, which soon resprout and form new plants. Mulching is a much less laborious way of killing them. (See box, What's A Weed?)

Annual weeds can be killed by digging them in. But it is rather a case of using a sledge hammer to crack a nut, and it can actually cause more of a weed problem than it cures. This is because many weed seeds only germinate when exposed to the light and digging does exactly that. Hoeing instead of digging reduces the amount of fresh seeds exposed to the light, and using a mulch to kill weeds avoids the problem altogether.

In any garden, whether dug or undug, the real key to dealing with weeds is to stop them becoming a problem in the first place. This means clearing perennial weeds before establishing the garden and never letting annuals set seed.

Seedbeds

Crops which are sown in situ rather than transplanted, such as roots, need a good tilth for germination. This can be made with a hoe rather than by digging, as long as the surface soil has good natural crumb structure.

WHAT'S A WEED?

Some permaculturists dislike the word weed. It suggests that certain plants are inherently good and others irredeemably bad.

Yet all of us who garden know what we mean by a weed: some of us have spent many a long hour removing them. In an attempt to soften our relationship with them people sometimes say that a weed is a plant in the wrong place. Myself I prefer to think of them as plants in the wrong quantity.

Having some wild plants in the garden can be beneficial. They cover bare ground, add to the organic matter in the system, act as hosts to useful insects and generally add to biodiversity. Some are directly useful, especially edible ones. I have actually introduced both chickweed and bittercress into my garden, and I always encourage fat hen. All three of them are edible. A problem only arises when there are more of them than we can eat. Then harvesting turns into weeding.

Some crop plants can be just as much of a problem as wild ones. Mints (especially applemint), coltsfoot and wild strawberries can all get out of hand, and I have seen the occasional garden that was more nasturtium than anything else. When they get to the point where there are more of them than we can use, once again harvesting turns into weeding.

Having said that, there are some plants which grow in quantities far greater than we can ever use. I understand there are some medicinal uses for couch grass. We could probably use one part in a million of what will grow if we let it have free rein. In plain English it's a weed.

A high humus content is the key to this. On soils which are prone to capping, which means those with a high silt content, protecting the surface from heavy rain with mulch or growing plants is also important.

It can be difficult to get a tilth on heavy soils in spring, especially if the soil is low in humus. The traditional method is to dig with a spade in autumn, leaving an uneven surface with plenty of edge for the frost to get at the soil. The frost shatters the soil into a fine structure which would be impossible to get by cultivating with fork, hoe and rake.

The alternative is to mulch the soil over winter, sheltering it from the impact of winter rain and allowing biological processes to carry on. When you pull back the mulch in the spring you should find a good natural crumb structure on the surface.

Deep cultivation is not necessary for good plant growth in a well-structured soil. The only exception to this as far as I know is carrots, which usually do much better on a dug soil than an undug one.

Incorporating Manures

This has already been discussed under compaction, above. It should not be necessary to dig in compost or manure once the soil is generally in good heart.

Mulching[35]

A mulch is any material laid on the surface of the soil. Bare soil is as rare in natural ecosystems as disturbed soil, in fact the two are usually one and the same. Keeping the soil covered, either with living plants or with mulch, is a direct imitation of nature. The specific functions of mulch are to:

- Kill weeds by denying them light.

- Conserve water in the soil by reducing evaporation from its surface.

- Protect the soil from capping and erosion.

- Encourage biological activity.

- Often, to add organic matter and plant nutrients to the soil.

There are three kinds of mulch: clearance mulch, grow-through mulch and maintenance mulch. Clearance and grow-through mulches are both used as a once-off measure to bring new ground into cultivation or to clear ground which has become very weedy. A maintenance mulch is used regularly in an established garden.

Clearance Mulch

A clearance mulch is used where there's a really severe weed problem. It covers the entire area, with no gaps, and has the sole function of killing the weeds. When the weeds are gone the mulch is removed and a crop is planted.

Mulch kills weeds by denying them light when they are trying to grow. They can't photosynthesise, so they use up the food reserves in their bodies. Once all the food reserves are gone the plants die.

This means that to kill weeds effectively the mulch has to be down during the growing season. To kill vigorous perennial weeds a whole growing season from early spring to late summer is needed. Some perennials, such as nettles and creeping thistle, need two years. It depends as much on the vigour of the weeds as their species. A one year mulch usually kills all the couch, but if it's very well established some ten percent of it may still be alive in the autumn.

A winter mulch in a mild winter may kill the annuals and less vigorous perennials, leaving you with, hopefully, small patches of perennial weeds to dig out in spring. This at least means that you don't have to dig the whole plot and it means you don't miss a growing season. The mulch is best removed a week or so before digging so that the parts which still need attention have time to reveal themselves by greening up.

One plant which is difficult to kill with a clearance mulch is bindweed. It has enormous reserves of food in its roots, and however large the area which is mulched it always seems to be able to get its shoots to the edge and put out leaves. Mulch can weaken it though, and the roots tend to grow between the soil surface and the mulch, so it's easy to scoop the majority of them up and dispose of them when you take the mulch up at the end of the season.

Anything which excludes light and is resistant to rotting can be used as a clearance mulch. Black plastic sheeting is ideal. It's a fossil fuel product and personally I won't buy it new, but used black plastic can be given another lease of life before it goes to landfill. Old silage clamp covers are ideal, though the thin film which is used to wrap silage bales needs to be put on several sheets thick and is a fiddle to work with. Plastic of any other colour is ineffective as it lets some light through.

Old carpets can be used, though couch grass shoots have been known to pierce them from below. Avoid carpets which have an artificial pile and a natural backing, as the backing can decompose, leaving you with masses of little tufts of artificial fibre in the soil. Rubber-backed carpets are as bad, leaving little bits of rubber in the soil. Large sheets of cardboard are usable but rot down quickly and may need replacing before the job is done.

Unsightly materials, like black plastic, can be covered with an organic material such as straw or bracken. If you want to re-use the black plastic this has a practical function as well as an aesthetic one. The ultra violet light in sunshine shortens the life of plastic and the covering layer will protect it.

Grow-Through Mulch

A grow-through mulch does the same job as a clearance mulch but in this case crop plants are grown through holes in the mulch. It's suitable for places where the weeds are not so severe, either when clearing a new garden or when weeds have got out of hand on an

THE THREE-LAYER GROW-THROUGH MULCH

1. The first two layers. Here a cordon of old carpet tiles has been put round the edge to prevent weeds creeping in from outside.

2. Making the hole.

3. Planting a potato.

4. The top layer goes on. In this case a mix of grass mowings and sweet chestnut leaves.

This kind of mulch combines all the functions of mulch: weed control, water conservation, soil protection and manuring. The method is as follows:

- Make sure the soil is moist before you start. If it's not, water it well. In spring it should have had time to warm up, too, as the mulch will insulate the soil and slow down the warming process.

- Cover the ground with a layer of cardboard or newspaper. Wherever two sheets meet, overlap them by about 20cm. Newspaper should be 15-20 sheets thick.

- Add a layer of manure, seaweed or other material which is heavy enough to hold the first layer in place and contains plant nutrients. Manure need not be very well-rotted but must have heated up enough to kill any weed seeds, ie at least partially rotted.

- At each planting station make a hole in the mulch with a spike, such as an old knife. Scrape the manure away from the spike and replace it with a double handful of topsoil or very well rotted compost.

- Plant the transplant into this. Don't try to plant it in the soil beneath – the roots will find their own way down through the little hole. If you're planting potatoes simply place the seed on top of the soil or compost. You will cover it later.

- Add the top layer of straw or other loose material. This is essential for potatoes but optional for other crops. It looks good and stops the manure layer from drying out. But in wet weather it can harbour slugs and should be withheld until the plants have grown enough to tolerate the occasional slug.

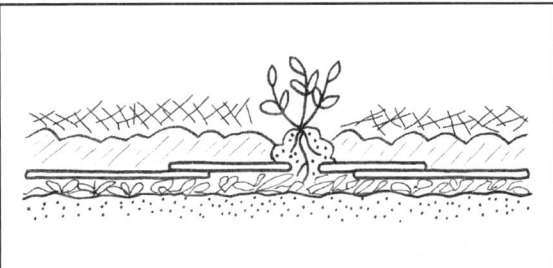

Layers of mulch.

existing plot. It's not suitable for a really heavy infestation because the weeds, especially creeping ones like couch, can grow out of the holes made for the crop plants. The advantage of a grow-through is that you don't have to lose a year's cropping while you clear the ground.

Not all crops are suitable for a grow-through mulch. Roots, which need to be sown into open ground are out of the question. Crops which are grown at a close spacing would reduce the effectiveness of the mulch because there would be too high a proportion of holes. Personally I find plants grown from seed don't do very well: there are usually more slugs in a mulch than on bare soil and the little plants tend to get eaten as they emerge.

Plants which are transplanted and grown at wide spacings, such as brassicas, sweetcorn and the pumpkin family, are ideal. Soft fruit and perennial vegetables can be started off this way, but you need to be sure that weed control will be total because this is your last chance to clear perennial weeds for several years.

Potatoes are a prime candidate for a grow-through mulch. Even where there's no need to kill weeds potatoes need to be covered, and the only way to do it on a no-dig system is with mulch. The seed potatoes are simply laid on the ground at the normal spacing and covered with a thick layer of straw or other organic material. Extra mulch is sometimes needed later on, when dug potatoes would need earthing up.

Where weed control is the main function, a grow-through can be done with the same materials as a clearance mulch with holes punched through it at the planting stations. Another option is the full three-layer grow-through, a permaculture classic. (See box on previous page.)

A biodegradable grow-through mulch suppresses weeds by a combination of the mulch itself and the growth of the crop plants. By the time the mulch has rotted away and any surviving weeds start to poke through, the vegetables will be growing strongly enough to shade them out. This is an important point. If the vegetables are planted long after the mulch is put down, or if they are too widely spaced, the mulch will not control weeds effectively.

Maintenance Mulch

A maintenance mulch can be used in the winter to protect the soil surface, or in summer to reduce evaporation and suppress weeds. In both cases any fibrous organic matter, such as autumn leaves, straw or bracken can be used. These materials are slow to rot, so they last longer than greener, more sappy materials.

In an annual vegetable garden it's best not to have mulch down in the spring, for two reasons. Firstly, it slows the warming of the soil. Most mulch materials are lighter in colour than the soil, so they reflect more heat; the mulch also insulates the soil. Secondly, it can harbour slugs. Slugs are at their peak in the spring in most gardens, and the plants are most vulnerable when they are little, which most of them are in spring.

If a winter mulch has been used it's a good idea to remove it about a week before sowing to let the ground warm up, and not put the summer mulch down till the plants are growing well and the worst of the slugs are over.

In wet years slugs can be a problem right through the year and it's better not to use mulch at all if the weather shows no signs of drying up as summer comes on. The moisture-retaining function of mulch will not be needed in such a summer anyway.

While mulch stops water escaping from the soil it can also stop it getting in. The relatively light rain of summer tends to wet only the mulch and then evaporate before it gets to the soil. So summer mulches should always be applied to a moist soil. If the soil's dry water it thoroughly before mulching. As long as the soil is moist when the mulch goes down the amount of moisture retained by the mulch will be much more than what it keeps out.

A summer mulch of straw is applied to a crop of garlic in late April. In a less sunny spring, or on a more slug-prone crop, this would be too early.

A mulch intended to feed the soil is a very different thing from one which suppresses weeds and keeps in moisture. Compost, manure, seaweed, comfrey or other high-nutrient materials can be used as a mulch to feed the soil. But they are all soft materials and will rot down and be absorbed into the soil relatively quickly. They can be placed under a more fibrous material to get the advantages of both kinds of mulch.

In my experience mulching saves a lot of work. I've grown potatoes by the digging method and I've grown them by the mulching method, and I'd never go back to digging now. I once started a new garden of five beds, four of them mulched and one dug. The ground had couch and creeping buttercup in it, and that one dug bed took more work than all the other four put together, including the time it took me to gather the mulch materials.

Having said that, getting hold of enough materials can be difficult, and it could mean that mulching ends up causing more work than it saves. A good way of making a little mulch go a long way is to slip a layer of newspaper under it. You can reduce the volume of material you need by half.

In the country straw or bracken can usually be had. (Bracken can be carcinogenic at certain times of the year, see Note 5 to Chapter 3, p45.) Beware of hay, even 'spoiled' hay, because it usually has viable grass seeds in it.

In the city, park maintenance people often have leaves and grass mowings which they want to get rid of. If they don't give them away to gardeners they have to pay to have them landfilled. Grass mowings on their own can only be used thinly, otherwise they heat up and go slimy. Leaves on their own tend to blow away. But a mixture of the two avoids both problems.

One way to become self-sufficient in mulch material is to grow your own reedbed using grey water from

the house. (See p97.) This is only possible in a relatively large garden, and much easier where the garden is downhill from the house so the grey water can flow to the reedbed by gravity. Reeds are much slower to decompose than the other materials mentioned here. They're rather stiff, almost like little bamboos, and need chopping before use. This is quite easy to do with a bundle of reeds under one arm and a good old-fashioned sickle in the other hand. (See Colour Section, 24.)

Perennial Vegetables

The most familiar edible perennials are fruits and the luxury vegetables, globe artichokes and asparagus. But there is also a modest range of less well known perennial vegetables, and there are some good reasons for growing these.

The fact that almost all the plants in natural ecosystems are perennials suggests intuitively that a garden of perennials may be more sustainable than one of annuals. But there are also some specific advantages to growing perennials. They are summarized in the mind map below.

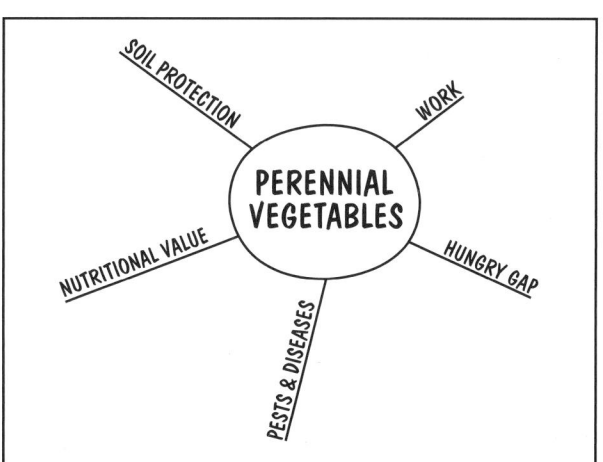

Work. Anyone who has grown both fruit and vegetables will know from experience how much more work there is in growing an annual crop than a perennial one, and this applies as much to perennial vegetables as it does to fruit.

Once they're established there's very little to do apart from a little mulching and picking the produce. There's no soil preparation, sowing, thinning or planting out and very little weeding. Once a closed canopy of perennials is established there's not much opportunity for weeds to grow. By comparison, in an annual vegetable garden there is always that time in spring and early summer when the plants are still small and there's space between them for weeds to get going.

Even those perennials which die down in the winter and regrow from their rootstock in the spring are well able to outcompete weeds. They start the growing season as a mature plants with plenty of food reserves

---

IS PAPER MULCH SAFE?

Some people are concerned about using paper and cardboard mulches because of the heavy metals in printer's inks.

This was a problem in the past but heavy metals have now been eliminated from inks and coloured paper is as safe to use as black and white or unprinted paper. This is dramatically illustrated by the analysis of recycled paper mill sludge. This sludge is the unusable residue left after the de-inking process, so it will have very high concentrations of any nasties which come from ink. Yet it turns out that the content of heavy metals in the sludge is lower than that in most natural soils![36]

Unfortunately that's not the end of the story. Recent experimental work in the USA has found that paper mulches can depress plant growth on some soils. A half-inch (1.3cm) mulch of ground paper pellets, made from 75% newsprint plus small amounts of magazines, caused severe reduction in the growth of a range of herbaceous ornamentals.

The problem appears to be not the inks but the paper itself, which has aluminium added to it during manufacture. This can leach out of the mulch and make phosphorous less available in the soil. The researchers recommend avoiding large amounts of newspaper around shallow-rooted plants growing in acidic soils, that is those with a pH of less than 6.[37]

This should not be a problem for gardens with a neutral to alkaline soil, nor where liming is used to keep the pH up to the recommended 6.5. In any case, the layer of paper granules used by the experimenters was thicker than the layer of flat newspaper which might be used in any of the mulching methods recommended here. But these findings do suggest that any product which has been through an industrial process must to some degree be suspect.

and can put on growth very much faster than annuals, which start off as seeds or little transplants.

The Hungry Gap. This also means that perennials start producing much earlier in the spring than annuals do. Spring is a lean time in an annual vegetable garden. The winter crops are over and the summer crops are not yet ready. But perennial vegetables have a head start. In fact many of them are at their most productive in spring time, just when the annuals are most scarce. So perennials are a good complement to annuals, especially for gardeners who aim for an all-year-round supply and who have a problem closing the hungry gap with annuals alone.

Pests and Diseases. On the whole perennials have not been as highly bred as annuals. Over the years a great deal of plant breeding effort has gone into annual vegetables, aimed largely at increasing yield and improving taste. Both of these make the vegetables more attractive to pests. Larger, faster-growing plants tend to be more soft and fleshy, and pests have a similar taste in vegetables to us. Although pest and disease resistance have also been bred for this has not compensated for the tendency of more highly bred plants to be more susceptible.

The established strength of perennials also helps them to resist pests. A couple of years ago I grew some purple sprouting broccoli in addition to my usual perennials. It was a terrible year for cabbage white butterflies and they feasted both on the broccoli and on the perennial kale. All the broccoli died but the kale, with all the stored energy reserves of a perennial plant, survived. I lost a few weeks' picking, but the plants soon recovered and are still there now, growing as strongly as ever.

The down side of this is that perennials can harbour pests which they don't mind too much but which annual and biennial relatives do. Mealy aphids on perennial kale is one example and leek rust on chives is another. But I haven't found these to be a problem myself.

When they're very young, some perennials can be as susceptible to slugs and snails as annuals are. But they're only young once in a lifetime of several years, whereas getting annuals through the slug-vulnerable stage is a yearly struggle in many gardens. Once established, most perennials are resistant to slugs, and many can self-seed in a relatively sluggy garden. Two exceptions are skirret and mitsuba. The skirret only needs some protection to get it going in spring, but mitsuba is impossible to grow in a very sluggy garden.

Nutritional Value. Wild plants tend to have a much greater content of vitamins, minerals and other life-enhancing substances than highly bred vegetables.[38] Nutritional content has never been bred for and it has been lost by default. To some extent breeding for yield and blandness of taste has actually selected against it. Large size may dilute the nutrients, and some of the stronger tastes are associated with specific nutrients. Vitamin C, for example, is what gives citrus fruits and sorrel their sharp taste.

Perennial plants have more opportunity to form effective mycorrhizas than annuals. Since mycorrhizas have such a central role in regulating mineral uptake, plants which have them well developed are likely to contain minerals in the appropriate balance for their nutrition. Since that is what we have evolved to live on, it's the appropriate balance for us too.

Growing perennials as well as annuals adds to the total repertoire of vegetables we can grow and thus adds diversity to our diet. Even if they didn't have a higher nutritional content than annuals, this alone would tend to improve our nutrition.

Soil Protection. Good organic cultivation of annual vegetables, especially on a no-dig system, does nothing but good to the soil. But on steep ground, which would be vulnerable to erosion if annuals were grown, growing perennials is an alternative to terracing.

The disadvantages of perennials really arise from the fact that they haven't received attention from plant breeders:

- there's a smaller range of perennials than annuals
- the yield is usually lower
- harvesting can take longer

Range. The ancestral plant of all the brassicas is wild cabbage, a mainly perennial plant which grows on cliff tops around our coast. From it have been bred all the annual and biennial brassicas, from brussels sprouts to kohl rabi, and many varieties of each. Yet there are only three perennial varieties of brassica, as far as I know: Daubenton's kale, tree collards and Nine-Star broccoli.

In the long term, if a greater range of varieties is bred, perennials could be a complete replacement for annual vegetables. There are certainly plenty of edible wild plants which could be developed for food. But in the short term it would make for a rather restricted diet, lacking in the tastes we know and love.

There is no real perennial equivalent of carrots, and if you absolutely must have lettuce – well, you absolutely must have lettuce. There are plenty of perennial and self-seeding vegetables which can take the place of lettuce in a salad as the main bulk ingredient with a mild taste, but none of them is exactly like lettuce, especially in texture.

Yield. As we have seen in Chapter 2, there's no reason why perennials should yield less than annuals. (See p15.) But as they haven't been intensively bred, they are in fact usually lower yielding per square metre. There are exceptions to this, such as the perennial brassicas and salad onions. But if space is your limiting factor and you want as much yield as possible you will probably concentrate on annuals. If, on the other hand, your limiting factor is the time you have available for gardening, you will find that perennials far outyield annuals in terms of produce per hour of work.

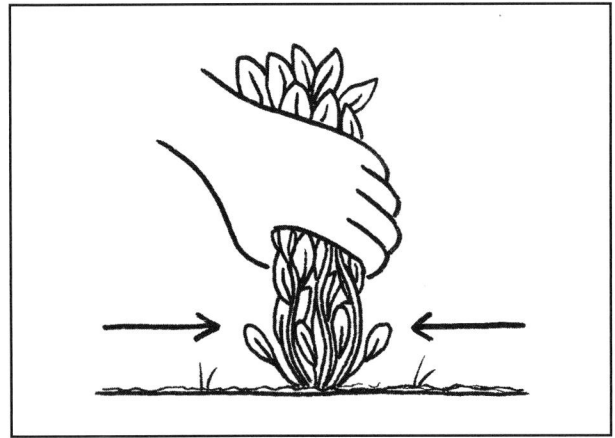

'Coppicing' perennial vegetables can save work and increase yield.

Harvesting. Many of the leafy perennials have smaller leaves than annuals, so picking can be tedious if you pick each leaf individually. It's much easier to harvest them by cutting the whole plant off, or 'coppicing' it, as shown in the illustration. The cut should be high enough to leave small leaves and buds so the plant can regrow.

As well as making harvesting easier and quicker, 'coppicing' helps to keep the plants young and producing leaves rather than flowers and seed. Thus it can both lengthen the picking season and increase the total yield of leaves.

A 'Minimalist' Vegetable Garden[39]

It's quite possible to grow a productive vegetable garden composed entirely of perennials, or a mix of perennial and self-seeding vegetables. Although it may not provide the full range of vegetables that a family will want it can put a salad or a pot of greens on your table any day of the year. Not all of us aim to be self-sufficient in vegetables, and the perennials and self-seeders are as good as any other vegetables for the portion of our needs we do grow.

The name minimalist refers mainly to the amount of work required by a garden of this kind. It's particularly suitable for anyone whose work or lifestyle often takes them away from home. You can leave a minimalist garden to its own devices at any time of the year and its productivity won't suffer from the neglect. Whereas if you left an annual vegetable garden alone for the month of May that would pretty well mean no crop at all for that year.

Alternatively, a minimalist plot can be grown in addition to a conventional vegetable plot. If so, it's best to keep the two separate. Mixing them would obviously make a rotation for the annuals very difficult. What's more the perennials make ideal slug habitat, and though they are on the whole resistant themselves any annuals planted out amongst them are likely to have a hard time. If there are self-seeders in the minimalist plot they can become invasive among the hand-sown annuals – although this need not be seen as a problem. (See box, What's a Weed?, p193.)

The annual plot may deserve first choice of site, both for soil fertility and zoning. It's a high-input/high-output garden so you need to be sure of a good return on the work you put in. Many perennials can stand some deciduous shade, because they do part of their growing before the trees leaf up. But this varies enormously between species, from ramsons which thrives in full shade, to Richard the Lionheart which is a Mediterranean plant and needs all the sunshine it can get.

The soil must be well-structured and in good heart before the garden is planted up, as there won't be a chance to improve it once the perennial plants are in. The same goes for removing perennial weeds. In an annual garden you get an opportunity to deal with them at least once a year, between crops. But once perennial vegetables are established you don't, so it's as well to start with a clean sheet. Having said that, most perennial vegetables are strong competitors, so a combination of nipping off the weeds when they show their heads above ground and the smothering effect of the crop is enough to keep them under control. I have bindweed in my minimalist garden and it really isn't a problem.

The key to weed suppression is having the crop plants at the right spacing. If they're planted too far apart weeds can grow between them. So if anything it's probably better to err on the side of planting them too close together. Typically perennials have a productive life of around five years, and the ground may well have acquired some weed burden by then. It's a good policy to remove the old plants at this stage, clean the ground and replant with other kinds.

PLANTING RATES

This table can help you calculate how many plants you need to get full ground cover with perennials. The recommended spacing for each vegetable is given in the table, Some Plants for a Minimalist Garden.

Spacing, cm	Plants per m²
100	1
60	3
30	11
25	16
20	25
15	44
10	100

This is like practising a rotation with five-year courses rather than annual ones. But rotation is much less important in this kind of garden than in a conventional vegetable plot. Perennials lend themselves to an intricate polyculture, which reduces pest and disease problems and, where roots mingle, can even out the demand for plant nutrients. Larger plants can

SOME PLANTS FOR A MINIMALIST GARDEN[40]						
Name	Habit	Height	Spacing cm	Main Use	Sun/ Shade	Notes
Nine-star broccoli	P	T	100	C	F	Not vigorous after first year
Daubenton's kale	P	M	60	G	F	
Fat hen*	S	M		G	F	
Good King Henry	P	M	30	G	P	Rather bitter taste
Sea beet*	P	M	30	G	F	Tolerates poor sandy soil
Swiss chard	S	M/T	30	G	F	
Tree collards	P	T	100	G	F	
Skirret	P	M	25	R	F	Susceptible to slugs
Alexanders*	S	T	50	S	F	
Sea kale*	P	M	90	S	F	Stems blanched
Chickweed*	S	L		SM	P	
Lamb's lettuce	S	L	10-15	SM	F/P	Winter salad
Mitsuba	P	L	15	SM	P	Susceptible to slugs
Musk mallow*	P	M	30	SM	F	
Pink purslane	S/P	L	15	SM	D	Prefers acid soil
Richard the Lionheart	P	L	20	SM	F	Prefers warm climate
Salad burnet*	P	L/M	20	SM	F	Prefers alkaline soil
Winter purslane	S	L	15	SM	P	Tolerates poor sandy soil
Bittercress*	S	L	10	ST	P/D	Tolerates wet soil
Chicory	P	Various	Various	ST	F	
Garlic cress	P	M	30	ST	P	
Land cress	S	L/M	15	ST	P	Prefers rich soil
Nasturtium	S	L	Various	ST	F/P	Tolerates poor soil
Ramsons*	P	L/M	10	ST	D	Does poorly in full light
Rocket	S	L/M	15	ST	P	Spring and autumn plant
Sorrel	P	L/T	25	ST	P	
Turkish Rocket	P	M/T	50	ST	P	
Watercress	P	L	15	ST	P	Open water is not essential
Everlasting onions	P	L	15	ST	P	
Tree onions	P	M	20	W	P	
Welsh onions	P	M	20	W	P	

Habit:　　　P = perennial　　　S = self-seeder

Height (approx):　L = up to 30cm　　M = 30-60cm　　T = over 60cm

Main Use:　　　C = curds　　　G = greens　　　R = roots　　　S = stems
　　　　　　　　SM = salad, mild　ST = salad, tasty　W = whole plant

Sun/Shade:　　F= full Sun　　　P = part shade　　D = deep shade

* Indicates a wild plant – seed available from wildflower specialists.

be planted individually in a polyculture, but smaller ones are best planted in drifts. Otherwise things get too fiddly.

If self-seeders are included in the garden the amount of work required is a little more than if only perennials are grown. They must have some bare ground to seed into, and where there is bare ground there can be annual weeds. Successful self-seeders often come up very thickly. In this case they can be treated as a cut-and-come-again crop (see p173), or they can be thinned to the recommended spacing and grown on to full size. Simply leaving them to grow in a crowded state will result in a poor crop.

Some perennials will also self seed, but those which do seem to vary from garden to garden and from year to year. Some perennials, such as chicory, make better eating in their first year than in subsequent years. These may be best grown as self-seeders, with one or two mother plants left to grow perennially and produce a new generation of young plants each year.

It's easier to establish perennials by transplanting than by sowing seed in the open ground. Though they are resistant to slugs when they're established, many kinds are susceptible when young. Self-seeders may also be easier to establish initially by planting than by sowing. They often seem to be more able to resist slugs when they self-seed than when hand sown.

There's an important place for herbs in a minimalist garden. As well as using them for cooking, I add them liberally to salads. One of the joys of minimalist gardening for me is experimenting with new salads. (See Colour Section, 23.) Even so there's a limit to the amount of herbs a family can use, and it's easy to allow herb plants to get over-large so they take up space which could be used more productively by other plants.

Herbs are also visually attractive, and there's no reason why some purely ornamental plants should not be included too.

Polyculture

The principles of diversity have been discussed in Chapter 2. (See pp19-21.) The main benefits of diversity in the garden are:

- a more diverse diet
- better plant health
- higher yield

The more diverse diet comes simply from growing a wider range of plants in the garden, which will also tend to improve plant health. But the biggest gain in plant health can be had from including plants which attract beneficial insects into the garden. The higher yield comes with polyculture in its fullest sense: carefully designed mixtures of plants, growing intimately together in the same soil at the same time. This is where a permaculture garden may differ from a standard organic garden,

where the emphasis on diversity is much more focused on a rotation of monocrops.

Plant Health

There are many insects whose larval stages feed on garden pests, such as caterpillars and aphids. The adult stages live on nectar and pollen from flowers, but as they only have small mouth parts they need small flowers to feed on. Plants with very small flowers include the daisy family, or composites, and the cow parsley family, or umbellifers.

More information on these insects and plants is given in Chapter 9. (See pp242-244.)[41] The plants discussed there are just as useful in a vegetable garden as in an orchard.

Many attractant plants are garden herbs, such as lovage, sweet cicely, tansy and chamomile, so they give a multiple output: food, pest control, and in many cases beauty too. Since a single herb plant usually produces more edible or medicinal produce than a family can use, their pest control function can be the thing which justifies their inclusion in a small garden where space is limited.

Productivity

A well-designed polyculture can outyield a monoculture only if each plant occupies a slightly different niche. For example they may have different rooting depths, nutrient requirements, height, shape, season of growth, speed of growth and so on. Thus they use different parts of the resources available for growth, including water, nutrients, light and time.

To put it another way, they are avoiding competition with each other. In some cases they can positively benefit each other as well. For example, legumes can benefit neighbouring plants by fixing nitrogen and taller plants can give shade to saladings in the heat of summer.

A simple example of sharing the resource of time is interplanting cabbage with lettuce. Cabbage is relatively slow growing compared to lettuce. Even when it's planted at equidistant spacing it takes quite some time before the plants have grown to cover all the space available. Lettuce, which is faster growing, can be planted between the cabbages to make use of this time and space which would otherwise be empty – or occupied by weeds.

Another example is sowing radishes and parsnips together in the same row. Parsnips take as much as three weeks to germinate, and if weeds come up in this time it's difficult to hoe them out for fear of killing a parsnip which is just about to emerge. The quick-germinating radishes show where the parsnips are, allowing you to hoe with confidence. The radishes arc harvested long before they compete for space with the slow-growing parsnips.

Mixing tall and short vegetables in a similar way is a form of stacking, and is covered in the next section.

The design of complex garden polycultures has been pioneered by Ianto Evans of the Aprovecho Institute in Oregon, USA, where the climate is not too dissimilar from that in northwest Europe.[42] An example of one of his designs is given in the box, A Nine-Plant Polyculture, below.

A complex mixture like that described in the box will keep the soil covered all year. This helps to ensure maximum use of space, protects the soil, reduces leaching of nutrients and keeps weeds down. It's not the easiest way to garden. Constant attention is needed to see that the population and vigour of all the different plants are kept in balance, so that one does not out-compete the other. But this extra care is well repaid by a very high output per square metre.

Every polyculture must be individually designed to fit both local conditions and the mix of produce required by the gardener. The example in the box should not be slavishly followed. But Ianto Evans gives some useful tips for polyculture design:

- Mix plant families, not just species.

- Sow several varieties of each species.

- Include many individuals of fast-producing, shallow-rooted vegetables like radish and buckwheat to provide cover rapidly. Spread the others along a calendar line to give overlapping harvests. There should never be a time when you're not harvesting, or growth will slow down.

- Don't mix your seed. Broadcast each kind separately, otherwise they'll separate out according to weight as you cast them, and come up in patches. Covering them with a thin layer of compost is better than raking in.

A NINE-PLANT POLYCULTURE

In early spring, broadcast a mixture of radish, pot marigolds, dill, parsnip and a selection of lettuce varieties. The radishes will grow fast, and help the germination of the other plants by shading the soil and keeping it moist. Harvest them as soon as they are ready and plant a selection of cabbages in the gaps left by them.

Start picking lettuce when the plants are still small, perhaps at six weeks from sowing. With a good selection of varieties you should have lettuce all summer from the one sowing. As the soil warms up plant French beans in the gaps left by the first lettuces.

All the other crops can be harvested as they come ready, with parsnips and late varieties of cabbage extending into the winter. As gaps appear in the autumn you can fill them with overwintering broad beans or garlic, or let them be filled by self-seeders.

- Don't overseed. If you have ten kinds use ten per cent of the normal seed rate for each. But increase this slightly for the later-maturing kinds, which will encounter more competition during growth, weakening some of them.

- Start harvesting quickly; don't wait for plants to mature.

- Pull out whole plants, taking care not to disturb root vegetables. Pull from the densest areas and choose thinnings to release slow developers.

- Take time every day to review your polycultures. Thin, harvest, observe. You'll probably need to pull plants every day to make space for new growth.

To this I would add: keep thinning even if you don't need the produce. Plants must be given space to grow, and crowding will lead to a lower yield in the end. In areas with high rainfall or low light conditions, too thick a growth may become rampant and leggy and more susceptible to fungal diseases.

Any polyculture which is working properly will be much more productive than a monoculture. Although this increased yield is partly accounted for by a more complete use of the resources present, it will still demand a high level of inputs. The soil must be well supplied with organic matter and nutrients, and regular watering is essential in dry weather.

Diversity can be overdone. If you want a reliable supply of food, it's usually best to stick to a moderate number of well tried crops and varieties rather than go overboard with many relatively unknown kinds. An experimental fringe around a core of old reliables is often a good compromise.

Indiscriminate diversity is not necessarily a good thing. Some plants have negative interactions. For example most of the aromatic herbs inhibit the germination of seeds. On the whole perennial herbs are best grown on their own, perhaps as a decorative border, or in a specially warm and sunny part of the garden.

Animals in the Garden
The garden polyculture need not be restricted to plants. Animals are part of the garden ecosystem whether we introduce them or not. Worms cultivate the soil. Birds forage in the garden, sometimes on pests and sometimes on our fruit and vegetables, and leave their nutrient-rich droppings in exchange. But if we introduce animals of our choice we can enrich the web of beneficial relationships in the garden.

The kinds of animals which may be useful in a garden are discussed in other parts of the book. They include:

- frogs for slug control (see p110)
- ducks for slug control and eggs (see p111)
- chickens for eggs, pest control and cultivation (see pp248-249)

- geese for weed control and eggs (see p247)
- bees for honey and pollination (see pp244-246)

In addition to these yields the poultry will help to make productive use of outputs which would otherwise be wasted, such as food scraps from the house.

Stacking

Stacking is a matter of working in three dimensions rather than two. In the garden it can take one of two forms:

- Combining tall and short plants in the same place to make the maximum use of the space available.

- Using vertical spaces, such as walls, fences, balconies and trellis, for growing food.

Stacking in the first sense is a form of polyculture. With vegetables it works best when a tall thin plant is combined with a short bushy one. Neither of these can cast much shade on the other, so competition for light is minimised.

An example is sweetcorn and the pumpkin family – squashes, courgettes and marrows. (See Colour Section, 1.) It takes a little more skill than growing each crop on its own, especially in ensuring that the squash crop doesn't out-compete the sweetcorn. This can usually be done by not planting out the squash till the corn is 50cm high. The combined crop will outyield the two monocultures and it needs a proportionately heavy dressing of manure or compost. In addition the weed-smothering effect of the squash saves the need to weed the sweetcorn crop.

Two other combinations which can work well are garlic with lettuce and leeks with celeriac. (See illustration, p266.)

On a somewhat larger scale, fruit trees can be combined with vegetables. Fan and espalier trees, trained in north-south rows (see p220), can be grown in a vegetable garden with little or no loss of vegetable yield. But where open grown trees are planted they may eventually cast too much shade, and then the vegetables must gradually give way to other more shade tolerant crops. This is the principle of succession in action. A possible sequence is described in Chapter 2. (See p24.)

Using vertical spaces to grow food is most valuable where horizontal space is lacking. Urban areas are usually short of horizontal space but they abound in vertical space. A good rule of thumb is: Wherever you feel you're short of space, look up!

As well as offering space, many walls offer favourable microclimates, so tender exotics like peaches and passionfruit can often be grown. At the same time there are plenty of shade-tolerant plants to make use of walls with less favourable aspects. (See p185.) High places, such as balconies and flat roofs often avoid shade problems which may limit growing in crowded urban areas.

Espalier trees in the vegetable garden at the Camphill community, Oaklands Park, in Gloucestershire.

Using window ledges, balconies and flat roofs for food growing has already been discussed. (See pp173-174.) Growing fruit on walls and fences is discussed in Chapter 9. (See pp231-233.)

Vegetables can also be grown on the vertical space offered by a wall, though the range to choose from is somewhat limited. Climbing vegetables include runner beans, climbing French beans, tall varieties of pea, some varieties of squash and marrow, climbing cucumbers, and nasturtiums – usually thought of as an ornamental, but also a delicious salad vegetable.

Tomatoes, although not climbers, can be grown to a considerable height against a vertical surface and they benefit greatly from the warmth of a south-facing wall. Sweetcorn is another tall, heat-loving plant, but it's not ideal for growing up against a wall, as pollination is much better when it's grown in a block rather than a row. Sunflowers are another possibility.

Some thought must be given to crop rotation. It's difficult to design a really satisfactory rotation with such a small number of species, most of which belong to only two families. So it may be necessary to grow shorter vegetables in the space in some years and forgo the advantages of stacking in favour of plant health.

The disadvantage of growing plants against walls is the dryness of the soil. (See p190.) This is particularly a problem with runner beans, which can fail altogether if the soil is not kept constantly moist. But any vegetable or fruit grown by a wall will need extra organic matter in the soil plus extra watering and mulch. Herbs and ornamentals are not usually so demanding.

Growing potatoes in towers made from old car tyres is another way of using vertical space. This method is intensive in both work and materials and is not worth it unless space is the main limiting factor. Many people would say that potatoes have no place in a very small garden anyway, as they are a low value, bulk vegetable. But if you are already growing all your other vegetables and have a little more space to spare it is worthwhile. Yields can be in the range of seven to ten kilos per tower, depending on the variety of potato. (See the box, Potato Towers, on next page.)

POTATO TOWERS, THE METHOD[43]

Use maincrop varieties – earlies will only produce tubers in the bottom tyre.

Sow as early as possible – mid February in southern England – using chitted (pre-sprouted) seed.

Allow three or four tyres for each tower, preferably of the same diameter. Cut away the walls with a craft knife.

Use compost, not soil. It needs a rough, open texture to allow for drainage but must be well enough composted not to heat up in the tyres. An example is equal parts of coarse garden compost, one year old horse manure and fresh straw.

Choose a sunny site. If there is soil there and you want to use the same site year after year, place a plastic bag under each tower to keep the roots out of the soil.

Put the first tyre of each tower in place. Put a 5cm layer of compost in it. Place three seed tubers on it. Cover with compost to the top of the tyre.

When the shoots show, put on the second tyre and cover it with a sheet of perspex.

When the shoots reach the top of the second tyre gently half-fill it with compost.

Keep adding and filling tyres till the end of July. During dry spells water at least once a week. Feed with urine, diluted 20:1, up to Midsummer's Day, then with comfrey liquid at 10:1.

Harvest when the haulm has died down in September or October. The used compost makes a good ingredient for potting compost.

Jerusalem artichokes, sweetcorn or sunflowers can be grown in the spaces between the tyres.

Forest Garden

Putting all these principles of Original Permaculture together – no-dig, perennials, mulching, polyculture and stacking – gives you a forest garden.

There can be little doubt that the most sustainable way to grow food in any part of the world is the way which most closely imitates the natural vegetation of the area. Here in Britain the natural vegetation is woodland, and a forest garden is closely modelled on a natural woodland. Fruit and nut trees, soft fruit and perennial and self-seeding vegetables take the place of native trees, shrubs and woodland herbs.

This kind of garden has long been grown in many tropical parts of the world, but here in Europe it was pioneered by Robert Hart of Shropshire. He created a forest garden beside his house on Wenlock Edge which has been an inspiration to many thousands of people.[44]

Because so many niches are filled the productivity of a forest garden is potentially very high, but the garden must be carefully designed to get the best out of it. The most important thing is to be sure that enough light reaches the lower layers.

The nature of the plants themselves goes some way towards ensuring this. Many of the self-seeding vegetables are winter salads, which do much of their growing when the trees and shrubs are bare of leaves. The perennial vegetables start growing early in the spring, the shrubs leaf up later and the trees last of all. Thus each layer of plants gets a crack at the sunshine. The winter twigs of the tree layer can actually enhance the growth of the lower layers in a cold winter, as they reduce radiation from the ground and thus keep it warmer.

Nevertheless a forest garden designer needs to make sure that enough light penetrates during the rest of the year. One way of doing this is to plant the trees slightly further apart than they would be in a normal orchard, another is to give the garden plenty of edge. Three concepts for creating edge in gardens of different shapes and orientations are shown in the illustrations on the facing page.

Another advantage of having plenty of edge is that it gives good access for picking. With so many different plants of different sizes growing together access can be difficult if it's not carefully planned from the start.

Succession in a forest garden. Standard fruit trees have been planted in this forest garden, but as yet they are small and take up little space. Redcurrants have been planted in a block between the fruit trees and covered with a cage. By the time the trees have grown enough to cast deep shade the redcurrants will have come to the end of their productive lives.

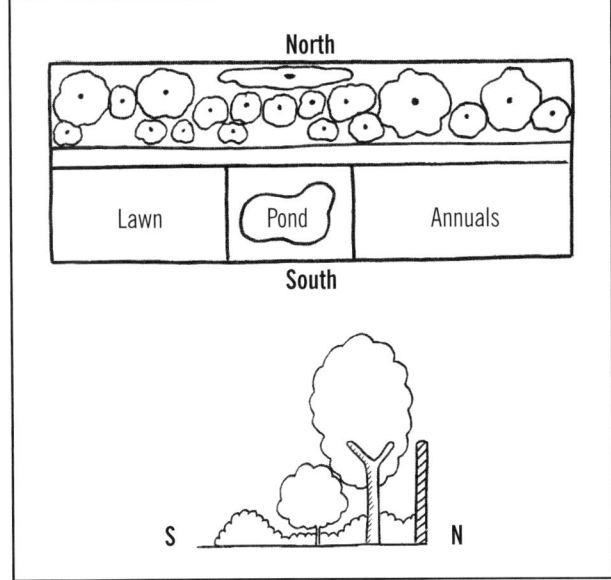

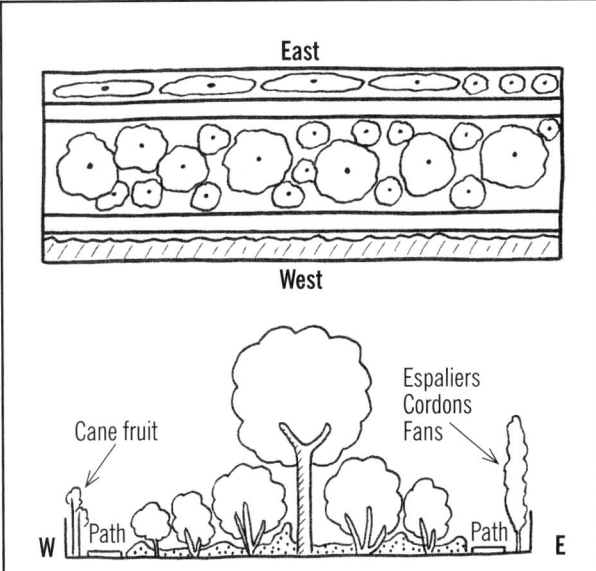

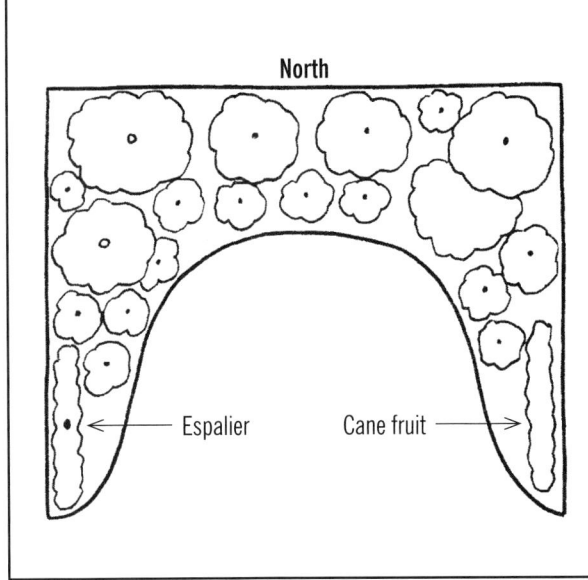

Three ways of getting edge in a forest garden.

As this is clearly a no-dig garden it's important not to step on the growing area, so a combination of edge, paths and possibly stepping stones is needed to make everything accessible.

There is a succession of yields in a forest garden. In the early years most produce will come from the lower layers, but as shade increases and the trees come into full bearing the balance will shift more to the upper layers.

As all the plants are perennials, a forest garden requires very little maintenance. Apart from a little pruning and mulching, the main job is to pick the produce. The key to keeping the maintenance requirement low is the design of the ground layer. This is very much like the minimalist vegetable garden described earlier in this chapter. If the perennial vegetables are planted at the right distances they will form a complete ground cover and no weeds will get a look in. If they are too far apart weeds will grow and the garden can become an endless source of labour.

Most people don't need as big an area of perennial vegetables as they do of fruit. Even those who do want the same area may not be able to get hold of enough perennial vegetables to plant the whole area up at once. The response to this is often to dot the vegetables around at very wide spacings, which is a recipe for disaster. A much better plan is to plant up part of the area with the vegetables and put the rest down to a simple ground cover.[45]

Is a Forest Garden For Me?
The forest garden has become an icon of permaculture. Many people feel attracted to the idea of growing one, but it's not the ideal solution to every person's needs. The main points to consider are summarised in the mind map below.

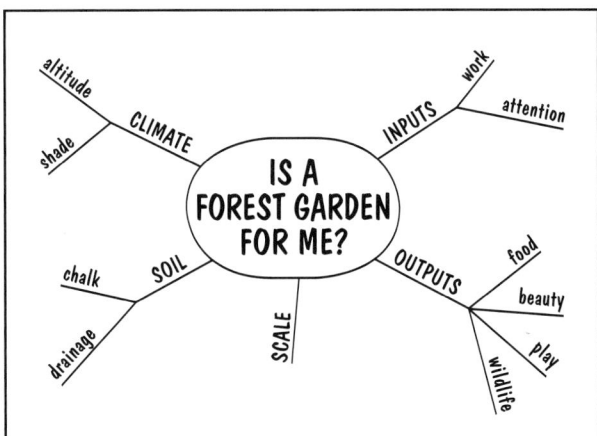

Inputs. If one of your main requirements is that your garden should need little maintenance, then a forest garden may be just the thing. Once established, it will be a lot less work than almost anything else occupying the same area, including a lawn.

On the other hand it does need regular attention, especially for picking the produce, much of which comes in little bits rather than one big harvest. The presence of people also discourages birds, which may take the soft fruit it it's not netted, and occasional weeding or mulching is needed to keep the ground layer in order.

Forest gardens are sometimes suggested as a community project. This may work well where there are people who will be in and out of the garden as part of their daily round. But where the work input is in the form of occasional work days attended by a group of enthusiastic volunteers, a community orchard is a better choice.

Outputs. The main outputs are fruit of all kinds and perennial vegetables. It can be possible to grow some annual vegetables between the trees when they're still young, but not in the mature garden. If annual vegetables are a priority a forest garden is not a good idea, at least not over the whole area you have available.

The non-food benefits of a forest garden are compatible with food production to varying degrees. People who want their garden to look really attractive can still plant a forest garden, but they may include some purely ornamental plants, or varieties of edible plants which are more pretty than productive, in which case the output of food is likely to be less. Children can have a drastic impact on the yield of a forest garden, but it can be an ideal place for a den or a bit of tree climbing. As a wildlife garden the forest garden is hard to beat. Native plants can be included, many of them edible (see p367), and whatever species are used the structure of the garden provides a rich habitat for a range of creatures.

Scale. A forest garden is intensive. As a polyculture it will potentially outyield any monoculture consisting of only one of its three layers: top fruit, soft fruit or perennial vegetables. But it is an intricate garden and requires a certain amount of purposeful pottering to give of its best. It works well on a domestic scale. People growing for sale, even on a relatively small scale, may find that splitting the three layers up into orchard, fruit cage and vegetable garden makes the work much simpler.

Soil. Generally speaking any reasonable garden soil is fine for a forest garden but there are two exceptions.

If there's one thing fruit trees and shrubs can't stand it's poor drainage. Vegetables can tolerate it much better, especially if grown on raised beds. Drainage can be improved, but the best approach to a piece of land which is too wet for fruit is to consider growing another kind of garden.

Most fruit does poorly on thin soils over chalk and limestone. The range of fruits which can be grown on such a soil is very limited. (See p228-229.) A forest garden is not impossible but once again another kind of garden may be a better solution.

Climate. As the upper layers of the garden shade the lower layers it's as well to start with a sunny site, or the shade will just be too much. On a shady site it may be wise to reduce the garden to two layers only, shrubs and vegetables, as soft fruit is more shade tolerant than top fruit.

There are some hardy fruits which can be grown in almost any climatic conditions, but at altitudes of over 150m it gets increasingly difficult. Again, soft fruit is more likely to succeed here than top fruit.

GOLDEN RULES

There is no single recipe for gardening success. In fact the mass of gardening advice on offer these days can be quite bewildering. But I find there are four simple rules which can act as signposts towards a garden which works, that is a garden which produces what we want from it while being a pleasure to work in. I recommend these to anyone who is starting out in gardening – and to any experienced gardener who feels things aren't going as well as they might.

They are:

- Start small.
- Start in the easiest place.
- Find your own style.
- Be kind to yourself.

Start Small

If, like me, you've spent time trying to keep up with a garden just a bit bigger than you could really manage you'll know the value of this one. I was always rushing about and never quite able to do anything to the standard I like. This took some of the enjoyment out of my gardening and left me feeling dissatisfied with what I'd done. I'm sure I would actually have produced more food if I'd been working a smaller area.

Starting small is probably the most important principle of all for a person starting gardening for the first time. Robert Kourik reckons nine square metres, that is three by three, is the *largest* size vegetable plot for a new gardener.[46] It looks too small when you pace it out. But when you get down there and start sowing and planting, thinning and weeding, it grows and grows. When you come to harvest the fresh produce it looks bigger again! The smaller the plot the easier it is to do it really well, and you usually grow more food per square metre than you would on a bigger plot. Nothing succeeds like success. In future years you can expand your area secure in the knowledge that you can make a good job of growing vegetables.

The area which is available for the plot to expand into in future years needs to be considered. A perennial green manure, such as white clover or lucerne, has much to commend it. It will only need mowing two or three times a year and it will be building up fertility all the time, even if you use the mowings to manure the area you already have in vegetables.

Gardening on this scale can be very productive but it does take a fair bit of work.

Experienced gardeners are not immune to the temptation to bite off more than they can chew. Often it comes from a reluctance to let go of land you, or some other gardener, have already invested a lot of work in. Trying to keep up with it is bad enough for one growing season but if it goes on for longer the store of weed seeds in the soil builds up. The old saying goes 'One year's seeds make seven years' weeds', and with each year that the weeds are able to set seed before they get pulled out the problem multiplies.

The worst burden of annual weeds I've ever seen was in the garden of a community where I was doing some teaching one time. They had four large plots which they ran on a rotation, but the total area was too big and they never quite got round it all. The weeds came up like a close-weave carpet, and the most difficult thing about weeding was being able to see the crop at all. The time spent on weeding was crippling, making it even more difficult to get round all the work. It was a vicious spiral.

When asked to suggest a solution I could only recommend that they took one plot out of cultivation and put it down to a perennial green manure. The weed burden would gradually decrease, as seeds lost viability and those which germinated were cut with the green manure. More labour would be released to get on top of the other plots, and the green manure plot could come back into cultivation sometime in the future, either if they felt they could take on four plots again, or to swap it with another.

They did not look happy at this suggestion, and I don't know whether they ever took it up. But one advantage of the plan I suggested is that you don't 'let go of' any land. It may make your rotation a bit complicated for a while, but the land doesn't deteriorate while you're not growing food on it. It improves.

Start In The Easiest Place

Usually the easiest place to start gardening is as near as possible to your own back door. This is the Kitchen Window Rule again, which states that the most productive ground in any garden is that which can be seen from the kitchen window. (See p170.)

It's also a good idea to start on the most fertile soil. I mean fertility in its widest sense, embracing not only the state of the soil but the microclimate, the weed burden and any other factor which makes one place easier to grow crops in than another. If there is one part of the garden which is markedly more fertile than the rest, this is the place to start. Once again, nothing succeeds like success and nothing is better for a new gardener than having a bountiful first season.

If this part of the garden is not visible from the kitchen window you will have to balance the pros and cons. Most of the factors which effect the fertility of a site can be improved. The exception is shade, which may be impossible to correct, and a very shady spot will never be good for food growing. As for the other factors: drainage can be installed, windbreaks erected, weeds removed, organic matter added to the soil and so on. But many of these things take time and during that time yields will be low. The best plan may be to start growing food in the fertile area while working on improving the bit by the back door.

It's a matter of weighing up the pros and cons. In a domestic garden I would say the value of having the productive garden near the house usually outweighs all but the worst deficiencies of fertility.

On a smallholding, where you probably plan to spend much of your time out on the land anyway, the zoning principle may carry less weight. If there's one patch of soil which is particularly fertile it would be foolish to ignore it through rigid adherence to the Kitchen Window Rule. Similarly, in a dry region the position of water sources may be the most important factor in siting the vegetable plot.

A vegetable garden on a smallholding becomes a centre of attention in its own right, especially if gardening is a major economic activity rather than a domestic sideline. The position of the garden may then even influence the placing of a building rather than vice versa. This is where the concept of network takes over from simple zoning. (See p28.)

When taking over an allotment which has been allowed to run down, the easiest place to start is usually where there are least weeds. This is often the bit which was abandoned last. A good plan is to start by cultivating this area intensively and put the rest down to mulch. A clearance mulch can be used on the most weedy part and maybe a grow-through mulch on an area which isn't so bad.

In fact this plan may be a good one to follow on a new allotment even if it isn't weedy, on the principle of starting small. A perennial green manure can be used instead of the clearance mulch on a non-weedy allotment, or once the clearance mulch has done its work on a weedy one.

Find Your Own Style

Finding your own style of gardening is as much a part of the permaculture design process as placing different physical things within the garden.

There are a host of different systems of gardening advocated by various experts and writers, often with the implication that theirs is *the* way to do it. They often present their ideas with such enthusiasm that it's hard not to get swept up with it and say "Yes! I must do this myself." Much the same can happen if you've learned all your gardening from one gardener.

But what works for one person doesn't necessarily work for another. We're all individuals with different temperaments, different gardens, lifestyles, families and so on. So how do you find out what's the right style for you?

One way of making sense of all the different styles is to see them as a spectrum, from the most intensive to the most laid back.

At one end of the scale is the bio-intensive method, as set out by John Jeavons in his classic book *How to Grow More Vegetables*.[47] It's a high-input/high-output method. There's little doubt that you can get far higher yields per square metre by this method than by any other. But it is very demanding. The double digging which is needed in the first few years is hard work; you need to be a pretty skilled gardener to start with; you have to give your garden constant attention throughout the growing season; and you have to be prepared to follow Jeavons' prescriptions closely.

If you have a strong back, a disciplined character and don't go away from home too much in the growing season this system is well worth considering – especially if space is your limiting factor and you want to grow as much as possible. Even if you're not limited by space you may find it the most efficient way to grow food. A small intensive plot of annual vegetables often yields more food per hour of work than a larger, more extensive one.

At the other end of the spectrum is the minimalist style. (See pp199-201.) If you're often away from home, not up to a lot of physical work, a touch slapdash by nature, and not concerned to get the highest yield or the widest range of produce possible, this may be for you.

Between these two extremes there are many other styles, or combinations of styles. Indeed some people may be happy with both an intensive and a minimalist plot in their garden. Trial and error inevitably plays a part in finding your own style. But some blind alleys can be avoided by sitting down and making a list of your requirements, lifestyle and personal characteristics, and those of your garden, before you start.

Be Kind To Yourself

We all have crop failures, even the most experienced of us. Sometimes the cause is wholly natural, like the weather or a pest attack, sometimes it's our own fault, but more often it's a combination of the two. Some years it seems as though more has gone wrong than right. Don't blame yourself. Don't be discouraged. I assure you it's perfectly normal. But unfortunately it's something that never gets illustrated in the gardening books.

Everything we do produce in our gardens is a cause for celebration. You put a seed in the soil, it grows into a plant, and you eat the produce of the plant. The fact that we've been doing this for ten thousand years easily blinds us to the miraculous nature of that chain of events. If you've grown something and eaten it, congratulate yourself. You deserve it.

Colin Leftly

What it's all about: John Walker with some of the vegetables from his no-dig mulched allotment.

CASE STUDY

A SMALL PRODUCTIVE GARDEN

photos by Michael Guerra

Growing some of our own food at home not only fulfills many of the principles of permaculture. It's also personally empowering because it's the most positive and active thing we can do to reduce our ecological impact. Many of us have difficulties standing in our way: too small a garden, too much shade, not enough time, no previous experience, or even a garden with no soil. Julia and Michael Guerra had all these problems.

They live in a small ground-floor flat in a commuter town north of London and grow much of their food in the equally small garden. When they moved in, in 1991, you would have thought it was an unpromising place to practice permaculture.

The housing estate was built on the site of a demolished factory and the 'soil' was a compacted mix of clay and builder's rubble. In the back garden this was paved over with patio slabs, while on the lawns at the side and front it was covered with a bare 15cm of infertile topsoil. Shade is also a problem. While the front garden is well lit, and the side reasonably so, the back garden lies to the north of the house and is shady for much of the year. A third limitation is space, with a total area of only 75m². Many people would have dismissed it out of hand as being too small for food production on anything but a nominal scale.

The social aspect was not promising either. Lawns and bedding plants are the gardening norm on the estate, and washing the car the main outdoor activity on weekends. In the early days their upstairs neighbours had to be dissuaded from taking a lawnmower to a patch of herbs, wildflowers and potatoes, which they considered to be unsightly and out of keeping with the neighbourhood.

Not surprisingly, Mike and Julia started their food growing in the secluded back garden. Mike had just been on a permaculture design course, which opened his mind to the possibilities of their little garden for food production. They immediately set to making raised beds by turning some of the paving slabs on their ends and filling the space behind them with soil.

This was mainly topsoil from a woodland which was to be clear-felled and built over, leavened with a mixture of equal parts of well-rotted manure, mushroom compost, sharp sand and coir compost. Normally they wouldn't consider importing topsoil, as they don't believe in impoverishing one part of the Earth in order to enrich another, but the wood was doomed anyway and they could make excellent use of the soil.

Neither of them was an experienced gardener. Julia's mother had been a keen flower gardener and some of her lore had rubbed off on Julia, while Mike had never sowed a seed in his life before. Nevertheless they grew plenty of food that first year. But they soon realised they could have grown even more, because far too much of the space in the garden was taken up by paths. So during the following winter they designed a layout which has no more than 15% path but with every piece of ground reachable without treading on the soil. They used railway sleepers to retain the beds, and it only took them a single weekend to change the garden to the new layout.

The amount of shade in the back garden varies from place to place, some parts hardly getting any sunlight all year and others getting a reasonable amount during the growing season. This meant that much of the design work consisted of placing things according to their light requirement.

The very darkest spot is occupied by a 1,000 litre water tank, which collects rainwater from the roof. The next darkest spot contains the compost bins and the area immediately adjacent to them has a selection of shade-tolerant edibles: a rhubarb plant, mints and a morello cherry fan-trained against the fence. At the spot which receives the most sunlight, the south-facing fence at the north end of the garden, they placed a little lean-to greenhouse. This was raised up on a plinth so as much sunlight as possible reaches the lower shelves. Sun-loving aromatic herbs were planted against the south-facing wall of the house, facing the street – the only productive plants they grew in the front garden at that stage.

On top of the compost bins they constructed a hot box, a growing space which captures the heat given off by the composting process. This technique was used by Victorian gardeners to grow exotic fruits, and the Guerras use it to bring on seedlings in the cooler times of the year and to grow winter salads. The roof of the hot box is perspex, to let in light, and the walls are lined on the inside with aluminium foil to reflect the light around inside the box and stop the plants from becoming too leggy.[48]

The main body of the garden is occupied by the raised vegetable beds, with fruit trees and soft fruit round the edges. "Apples have very little footprint," says Michael, "so in a small space you think about volume rather than area." They grow potatoes in tyre towers, also placed on the edge of the garden. (See p204.) This gives them a much higher yield of potatoes than they could get on the flat in the same area, but the main reason for growing their own potatoes is not bulk but freshness. They grow five towers, each with a different variety, and look forward with relish to the first meal of each, so fresh that you can put the water on to boil before you pick the potatoes!

The vegetable beds are run on a normal organic rotation, with a much smaller area set aside for perennial salad plants. The different crops are planted in as intimate a mixture as is practicable so as to get the full benefits of polyculture. Weeds don't get a look in: in such an intensive garden the ground is almost always covered with crop plants and any bare areas are small and temporary. Thanks to the raised bed design no digging is necessary.

They were a little apprehensive about extending cultivation to the area beside the house, which is in full view from the road. Pollution is not a problem, this being a quiet cul-de-sac. But how would the neighbours react to an edible jungle advancing round the side of the house? In the event they really liked it. Mike and Julia had been there long enough by now that people knew them and trusted them. No-one felt threatened by runner beans and squash spilling out into a previously sterile area.

The final step was the front lawn. In addition to the herbs they had planted by the front wall, they had harvested crops of berries from the existing rowan tree to make rowan and apple jelly. Now they converted the area to a small forest garden, including two additional fruit trees. So as not to cast too much shade on the house and herb bed these are dwarf trees, and planted as far from the house as space allows.

Despite the emphasis on growing as much food as possible there are some ornamental and wildlife plants in the garden, and even a tiny pond made out of a sunken plant pot with the drainage holes filled. Climbers, including a clematis and a fragrant rose, give maximum ornamental value in the minimum space.

The wildlife plants add indirectly to the yield of food by attracting pest predators. A small buddleia is credited with ridding the garden of more than one plague of greenfly.

With the garden in full swing, they grow virtually all their fruit and vegetables. Before they had their first child their weekly food bill was a mere £3 a week during summer, and averaged only £15-20 a week through the year after he was born. The gross yield of food averages out at 200kg per year, a fifth of a tonne. This is equivalent to 36 tonnes per hectare.

The inputs needed to produce this yield are modest: around two hours' work a week, £30 a year for seeds, and a few barrow-loads of manure from the riding stables ten minures walk away. The latter is a specially useful resource, as they don't own a car.

"Nothing's easier than a small garden," says Mike. "There's no pulling on of wellies. We garden in our stockinged feet." Sometimes it can almost be too easy. Julia has known times when she came home from work looking forward to unwinding with a bit of gardening, only to find that there was nothing that needed doing.

Now they have children the question of play space arises. "Although an intensive edible garden is no place for football," says Mike, "the maze of paths and food provides a fascinating environment for play, hiding and eating. Children soon lose their fear of fresh food, and soft fruit rarely makes it onto the table."

The only thing which does need modifying because of the children is the little greenhouse at the end of the garden. The three-year-old has started breaking the panes, so it has been given away and will be replaced with a Wendy house with an edible turf roof and multiwall polycarbonate walls. This will be used for summer play and over-wintering plants. In any case the greenhouse had proved too hot for summer crops, and heat-loving plants such as chillis do better in the hot box.

The biggest influence of the children is to temporarily take Julia, who does most of the gardening, away from the garden. With three under-fives her hands are

Julia prepares one of the new mulch beds at the side of the house. Note the general character of the neighbourhood.

The same bed a few months later.

full and the garden is running on minimal maintenance mode. Production is down to fruit, herbaceous
perennials and a few easily grown annual vegetables.
These are grown at wider spacings than usual to
reduce the need for manuring. Total output this year
is likely to be 50kg, a quarter of the usual yield.
When all the children are at school or playgroup for
at least one morning a week the garden will return
to full productivity.

*The back garden. Making full use of the vertical
dimension is one of the keys to high production in
small spaces.*

CASE STUDY

COMMUNITY GARDENING
IN THE CITY

Permaculture emphasises change from the bottom up. It's more about doing things for ourselves than about telling the powers that be what they ought to be doing on our behalf. There's nowhere where this approach is more important than in urban food production: there's little point in telling the authorities about the value of allotments when so many of them lie unused.

Taking on an allotment on your own can be a daunting prospect for someone with no skills or experience, and this is where community action can be very valuable. But community action doesn't spring from nowhere. It's usually the result of dedication and hard work on the part of a few individuals, as is the case in Sheffield.

The Sheffield Organic Food Initiative (SOFI) is the creation of a few individual people who are personally committed to growing food and greening their community. Now that the initiative is up and running official bodies have started to co-operate with it. The City Council and Health Authority both make use of the expertise it can offer and help to make grant money available. But if the people had waited for officialdom to give a helping hand to get started they would still be waiting now.

Richard Clare is one of those committed individuals. In 1988 he realised that growing food could help to solve many problems, ecological, social and personal. He had no experience of growing and there were no relevant courses on offer in the area, so he taught himself, partly from books and partly by learning from his own mistakes. Like most people in central Sheffield he had no garden or nearby allotment site, so he took an allotment on the outskirts, a two mile walk away.

Bit by bit he developed both his gardening skills and his allotment, starting with a few easy-to-grow vegetables. Right from the start he recognised the central importance of organic matter, and with his friend Darrell Maryon started collecting unwanted organic materials and composting them. At the time Darrell worked in a wholefood shop, and the unsold vegetables were an obvious resource to start with. Soon they were collecting from six local shops – and saving the shopkeepers a lot of money in disposal charges. Strawy horse manure from a local stables helped to balance the sloppy vegetable waste, and they started making leaf mould from the huge bulk of autumn leaves produced in the city.

At this point Richard compiled a report outlining the potential for composting in Sheffield, which he submitted to the council's recycling officer, but it was not acted upon.

As their skills and confidence increased Richard and Darrell found that a single allotment was not big enough for everything they wanted to grow, and they started taking on more plots. They have become adepts at urban soil improvement and have learned how to grow dozens of varieties of fruits and nuts and more than a hundred vegetable maincrops. "I treated this exercise as an experiment or research project," says Richard. Over the years he developed a set of gardening techniques which are specially relevant to urban allotments, and kept careful records of his work.

They started to encourage their friends and neighbours to grow some of their own food. People who hadn't the time to take on a whole plot could help on Richard and Darrell's established ones in return for produce, while those who were able to take a plot received help and advice to get going.

The permaculture element in their work is evident in the different design approaches taken on different allotments, according to the needs of the plot holders and the characteristics of the individual plot. Examples include: a fruit-only allotment for a busy wage-earner with little time to spare; incorporating a play area for a family with children; wide concrete paths for wheelchair users; terracing for steep slopes; and double digging plus plenty of compost for an abused soil.

Three Allotmenteers celebrate abundance with a snack of super-fresh organic strawberries.

Richard and Darrell are keen to explain everything they've learnt to volunteers, especially beginners. "There's a desperate need for basic, practical advice," says Richard. To help fill this gap he has written some brief, accessible guides to some of the key techniques, such as composting.

The number of people actively involved in the project has grown steadily, to around a dozen people actively and regularly growing food with several dozen non-active supporters. SOFI now operates on more

Richard shows visitors round Unstone Gardens on an open day.

than twenty sites in and around Sheffield, cultivating four acres of intensive food crops. It also has links with other groups doing similar work in the area.

One of these is Beanies' Wholefood Co-op, an indirect organic vegetable box scheme. (See p291.) The co-op is an important supplier of vegetable waste to SOFI's composting operation. They would like to sell locally grown vegetables rather than the ones they buy in from the other side of the country at present, and SOFI is building up the skill base among local people which can make this possible.

Another is the Ponderosa Environment Group (PEG), a community group dedicated to improving a six hectare urban open space. SOFI members have been at the core of the group from the start, and the work has included establishing a community orchard and planting thousands of native trees. PEG has been invited by the City Council to take part in a major refurbishment programme involving over 2,500 council homes. They have also been commissioned by Hallam Univeristy to undertake an environmental assessment of north-west inner-city Sheffield.

The Allotmenteers is a scheme operated by SOFI to introduce people to the basic methods of organic crop production. They meet once a week to work on renovating derelict allotments and maintaining those already in production. In return for their work the volunteers get instruction, camaraderie and a share in the produce. "Working with a wide range of people has given us realistic expectations about the level of skill and understanding you can expect today," Richard reflects. "Few people have the commitment to grow food in the face of so many other priorities. Anyone who manages to grow some of their own food organically deserves to be celebrated."

In 1995 SOFI members initiated a project to restore a large antique kitchen garden and orchard at Unstone Grange, five miles south of Sheffield. The gardens are owned by a charitable trust, which operates a conference centre on the site, and are let to SOFI in exchange for maintenance and a share of the produce. In a few years the gardens have been transformed from their moribund state and now form a large part of SOFI's total growing area. Peaches, apricots and other tender fruits have been planted along a south-facing wall and a huge polytunnel has been installed.

Over the past three years Unstone Gardens have welcomed thousands of visitors to their organic open days and hosted the annual conferences of various national gardening organisations. It has recently been certified organic by the Soil Association.

SOFI now runs the kind of courses which Richard found so woefully lacking in his early days. It also publishes documents on composting and urban growing and acts as a consultant to other community groups which want to develop these activities. As it has grown in size, expertise and local renown it has increasingly been invited to participate in officially-sponsored bodies. An example is the Sheffield Healthy Gardening Group, which seeks to encourage gardening for its combined benefits of nutrition, exercise, community activity and emotional wellbeing.

SOFI's links are not confined to voluntary and official bodies. One private business which has become intimately involved with the project is the Ecology Company, a shop specialising in books and magazines relevant to permaculture and organics and a range of green products. It acts as a drop-off point for compostable materials and a distribution point for free seaweed fertiliser, made available through PEG and the Healthy Gardening Group to help budding organic gardeners. The shop acts as a focus for many people involved in organics and other green activities.

SOFI is now registered as a charity. With official recognition has come some funding, at first on a very small scale but now growing. This includes help with costs of printing, putting on courses and paying key people for the work they have done for so long for free. Until recently SOFI was run on a completely voluntary basis by people who are unemployed or full-time carers. Now, after a decade of unpaid work, Richard is paid a part-time salary by the council as Sustainable Food Development Worker. Darrell, in addition to his teaching work, is employed part-time to manage Heeley City Farm's food-growing project.

The lack of official support in its early years is the proof of SOFI's strength, and its relevance. It has succeeded simply because a small group of people wanted it to and had the commitment and perseverence to make it happen. Although it has expanded into the fields of education, consultancy and municipal activities, it remains firmly rooted in the soil. The central activity is still the actual growing of food in the city.

A young harvester.

1. An example of the permaculture principle of stacking, this two-layer crop of sweet corn and pumpkins makes more complete use of the available space and light than either could alone, and thus it can yield more than a monocrop of either. (See pp22-24.)

2. An example of the Key Planning Tools in action. (See pp27-30.) This orchard is sheltered from the prevailing south-westerly winds by the rising ground to the south west, from which the picture is taken – a sectoring consideration. It's also in a key area, above the frost-prone low-lying land in the distance – an elevation consideration. The two ponds either side of the orchard are also in the key area.

Howard Johns

3. Although making the best use of existing microclimates is the priority in permaculture, creating a totally new one can be worthwhile. A polytunnel can transform the cropping potential of a domestic or market garden by extending the season and allowing a wider range of crops to be grown, as here in Pippa James' small market garden. (See pp74-75.)

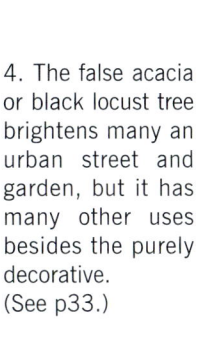

4. The false acacia or black locust tree brightens many an urban street and garden, but it has many other uses besides the purely decorative. (See p33.)

5 and 6. Welsh onions, grown on the sunny and shady sides of the same garden, show that a shady microclimate can be an advantage if you want to grow succulent leaves for salad. On the sunny side (*left*) the onion has already flowered and now has nothing but dry, fibrous leaves to offer. In the cool, moist conditions of the shady side (*right*) the onions are producing an abundance of juicy leaves. Photos taken on the same day.

7. This reedbed sewage system at Oaklands Park in Gloucestershire treats the output from a community of 90 people, with vertical and horizontal beds and a pond. (See p103.)

8. Timber can be a highly ecological building material, and a creative one, as in the London Wildlife Trust headquarters building, designed by Architype architects. (See pp161-162.)

9. The vertical flow beds in the Oaklands Park system, one planted with common reed the other with yellow flag iris. The flowers make an attractive border but are not essential.

10. This massive fireplace stores heat and helps to keep the room warm after the fire has gone out, though the room may take a little longer to heat up when the stove is first lit.

11 and 12. Over-wintering broad beans were planted by a south-facing wall (*left*) and in the open garden (*below*). Their growth the following spring, and their subsequent yield, shows the value of the warmer microclimate by the wall. The hand fork gives the scale.

13. Richard Fish makes a modest living from his herb garden, at the same time it is a delight to both eye and soul.

14. A border of flowers round the vegetables can set them off beautifully.

15. Pot marigolds and musk mallow both have edible leaves and flowers.

16 and 17. Lollo rosso (*left*) and salad bowl (*right*) lettuces can give an accent of colour to the garden. As they are pick and pluck varieties they will stay in place all summer long. (See p173.)

18. Container growing. These nasturtiums, all self-sown and all of the same variety, illustrate the difference between growing in containers and the open soil. Although the ones in the pot are regularly watered they don't have the same access to either moisture or nutrients as the ones in the ground, which have received no attention at all, and the difference is plain to see. (See pp175-176.)

Alice Naish

19. A community garden can give children contact with nature right outside their own back door. (See p184.)

Beryl Smith

20. Making a community garden in Bromley-by-Bow, London: Steve Pickup shows local residents how to make a structure out of green willow.

Steve Pickup

21. A year and a half later the willow structure (*being built in photo 20*) is in full leaf, and the intervening space planted up with edible and medicinal herbs.

Howard Johns

22. Allotments have great value in urban areas, both socially and ecologically. Saving them from building development is an urgent priority. (See p144.)

23. A salad from my 'minimalist' garden. A mix of green leaves is topped off with leaves of red orach and flowers of land cress. (See p199.)

24. Chopped reeds make an effective mulch which is slow to rot down. (See p197.)

25. In autumn both the leaves and the berries of guelder rose give a striking display of colour. (See p182.)

26. Native shrubs can be as beautiful as exotic ornamentals, but are much more beneficial to wildlife and if well chosen need less care. Spindle is exquisite in autumn.

27. Guelder rose berries give a splash of red in summer to this planting of native shrubs, but the contrast in leaf colour, shape and texture gives an equal, if subtler, delight to the eye.

28. An abundance of produce on farmer Phil Collins' stall at the farmers' market in Bath. Direct sales from producer to consumer have great benefits to both parties, and to the Earth. (See pp288-289.)

Carlo Chinca

29. This walnut tree is well placed in the middle of the shopping centre in Street, Somerset. It's a survival of a former orchard on the site, but there's every reason to plant more fruit and nut trees in places like this, where the present fashion is for purely ornamental trees.

Horticultural Research International, East Malling

30. Plant diversity is a real alternative to chemical use. This pear orchard has been underplanted with flowers to attract pest predators and pollinators. (See pp242-244.)

31. These damson trees grow happily on a woodland edge, and regularly fruit well on the side exposed to the Sun. Ragmans Lane Farm in Gloucestershire. (See p234.)

32. My traditional hay meadow is a polyculture of herbaceous perennials. This kind of grassland is the closest model for the 'domestic prairie' concept of grain growing to be found in this country. (See p272.)

33. This woodland, at Oaklands Park in Gloucestershire, shows the multi-layer structure typical of group selection forestry. (See p319.) It is certified by the Forest Stewardship Council. (See p134.)

34. This timber-seasoning shed is a key element on Neil and Maggie Sutherland's croft in Wester Ross. (See box, Local Timber for Local Houses, p302.)

35. A Norwegian wheel sledge. Extracting timber with horses rather than heavy machinery can save the woodland soil from compaction.

36. A traditional coppice-with-standards woodland, coppiced the previous winter. For optimum coppice growth it would have been better to have left fewer standard trees.

37. Stuart Norgrove's Grandfather planted these variegated hollies for the Christmas trade in the hedgerows of their small farm in Worcestershire. This kind of imaginative use of space is a great aid to making a living on a small acreage.

38. Gardens can provide immense enjoyment to people of all ages.

39. Creative park keeping: the formal style in the foreground rapidly grades to a large naturalistic area in the background, which is both good for wildlife and easy on the park keeper's budget. Abbey Park, Glastonbury, Somerset. (See p352.)

40. Orchids survive on this slope because it's too steep to drive a tractor on but is still regularly grazed. (See p358.)

41. Part of Ashley Vale open space in Bristol was saved from building development by a determined local campaign. Local school children made this pictorial map of the site, placed at the main access point.

42. With a little benign neglect, cemeteries can develop into a uniquely urban ecosystem with a mix of native and exotic species. Hampstead Cemetery, London. (See p352.)

Chapter 9

FRUIT, NUTS & POULTRY

This chapter is intended both for home gardeners and for people who want to make all or part of their living from tree crops or poultry. Much of the information applies equally to both, and where it applies to one or the other this is made plain in the text.

Nuts are not widely grown in Britain at present, and sustainable fruit growing, whether we call it organic or permaculture, is only just starting. All the information in this chapter is carefully researched and represents the state of the art at the time of writing, but the art is not highly developed. Anyone who sets out to grow fruit and nuts sustainably on a commercial scale in this country must consider themselves a pioneer, taking part in something which is at least in part experimental.

Nuts and fruit are dealt with separately. The big difference between the two is that nuts have the potential to become a staple food, whereas fruit, though a valuable part of our diet, can never be a source of bulk carbohydrate and protein. Nevertheless the two are grown in a similar way, and may well be grown together in the same orchard. Thus some of the information given here under Nuts is relevant to fruit growing, and vice versa.

Poultry are discussed in terms of their versatility as part of an integrated system, and we also look at chicken forage systems, which may provide an alternative to feeding them grain.

Tree crops are the ultimate permaculture food. As long-lived perennials they protect the soil from erosion and allow the natural soil ecosystem to flourish. Once they are established they require little work or energy input apart from harvesting and perhaps pruning. As woody plants they lock up carbon dioxide which would otherwise contribute to the greenhouse effect. They also lend themselves to stacking with shorter crops, and have the potential of yielding a multiple output – wood and timber as well as food. If sensitively grown they can have the same positive effects on the water cycle as woodland does.

In general there is no doubt that where woodland is the natural vegetation, as it is here in northwest Europe, the most sustainable way to grow food is on trees. Potentially it's also the most productive. As we saw in Chapter 2, the biological productivity of the land tends to increase with increasing tree cover. (See p22-23.) If we can increase the proportion of that biomass which is edible to humans then trees can outyield annual cereals.

Increasing this proportion (known as the harvest index) has been one of the main aims of cereal breeding over the ages. The success of generations of farmers, and more recently professional plant breeders, in doing

this is one of the main reasons why cereals yield as much as they do. At first sight it may seem difficult to increase the harvest index of trees in the same way because they have all that wood to make while cereals only have straw. But trees only add a small increment to their wood each year and this is similar to the straw which a cereal crop produces each year. Annuals are not inherently more productive than perennials, they've just received more attention from plant breeders. This is also discussed in Chapter 2. (See p15.)

In the long term there's no theoretical reason why tree crops should not replace annual field crops as our main food source. That would revolutionise not only our agriculture but our landscape too. Gone would be the open fields of cereals and grass, replaced by orchards or edible woodlands, with perhaps as many clearings as we now have woods. Our landscape would cease to be an imitation of the prairie or steppe, and become an imitation of what it was before we turned it upside down: woodland.

That's very much a vision for the long term, and it may or may not turn out to be the ideal solution. Far be it from me to try to predict the future. But there can be little doubt that the further we go down the route towards perennial crops, especially tree crops, the more sustainable our agriculture will become.

NUTS

As well as holding the long-term prospect of a sustainable food supply, nuts are a valuable crop in the short term. Virtually all the nuts consumed in Britain are presently imported, and there is certainly potential for expanding home-grown production with the present market base. Discerning consumers are becoming more interested in eating locally produced food, and more varieties of nut trees suitable for our climate are becoming available.

One of the very few walnut orchards in Britain, at Boxted in Essex. Note the form of the trees, with short trunks and wide-spreading branches. (See p220.)

Whether nuts are a practical proposition as a staple food in this country has to be an open question at this stage, though I am personally convinced that they will prove to be so. Much depends on answering these questions:

- How much food can they yield?
- What's the quality of that food?
- How easily can we grow them?
- How easily can we breed improved varieties?

Yields. The three kinds of nut with the most immediate potential for growing in Britain are hazels (including cobs and filberts), the common walnut, and the sweet chestnut.[1] Yields from mature trees of the best modern varieties may be in the region of:

	Tonnes/ha	Years from planting to full yield
Walnut	7.5	20
Chestnut	5	10-12
Hazels	3	15

The yield of walnuts is comparable with that of conventionally grown cereals, and that of chestnuts with organically grown cereals. Whether the low yield of hazels is due to the intrinsically lower yield of a smaller tree or because they have received less interest from plant breeders is unclear. Either way there is no

doubt that yields of all three can be raised considerably if the amount of plant breeding effort which has gone into annual crops is put into nuts.

Because the trees take some years to work up to their full yield, the average yield over the lifetime of the trees is less than the figures given above. But field crops can be grown between the young trees as they grow, so the total output of food can be similar. In fact since polycultures usually yield more than monocultures it's likely to be greater. (See p19-20.)

Nut yields vary widely from year to year according to the weather. Nevertheless nuts have a yield advantage over cereals in that cereals, if grown organically, must be grown on rotation. This means that each piece of ground will only produce a cereal yield, say, two years out of four. The main produce in the other years consists of meat and milk, which give us far fewer calories per hectare than either cereals or nuts. (See pp119-120.) In the long term nuts must be grown on rotation too. You can't follow walnuts with walnuts. But you can follow them with another kind of nut and thus get a continuous yield of nuts through time.

In addition to the yield of nuts there is always a yield of wood. In practice it may be difficult to get both nuts and high quality timber from the same tree, but even if it's only firewood the increment of wood each year on a hectare of nut trees is a useful yield.

Food quality. Chestnuts are comparable to cereals in the levels of carbohydrate and protein they contain. Hazels and walnuts are both very high in oil, which means that weight for weight they have a higher calorific value than cereals. Walnuts have a particularly high protein content: 18% in the common walnut and 30% in the black walnut.

Although nuts are considered a luxury food by many, adopting them as a staple would involve some cultural change. In particular they lack gluten, the protein in wheat which makes bread rise. Life without bread would be different, though probably more healthy.

Ease of growing. There can be no doubt that the cradle-to-grave energy costs of growing nuts is less than that of growing annual cereals. Much of the work which is needed, including establishment, pruning and harvesting, is hand work. The low price we pay for grains compared to nuts is closely linked to the low price we pay for fossil fuel.

The main work involved is harvesting. This takes more work than harvesting field crops, both because trees are more difficult to get at, and because the nuts ripen over a period, so harvesting must be done repeatedly rather than all at once.

Harvesting can be done in one of three ways: by hand, by hand assisted by nets, or by machine. Net harvesting is unsuitable for hazels as the trees are too low. There are two methods of mechanical harvesting: shaking the tree with a machine and collecting the falling nuts in a net, or shaking the tree and picking the nuts up off the

We may think of chestnuts as a treat for Christmas, but there's no reason why they shouldn't become a staple.

ground with another machine. Efficient machinery for these tasks has been developed and is in regular use in countries where nuts are grown commercially.

With currently available machinery mechanical harvesting is difficult or impossible either on a steep slope or where the trees are underplanted with another crop. This is unfortunate. Tree crops are particularly suited to sloping land as they protect it from erosion, and growing field crops between the trees in their early years is essential if the full yield potential of the land is to be realised. No doubt machinery can be developed which overcomes both of these limitations.

The British climate is not ideal for nuts. Hazels are native all over the country, and chestnuts are naturalised in the south. Nevertheless, for maximum yield all three kinds really prefer a more continental climate, with warmer summers, and a more marked contrast between summer and winter. Yields here tend to be erratic, with big variations from year to year according to the weather. A total crop failure is always possible if the weather is bad at pollination time. In fact chestnuts do not normally yield every year in Britain away from south-east England.

In the long term there's no doubt that more suitable varieties of nuts can be bred. After all, wheat and barley originally come from south-west Asia, and both are now successfully grown in Scotland. New nut varieties are being bred at present, but mostly in North America, where the continental climate is very different from ours. Nevertheless many of the new varieties may turn out to be suited to our conditions, and there is plenty of scope for small-scale trials of these varieties by anyone with a patch of land available.

How our climate will change in response to the greenhouse effect is an unknown quantity. (See p71.) If it gets warmer it could mean that trees planted today will enjoy a more favourable climate when they mature. If it cools down, as well it may, the outlook for nuts is not so good.

Nuts are easy to grow organically, having little in the way of pest and disease problems. Chestnuts and walnuts have high nutrient requirements, and these could be met with human manure. Tree crops are ideal for human manure because, since the food is borne up in the air, the risk of contamination by manure applied to the ground is very slight, especially if the crop is harvested by net.

Nut breeding. One of the great advantages of annual crops is that the process of breeding new varieties is relatively quick, because a new generation can be bred each year. When new strains of pathogens arise resistant strains of crop plants can be bred relatively quickly. With tree crops, which may take several years before they start to bear and more years before an assessment of their life-time performance can be made, plant breeding must be a slower process.

The process can be greatly speeded up by selecting strains of nuts which flower and set seed much earlier in life. Phil Rutter of Badgersett Farm in Minnesota USA, nut breeder extraordinaire, has already bred a chestnut which produces pollen in its first year of growth. He predicts that breeding and propagating a new nut variety need take no more than two to three years longer than producing a new cereal variety.[2]

This early-bearing characteristic could be a useful trait to breed for in any case, as it cuts to a minimum the early unproductive years of the tree. Indeed this is one of the reasons why most apples are grown on dwarfing rootstocks these days. Early bearing is usually linked to a short overall life of the tree, but this may be no bad thing either. A tree which bears for 50 years rather than 200 may be ready for replacement just when new improved varieties are rendering it obsolete, and the timber may be in its prime rather than over-mature.

Designing a Nuttery

Choosing the ideal site for the species to be grown is often the key to success with trees. But in a way it's less important with high-value trees like nuts than it is with relatively low-value timber trees. A timber tree is usually notch-planted straight into whatever soil is there and largely left to get on with life. But for a nut tree it can be worth digging a large hole and creating the soil conditions the tree likes, and possibly modifying the microclimate with a windbreak.

The preferred soil, climate and microclimate conditions for each of the main nut species is given below. (See pp222-225.) If your site is completely unsuitable it's best to consider growing something else. But if conditions are marginal and the site is close by your house the care and attention you are able to give your trees may compensate for poor soil and microclimate.

Windbreaks are not as crucial for nuts as for fruit, but they should be seriously considered wherever conditions are less than ideal. (See box, Breaking the Rules, p226.) In areas of mild winds nut trees can be used to shelter more wind-sensitive fruits. But the yield of nuts will always be less than if they were planted in shelter themselves, and they are not the most effective trees at giving shelter, so the idea should be used with caution.

Choice of varieties is particularly important with nuts, as most of them are at their extreme climatic range in Britain. Most nuts are at least to some extent self-sterile, so more than one variety must be planted. Usually a distinction is made between the main

ALLELOPATHY[3]

Allelopathy is where one plant affects another plant by means of chemicals which it releases. Any plant which produces a chemical which has an effect on another plant is called allelopathic, or an allelopath.

Strictly speaking allelopathy includes all biochemical interactions, both beneficial and detrimental, that occur between plants. But in practice most of the allelopathic interactions which can be easily recognised are detrimental ones, and the term is commonly used to imply a detrimental effect by the allelopath on its neighbour.

Aspects of plant growth and function which may be affected include: germination, root growth, shoot growth and uptake of minerals. Effects can be complex. For example, plant A may enhance the germination of plant B, but decrease its root and shoot growth.

The chemicals can be moved from plant to plant in four different ways. They can be washed from the leaves by rain onto surrounding soil or plants, exuded from the roots, released from decomposing plant debris, or released as volatile chemicals from the living plant into the air.

Research suggests that almost all plant-plant interactions involve some allelopathy. But it's often such a small effect that it's masked by greater ones, such as competition. Allelopaths can affect plants of other species or other members of their own species. The advantage to be gained from inhibiting the growth of other species is fairly obvious. Inhibiting other members of its own kind often seems to be a natural form of crop rotation.

For example, replant disease, a disorder which effects fruit trees when they're planted where another related species grew before them, is partly an allelopathic effect. (See p227.) So is clover sickness, a disorder which affects pasture legumes when they're grown on the same ground for a number of years. There's also an element of disease build-up in both cases, and the allelopathic effect is there to discourage plants from growing in soil where the pathogens which affect them have built up to a high level. In mixed woodland there's a tendency for trees to regenerate more easily under trees of a different species. No doubt this is also an allelopathic effect.

Grasses, bracken and heather all use allelopathy as well as competition to inhibit the growth of trees and shrubs and thus slow down the regeneration of woodland, which will eventually shade them out. The effect is sometimes indirect, as when heather prevents the establishment of both birch and Norway spruce by inhibiting the formation of mycorrhiza.

The inhibiting nature of grass on trees has long been known. The fact that allelopathy, and not just competition, was involved was confirmed by an experiment with apple trees. Water which had been used to irrigate grass plants in a tray was drained off and used to irrigate a tray of apple seedlings. The apple seedlings grew less than a control tray irrigated with plain water. Some other examples include:

Brassicas are believed to suppress weeds both through root exudates and from the decomposition of plant residues.

Vetch, *fodder radish*, *mustard* and *buckwheat* are green manures which inhibit weeds allelopathically, as well as by competition.

Potatoes have an inhibiting effect on apples, so are unsuitable as an intercrop.

Elder has a negative effect on many plants, especially brassicas.

Corncockle, a once common cornfield weed, increases the yield and quality of wheat.

Barley inhibits the growth of chickweed and shepherd's purse, but wheat does not.

Aromatic herbs, including members of the labiate, umbellifer and composite families, inhibit the germination of annual plants in their vicinity. The chemicals involved are apparently the same ones which give the herbs their medicinal and culinary uses, and which are important to the health of the herbs themselves.

Marigolds of the *Tagetes* genus have an allelopathic effect on various weeds, including bindweed, couch and ground elder, weakening or even killing them. *Tagetes minuta*, the Mexican marigold is the most potent species.

Walnuts – see facing page.

productive variety or varieties and the pollinator or pollinators. Of course the pollinators will also bear nuts, but the yield and quality may not be so good since they're primarily chosen on their ability to pollinate the main varieties. Details of pollination requirements for each species are given below. (See pp222-225.)

Nut Polycultures

The potential for polycultures is somewhat different for each of the three main species.

Chestnuts are very large trees and cast a heavy shade. It's not possible to grow anything underneath them, and anything grown between them will suffer from considerable side shade and root competition. The fallen leaves are very persistent; they can collect in natural traps and kill off the ground layer. Chestnuts are probably best grown alone.

All members of the walnut family are allelopathic, but the strength of the effect varies. (See box, Allelopathy.) It's strong in the black walnut, which is said to stunt or kill many plants, including: apples, blackberries, tomatoes, potatoes, grains and lucerne; but not to effect others, including: pears, plums, peaches and clovers.[4]

However in the common walnut the effect is negligible. It happily coexists with apples in English orchards, and is unlikely to have a significant allelopathic effect in a polyculture. But before planting an extensive polyculture based on walnuts it might be wise to experiment by planting a little of the proposed companion species by or under an existing mature walnut.

Walnuts can cast a heavy shade, though a clear stem to two metres minimises shade at ground level. But they come into leaf late, in May, so they combine well with plants which do much of their growing earlier in the year, such as winter cereals and currant bushes. Taller companion crops may restrict air movement around the trees and thus increase the likelihood of walnut blight.

Hazels are tolerant of light shade, and can be planted as an understorey beneath taller trees. But nut production is much less in shady conditions, and nutting hazels are invariably grown as the canopy crop. They cast a fairly heavy shade themselves, but in Kentish plats, as hazel nutteries are known, they were sometimes interplanted with currants and gooseberries in the past.

In all nut polycultures, the growth habit of the companion crop or crops has an effect on the ease of harvesting. An intercrop of similar sized trees should not pose a great problem, nor should an undercrop like grass or cereals which can be grazed, mown or harvested before the nut harvest. But an undercrop of shrubs or of herbaceous vegetation which persists into the autumn will make harvesting by hand more difficult, and harvesting by machine difficult or impossible.

This means that a much wider range of undercrops can be considered for a small-scale or domestic nuttery, which will be harvested by hand, than for a large-scale commercial operation where machinery will be used.

Nitrogen-fixing trees may be a useful interplant. Walnuts and chestnuts have a high nitrogen requirement and hazels a moderately high one. More nitrogen is needed in the early stages of development, when vigorous vegetative growth is required and nut production has not started. But when the trees are mature abundant nitrogen may encourage vegetative growth at the expense of nutting. One way to provide this sequence is to plant the nut trees at the normal spacing with the nitrogen fixers between them. There is plenty of room for the extra trees in the early years when they and the crop trees are still small, and when the trees begin to crowd each other the nitrogen fixers can be removed, leaving the nuts to mature with less nitrogen in the soil.

One experiment in North America gave dramatic results. Black walnuts were interplanted with nitrogen fixing trees and shrubs: Russian olive, autumn olive (both species of *Eleagnus*), common alder and Siberian pea shrub. During the first 12-14 years after planting the walnuts showed 50% more growth in height and stem diameter than all-walnut control plots, while the difference in crown growth was double or triple that of the control plots. The population of walnut crop trees was the same in both plots. It's interesting to note that by far the greatest increase in growth was in the crowns, where the nuts are borne. Unfortunately nut yields were not recorded in this experiment.[5]

A selection of nitrogen-fixing plants is given in Chapter 3. (See p49.)[6]

Layout

Three possible layout patterns for a nuttery (or fruit orchard) are: in rows, on the square, and hexagonal.

Choosing which pattern to adopt is often a matter of balancing yield with convenience. The row layout is likely to give the lowest yield of the three, as trees don't make full use of the space between the rows. The square pattern is intermediate, and the hexagonal pattern yields the most, as each tree is at its ideal spacing

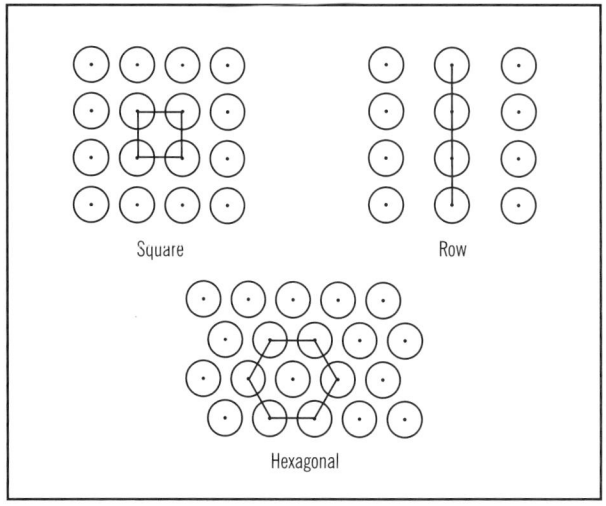

The three basic layouts.

relative to every one of its neighbours, resulting in 15% more trees per hectare than on the square pattern.

The advantage of the row pattern is that it allows for mechanical harvesting with smaller trees such as hazel. With larger trees such as walnuts there may be enough space for machine access on the square or hexagonal patterns, especially the square. The row pattern also leaves a decent amount of space for an intercrop and allows more light to get to it.

Wherever nuts are grown in rows the ideal alignment for the rows in our latitude is north-south. This means that each side of the row, and the intercrop if any, gets direct sunlight either in the morning or in the afternoon. If the rows run east-west the northern side will be in the shade almost all the time, as will the intercrop to the north of it.

If the nuttery is on sloping land it's best to have the rows running up and down the slope in order to allow cold air to drain away and thus minimise the chances of spring frosts. It's also easier to use machinery up and down the slope than across it. On the other hand, if bare soil crops are grown between the tree rows it's best to align them across the slope to minimise soil erosion.

Obviously these requirements will conflict in some cases, and if one requirement is particularly strong it will over-ride the others. Sometimes it may be possible to minimise the frost risk by other means, such as by planting a dense tree belt which deflects cold air away from the orchard. (See illustration, p78.)

Dual-Purpose Trees

Taking a yield of timber or wood from a tree as well as one of nuts or fruit is an attractive idea. Hazel and chestnut trees can both be coppiced for craft materials, while both chestnut and walnut produce timber, the latter of very high quality. Oak and beech, both well known as timber trees, also produce edible nuts. Many fruit trees, such as apples, pears and cherries, have timber which is valuable for carving and high-quality furniture.

Some sort of wood yield is always possible, both in prunings and in the whole tree when it reaches the end of its life. But combining a nut or fruit yield with one of high quality timber or wood is problematic. There are three of ways in which the two kinds of yield conflict with each other:

- the shape of the tree
- the length of the rotation
- establishment costs

Tree shape. Most timber and coppice trees are grown close together, partly to get the maximum yield of timber or wood per hectare, and partly to get stems of a usable shape. For timber, and coppice wood destined

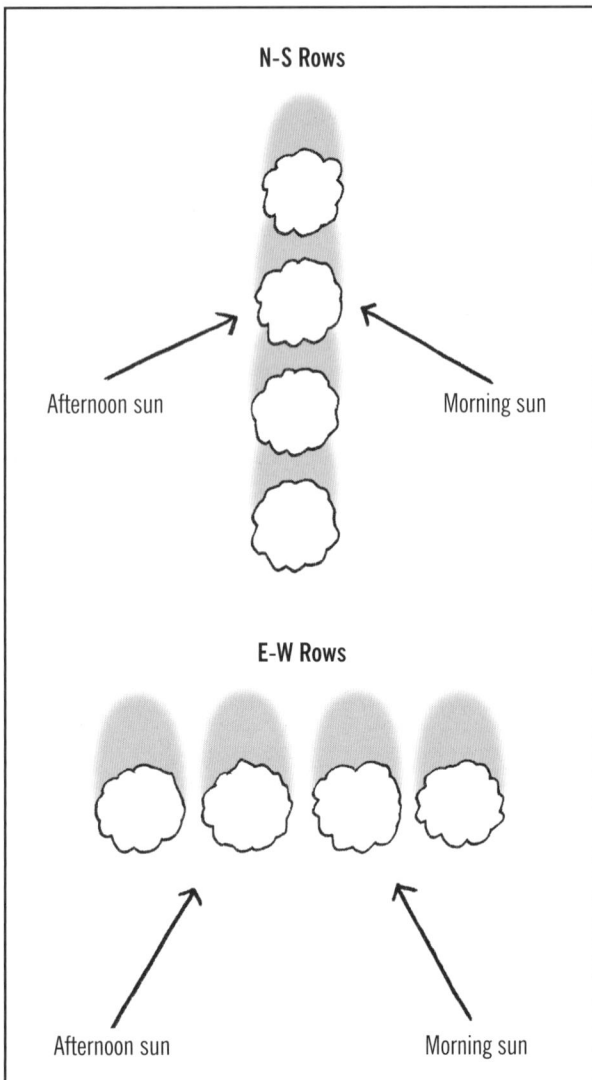

North-south rows reduce shading.

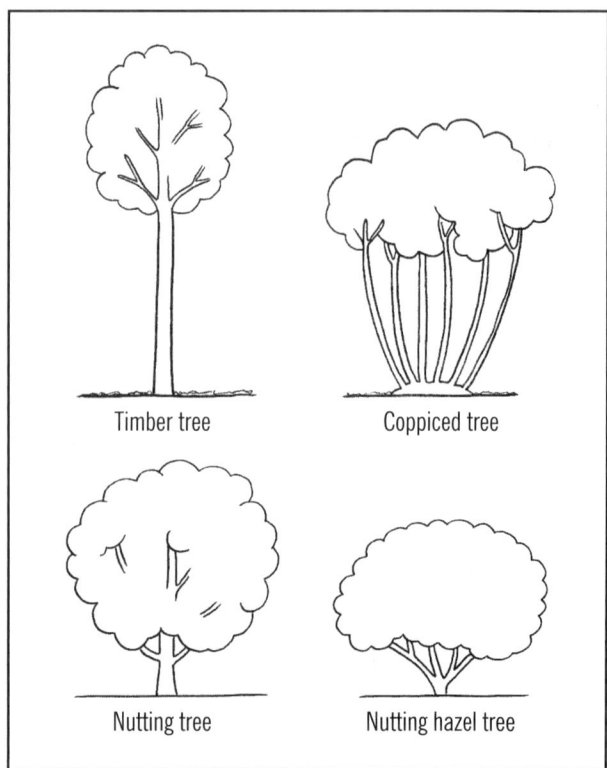

Tree form varies according to the kind of produce required.

for craft uses rather than firewood, usable shape means tall, straight and branch-free. Growing them relatively close together achieves this, as each stem strives to grow upwards towards the light, and the closed canopy of the woodland casts shade which suppresses lower branches.

The ideal shape for a fruit or nut tree is quite the opposite. Fruit and nuts are borne on twigs, and in most cases only on twigs which are exposed to the Sun. So the trees need to have many branches, with a large well-developed crown, and enough space left between them so the crown of one does not touch the crown of another.

Rotation length. The rotation is the number of years from planting to felling, and it's shorter for both timber and coppice than it is for nutting trees.

Timber trees are felled when they're middle aged, when the fast growth of youth is beginning to slow down, and before the possible rots and splits of maturity start to affect timber quality. At this time of life, a nut-bearing tree still has many years of prolific nut production ahead of it, and it would be wasteful to fell it. By the time nut production tails off in old age the quality of the timber may have deteriorated. Walnuts, for example, are usually felled at 60 years for timber, but they can go on producing nuts for a century or more.

Coppiced chestnut and hazel trees don't produce much in the way of nuts for a few years after coppicing. Hazel is normally coppiced on a rotation of seven years, but for the first five years after coppicing it produces few nuts. It may be just about getting back to full production by the time it's cut again, so the average nut production over the coppice rotation is very small. The chestnut coppice rotation is longer, but they take proportionately longer to come back into bearing after coppicing.

Having said that, Phil Rutter in Minnesota reports harvesting the equivalent of 764kg of kernels per acre off one of his hazel trees after only two years of growth from coppicing. This was an exceptional strain amongst thousands on his experimental nut breeding farm. Maybe some day varieties like this will be available commercially, but they are not now.[7]

The possibilities of coppicing fruit trees, as opposed to nuts, is discussed in the guest essay at the end of this chapter. (See p254.)

Cost of establishment. Timber and coppice crops are grown from unimproved strains of trees, and small, cheap transplants are used. They are usually notch planted, a process taking a few seconds per tree. The entire cost of establishment may be counted in pence per tree. In the case of timber most of these trees will be removed as thinnings during the life of the plantation, leaving about ten percent of them as the final crop.

Fruit and nuts, on the other hand, are grown from cultivated varieties, usually grafted onto a rootstock, carefully pit-planted, and given intensive after-care. The cost of establishing each tree may be tens of pounds, especially if labour is included. It would be a great waste to fell or coppice such a tree before it had yielded its full potential of nuts, as both timber and wood are much lower value products.

It would be equally wasteful to leave a timber tree to grow on after the optimum felling age for the sake of the nuts it will produce in the ensuing years. Since it has not been bred for nuts the crop will be light and infrequent and the nuts small. Neither is there much scope for lengthening coppice rotations in order to get more nuts from the trees. Both chestnut and hazel are grown for craft purposes rather than firewood, and the quality of the wood would decrease if they were grown on beyond the optimum coppice rotation.

So much for the case against dual-purpose trees. But in practice there is some scope for compromise, and dual purpose trees have often been grown in the past. In parts of Switzerland, for example, apples are traditionally grown as very tall standards so that the trunk can be used for timber.

The most promising nut tree for dual purpose cultivation is the walnut. It's one of the few timber species which is not grown close together and progressively thinned, but planted at an orchard-style spacing, with the canopy of one tree never touching that of another. This is called free growth, as the trees do not compete with each other. It leads to rapid growth in stem diameter, which is a key to high timber quality, and is used for walnuts because they are grown for quality rather than quantity. Total production of timber would be greater if the trees were close grown. But the high value of walnut timber justifies both the loss of bulk and the expense of stem pruning, which is necessary to get knot-free timber when it's free grown.

It's certainly worthwhile to collect nuts from walnuts being grown primarily for timber, and there is often valuable timber in the trunk of an old nutting tree when it comes to the end of its productive life. Trees grown

The contrast between sweet chestnut trees grown for coppice, foreground, and for nuts, background, could hardly be more extreme. Prickly Nut Wood, Sussex. (See case study, Living In The Woods, p336.)

for nuts can be useful for timber if they have a height of 2.4m to the first branch, and this should guide the pruning of a dual-purpose tree. The key to success with dual-purpose trees, whether fruit or walnut, is probably to prioritise either nuts or timber as the main output, manage the plantation with that in mind, and accept the other as a useful by-product.

There may be some potential for growing other tree species for a dual harvest of nuts and timber. There have been some experiments with free growing of oak in Britain.[8] The method here has been to plant oak at a normal forestry spacing, choose the final crop trees early on, and thin back the other trees so that the chosen ones always have free space around them throughout their lives. This means that the final crop trees grow faster and can be harvested sooner than conventionally grown trees, and more veneer-quality logs can be expected, both of which increase the profitability of the timber operation.

It will also mean that the trees have a much larger canopy exposed to the light, and acorn production will be much greater. The plantation as a whole develops a crenalated profile, as opposed to the flatter profile of a conventional plantation. In effect the amount of edge is greatly increased. If mechanical harvesting is used the forest floor would need to be kept free of undergrowth.

Standard trees in coppice-with-standards woodland can also yield a good crop of nuts, but mechanical harvesting is pretty well impossible among the coppice stools. Although yields of nuts from close-grown high forest or coppiced chestnut or hazel are usually light and erratic, they can be worth gathering for home consumption or for retail sale as an occasional sideline.

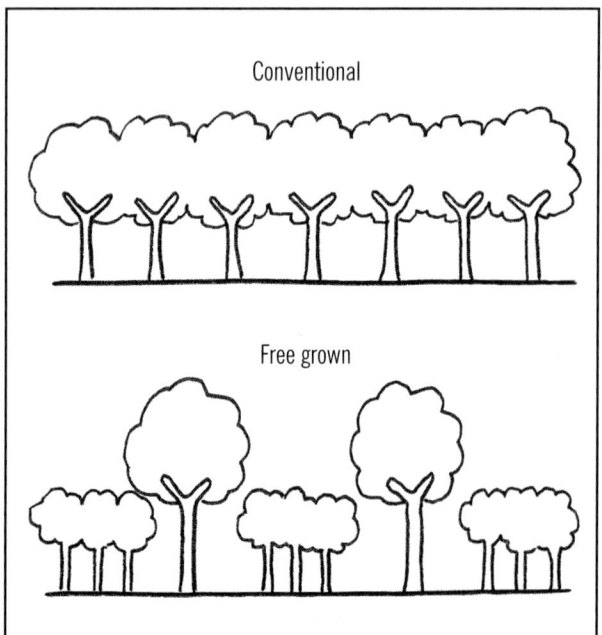

Comparison of potential nutting surface in conventional and free grown timber plantations

Nuts One by One

The Obvious Ones

When buying nut trees it's essential to buy named nutting varieties. Trees sold under the species name only will probably yield very little.

These notes are a supplement to the general discussion of nut growing above. (For yield per hectare see p216.)

Hazels, including both cobs and filberts.
Soil. Of moderate fertility; too fertile a soil produces vegetative growth rather than nuts; best on moist soil with plenty of calcium; do badly on dry soils; pH 5.5-7.5.
Climatic range. Should crop well wherever apples will.
Microclimate. Fairly frost hardy; need some shelter to crop well; prefer full Sun but give some yield in part shade.
Spacing. Traditionally 4.5-5m each way, or 6 x 3.5m to allow mechanical access; there's a current trend towards closer spacings of 4.25 x 2.25-3.5m
Height. Usually kept to no more than 2m for ease of pruning and picking.
Pollination. Pollinators required; a minimum of three varieties recommended including main varieties, but wild hazels are effective if present; pollinators effective within 45m.
Pests. Grey squirrels can strip whole trees before they ripen; if a few trees are grown in an area where there are squirrels they will take all the nuts; in large plantings the squirrels will only take a proportion of what's there; hazels can only be a domestic scale crop in areas with no grey squirrels; netting is not effective.[9]
Yield per tree. Typically 3-5kg, exceptionally 11kg; maximum from 15 years after planting; often erratic and biennial.

Walnut, the common walnut, also known as English or Persian walnut.
Soil. Deep, moist, well-drained, medium-textured loam with good nutrient status and pH 4.5-8.3, ideally between 6 and 7.
Climatic range. As far north as South Yorkshire, and probably in favourable microclimates further north.
Microclimate. Very vulnerable to late frost in early leafing stage; if caught by late frost young trees can be killed and the year's crop lost from mature trees; mid slope on a gentle south or south-west slope is ideal, easterly aspects are best avoided (see p238); late-flowering varieties can be grown where late frosts are likely; less heat-demanding than chestnut; light-demanding.
Spacing. 9-15m, depending on variety; dwarf walnuts are now available which grow to 4-6m.[10]
Pollination. Pollinators required, though some varieties are partially self-pollinating; pollinators effective within 200m, but only 80m in plantation conditions.
Allelopathy. Very mild; perhaps avoid underplanting with potatoes or tomatoes.

Pests. Grey squirrels can damage young trees, and may take nuts, especially where a few trees grow near woodland; squirrel predation can be avoided by picking the nuts early for pickling, for which there is a good market at present.

Disease. Blight can be severe if weather is cool and wet at flowering time; avoid low pH (<6) and excessive nitrogen, allow some air circulation around trees; resistant varieties to both blight and anthracnose are available.

Yield per tree. 3-5yrs 5kg; 10-15yrs 50kg; 20yrs+ 75kg.

Chestnut, the sweet, or Spanish chestnut; other species and hybrids may be suitable for British conditions (See Appendix E, Suppliers of Permaculture Requisites.)

Soil. Prefers light, well-drained loam, acid to neutral; tolerates all textures except heavy clay; tolerates very acid soil, but usually not limestone or chalk; tolerates drought.

Climatic range. Best in south-east England; size of nuts, yield, and regularity of yield all decrease further north or west; not above 100m altitude in Britain; requires average annual temperature of 8-15°C, and September and October averages of 14.5 and 8.5°C respectively.

Microclimate. Young trees can be damaged by frost; flowering takes place in June, so it is not affected by frost, but the late flowering date means that there is less accumulated heat during fruit swelling time, so a warm site is essential.

Spacing. Usually 10x10m or 6x8m in intensive plantations; mature tree grows to 20-30m; dwarf chestnuts should be available soon.[10]

Pollination. Many nutting varieties are effectively self-sterile, so pollinators are needed; they are effective within 40m, therefore must be planted every fourth tree.

Yield per tree. For hybrid varieties planted at 10x10m, 50kg when in full yield, from 12 years after planting; indigenous varieties at same spacing, 30kg; a mature full-sized tree can yield hundreds of kilos. (See case study, Living in The Woods, p336.)

The Less Obvious Ones
Oaks.[12] Acorns have been eaten with relish in many parts of the world, and still are in some countries, notably Korea. All acorns are edible, but most have to have the tannins leached out of them first. Leaching the tannins may also remove vitamins and minerals, leaving a basic staple which is mostly carbohydrate. But what an easy one to grow!

The level of tannins varies both between species and between individuals. Both of our native species, the pedunculate and sessile oaks, have fairly high levels. Since the tannins are a protective measure against herbivores, there is a natural selection pressure towards high levels. But there is a wide variety of characteristics in any native population, including tannin levels in acorns. Individual oaks with unusually low levels can be found and selectively bred for yet lower levels till the acorns are edible without leaching. Then these can become the parents of oak orchards. There may even be some individuals out there which are already sweet enough. Our cultivated varieties of sweet cherry have been bred like this, by selecting from a wild population whose fruit varies from the completely inedible to the delicious.

Some of the oak species from warmer climates have lower levels of tannins, though again this varies from one individual to another. The Mediterranean ones tend to do better here than the North American ones. An example is the holm oak, an evergreen sometimes grown in the south of England as an ornamental. It's tolerant of coastal conditions but should be planted with caution as it can become invasive in semi-natural woods on coastal sites.

Oaks are irregular croppers, yielding heavily after a warm summer and little or nothing after a cool one. A year when they produce a full crop is known as a mast year, and these come at intervals of anything from two to seven years. Again there is variation between individuals, some cropping more regularly than others, and some giving at least a moderate crop

---

PREPARING ACORNS FOR EATING

It may not be worth planting oaks for food till we have improved varieties available, but in the meantime we can pick and eat acorns from wild trees. The tannin level is lowest when they're fully ripe, and the best way to ensure they are ripe is to store them for at least two weeks before eating.

There are a number of ways of leaching the tannins out of acorns. The most practical include:

- Take dried acorns, shell them and grind to a flour. A coffee mill or nut grinder will do the job. Put the flour in a cloth bag and pour boiling water over it. Use the resulting paste immediately.

- Put shelled acorns in a sack in constant running water for two to three hours. Roast to remove other tannins.

- Soak shelled acorns in water for 8 hours. Renew the water and soak for 4 hours. Roast to remove other tannins.

The roasted acorns will keep for many months. Acorn flour resulting from any of these three methods, mixed half-and-half with wheat flour, gives a rich, nutty flavour to breads, biscuits, cakes etc.[13]

in their off years. These trees could be bred from, and crossed with sweet individuals, with the ultimate aim of producing annually-cropping sweet oaks.

In good mast years yields can reach 7.5 tonnes/ha or 90kg per tree under woodland conditions. This yield represents the accumulated energy of several years, so annually-bearing trees would be unlikely to average this until they had been selected for high yield over a generation or two. Trees start to bear at between five and fifteen years. Peak yield is between 50 and 140 years, with useful yields continuing till 250 years.

It's always possible to breed for an earlier maximum yield, but this is usually linked to a shorter lifespan, so the gain of one generation would be the loss of the next.[14] It may be better to see oak as the long-term climax producer in a polyculture which also contains shorter-lived trees which will give their maximum yield while the oaks are still immature.

The value to biodiversity of using oak as a staple food crop cannot be overstated. Not only is it a native tree, it's one with a remarkably high number of associated invertebrate species. (See p368.) What's more, many of these species only live on an oak when it's well past timber age, but nutting trees would have a much longer life.

Monkey Puzzle. Although usually thought of as an ornamental, the monkey puzzle has great potential as a nut tree in Britain. Its home is the south of Chile, where it has long provided the staple food of the native people. Southern Chile is one of the few parts of the world which has a cool maritime climate like ours, and monkey puzzle is perhaps unique among nut trees in preferring a maritime climate to a continental one.

In Britain it prefers the wetter west of the country, and tolerates coastal conditions. In fact it can be killed by exceptionally severe winters if grown inland. It likes a deep, well-drained soil with the pH on the acid side.

Male and female flowers are normally borne on different trees, though individuals with both sexes occasionally occur. At least one male tree to five females is required. Seedlings can't be sexed till they flower, but vegetative reproduction is possible.[15] They are big trees, up to 30m high and 15m spread, and although evergreen cast a light shade. They could easily be underplanted with other productive trees, perhaps hazel.

The big disadvantage is that they don't start producing till they're 30 to 40 years old, so planting them is very much an act of faith in the future. Also, though monkeys may find them a puzzle, grey squirrels apparently don't, and have been known to take the nuts.

Plum Yews. These trees are similar to the native yew, and have two properties which make them particularly attractive. Firstly, they are shade-lovers, and thus ideal for stacking. Some kinds will even grow under conifers. Secondly, the nut is surrounded by a foul-tasting fleshy aril which even grey squirrels can't stand, making them one of the few truly squirrel-proof nuts which can be grown in Britain.

Male and female flowers are borne on different plants, though there are occasional individuals with both sexes. I can find no information on how many males are needed for pollination. They tolerate a wide range of soils. The crop is said to be heavy and easy to harvest. The nuts are sometimes bitter, in which case they need leaching, and can be eaten raw or cooked.

Cephalotaxus harringtonia is probably the most suitable species for Britain. It grows to around 5m tall and wide but can be kept smaller in cultivation. It must be grown in shade in the south but could succeed in the open in the north. The variety *drupacea* is tolerant of chalky soils. The variety *nana* is a dwarf shrub 0.3-1.8m high which suckers and forms thickets. It could

The monkey puzzle, grown here as an ornamental on the Gower Peninsula in South Wales, could be an important food producer in the west of Britain, where the climate is less suitable for other nuts.

Plum yews have the great advantage of being unattractive to grey squirrels.

make a useful understorey in a forest garden. These two varieties are also said to have the best flavoured nuts.

Black Walnut. The nuts are similar to common walnuts, but with a stronger flavour, a higher protein content, and a much stronger shell, which needs a special nutcracker. The trees have similar requirements to common walnuts except that they need a higher summer temperature than we normally get in Britain. This is probably why they are rarely grown here for nuts. They are strongly allelopathic towards certain other plants. (See p218.)

Other members of the family include: the heartnut, the butternut or white walnut, and the buartnut, a hybrid of the two. They are said to be hardier than the common walnut and may be worth trying.

Almond. Almonds are said to be the most health-giving of the nuts. In the right climate they are tough and undemanding trees. But they blossom in February and March, when the possibility of frost is high and there are few insects on the wing to effect pollination. They can also suffer from peach leaf curl in our rainy climate. They can be open grown as a garden tree in the extreme south, especially on chalk or limestone, and against a wall further north. But they have no potential for large-scale production here under present climatic conditions.

Pines.[16] All pine nuts are edible but only some kinds are big enough to be worth bothering with. Some of the species which have large nuts will grow in Britain, but yields are very low. The stone pine, which is cropped commercially in Mediterranean countries, gives only 5-15kg per tree, which is 1.5 tonnes/ha. They surely aren't worth growing purely as a food crop. It may be worth collecting nuts from trees grown primarily for timber, but none of the large-nut pines are good timber trees here. It's doubtful whether the bonus of nuts would compensate for the loss of timber yield due to choosing a poor timber species.

Beech. Beech nuts are tiny and not worth picking and shelling by hand. But if they can be harvested economically a high quality oil can be pressed out of them. Mast years are very irregular, occurring at anything between 4 and 15 year intervals in Britain.

FRUIT

There are many reasons for growing fruit. From a home gardener's point of view, it's delicious to eat, and being perennial takes much less work than annual vegetables. Organic fruit is expensive to buy, and when conventionally grown it's sprayed with more pesticides than any other kind of food, many of which are still there when we eat it.

Some people may be reluctant to plant fruit in their gardens if they think they might move house in a few years.

But trees on the really dwarfing rootstocks will start to bear within two to three years, and soft fruit in the second year for sure. Anyway we don't have to take such a narrowly self-interested view of things. The next occupants of the house may appreciate the fruit too. I like the old saying, Live as though you're going to die tomorrow but farm as though you're going to live forever. In a way that's what permaculture is all about.

As a commercial permaculture enterprise organic fruit has much to commend it, being a perennial food crop, greatly in demand and under-supplied at the moment, which means it fetches a good price. But the reason why it's in short supply is that it's very difficult to grow fruit organically, top fruit at least. It has been described as 'organic's final frontier'. Of course fruit was grown organically in ages gone by, but it was more of a luxury product in those days. Chemical control of pests, diseases and weeds, and imports from countries where the climate is ideal for top fruit, have brought down the price and raised the public's expectations of visual perfection. There's only so much extra you can charge for organic produce before people stop buying it.

Major research into organic fruit growing is now under way,[17] and in the future there should be more solid recommendations available on suitable varieties and pest and disease control methods. Meanwhile, it would be rash to try making your living at organic fruit unless you already have an in-depth knowledge of fruit growing, a site which is well suited to fruit, and plenty of capital. One authority on organic top fruit reckons that the only way to do it without risking your shirt is to start the trees off conventionally and convert to organic in the third year, when they're past their most vulnerable stage.[18]

On the other hand there are more imaginative approaches. One is to combine a fruit orchard with free range chickens on the same ground. The chickens will help protect the trees from pests and diseases and provide a parallel source of income, especially before the trees start to bear. (See pp246-247.)

Growing fruit for juicing rather than for sale as fresh fruit is another possibility. The need for general tree health and a good yield of fruit will be the same, but fruit which is misshapen or munched by pests will still be saleable. It may be practicable to sell the cream of the crop as fresh fruit and to juice the rest.

Another approach is to move away from the idea of an orchard and to tend and harvest the neglected trees many people have in their urban back gardens. Being relatively isolated from other fruit trees these don't have the pest and disease problems which orchard trees do.

The saving in pest and disease control will be partly offset by a lot of unproductive travelling time. But the big advantage of working like this could be in marketing. Selling direct to the consumer is often the key to making small-scale organic production of any sort profitable (see Food Links, pp285-293), and towns are where the consumers are. This idea has already been used successfully with chestnuts in Australia. (See box, A Successful Niche, p235.)

Choosing What to Grow

The first key to success with fruit is choosing the right kinds for the local climate and microclimate. This is especially true if you grow them organically, as you're depending on the general health of the plants to keep them free from disease rather than curing problems with chemicals as they come up.

It's no coincidence that Kent is the centre of top fruit growing in Britain. Most fruits like a hot summer without too much rain, and the further north and west you go the less likely you are to get that. Height above sea level also effects both temperature and humidity. (See p69). The rule of thumb for most of the country is that commercial fruit is best grown below 120m but some fruits can be grown up to 180m.[19] Above that a very favoured microclimate is needed.

The best way to assess what can be grown on a particular site is to look at what's growing in the same locality at the same altitude. But microclimate can make a big difference. A sheltered, south-facing suntrap is a very different matter to an exposed north slope. In fact even quite minor differences in microclimate can make a big difference to the range of fruits which can be grown.

This is especially true of late spring frosts, which are a particular problem of our maritime climate. (See pp67-68.) Late frosts can be a problem in all parts of the country and at all altitudes. Whether they will be a problem on a specific site is more a matter of microclimate, and small local changes in topography are significant here. (See pp78-79.)

As a rough guide, the following list places the most commonly grown fruits in order of climatic tolerance, starting with the toughest:[20]

- wildings, eg bramble, rowan, wild crab
- soft fruit
- cooking apples, damsons

Pears are a good choice of fruit for a warm, sunny wall.

- dessert apples, cooking plums
- pears, desert plums
- gages, tender pears
- peaches, apricots, figs etc (only worth growing on a domestic scale in Britain)

There's also a great range of toughness among the different varieties of these fruits. Much can be learned from nursery catalogues, particularly those which specialise in varieties suitable for their own part of the country. The directories of varieties published by the Agroforestry Research Trust are especially useful in identifying disease-resistant varieties and those suitable for organic growing. They are also more comprehensive than any nursery catalogue, covering all known varieties and listing all the available information on each.[21]

Fruit needs good soil as well as the right climate. Poor drainage and high pH are the two big problems. Where the only available orchard site is poorly drained some kind of artificial drainage is essential. There is really no leeway with bad drainage for fruit.

BREAKING THE RULES

I know a man who grows successful raspberries on a thin chalky soil, and I'm sure there are others who grow healthy top fruit more than 180m above sea level. If all you want is to grow some fruit for yourself and your family, it's always possible to break these rules of thumb – if you keep to two others.

The first is to find the right varieties. Varieties which can tolerate extreme conditions, such as high alkalinity, may not be the popular ones which feature in the average nursery catalogue. It may take quite a search to find them. In fact the best way is often to seek out someone who's growing what you want to grow in a similar environment to yours, and find out what varieties they're growing.

The second is to do them well. Whenever a plant is stressed for any reason, reducing other sources of stress to the absolute minimum will help it gain the strength to overcome that stress. For example, providing fruit with a windbreak won't make the Sun shine more often, but it may help the trees' general health improve enough to cope with a certain lack of sunshine.

If you're a skilled grower, prepared to give your plants all the attention they need and willing to take a bit of a risk, you have a good chance of breaking the rules and getting away with it. If you just want a reliable yield without too much effort, you'd be well advised to stick to the rules of thumb.

On a thin soil over chalk or limestone the best options are apricot, elder, bramble and fig, though hazel and walnut may also do well there. Where there's a good spade's depth of soil above chalk or limestone you may be able to get away with growing other kinds of fruit if you add plenty of organic matter to the soil, as long as you avoid raspberries and blueberries.

Another thing to note is what was growing there before. Replant disease is a condition which affects fruit trees planted where closely related trees were growing before. Apples and pears can follow stone fruits and vice versa, but neither group can follow itself, at least not for fifteen years.[22] This is the traditional advice, however, recent work in the USA suggests that growing wheat between grubbing out the old trees and planting the new can reduce the effect of replant disease.[23]

See tables, Fruit Trees and Shrubs, and Yield Guide to the Common Fruits on pp228-230.

Growing Fruit For A Living
If you're thinking of growing fruit commercially it's more a matter of choosing a site which is suited to fruit than choosing fruit to fit an existing site. There's all the difference in the world between growing fruit for domestic consumption with perhaps a few sales of surplus on the side, and relying on it for a major part of your income. A crop failure or chronically low yields may be an inconvenience on a domestic scale, but could mean ruin for a commercial grower. Also, if you're only growing a few trees of each kind you can always find an ideal site for them, perhaps up against a south-facing wall. But if you're growing dozens or hundreds of them the site must be good overall.

The decision to grow fruit commercially should come before acquiring the land, not after. It may be possible to turn a somewhat poor fruit-growing site into an adequate one by investing a great deal of energy in it, for example by installing land drainage. But the extra investment will be a burden on the profitability of the enterprise, and the yield may still be less than could be had on a naturally good site. This approach is working in spite of the land rather than with it, and that is the antithesis of permaculture.

The checklist in the next column summarises the main characteristics to look out for when choosing a site for an orchard. Each kind of fruit has somewhat different requirements, for example, pears can tolerate a slightly wetter soil than apples. So the requirements given are broad indications only. Nut crops are less exacting in some respects, but the same questions need to be asked when selecting a site for a nuttery. Again, different nuts have different requirements, as indicated in the nut profiles above. (See pp222-225.)

No site is perfect, and some compromises will have to be made. A site in the north or west of Britain, for example, will have to score better all round than one in the south-east. But some factors, such as being in a serious frost pocket, will rule the site out altogether.

THINGS TO CONSIDER WHEN LOOKING FOR A SITE		
Factor	Characteristic	Requirements
Soil	depth	at least 60cm for top fruit
	drainage	good drainage essential
	pH	ideally 6.6-6.7; high pH a problem for many fruits
	texture	medium loam is ideal
Climate	region	south-east of Britain better than north or west
	altitude	ideally below 120m
Microclimate	frost	avoid frost pockets
	wind	shelter needed
	humidity	avoid excessive shelter
Vegetation	previous	beware replant disease
	other orchards	isolation reduces pest and disease problems
	nearby woods	honey fungus, birds and grey squirrels can be a problem (see p234)
Markets	local population	enough for direct sales (see p286)

FRUIT TREES & SHRUBS

This table can only give approximate information on the characteristics of the different fruits. For more detailed information see How To Make A Forest Garden. (See Appendix A, Further Reading.)

Fruit	Season	Pollin-ation	Dis-ease	Hardi-ness	Frost	Shade	Aspect	Coastal	Drainage	pH	Notes
Apple, cooker	Au-My	I		H		HS	(N)		T		
Apple, desert	Au-My	I					W				
Apricot	Jn-S	F		T			S		T	Ch	Very susceptible to frost
Asian pear	S-D	I		T	T		S				Mildest areas only, not commonly grown
Beach Plum	Au	F	R					C			Very tolerant of coastal conditions
Berries, hybrid	Jl-S	F			T	SH+Sh	(N)+N				Many different hybrids with a wide range of characteristics
Blackberry/Bramble	Au-O	F		H	T	HS	N		T	Ch	
Blackcurrant	Jl-Au	F				HS			T		
Blueberries	Jl-S	I	R		T	HS				A	Needs acid soil, maximum pH 5.5
Cherry Plum	Jl-Au	F			T						Good windbreak tree, irregular yielder
Cherry, sour	Au-S	F				Sh	N				
Cherry, sweet	Jn-Au	I					W				Extremely susceptible to predation by birds
Crab Apple	S-Mr	F	R	H		HS	(N)		T		Some varieties good pollinators for domestic apples
Currants, red & white	Jl-Au	F				Sh	N				
Damson	S-O	F		H		HS	(N)				Good windbreak tree
Elders	Au-S	F+I	R	H	T	HS	N	C	T	Ch	American species (*Sambucus canadensis*) yields best
Eleagnus, evergreen	Ap-My	F	R	H		Sh	N	C			Many species†
Eleagnus, deciduous	O-N	F	R	H			E				Many species, some very high yielding†
Fig	Au-S	F	R	T			S	C		Ch	Susceptible to winter frost
Gooseberry	Jn-Au	F				Sh	N				
Grape, outdoor	O-N	F		T			S				
Hawthorns	S-O	F	R		T			C	T	Ch	Several exotic species with very edible fruit†

	Season	Pollination	Disease	Hardiness/Frost	Shade	Aspect	Coastal	pH	Notes
Japanese Wineberry	Au	F				N			A very decorative plant
Juneberries	Jn-Jl	F	R		HS	W			Very susceptible to predation by birds
Kiwi	O-N	D	R	T		S			Very susceptible to frost
Kiwi, hardy *A. arguta*	Au-O	D	R			W			Extremely vigorous, can smother trees if not controlled
Kiwi, hardy *A. kolomikta*		D	R		HS	E			More tolerant of frost than both other kiwis, not commonly grown for fruit
Medlar	N-D	F	R	T	HS	(N)			Fruit edible when bletted
Mulberry	Au-S	F	R	T		S			Vulnerable to autumn frosts
Oregon Grape	Jl-S	F	R	T	Sh	N			Very shade-tolerant, flowers in winter
Peach	Jl-S	F		T		S			Very susceptible to frost, ripens well outdoors
Pear, cooker	Au-Ap	I			HS	(N)			
Pear, desert	Au-Ja	I				S			
Plum	Jl-O	F+I				E			
Quince	N-D	F	R	T		S			
Raspberry, autumn	Au-O	F						A	
Raspberry, summer	Jn-Au	F			HS			A	
Whitebeam	Jl-S	F	R	H			C	Ch	Fruit usually needs bletting
Worcesterberry	Jl-Au	F	R	H	Sh	N			A tougher but lower-yielding alternative to the gooseberry

NOTES

Season — The months during which the fruit is ripe, including keeping time for fruits which can be kept without bottling. For most fruits the whole season can only be covered by growing more than one variety.

Hardiness — Tender fruits may be grown in cooler climates against a south-facing wall.

Shade — All fruits prefer full Sun unless stated otherwise.

Aspect — This refers to the aspect of a wall on which the fruit can be grown. All fruits can be grown on warmer aspects than that indicated. The key gives aspects in decreasing order of warmth.

Drainage — All fruits prefer perfect drainage. The degree of imperfect drainage tolerated by the species marked 'T' varies from one to another, and is not great in any case.

† For further information on these fruits, and many other unusual species, see *Plants for a Future*. (See Appendix A, Further Reading.)

KEY

Pollination
S = most varieties self-fertile
I = most varieties self-infertile
D = dioecous: male & female plants needed

Disease
R = generally resistant to diseases

Hardiness
T = tender, needs warm climate or microclimate
H = more than average hardiness

Frost
T = tolerates spring frosts

Shade
HS = needs Sun for at least half the day
Sh = tolerates more than half the day in shade

Aspect
S = south, south-east or south-west
W = west
E = east
(N) = north-west or north-east
N = north

Coastal
C = some tolerance to salty winds

Drainage
T = tolerates imperfect drainage

pH
Ch = can be grown on shallow soil over chalk or limestone
A = most intolerant of chalky soil

YIELD GUIDE TO THE COMMON FRUITS				
Fruit	Tree or Bush Type	Size, m	Yield, kg	Notes
Apple	Standard	6-8	45-180	
	Bush on M2 rootstock	6-8	90	
	Bush on M26	2.5-4	27-55	
	Dwarf bush on M9	2-3	18-27	
	Cordon	0.75-0.9	1-4	depending on rootstock
	Espalier, per 2 tiers	3-5		half the yield of a bush on the same rootstock
Pear				60-80% of apple of similar size and form; more variable from season to season
Plum	Bush on St Julien A	3.5-4.5	15-25	first 10 years of fruiting
	Same		40-50	full bearing
	Fan on St Julien A	3-5	7-14	
	On Pixy	2.5-3 (bush)		60% of St Julien A in same form
Gage	All types			most varieties – 60% of common plum variety of similar size and form
Berries, hybrid	Fence-trained	1.8-2.5	3-8	
Blackberry	Fence-trained	3-4.5	5-15	
Blackcurrant	Bush	1.5-1.8	4	
Blueberry, high bush	Bush	1.5-2	5	slow to come into full bearing; yield 2.5kg in fifth year after planting
Currants, red & white	Bush	1.5	4	
	Cordon	0.4	1	
	Fan	2	5	
Gooseberry				as red and white currants
Raspberry	Yield per metre of row	0.45	2-3	

NOTES
Size indicates diameter of mature tree or shrub, or width of restricted form tree. Planting distances should be wider than the diameter to allow for all-round sunlight.
Yields are very variable, depending on soil, microclimate and season.

Where and How?

There's a whole spectrum of ways to grow fruit, from the most intensive to the least. This is partly a matter of species choice, as discussed in the previous section, and partly one of the style of growing. (See boxes, below.)

The most intensively-grown fruits are best placed where they will get the attention they need, either in a domestic garden or an orchard. Trees grown in more isolated situations need to be more self-reliant, and will inevitably be grown in a less intensive manner. Thus to a great extent the place and method of growing are complementary and can be considered together.

At one extreme are dwarf or restricted form trees, as used in a modern commercial orchard or an intensive home garden. In the middle is the traditional orchard of standard or half-standard trees with pasture or lawn underneath. At the other extreme there are occasional trees of hardy varieties dotted around the rural or urban landscape, possibly grown primarily as ornamental or timber trees with fruit as a by-product.

The least intensive option may seem like the permacultural ideal. Mixing fruit with other elements in the landscape increases diversity, allows for the use of existing favourable microclimates wherever they occur, and encourages stacking. On the other hand there is much to be said for growing fruit intensively. The output is potentially high, and it will only be realised if the trees get the attention they need. It's much easier to give them this when they're all in one place, especially if that place is near to home.

Having said that, there's a place for every style of fruit growing. Which level of intensity is most suitable in any particular situation depends as much on the aims of the grower as on the characteristics of the site.

OPEN-GROWN & RESTRICTED FORM FRUIT TREES

The open grown forms of fruit trees are on the whole the 'tree-shaped' ones. They're normally pruned in the winter, and can be grown more or less intensively. Regular pruning is not essential, and they can yield well with little attention once they're well established. The open-grown forms are illustrated in the top picture.

The restricted forms are mostly two-dimensional, grown on wires, either against a wall or on a free-standing framework. Although each tree has a relatively low yield they take up little space and the yield per square metre is very high. They are pruned in the summer, usually more than once, and if pruning is neglected they yield very little. In general they need much more care and attention than open-grown trees. See the lower picture.

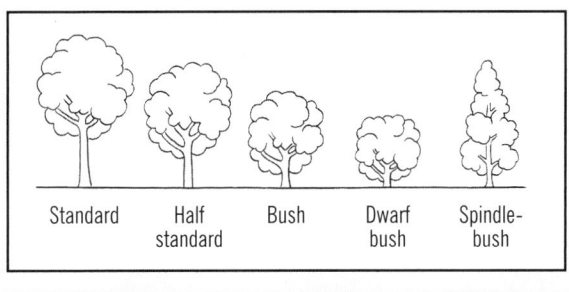

Standard   Half standard   Bush   Dwarf bush   Spindle-bush

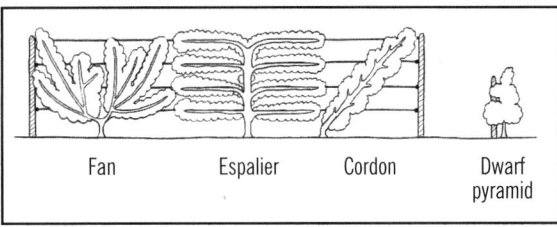

Fan   Espalier   Cordon   Dwarf pyramid

DWARF AND VIGOROUS FRUIT TREES[24]

Almost all fruit trees are grafted. The roots and the base of the stem are of one variety, known as the rootstock, and the upper parts of the tree are of another, either known simply as the variety or as the fruiting variety.

The fruiting variety determines what the fruit will be like, when it will ripen, the time of blossoming and so on. The rootstock determines the size the tree will grow to and the speed with which it matures; the smaller the tree the sooner it starts fruiting and the shorter its overall lifespan. Along with size goes vigour. In fact the largest trees, and the rootstocks on which they are grafted, are known as vigorous while the smallest ones are called dwarf.

Dwarf trees have much in common with restricted form trees. They can produce a great deal of fruit per square metre or per hectare. Their early fruiting and small size can be advantages: small trees are easier to tend, and a small quantity of several fruit varieties is more useful in a domestic garden than larger amounts of one or two.

But these advantages need to be balanced against the fact that dwarf trees are in effect infants throughout their lives and need a great deal of care and attention, including constant staking and weeding or mulching. Vigorous trees, on the other hand, are much more able to look after themselves once they are past the first few years.

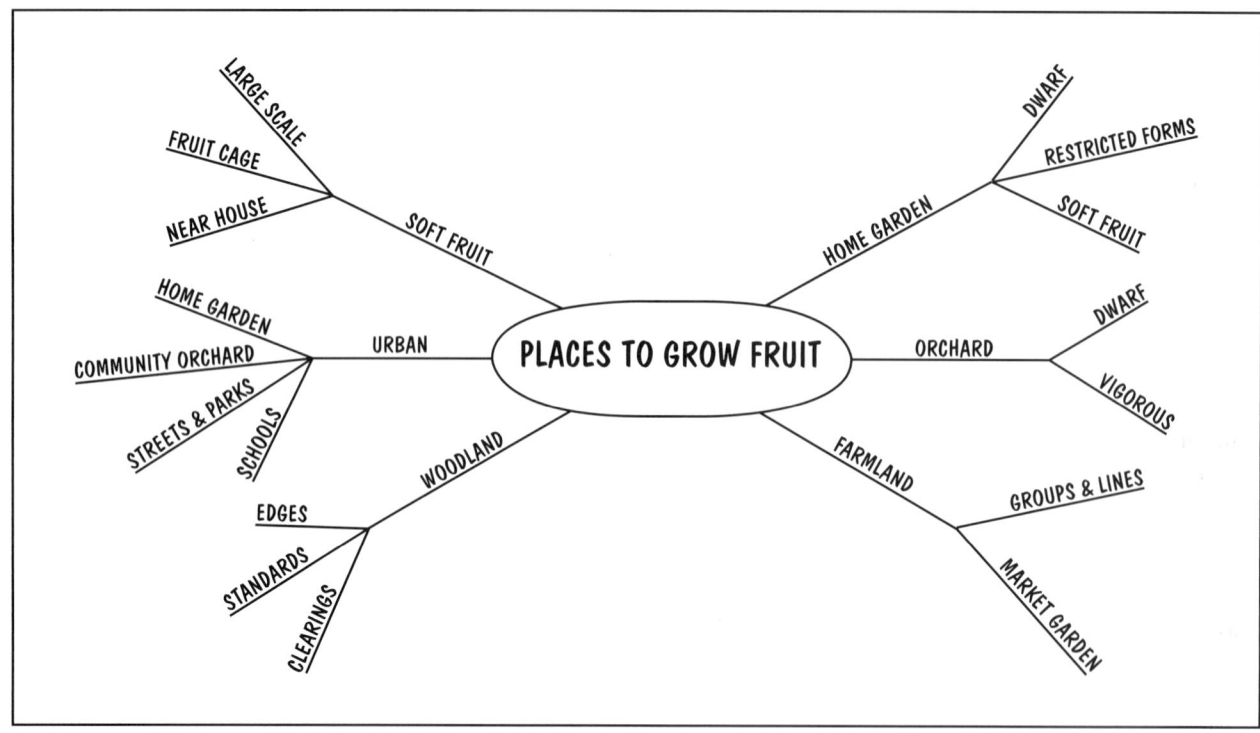

LARGE SCALE
FRUIT CAGE
NEAR HOUSE — SOFT FRUIT
HOME GARDEN
COMMUNITY ORCHARD — URBAN
STREETS & PARKS
SCHOOLS
EDGES — WOODLAND
STANDARDS
CLEARINGS

PLACES TO GROW FRUIT

DWARF
RESTRICTED FORMS
HOME GARDEN — SOFT FRUIT
DWARF
ORCHARD — VIGOROUS
FARMLAND — GROUPS & LINES
MARKET GARDEN

Home Garden

Although both restricted form and dwarf trees need more attention than vigorous open-grown ones, for most home gardeners the advantage of the extra yield from a limited space will outweigh the disadvantage of the extra care needed.

But where low maintenance is a priority, choosing what size of fruit trees to grow can be a bit of a compromise, because the bigger trees, although easier to grow, can easily produce more fruit than the family can eat.

This is not often the case with late-season apple varieties, which may keep for several months. But with non-keeping fruit, such as plums, most pears and early-season apples it can be difficult to make use of all the fruit before it goes off. There's no point in having 50kg of Worcester Pearmain apples, which keep for two or three weeks, if the family eats a maximum of 5kg of apples a week – unless you plan to sell, barter or give away the surplus. It would be much better to have two small trees of different varieties which ripen their fruit at different times.

All stone fruit and soft fruit must either be eaten soon after picking or preserved by bottling, freezing or jam-making. Growing an abundance of fresh fruit in season is one thing. Committing yourself to hours of work over the preserving pan each year is another. It's easy enough to match soft fruit plantings to requirements because the individual plants are small. But with non-keeping top fruit there is a choice to be made between large vigorous trees which need little maintenance and smaller trees which don't land you with a glut of perishable fruit.

The ideal is a succession of different fruits, and different varieties of each fruit, to give a continuous supply over as long a period as possible, and this means a large number of small trees. If there's space for one or two large trees it's probably best to make them keeping varieties of apples.

Another advantage of growing a larger number of small trees rather than fewer larger ones is pollination. Most, though not all, fruit trees are self-infertile. This means they can't set a crop of fruit without another tree of the same species but a different variety growing nearby to pollinate them. Different varieties blossom at different times during the spring, and two varieties must blossom at the same time in order to pollinate each other. If you only plan to have two apple trees and your two favourite varieties blossom at different times you could have a problem.[25]

Restricted Forms. In small gardens there's much to be said for restricted form trees rather than open-grown dwarfs. They have the same advantages of size, but

Cordons take up very little space and allow a range of varieties to be grown in a home garden, but they do need careful nurturing. Note that one of these has died.

they're ideal for stacking, and enable you to take full advantage of the special microclimates offered by walls. The disadvantage of growing fruit by a wall is that the soil is almost always drier here than in a more open position. (See p190.) Extra organic matter must be dug into the soil before planting, and the trees will benefit from plenty of mulch. The problem can be minimised by planting 30-50cm away from the wall and training the stem back to it.

Soft Fruit. Soft fruit comprises the bush fruits, including currants and gooseberries, and cane fruits, including raspberries, blackberries and their hybrids. Most soft fruits are admirably suited to the intensive conditions of a domestic garden. All the commonly grown kinds except raspberries and blackcurrants can be trained up a wall or fence. Gooseberries and red and white currants can be fan trained. Most kinds of soft fruit can manage with half a day's sunshine, so they do well on the more northerly walls in a garden.

One group which deserves particular mention is the hybrid berries, which comprises most of the kinds of cane fruits which can be grown. Most of them are similar to blackberries in fruit and growth habit. They make excellent use of vertical space in gardens of all sizes, including the smallest. They have some effect as a windbreak when grown up a fence or free-standing trellis. Unfortunately they aren't dense enough to give privacy, at least not if they are kept pruned for fruiting. Non-pruning is not really an option for more than a few years, as they will end up like wild bramble bushes, and few of us have enough space to accommodate these in our gardens.

Orchard

The dwarf option is the one most often taken by commercial growers. Not only are the trees easier to manage, but they come into bearing sooner, which means an earlier return on capital – a crucial point to anyone starting up with limited funds. On the other hand they have a much shorter life, perhaps 35 years for a dwarf apple compared to 120 for one on a vigorous rootstock.

Dwarf trees do need careful looking after. They need constant staking, are unable to compete with grass, are vulnerable to pests, diseases and inclement weather, and since their roots are restricted they can't forage far, so they need feeding.

The larger, vigorous or semi-vigorous rootstocks are perhaps more suitable for organic growing. It's also much cheaper to establish an orchard of vigorous trees, as the number of trees required may be a tenth or less of the number needed for a dwarf orchard. That means a tenth of the number of holes to dig too. In an orchard of large trees there's plenty of space and light to grow an intercrop in the early years before the trees start to bear, and this can compensate for the lateness of first bearing.

It's even possible to interplant with dwarf fruit trees, so that the produce is apples right from the start. A row layout is best for this, with the dwarf trees planted down the middle of the inter-row as far from the vigorous ones as possible. Eventually the vigorous trees will outcompete their little neighbours and render them completely unproductive. But by then the dwarfs should have paid for themselves.

Trees on the more vigorous rootstocks may be grown as a bush, half standard or standard. The advantage of the bush is that it's relatively easy to reach for harvesting and pruning, though the upper branches will still be out of reach of a person on the ground. The traditional advantage of the half standard and standard forms is that sheep and cattle respectively can graze under them. This also means that they lend themselves to underplanting, as long as the undercrop receives enough light, and this may be an important advantage in a permaculture design.

Farmland

Standard fruit trees were traditionally grown in hedgerows in Worcestershire, and a few still remain, scattering their fruit forlornly on the ground to feed the redwings and fieldfares. These days few people have the time to make a trip out with ladder and basket to pick a single tree, especially one of an old variety with an unfashionable taste. But on a smallholding, where space is limited and you want to make the most of the resources the land offers, it can be worthwhile.

Small groups or lines of trees, perhaps positioned to take advantage of a special microclimate, may be a better idea than single trees. A group is more worth the trip to harvest or prune than a single tree. A line of trees along a regularly used track or path can work well, as the plants get the attention they need as part of the natural course of things.

Tough trees are needed for these kinds of situation, where they will get a minimum of care: standards on vigorous rootstocks, cookers rather than eaters, and traditional varieties, or at least those recommended for organic growing in your area. Cider apples and perry pears will do well enough in this kind of situation, but there should not be dense vegetation at their feet, as they are normally harvested by shaking and picking off the ground. In the Lake District, damsons are traditionally planted in odd corners and along by walls. They are the toughest of all commonly grown top fruits and the only one which is worth planting in the local climate.

Market Garden. In a market garden fruit trees can be integrated with field crops in a more intensive way. Lines of espalier and fan fruit can be grown between the vegetables without interfering with the work or unduly competing with the vegetable crop, especially if the rows run north-south. (See p220 and illustration on p203.)

On one farm I know where field-scale vegetables are grown, lines of espalier trees are used to mark the boundaries of the different courses in the rotation.

Before the trees were planted it was difficult to see in the spring where the boundary between different crops had been the previous year.[26] Where vegetables are grown in permanent beds this is not necessary, but in either case it makes for intensive use of the land, and is a good example of the stacking principle.

Woodland

As we have seen there is a conflict of aims between growing timber and fruit on the same tree. (See pp220-221.) Here we look at the possibilities for growing some trees purely for fruit in woods and plantations where the majority of the trees are grown for timber or wood.

Edges. Edge trees have at least one side exposed to the light, and may yield a moderately good crop of fruit. Timber quality is usually poor on an edge, as the lower branches on the exposed side of the trees are not suppressed by a closed canopy. Some foresters plant an edge strip of non-timber native trees for wildlife and aesthetic value, but hardy fruit trees can be used too. Edges facing east, west or south are all suitable.

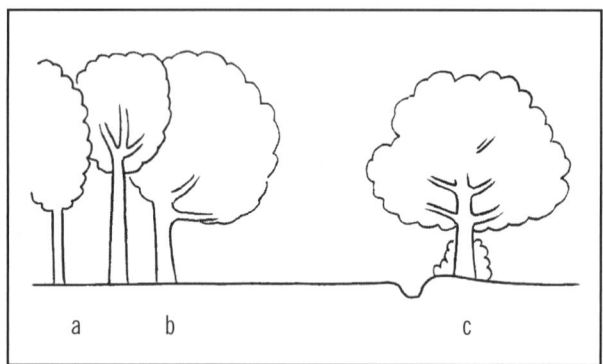

Typical tree forms in different situations: a) in a plantation, b) on an edge, and c) in a hedgerow.

They will be exposed to root competition from the timber trees, which are invariably more vigorous than them. There is also the possibility of infection with honey fungus, a parasite of trees which is very often present in a woodland. The woodland trees tolerate it but fruiting trees are more susceptible.

Semi-wild fruits and nuts grown on their own roots, including bullaces, damsons, cherry plums, mirabelles, chestnuts, crab apples, elder, wild service and blackberries can all handle these conditions. The first four of these are members of the plum family and will sucker, a habit which may enable them to move around over time as light conditions change. Sweet cherries and hazels can also compete, but the fruit and nuts will all be taken by birds and grey squirrels in this situation. (See the illustration of damsons growing on a woodland edge in the Colour Section, 31.)

It may be worth trying a few vigorous grafted trees of other kinds, like those recommended for farmland above. But because the yield potential is limited to one

side of the tree it's probably not worth spending much money on establishing them. Cheap trees planted on a sink-or-swim basis are probably the best bet. Soft fruit may do well, as most kinds are pretty shade tolerant, but there is the problem of birds. (See p237.)

As well as the exterior edge of the wood, an edge on a ride or a glade can be used, but there's the added problem of shading by the trees on the other side of the ride or glade. Only a south-facing edge is really worth considering, and at a very minimum the width of the glade should be twice the height of the trees on the south side, that is the north-facing edge. Of course the height of the trees will vary as they grow and are felled, and the measurement should be based on their maximum height through the rotation. Treating the woodland to the south of the glade as short rotation coppice would help, though it may not be worth it if the value of fruit is low and short rotation coppice is not wanted for its own sake.

Clearings. It is possible to grow an orchard within a woodland glade or clearing, as opposed to placing individual trees on an edge. Conventional wisdom regards woods as nothing but trouble for orchards as they harbour so many pests and diseases. A minimum distance of 25-30m is recommended between woodland and orchard, and this should be free of woody plants so as to keep squirrels and honey fungus at bay. This would also avoid possible problems with shading and root competition.

I was somewhat sceptical about orchards in woods until I visited Ben Law and saw his woodland orchard. Sited in an existing clearing in a chestnut coppice, it yields abundant fruit, with some trees only 3m from the nearest coppice stool. (See case study, Living in the Woods, p336.)

It would not be safe to extrapolate from Ben's experience and say that anyone could do the same in any wood. Conditions vary from place to place. For example, where there are mature ash trees and dense shrubby cover bullfinches may be a significant problem. (See p248.) Where there is no existing clearing and one has to be made, there is the possibility of carry-over of soil-borne pests and disease from the woodland trees, especially honey fungus.

Honey fungus is the big unknown. It's present in most woods, but there are many strains of varying virulence. Fruit trees and walnuts are among the most susceptible of trees, but they may be able to cope with an infection by one of the weaker strains. Trees can sometimes grow quite happily for many years and succumb when middle-aged, so there is always some risk, even when initial plantings have got off to a good start. Although a recent clearing carries the highest risk of honey fungus it can spread underground from nearby trees.

Two rootstocks which have some immunity are pear and hawthorn. Pears grown on pear rootstocks make huge trees which take decades to come into fruit, but this may be appropriate in a woodland.

Hawthorn is only used commercially with medlars at present, but apples and pears can be grafted onto it.[27]

Shading and root competition can both be controlled in a coppice wood by adjusting the length of the coppice rotation in the coupes immediately adjacent to the orchard. Coppicing not only reduces the height of the trees but causes some die-back in the feeding roots when the upper parts are cut. In high forest grown for timber this is not possible, and I really don't think an orchard would work in high forest.

Birds are much less of a problem when there are people around. A coppice where value is added to the wood on site – charcoal-making, hurdle-making and so on – is much more inhabited than one which is just a source of raw materials. In fact an orchard will really work best when the coppicer actually lives on site, not just because it will keep the birds away but also because it means the fruit trees are more likely to get the attention they need.

Glades and clearings can be frost pockets. Cold air is not as 'liquid' as water and will flow along the top of woodland rather than sink down into it. But it will flow down into any gap in the trees, and if it can't flow away from there a frost pocket will form. A flat clearing with no outlet for cold air is likely to be a frost pocket, whereas a clearing on a slope with a ride at the bottom to allow cold air to flow away is unlikely to be one. But it's always better to find frost pockets by observation than by trying to predict them. The great advantage of a woodland clearing from a microclimate point of view is shelter from wind. But some air movement is desirable to prevent fungus diseases.

and pears on the most vigorous rootstocks, preferably cooking or cider and perry varieties are also a possibility. Harvesting will be difficult except in years when the coppice is cut.

This is something else which Ben Law is trying in his woodland.

Urban

Home Garden. The biggest potential for fruit growing in towns is certainly in people's private gardens. Although many people nowadays have neither the time nor inclination to grow vegetables, fruit takes hardly more work than perennial ornamentals, and is much more fun. A couple of days' work a year, apart from picking, is enough to keep a family in fruit for almost the whole year.

Sadly all too often the few fruit trees which remain in back gardens are ignored, the fruit left to fall and swept into a corner to rot down. There are a number of reasons for this: the fruit varieties are often old and unpalatable to the modern taste, people eat less fresh unprocessed food anyway, and food is now so cheap compared to wages that people can't be bothered. But above all it's because we've lost the culture of producing things for ourselves.

As I noted at the beginning of the Fruit part of this chapter, this unused resource presents a business opportunity, and is possibly one of the most realistic ways of growing organic fruit, especially if the emphasis is on juicing rather than selling fresh fruit. This is an example of niche management. (See p261.)

Ben Law in his woodland orchard. The tall trees in the background are chestnuts left as standards for nut production in the adjacent coppice.

Standards. It may also be possible to grow fruit and nut trees as standards over coppice. They must be very vigorous kinds, and the coppice must be cut on a short rotation. Pear, chestnut and walnut seedlings, as opposed to grafted trees, are the best bet, but they are unlikely to yield as much as grafted varieties and may take many years before they start to yield. Apples

---

A SUCCESSFUL NICHE

I heard this story about a man in Melbourne, Australia. He noticed that many of his neighbours had chestnut trees in their gardens which they were not harvesting. He approached the families and asked them how many chestnuts they ate in a year. In all cases the amount was a small fraction of what the trees were yielding. "How would you like it," he asked them, "if I was to pick and husk that amount of chestnuts for you and keep the ones you don't want in return?" Most of them said yes.

This gave him a good enough supply of chestnuts to start a wholesale business, and it grew till there was hardly a chestnut tree in the city left unharvested. After a year or two he started approaching people who didn't have chestnuts in their garden and offering to plant one free, on the agreement that he would do the harvesting and keep any surplus for himself. His business expanded even more, and now they say he's driving a Mercedes.

Community Orchard. A community orchard, whether in town, suburb or village, is perhaps the best way people can get together to grow food. (See Community Gardens, p183-184.) Fruit and nut trees have the advantage that they don't need the frequent attention which annual vegetables or a forest garden do. The work, mostly pruning, grass cutting and harvesting, comes infrequently, but when it does it's on a big enough scale to make a communal work day.

A traditional orchard of standard trees and grass also makes a meeting place for celebrations, open air parties and picnics. In an orchard of bush trees there's less space, at least once the trees have grown, but an area can be kept open for events and gatherings.

Many old orchards, in both rural and urban areas, are threatened by development, and they can often be saved by determined action on the part of the local community. This is not only a gain for sustainability, biodiversity and landscape amenity, but the action can play an important part in bringing people together and building community.[28]

Vigorous trees are probably better for communal orchards, as they may not get as much regular attention as private ones, but they don't start producing fruit for perhaps eight or ten years after planting. A good compromise is to use the moderately dwarfing root-stocks (such as MM106 for apples and Quince A for pears) which can start to bear in three to four years.

In many areas security must be considered. It's very disappointing to grow a fine crop and find it vanish a few days before harvest. But the most important consideration of all is clarity about how the orchard is to be run. Planning for continuity is especially important, as trees are a long-term commitment, and they will probably outlast successions of different people involved in the project. It's a good idea to have everything agreed and written down from the start of the project.

Streets and Parks. There are limitations on growing fruit in public urban spaces. Pollution from road traffic is one. (See p190.) This will vary greatly, from busy roads where any food growing is out of the question to quiet suburban streets which may be less polluted than some country areas. Considering the pesticide residues in the average apple, even when the pesticides have been applied strictly according to the manufacturers' instructions, an apple grown organically in a city location may not be a bad bet. Nevertheless it would be wise to monitor air pollution levels over a period of time before deciding to plant trees in any open space where traffic levels are high. Soil pollution may be a problem on some urban sites, especially where there has been industry in the past, and this may need testing too.

The other big question about growing fruit in public places is Who's going to look after it, and who's going to eat it? The answer depends to a great extent on the situation.

Village Homes, a suburban development in the town of Davis, California, was designed and built to be as ecologically sound as possible. The roadsides and the spaces which would otherwise be down to sterile lawn and ornamentals are planted with fruit trees. The trees are mostly half standards, so there's still plenty of play space for the children. The work is done by a maintenance team, part of whose wages are paid out of sales of organic fruit.

Not many residential areas are as tight-knit and organised as that, and the Village Homes model wouldn't work where theft is a real problem. But it's impossible to steal something which is freely given, so the best approach in areas where security is impossible is to put up a sign saying, 'Please pick the fruit'. Why not? Any fruit not picked by the public could be picked by the council workforce and juiced.

The cost of planting and tending the trees would hardly by met by sales, but the costs need not be much more than those of non-edible trees. Tough, self-reliant kinds of fruit can be used, especially tip-bearing varieties which need little or no pruning once the basic shape of the tree has been determined. Walnuts, almonds, cherry plums, damsons, cooking apples and sour cherries are some possibilities. There's no reason why these kinds of trees should not become the norm for parks and street trees in many areas.

Where the cost of establishment *is* higher than for purely ornamental trees it would be an excellent use for lottery money. Our descendants, in a world less replete with under-priced fossil fuels, would thank us for it.

However the biggest problem with growing fruit and nut trees in towns is prejudice. In their otherwise excellent book *Planting Native Trees and Shrubs*, Kenneth and Gillian Beckett make the bizarre statement "...[hazel's] very edible nuts perhaps reduce its usefulness as a planted tree in heavily populated areas." No reason is given, and it's hard to think what it could be, other than an ingrained cultural attitude.

These apple trees bring life to a corner of the urban landscape.

Perhaps the most important benefit from growing fruit and nut trees in urban situations is to help recreate the link between plants and food which has to such an extent been lost in our culture. Everyone deserves to have the experience of gathering at least a little of their own food from a living plant.

Schools. Schools provide perhaps the best opportunities for growing fruit in public, or at least semi-public places. Not only can they be made secure if necessary, but the educational opportunity is really valuable. If attitudes are to change in future this is where the change must start. An orchard has the advantage over a vegetable garden that it can cope better with neglect over the school holidays, and most children are more keen on fruit than vegetables.

Soft Fruit

On the whole soft fruit is best grown intensively. Small plantings scattered around the landscape are likely to be taken by birds, though the different kinds vary in their susceptibility, from redcurrants which almost always get stripped by birds to blackberries which are relatively immune.

There are three ways to get round the bird problem. The first is to grow the bushes close enough to your house that the birds are frightened off by the constant possibility of human presence. How effective this is will depend on the size, tastes and timidity of the local bird population. In inner city areas, where bird populations are low and cat populations high, soft fruit may not be much troubled even in places which are relatively unfrequented by people.

The second way is netting. A full-scale walk-in fruit cage can be quite an investment, but it will be repaid many times over by the increase in the weight of crop actually harvested. Smaller plantings can simply be covered with nylon netting which is removed to get at the bushes. This is easy to do on fruit which is trained up a wall or fence.

The third way is to grow them on a really large scale. Where there are only a few fruit bushes in one place the local birds can take all the fruit with ease. But where there are acres of them the birds can eat their fill and still only take perhaps five or ten percent of what's there.

Another reason why soft fruit is best concentrated in one place is weeds. Unlike trees, the bushes can easily be swamped by herbaceous growth, and the closer they are to the eye of the gardener the more likely they are to get the attention they need.

Microclimate

Choosing the right microclimate and improving the one you've got are both important to successful fruit growing. A general introduction to microclimates is given in Chapter 4. The requirements of most fruits are:

- Freedom from frost at blossom time – late April to early May for most common fruits.

- Shelter from wind in winter to protect fruit buds.

- Shelter at blossom time, both to protect blossom and to provide the right conditions for pollinating insects.

- Shelter in summer and early autumn to prevent the fruit being damaged or blown off the trees. Shelter can also give earlier ripening.

- Warmth through the summer.

- Full sunshine for dessert varieties of top fruit.

- At least half the day's sunshine for most cooking and soft fruits.

In a home garden there's often a range of microclimates, each suitable for a different fruit. If the garden is in an area which is marginal for fruit, most of the microclimate factors can be enhanced by growing fruit against walls, especially those with a southerly or westerly aspect. The extra care required by restricted forms is well worth it if it means that delicious tender fruits can be grown.[29]

When choosing a site for a larger orchard, it's well to remember that fruit is a relatively high value crop and a very demanding one as regards microclimate, so it has a better claim to a favourable microclimate than does arable or pasture. Vegetables are equally high value but less sensitive to frost, so there may not be competition for the best site between them and an orchard. (See case study, A Permaculture Farm, p294.)

Shade, which can be the dominant microclimate factor in home gardens, is usually less important on a field scale. Fruit needs plenty of sunshine for ripening, which means in autumn for top fruit, and the Sun is lower then than it is in summer. If there are tall trees or buildings to the south of a possible orchard site the amount of shade they cast should be checked during September or early October. Any area which receives less than half the day's sunshine in this season should be planted to cooking varieties, or ideally to soft fruit, most of which ripens in summer.

Ways of minimising frost have been described in Chapter Four. (See p78-79.)

Aspect

Aspect is a complex subject, as many factors come into play. One grower who had a complete hill under apples didn't notice a great difference between aspects over the years. In a wet year the south-facing slope did well, and in a drought the north-facing one. But much depends on the part of the country you're in and the range of fruit to be grown. For orchards of our traditional fruits – apples, pears and plums – the following points are significant:[30]

- North slopes are often reckoned to give the most reliable and regular cropping in warmer parts of the country. They're sheltered from damaging south-west gales and from extreme heat and drought. They may escape frost damage when fruit on warmer aspects is encouraged to blossom early by a precocious warm spell only to be caught by a later frost.

- East slopes are sheltered from south-westerly gales, and may be as reliable as any other slope. In eastern areas, cold east winds may cause bud death in winter, and this can narrow the choice of varieties. Frost damage may be greater on east-facing slopes. (See below.)

- South-west slopes may lose crop from the wind, even relatively moderate winds. They are usually the driest aspect, firstly because the afternoon Sun is hotter than the morning, and secondly because of the drying effect of the prevailing wind. These slopes are good for later ripeners, though they may be caught by autumn gales.

- South slopes are even hotter and most in need of irrigation during drought. They are good for late ripeners. In northern areas, where summer warmth, light, and the length of the growing season are limited, south-facing slopes are usually best. The same may be true in cloudy, wet areas in the west.

Tender fruits have their own specific requirements. Some, such as grapes, don't mind a summer drought, and most are more susceptible to frost than the traditional fruits. In general south and south-west slopes are best for them.

Aspect and Frost. The relationship between aspect and frost is particularly important for wall-trained fruit. Much of the damage caused by frost to blossom and young shoots occurs not when they are frozen but as they are thawing out. The faster the thaw the worse the damage, so fruit situated in a position which warms up early in the morning is more vulnerable than that in a spot which warms up more slowly. In general an easterly-facing aspect will be more vulnerable than a westerly one because the Sun rises in the east.

For apples, pears and plums grown in an open orchard this is hardly a significant effect, but because a wall concentrates the heat, wall-trained fruit may be more susceptible. The time of blossoming is significant here. In April the first heat of the morning Sun comes approximately from the east, but by the end of May the Sun is rising in the north-east, and northerly and north-easterly aspects are vulnerable.

Species which have tender spring foliage, such as walnuts, grapes and kiwis, including the hardy kiwis, are more susceptible. If there's a choice, easterly aspects are best avoided for these fruits even when they are grown in the open.

Shade cast by trees and buildings can slow down the thaw on a vulnerable aspect. If you're thinking of growing fruit, especially tender fruits, on an easterly wall or slope the best thing is to observe it through the springtime before making a planting plan. If possible this should be done on frosty mornings, but if the opportunity doesn't arise any sunny morning will do because the key thing to observe is how quickly it warms up.

Orchard Windbreaks[31]

The vast majority of commercial orchards in Britain have windbreaks, and they are just about the only crop which usually does have them. This is proof enough of how important windbreaks are to the success of fruit trees on dwarfing rootstocks, which all modern orchards consist of. Old orchards of standard trees often don't have windbreaks because the vigorous trees are that much hardier. But even where they're not essential orchard windbreaks are usually worthwhile.

Unless your orchard is unusually well sheltered, the first thing to decide is what kind of windbreak you need or want. This depends to a great extent on what you're protecting and on how strong the winds are. If you're planning an orchard of tender fruits on dwarf rootstocks and/or your climate is marginal, you need to get serious and design a windbreak for maximum effect. If you're planning on tougher fruits and/or have an ideal climate for fruit you have much more latitude.

Broadly speaking there are three styles of windbreak:

- Business-like, designed for maximum effect where maximum protection is the aim.

- Informal, low-input / low-output using locally native trees and shrubs.

- Edible.

The principles of windbreak design are given in Chapter Four, (see pp83-84) and the following points are specific to orchard windbreaks. They apply more to the conventional 'business-like' style, because this is the kind which needs the most precise design. The other two can be seen as more relaxed versions of the conventional type, and their design is perhaps more up to the preferences of the individual.

Layout. Because of the high value of shelter to orchards, windbreaks are usually placed closer together than they would be on farmland. Distances as close as six times the height of the windbreak for apples and four times for pears have been used profitably – that is 5m windbreaks at 30m and 20m apart respectively. This is not universal, and 20-25 times the height is more common.

There is a qualitative difference between these two extremes of spacing. Windbreaks at the wider spacings act independently of each other, with windspeed at ground level recovering between them. Those at closer spacings, around ten times the height or less, act together to lift part of the airstream right above the area over which they form a network. On this basis a spacing of ten times the height may be the best compromise between effectiveness and economy.

Single lines of trees are normally used. This is partly because fruit is a high value crop, so the land is valuable. It's also because letting part of the airflow through the windbreak is particularly important for fruit, firstly to reduce the likelihood of frost (see p79), and secondly to allow the trees to dry out after rain to reduce the incidence of disease. It's also relatively easy to prune a single row to keep it at the desired height.

Where winds are stronger a multiple-row shelterbelt may be preferable because gaps are more likely in a single line. A belt may also grow faster than a single row and be taller when fully mature. But it should not be dense, and a total width of 6m should be sufficient. Another option for windy areas is to plant single-row windbreaks more frequently across the orchard and make them shorter – hedges rather than tree rows. This may give less shading and root competition.

Whatever the layout, shading and root competition can be reduced by placing roadways next to windbreaks, or using the space to plant a strip of herbaceous attractant plants. (See pp242-244.) One situation in which shade can be an advantage is on the east side of an orchard to shade the trees from the morning Sun after a frosty night. (See Aspect and Frost, opposite.)

Tree spacing within the windbreak is commonly 1.5-2m in a single line windbreak, but with poplars it varies according to the variety used, and may be up to 5m with very vigorous ones. Advice should be sought from the supplier of the poplars.

Tree Species. Hybrid black poplars are the most commonly used trees, mainly because they grow so fast. They can grow as much as two metres a year in ideal conditions, and many of them have the tall, narrow shape required for a windbreak. Because they take from cuttings establishment can be quick and cheap. They're thirsty trees, and should be avoided both in low rainfall areas, where they might compete with the crop for water, and near buildings on clay soils, where they could damage the foundations. The best varieties for windbreaks are those which keep their lower branches as they mature, developing a columnar rather than a lollipop shape.

The white and grey poplars are suckering trees. This can be an advantage in that the suckers often thicken up the bottom of the windbreak, but a disadvantage in that they can come up among the crop trees. In a way this is an advantage too, as the suckers are visible evidence of how far the roots of the windbreak trees have reached. Black poplar roots are no less competitive for being invisible, and they can reach as far as 30m away from the trees.

Root pruning is necessary for all poplars. This can be done with any single-tined implement capable of cutting the roots, drawn by a tractor parallel to the windbreak and a few metres away. If it's done every year or two the roots will not grow too thick to be easily cut.

Alders can grow almost as fast as poplars, especially in the first few years, and like poplars have a relatively tall, narrow habit of growth. They are often used, and may be the best choice for an organic or permaculture orchard as they provide habitat for some pest predator insects.[32]

This windbreak of poplars has been underplanted with shade-tolerant conifers.

The roots tend to grow downwards rather than horizontally, which means there's less problem with competition, and hopefully less need to root prune. The native alder requires a constant supply of moisture during the summer to achieve good growth, but it doesn't need the very wet conditions which it favours in the wild. On drier soils and in drier parts of the country the Italian alder is preferred. All alders fix nitrogen, and they make a good windbreak when alternated with poplars or other trees, which can benefit from their nitrogen.

The trees may get thin at the bottom and require underplanting with shade tolerant species. Conifers such as Lawson or Leyland cypress or western red cedar are sometimes used. These will eventually become as tall as the poplars, and can be used to replace an ageing windbreak. If they are kept well thinned they can perform quite well, though the overall permeability of the broadleaves is preferable to conifers with gaps between them.

Alternatively more shrubby species may be used as an understorey. Hazel and *Eleagnus* are two options with edible outputs, and the latter fixes nitrogen. Holly, although slow growing, is very shade tolerant and has a saleable product at Christmas time. Note that the variegated hollies are less shade tolerant, and better grown in an open hedge than as a windbreak understorey.

Spring-flowering willows, although useful as pollinator attractants, are neither tall enough to act as main trees nor sufficiently shade tolerant to go in the understorey, unless they are on the south side of the windbreak. They could fit well if the option of more frequent, shorter windbreaks is preferred.

Trees to avoid in an orchard windbreak are those which may suffer from the same pests and diseases as fruit trees. Fireblight is the most important disease to consider here, as it affects all members of the Rosaceae family, which includes both fruit and potential windbreak species. The family includes: all common top and cane fruits, blackthorn, hawthorns, roses, rowan, whitebeams, service, *Amelanchier*, *Spirea* and *Sorbus*. Pears are the most susceptible fruit and hawthorn can be a serious carrier of the disease, so these two should definitely not be combined. Ash is host to the same canker as apples and pears.

There are other pests and diseases common to both fruit and potential windbreak trees. But it's probably not worth worrying too much about them as long as the orchard enjoys both a good microclimate and the kind of healthy diversity described in the sections below, and resistant varieties of fruit have been chosen. If the soil or microclimate is marginal for fruit then more care should be taken. (See box, Breaking the Rules, p226.)

The Long Term. It's important not to let the bottom of the windbreak get too thick or the fruit trees will suffer from lack of ventilation. This may mean occasional selective coppicing of shrubs.

The longevity of the windbreak must be considered at the design stage. In humid areas with little wind, ventilation may be more important than shelter, and it may be best to remove windbreaks when the fruit trees are ten years old. But usually they will remain for the lifetime of the orchard. How long this will be depends firstly on the rootstocks on which the fruit trees are grown, and secondly on whether the orchard is to remain in situ for more than one generation of crop trees.

In practice it's difficult to extend the life of a single-row windbreak beyond two generations of windbreak trees. This is because trees which cast little shade are generally not tolerant of shade themselves while those which are tolerant of shade cast a dense shade themselves. The initial planting needs to be a light-shading tree to allow for underplanting, and the second generation needs to be tolerant of shade, which means it will cast too much shade itself to allow a third generation to be planted under it.

Informal and Edible Windbreaks. In a less intensive orchard a serviceable windbreak can be grown simply using the trees which grow wild in the locality. They may not be perfectly suited to the job but they will probably grow well, especially if they are raised from local seed or cuttings. Birch is tough and fairly fast-growing, but has a short life and may lose its lower branches rather quickly. Aspen, a white poplar, is tough and easy to propagate from suckers. The suckers, as noted above, are a mixed blessing.

An edible windbreak, composed partly or wholly of fruit and nut trees, may be appropriate where space is limited and you don't want to devote any land to plants which don't produce food. Edible windbreaks are discussed in Chapter 8. (See box, Edible Hedges, p187.) A sloping profile on the downwind side of the windbreak gives more edge and may make it more productive. But the upwind side should be kept as nearly vertical as possible for maximum effectiveness in reducing wind. (See p85.)

Plant Diversity

Crop Diversity

Many conventional commercial orchards contain only one variety of one fruit species plus a pollinator. Organic or permaculture orchards contain a mix of varieties, and often of species. This diversity helps plant health and insures against a total crop failure if things go wrong. A monocultural orchard is designed to supply in bulk to a wholesaler or supermarket, who want large quantities of each variety. A polycultural orchard, on the other hand, suits direct local marketing. Retail customers want diversity, and they want to be able to buy fruit through as long a season as possible, which can only be done with a range of species and varieties.

LAYOUT FOR POLLINATION

The trees in an orchard can be laid out in rows, on the square or on the hexagonal pattern, as described above for nuts. (See p219.)

Where honey bees are the main pollinator the arrangement of varieties within these overall layouts can have an effect on the thoroughness of pollination. The bees economise on energy by moving from one flower to another close by. In an orchard laid out in rows they tend to work up and down the rows rather than crossing the wide gap to an adjacent one. Thus trees which pollinate each other – trees of the same species and of varieties which blossom at the same time – are best placed in the same row. Where the square or hexagonal layouts are used it's best to group trees which pollinate each other in the same part of the orchard.

Fruit trees can be combined with nuts. One possibility is to place the nuts so as to shelter the more tender fruit trees. Another is to grow low-pruned hazels under standard apples or pears. But nut trees have a very similar niche to fruit trees, and a more productive polyculture can be had by interplanting quicker-growing crops while the trees are young.

Soft fruit is an obvious choice. Most soft fruit is productive for about a dozen years, though gooseberries can last much longer. Top fruit on vigorous rootstocks starts bearing, lightly at first, at around eight to ten years. When the soft fruit is finished the trees will be just starting to crowd it and shade it, and just ready to take over the productive role.

It's quite possible to net the soft fruit between the trees. (See illustration, p204.) On a large scale a single overall net could be more economical than a series of small ones between the orchard rows, but then on a large scale netting may not be needed at all. (See p237.)

The same can be done with vegetables. Planting young fruit trees in a vegetable garden means that you don't have to make any effort to weed, water and manure them because they're already in an environment which is kept weed-free, moist and well manured. The vegetables are not vigorous enough to significantly compete with the trees, though very competitive plants, such as mints, should be avoided at the base of the trees.

Nor, for a number of years, will the trees significantly affect the vegetables, either by root competition or by shading. Just how many years will depend on the rootstock chosen. Dwarf and vigorous trees are much the same size when they're young, but the dwarfs are planted much closer together and so will start to affect the vegetables sooner.

Of course you end up losing your vegetable garden, which becomes an orchard. Charles Dowding of Shepton Montague in Somerset planted standard apple trees in his raised-bed market garden. When I asked him why he'd put them there he told me he would want to retire some day and when he did he would have an orchard instead of a garden. This is the permaculture principle of succession in practice.

Standards or half standards are a better shape than bush trees for this, as they grow upwards and leave space beneath for growing the vegetable crop. Charles also told me that he wished he'd pruned his trees so that the main branches lie along the beds rather than across them. The ones which came out at right angles stretched across the paths and he kept hitting his head on them! Espalier and fan trees are also a possibility, and would completely remove this problem, but they require much more work both to establish and to maintain. They are not often used on an orchard scale these days as dwarf trees provide an easier way to grow fruit intensively in the open.

The amount of light reaching the ground of this mature apple orchard at noon on midsummer's day suggests that an understorey would be possible here.

But the same scene at 6pm on the same day suggests that the average light level through the day and through the growing season would not be enough. Either the trees must be at a wider spacing, or an understorey could be grown in the early years of the orchard when the trees are smaller.

A possible cash crop for established orchards is spring flowers, grown directly under the trees. This may be specially suitable for dwarf trees. Dwarfs have to have an area beneath them kept grass-free throughout their lives, but they could tolerate the milder competition of bulbs and other flowering perennials. (See Ground Covers, below.) The flower crop would give an additional economic return for the task of keeping the area weeded. But the idea that the flowers could outcompete grass without any help from human hand is a bit optimistic.

One potential problem with any orchard understorey is control of apple and pear scab. Conventional growers spray for it, up to 15 times a year. Organic growers use resistant varieties, and break the disease's life cycle by getting rid of the leaves in winter. This is done either by collecting them up and composting them or by shredding them with a mower, so the small pieces get taken underground by earthworms. Either of these is only possible if the ground is clear in the winter. Another possibility for scab control is to run chickens in the orchard in winter. They will make a good job of shredding the leaves, and reduce other pests. But they will limit which undercrops can be grown. Crops which are green in the winter, such as leeks or winter cabbages would be eaten by the chickens.

Non-Crop Diversity

Non-crop plants which can be useful in an orchard include:

- Ground covers.
- Nitrogen fixers.
- Pest-predator attractant plants.
- Pollinator attractant plants.

Ground covers & nitrogen fixers. It's normally recommended that young trees, and very dwarf trees throughout their lives, should be grown in bare soil or kept constantly mulched. Keeping soil bare is obviously undesirable, and can only be done economically with herbicides, while mulching involves a great deal of hard work, unless woven plastic mulch is used.

But it's not plants in general that young trees can't bear competition from, it's grass. (See p327.) The bare earth policy has arisen because grasses are the most likely plants to grow on ground where the only maintenance is cutting.

Young trees can thrive with a ground cover of clover. In fact they can do better with clover than with mulch. I have seen a young orchard, half of which was heavily mulched with straw and half underplanted with red clover. The underplanted trees were doing noticeably better than the mulched ones, no doubt due to the nitrogen provided by the clover. Mulch could perform better than clover where something other than nitrogen was the limiting factor to tree growth, either moisture or other nutrients. But mulching is very hard work, unless plastic mulches are used, and on anything but the smallest scale the mulch material must be bought in.

The clover can easily be maintained by cutting, with the cut material left on the surface as a feeding mulch. Compost can be added as a top dressing as needed. Any perennial kind of clover is suitable, with white and alsike clovers lasting longer than red. Lucerne is less suitable as it may compete excessively for water in a dry year.

Clover does not persist for ever, especially in a pure stand. It's likely that grass will eventually establish itself in the sward, even if some clover remains. This suits trees on vigorous stocks, especially if it coincides with the onset of fruiting. They need nitrogen when they're young and growing vegetatively, but too much of it can suppress fruiting. In fact one reason for putting orchards of trees on vigorous stocks down to grass is to depress vegetative growth and thus encourage fruiting. If the clover does persist once fruiting has started, a high-potassium fertiliser, such as comfrey liquid or wood ash, will redress the nutrient balance and encourage fruiting.

Nitrogen fixing nurse trees may also be used, in the same way as suggested for nuts. (See p219.)

There is a wide range of non-nitrogen-fixing ground cover plants which may be appropriate for dwarf trees during their fruiting years.[33]

Attractant plants. Pest-predator attractants are the main means of controlling pests in an organic orchard, just as resistant varieties are the main means of controlling diseases.

The two most important groups of pest-predator insects are ichneumon wasps, whose larvae parasitise caterpillars, and hoverflies and lacewings, whose larvae eat aphids at a prodigious rate. The adults of these insects eat nectar and pollen (though adult lacewings are omnivorous) and because they have short mouth parts they find flat, open flowers the most accessible.[34]

Two families in particular have flowers like this: the umbellifers, or cow parsley family, which have many small flowers gathered together in umbrella-like

A hoverfly on a pot marigold flower. Although an annual, pot marigold can self-seed prolifically in garden conditions.

flower-heads, and the composite, or daisy family, whose flowers are really a mass of tiny florets. Perennial species are clearly most useful in an orchard, and a selection is listed in the box.

In experiments at East Malling Research Centre corn marigold was found to be the most effective of the perennial attractant plants, along with the annuals corn chamomile and cornflower.[36] Plants should be selected which are suited to the soil of the orchard, especially its pH, and plants of native provenance are ideal. Unsuitable plants will not thrive and may die out. Ideally the level of nutrients in the places where they are planted should not be too high or they will be superseded by grasses and other coarse plants. Mowing once a year and removing the mowings will help to keep the level of nutrients low. A sunny microclimate is best both for flowering and for the insects.

Professor Bob Crowder, working in New Zealand, has found that this kind of planting has given complete pest control without the need for any other measures.[37] But these results should not necessarily be extrapolated to other temperate countries, as the pest burden can vary greatly from region to region.[38]

Predator levels can be built up by encouraging populations of alternative prey. This can mean nothing more than allowing some patches of weedy vegetation to grow around the edges of the orchard. Many weeds are host to species of aphids and caterpillars which do not trouble crop plants.

Nettles, for example, support many species of caterpillars, and an aphid which feeds exclusively on them and is present in the spring. This enables aphid-eating predators to build up their numbers in time to go to work on the fruit trees, without increasing the population of fruit tree pests. When the fruit trees are troubled by pests the nettles can be mown to encourage the predators to move on to the trees. Nettles can be equally useful in this way in a vegetable garden.

In addition to food the predator species need habitat, both during the growing season and over winter. To a great extent providing this is a matter of not being too tidy. Mulches, log piles, stone piles or a dry stone wall, piles of leaves, and tussock-forming grasses such as Yorkshire fog and cocksfoot – both of which are visually attractive when in flower – are all useful. Ladybirds and other predators often overwinter in the hollow stems of herbaceous plants, so at least some of these should be left standing in the autumn. Nettles are ideal.

Many of the plants which attract pest-predators also attract pollinators. Providing these pollinating insects with food at those times of the year when the fruit is not in blossom can make pollination more certain in a difficult season and more complete.

The spring flowering, or pussy, willows are particularly useful here as they produce abundant pollen early in the spring when little else is available. They are valuable to bumble bees, which make better pollinators than honey bees because they work faster,

for longer hours and in worse weather than honey bees. The pussy willows give the bumble queens the food they need to raise the first batch of workers, who should be ready in time to pollinate the fruit, especially the earlier flowering fruits such as pears and plums.

Bumble bees also need suitable nesting habitat. Each species has its own preference, such as raised hedgebanks or old birds nests, and many of them can be accommodated in a good hedgerow. Straw bales and nest boxes similar to bird boxes can provide nest sites for others.[39]

Whether attractant plants are primarily for predators or pollinators, a selection should be planted to give successional flowering through the season. Unlike domestic bees, bumble bees and other wild insects don't store food in the form of honey, so they need constantly available food. (See box, Wild Flowers Attractive to Bumble Bees overleaf.)

If possible there should be a gap in the succession of pollinator attractant plants at fruit blossom time, otherwise the pollinating insects will be attracted away from the blossom. Fruit tree blossom is relatively unattractive to bees of all kinds because the nectar is low in sugar, so the bees will take other flowers in preference. It's hard to achieve this gap by species choice alone, because the relative flowering times of different plants is not that predictable. Mowing herbaceous flowering plants just when the fruit trees come into blossom is more reliable.

SOME PERENNIAL PEST-PREDATOR ATTRACTANT PLANTS[35]		
Umbellifers	Composites	Others
Angelica	Cornflower, perennial	Alder*
Chervil		Currant, flowering
	Corn marigold*	
Dill		Heathers, winter flowering†
	Fleabane*	
Fennel		
	Golden rod*	
Hogweed*		Hazel*
	Michaelmas daisy	
Lovage		Ivy*
Sea holly*	Ox-eye daisy*	Marjoram, wild*
Sweet cicely	Shasta daisy	
		Willows, spring flowering*
	Yarrow*	
		Vipers bugloss*
* Native plants † Good if there's a sudden warm spell in early spring with few natives flowering.		

Whether to place the attractant plants around the edge of the orchard or to interplant them depends both on the size of the orchard and the size of the plants themselves.

In a small home orchard, where nowhere is very far from the edge, planting round the edge may be sufficient, although interplanting may give better results. But in larger orchards interplanting is necessary, both to get a high enough population of insects and to get them in among the trees. It may be necessary to leave a mown access way between each row of trees, but all the space beneath and between the trees can be taken up with attractant plants. Many kinds will die back by harvest time, but some cutting down, mostly of dead stalks, may be necessary.

The shorter attractant plants can be grown under the trees, as long as there's enough light there. Many of the umbellifers are tall plants, often as tall as dwarf fruit trees, and these are best placed round the edges of the orchard.

WILD FLOWERS ATTRACTIVE TO BEES[40]			
Flowers	Spring	Early Summer	Late Summer
Birdsfoot trefoil	✓	✓	✓
Deadnettle, red & white	✓	✓	✓
Red clover	✓	✓	✓
White clover	✓	✓	✓
Dandelion	✓		
Pussy willows	✓		
Hedge woundwort		✓	
Honeysuckle		✓	
Roses, dog & field		✓	
Vetches		✓	
Bramble		✓	✓
Foxglove		✓	✓
Black horehound			✓
Black knapweed			✓
Common mallow			✓
Scabious			✓
Teasel			✓

Woody attractant plants, such as willows and hazel, can be included in the windbreak.

Some attractant plants have another useful output, and this can be harvested as long as that doesn't interfere with flowering. Many of the umbellifers are herbs, and moderate picking of the leaves does not affect flower production unduly. Some, such as fennel produce useful seed, and this may be harvested later in the year as long as the plants don't need to be cut down to facilitate fruit harvesting before the seed is ripe.

In general, attractant plants will provide food and habitat for a wide range of wildlife. Birds will feed on the increased insect life and the gone-to seed flowers, and insects will overwinter in dead stalks of nettles and other plants. Many of these creatures are also pest predators, and indeed the predators mentioned above are wildlife in their own right. In a really well designed orchard the interests of humans and those of wild species can be hard to separate.

Studies of wildlife on organic farms have shown that while general biodiversity is higher than on conventional farms populations of pest species are the same or lower.[41] Where an orchard has been specifically designed to attract beneficial wildlife we can expect that this favourable balance will be increased.

Animals In The Orchard

The traditional orchard animals are cattle and sheep, grazing the pasture beneath the standard trees. Other domestic animals which can have a beneficial interaction with orchards are bees, chickens, geese and pigs.

Bees[42]

Fruit and honey bees are natural companions: the fruit trees and shrubs get pollinated, and the bees get nectar and pollen from the trees. Although fruit trees don't provide the richest nectar, and a single honey bee is not as effective a pollinator as a single bumble bee, the volume of blossom in an orchard at one time amounts to a major resource for the bees, and honey bees can be brought to bear on an orchard in huge numbers at the crucial time. In a large orchard it would be hard to get such a force of pollinators simply by encouraging wild species. Honey bees can also be valuable pollinators for other large-scale crops, especially peas and beans.

Small orchards and domestic fruit can manage quite well with natural pollinators, especially if they are provided with year-round food and nesting sites. (See Attractant Plants, above, and Wild Animals, below.) Also, honey bees do to some extent replace wild, or bumble bees in the ecosystem, and populations of wild bees are declining drastically. So people who place a high priority on preserving and enhancing biodiversity may choose not to keep honey bees for that reason.

In any case it's not worth keeping bees purely for pollination, though it certainly is worthwhile for an orchardist to get a local beekeeper to bring their hives in at blossom time. Bees require regular attention, and beekeeping is a skilled craft, and it's only worth investing your time and energy in them if you're seriously interested in the yield of honey. It's possible to make a living at beekeeping, whether you have land or not, and it's equally viable as part of a polycultural income or as a productive hobby. Bees integrate well into a smallholding, and are an excellent way of increasing output where space is limited.

They can be kept in either town or country. In fact these days there is often more bee forage to be had in urban and suburban areas than in the country. Fifty years of intensive chemical farming have left the countryside sadly depleted of wild flowers, while the flowers in urban gardens bloom unabated.

The best way to start beekeeping is to contact established beekeepers in your area, who are usually happy to provide you with advice, and often with bees. The British Beekeepers Association will help you locate them. (See Appendix D, Organisations.)

Siting an Apiary. The key to keeping bees in towns is to avoid any problems with people being stung, or their fear of being stung. This can be done by two simple measures:

- Avoid siting bees where they can see people moving about in front of the hives.

- Make sure their flight path is up above people's heads.

Both of these can be achieved either by surrounding the apiary with a tall hedge or by siting it in a high place, such as on a flat roof. A hedge may also help to provide the ideal microclimate, though putting them in a high place may make it more difficult to provide the necessary shelter.

The microclimate needs for an apiary, whether in town or country, are:

This flat turf roof is an ideal site for urban bee keeping.

- All-round shelter from winds; this includes winter winds, so some evergreens are desirable in the windbreak.

- Good ventilation, to reduce excessive humidity and frost.

- Avoid frost pockets.

- Plenty of sunshine.

A site which fulfils all these requirements is ideal, but it is quite possible to keep bees in a site which falls short of the ideal, and bees are often kept in unlikely places with good results. A trial with one or two hives is a good idea for an unpromising site.

The third requirement for an apiary site is good access. Best of all is the beekeeper's own home garden, but if that's not possible it should be easy both for regular inspection and for hauling out heavy loads – 'supers' full of honey!

These requirements are for a home apiary where the bees spend most of their time, including the winter. In summer they may go to an out-apiary where there is a large seasonal supply of forage, perhaps an orchard or a field of oilseed rape, or simply a tract of country where there is generally good forage. This is usually on someone else's land. Access and security are just as important for an out-apiary as a home apiary, but the microclimate requirements can be relaxed somewhat, in line with the weather you can expect in summer.

The rule of thumb for moving bees is, 'Less than three feet or more than three miles'. This is because foraging bees find their way home by learning the landscape. If the hive is moved more than a very little they will return to the place where it was before, and when they fail to find it they will not go and look for it. But if the hive, with the bees in it, is moved right out of their foraging range they will not recognise the landscape and will immediately learn to return to their now home.

Providing Forage. The aim is to provide nectar and pollen over as long a season as possible, and, if pollination of an orchard or other crop is a priority, to avoid alternative sources when this crop is in flower. A domestic beekeeper doesn't have control over a wide enough area of land to do much about this other than to move the bees to suitable out-apiaries at different times of the year. But farmers, smallholders and landscape architects may consider planting bee fodder plants.

Often these will be plants with another main function and bee fodder will be a supplementary output. It's always worth considering bee fodder where there is a choice of species to meet a specific need, whether ornamental, food, timber or windbreak, even if there is no beekeeper in the area at the time of planting. If two species are equally suitable for the main purpose and one of them also feeds bees, why not choose that one?

IMPORTANT BEE FORAGE PLANTS		
Plant	Season	Notes
Willow, pussy	March - May	Important pollen source in spring (male plants) – a must for the beekeeper's garden
Soft fruit	end March - mid April	Nectar and pollen
Top fruit	end March - May	More pollen than nectar
Dandelion	April - June	High yield of nectar and pollen
Sycamore	May	
Horse chestnut	May	Only white-flowered varieties
Beans, broad & field	May	Autumn-sown
Oilseed rape	May - early July	Poor quality honey; danger from pesticides
White clover	mid June - end July	High yield and good quality honey, the bee plant par excellence
Other clovers & lucerne	if & when allowed to flower	Red clover yields heavily
Limes	end June	Seem to be taken less by bees than formerly
Blackberry	June - August	Worked by bees even at low temperatures
Oaks	mid & late season	Yield honeydew rather than nectar & pollen
Willowherb, rosebay	July - August	Good quantity and quality
Heathers	end July - August	Not every year, but high yield when it comes
Ling (common heather)	mid August - early September	Good yields every 3-5 years
Michaelmas daisy	August - November	All the daisy family are useful
Ivy	end September - October	Important for winter stores

Bees take nectar and pollen from a wide range of flowers, but some of the most important ones in Britain are given in the table, above.

Chickens

Chickens can be let into an orchard in the autumn and winter to control pests and diseases which overwinter either in the soil or on windfalls or fallen leaves. These include:

Crop	Pests & Diseases
Apples	codlin moth, apple sawfly, winter moth, scab
Pears	pear midge, slugworm, winter moth, scab
Plums	plum sawfly, winter moth
Cherries	slugworm
Cane fruits	raspberry beetle, cane midge
Gooseberry	sawfly

Of course there are other pests and diseases of fruit, and other control measures will be needed for them. But in many areas a combination of chickens in winter and pest predator attractant plants which bloom in summer can make the need for other interventions rare. As far as apples are concerned, codlin moth is the most important pest and scab the most important disease, and chickens can have an impact on both.

Chickens require daily attention, and on a domestic scale it would never be worth keeping them purely for their pest control abilities. But anyone who has both chickens and fruit would be well advised to integrate the two.

On a commercial scale, a free range chicken enterprise may be the best vehicle to get an organic orchard up and running. The trees can be established in the chicken runs, where they will stand a chance of getting through their vulnerable early years without recourse to pesticides and herbicides. Meanwhile the chickens can give an income during the unproductive years. The daily attention needed by the chickens can also mean that the trees get inspected more often than they might otherwise, and any problems can be spotted early.

120-180 chickens per hectare is the recommended density for an established orchard, though smaller numbers will still have a beneficial effect. More can lead to excess nitrogen in the soil, and, if they are there all year round, they will start to kill out the ground cover, leading to the familiar bare-earth-plus-nettles of the over-stocked hen run. Keeping them in the orchard year-round is probably not a good idea anyway, as it could lead to a buildup of chicken pests and diseases.

A possible exception to this is a young orchard, before the trees have started fruiting. Here a high level of nitrogen and the absence of ground vegetation may be just what the young trees need to get them going. But this is best not left as a permanent arrangement.

Chickens will not get a significant proportion of their energy input from foraging in an orchard. In fact the energy they use up in foraging may even be more than they gain from the food they find. But the diversity of creatures and herbs they find will be good for their health, and will certainly improve the quality of the eggs and/or meat produced. Chickens often range farther in an orchard than on open pasture, due to a greater feeling of security, and if they are getting a net gain of energy from foraging this effect will increase the benefit.

Alternatively, it may be possible to combine chicken forage trees with the fruit trees, so that the birds do get a significant amount of food from ranging in the orchard. This is discussed under Chicken Forage Systems. (See pp251-253.)

In either case the chickens will have the benefit of an improved quality of life. Specifically, the trees can give shade and shelter, and reduce aggression by keeping individual chickens apart from each other and providing cover for the weaker ones. The overall effects can show up in improved health and increased egg production.

On the down side, the trees will attract wild birds which may carry diseases that the chickens are susceptible to, and possibly provide cover for predators. If there is cover at ground level the hens may take to laying in it, and you will have difficulty finding the eggs. If the trees are mulched the chickens will make a mess of the mulch by scratching in it for insects. They will eat any vegetables which are around when they are in the orchard and they can also defoliate any branches within reach, so they should not forage among soft fruit or dwarf trees in summer. Most of these problems can be avoided by keeping chickens in the orchard during the winter only.

The selection of suitable breeds of chickens is discussed below. (See box, Choosing Poultry Breeds, p251.)

Geese

Geese are natural grazers and can make good use of the grass and other herbage in an orchard which is not underplanted with more intensive crops. The stocking rate needed to keep the grass down through the spring and summer will depend on the ability of the local soil and climate to grow grass. 10-15 per hectare is likely to be the minimum required to avoid any need for mowing in a British summer.

Although they can feed themselves through the summer it's as well to give them some food each evening to keep them in the habit of coming in. Foxes can kill them if they stay out. Like ducks they are water fowl and must have access to clean water to keep healthy, at least enough to fully submerge their heads in, but ideally a pond they can swim in.

When the grass stops growing in autumn the geese will need feeding, and this point comes well before the Christmas market. This means that totally self-foraging geese are not a realistic proposition, and it's probably not worth keeping them purely as a way of keeping the orchard weeded. But if you want to keep geese anyway, grazing them in an orchard over the growing season is a good way of getting a multiple output from them.

They can be trained to eat grass and weeds rather than the crops by making sure that these are the only green plants they come into contact with when they first start grazing as young goslings. But they do have their preferences, for example they're not fond of docks and dandelions. They sometimes strip the bark off young fruit trees, so they should be kept out of an orchard till it's well established unless the trees are individually protected.

They are noisy animals and may not be popular with neighbours – or with orchard keepers who want a quiet life! They are said to make good 'guard dogs' but in practice they honk at almost anything, even a blackbird, so you end up not paying much attention to them.

In California and Oregon weeder geese have been used in orchards, vineyards and field crops, including garlic, tomatoes, cucumbers, cotton and mint, at a stocking rate of five or six per hectare. They can only be introduced when the crop is big enough not to be eaten.[43] The low stocking rate may be in part because of slower weed growth in the dry summers of California, and also because some of these crops give more ground cover than fruit trees and so leave less space for weeds.

Pigs

In the 1920s, before farming and horticulture became as specialised as they are now, and before fruit growing became more a matter of chemistry than of biology, the benefits of keeping pigs in orchards were well appreciated.

> Spencer [writing in 1921] was notably expansive in his praise of pig/orchard systems, and suggested that they were perhaps the most profitable form of outdoor pigkeeping. He stated that the pigs benefited from shade, shelter and fallen fruit, while the trees benefited from manure and the clearance of insect contaminated windfalls. He cited two fruit growers who claimed that the size and quality of their apples and cherries, respectively, were improved by the presence of pigs under the fruit trees.[44]

Pigs are not compatible with dwarf trees, and no pigs should be put into an orchard of vigorous trees till the trees are big and sturdy enough to be used as scratching posts without suffering damage. It's hard to say when this will be as there are so many variables, including how long the pigs are in the orchard at one time, and the itchiness of individual pigs. But it's best to think in terms of ten year old trees at the least.

The most suitable breed of pig for orchards is the Gloucester Old Spot, which actually prefers to graze rather than root, which is unusual for a pig. Most breeds will start to root as soon as they're let onto new land, but Old Spots don't usually start to root till they've eaten off what they can find on the surface. Rooting is very damaging to fruit tree roots, and other breeds of pigs need to have rings put in their noses. Old Spots will start to root once they've finished whatever grazing and windfalls are there, so if they're not rung they have to be taken out of the orchard promptly when they reach this stage.

Wild Animals

Wild birds can be either friend or foe in an orchard. Tits feed their young on caterpillars, including those of the winter moth and codlin moth, so they are friends. Providing nest boxes for them is worthwhile.

Bullfinches, on the other hand, can be a serious pest. Their staple winter food is ash keys, as the fruit of the ash tree is known. In years when the ash trees produce no keys, which can be every other year, they turn to the fruiting buds of fruit trees. Pears and plums are the worst affected, followed by gooseberries and currants. Apples and cherries are among the least affected, but they will be taken where the other fruits are not present. Losing the fruit buds means no fruit that year, and plums and gooseberries can suffer a permanent loss in their fruit-bearing potential.

There's little to be done to prevent them other than netting the trees and shrubs. But bullfinches are shy little birds and never stray far from dense cover, so an orchard which is not too close to dense woodland or other thick cover should be safe. It's impossible to be precise about the distance as it depends on a number of factors, including the individual habits of the local bullfinches and how often people are present.

A barrier of unattractive fruit may be effective in some cases. One large orchard, planted near woodland, was protected by two rows of quince trees round the edge. Neither the birds nor the squirrels like quince, and none of them penetrated further into the orchard than the two unappetising rows. But I wouldn't count on this to work everywhere.

POULTRY

Chickens and other poultry are perhaps unique in the number of beneficial relationships which can be set up between them and other elements of a household,

smallholding or farm. Some of these links are more applicable in a home garden, others on a commercial holding, but in general they are easier to make on a small scale than on a larger one.

Garden. The idea of the 'chicken tractor' – allowing chickens to cultivate the ground for us – can be used on different scales.

On the scale of a domestic garden it's best done by means of a small ark, with a base the same width as that of the vegetable beds. Two to four chickens, according to the size of the ark, can be taken out of the flock and put into the ark, which is placed over the bed to be cleared. In a day or two they will have removed all crop debris and weeds, eaten any weed seeds in the surface layer, deposited a little manure, and eaten all the slugs plus a good deal of any other pests which are present. Earthworms tend to head for deeper soil under these conditions so not too many of them will be taken. Then the ark can be moved onto the next section of the bed, and when the whole area is tractored the chickens can rejoin the main flock. The soil surface may be slightly compacted, but this can easily be remedied with a quick shallow forking.

Chicken tractor ark.

Chicken tractor with movable electric fence.

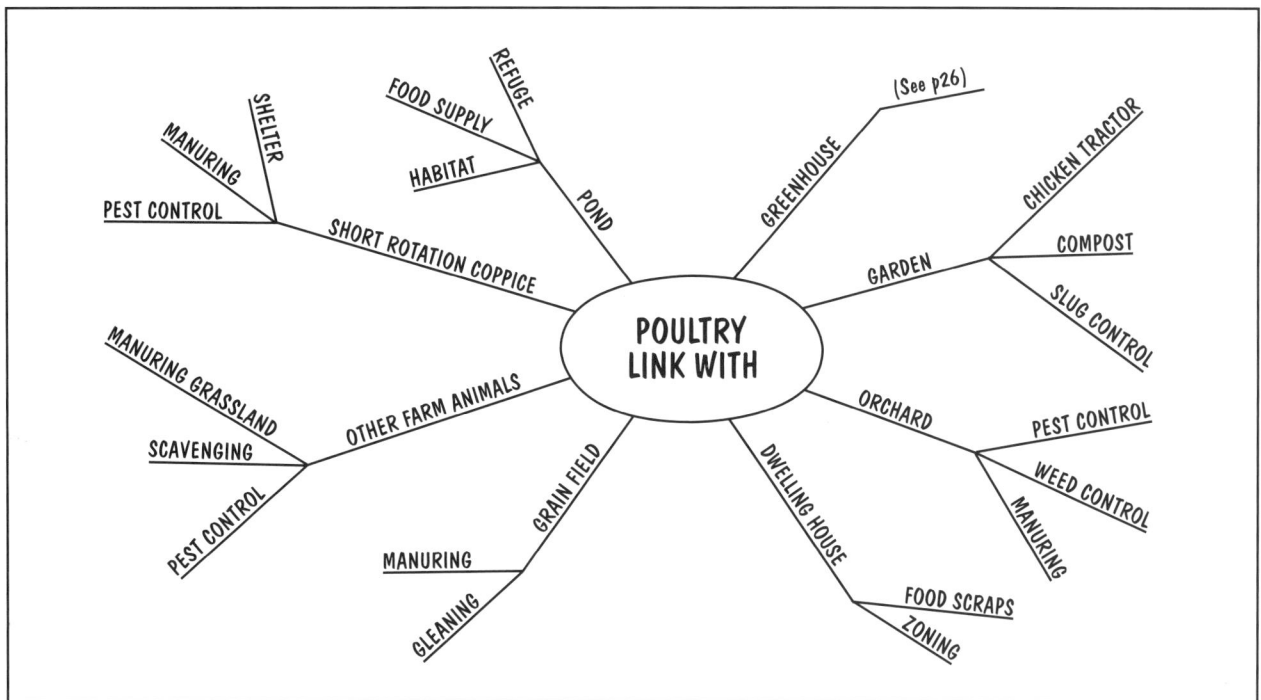

On a larger scale the whole flock can be fenced onto an area of land with movable electric fencing. This kind of fencing is easily moved and is a useful tool for free range chicken keeping whether tractoring is part of their lifestyle or not. A mobile chicken house is needed, unless the garden is arranged so that all parts of it are accessible from a permanent house site.

An ark can be easily fitted into any garden, especially if it's laid out on the bed system. Fenced chickens will fit in better if the garden is designed from the start with them in mind. This means giving some thought to where fences can go, and to where a permanent house should be sited or how a mobile one can be moved around the garden. The chickens can also be designed into the rotation. This applies particularly to the use of green manures, which can be cleared by the chickens instead of being dug in. The soil will still benefit from the green manure, as the nutrients in the top growth will be returned to the soil in the chickens' manure, while the chickens get the benefit of valuable greenstuff, and the need for digging is avoided.

Whichever method is used, the chickens should not be left on the land any longer than necessary, or the accumulation of raw manure may be too much for the following crop. It's always wise to have a non-garden area where the chickens can range when they're not needed for tractoring. One option is to keep them in a comfrey patch. Comfrey is about the only plant which can handle raw chicken manure in quantity, and make good use of it. Chickens tend not to eat it unless they have been taught to at an early age and it will appreciate being weeded.

Mature compost builds up a population of brandling worms, which are an excellent source of protein for chickens. If it's dumped in the tractoring area just before they move on they will feast on the worms (which are not the same as earthworms, and have no function in the soil) and spread the compost around. They should be moved on as soon as they've finished the worms or they will eat up the compost too and deprive the soil of valuable organic matter.

The use of ducks for slug control is described in Chapter 5. (See p111.)

Orchard. The use of chickens and geese in orchards is described in the Fruit section. (See pp246-247.)

Dwelling House. The distance between the poultry house and the dwelling house needs to be thought about carefully. As well as feeding and collecting the eggs, the birds need to be shut up at night and let out in the morning, or sooner or later they will be had by a fox. That twice-daily trip can mount up to a great deal of walking. Every extra ten metres between house and poultry house means another eight miles a year walked by the poultry keeper. Why not place the poultry a little closer and go for a good walk in the country once a year? But perhaps you have a good reason not to. (See box, No Rules, overleaf.)

On a domestic scale chickens can make good use of food scraps which are unsuitable for the compost bin. Broadly speaking this means cooked food, which will attract mice and rats. This is one of the great advantages of small-scale domestic production: useful materials which come in small, dispersed packages can be easily utilised. But scraps should not be seen as a substitute for a properly balanced diet containing the energy and protein which poultry need. They should not amount to more than a third of the birds' total intake.

benefit the coppice by manuring the soil, controlling weeds and eating the main pest of the coppice crop, the brassy beetle, which overwinters on leaf litter.

This would not be a good situation for laying birds as they would tend to lay among the coppice and finding the eggs would be a problem. But the system has recently been tried with meat birds with good results. The breed used is the Ixworth Cross, which combines the hardiness and outdoor characteristics of traditional breeds with the high production levels of modern hybrids. They can be put into the coppice as soon as it's tall enough for the leaves to be out of their reach, usually half way through the first growing season. The stocking rate can be as high as good poultry management allows. The Soil Association standards are a good guide to this and they presently give an upper limit of 600 birds per hectare.

The pioneer of the system, Derek Jackson of Devon, envisages a more integrated system in the future, with north-south strips of short rotation coppice alternating with strips of vegetables. The north and south ends of the plot can be closed off with shelterbelts of elephant grass, a huge grass which can also be grown as an energy crop.

Under this system the vegetables are grown in rotation with green manures which are grazed off by the chickens. The vegetables benefit from the shelter given by the strips of coppice and the coppice yields more than it would if planted in a solid block, due to the edge effect. Access paths are best placed on the edge between the coppice and the vegetables to avoid undue root competition and shading. Manure from the poultry arks can be applied to the vegetables, and the only input of nutrients to the system are those contained in the chicken feed.[45]

NO RULES

On permaculture courses I usually mention the advantages of siting a chicken house near to the dwelling house. But people don't always agree.

One student said that the walk to the chickens was her favourite part of the day. It took her through her vegetable garden, a trip which gave her both pleasure and an opportunity to see what was going on there. For her it was a positive advantage to have the chickens that distance away from the house.

Another student, in Croatia this time, said "That may be very well in England, but here in Croatia it gets hot in the summer and the chicken house starts to smell. You wouldn't want it too close to the house if you lived here." To which I could only reply, "Exactly."

These two responses illustrate how there are no hard and fast rules in permaculture, no off-the-peg designs for all situations. Every situation is unique. Perhaps the only rule is that every factor needs to be considered, and even as powerful a design tool as zoning will often be superseded when everything is taken into account.

Grain Field. Poultry can be ranged over grain fields in autumn and winter to make use of the fallen grains missed by the combine. This is more practicable where the following crop is to be spring sown, as there is not much time between harvest and autumn sowing. This is a good way of making use of an otherwise wasted resource, and the field will receive a manuring in return. But some grain should be left for seed-eating wild birds, which have greatly declined with the modern fashion for growing only autumn sown crops.

Other Farm Animals. A mature cowpat hosts a thriving population of grubs, which are a protein feast for chickens. Ranging chickens over grassland after cattle will make use of this food source, spread the cattle manure more evenly, and add the chickens' own manure to the grassland.

Pigs are not the tidiest of eaters and a small family of chickens can live off what three or four sows lose among their bedding or leave in the corner of the trough. They will also pick the lice off the pigs, making life more pleasant for their neighbours at the same time as feeding themselves.

Short Rotation Coppice. (See p325.) Keeping chickens in a short rotation coppice plantation has benefits for both trees and birds. The coppice provides the chickens with ideal shelter, and if the rods are chipped on site some of the chippings can be kept for bedding in the chicken ark and fuel to heat the brooder. The chickens

Pond. Ducks are aquatic animals and it's unkind to keep them without access to a pond. But if the pond is small they will denude it of all plant and animal life, so where space is limited it may be a choice between ducks or a pond for other purposes. When a small duck pond becomes so dirty that it needs cleaning out, the water can be used as a liquid manure in the garden. On a larger pond an island can provide them with a fox-proof refuge, which saves their keeper the task of shutting them up and letting them out each day. Of course you do need a pair of waders, a boat, or a drawbridge to collect the eggs!

Chickens will avidly take tadpoles from the shallows at the edge of a pond, and this can be a significant source of protein in spring and early summer for a smallish flock with access to a large pond. All amphibians are in drastic population decline, so this resource must be used wisely. But one adult frog lays so many eggs that the vast majority of tadpoles hatched never make it to adulthood, even if the population is increasing. There's little harm done if some of that wastage goes to feed chickens rather than other predators. Indeed, if the chicken keeper has provided new frog habitat by digging the pond in the first place, there is a net gain to wildlife.

Chicken Forage Systems

Although chickens can be very versatile members of the farm or garden community, the fact remains that they compete with us for food. They need a diet of grain plus high quality protein, both of which we can eat directly ourselves, and we would gain far more nutrition from them if we did that. (See pp119-120.) It takes both energy and fertile soil to produce grain. Thus if we can substitute some of it with self-forage plants which can grow with lower inputs on less favoured soils we can increase the ecological viability of keeping chickens – and greatly reduce the work involved.

Chickens can get a limited amount of nourishment from grazing, just as we can from a salad, but probably not more than around ten percent of their grain ration can be saved in this way. Ducks can get a larger proportion of their food from green stuff, and geese can more or less survive on grazing alone during the summer. But if we want to keep chickens without feeding them on grain we need to find another source of high quality carbohydrate and protein.

Much of their protein needs can be met from the kind of scavenging described in the previous section. But even if all the resources mentioned could be tapped – and it would be a rare situation where that was possible – most of their energy needs would still have to come from grain, and some protein supplement would be necessary at certain times of the year.

An answer to this problem is to plant up an area of land to trees and shrubs which yield seed or fruits which are edible to chickens and which are shed onto the ground where the chickens can pick them up. This kind of planting is called a chicken forage system.

Tree crops like this can be grown on land which is too steep for growing cereals, and some of the most suitable chicken forage plants do well on poor sandy soils (see box, Chicken Forage Shrubs, overleaf). What's more, these perennial plants avoid all the ecological costs associated with growing annual crops and the energy involved in transporting and processing bought-in feed. Even the labour of actually feeding them is saved, as once the trees and shrubs are established the food simply falls on the chickens from above!

In tropical countries it's possible to design a chicken forage system which meets all their needs all year round. Here in Britain most seeds and fruit are available between late summer and early winter, and one of the main challenges facing the designer of a chicken forage system is to extend the season as far as possible by including early and late season trees and shrubs in the planting plan.

The species chosen must have a heavy enough yield to justify devoting the land to them and that yield must be reasonably reliable from year to year. They should also shed their seed or fruit rather than retaining it on the plant. Otherwise it has to be harvested by hand, a job which will take longer than simply dishing out purchased feed, and so reduce the attractiveness of the system.

Since the most promising chicken forage plants are legumes, which fix nitrogen, they could be a useful interplant in a young orchard of vigorous fruit trees. The fruit trees could be planted at normal spacings with chicken forage shrubs between them. The shrubs could be removed when the growing trees start to crowd them, at perhaps ten or twenty years. This would give nitrogen and pest control during the early years of the trees' growth when both of them are most needed.

CHOOSING POULTRY BREEDS

It's important to get the right breeds of chicken for permaculture systems. Whether they're wanted for controlling pests in an orchard, tractoring or making use of a chicken forage system, they do need to be good foragers with all the instincts that make them go out and search for food. Modern hybrid chickens can produce very high yields of eggs or meat in a short time, but they've had most of the foraging instinct bred out of them. When kept on free range they are often reluctant to go outdoors at all, and if they do they do little more than hang about near the house waiting to be fed.

Even among the traditional breeds some are better foragers than others, and it's a good idea to get advice from someone with experience of a number of breeds before making a selection. Beware of the single breed enthusiast, who will tell you that their breed is the best for every purpose. Also, many traditional breeds are now kept more for show than for production. They may look perfect but perform poorly. It's always best to get birds from someone who keeps them for food rather than for fancy.

In general light breeds tend to have more of the foraging instinct than heavy ones. Bantams may be the most suitable breeds in gardens where there is mulch, as they don't make such a mess of it as full-sized breeds, but then they won't get to the pest organisms either if they don't scratch the mulch back.

Vegetarians may have another reason for choosing traditional laying breeds over hybrids. Although hybrids produce more eggs than the old breeds in the first year or two, they fall off rather sharply after that, while the traditional breeds will carry on laying at a moderate rate for a number of years. This means you don't have to kill and replace your birds nearly as often. Having said that, all chickens can live for years after they stop laying.

Alternatively the fruit trees could be planted at a wider spacing and the chicken forage shrubs allowed to complete their life cycle. In either case the chickens would be present in the summer, as the forage season begins in July, and this may not be ideal for a number of reasons. (See p247.) The shrubs should not need inoculating with nitrogen-fixing bacteria, as they belong to the same group as some native clovers, so their bacteria should be present in the soil. (See p48.)

The requirements for ground layer vegetation in a chicken forage system are less clear cut. One option is to plant it up with herbaceous perennials which provide edible seed and greens for the chickens. In order to prevent the chickens destroying these plants the stocking rate and length of time spent in any one part of the forage area would have to be carefully controlled. This would require a network of internal fencing, temporary or permanent. It could sometimes mean that the birds had to be moved on before they had finished eating the tree produce. An edible ground layer may be of most use in the early years of the system, before the trees and shrubs are yielding fully. Or it could be provided in separate enclosures without trees.

A simpler and more flexible approach is to accept whatever ground cover there is, and just move the chickens on to another area sufficiently often to prevent them from reducing the plant diversity by overgrazing.

If a chicken forage system is situated next to a vegetable garden it can be combined with a chicken tractor system. But during the time the chickens are confined on the vegetable beds for tractoring they would not have access to seed from the forage system and would need to be hand fed with a grain ration.

Meat chickens may be more suitable than layers for a forage system for two reasons. Firstly, batches can be bought in at specific times of year to make use of seasonal yields of forage. This may be easier than trying to provide a constant supply through the year for layers, and would avoid the problem of overgrazing the ground vegetation. Secondly, hens may take to laying their eggs out in the forage area rather than in the house.

CHICKEN FORAGE SHRUBS

Perhaps the most promising chicken forage plants for temperate climates are the Siberian pea shrubs and the closely related bladder senna. They are sometimes grown here as ornamentals, for their display of flowers in the spring and their attractive foliage, but in Siberia they're used to feed chickens. The seed is said to be edible for humans, rather like lentils, though I haven't tried it myself.

They are legumes, and like all members of that family the fruit is a pod which splits open when ripe, scattering the seed on the ground. The seed is released gradually over a number of months. In fact the chickens like the seed so much that they often can't wait for it to drop and do their best to fly up and pick it out of the pods.

The Siberian pea shrubs grow to some 4m height and spread, but can be kept smaller by pruning. They are extremely tough and have been used in windbreaks on the Canadian prairies. They tolerate cold, salt winds, drought, poor soil, in fact almost anything other than waterlogging. For good seed production they prefer a hot summer and a fairly dry winter. This means they do better in the east of Britain than in the west, where they may not set seed every year. They also prefer a light, sandy soil.

The bladder senna is somewhat more tolerant of heavier soils, and may do better than the Siberian pea in western climates. But there is certainly scope for selecting varieties of Siberian pea which are more suited to the maritime climate. Until we have such varieties the best the western grower can do

is to choose the planting site carefully. A sheltered, unshaded, south-facing slope will give the pea shrubs every chance to ripen their seed, and if a light soil is not available it must at least be well drained.

Since they are mainly grown as ornamentals in Britain, the most commonly available varieties will have been selected for appearance rather than yield of seed. Anyone wanting to grow them for chicken forage, or human food, may get a better yielding plant by buying from a specialist permaculture supplier, such as the Agroforestry Research Trust (See Appendix E, List of Suppliers.) I'm unable to find any information of the yield which can be expected, either per plant or per hectare.

Bladder senna in flower. Inset shows close-up of seed pod.

As far as I'm aware there is no intentionally designed chicken forage system at present in Britain, though some people have the occasional Siberian pea or bladder senna to supplement the chicken's food. So it remains to be seen what proportion of chickens' diet can be met by this means, and for how much of the year. But to the extent to which it reduces bought-in food it will reduce the ecological impact of chicken keeping by that much. There's no reason why a system which meets half the chickens' food needs should not be well worth while. But in practice it would be necessary to calculate how much that land could produce if put to another use. This will depend very much on the individual circumstances of the particular farm or smallholding.

On the domestic scale, a single pea shrub in a suburban hen-run may not make much impact on the annual feed bill. But it will provide the hens with some food and summer shade, and the people with an attractive shrub to look at, so why not?

GUEST ESSAY

OWN-ROOT FRUIT TREES

by Phil Corbett

Phil Corbett is an experienced horticulturist with a deep knowledge of and feeling for plants, and is a leading light in the British permaculture scene. Inspired by the work of Hugh Ermen, formerly of Brogdale Horticultural Experimental Station, Phil has set up the Own-Root Fruit Tree Project to experiment with the idea of growing fruit trees on their own roots instead of grafting them onto a rootstock.

Hugh discovered that there are several advantages in growing apples on their own roots. The graft union, which is a union between two genetically different individuals, always creates a degree of incompatibility. The advantages of own-root trees are:

- Better health, although not altering the basic susceptibility of the variety to disease.

- Fruit development is typical of the variety, giving:
    - best possible flavour;
    - best storage life;
    - typical fruit size for the variety;
    - best overall fruit quality.

- Best fruit set, given adequate pollination; fruit from own-root trees have more seeds, indicating increased fertility.

- It is highly likely that the degree of self-fertility is increased.

The only disadvantage of own-root trees is that most varieties are more vigorous than is usually wanted. This means that trees may make a lot of wood at the expense of fruit bud production, giving big trees that take a long time to come into crop. Conventionally this vigour is controlled by grafting onto a dwarfing rootstock.

With own-root trees a number of traditional techniques are used in the first few years of growth to induce early cropping. These are:

- Withholding nitrogen, which stimulates growth, and withholding irrigation, except in serious drought.

- Tying down 1 and 2 year old branches to the horizontal, which induces fruit bud formation.

- Summer pruning, which induces fruit buds, and avoiding winter pruning, which stimulates regrowth.

Once cropping begins the tree's energies are channelled into fruit production, and growth slows down to a controllable level, so a normal feeding and watering regime can begin. The average cropping own-root apple tree can be maintained at a size very similar to the same variety on MM106 rootstock.

Unfortunately Hugh's own-root trees at Brogdale have now been destroyed for redevelopment of the site. But the Own-Root Fruit Tree Project is in the process of setting up new trial grounds to continue his work.

The Coppice Orchard

An extra advantage of own-root trees is that if they are cut down to ground level the regrowth will be of the fruiting variety, whereas with a grafted tree it is the rootstock which will regrow. This may be useful where damage from gales, animals or vandals is likely. It may also mean that own-root fruit trees can be coppiced, and this is the basis of our coppice orchard which we are now establishing.

This consists of an acre of own-root trees planted in rows running north-south. When the canopy of the orchard closes, a row will be coppiced and the land in the row used for light demanding crops, such as vegetables on a no-dig system, while the trees regrow. The trees either side of the glade will have higher light levels on their sides and produce more fruit buds. The next year another row is cut, but not the immediate neighbours as these will have the extra buds, so the next row for coppicing will be next-door-but-one.

In other words this will be alternate row coppicing. This process is repeated every year, creating a series of parallel, sheltered glades. Eventually the rows of trees forming the avenues between the glades will also be coppiced in turn, but by then the 'glade' trees will have regrown to form the avenues.

As the trees regrow there will be glades at all stages of regrowth until the cycle repeats itself, and niches for plants suited to full light, semi-shade or heavy shade, creating opportunities for different types of land use. The number of years before re-coppicing, and so the length of the coppicing cycle, is one of the many aspects of the project that we will only learn by doing it. The exact timing of coppicing can be adjusted to suit the type of produce that is wanted most, either more coppicing for more vegetables or less coppicing for more fruit.

After coppicing the trees will probably return to fruiting after about three years, the normal shoot-maturing cycle for apples. This will vary with the variety. If regrowth is kept to a single stem it will closely resemble a maiden tree. Multi-stem regrowth will fill space more quickly, but a new pruning and management system will need to be devised for it.

Apart from apples, the main planting sites of the orchard will also have own-root pears and plums, hazelnuts, and nitrogen fixing trees and shrubs. Instead of just producing top fruit the coppice orchard can produce a wide range of crops: small wood, soft fruit, vegetables, possibly cereals, fungi and the more traditional bees and poultry.

An exciting possibility is the production of shredded woody wastes for compost heat production using the Jean Pain system. (See pp315-316.) This needs a minimum of 30 tonnes of material to generate useful heat for 15 months, ie over two winters. It will be very interesting to work out how much of this bulk can be sustainably produced from a coppice orchard system, using coppice wood plus the various intercrops that can be grown.

One of the great virtues of the Jean Pain system is that at the end of the heat cycle you still have virtually the same bulk of compost that you started with, and this can be returned to the system, perhaps as mulch in the vegetable areas. It could be that as well as producing most of our crop needs the coppice orchard could also provide much of our domestic and horticultural heating, and instead of a residue of wood ashes, leave us with a heap of valuable compost.

There is an old Chinese proverb that says 'fertility follows in the footsteps of the farmer' which is reworked as the permaculture principle: fertility follows attention. In the coppice orchard there is the potential for producing a great range of our needs in a single system, and productivity should benefit from our attention not being divided between vegetable plot, orchard, woodland, and so on.

The rotation of crops avoids disease build-ups, and if all residues are returned to the site there should be a build-up of fertility. The plan is to include nitrogen fixing and soil conditioning plants, insectary plants to support useful insects, bird and bumble bee nest boxes, small ponds for amphibians and hedgehogs and generally to maximize the natural diversity, and yields, of the site.

Chapter 10

FARMS & FOOD LINKS

Farming faces two serious crises. Economically it's in the worst state it has been in since the 1930s. World over-production of food has brought prices ever lower, to the point where many perfectly competent farmers are unable to make a living on the land. Ecologically, farming is skating onto ever thinner ice, as the natural capital on which the health of agroecosystems depends is increasingly depleted by methods which are geared to the short term. The two crises are connected, as the vicious spiral of lowering prices is one factor which has led to unsustainable farming methods.

The practices described in the first three parts of this chapter, Animals, Arable & Vegetable Crops, and Agroforestry, can go a long way towards bringing farming back into ecological balance with the land. The fourth part, Food Links, offers hope on the economic front. Making direct links between producers and consumers is not a panacea but it is an alternative to the conventional 'solution' of bigger, more specialised farms with more machinery and fewer people on the land. Direct food links offer the prospect of financial viability to small farms and smallholdings, while actually rewarding diversity on the farm.

Part of the vision of permaculture is that we can feed ourselves much more efficiently from gardens than from farms. (See pp26-27 and 31-32.) But that vision presupposes enormous changes in settlement patterns, the structure of society and people's values. It's our best bet for a sustainable future, but it won't happen overnight. At present almost all of our food comes from farms and it will continue to do so for some time to come.

Given that farms are not the permacultural ideal, how can permaculture bring them closer to that ideal? The answer is, not surprisingly, something of a compromise.

The degree of diversity and intricacy which is possible in a home garden is not possible on a farm scale. But relatively simple forms of polyculture and stacking, together with no-till methods and perennial crops such as nuts can all be introduced on farms and make them much more sustainable – and in many cases more productive. Some of the techniques described here involve only small changes to present farming methods, others are much greater departures from present practice, and closer to the permaculture ideal. Some are tried and tested, others more experimental. But all represent a step, however large or small, towards farms which work like natural ecosystems.

Good design of the farm layout is an aspect of permaculture which may not be apparent to the casual observer. A permaculture farm may have all the same elements on it as another farm. But if these elements are arranged in such a way that they work well together with a minimum of external inputs and maximum cycling of on-site resources, both sustainability and profitability can be increased. (See case study, A Permaculture Farm, p294.)

Organic Farming

Permaculture and organics are complementary. Permaculture is very much about spatial design, while organics is more about method. The two together make a powerful combination, but it's not essential to use organic methods in order to practice permaculture. There are no rules in permaculture and if non-organic farmers and growers want to make use of some permaculture design ideas there's nothing to stop them.

On the other hand, since the aim of permaculture is sustainability it can hardly be compatible with the unrestrained use of non-renewable chemical fertilisers and synthetic pesticides. But is there a middle way? Is it possible to farm in a way which minimises the use of harmful chemicals without eschewing them altogether?

Yes, there is such a middle way. It usually goes under the name of Integrated Crop Management, or ICM. The aim of ICM is to reduce inputs of pesticides and chemical fertilisers while maintaining farm income. Some of the methods used to achieve this are similar to those used on organic farms: crop rotation, disease-resistant varieties and nitrogen-fixing legumes. Chemical fertilisers and pesticides are accurately targeted to avoid over-use, and the pesticides used are carefully chosen to be as selective as possible, so they kill the target species only and not other organisms.[1]

It's an attractive idea. Indeed reductions in nitrogen fertiliser and pesticide inputs have been achieved on trial farms with little or no reduction in profitability. It looks like a way forward. But in fact it's a dead end.

It's a dead end in the sense that it can take us so far down the road to sustainability and no further. There is not a steady continuum from the all-out chemical farmer at one end to the purely organic farmer at the other. There's a qualitative difference between using chemicals and not using them. This difference arises because a farm is an ecosystem and in an ecosystem you can't change one thing without changing a lot of others.

An organic farm relies on biological processes for the supply of plant nutrients and protection from pests and diseases. These processes run at their own pace, and are rarely as immediate in their effect as applying a chemical. Suppose we add just one chemical to an organic system, say nitrogen fertiliser to give a crop a quick burst of growth in the spring. What happens?

The crop plants become lush and more vulnerable to pests and diseases, so we are drawn towards using pesticides and fungicides to control them. The pesticides are likely to have a longer lasting effect on pest predator insects than on the pests themselves, so natural controls are reduced and we are more likely to have to use pesticides again. Certain fungicides, and indeed some kinds of nitrogen fertilisers, kill earthworms, thus reducing the ability of the soil ecosystem to provide its own fertility. The presence of soluble nitrates reduces the activity of nitrogen-fixing bacteria, increasing the crop's dependency on artificial nitrogen. The ethylene cycle is disrupted by the temporary excess of nitrate, causing loss of organic matter in the soil, reduced availability of nutrients and increased activity of soil-dwelling pathogens.

In short, the whole nature of the system has been altered by that one action. Of course it won't be irrevocably changed by one modest application of fertiliser. But the example illustrates that there is a qualitative difference between organic farming and chemical farming. It's not a matter of degree. Either you trust biological processes or you feel you need to add something; and when you add one thing you find you have to add a whole lot of others. Thus you become dependent.

The dependency can be modified a bit, as it is in ICM. But it can't be reduced beyond a certain point unless you take the plunge and decide to trust biology rather than chemistry.

Any discussion of organic farming raises the question, can we feed ourselves by organic methods?

The assumption that organic yields are lower doesn't always hold true. Yields of grassland are usually similar before and after conversion to organic farming, though there is a dip in yields during the conversion process.[2] Yields of arable crops are usually lower, though not drastically so. But the real answer to the question lies not in yield comparisons so much as in the place animals have to play in both conventional and organic farming.

ANIMALS

Omnivorous or Vegan?

The decision whether or not to eat meat and other animal products is both a practical matter and an ethical one. Clearly it's the practical side which affects the answer to the question which was posed at the end of the last section, but let's look at the ethical dimension first.

The vegan point of view, that eating any food of animal origin is unacceptable, is one I respect though I don't share it. The vegetarian stance, that it's alright to eat milk and eggs but not meat, is more problematic.

What happens to the bull calf or male chick which cannot be used for milk or egg production, or the cow or hen when they grow too old to produce an economic yield? Of course they are fattened up, killed and eaten. It's impossible to eat animal products without contributing to the death of animals.

Personally I'd rather eat a lamb chop than a battery egg. What matters to me is how much the animal has suffered. The amount of stress involved in humane slaughter is tiny compared to the suffering borne during their whole lives by the majority of pigs and poultry, and many dairy cows.

People who want to eat animal products but don't want to inflict a lot of suffering on animals can eat certified Organic food with confidence. The Soil Association rules are as strong on animal welfare as they are on the non-use of chemicals. Calves born into a conventional dairy herd are separated from their mothers at birth, causing great stress to both of them. Cows in a conventional herd are often pushed so hard to produce more milk that their udders swell to a point where their hind legs are skewed and they suffer from lameness. The Organic rules forbid these things, and purely from an animal welfare point of view I'd rather eat an organic steak than drink a pint of conventional milk.

The alternative to buying organic is to buy direct from a farm you know and can visit personally. (See p285.)

The practical question is, 'How important is the keeping of animals to sustainable farming?' Here there are two conflicting principles to be balanced. On the one hand we can get a far higher yield of food from the land by eating plants rather than animals. On the other hand animals have an important part to play in maintaining soil fertility.

When an animal eats a plant, the majority of the energy in the plant is lost as low grade heat or goes to maintaining the animal's everyday activity. Only a minority of it goes towards increasing the animal's body weight. Typically only 10% of the food consumed by farm animals is available as edible meat when they are killed.[3] The ratios for milk and eggs are somewhat better, but similar. (See The Trophic Pyramid, pp119-120.)

This is obviously inefficient. A hectare of land can support ten times as many people on a vegan diet than on a purely carnivorous one. Meat and other animal products do contain very high quality protein and some B vitamins which are hard to get from plant foods. But with a little knowledge of nutrition a vegan diet can be perfectly healthy. We don't actually need to eat meat.

British farmland can broadly be divided into two kinds, the upland and the lowland. By and large the upland is not suitable for arable farming. It can grow plenty of grass, but relatively little food for direct human consumption – unless we can eat the grass. (See box, Leaf Curd, overleaf.) Farming in the uplands is important to the people who live there, but in terms of putting calories on the plates of the nation it's contribution is not great.

The question of whether we can feed ourselves has to be answered on the lowlands, where we can grow crops for direct human consumption. But by and large we don't do this. We use most of the land to feed animals. This is partly because there is considerable grassland in the lowlands, but also because 60% of the grain grown in Britain or imported here is fed to animals. In fact only 25% of the good agricultural land in the country is used to grow food for direct human consumption. If you include rough grazing land in the total agricultural area, only 15% of the total area produces direct human food.[5]

So how important are the animals which graze the lowland grass and eat the lowland grain to the maintenance of soil fertility? There's a big difference between grass-fed animals and grain-fed. The grain fed are eating what we could eat – and losing 90 per cent of it in the process – whereas the grass-fed are central to providing soil fertility in an organic system. (See box, Ruminants and Non-ruminants.)

The most important requirement for maintaining true soil fertility is not the supply of mineral nutrients but maintenance of the soil organic matter. (See p40 and p42.) The main source of organic matter on an organic grain farm is not manure but the ley. A ley is temporary grassland, a mixture of grasses and clovers grazed by animals or mown for hay or silage. In an organic rotation a number of years of arable crops are alternated with years of ley.

As the grasses and clovers in the ley are regularly defoliated, either by grazing or by mowing, their roots die and thus organic matter and nitrogen are added to the soil, and structure is improved. On a farm scale it's impossible to add the same quantity of organic matter to the soil in the form of composted manure, although this is possible on a market garden scale.

In theory it's possible to grow a ley and get the soil fertility from it without involving animals. The herbage can be mown and left in situ as a mulch, or taken away and composted. Where high-value crops such as vegetables are grown this is worthwhile, but on a grain farm it would mean that about half the farm would not be producing any food at all, and it would be impossible to sustain economically.

There are some 'stockless' organic rotations being developed at the moment.[6] These are based on growing green manures as catch crops, that is crops of a few months' duration fitted in between the main crops. These rotations may be effective in providing weed, pest and disease control and biologically fixed nitrogen. But quick green manures do not build up the long-term organic matter in the soil in the way a ley does. Some of the organic matter is lost every time the soil is cultivated

RUMINANTS & NON-RUMINANTS

The main constituent of grasses and other leaves, apart from water, is cellulose, a complex carbohydrate which cannot be digested by mammals. But certain mammals, the ruminants, have bacteria living in their guts which can digest it and thus make it available to the ruminants themselves. The ruminants include cattle, sheep and goats.

Pigs, chickens and humans do not have bacteria of this kind in our guts, so we can't digest cellulose. We can only digest the more soluble kinds of carbohydrate such as those found in grains. Ruminants can also digest these more soluble carbohydrates, and under some farming systems they may be fed a high proportion of grain. But it's not healthy for them, and under organic systems the grain in ruminant diets is kept to a minimum. Likewise, pigs and poultry can make use of a small proportion of leaves in their diets, but the bulk of it must be grain or other soluble carbohydrates.

Much the same is true of protein as of carbo-

hydrate. Pigs, poultry and humans need high-quality protein in our diet, while ruminants can make use of any organic substance which contains nitrogen and turn it into protein – at least the bacteria in their guts can. These non-protein-nitrogen substances are found in grasses and other leaves.

All this means that the non-ruminant animals must be fed foods which could equally well be eaten directly by humans, whereas ruminants can make use of foods which we cannot.

The down side of the ruminant digestive system is that it gives off methane as a waste product. Methane is a greenhouse gas, some 11 times more potent than CO_2. A sustainable population of ruminant farm animals would certainly be much smaller than the present number worldwide. Although they may be useful in sustainable agriculture, we should not keep any more than are needed to maintain soil fertility, and certainly not feed them a grain-based diet.

LEAF CURD

Small-scale production of leaf curd for sale at Coombe Farm near Tiverton in Devon.

One problem with a vegan or vegetarian diet in our part of the world is that most of the pulses which provide the protein part of the diet don't grow in our climate. The great majority are imported. One thing we really can grow well, especially in western areas, is grass. Making leaf curd is a way of getting a high-protein food directly from leaves, including grass, rather than feeding the leaves to animals and then eating the animals. Leaf curd is high in protein, vitamins and minerals, but low in carbohydrate and bulk. The process is simple. (See illustration right.)

The taste of the curd varies according to the kind of leaves used, but it's usually inoffensive rather than attractive. It's best used as a supplement to dishes rather than as a central ingredient. As it's highly concentrated only a small amount is needed. It goes well in everything from soups and stews to cakes and sweets.

Any leaves can be used as long as they don't contain poisons. Potatoes and tomatoes are obvious ones to avoid for this reason. Lucerne has often been used because of its high yield of fresh leaves, but it does have a slightly bitter taste. Grasses are an alternative, as long as they are taken young, when they are still juicy. Nettles, also best taken young, are very easy to grow and yield well.

Organic market gardeners may find leaf curd is a good use for the unsaleable parts of vegetables, such as carrot tops or the tatty outer leaves of cabbages. Weeds such as fat hen and chickweed can go in too. Many tree leaves are suitable, some, like beech, only when they are young, others, like lime and ash, all summer long. Trees like oak and chestnut with their high tannin contents are not suitable.

Growing leaves for curd can't compete in terms of sheer bulk with growing a grain crop or potatoes. But the yield of protein can be very high, as the following figures illustrate:

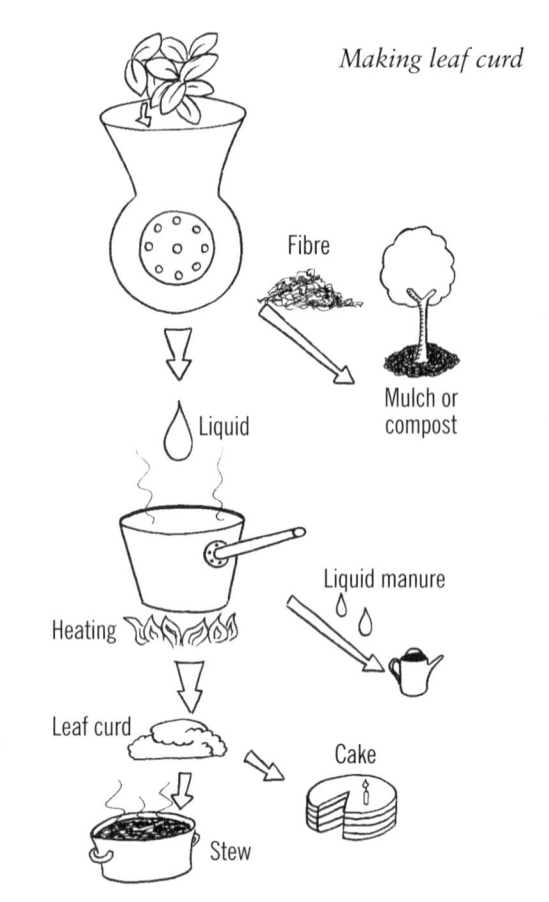

Making leaf curd

1. Crush the fresh leaves in a meat-mincer or liquidiser. (A food blender won't do.) If you use a mincer, separate the liquid from the solid with a sieve. Keep the liquid.

2. The solid can be used for mulch or compost.

3. Heat the liquid to boiling point, and the curd will solidify immediately.

4. The remaining liquid can be used as a liquid fertiliser, high in nitrogen and potassium.

5. The curd can be used immediately, or dried, frozen, pickled etc for future use.

Crop	kg Edible Protein per ha in 6 months[4]
Lucerne made into curd	1070
Soya	660
Wheat	475
Grain fed to pigs	100
Pasture grazed by sheep	85

for an arable crop, and there can be a net loss of organic matter from the soil over a stockless rotation on an organic farm.[7] This is hardly more sustainable than chemical farming.

Bicropping is an alternative to the ley/cereal rotation. In a bicrop grain and clover are grown at the same time in the same field. (See pp268-270.) The relationship between grain and ley are fundamentally the same as in a rotation except that both are present at the same time rather than in sequence. There's a grain crop every year, but it yields less than it would if grown on its own because it shares the available resources with the clover. This is not a loss if the clover contributes to the farm output by being fed to ruminant animals.

A bicrop would be possible without animals, but at the yield levels this system has achieved so far the total output would be low. If the yields of a bicrop can be brought up to something approaching that of a monocrop, the system could be viable without animals, with the clover yield used simply as a source of fertility. Whether this will be possible remains to be seen.

What all this means is that there's a fundamental difference between animals which eat grain and those which eat grass.

Grain-fed animals, including both non-ruminants and ruminants fed a high-grain diet, compete directly with us by eating food which we could eat ourselves. This kind of animal farming represents a net loss to us in food terms.

Grazing animals can make use of land which is unsuitable for crops for direct human food, but when they graze a ley in an arable rotation they are competing with us indirectly because we're unable to grow a grain crop on that land in that year. The food value in the meat or milk is a fraction of what we could get with a direct food crop. But the ley is so important to the fertility of the soil, and thus to the sustainability of the system, that their presence is a net benefit.

The total output of grain from a wholly organic agriculture would be less than at present, partly because of lower yields per hectare, and partly because a proportion of the arable land must be down to ley at any time. We would have plenty of grain for ourselves but not much to spare for feeding animals. So production of pigs and poultry would have to be reduced. But grass-eating ruminants – sheep, cattle and goats – would probably have a central place.

Choosing Animals

On very small holdings, where vegetables are the main output, the question may not be so much which animals to keep but whether to keep any at all.

While the fertility of the broadscale soil is bound up with the dynamics of grass and clover growth, intensive vegetable growing is usually dependent on brought-in manure. This can be seen as a tithe on the fertility-building capacity of the surrounding landscape, invested in the most intensively productive areas. There's nothing inherently unsustainable about this, and it matters little whether the area from which the fertility is gathered is on the same holding as the vegetable garden or a neighbouring one. Self-sufficiency in soil fertility is an essential component of sustainability, but it is best measured on the scale of the local landscape rather than the individual property boundary.

A few chickens are almost always useful for a bit of tractoring and eggs for home consumption. But looking after animals does take time. Vegetable growing is a hard way to make a living, with long hours. Having to go off and milk the goat or feed the pig on a busy day can be an unwelcome distraction, out of all proportion to the benefits they may bring to the agroecosystem by scavenging unused food or adding activator to the compost heap. In other words, you can have too much diversity, at least from a work management point of view.

Wherever animals are kept, a mixture of species is always desirable from a biological point of view. A polyculture of animals has much the same benefits as a polyculture of plants: it's important to animal health, and can give a higher yield than a monoculture.

Sheep in particular are very prone to internal parasites, which spend part of their life cycle in the grass sward and reinfect the sheep when they are ingested. The only non-chemical way to control these parasites is to practice rotational grazing, alternating sheep with cattle, which are not prone to the same parasites. An organic monoculture of sheep is impossible.

NICHE MANAGEMENT

An alternative to keeping your own animals, both on smallholdings and on larger farms, is niche management. Niche management is an agreement between a farmer or landowner and a person who wants to carry out some kind of productive activity on the land which doesn't involve having full use of it. This is how beekeepers often operate, by taking up the beekeeping niche on a number of different holdings. It can also be done with grazing animals such as sheep on farms where the amount of grassland is not enough to justify keeping a flock full time. Usually the niche manager pays a fee for using the land in a specific way for a specific time.

A variant on niche management is where a person takes on responsibility for a particular task in return for receiving something which the farm has to offer, for example feeding the pigs in return for space to pitch a caravan or bender. In these ways the farm can get some of the advantages of diversity which animals bring without making life too complicated for the farmer.

Cattle and sheep eat grass at different lengths, so they make fuller use of the available food. In fact, for each cow grazing on a farm, one ewe can be kept without any increase in the grazing area.

Goats can be useful where brambles or other scrub have taken over grazing land and the intention is to bring it back to productive pasture. Since they prefer woody plants to herbaceous ones, they will reduce the scrub and thus make the land more productive for other grazing animals while feeding themselves in the process. On the other hand, where you want trees and shrubs to grow among the pasture, which is often the case in permaculture systems, goats need very strict control, which must be 100% effective. If they get in among the trees and shrubs they can cause untold damage in a few minutes.

They can be very productive, giving a high yield of milk in relation to their bodyweight compared to cows. But they are somewhat tender animals, needing more care and attention than most, and housing throughout the year.

Horses are bad grazers. They will eat some parts of the pasture down as flat as a billiard table and leave other areas, where they dung, untouched. They also cause a fair bit of compaction with their feet. But they only do serious damage in a monoculture. If they're rotated with other grazers they do little harm.

Since they compete directly with us for food, pigs are hard to justify on ecological grounds unless they can make use of resources which would otherwise not be used. An example is the mast (seed) of woodland trees. But mast is not produced every year, and is only a seasonal food when it does come. (See pp223-225.) The pigs have to be fed for the rest of the time. Another example is picking up windfall apples in an orchard. But fruit can only be a supplement to the diet, and the practice is probably of more benefit to the health of the orchard, by removing pests and disease spores with the fallen fruit, than to the nutrition of the pigs.

The idea of a pig tractor – pigs cultivating the ground by rooting – is attractive. They are particularly useful for bringing new land into cultivation or reinstating temporarily abandoned land where there is a mass of herbaceous vegetation with persistent roots or rhizomes, such as bracken or bindweed. They can remove roots to a depth of more than 75cm at a density of 25 pigs per hectare.[8] Where brambles and other shrubs need removing a density of 50 pigs per hectare is recommended.[9] Traditional breeds tend to be the most effective rooters, especially the athletically built Tamworth.

On heavy soils pigs can cause severe compaction if they are left on the area they are clearing for too long, or if the weather turns wet. They need to have permanent quarters which they can be returned to as soon as their presence on the tractoring area threatens to do more harm than good.

However on light, well-drained soils a pig tractor can form part of the rotation, with the pigs spending a pre-set period of time in the field regardless of the weather and with nowhere else to go. Some soil compaction is likely under these conditions, but the farmer may be prepared to accept this in return for weed control and the nutrients the pigs leave behind in their dung. These nutrients are in effect an external input, as they will have come mainly from the grain-based pig food which forms the bulk of their diet. The plants they eat in the course of tractoring are only a supplement to this.

As well as choosing the appropriate species of animals, care needs to go into the choice of breeds. Modern breeds are mostly in the high-input/high-output mould. They produce a great deal of milk or meat, but in order to do it they need large quantities of food, veterinary attention, shelter and so on. Old-fashioned native breeds may not produce so much in total, but they tend to be thrifty, hardy, good mothers and have easy births. In the case of meat

A group of young pigs clear a vegetable plot on Ragmans Lane Farm, Gloucestershire. Shortly after this picture was taken the weather turned wet and they were quickly moved to a strawed yard.

These pigs on Hen and George Curtis' farm in Powys live in the fields full time. After six months in one field they move on and the field is sown to grass or kale for the cattle and sheep, with excellent results.

animals, the taste is often much better too. This can be especially important when making a living on a small farm, where high volume production is impossible and going for high quality is the only way to make a living.

Grazing Animals

Foggage

The foggage system is a way of feeding grazing animals which has received much attention from permaculturists in Britain. Foggage is grass which has been neither grazed nor cut, but left standing from summer into winter. On a farm which is run on a total foggage system no hay or silage is made. The animals simply eat the foggage where it stands.

No work or machinery is needed to provide the animals with their winter feed, and because they live outside all year, no buildings are needed, no bedding down and no hauling out of manure. Foggage is less nutritious than hay or silage, and there is more waste, so the gross output is less than on a conventional farm, but the net output in energy terms is more.

This is a traditional method but these days it's very rarely used. It's principal champion is Arthur Hollins of Market Drayton in Shropshire, who has run his whole farm on a foggage system for several decades now.

The entire 60 ha is down to permanent pasture. He runs a multiple suckling herd of 100 beef cattle plus 100 sheep. During spring and early summer the cattle range the whole farm. But from mid-August one third of the farm is closed off to provide foggage for the first half of the winter, and from the end of August another third is closed to provide late winter foggage. The latter receives any available manures, often sewage sludge. The remaining third is left for late summer and autumn grazing, and the three different treatments are rotated round the farm over three years.

It's important to have a good proportion of tussock-forming grasses in the sward to keep it standing well, and cocksfoot is the most important of these. But a wide diversity of grasses, legumes and herbs is desirable. New seed could be introduced with a strip-seeder, as used in the clover-cereal bicrop (see p269), but Arthur Hollins sows it by hand into the abundant molehills!

A mature sward is also needed. The reason for keeping cattle indoors in the winter in lowland areas is more to protect the soil from their feet than to protect the cattle from the weather. Only when a strong mat of roots and stems has formed does the sward have a chance of surviving a winter of grazing. Foggage farming provides ideal conditions for a mat of this kind to develop.

The stocking rate is some two thirds of what it would be on a more intensive farm and the growth rate of the animals is slower, giving a gross output of perhaps half what it could be. But with the lower costs net profit is similar, and with virtually no fossil fuel

Cattle on Arthur Hollins' farm. Note the prolific molehills.

inputs the net energy output is probably higher than in any other meat production system in Britain. Arthur Hollins only keeps a tractor to cut the thistles once a year. There's virtually no work involved other than caring for the animals.

However foggage farming does have some limitations. The first is climatic. When the ground is covered in snow the animals can't get at the foggage, and if the whole farm is down to foggage fodder must be brought in from elsewhere. The second is soil type. Arthur Hollins' farm is on a very light sandy loam. Given a medium loam and a wet winter a foggage field can become a mudbath. This was the experience of bio-dynamic farmer Richard Smith when he tried wintering his beef herd on foggage on his farm at Ashprington in Devon. He had to abandon the experiment and bring the cattle in to save the pasture.

Even where the sward is not destroyed the compaction caused by keeping the cattle out over winter can reduce yields. My own field, also a medium loam, is usually cut for hay, but one year it was fed as foggage and the following year the hay yield was down some 40% from usual. The cattle were in the field no longer than it took them to eat the foggage, but it was a wet time, and the yield drop was almost certainly due to compaction.

The third limitation is the condition of the cattle: they become very lean during the winter and spring. Although allowing an animal to stop growing is against conventional wisdom, the situation is retrievable with beef cattle. With dairy cattle it's not. A milking cow would dry up immediately if she had to survive the winter on foggage. I have heard of a dairy farmer who keeps his dry cows on a foggage system while feeding his milking cows normally, and this is probably the only way foggage can be used on a dairy farm.

However the biggest drawback of the foggage system is that it is a purely animal farming system. As we have seen, the main ecological function of grazing animals in sustainable agriculture in the lowlands is to make use of the leys which provide soil fertility for direct human food crops. Foggage can't do that because it can only operate on permanent pasture. The second function of grazing animals is to make use of land which can't grow direct human food. Much of this land is in hill and mountain areas and is unsuitable for foggage because the climate is too harsh to allow cattle to winter outside.

The best place for foggage is probably on lowland farms where there is land which is too steep for arable farming. This land is often also too steep for hay and silage making and could be ideal for foggage, especially if combined with tree crops in an agroforestry system.

ARABLE & VEGETABLE CROPS

Diversity in the Field

Rotation, or diversity through time, is the backbone of organic farming and growing. But in permaculture we emphasise simultaneous diversity, with more than one kind of plant growing in the same field at the same time. Broadly speaking there are three ways of increasing diversity in the field:

• Mixing crop varieties.
• Mixing crop species.
• Mixing crop and non-crop plants.

Much of the world's food is grown using these practices. They are all very common among small-scale farmers in the less developed countries. Once they were common here too but the trend of industrial agriculture has been towards uniformity. Monocrops are suited to high levels of mechanisation and pesticide use. But given the same level of inputs well-designed polycultures usually yield more than monocrops. (See pp19-20.) They also usually have a more stable yield. In a year when one component of a mix does badly another component is likely to yield better than average.

Varieties[10]
Interest in variety mixtures is currently centred around disease control. At the simplest level, a mixture of two varieties with different levels of disease resistance will become diseased to a level similar to that of the more resistant variety.

In addition, each variety of a crop species has resistance to some strains of the disease-causing pathogen and is susceptible to others. When a large area of a single variety is grown in one place any pathogen strains to which it's susceptible multiply rapidly and become the dominant strains. Thus the variety's resistance to disease is soon lost and serious outbreaks of disease are likely. Plant breeders respond to this by breeding new varieties but these soon succumb in turn to new strains of the pathogen. This process develops into a race between plant breeders and pathogens.

When mixtures of varieties are grown, different plants in the field have resistance to different strains of a pathogen. This means that serious outbreaks are unlikely. Plants will be affected here and there, but it's hard for the disease to spread quickly across the field as every pathogen is restricted by a barrier of resistant plants. It also slows down the rate at which new strains of pathogens can evolve to overcome the resistance of any one variety.

The effectiveness of varietal mixtures was demonstrated in East Germany in the years before reunification. In order to save the hard currency needed to buy fungicide, the nation's entire spring barley crop was put down to mixtures. This was so successful that fungicide use was cut by 80%. After reunification monoculture and fungicides were reimposed. This was partly at the insistence of the West German brewers, who would only buy single varieties, though the East German brewers had brewed perfectly good beer with the mixtures.

These mixtures were based on the experimental work of Professor Martin Wolfe and others at the Plant Breeding Institute at Cambridge. Martin Wolfe is one of the leading practitioners of sustainable farming in Britain, and now combines variety mixtures, mixed-species crops and agroforestry on his Wakelyns Farm in Suffolk.

Cereal variety mixtures are grown in a number of other European countries, and in the USA 100,000ha of mixed variety wheat is grown, and accepted by the grain buyers. Many other crops are suitable for mixtures, including vegetables and fruit, although with top fruit the large size of individual plants reduces the intimacy of the mixture.

Mixtures usually yield at least as much as the average of the component varieties when grown as monocultures, and often more. In any one year there is usually one component variety in a mixture which will yield more than the mixture. But the top-yielding variety varies from year to year and at sowing time you can't predict which one it will be.

The more varieties the more resistant the mixture is to disease, though above three or four components you may get diminishing returns. When choosing varieties to put into a mix the first thing to consider is whether they will grow well together. They need to have the same sowing and harvesting times and so on. In a crop which cannot be separated after harvest, like cereals, they also need to have the same end-use qualities. With vegetable crops it's not too difficult to separate the varieties at harvest time for separate marketing.

The varieties should have some resistance to disease, and ideally have resistance to different strains of disease. The latter is not a foregone conclusion, as many varieties have common ancestors and thus common genes for resistance.

As regards layout of plants in the field, the most intimate mix of varieties gives the best disease control. But if the varieties need to be harvested separately, growing in alternate strips can be worthwhile. If vegetables are grown on a bed system this can be done by planting different varieties in adjacent beds. Although this is less diverse than either mixing varieties within the bed or growing different crops in adjacent beds, it makes management simpler and may be a useful compromise.

There is a danger of a single really good mixture being over-used. For variety mixtures to work properly farmers and growers need to use different mixtures in different fields and change them from time to time.

Another advantage of variety mixtures, apart from disease resistance, is that they may be able to yield better where soil conditions vary in different parts of the field. An intimate mixture would probably be most effective, so that the most successful variety in any particular spot can easily take up the space left by its less successful neighbour. Uneven soil conditions are more likely to occur on organic farms than on conventional ones where blanket applications of chemical fertiliser are used, though in-field yield variations of up to 50% have been recorded in conventional potato crops. This hypothesis has not yet been tested experimentally.[11]

Species[12]
There are a number of ways in which species diversity can be increased in the field or market garden:

- Border cropping – a border of a different crop is grown around the main crop.

- Strip cropping – alternate strips of different crops are grown.

- Intercropping – alternate rows of different crops are grown.

- Mixed cropping – two crops are mixed at random.

An example of border cropping is sowing wild mustard around brassicas to draw off flea beetles, which are attracted by the stronger scent of the mustard. Success with this particular combination has been mixed. Often it reduces the number of flea beetles on the crop, but it has been known to increase it. The key to success is an intimate knowledge of the ecology of your own locality. (See box, Skill or Energy?)

Strip, inter- and mixed cropping can all have a positive effect on pest and disease levels in just the same way as variety mixtures. While variety mixtures have an effect on diseases, species mixtures affect both pests and diseases. This is because, while most pests only affect one or a few species, there is usually no difference in pest resistance between different varieties. Pest predators are usually more numerous in these diverse crops and this has an added impact on pest levels.

Microclimate can be improved, particularly by strip cropping. Tall crops, alternating with low-growing high-value crops, can act as windbreaks. Winter rye has often been used. In Russia, rye, sunflower, fodder cabbage and maize, sown in strips 0.6 to 1.8m wide and 5 to 7m apart have raised yields of tomatoes, potatoes and cucumbers growing between them. In the Fens of eastern England bulbs have been strip cropped with wire-trained blackberries in the same way.[13] Care must be taken to align the rows across the prevailing wind, otherwise this arrangement will actually increase wind speed.

The components of a strip crop can be harvested separately. But inter- and mixed crops can't unless they're vegetables which will be harvested by hand. This means the different components must ripen at the same time. Examples of grain bicultures which do this are winter wheat with winter field beans, and spring oats or barley with spring peas.

SKILL OR ENERGY?

Increasing diversity of all kinds is a realistic alternative to chemical use, but it has its down side. Although there are plenty of reliable crop combinations which can be used anywhere with confidence, others are less widely applicable. Very often a combination which works in one place is not effective in another, and some can work better in one season than another.

The interactions between plant and plant and between plants, pests and pathogens are complex and we are far from understanding them completely. The key to success is an intimate knowledge of the ecology of a specific locality. This is a very different kind of skill to that required for conventional farming and growing, which deals with simple interactions: this pesticide kills this pest. The skills required for sustainable growing take longer to acquire and require a more observant and reflective frame of mind.

Because more diverse systems are more complex they often also require more day-to-day management and more labour than monocultures. This is a general characteristic of sustainable systems. They substitute human input, mainly in the form of skill, for heavy inputs of chemicals and machinery.

The economics of the sustainable approach are very much dependent on the relative prices we pay for fossil fuel and people's time. At present the price of fossil fuel is absurdly low and sustainable farming and growing are only viable if the produce receives a premium price, either by Organic certification or by direct marketing to consumers who value it. (See Food Links, p285.)

In the barley-peas biculture the main function of the barley is to act as a support for the peas, and not much of it is needed. In one trial in south-east England the barley was sown at only 3% of its usual seed-rate, giving a land equivalent ratio (LER) of 1.1.[14] An LER of 1.85 has been recorded for a biculture of barley and field beans, and in general mixtures of legumes and non-legumes have been found to have a positive LER.[15]

Separating the different components after harvest can be done with existing seed-cleaning equipment. This is easier if there is a marked difference between the components in seed size, density or shape, such as there is between cereals and legumes. Alternatively the mixture may be suitable for animal feed as it is.

Crop mixtures can give greater yield stability. If one component of the mixture fails for any reason the other may increase its yield to compensate. In one example, the oats in an oat-pea mixture were attacked by wireworm. The yield of oats was reduced by half, but the pea yield increased fourfold.[16]

These legume/non-legume mixes are based on complementary use of nutrients by the two components. Mixtures can also be based on the different life-cycles of the components, making complementary use of the resource of time. An example of this is a biculture of brussels sprouts and summer cabbage. Both are sown at the same time and require much the same cultural treatment, but the cabbage matures much more quickly than the sprouts. They can be planted together, more densely than either would be alone, and the cabbage removed when mature, allowing the sprouts to continue growing.

In an experiment at the Wellesbourne research station in Warwickshire, the LER of this combination ranged between 1.11 and 1.34. The cost per tonne of produce fell by between 10 and 20%, making this system significantly more profitable than the equivalent monocrops. The fact that both plants were brassicas was considered an advantage under a conventional system, as the same fertilisers and spraying regime were appropriate to both crops. Under an organic regime

A strip crop of barley, wheat and oats on Martin and Ann Wolfe's Wakelyns Farm in Suffolk.

Celeriac and leeks make a good combination, a tall thin crop and a short bushy one.

crops of different families would be favoured to reduce pest and disease problems. But the experiment does give hard evidence that the principle of mixing fast- and slow-maturing crops can give higher yields.[17]

Non-Crop

Non-crop species which may add to the diversity of the field ecosystem include both non-crop plants within the field, commonly known as weeds, and field edge vegetation, such as hedgerows. Both can have both positive and negative effects on the crop plants, and it takes some skill on the part of the farmer or grower to maximise the positive and minimise the negative.

Weeds compete with crops, but after the crop is harvested they can protect the soil from erosion and take up nutrients which would otherwise be lost by leaching. A few species of weeds actually benefit the crop while it's growing. One such is corncockle, which can increase both the yield and quality of wheat by means of a chemical it gives off, an example of positive allelopathy. (See p218.)

Pest levels can be both increased and decreased by weeds and field edge plants. If the non-crop plants provide alternative hosts for pests they can increase the problem. But if they have their own specific pests which do not feed on the crop species they can help to increase the population of pest predators without increasing crop losses. Many flowering plants, including members of the umbellifer and composite families, feed the adult stages of insects whose larvae predate on pests. (See pp242-244.)

A strip of perennial grasses and wild flowers around the edge of a field can have multiple benefits. It can stop annual weeds such as cleavers spreading into the crop, the flowers can attract pest predator insects, and it provides habitat for other predator insects such as beetles. It also provides habitat for small mammals which are prey for predatory birds such as the barn owl. A width of 1-2m is usually appropriate, and this is currently eligible for grant aid under the Countryside Stewardship Scheme.[18]

Grain

Five different ways of growing grain are described below. The first is the 'conventional' way, the way the vast majority of cereals are presently grown in Britain. This is our starting point and other methods can be seen in comparison to it. The organic method is practised by a very small but growing number of farmers and is an established method with a track record.

The other methods described here are to varying degrees experimental. The bicrop method is being developed by scientists at mainstream agricultural research stations, and could well be adopted by some farmers in the near future. The Bon Fils method is more unconventional and has not even been developed on an experimental level. It's essentially a small-scale way of growing grain and would seem quite alien to a contemporary European or North American farmer. The domestic prairie system is being developed experimentally, but it's such a radical departure from what is now done that firm results cannot be expected for decades. But there's little doubt that in the long run it's the most sustainable way to grow grain.

Conventional

Most cereals are grown according to the yearly cycle shown in the illustration below. The land is ploughed and cultivated ready for seeding in September or October. The plants are still quite small when growth stops with the falling temperatures of late autumn.

This has two important effects. Firstly, the root systems are not sufficiently developed to take up all the nutrients in the soil, and leaching takes place. Secondly, there is much bare soil between the plants, allowing considerable erosion to take place.

The erosion potential is also increased by the

Tramlines in a cereal crop.

degree of mechanisation and chemical use. It's difficult to operate heavy machinery across the slope, so all operations are carried out up and down the slope. At regular intervals across the field pairs of rows are left unplanted to take the wheels of the tractors which will apply pesticides and fertiliser as the crop grows. These tramlines, as they're called, become compacted and act as channels for water, which can carry away the soil.

In the spring the plants start to grow again, but they are still small, with an undeveloped root system. This means they're little able to forage nutrients for themselves, so they're given a dressing of nitrogen fertiliser. This makes them grow fast but it also makes them fleshy and vulnerable to pests and diseases, increasing the need for pesticides and fungicides. The bare soil between the plants allows weeds to grow and these are controlled with herbicides. Any nitrogen not used by the plants can be leached to the water table in wet weather, where it can become a serious pollutant of drinking water.

The crop is harvested in late summer. Then the land is ploughed, cultivated and reseeded. On a specialised arable farm cereal crops may be grown continuously on the same land, or they may be interspersed with arable break crops, such as oilseed rape, which are grown in the same way as the cereals. Conventional farms running a rotation where a ley alternates with arable crops are becoming increasingly less common as the economies of scale drive farmers to specialise in either arable or animal farming.

Organic

On organic farms the annual cycle is similar, without the applications of chemicals of course, but the year-on-year cycle is very different. Most organic farms which grow grain are mixed farms. A rotation is practised, usually with a number of years of cereal crops alternating with a number of years of ley. Thus a series of annual monocultures alternates with a perennial polyculture.

The illustration overleaf shows a typical rotation. During the years of ley soil fertility is increased in a number of ways:

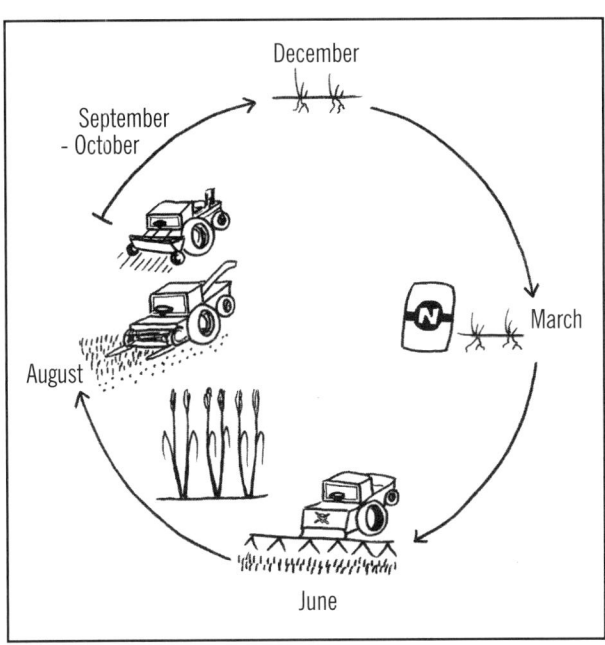

Conventional grain growing, the annual cycle.

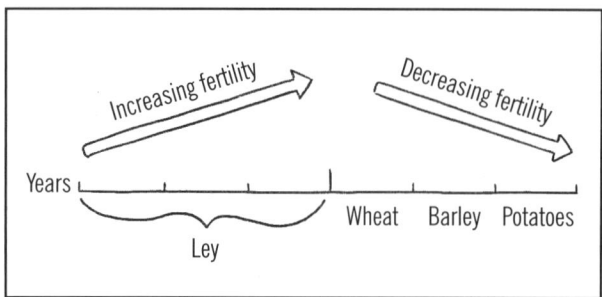

The cycle of fertility in a typical organic rotation.

- Organic matter levels, earthworm populations and soil structure are all built up.

- Nitrogen is fixed by the clover in the ley and retained in an organic form which is resistant to leaching.

- The perennial ground cover protects the soil from erosion.

- Both annual weeds and the soil-borne diseases of cereals are controlled.

All stages of the rotation are present on the farm at any one time, giving a spatial diversity which helps to control other pests and diseases.

When the ley is ploughed up the arable crops make use of the accumulated fertility. The first grain crop after the ley gains the most. Soil fertility and yields usually decline in subsequent arable crops, but the level of organic matter in the soil is unlikely to fall to the low levels found in continuous arable soils.

Soil erosion during the arable part of the rotation is less than on a conventional arable farm, mainly because the soil contains more organic matter and has a better structure. (See pp54-55.) The lack of chemical fertilisers means lower yields, but it also means less leaching. There is some leaching of nitrogen from an organic rotation at the time when the ley is ploughed up for the first arable crop. This brings an influx of oxygen into the soil and much of the organic matter is oxidised, converting the nitrogen to a mineral form which can be leached. But average yearly leaching from organic farms is typically much less than from conventional ones.[19]

Energy consumption on organic farms is far less than on conventional ones, 65% less in one study.[20] This is mainly due to the high embodied energy of nitrogen fertiliser, which is the biggest single component of the energy budget on conventional farms.

A significant difference between conventional and organic farming is the level of skill required. If things start to go wrong on a conventional farm it's often possible to put them right by a quick application of fertiliser or spray. On an organic farm, plant nutrients and plant health must be supplied by careful husbandry over the whole rotation. By the time things go wrong it's often too late to do anything about it.

Bicrop[21]

A bicrop or biculture is any combination of two crop plants grown on the same land at the same time. Here we are looking specifically at growing cereals, through a perennial sward of white clover. This is simultaneous diversity in the field, compared to the diversity through time of a traditional rotation of ley and cereals. The clover element of the bicrop has similar functions with respect to soil fertility as the ley in an organic rotation.

Experiments on this system have been carried out over the past few years at mainstream research stations in Denmark, Ireland and Britain, by Dr Bob Clements and others. This work is particularly interesting to permaculturists, as it amounts to scientific investigation of a number of permaculture principles in action: stacking, polyculture, covered soil and minimum tillage. (See box below.)

The system is not primarily an organic one, although trials on organic farms have started at the time of writing, notably on Martin Wolfe's Wakelyns farm. The mainstream experimenters would use any chemicals they thought necessary, but the nature of the system itself has meant that both pesticide sprays and artificial nitrogen have been drastically reduced, though phosphate and potash fertilisers are used as normal.

The annual cycle of growth is much the same as that shown for conventional cereals in the illustration on the previous page, but with some important differences.

First a sward of white clover is established. This is the most suitable legume, as it has a prostrate, mainly horizontal growth habit, which means it will offer minimum competition to the upright-growing cereal, and it's a persistent perennial.

MINIMUM TILLAGE

In recent years minimum tillage has been increasingly used in North America as a way of reducing soil erosion. There are a number of different techniques, but all aim to avoid inverting the soil, and to leave as much residue from the previous crop on the soil surface as possible. The seed is sown through this layer of debris, which protects the soil from erosion.

Soil erosion certainly has been reduced. But dependence on herbicides has increased, as the plough is no longer used to control weeds. Using a cover crop such as the clover in the wheat-clover bicrop is one way to solve this dilemma. But whether bicropping could work as well on the great plains of North America as can here in Europe is an open question. Soil moisture is much more of a limiting factor to growth there than it is here, and competition for water between cereal and cover crop may be too severe.

Young cereal plants growing through a sward of white clover.

In autumn, just before seeding, sheep are introduced at a high stocking rate. They graze down the clover so that it will not compete with the young wheat plants. In addition their urine provides a boost of nitrogen and they reduce the slug population, which can be a big problem in minimum- or no-till crops. Trials in New Zealand have shown that stocking rates of 1,500 sheep days/ha or more, over not more than eight days, can reduce slugs by 90% – compared to methiocarb slug killer which only achieved 50%. The actual mechanism by which sheep kill slugs is not known but it may be due to trampling effects.[22]

Alternatively the clover can be cut and used as fodder, mulch or compost material. This would be as effective in controlling the clover, but would not provide nitrogen or slug control, which could be crucial.

The wheat seed is sown with a strip seeder, which cultivates a series of narrow strips and drops the seed into these. Strips of 5cm at 22cm centres have been used in the non-organic trials. But cultivated strips of 20cm with 30cm uncultivated between them may work better in organic systems as this allows for mechanical control of the clover. (See pp274-275.) The wider spacing also gives the cereal a better start, with less competition from the clover, and the clover soon recolonises any available space. Either way most of the soil remains undisturbed.

The vast majority of the nitrogen for the wheat comes from the clover. But this is not available in any quantity till the summer, so a small dose of artificial nitrogen or slurry is normally applied in spring to get the cereal crop going.

In autumn the crop is harvested in the normal way, but with the combine set high so as to avoid the clover. The straw and clover are then grazed off or mown for fodder, and another cereal crop can be sown as before. Alternatively, grain and clover can be harvested together earlier in the season as whole-crop silage.

The key to success in bicropping is to balance the growth of cereal and clover so that neither one takes over and smothers the other. Early experiments found that spring-sown cereals were not vigorous enough and got smothered by the clover, while winter cereals other than wheat were too vigorous and killed out the clover, so only winter wheat was recommended for bicropping. But results at Wakelyns Farm have shown winter oats to be the best cereal for bicropping, followed by wheat and triticale. This may be in part because a mix of five or six clover varieties is used, including early and late season ones, and the mixture is surely more vigorous and persistent than a single variety. It may also be related to the fact that these trials, unlike the earlier ones, are on an organic system.

Breeding cereal varieties specifically for bicropping would clearly be a great help in balancing the vigour of the two components and thus increasing the range of cereals which can be grown.

The advantages of this system of bicropping are numerous:

- The lack of soil disturbance allows soil fertility, and thus yields, to rise year by year, instead of falling as they do in the years following the ploughing up of a ley.[23] Earthworm populations, organic matter levels, soil structure and soil nitrogen levels can all rise to a high level and stay there.

- Soil erosion is virtually eliminated by the complete ground cover.

- Leaching of nitrogen is reduced because there is a mature root system present throughout the year, and there is no need to plough up the ley before sowing cereals.

- Fossil fuel consumption is around 50% of the conventional level.[24] Less machinery is used, and less artificial nitrogen.

- Pests are reduced to below the problem threshold. Aphids are the major pest of wheat, and four reasons probably contribute to their control:
    - aphids are thought to recognise their host by the contrast of green plants on a brown background, but in a bicrop it's all green
    - the perennial understorey allows the population of predators, such as beetles and spiders, to build up
    - the aphids frequently fall off the plants, and the understorey makes it difficult for them to find their way back
    - the excess of soluble nitrogen which makes plants in a conventional system so vulnerable to pests such as aphids does not occur.

- Diseases are so reduced that sprays are rarely used, and when they are they are applied at half dose. The reasons for this include:
    - virus diseases are mainly spread by aphids
    - *Septoria* is spread by rain splash, which is greatly reduced by the ground cover

- – mildew is reduced both by the moderate nitrogen content and by the buffering effect of the clover
- – the high earthworm population quickly clears up debris from the previous crop.

This experiment at Long Ashton Research Station shows how effective the clover in a bicrop can be in controlling weeds. The plot on the right has been treated exactly the same as the one on the left except that the clover has been left out, and it's full of volunteer oilseed rape.

- • Weeds are largely suppressed by the perennial ground cover, reducing the need for herbicides.

- • There's a general increase in biodiversity, including wildflowers, insects and so on up the food chain.

Many, if not all, of these advantages flow from the fact that it's a system of simultaneous diversity, as opposed to a rotation, which is one of diversity through time.

The disadvantages of the system are that:

- • Inputs of phosphate and potash are relatively high in the system as currently practised.

- • Slugs have occasionally been a problem.

- • If a weed problem does arise it may be difficult to control it without using herbicides. Normal weeding cultivations are not an option.

- • An infection of clover rot would be a serious problem, but there are resistant varieties, and a mix of clover varieties can minimise the risk.

- • Extra skill is required, especially in balancing the vigour of the two components so that neither out-competes the other.

- • Yields of grain are lower than in a monocrop, mainly due to competition for water from the clover from late May onwards.

Results from the first three years' trials with non-organic bicropping showed grain yields averaging 60% of conventional crops. After allowing for the fodder value of the clover and the lower input costs, profitability was 80% of conventional.[25] When the bicrop was harvested as whole-crop silage rather than grain, yields were 78% of conventional whole-crop silage, and the profit was greater.

There are indications that the grain yield can be increased, perhaps to the same level as that of monocrop cereals, by reducing the vigour of the clover during the summer period when it competes with the cereal for water. This can be done with a selective herbicide on a conventional crop and by means of an inter-row mower on an organic crop sown on the wider spacing. When this is done, bicropping could be more profitable than monocropping, as the costs are so much less.

Since the aim of bicropping is to grow grain in a more ecologically sound way, it deserves the premium price which consumers pay for organic produce. It's unlikely that consumers will accept a new category of non-organic but bicropped food and pay a premium for it, so farmers are only likely to get the full financial reward from bicropping if it is also organic. The advantages of bicropping listed above are just the features needed in an organic system. In addition bicropping may mean that wheat can be grown more times in an organic rotation than it can with a series of monocrops. This could go a long way towards bringing the overall grain yield of organic farming closer to that of conventional. (See p261.)

In the long term, when we start paying something closer to the true price of energy – which we surely must if we are to survive at all – the lower energy cost of bicropping will make it competitive with monocropping even if yields are somewhat lower.

Clover bicropping can also be used with other crops, such as vegetables (see pp273-275) and maize. Maize, which is widely grown for silage in this country, is a serious source of both soil erosion and persistent pesticides in the environment. A maize-ley bicrop, where maize is drilled into a ley without tillage, is being promoted in Switzerland to reduce erosion and leaching. It has also been shown to reduce pests and diseases.[26]

Bon Fils[28]

The organic and bicrop systems are not so far removed from what conventional farmers do today that they couldn't imagine themselves farming that way. But with the Bon Fils system we move into the realm of radical experiments. It's a cereal-clover bicropping system, and it shares most of the characteristics of the bicropping system described above, but the annual cycle is radically different. The cereal plants are allowed to complete their full development, which means they are in the ground for more than twelve months, each year's crop being sown when the previous crop is still present. This makes mechanisation problematic, so it's essentially a hand method.

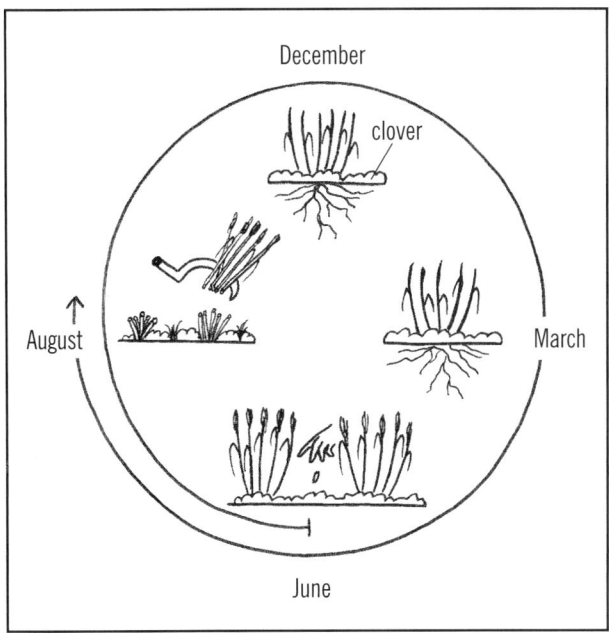

The annual cycle of the Bon Fils system.

When Marc Bon Fils devised his system for growing grain in France he hadn't heard about Fukuoka's work. But his system could be seen as an adaptation of Fukuoka's to European conditions. Unlike Fukuoka he hasn't used his system over a long period of years, so as yet it must be regarded as interesting but untried.

A perennial stand of white clover is established first. Cereal seed is sown at around the Summer Solstice, and germinates as soon as the nights start getting longer. Plant spacing is very wide, 50-60cm each way, giving a plant population around one tenth that in a conventional field. Thus the plants have the space as well as the time to grow ten times as big as conventional ones, with ten times the number of tillers (stems) by the

A single wheat plant grown on the Bon Fils system.

time growth stops in autumn. Each tiller can potentially carry an ear of grain.

In spring these plants have all the resources they need within their bodies to start growing rapidly. They don't need the quick fix of soluble nitrogen which conventional plants get. Their root systems are very much more extensive than those of conventional cereals, and are able to forage effectively for nutrients and water in a much greater volume of soil. The plants grow vigorously through the summer and are ready for harvest at the usual time in late summer.

It's essential to use obligate winter varieties of cereals, that is varieties which can only flower and set seed after they have been through the cold of a winter. Otherwise they will try to complete their whole life cycle between June and the end of autumn, and the crop will come to nothing. No modern varieties of wheat are obligate winter wheats. It's said you need to get varieties which go back to before 1826, when a

---

FUKUOKA[27]

The work of the Japanese farmer, scientist and sage Masanobu Fukuoka has been very influential in the permaculture movement worldwide. But since the Japanese climate is so different from the European, his work is more relevant to us as an inspiration than as a model.

He evolved his system of Natural Farming over a period of 25 years, and with it achieved yields which are comparable with those of conventional farmers in his locality. It is based on four principles: no ploughing, no chemical fertilisers or prepared compost, no weeding by tillage or herbicides, and no dependence on chemicals. It is a bicrop system, with grains grown in a permanent sward of white clover.

In the climate of southern Japan it's possible to grow two crops a year, and the Fukuoka system follows the traditional rotation of summer rice alternating with winter barley. Each crop is established by sowing pelleted seed on the soil surface before the previous crop is harvested. Fertility is maintained by the clover, and by returning all the straw as a mulch, with a light dressing of chicken manure to help it decompose.

Another practice he eschewed was the constant flooding of the rice field. He observed that rice grows quite well on unflooded land as a volunteer, so why shouldn't it do the same in his fields? But he did flood once a year, deep enough to cover the clover but not the rice and for long enough to weaken the clover but not to kill it. This is necessary to stop the clover becoming too strong and out-competing the cereals.

non-obligate wheat variety was bred which has parented all subsequent varieties.

In order to grow a crop each year, one year's crop must be sown before the preceding year's has been harvested. It's perfectly possible to sow seed between the large maturing plants. Seed can be broadcast on the surface, as long as it's pelleted in clay to protect it from birds, or it can be dibbed in, which is a more secure way of getting the right spacing. Trampling of the young plants when the previous crop is harvested is not harmful, and may even be beneficial by encouraging the production of more tillers.

The need to control the vigour of the clover while the cereal crop is growing is more acute with the Bon Fils method than with the conventional bicropping described above, as there is no gap between harvest and seeding when the clover can be grazed or mown. Fukuoka was able to use flooding to weaken the clover, but with the Bon Fils method some mechanical means would be needed, either mowing or cultivating between the cereal plants. The ideal time to do this would be late spring to early summer, when the clover may start to compete with the cereal and before the new cereal crop is sown.

Domestic Prairie[29]

All the systems of grain growing we have looked at so far are based on annual cereals, grown as a monocrop or bicrop. Domestic prairie is based on perennial grain plants grown as a polyculture. As such it's much closer to the permaculture ideal. (See Colour Section, 32.)

Wes Jackson, an agricultural scientist from Kansas, USA, became horrified by the loss of soil caused by growing annual grains in his home country. (See p53.) He looked at the natural vegetation of the North American Great Plains, which is prairie. He saw how it preserves the soil and goes on producing year after year without any of the fossil fuels or chemicals used on farms. He realised that an agriculture which imitates the natural vegetation of the locality as closely as possible must be the most sustainable way to grow food.

In order to start the long task of turning this vision into reality he set up the Land Institute at Salina, Kansas.[30] The first task of the workers at the Land Institute is to learn all they can from the remnants of native prairie.

Prairie is a polyculture of herbaceous perennials, and domestic prairie is just the same thing but consisting of plants which produce seed edible to humans. The word 'domestic' is used in the same way that we describe sheep and cattle as 'domestic animals'.[31]

The plant composition of the prairie varies greatly with climate and soil type, both over broad regions and over quite short distances. This means that a polyculture designed for Kansas may not do well in Saskatchewan. Equally, one which did well on a sandy loam on a hilltop may be a failure on a heavy soil or in a valley bottom. A perennial polyculture is much less under the control of the farmer than a rotation of annuals. The opportunities for intervention are less. So the match between plants and environment must be all

the closer. Individual mixtures will have to be designed for individual places.

The composition of prairie also changes from year to year. The list of species present may not change, but the relative growth of each may change dramatically in response to weather. For example legumes may be less prominent in a dry year, as water rather than nitrogen becomes the limiting factor to plant growth. A domestic prairie needs to be designed so that it can maintain its functional diversity over a number of seasons. The relationships between species must be such that plants do not die out in years which are unfavourable to them.

Whether a domestic prairie can work in practice depends on answering four questions:

- Can perennial grain crops yield as much as annual ones?

- Can a perennial polyculture yield more than a monoculture?

- To what extent can a perennial polyculture meet its own need for nutrients?

- Can a perennial polyculture manage weeds, pests and diseases?

Perennial vs Annual. To answer this question, workers at the Land Institute first studied the agronomic characteristics of thousands of species and varieties of prairie plants. From these they have selected a handful which they are breeding so as to enhance their desirable characteristics.

The selected plants come from three families: the grasses, which includes the domestic cereals; the legumes, which includes peas and beans; and the composites or daisy family, which includes sunflowers. The two main approaches to breeding perennial grains are, firstly, to breed up wild perennials which have not previously been grown for grain, and secondly, to cross domestic annual grain species with wild perennial relatives.

Critics of the perennial grain idea theorise that perennials should yield less than annuals because they need to devote part of their energy to overwintering organs. (See p15.) But results so far have been encouraging. Two of the wild legumes under investigation have yielded around a tonne to the hectare. This is a low yield by modern standards, but similar to what medieval farmers achieved with annual cereals before the start of modern plant breeding. Meanwhile another North American plant breeder has produced a perennial wheat with 70% of the yield of annual wheat.[32]

There is a general tendency for yields of perennial grains to be highest in the first year and then decline. While this is true of some species there are others which yield best in subsequent years, and the problem can be minimised by the appropriate species mix. Reliability of yield is also important, and here perennials have an advantage. In drought years, when annuals may fail to produce any grain at all, the perennials with their deep, established root systems can get enough water to produce a crop.

An important characteristic of any grain crop is that the grain should stay on the plant when ripe and not fall to the ground before it can be harvested. In a polyculture this requirement is compounded by the fact that different component species may ripen at different times. Resistance to shattering, as this characteristic is called, is one of the traits which plant breeders at the Land Institute are selecting for.

Polyculture vs Monoculture. Trials with bicultures of perennial grains at the Land Institute suggest that polycultures of these plants will yield more than monocultures. Land equivalent ratios of up to 1.61 have been recorded. The reasons for this include the fact that the different species take water and nutrients from different layers of the soil and at different seasons, and that legumes are included in the mix. It is also due to the fact that the decline in yield after the first year has been less in bicultures than in monocultures.

There are competitive and allelopathic relationships between the different plants, as well as positive interactions, and research has shown big differences in the yield of individual plants according to which species is growing next to them.[33] Thus polycultures need to be designed to place good neighbours next to each other. The design need not be intricate, and research suggests that an intercrop with a single species in each row may be sufficiently diverse. This can easily be sown with a maize planter with a different kind of seed in each hopper.

Nutrients. A grain polyculture can meet its nitrogen needs from the legume component, just as a ley can. Mycorrhiza have an important role to play. (See pp56-57.) One mycorrhizal fungus can simultaneously combine with the roots of plants of different species. Thus nitrogen can pass directly from legumes to other plants through them rather than via the indirect pathway through the soil, which only operates at certain times of the year and is vulnerable to leaching.

Mycorrhizal transfer can happen in polycultures of annuals, or in an annual/perennial bicrop. But mycorrhiza are sensitive to the disturbance caused by tilling and the undisturbed conditions of a perennial agroecosystem give ideal conditions for the development of a really effective mycorrhizal network.

The most significant role of mycorrhiza in plant nutrition is that of making phosphorous more available to plants, and here the perennial nature of domestic prairie may give it a distinct advantage over other ways of growing grain.

Weeds, Pests and Diseases. Most weed growth in annual systems takes place during the spring, when the crop plants are small and don't cover all the ground. Herbaceous perennial crops can be much more effective competitors with annual weeds, because they grow quickly in spring from a large established rootstock.

But perennial systems do offer less opportunity to control perennial weeds than do annuals, as there is not that gap in time between one crop and another when weed control cultivations can take place. The inclusion of allelopathic plants in the polyculture can help to control perennial weeds. At the Land Institute a perennial sunflower has shown itself capable of taking on this role, as well as yielding edible seed.

We can expect the incidence of pests and diseases to be lower in a polyculture than a monoculture, and trials at the Land Institute have shown this to be true for virus diseases in the mixtures they have tried.[34] So far there are no results for other diseases and pests.

Developing domestic prairie is a long-term prospect. Land Institute researchers say they think in a 50 to 100 year time frame. There's not even any certainty that in the end it will work. What is certain is that if it does it will be the most sustainable way to grow grain.

Bicropping Vegetables

Most permaculturists who grow vegetables do so in the standard organic way. A distinction is usually made between intensive vegetables, grown mainly by hand in a dedicated vegetable garden, and field vegetables, grown on a larger scale with tractor power as part of an arable rotation. The former often fit better into permacultural systems, especially when combined with some form of direct marketing such as a box scheme, which can make this scale of operation financially viable. (See Food Links, pp285-293.)

The polyculture of perennial vegetables as described in Chapter 8 may be the permacultural ideal. (See pp199-201.) But it's probably not a commercial proposition today, especially with the limited range of perennial varieties available. Bicropping field vegetables with clover, on the other hand, has been well developed experimentally and could be used to advantage both by commercial vegetable growers and home gardeners.

A crop of cabbage grown through a sward of white clover. Flevoland, Netherlands.

Pest and Weed Control[35]

This technique has been pioneered mainly in the Netherlands, mostly on leeks and brassicas. In contrast to the work on wheat, the aim has been primarily to reduce pesticide use, and the other benefits of the legume have been ignored by the mainstream experimenters. It has also been mainly an annual system, with the clover only persisting for the life of a single crop, in contrast to the perennial approach of the cereal bicropping experiments. (See pp268-270.)

However there is some evidence that a perennial stand of clover may be an advantage over an annual one even when the aim is only pest control. This comes from a trial in Scotland on the control of cabbage root fly in swedes. A perennial stand of clover appeared to be necessary for good results, as many eggs are laid early in the spring before it's possible to establish a good ground cover of clover. However, trials with annual clover swards have recorded good control of aphids and moderate control of butterflies on cabbage, and successful control of thrips on leeks.

Although the mechanisms of control are not fully understood, it has been observed that they don't operate unless all the soil is covered, and with that much clover present there is inevitable competition between the clover and the main crop. Competition effects can be reduced by minimising the check to the main crop on transplanting. With cabbage, growing seedlings in larger modules has been shown to help. Plants raised in 90ml modules did better than both those in 15ml modules and bare rooted plants.

Yields of 60-70% of conventional crops have been recorded with both cabbages and leeks. But in one trial a bicrop of leeks, although two weeks behind a comparable conventional crop, caught up and yielded the same when harvest was delayed.

In fact the reduction in yield may be more apparent than real, at least on an organic system. Most organic field-scale vegetable rotations include a ley or green manure to rebuild fertility. A bicrop simply incorporates the ley simultaneously rather than in series and any yield comparison should be between whole rotations rather than single crops.

Choosing the right kind of clover is important in reducing competition. The annual subterranean clover is less competitive than white clover and may cause less yield loss. The varieties Geraldton subterranean and Asterix white are recommended, though these are both hard to obtain in Britain. Fenugreek may be a possible annual substitute.

Another way to reduce competition from the clover is to knock it back with an inter-row mower. This may also have the added benefit of making more of the nitrogen in the clover roots available to the vegetable crop as clover roots die off in response to cutting back the top growth. With a perennial stand of clover it's also necessary to mow immediately before planting out the vegetables to prevent the young transplants being physically swamped by the clover.

As the season progresses heavy-shading crops such as cabbage become more competitive than the clover and they can kill it out. One way to prevent this may be to plant alternate rows of light and heavy shading crops so that the clover can survive and provide complete ground cover for the next year.

Experimenters at the University of Connecticut, USA, have used a cover crop of summer purslane with broccoli, not with the intention of reducing insect attack, but to suppress weeds. Although some hoeing and hand weeding were also needed at the beginning of the season, they found costs were similar to chemical control and cheaper than black plastic. They found that yields of broccoli were similar under all three treatments, presumably because the purslane roots occupy a shallower level of soil than those of the brassica.[37]

A Complete System

In contrast to the limited aims of the experiments described above, a complete organic system for bicropping vegetables is currently being developed by Martin Wolfe and his associates at Wakelyns Farm. It's based on an agroforestry layout with rows of fruit and nut trees alternating with cropping alleys 18m wide. (See Alley Cropping, pp280-281.) The perennial sward of mixed-variety white clover covers the whole field, tree rows included, and this helps prevent problems with weed growth in the tree rows.

The vegetables are grown in 20cm cultivated rows with 30cm of undisturbed clover between them, the same design as is used for cereals on the farm. Every three rows make up a bed, with tractor wheelings on the clover either side of each bed. A set of simple

Onions growing in clover on Andy Waterman's allotment in Hampshire. He has integrated bicropping into his standard organic rotation.[35]

Young vegetable crops receive target irrigation in the bicrop on Wakelyns Farm.

tractor-mounted machinery has been developed on the farm to cultivate, sow, weed and irrigate in these rows, and to mow the clover between them. Once the crops are well established the clover recolonises any remaining bare ground in the vegetable rows.

A seven-year rotation is practised, with two beds each of lettuce, onion family, brassicas, beet family and carrot family, and one bed of beans, in each alley. As the rotation proceeds each crop is moved not to the adjacent two beds but to the next two. This is to avoid the possibility of buildup of root borne diseases, as the roots of one crop could easily penetrate the adjacent bed. In the final year the land is ploughed and potatoes are grown, then the clover is re-established and the rotation starts again. Clover and vegetable crops can be established simultaneously in the first spring of the rotation.

No direct yield comparisons have been made with similar crops grown without clover, but Martin Wolfe is convinced that any yield loss is more than compensated for by the lack of non-productive fertility-building courses in the rotation. He describes the impact on pests and diseases as "phenomenal." A further advantage has

been a great reduction in soil compaction, as machinery runs on the clover sward rather than on bare soil. This also means that it's easier to get on the land when weather conditions are on the wet side, an important advantage on the farm's heavy soil.

One key to the good yield level of the Wakelyns system is undoubtedly the 20cm wide cultivated strips that the vegetables are sown or planted into, giving them a good start with little or no competition from the clover. Inter-row mowing is also essential. Trials suggest that monthly mowing is better for the crop than weekly. Both these tasks require highly specialised machinery which is not available on the market, and this has been designed and constructed on the farm by Paul and Mark Ward.

The system could easily be adapted to intensive vegetables, perhaps using an adapted garden rotavator, or to a home garden using hand tools.

AGROFORESTRY[38]

Agroforestry means growing tree crops and field crops on the same piece of land at the same time. Tree crops include both timber and fruit, while field crops include arable, vegetables and grassland. A traditional orchard, grazed by cattle and sheep, is one example of agroforestry.

There are many variations on this theme, some of which are described in other parts of this book. Page references to these are given in the mind map below. In this chapter the general principles of agroforestry are given, with examples of how it can be applied on farmland with arable crops and grass.

Principles

A mixture of trees and field crops is one step nearer to the natural vegetation of this county than simple fields

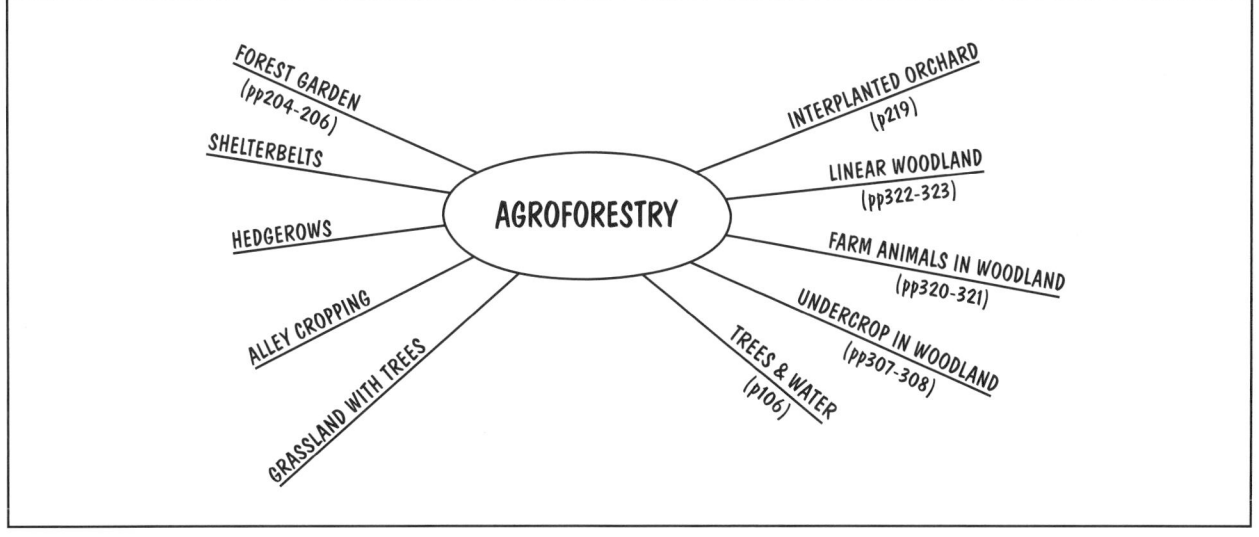

of arable or grassland. Intuition tells us that this is a more sustainable way to work with the land. But we can also point to a number of specific advantages of agroforestry over simpler systems.

Advantages

There are a number of ways in which agroforestry leads to higher yields, and there are other advantages which are not directly related to yield. Let's look at the yield advantages first. See the mind map below.

normal forestry spacing. They may not increase height so fast but total growth is greater.

Although they can reduce the growth of the field layer by intercepting some of the light, the conflict between the layers is minimised by the different annual growth patterns of trees and field crops. A good example is ash trees grown in pasture. Ash is a late leafing tree, not coming into full leaf till the end of May. In one study it was found that pasture grasses produce 60% of their growth during the months when ash is not in

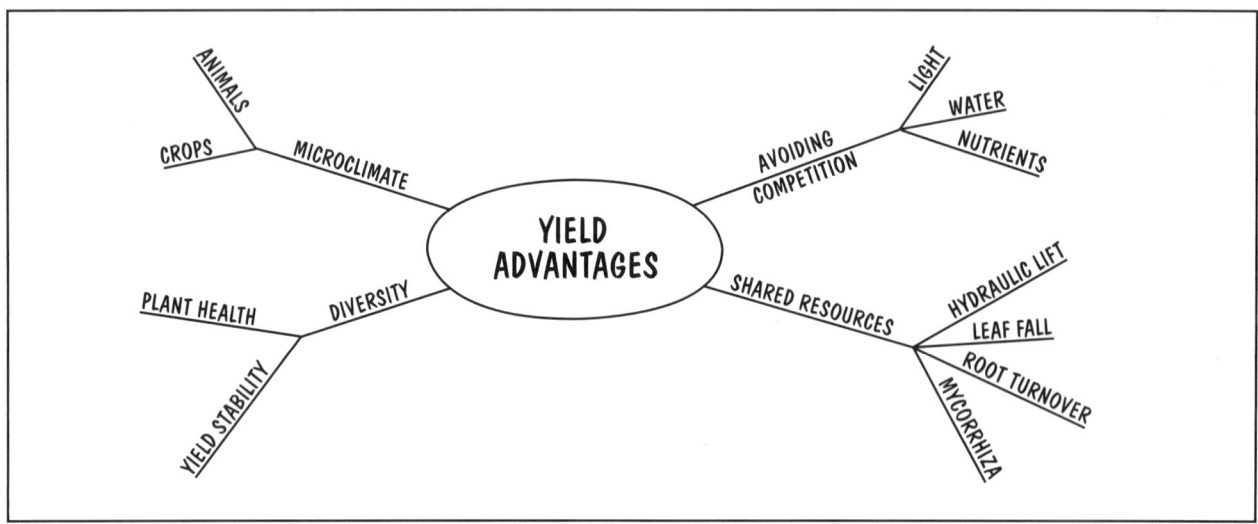

Avoiding Competition. Plants compete for space, light, water and nutrients. If every plant in the field is the same size and shape and has the same annual cycle they all compete for these resources in the same space and at the same time. But trees and field crops differ both in shape and in annual cycles, so a mixture of the two reduces competition and makes more complete use of the available resources.

Trees grown at a wide spacing, as in an agroforestry situation, have space all around them and receive light from all around, so they grow faster than trees grown at

Walnuts and winter wheat are a good agroforestry combination as each does some of its photosynthesis when the other is not in leaf.

leaf, mainly in the spring.[39] Autumn sown arable crops, including cereals, pulses and oilseeds, also benefit by doing much of their photosynthesis before the trees come into leaf. At the other end of the year, in the autumn, the trees stay in leaf long after the field crops have been harvested.

What's more the trees are relatively small for much of their life. At the beginning of the tree rotation they have no measurable effect on the productivity of the field layer. (See p278.)

Water and mineral nutrients are also used more effectively by a system like this which is actively growing for the longest possible period of the year. When there is no crop growing water can be lost by deep drainage or evaporation and nutrients by leaching. Tree roots and their associated micro-organisms are even active in the winter and can take up nutrients then, when most leaching takes place.[40]

Water and nutrients can also be gathered from a greater volume of soil as, in general, tree roots reach deeper into the soil than those of herbaceous plants. In fact the presence of the herbaceous roots can make the tree roots develop more strongly in the lower layers than they would if grown alone. This not only increases the volume of soil being actively used, it also means that nutrients leached below the root zone of the field crop can be captured by the trees.[41]

Shared Resources. The water gathered by deep tree roots is not only a resource which would otherwise be lost to the system, it may actually be made available to the herbaceous plants by a process called hydraulic lift. In hydraulic lift water is extracted from a relatively wet subsoil by tree roots and released by them into a relatively dry topsoil, where it may be taken up by some herbaceous plants. This process is still poorly understood and no-one yet knows for sure how significant it may be in practical agroforestry.[42]

It's not just the size and depth of tree roots which enables them to take up more water and mineral nutrients, but the fact that they live longer and have greater opportunity to form mycorrhizas than herbaceous plants. (See pp56-57.) Mycorrhizas are particularly important in the uptake of phosphorous.

These nutrients can be shared with the field crops in three ways: firstly by leaf fall; secondly through the constant turnover of roots, as older roots die and are replaced by new ones; and thirdly by shared mycorrhiza. Mycorrhizal fungi are not species-specific, and it's possible for a single fungus to be associated simultaneously with the roots of two unrelated plants, such as a tree and a cereal. Nutrients can then pass in any direction. If either the tree or field layer contains nitrogen fixers there is an extra resource available which may be shared with the other components of the system.

Leaf fall and root turnover can also contribute significant amounts of organic matter to the soil.

Diversity. The diversity of structure in agroforestry systems can lead to higher biodiversity and thus to reduced pest problems.

In an experiment by Leeds University aphid populations have been found to be consistently lower in arable agroforestry compared to the same crops grown in open fields. This may be due to increased habitat for aphid predators such as ground beetles and hover-flies in the non-crop vegetation at the base of the trees. On the other hand, this same vegetation harbours slugs, which have caused some crop losses.

As with any polyculture, the greater diversity gives greater stability of yield. In a year when the field crops do badly the trees may yield well and vice versa. If the trees are being grown purely for timber this may not be much help to the farmer in the short term, but if they have an annual yield, such as fruit, nuts or animal fodder, it could help in a difficult year.

Microclimate. The shelter given by the trees to both crops and farm animals can be beneficial, in much the same way as windbreaks and shelterbelts can. (See p282.) In particular shelter from the wind can slow down the loss of soil moisture during the summer, which can increase the growth of both grass and arable crops, especially in dry areas.

In grassland with trees the shelter is spread over the field rather than concentrated near a shelterbelt. This encourages the animals to make better use of the grazing over the whole field rather than staying huddled in one place. Both this and the direct effects of shelter can increase animal growth and productivity.

Other Advantages. Trees are the best way to prevent wind erosion. Those parts of the country which are prone to wind erosion are on the whole fertile in other respects, too fertile to be turned over to simple forestry, and agroforestry is probably the best land use system for them.

Tree roots and leaf litter keep the soil open and allow water to infiltrate rather than run off. Thus lines of trees on the contour can prevent water erosion and reduce runoff without the construction of a bank or swale. This is an example of the use of a biological resource, where planting trees replaces the use of heavy earth-moving equipment.

Any actively growing trees are taking CO_2 out of the atmosphere and storing it. Although planting trees is no substitute for reducing CO_2 emissions if we want to check global climate change, it will help. (See p74.)

From an economic point of view, agroforestry adds diversity to the farm business. In times when demand for mainstream crops and animal products is low,

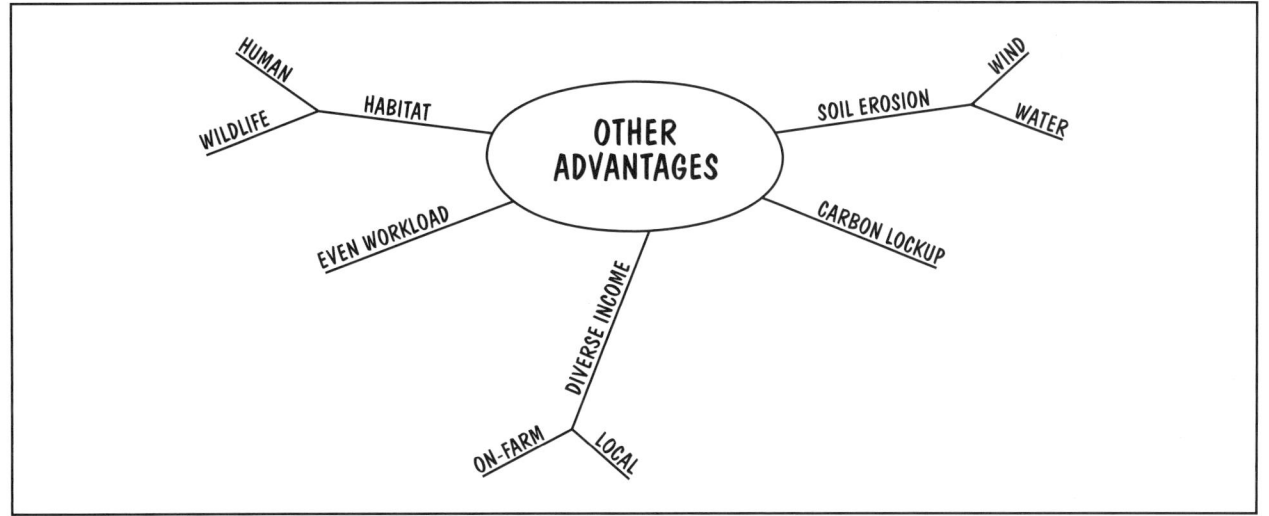

tree produce, such as fruit and nuts, may keep its value. Timber is a long term investment, yielding a lump sum at the end, which can complement the annual income from agricultural crops. It may be seen as the farmer's pension, or as insurance against hard times. As I write this there is a drastic recession in farming in Britain, and many small farmers are losing their livelihoods. Having timber to sell could make the difference between surviving till conditions improve or going under.

Timber production is most worthwhile if value is added to it before sale. Planked and seasoned timber fetches many times the price of a newly-felled log (see p301), and finished furniture an even higher price. This could mean an extra enterprise on the farm, or a new business in the locality. Thus agroforestry can bring employment into rural areas and help to bring life back to the local economy.

Most tree work, such as planting, pruning and harvesting, is done in the winter, which is a quiet time on most farms, especially arable ones. So agroforestry can help to even out workload over the year.

Many wild animals and plants can benefit from the increased diversity of structure which agroforestry brings. This increased biodiversity is not only intrinsically valuable but can make the place more pleasant for humans to live in. Bringing trees back into landscapes which have become increasingly barren in recent decades also makes them more pleasant places to live in.

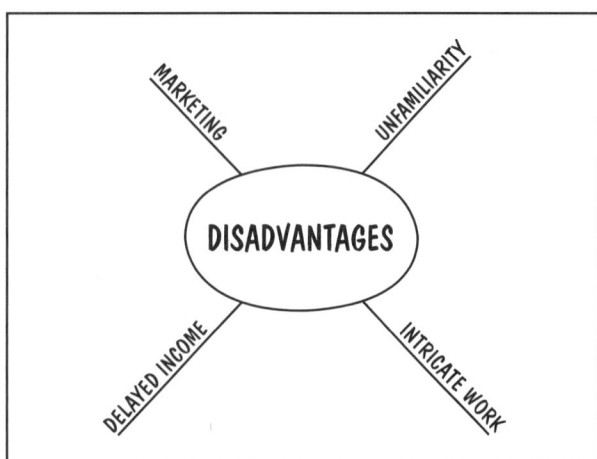

Perhaps the biggest obstacle to the adoption of agroforestry in Britain, and many other European countries, is the estrangement between farming and forestry which has left most farmers with little knowledge about or feeling for trees. (See pp300-301.) This doesn't mean that farmers don't like and appreciate trees. But they more often think of them as an amenity than as a productive component of the farm, and usually lack the skills to work with them.

There's no doubt that having trees in the fields makes field work more intricate and time-consuming, especially for farmers with big machinery and large fields. This is a cost which must be set against the overall yield benefit of agroforestry.

Growing timber does not directly replace any income forgone from the field layer, because it won't be realised for many years to come. This could be a problem for a farmer hard pressed to make a living or repay borrowings. Trees grown in the relatively open conditions of an agroforest can grow much faster than in the crowded conditions of a forestry plantation. Ash, for example, can be ready to fell in 30 years rather than the 70 which is more typical in forestry. Poplar can be ready in 15. But this is a long time for a farmer used to annual crops. Fruit and fodder trees will start to give a return much sooner.

However, there is little or no loss of yield from the field crops in the first few years, or even decades, after planting the trees, and the cost of establishment can be spread by planting up a small area each year. This is sometimes known as 'rolling permaculture'. At present grants are available under the Woodland Grant Scheme for planting timber trees in agroforestry, though not for fruit and nut trees.

The tree component may also require new marketing skills. In the case of timber crops the return may not be worthwhile if there is no value added on the farm.

Grassland with Trees

Combining trees with farm animals is known as silvo-pastoral agroforestry. There is a network of silvo-pastoral experimental sites in Britain and Northern Ireland in a variety of upland and lowland situations. They were planted in 1987-88, each with trees at both 100/ha (spaced evenly at 10m x 10m) and 400/ha (5m x 5m), in grassland grazed by sheep.

More than a decade later there is still no reduction in sheep output due to the presence of the trees. The total loss of grassland production through the whole of the tree rotation is estimated to be between 10 and 25%, with trees planted at 400/ha.[43]

The shelter given to the sheep by the trees certainly compensates for at least part of any drop in grass

Groups of trees are cheaper to protect than individual ones, but there's a less intimate mixture of trees and grassland.

production. Improved animal welfare was seen as one of the biggest advantages of agroforestry by farmers in a survey conducted in Northern Ireland.

The animals can damage the trees, both directly and by soil compaction in the rooting area. Because the sheep are attracted to the trees for shelter, compaction is greatest just where it can do most damage to the trees, and it can cause losses of trees in the first few years. The effect is greater when the trees are at the wider spacing, as the same number of sheep are gathered round a smaller number of trees. It is not a significant problem when they are planted at 400/ha.

Direct damage to the trees by browsing and using them as scratching posts is perhaps the greatest limitation on grassland agroforestry, because protecting the trees is expensive. Options for reducing the cost include:

• Grouping the trees together. This means that individual protectors can be replaced by a ring fence, which is cheaper. But this is a less intimate mixture of grass and trees, so some of the benefits of diversity are lost.

• Planting in rows. This has the same advantage. It's also said to make herding the animals easier than either even-spaced or group layouts.[44]

• To graze only sheep in the early years when the trees are most vulnerable. It's still necessary to protect each tree but the protection can be less robust and cheaper than that needed for cattle. However this would not be possible on an organic system because without routine use of chemicals parasites can only be controlled by rotational grazing.

• Geese are another option. But they too can damage trees, and the market for them is limited.

• To take only hay and silage in the early years and introduce animals later. But all mowing with no grazing can lead to deterioration of the sward, perhaps to the point where it would need to be reseeded.

• To grow arable crops in the early years and grass the land down when the trees are big enough.

PROTECTING TREES FROM CATTLE

The amount of hardware needed to protect trees from cattle is less than many people think.

• The kind of structure in the photograph below is 100% effective but uses an excessive amount of timber and steel.

Almost as effective, and at a realistic cost.

Effective but prohibitively expensive.

• In a trial of different levels of protection, a tree shelter, two round stakes and a spiral of barbed wire, as in the photograph above right, was found to be 90% effective. Where grazing pressure is not intensive this can rise to 100%. The tree protector needs to be at least 2m tall, and plastic mesh may give better tree growth than a rigid tube.

• In the same trial using three stakes instead of two, arranged in a triangle round the tree, was 100% effective.

Anything less robust than these is likely to give an unacceptable level of damage to trees where cattle are grazed. All kinds of tree protection need occasional checking so that damage to the protector can be remedied before any damage is done to the tree. The kind of protection shown in the photograph above needs to be expanded as the tree grows.

This is perhaps the best solution where the land is suitable for arable. It would also help to bring back mixed farming to areas which have become a grass monoculture.

Avoiding damage to trees is an art rather than a science. The age of the trees is only one factor. Animals will often ignore the trees when they're first let into a pasture but start to take an interest in them when they've been there for a while, either when they've eaten off all the more palatable grasses or simply when they get bored. There's also a great deal of difference between breeds and between individuals. Some are always more interested in browsing than others and it's not unknown for a whole herd to become keen on eating trees through the example of one animal.

Apart from growing timber or fruit, trees in pasture can be used to produce fodder for the grazing animals, which can be fed to them simply by pollarding the trees. This is a traditional practice and is especially useful in high summer, when grass growth is reduced. (See graph below.) Cattle and sheep will eat the leaves, twigs and bark of palatable tree species, leaving the sticks as ideal fuel for a masonry stove. (See pp152-153.) In autumn, sheep in areas of rough grazing will often seek out ash trees to eat the fallen leaves. Pollarded branches can also be dried as 'leaf hay' and fed in the winter.

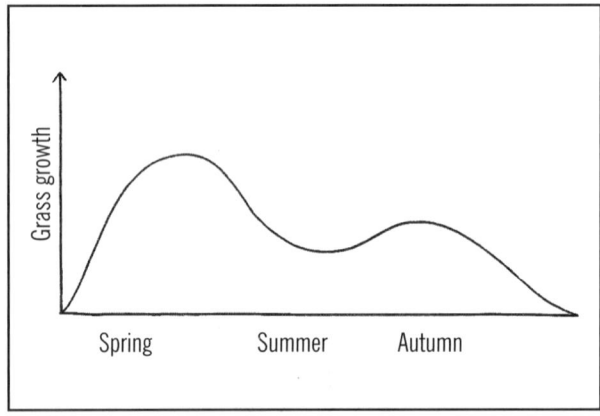

The annual growth curve of grasses.

Palatable trees include ash, limes, willows, poplars and alders. Beech, oaks and sweet chestnut have more tannins in the leaves and are not suitable. For winter fodder holly has often been used in the past. Although the lower leaves are prickly the leaves on the upper branches are not, and they are quite palatable to grazing animals. (This only applies to trees grown from seed, a cutting taken from the prickly part of a holly tree will grow into a wholly prickly tree.) Holly leaves are said to have one of the highest calorific contents of any tree browsed by animals, and are rich in nutrients.[45]

To grow high quality timber in any agroforestry system the trees must be pruned. (See Alley Cropping, next column.) The prunings make a useful addition to the grazing animals' diet.

Alley Cropping

Combining trees with arable crops is known as silvoarable agroforestry. By far the commonest layout is parallel lines of trees with field crops grown in the space between them. The lines of trees are known as production hedges and the spaces between as alleys, hence the name alley cropping for the system as a whole.

In our latitudes the best alignment is north-south (see p220), although where waterborne soil erosion is a possibility hedges on the contour should be considered. The alley width is determined by the width of the agricultural machinery used. Typically 12, 18 or 24m of field crop lie between hedges 2m wide, giving an overall width of 14, 20 or 26m. If the field crops are grown on a no-till system it's necessary to run along the sides of the hedges with a deep tine periodically to prune the tree roots and prevent them competing unduly with the field crops.

With the narrower alley widths shade may become too much for arable crops towards the end of the tree rotation and the alleys may best be put down to grass. The trees will not need protection from the animals by this stage. With wider alleys arable can be maintained throughout the rotation.

Experimental work on alley cropping in Britain shows less encouraging results than the grassland work. Yield trends of field crops in the alleys have not been consistent, being sometimes higher than simple arable and sometimes lower. No doubt a clearer picture will emerge as the present plantings get further into their rotation, but researchers are confident that the outcome will be positive.[46]

If the tree yield is in the form of timber, the system will only be viable if this is high quality timber of furniture grade. In a forestry plantation quality is achieved by planting the trees close together and progressively thinning them. This effects quality in two ways. Firstly, the close planting encourages the trees to grow straight and suppresses branches on the lower trunk. (See illustration on p220.) Secondly, the thinning process allows the forester to choose the best trees to grow on for the final crop, often less than ten percent of the trees that were originally planted.

In an agroforest trees are either grown at the final density from the start, giving no opportunity to select the best ones, or at two to four times the final density, giving little opportunity for it. In either case the trees are not 'drawn up' by close neighbours so they must be pruned.

Pruning is an extra expense. It also makes the tree rows excessively permeable to wind at ground level, which can actually increase windspeed. (See p85.) But the closeness of the tree rows means that overall windspeed is reduced, and this is not the problem it would be with a single windbreak. Where shelter is a priority an understorey of shrubs can be grown between the trees.

On the other hand, pruning reduces shading and thus prolongs the time arable crops can be grown in the relatively narrow alleys.

If trees in agroforestry are to yield quality timber they need to be regularly pruned, as these roadside poplars have been.

Because foresters can select the best trees they can get quality trees for the final crop even if the initial planting stock is highly variable, so plant breeding has been largely ignored in forestry. But if agroforestry with timber is to be viable, uniform high quality cultivars must be bred. This is a long-term prospect, but in the short term the best individual trees can be multiplied by vegetative reproduction. In the case of trees which reproduce by suckers, such as wild cherry and some poplars, this is relatively straightforward. Other trees can be reproduced by micropropagation.

This is now being done, and selected strains of poplar and cherry are already available. Work is underway on ash, oak, beech and sycamore.

Hedgerows

The pattern of small fields and hedgerows which is traditional over most of Britain can be seen as a form of agroforestry. In the past, in areas where the fields were really small and hedgerow trees frequent, there were often more standard trees in farmland than in coppice-with-standards woodland.

The main functions of hedgerows are containing animals, giving shelter, and checking soil erosion. Additional outputs can include:

- timber from standard trees
- fuel, tool handles etc from pollards and coppice
- fruits and nuts
- edible and medicinal herbs
- supplementary food for animals, including health giving herbs
- decorative foliage
- habitat for beneficial wildlife

Where ancient hedgerows exist many of the plants which can provide these outputs are present, and only need encouraging.

For example, many potential timber trees get cut off every year when hedges are trimmed by machine. With a little care they can be marked with a tape so that the driver of the hedge trimmer can see where they are, and thus the trees can be preserved to grow on. The hedge shrubs in the immediate vicinity of the young tree will need to be trimmed by hand. Trees in north-south hedges cast less shade on the fields than those in east-west hedges and these may be the best ones to leave.

They will need pruning if they are ever to form long, valuable trunks. All too often trees have been used as fenceposts by farmers. Sawmills always reject the bottom 1.2m of a hedgerow tree, even if no wire is visible, to protect their saws. Unless the tree has been pruned this may leave little usable timber.

Where new hedges are planted the appropriate plants can be included. Fruit and herbs for human consumption can be included, but this is more likely to be worthwhile for domestic use than as a commercial proposition. (See p233.) Medicinal herbs for animals to self-select may be very worthwhile on an organic livestock farm. The edge habitat of a hedgerow is ideal for a wide range of herbs, both woodland and grassland species. Hedgerows are of course an ideal place for fodder trees for pollarding. Holly trees for Christmas foliage are an imaginative use of hedgerow space. (See Colour Section, 37.)

Ash makes a good hedgerow pollard, yielding tool handles, firewood or forage, all of high quality.

A hedge bank gives overwintering habitat for beetles which migrate out into the fields in summer where they eat aphids and other pests. It is also nesting habitat for wild bees, and the shrub layer of a well-grown hedge gives shelter to bees as they move around the landscape. It's estimated that a third of our food comes from insect-pollinated plants, including fruits, oilseeds, peas and beans and animals fed on clovers.

Shelterbelts

The principles of windbreak and shelterbelt design are given in Chapter 4. (See pp83-88.) Here we look at how they may be used on farms.

The use of shelterbelts has been neglected on British farms during recent decades. One reason for this must be the cultural split between agriculture and forestry. Another is the chemical approach, which has largely ignored factors such as microclimate. 'Why wait years for trees to grow when you can put on an extra bag of fertiliser and get your money back right away?' is a message which has had the full force of the advertising industry behind it.

It's easy to get the impression that shelterbelts cause a net loss of yield, because the yield loss immediately beside the belt, mainly due to shading, is much more visible than the gentler but more widespread yield increase over the field as a whole. (See illustration below.) Although there is no direct evidence to show that shelterbelts give a net increase in yield in this country, a study of other temperate countries recorded an average yield increase in the order of 10-20% for arable crops and grass.[47] A Danish study showed a yield increase four times as big as the yield loss due to land take for the shelterbelt and negative edge effects combined.[48]

A good use for the narrow strip of land right by the windbreak where yield is reduced is to plant it up with a strip of wildflower meadow to attract beneficial insects and control the ingress of annual weeds. (See p242.)

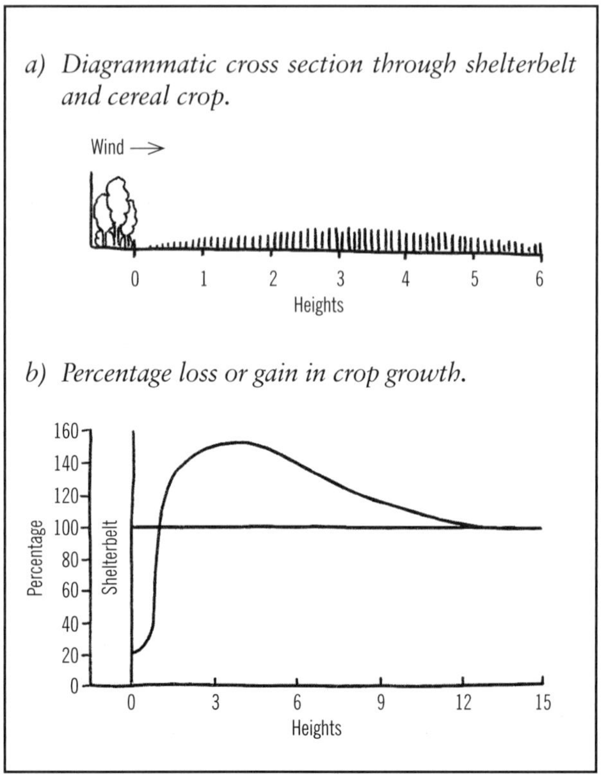

a) *Diagrammatic cross section through shelterbelt and cereal crop.*

Wind →

b) *Percentage loss or gain in crop growth.*

Effect of a shelterbelt on crop yields. The figures on the horizontal axes of both diagrams are multiples of shelterbelt height.[49]

At present this can be grant aided under the Countryside Stewardship Scheme.

In addition to boosting crop yields shelter can:

- Improve the quality of crops and grassland.

- Extend the growing season, giving:
  - earlier ripening of crops
  - an early bite for lambing ewes and other grazing animals
  - a longer grazing season on upland pastures, allowing more hay and silage to be made on in-bye land.

- Increase animal growth rate and milk production, while reducing food intake.

- Improve health and disease resistance of both crops and animals.

- Reduce losses due to unusually bad weather.

- Reduce soil erosion.

The main disadvantage is that the reduction in drying winds can delay cereal harvest or haymaking. Some farmers also see shelterbelts as refuges for crows and foxes, which can predate on lambs.

A recent study calculated the possible economic benefits of shelter on British farms. The calculation was based on shelterbelts which occupy 5% of the land area, and are 5-10m wide and spaced 300m apart, that is approximately 15-20 times mature tree height. It found that on lowland farms – whether arable, livestock or fruit – the net yield increase would be around double the yield forgone due to the presence of the shelterbelts, whereas on upland and hill farms there was little gain or even a loss.[50]

The message is clear: shelter is a priority not so much where the wind is strongest but where the produce of the land is most valuable. As well as indicating which farms will benefit most from shelter, this principle can be used to guide the layout of shelterbelts within an individual farm. Houses, farmyards, gardens, orchards, lambing areas, dairy pastures and intensive cropland are the places most worth sheltering.

On farms where there is no great difference between the value of activities in different areas, the positioning of shelterbelts will be determined by the windiness of different parts of the farm and the direction of the winds. In either case the existing layout of woods and hedges will also have an influence. Often linking up and enhancing existing shelter can yield a great improvement in microclimate for a comparatively small input of planting. Small, isolated blocks of woodland can actually increase the intensity of winds, and joining them up with new planting should be a priority. (See illustration, p88.)

The shelterbelts on this hill farm in Mid Wales concentrate on the high-value areas – the house, yard and in-bye land where lambing takes place – rather than the windy tops, which are mainly grazed in summer. This reflects the permaculture principle of zoning.

Because shelterbelts have so much edge, fencing is one of the biggest costs of establishing them on farms with animals. This can be kept to a minimum by following existing fence and hedge lines. In some cases a hedge can be thickened to a shelterbelt when the fence that protects it needs replacing anyway. This approach is also aesthetically pleasing as it blends with the existing grain of the landscape.

Internal Design

A multiple row shelterbelt is both more forgiving and more flexible than a single row windbreak. The death of one tree doesn't form a gap; crops of wood and timber can be taken from time to time; and long-term replacement can be designed in from the start. The advantage of single row windbreaks is that they take up less space. They are more suitable where land is precious, either on small farms or where high value produce is grown. They are described in Chapter 9. (See pp238-240.) Here we consider the design of a multiple row shelterbelt.

Species choice is best guided by what is growing well locally in positions of similar exposure and soil to the shelterbelt site. There are four things which shelterbelt trees must provide:

- quick growth for early height
- bottom growth, especially important as the main trees grow tall
- wind-firmness on exposed sites
- long life

This mix of functions is more likely to be had from a mixture of trees than from a single species. For example a fast-growing species, like poplar, can give early height and then be removed to favour a long-lived species, such as beech or oak. There is usually a distinction between the tall, main tree species and the bottom

species, which can be either shrubs or coppice of shade tolerant trees. Coppice is of course only suitable where the width of the shelterbelt is enough for at least two lines of understorey, so they can be cut alternately to give permanent shelter. A mix of main tree species is desirable, not least because it gives a crenellated profile to the shelterbelt, which reduces eddying. (See p84.)

The more rows the shelterbelt has the more opportunities there are to share the different functions out between different species.

The amount of land taken up by shelterbelts with different numbers of rows is approximately as follows:

Rows of trees	1	2	3	4	5
Width of belt in metres	3	4.5	6	7.5	9

The next three illustrations show examples of layouts.

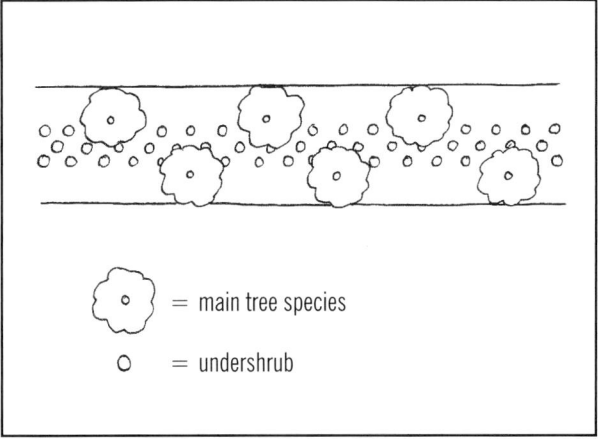

A simple design for an area fertile enough to grow good poplars. One row of poplars is felled and replanted at 15 years, the other at 30 years, and so on to give permanent cover. Where poplar follows poplar the new plants should be planted as far as possible from the stumps of the old in the rows.

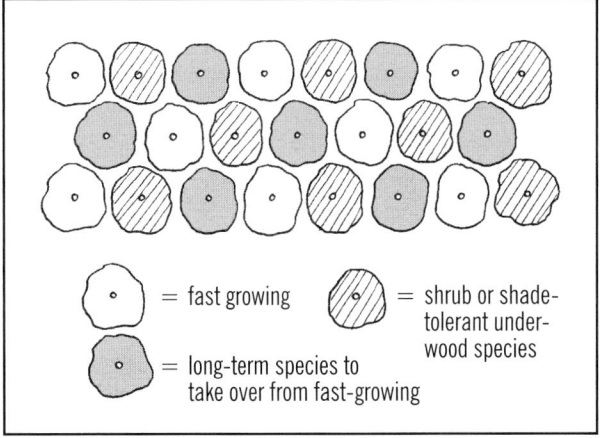

A tree-by-tree mix, which gives a high degree of diversity in a relatively narrow belt. But close attention is necessary to see that no individuals are suppressed by competition.

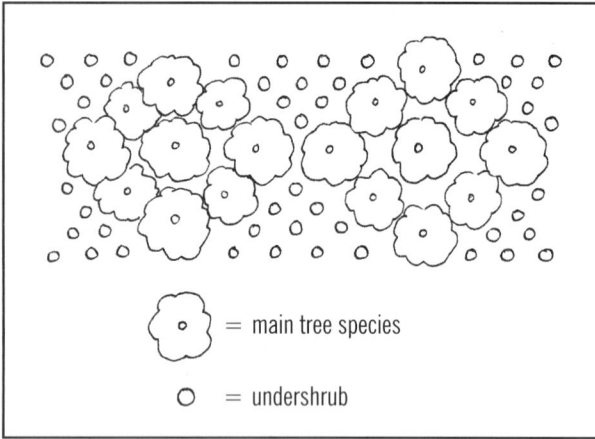

= main tree species

= undershrub

A group mix. This requires a wider belt than a tree-by-tree mix but needs less attention. Each group of main trees will eventually be thinned down to one or two individuals. A mix of main tree species can be used, but each group of main trees consists of only one species.

It is possible to grow timber in a shelterbelt but quality stems will only be produced in inner lines. The outer lines will have many branches on their exposed sides, and these can't be pruned because it's the branches which give the shelter.

Thinning must not be neglected, both to keep the trees in good heart and to keep the right level of permeability. This is hard to measure precisely, but there are two rules of thumb. One is that when standing close to the belt when the trees are in leaf it should be possible to see the colour of the field on the other side but not to tell what the crop is. The other is that you should be able to tell if an animal or machine is moving on the other side, but not to make out its outline.

One way of getting continuity is to fell each side of the belt alternately, as in the first illustration. Another is to fell individual trees from time to time and replant in gaps to give an uneven-aged stand similar to that in continuous cover forestry. (See p318.) This can only be done with a relatively wide belt, and trees planted into the existing stand can only be shade-tolerant kinds. Fortunately several shade-tolerant kinds, such as beech and sycamore, are good shelter trees.

A third option is to plant a new belt alongside the old when the latter is half way through its rotation. Space for this must be allocated in the initial design.

Hill Farms

Although a complete network of shelterbelts may be hard to justify in upland areas, refuges for animals grazing on exposed moorland are certainly worthwhile. This is one situation where a permeable windbreak is not appropriate. What is wanted is maximum wind reduction over a comparatively small area where animals can gather in rough weather, and this is best provided by a dense barrier.

A wide block of trees is always impervious to wind, and an approximately square timber plantation makes

a good refuge. This is much cheaper to fence than a long narrow belt. It's also a better shape for timber production as less edge means fewer trees with poor timber form.

In windy upland areas there may be no alternative to clear felling for renewing the plantation. (See p318.) So if continuity of shelter is required a second plantation should be made when the first is half-way through its life. Another option may be to maintain a margin of hardy, wind-firm and long-lived trees, on the windward side at least, to protect two or even three rotations of timber trees.

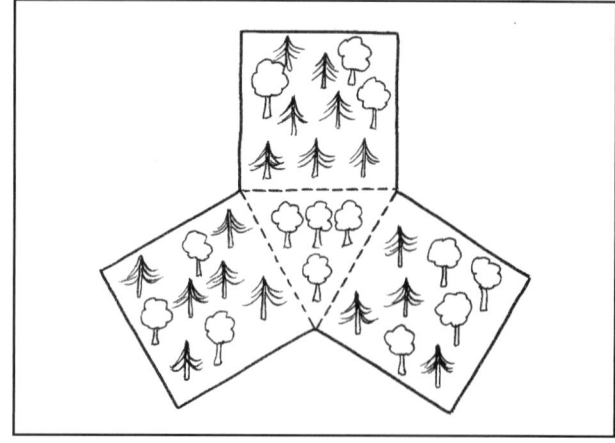

The Manx leg is an ideal layout for a shelter plantation. It gives shelter from every direction with little more fencing required than if the three squares were laid out end to end to make a belt.

The best guide to species choice is usually what is growing locally – if there are any trees at all. But generally the hardiest species for exposed sites include: sycamore, beech, whitebeam, rowan, grey alder and lodgepole pine. It's essential to match the trees to soil conditions as well as microclimate.

When establishing trees in exposed areas it's worth making use of every bit of shelter the existing topography offers. Trees which start life in a sheltered spot can soon overtake their neighbours in more exposed positions. An existing wall or bank can give this shelter.

As well as providing shelter it's worth planting up steep gullies. Sheep like to take refuge from harsh weather in such places, and if they're there while heavy snow falls they can get buried en masse. Planting and fencing such gullies keeps the sheep out of them.

Existing woods, both native and plantations, can provide useful shelter in the uplands. Most of these are presently unfenced and animals have free access to them, but continuous grazing will prevent regeneration. In order to keep these woods alive for future generations it's necessary to keep animals out of them periodically. The animal-free period must last long enough to allow the new generation time to germinate and grow to a size at which they can resist browsing.

FOOD LINKS

Permaculture is all about setting up beneficial relationships, and one of the most important relationships is that between the people who produce food and the people who eat it. The present food system operates on a very large scale and works in the interest of corporate profit, not of the producers or consumers of food. Direct marketing links between producers and consumers can benefit both parties, and reduce the ecological costs of distributing food.

The money which normally goes to the food marketing business can be split between producer and consumer, with the one receiving a somewhat higher price and other paying somewhat less. This means that organic produce can be sold at a price which ordinary people can afford. It also means that organic farmers and growers can get a price which makes their livelihood that bit more viable.

Organic food can be expensive in the shops, but it doesn't cost very much more to produce on the farm. Most of the price difference is generated in the distribution of the food.

Out of every pound spent on food in the shops, on average only ten or twenty pence goes to the farmers who grew it.[51] All the rest goes to processors, distributors and retailers. Thus even if organic food cost twice as much to produce as conventional food this would only add five or ten percent to the retail price. But the food industry charges more for handling organic food than conventional because it's sold in smaller volumes. It's the economies of scale that are largely responsible for the price difference in the shops.

The price gap is often exaggerated by a difference in marketing attitude. Conventional lines are sold on the principle of 'pile it high and sell it quick', with profit coming from very small markups on a very large turnover. Organic lines are usually regarded as a niche market, one where volume is low so that markups must be high – a self-fulfilling prophecy if ever there was one.

At the time of writing there is a catastrophic slump in the price of many kinds of farm produce. Sheep farmers are getting less than five percent of the retail price for their lambs. That's a mark-up of two thousand percent. As prices have fallen and many farmers have lost their livelihoods, the price of lamb in the supermarkets has not gone down one penny.

Over the past 50 years farmers and growers have been increasingly squeezed. The 5-20% share of the retail price which they get has dropped from 50-75% in the 1950s.[52] The price they are paid for cereals, for example, is around the same today as it was when Britain joined the EEC in the early 1970s. This price squeeze has forced farmers to strive for an ever higher gross output just to stay in the same place. In order to stay economically viable farms have become bigger, more specialised, more mechanised and more chemicalised. The majority of small farmers have been forced off the land.

Direct food links offer an alternative, especially to the few remaining small farmers and those just starting up. By taking on the job of distributing the food as well as growing it farmers are not just adding value to their produce, they are also taking back control of their lives from the big retailing companies. It does require more work and some marketing skills. But the added value means that farmers don't have to go for maximum volume at all costs and cut corners in order to keep the cost of production as low as possible. It becomes possible to make a living on a relatively small volume of output.

It's more than just a financial arrangement. Direct food links can also enable a personal relationship between producers and consumers to develop. This varies greatly between the different kinds of food links which are described below, but in many cases the people who eat the food can visit the farm or garden where it's grown to see for themselves how it's done. Then they can decide for themselves whether the food is grown in the way they would like it to be, and if they don't they can ask for things to be done differently.

Direct food links are an alternative to the certification system for organic food. For people who buy their food in shops the organic symbol is a reliable guarantee that the food has been produced in a way which is ecologically responsible and humane.[53] But it's inevitably remote, somewhat bureaucratic and generalised. Valuable as it is in present circumstances, it's second best to a direct personal relationship. (See case study, A Permaculture Farm, p294.)

From an ecological point of view the great benefit of direct links is the reduction in food miles, the distance that food travels between field and plate.

A classic example is the farmer from just outside Leominster in Herefordshire who asked a supermarket in the town if they would like to buy his organic potatoes. "Certainly," they said. "But they'll have to go through the company buying system." This meant they would have to go to the organic packing house

Ashurst Organics in Sussex is one of many small organic holdings which are made viable by direct marketing, in this case a box scheme.

at Lampeter in Carmarthenshire, from there to the company distribution centre in Cheshire, and then back to Leominster, a total distance of over 200 miles. The farm is some six miles from the shop.

These food miles pale into insignificance beside importing food from other continents. Green beans come from Kenya, 3,600 miles away, apples come 12,000 miles from New Zealand. Much of this long range transport is of out-of-season fresh produce and is air-freighted. Air freight uses four times as much energy per tonne-kilometre as road and forty times as much as water.[54] All this energy is expended merely to escape from the natural rhythm of eating with the seasons, something which can be a real pleasure. It is still perfectly legal to describe such produce as Organic!

Another advantage of direct food links is the reduction in waste. Supermarkets sell food by its visual appearance. Customers choose their fruit and vegetables from the shelves and anything which looks less than perfect is left unsold. Conventional growers achieve visual perfection by spraying – a large proportion of the pesticides used on fruit and vegetables are purely for cosmetic reasons. Organic produce is brought up to the same visual standard by rejecting anything which looks imperfect. Outgrades can amount to a quarter or even a half of the whole crop.

Growers and farmers who sell direct are selling genuine quality, not visual appearance. People buy from them because they know how the food has been grown. The odd hole in a leaf is not important. Nor is eye-catching packaging, and the reduction in packaging is another ecological advantage of direct marketing.

Direct food links are not a panacea for every farm or market garden. In particular those which are situated far away from centres of population will find it hard to sell all their produce directly. In practice you need to be within one and a half hours' drive of your customers in order to make a reasonable day trip out of delivering boxes or attending a market. In areas of modest population it may only be possible to sell part of the farm's output directly and in remote areas it may not be possible at all.

There are many ways of linking up producers and consumers of food. Each individual system is designed to meet the specific needs of a particular group of consumers and/or a particular farm or market garden. This makes it hard to classify them, but there are some broad categories and these are given in the mind map below.

Not all of them share all of the characteristics of a full-blown food links scheme. For example, food co-ops don't involve direct contact between producers and consumers, and the food in farmers' markets is not necessarily cheaper than that in shops. But all represent a move towards closer links between the people who grow food and the people who eat it.

Food Co-op

A food co-operative is an initiative taken by consumers. It's a voluntary organisation which buys food wholesale and sells it to its members at cost price. Despite the lack of direct contact between producers and consumers, co-ops can do valuable work by making healthy food more affordable.

They are particularly valuable in poor urban areas where there are few shops and what shops there are may not sell fresh unprocessed food. Shopping for fresh fruit and vegetables may involve the expense and inconvenience of a bus ride, and people often subsist on processed foods with little nutritional value. Even where fresh food is available, calorie for calorie, junk foods are much cheaper.

A food co-op alone will not solve the complex of causes which lead to poor nutrition in deprived areas. But they can at least make decent food available at prices which people can afford. Fruit and vegetables are a good place to start, and once a co-op is up and running other staple foods can be added to its repertoire. A successful food co-op can give people a sense of achievement and may be the nucleus which leads to other community action.

In some cases they can link up with urban food

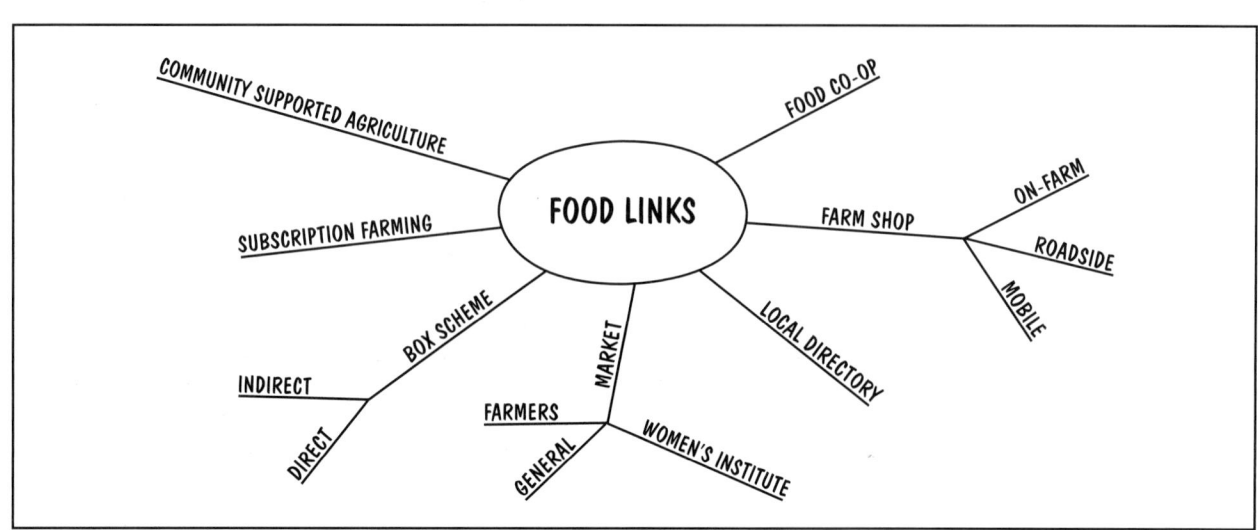

growers who have a surplus. This has happened in the Cuthelton Lilybank area of Glasgow, where the local community worker, Jim Lister, put the local allotment association in touch with the food co-op to the benefit of both.[55]

Co-ops also have their place in less deprived areas. They can provide the means for people on modest incomes to buy wholefoods and organic vegetables at lower prices. In particular they can put organic food within reach of people who would not otherwise be able to afford it. My own home town offers an example. (See box, below.)

Every co-op is different and must be designed to address the real needs of the people in a particular area. The three basic requirements to get a co-op going are:

- A core group of volunteers, or even a single person, with the determination and commitment to make it happen.

- Enough local people who are sufficiently interested to get the co-op going in the first place; once it's seen to offer benefits more people will join.

- Premises to operate from.

It's usually a good idea to set up a formal co-operative structure and register as a co-op. This gives credibility to official bodies, and demonstrates that the co-op is not someone's money-making enterprise in disguise.

Also, some wholesalers will not supply unregistered co-ops, because their retailer customers suspect these are simply middle-class families who want to get their food cheap.

Farm Shop

A farm shop starts from the other end of the chain, with the producers. The main emphasis is on making their livelihood more viable, though there is usually a cost advantage to the customers too.

A full-blown farm shop, open for normal shop hours with a person serving, is not often worthwhile. On most farms it's hard to generate enough business without stocking at least some food which was not grown on the farm, and it's much more difficult to get planning permission for the shop if this is the case. An on-farm shop does not come out too well on the ecological balance either, as each customer goes there in their own car to buy a few items.

A roadside stall with an honesty box can work for smaller producers in quiet rural areas. If it's well placed people can stop by as part of their normal daily round or walk from their homes. My friend Carol Jacobs, who has a semi-commercial organic garden in Wiltshire, makes most of her sales from a stall like this and has done so for many years. Since no-one has to tend the stall full time it's viable for even the lowest volume of sales. Of course the possibility of

---

A SMALL-TOWN FOOD CO-OP

The Glastonbury Wholefood Co-op was started by a young couple operating out of their own home, but it soon moved to a small room in the basement of the town's community building, the Assembly Rooms, which was available at a very low rent. Having this room was crucial as the co-op could not have paid a commercial rent.

Each member, either an individual or a family, paid £10 to join. This made up the working capital needed to buy the initial stock of food. Bins for the food, shelves, scales and a table and chair were donated. An annual renewal fee of £2 was also paid, and a markup of 10% on loose food and 5% on packaged items was enough to cover the rent and any losses from spillage and so on. This compares to a typical markup of 30% or more in a wholefood shop and helped many people on low incomes to afford organic wholefoods.

Registering as a co-op was easy as there was an umbrella co-operative in the town which any new initiative could affiliate to.

Everyone committed themselves to do some work. Most members took a turn on the atten-

dance rota. This meant being there for the four hours a week that the co-op was open, taking the money, keeping an eye on things, and making up a list of what needed to be ordered for the next week. Four or five people did the banking and ordering. Because there were fewer of them these people did more work than other members, and the venture would not have succeeded if there had not been some people willing to do more than their minimum share.

For some 13 years the co-op worked well. There were some freeloaders who didn't take their turn on the rota and the occasional spate of minor theft, but the general will to make it work was strong enough to cope with these. Then, as is so often the case, the impetus faded and the co-op was wound up.

Recently a new co-op has started up, selling organic fruit and vegetables as well as wholefoods. This has a different structure. It's part of a child-care and educational centre, and most of the work is done by volunteers involved with other work at the centre.

theft means that it won't work in every area, but the general experience is that theft is much less of a problem than most people expect.

A mobile shop, or delivery service, saves a lot of energy by replacing the many car trips people would make to get to the shops with one trip by a van. This needs no planning permission and no investment in a shop building which must be up to food hygiene standards. But it can be time-consuming for the producer.

The time taken can be kept to a minimum if it is organised in a similar way to a box scheme. (See below.) The food is delivered weekly in boxes to a number of drop-off points and customers pick up their box from their nearest one. This kind of service differs from a box scheme in that customers can choose each week exactly what they will buy the following week. This means prices will be a little higher than in a true box scheme. Radford Mill Farm in North Somerset sell most of their organic produce in the nearby city of Bath through a delivery service of this kind.

One advantage of this arrangement is that less food is wasted. Because people order what they want in advance, exactly the right amount of each vegetable can be picked. A normal farm shop, whether mobile or farm-based, must always be stocked with a range of produce, and anything which is not sold while it's still fresh has to be thrown out.

Local Directory

Small food producers often produce a high-quality product in terms of taste, wholesomeness, animal welfare and ecological impact. In order to get a financial reward for this they often have to sell to the luxury market in distant big cities, something which

many of them would rather not do.

At the same time many rural and small-town consumers would like to buy local food, not just because of quality but also because they want to support the local economy, especially small farmers. Perhaps few of them are aware of the ecological importance of reducing food miles, but, as in so many aspects of life, what's good socially is good ecologically and vice versa.

In many places all that's needed is to let the public know what is available locally, and direct sales to local customers can rise significantly. One way of doing this is to publish a local food directory, listing all producers in the area who sell their own food locally. The Forest Food Directory was the first of its kind in Britain. (See box.) Some two years on there are fifty local food directories in the country, most of them to a greater or lesser extent promoted by local councils.

Markets

Farmers' markets are stall markets where producers sell their own food direct to the public. They have been running in the USA for decades, and have recently been introduced to Britain where they're spreading at an astonishing rate. They are in fact a formal re-creation of the kind of market which has never died out in countries like France.

There are two stipulations about what can be sold at a farmers' market:

- Everything offered for sale must be produced by the stall-holder.

- All stall-holders, and thus the food they sell, must be local.

THE FOREST FOOD DIRECTORY

This is a pioneering food links project in the Forest of Dean, initiated in 1998 by farmer Matt Dunwell and author Kate de Selincourt. (See case study, A Permaculture Farm, p294, and Further Reading, p442.) It lists 33 individuals or companies in the district who produce food for local sale.

Produce includes: vegetables, fruit, eggs, cheese, milk, honey, ice cream, meat, fish, bread, cakes, preserves, wine, cider and beer – yes, even the local brewery is in there. Most of the items are actually grown by the people who sell them, but food and drink made by the seller from bought ingredients are included too. There are also advertisements for shops and restaurants which sell or serve local food, and articles on subjects like food miles and traditional breeds.

Much of the food listed in the Directory is organic, but by no means all of it. The only ethical

criterion is that any producer who is listed must be willing for customers to visit and see for themselves how the food is produced.

The cost of producing the booklet is met partly by the producers themselves, and partly by various sponsoring organisations, including the District and County Councils. It's distributed in a number of ways. The Council includes a flyer about it in general mailings to people who live in the district, and the producers make it available to their customers, so once someone has bought from one of them they get to know about all the others.

Now the directory is also available as a single-page fold-out map, illustrated in full colour by school children. Although this contains less detail than the booklet, it's much more accessible and it still contains all the key information about who sells what and how to contact them.

Local Produce from Radford Mill Farm at the farmers' market in Bath. (See also Colour Section, 28.)

Produce made by the stall-holder from bought-in ingredients, such as home-baked bread, is allowed, but in practice most of the sellers are farmers, gardeners, orchardists and beekeepers. The range of produce at our local farmers' market is very similar to that listed in the Forest Food Directory. (See box, opposite.)

Although there are some farmers' markets which are not strictly local, in most cases stall-holders must live within a certain radius of the market. When a market first starts up the radius is often quite wide to make sure there are enough stalls to get things off to a good start. Later on it can be drawn in as more local producers take stalls and more markets start up in adjacent towns. 40 miles is a typical starting radius.

These markets are growing so fast because they give both consumers and producers something they can't get anywhere else. A survey of customers at the first farmer's market to be run in Britain, at Bath in 1997, asked the question Why did you buy from the farmers' market? The most frequently given answers were:[56]

It was freshly made or harvested	47%
It was good quality	45%
It was directly from the producer	38%
I like to support local business	37%
It was not available elsewhere	15%
It was organic	14%
It was cheap	9%

It's interesting to see how many respondents valued the opportunity to buy direct from the person who grows the food and to support local food producers. 65% said they normally got the items they had bought from a supermarket.

The community building aspect is also much appreciated. A farmers' market is a good place to meet friends you don't see often, and they tend to have a distinctly warmer and friendlier atmosphere than a general stall market selling miscellaneous goods. Our elderly neighbour said of the first farmers' market in our town, "It's just like Glastonbury used to be – and how it should be now."

From the producers' point of view, a farmers' market is more than just an opportunity to sell some of their produce at a retail price, though this is valuable in itself. For many of them it's the only time they come face to face with their customers. They can get direct feedback on what they're selling, find out what people would like, and explain how they grow the food and why they choose to grow it like that. Some put up small displays with photographs of their farm or garden and a brief account of what they do there. Others have leaflets which they hand out.

Some markets have been set up by local councils, usually by the Local Agenda 21 Officer. In many cases the aim is to hand over to a body composed of stall-holders in the long run. Others have actually been set up by supermarkets, sited in their own car parks. To begin with most farmers' markets are held monthly, but many are aiming to become weekly when sufficient impetus has built up.

Of much longer standing are the weekly markets run by the Women's Institute (WI) in many rural areas. You don't have to be a woman to sell in a WI market, but you do have to have grown or made all the produce yourself. Some small vegetable growers and poultry keepers have made a significant part of their income from these markets over the years.

A stall in a general market also has its advantages. They reach a somewhat different clientele, and there is no stipulation that everything offered for sale is produced by the seller. This can be an important advantage for fruit and vegetable growers who are not able to grow a complete range of produce throughout the year. Customers value being able to get all their needs from one stall, and a few organic dairy products to go along with the vegetables can make the stall that much more attractive, and thus more viable.

Box Scheme

With the box scheme we move beyond the realm of simple marketing to one where there is a relationship between producer and consumer which involves a degree of commitment.

The great majority of box schemes at present are for vegetables. The grower commits to providing a box of assorted organic vegetables every week to each customer, and the customers in return commit themselves to accept a mix of whatever vegetables are in season. The price of each box is the same every week, set at what a similar mix of conventional vegetables would cost.

I've not heard of a single case of a successful box scheme selling conventional vegetables. The commitment of accepting what is in season does limit the consumers' choice, and people are only prepared to accept this for the sake of getting organic vegetables at a price they can afford.

Howard Johns

Pippa James, box scheme grower, with one of her vegetable boxes.

In most cases the staples – potatoes, carrots and onions – are included every week, but the rest of the selection depends on what is ready to pick. Some schemes allow customers to ask for certain vegetables to be left out of their box, either things which they don't like or which they grow themselves. But giving more choice than that makes packing increasingly complex and time consuming, and the price has to go up. When you get to the stage where people put in an order of exactly what they want next week, the box scheme begins to turn into a mobile farm shop.

Limitation of consumer choice is anathema to the market economy. You could say that box schemes extend consumer choice to people who couldn't afford organic vegetables otherwise, but that is really to miss the point. Box schemes actually replace the idea of having exactly what you fancy when you fancy it with the pleasure of eating with the seasons. Not only is there the anticipation of wondering what will be there this week, but there's the pure joy of eating a favourite food for the first time since last year. What can compare with first taste of young broad beans, lightly boiled with a knob of butter!

On a deeper level, eating with the seasons is one way town dwellers can reconnect with the rhythms of the Earth. The variation from week to week reminds us that there's a whole world out there which responds, not to the ups and downs of numbers on computers in New York and Tokyo, but to the annual wheel of real time and the coming and going of the weather.

Receiving a box of fresh vegetables every week has other consequences. When the box contains something they've not eaten before, it can lead people down new and unexpected culinary paths. Many box schemes put an occasional newsletter in the boxes, with suggested recipes for both familiar and unfamiliar vegetables. Some families have found that cooking more real food has led to them having more sit-down meals together, rather than grabbing individual snacks. This brings the family together as a whole more often and strengthens the family community.

It has consequences in the market garden, too, apart from the most obvious one of bringing in more money. Large-scale markets require large quantities, so growers who sell to supermarkets have to concentrate on a very few vegetables and perhaps only one variety of each. But families want a wide diversity of vegetables, and they want to eat each vegetable over as long a season as possible. So a garden growing for a box scheme must grow a wide range of vegetables and a wide range of varieties of each vegetable, both early and late maturing. This makes for a very diverse garden. Mandy Pullen grows some 150 different varieties of vegetables each year on half a hectare. (See case study, A Permaculture Farm, p294.) A grower selling to a supermarket might grow five.

FOOD LINKS IN JAPAN[57]

The movement towards direct links between farmers and consumers has been going on for much longer and is much more developed in Japan than here in Europe and North America.

It all started in 1965, when a Tokyo woman formed the Seikatsu Club. The initial purpose of the club was to avoid the high price of milk by buying direct from the wholesalers, but it soon expanded to cover other produce, both food and non-food, and by 1971 members were buying directly from farmers. By the early 1990s some 20% of the Japanese population were members of consumer co-operatives, accounting for 2.4% of total retail sales. Over time the emphasis has shifted from buying cheaply to providing safe food through the direct connection with producers.

In contrast to the large-scale mainstream co-ops such as Seikatsu, a wave of smaller groups known as teikei, which means 'tie-up' or 'co-partnership', have grown up. These mainly involve organic produce and emphasise the importance of face-to-face contact, often as an alternative to the costly process of organic registration. This means that they are smaller individually, but the total number of farms involved has been estimated at between 10,000 and 25,000. Like the biodynamic movement in the West, many teikei groups have a spiritual aspect, treating the relationship between consumer, producer and the land itself with a reverence which transcends physical nutrition.

Of course this diversity is good for the health of the garden ecosystem. The primary aim of a box scheme is the mutual benefit of the two groups of people involved, but it ends up benefiting the land as well.

Some growers find it hard to grow a wide enough range on their own, and they may co-operate with others to supply a single box scheme. This requires a little more organisation, but it can mean that very small, part-time growers can get a reasonable return on their labour without having to grow tiny quantities of many different vegetables. A box scheme of this kind was started up at Exmouth in Devon, and it meant that one or two small-scale organic growers who were on the point of giving up were able to stay in business.

Other growers may need to buy in produce to complete their range. This can be necessary in the case of crop failure of some really popular vegetable, or simply because their land is unsuitable for one crop or another.

During the 'hungry gap' in the spring, when the winter crops are over and the summer ones have not yet started, most growers would have to fill the boxes almost entirely with bought in produce. Many of them don't supply boxes for a month or two at this time, which is also the busiest time in the garden.

WHAT ABOUT THE LOCAL SHOP?

Box schemes can be seen as a threat to local food shops. Most small food shops have been forced out of business in recent decades with the rise of supermarkets, and those which remain often struggle to keep going. If a proportion of their business is taken from them by a box scheme it could make the difference between survival and closing down.

In fact box schemes can have the opposite effect. Receiving a vegetable box means that a family has less to buy at the supermarket and it makes that weekly trip less worthwhile. If the box also changes the family's eating habits it will mean they eat less processed food, further reducing the need for the big weekly shop. Buying what remains from a small local shop then becomes more attractive.

In some places it may be possible for a shop to become a drop-off point for the box scheme, which can draw more people into the shop and increase trade.

Taking a broader view, box schemes can only be seen as a threat to small shops if they are considered in isolation. Some people may join a box scheme and leave the rest of their life unchanged, but in most cases it's part of a general awareness of ecological and social issues. The importance of supporting local traders is part of that awareness.

In most schemes the work of delivering the boxes is shared by the growers and the customers. The customers are grouped into clusters of about ten households, and one household in each group acts as a drop-off point. The growers make up the boxes on the farm and deliver them to these families only, which saves them the big job of delivering to every single one. The other families pick up their boxes from the drop-off family and pay the money through them. The drop-off people are usually paid for their services with a discount on their own box, or even a free one.

If people live within walking distance of the drop-off point, the need for motor transport is reduced to a minimum. Empty boxes can be returned for re-use, so packaging is also minimised.

Box schemes can be initiated by groups of consumers or by growers. The Soil Association has taken a prominent role in encouraging the growth of box schemes, and most new organic vegetable growers in the past few years have started up with a box scheme as their main or only marketing outlet. The majority of box scheme growers are certified Organic, but in some cases the direct relationship with their customers takes the place of certification. (See case study, p294.) Open days and farm walks for the customers are essential in that case, though many growers who are certified Organic also organise these.

A key part of the box scheme concept is that it should be local. The ideal set-up is a farm or garden, or a group of small gardens, feeding people in the immediate area. But with our present pattern of settlement that's not always possible. Most people live in big conurbations and organic growers are often located in remote areas with low population.

It is possible for an organic farm or garden to set up a box scheme with a group of customers in a city at some distance from the farm. But at present much of the demand for vegetable boxes in cities is being met by indirect box schemes. An indirect box scheme is a business which doesn't grow any food itself, but buys organic produce from the growers and sells boxes to the consumers. They perform a useful function for consumers, and for producers who can't or don't want to sell their produce directly. But they are not a form of direct food links.

Subscription Farming

Subscription farming involves a higher level of commitment on the part of the consumers than a direct box scheme. The food is distributed in the same way, but the consumers pay for a whole season's produce at the beginning of the season.

People who can't raise all the money at once can usually pay in instalments, but however they pay they are buying a share in whatever the year will bring, not a fixed amount of food. The growers usually give a detailed list at the beginning of the year of the kinds and quantities of produce one share is expected to include. But in a good year it will be more and in a bad year less, and the consumers agree to share that risk.

The cost of a share is based on the costs of production, including a living wage for the farmers or growers themselves. Of course it does also have to be related to the value of the expected yield of food, but the emphasis is on meeting needs rather than on market prices.

Subscription farming has developed strongly in the USA over the past twenty years, and there are now an estimated 1,000 subscription farms there. The pay-as-you-go box scheme does not seem to exist there, while here in Britain subscription farming schemes, as defined above, are rare.

One great advantage of subscription farming is that the growers get most of their money at the beginning of the year, when they need it for seeds and other expenses. This takes them out of the hands of the bank, which can be crucial for small producers starting up without much capital. The profit margin on growing food is paper thin, and if interest charges have to be paid they can swallow it all up leaving the growers with nothing to live on.

A somewhat different model of subscription farming is based on consumers taking a share of the capital cost rather than the annual running cost. An example of this is Sharpham Barton Farm in Devon, a small bio-dynamic farm, producing eggs, meat and grain.

Some years ago the farm needed an injection of capital in order to develop the business and keep it at a level where it would provide a living for the farmer and his family. The interest payments from a bank loan would have been crippling, so the farm decided to ask the local community to provide the needed finance. Shares were offered for a few hundred pounds each. Instead of interest the shareholder gets a discount on food they buy from the farm for as long as they hold the share.

The financial return on a share may not be competitive with a more conventional investment, but that's not why people bought them. They wanted to support a local farm which is being run in a sustainable and wholesome way. That they have done, and the farm continues to thrive.

Community Supported Agriculture

There's a fuzzy edge between subscription farming and community supported agriculture (CSA), in fact the majority of the subscription farms in the USA consider themselves to be CSAs. The level of commitment which is inherent in subscription farming tends to grow and lead to a greater degree of involvement until the distinction between producer and consumer begins to fade.

An example of this process comes from Brookfield Farm (sic), a subscription farm in Massachusetts, USA.

The consumers pay for their shares of the year's produce each March and the amount paid usually works out about right by the end of the season. But in 1988 it became clear as the season progressed that there would be a deficit of money. The community of shareholders saw it not as a problem for the farmers but for the community as a whole. They set to and fund-raised by various means until the deficit was made up. The people who were eating the food cared as much about the future of the farm as the farmers themselves did. Calling themselves 'the community' rather than 'customers' is not a bit of hopeful semantics; it reflects the reality.[58]

Elizabeth Henderson (see Appendix A, Further Reading) bases the distinction between subscription farming and CSA on whether the 'sharers' do any work on the farm or not. This work can be on the land, in distributing the food or in administration. Either way it represents a further level of commitment and begins to chip away at the distinction between producers and consumers.

But it would be wrong to see CSA as purely a development of subscription farming. Often it's the other way round: the subscription model is selected as the best way to organise a farm by people who have a strong desire to grow food in a radically different way, both ecologically and socially. (See box, Buschberghof CSA, opposite.)

Most of the American CSAs are biodynamic rather than simply organic. The biodynamic people see a farm as an organism, and extending that organism to include the people who eat the food is a natural step. The sharers become part of the farm community. The support they give the farm is an alternative to government support. While government subsidies encourage large-scale, chemically-grown monoculture, community support encourages what the community members want: healthy food, a diverse diet, and well cared-for land.

It's much more than a way of marketing. The community members are not buying food. They're providing the financial input needed by the farm organism, of which the soil, plants, animals and people are all a part. The farm provides the food needed by the members, which is distributed to them without direct payment. Just as the yield of produce can vary with the season, on some farms the contribution of each family varies according to what they can pay, with richer ones paying more than the standard share and poorer ones less.

This needs to be arranged with some sensitivity, and only the individual families themselves can decide how much is appropriate for them. The fear that everyone will offer less than the standard share can be laid to rest by the following method. At the pre-season meeting for the whole community, when the plan for the year is decided, the farmers tell the other members how much the average share needs to be. Each family is asked to write down on a piece of paper how much they would like to contribute. The sums are quickly added up, and if the total is not enough the pieces of paper go round again. In practice the required total usually comes up on the first or second round.

Many of the American CSAs go further than this and find ways of making their food available to the urban poor, often by linking with food co-ops, which takes us a full circle to the first kind of food links we looked at.

BUSCHBERGHOF CSA[59]

Buschberghof near Hamburg in Germany is a mixed biodynamic farm, with arable land, vegetables, permanent pasture, woodland, a bakery and a dairy producing cheese, yoghurt and liquid milk.

The change to CSA started with a desire on the part of the original farmer, Carl-August Loss, to "eliminate personal profit as the principal motive behind the farm and to use the land instead as a healthy base for people, animals and plants". So, in 1968 he took the huge step of setting up a land trust and donating his farm to it. He gathered around him a group of people who wanted to farm with him and they formed a farmers' association. The trust gives them the use of land, buildings, animals and equipment in exchange for a small fee to cover depreciation. The farmers retain the right to choose their successors.

Many different ways of selling the food were tried, including a farm shop, wholesaling and a weekly delivery round. With such a diversity of produce marketing by any means was very time-consuming. So in 1988 they started subscription farming, a move that was very much in tune with the farm's ethos. This led to the creation of a third body in addition to the trust and the farmers' association, the community of consumers.

The food is distributed in much the same way as in a box scheme, with families divided into groups around a number of drop-off points, and those who do the work of distribution being paid in food. There is no set share: each family takes as much food as it needs. The diversity of food grown on the farm is so great that community members get almost everything they need from it. Only things like tea, coffee, alcohol and salt need to be bought.

Once a year there's a meeting of the whole community, including both farmers and consumers. The farmers present the budget for the year for discussion and approval, and each family makes a contribution according to what they can pay. There are also regular monthly meetings, at which each group must be represented by at least one member. This is a time to discuss both general and specific issues about running the farm and for feedback from the wider community to the farmers.

Some members of the wider community work on the farm, either regularly each week or for longer spells in the summer. There are regular farm walks so that members can learn and understand as much as possible about the farm. They also have access to the farm for informal walks or picnics whenever they want.

One of the farmers at Buschberghof, Christina Groh, sums up what the farm means to her: "This farm is now an integral part of all of our lives. It exists not chiefly to provide us with a living, nor to earn us fame by dramatically changing the world, but to recreate a bond between ourselves and the land. Each person has to create his or her own individual bond both with the land and also with the other members of the community."

CASE STUDY

A PERMACULTURE FARM

One of the aims of permaculture is to reverse the present trend towards ever bigger farms employing ever fewer people, and to create a repopulated countryside with many more people making a living on the land. Ragmans Lane Farm is one place where this has already happened.

The chief permaculture principles which have been applied there are those of linking and diversity. Direct links with the consumers of the food are vital to the economics of the farm. But links with local resources and people, and the internal links within the farm, also play an important part in its success. As for diversity, the farm business consists of a number of small enterprises, making use of the multiple niches and resources available, rather than aiming for bulk output of a single product.

When Matt and Jan Dunwell took over Ragmans Lane in 1990 the valuer commented that its 24ha of grassland could only provide half of one salary. Yet today three people make their living there. They are modest livings by material standards, but rich in quality of life.

The farm nestles in a side valley overlooking the river Wye on the edge of the Forest of Dean. At a casual glance it may still look like an ordinary grassland farm. But a closer look reveals a hive of intensive activity on a relatively small area of land round the farmhouse. In fact some of the permaculture features which make the farm so productive are invisible. These are the links the farm has with its customers and with other farms and businesses in the locality, and the links between the different enterprises on the farm itself.

Three Small Businesses

The main enterprise on the farm, at least in terms of the area it occupies, is its traditional one of meat production from the permanent pastures. Almost all of the farm is too steep for arable cultivation of any kind.

Beef cattle, sheep, pigs, meat chickens and layers are kept on the farm, and all the produce is sold direct to groups of consumers within a 30 mile radius on a system similar to a box scheme. (See pp289-291.) The animals are slaughtered at an abattoir just ten minutes journey from the farm and the meat is prepared by a village butcher. The extra work he gets from processing the Ragmans meat helps to keep this small local business going, at a time when so many rural shops are closing down.

Matt and Jan have now taken a back seat on the farm to pursue other interests, and the animal side is run by Peter Kramer on a share-farming arrangement. The Dunwells provide the land, breeding stock and equipment and Peter provides the skilled labour. This arrangement allows him, a young man with no capital,

to start up in farming as his own boss. But he couldn't make a living on such a small acreage without the direct link to the customers, which brings all the profits from the meat back to the farm.

Mandy Pullen's garden.

Ragmans Lane is run organically but it's not certified as Organic by any official body. Some of the rules which farmers have to follow in order to be certified Organic are debatable, for example whether or not to put rings in the pigs' noses to prevent them tearing up the pasture. "I have great respect for the work of the Soil Association," says Matt, "But here at Ragmans we're able to have that debate direct with our customers." The customers are encouraged to come to the farm on open days, when they can see for themselves how their food is produced, and ask for changes if they want to.

Not being certified Organic also means that permaculture principles can sometimes be put before organic ones. An example of this is the pig feed. Is it better to feed them on Organic feed which has been trucked in from East Anglia or on a mix of locally grown non-organic feed? Both have been done at Ragmans from time to time. The responsibility lies with the farmer – and the people who eat the food.

At the heart of the farm lies Mandy Pullen's vegetable garden. On half a hectare of former grassland she grows vegetables for a box scheme which supplies 60 families. Working almost entirely by hand, with occasional careful use of a two-wheel rotavator, she grows some 150 varieties of vegetables. "My income is very small," she says, "And that's because I'm not prepared to compromise what I'm doing, or to get bored. Ten acres of cabbage for a supermarket would be more profitable, but a much greater departure from nature – and boring."

Mandy with produce.

She rents the land from the Dunwells and runs the garden as an independent business, but there is plenty of room for co-operation. For example, when there are deliveries of meat and vegetables to be made at the same time, one of the two businesses can deliver for the other, and a new customer for one may mean a new customer for both. There are on-farm links too, such as when the pigs 'tractor' a bit of ground prior to planting vegetables. (See p262.)

Each year Mandy takes on an apprentice who helps her in return for the skills she can pass on plus a small wage. Some of these apprentices have gone on to start up their own box schemes in other parts of the country. She herself learnt her trade almost entirely on the job, having gardened on a domestic scale for just one season before starting up her box-scheme garden.

There are also a number of smaller enterprises on the farm. Matt manages these directly, assisted by the third person to make a living on the farm, 'Josh' Joshua. They include: cider and perry, fruit juice, charcoal, shii-take mushrooms and comfrey for both liquid manure and root cuttings. Some of these are quite experimental, for example the mushrooms, which take up a great deal of time but whose output is still an unknown quantity. Nevertheless the combined sales of these products more than pay Josh's wages.

All of these minor enterprises are based on the use of perennial plants, but perhaps the most important permaculture characteristic they have is their links with other enterprises, both on and off the farm.

The on-farm links are similar to those between the meat and vegetable enterprises. A £20 stall fee at the local farmers' market would be prohibitive for Mandy, but when it's shared between all the businesses on the farm it's not, and the job of actually attending the market can be shared too. The people who live and work on the farm can also cover for each other when one of them

wants a holiday, something which is often impossible for people making a small living on the land. Liquid manure from the cattle yard fertilises the comfrey, and the pulp left over after juice-making is fed to the pigs and cattle.

The off-farm links are mainly a matter of adding value to unused or under-used outputs from the locality. Many fruit trees have been planted on the farm, but most of these have yet to come into full bearing, and the main supply for the cider, perry and juice comes from local orchards off the farm. Cider and perry are traditional in the area, and many old, unsprayed orchards survive. Those on flat land are harvested mechanically and the fruit sold to large commercial cider makers, but those on steep land are no longer considered economic.

Matt pressing apple juice.

Matt and Josh buy the fruit from these orchards and harvest it by hand. This is more expensive than machine harvesting, but it pays because they sell the produce direct to the consumers through the marketing systems already in place for meat and vegetables. This means that orchards which might otherwise be grubbed out are preserved in the landscape.

Mushroom logs make good use of a moist, shady microclimate.

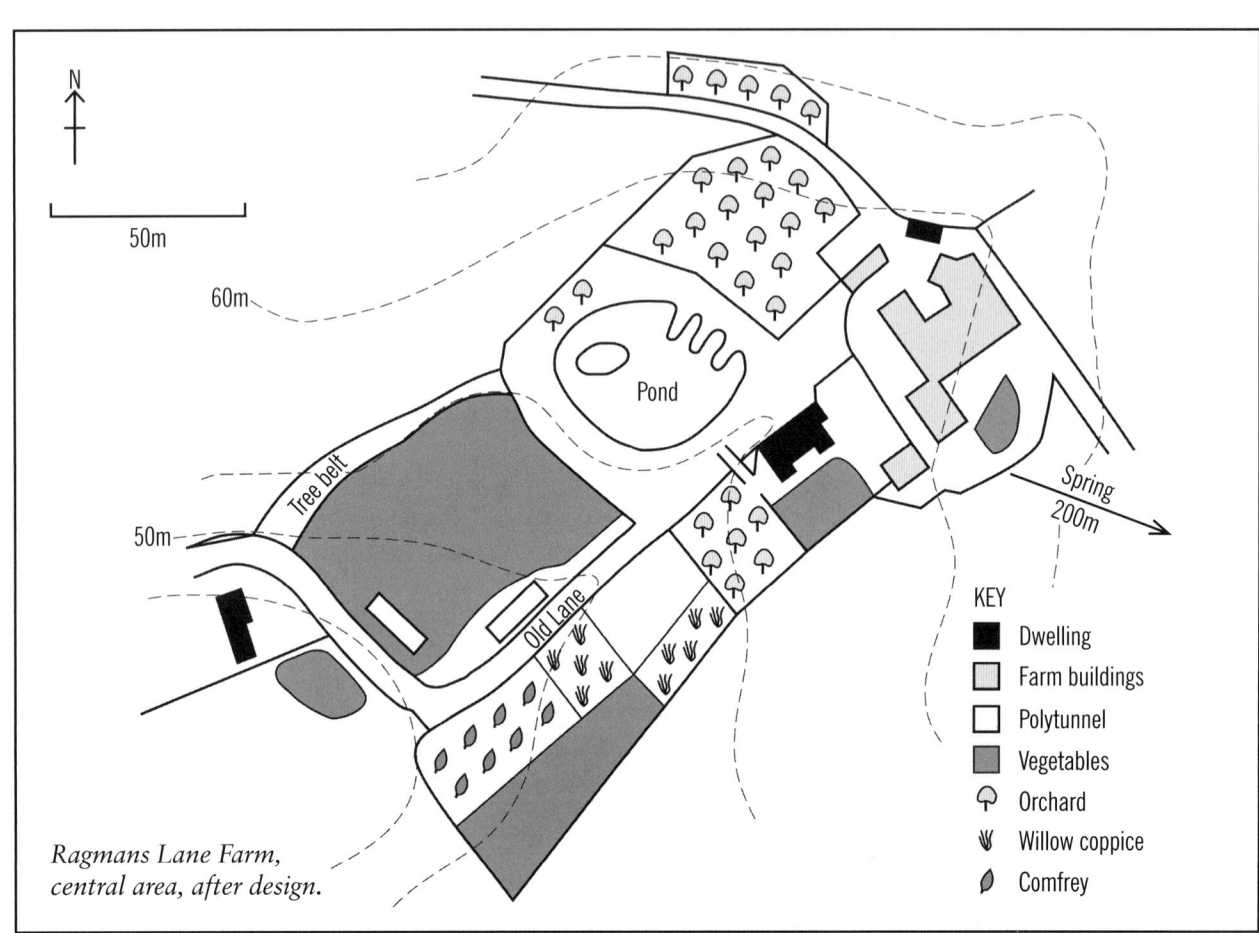

*Ragmans Lane Farm,
central area, before design.*

KEY
as on map below

*Ragmans Lane Farm,
central area, after design.*

KEY

- ■ Dwelling
- ▨ Farm buildings
- □ Polytunnel
- ▨ Vegetables
- ♀ Orchard
- ⚘ Willow coppice
- ◍ Comfrey

The charcoal also makes use of a local product which is not highly valued. This is slab wood, the outer parts of logs which are left over after planking, from local sawmills. There is woodland on the farm, both mature and newly planted, but this is not enough to support a charcoal enterprise. Thus a low value product becomes a high value one, and more money is kept in the local economy.

The mushrooms are grown on logs from a coppice which lies just over the boundary fence. This coppice would certainly remain uncut if the Ragmans Lane people hadn't found this high value use for it. Once they are inoculated with mushroom spawn the logs are stacked in the old green lane from which the farm takes its name. This is a very shady place and would normally be written off as totally unproductive, but it's ideal for mushrooms.

Design Layout

Most farmers inherit the basic layout of the farm from previous generations, and at Ragmans this was well designed. The farmhouse and buildings are near the centre of the holding, at a keypoint (see p29), just above the frost pocket, well sheltered by the surrounding hills, and downhill from the spring which is the farm's main water supply. (See top illustration, opposite.)

The existing woodland is on a steep patch of land on the boundary, and so is the new woodland planted since Matt and Jan took the farm. The site of the new wood has unfortunately turned out to be very dry, facing both the Sun and a drying wind, and tree growth has been patchy. But even with hindsight it would be hard to think of a better use for this piece of land, and bit by bit the trees are growing.

Other steep and bumpy areas have been planted to new perry orchards. It's hard to make hay and silage in an orchard, but these areas were unsuitable for tractor work anyway, and they can still be grazed. Thus there is no loss to the animal enterprise from the addition of the new trees. Some cider apple trees have been planted along hedge and fence lines, where they take up little space.

In the intensive area at the centre of the farm a few crucial placement decisions have helped to make the farm work more efficiently, especially with respect to the linking of resources. (See bottom illustration, opposite.)

Mandy's vegetable garden lies in a hollow below the farmhouse. This is a frost pocket, but frost is not as crucial for vegetables as it is for fruit, and in summer the garden is a sheltered suntrap. Being low in the landscape it has a deeper, more fertile soil than other parts of the farm and retains moisture well in the summer. It's also the largest area of relatively level land on the farm. On balance it's the best place to grow vegetables.

The main part of the garden is laid out on a bed system, with the beds running along the contour to prevent soil erosion. (See illustration on p31.) The wisdom of this layout was not really apparent till some ten years after establishment, when there was a tremendous once-in-a-decade rainstorm. This caused serious erosion on a subsidiary garden, which was not laid out on the contour, but left the main garden untouched.

A pond has been installed on the keypoint just above the garden. This goes against conventional wisdom, which usually places a pond at the lowest point in the landscape, but it allows the garden to be irrigated from the pond entirely by the force of gravity. Three small orchards have been planted just above the pond, well out of the frost pocket. As these grow they will give the pond some shelter from the north and north-east and help to prevent any silt from being washed into it.

The comfrey patch lies beside the vegetable garden, downhill from the main nutrient outflows on the farm. These include the liquid effluents from the composting of the cattle manure and from the septic tank, both of which are easily led into the comfrey patch. This keeps valuable nutrients on the farm, and out of the river, where they would become pollutants. The fact that the comfrey is used for liquid manure rather than human consumption puts an extra step between people and their own effluent, which is just as well from a hygiene point of view.

As well as being a working farm Ragmans is also a venue for courses on permaculture and other related subjects, with a converted stone barn acting as residential and teaching accommodation. An important part of the farm policy is to keep the teaching side quite separate financially. The farm does not depend on it for its economic viability.

One of the aims of the farm is to show that permaculture can be economically viable and it's run as a commercial operation. It stands as an example of how the countryside can be repopulated and how people can make a living from the land by applying permaculture principles and practice to a traditional landscape.

Note

Since I wrote this case study in 1999 much has changed at Ragmans Lane. Some of the people who were involved then have left and new people have arrived, and there is some change in the mix of enterprises. The farm is still in a state of change, and it seems more useful to let this description stand than to attempt a snapshot of the present changing situation.

Chapter 11

WOODLAND

The vast majority of tree-covered land in Britain is large-scale coniferous plantation managed on the clear felling system. In fact the commonest tree in the UK is not oak or ash but Sitka spruce. But practical permaculturists are more likely to be involved with a very different kind of woodland, usually small-scale, often broadleaved and frequently semi-natural rather than planted. So the emphasis of this chapter is more on this kind of woodland than on big plantations.

While the previous chapters, from Buildings through to Farming, have looked mainly at the distinctive contribution which permaculture can make to their subjects, this one aims to give a more general view of sustainable woodland practice. I've taken this approach because there is not the same general information in print on the sustainable approach to woodland as there is on, say, organic growing, so I feel a concise summary is worthwhile.

Perhaps the most distinctive idea which permaculture brings to the subject of woodland is the principle of multiple outputs, and this is the theme of the first part of the chapter. Working with existing woods and designing new ones are the subjects of the two central parts. This general information on woodland is applicable to woods in all situations, urban and rural, in private or public ownership. Thus the final part, Urban and Community Woodland, concentrates only on those aspects which are specific to urban locations and community action.

I hope this chapter will be equally valuable as a beginners' guide and as a new perspective for conventionally trained foresters.

Any discussion of productive woodland in Britain must be seen in the context of our consumption of timber. At present we import 90% of what we consume, so if we're talking about sustainability it's premature to look at domestic production until we've firmly grasped the nettle of reducing consumption. This is discussed in Chapter 6. (See pp133-134.)

Trees live for a very long time, so good initial design is perhaps more important with woodland than with any other part of the landscape, whether we're designing a new woodland or planning how to work with an existing one. These two tasks obviously require considerable differences in approach, but in both cases a clear idea of the potential outputs is needed from the start.

OUTPUTS OF TREES & WOODLAND

Ecological Functions

It is possible to get over-enthusiastic about trees. They are not 'better' than other plants, nor are woody

ecosystems necessarily 'better' than those without trees or shrubs. Peat bogs, for example, keep much more CO_2 out of the atmosphere than any plantations of trees which might replace them (see p72), and in some cases grassland ecosystems can be more diverse than a woodland which might succeed them.

Nevertheless there are some ecological functions that trees perform which, if not unique to them, can usually be done more effectively by them than by any other life form. To a great extent this is because they're so much bigger than anything else which can grow on land, so any ecological effects associated with plants are usually much greater if the plants are trees.

The mind map overleaf summarises the main ecological functions of trees in a western European context, many of which have been discussed in other chapters of this book. In other parts of the world they have other functions which don't apply here or are less important because of differences in climate and geography. Examples include: rehumidifying the atmosphere to produce more rain, which is important in the interiors of large continents; condensing fog to liquid water, which is important in coastal areas which receive little

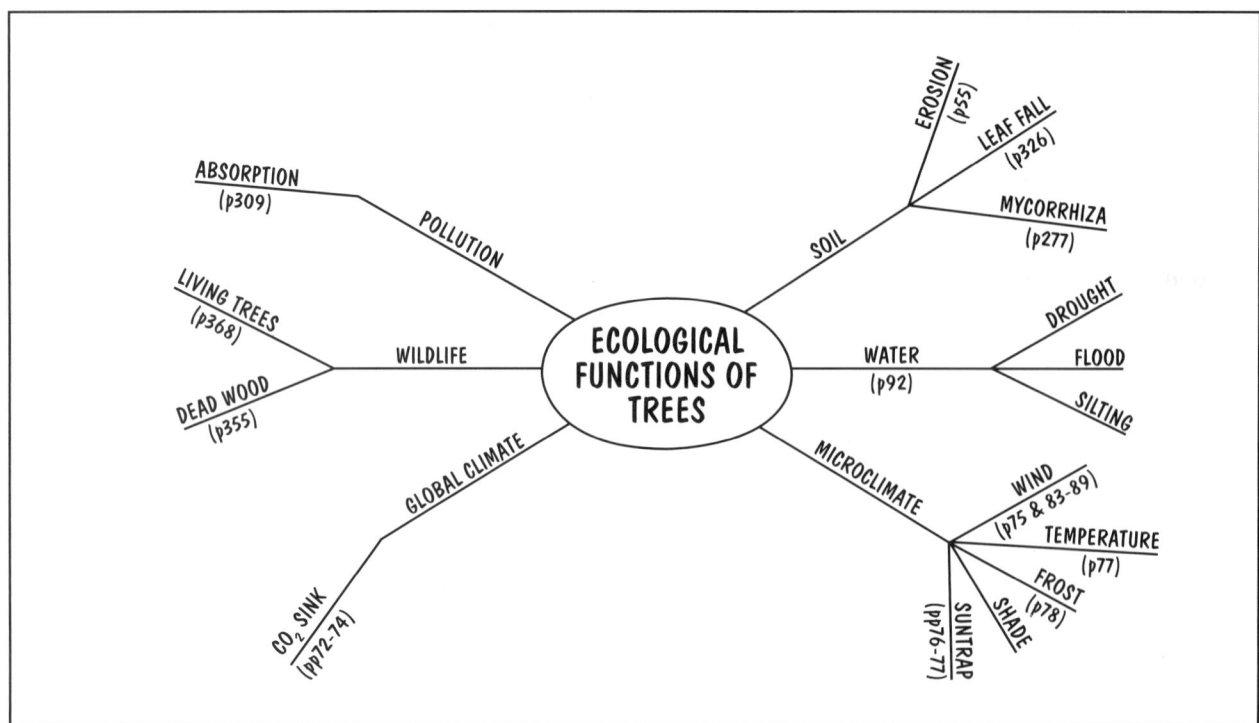

ECOLOGICAL FUNCTIONS OF TREES

- ABSORPTION (p309)
- POLLUTION
- LIVING TREES (p368)
- WILDLIFE
- DEAD WOOD (p355)
- GLOBAL CLIMATE
- CO₂ SINK (pp72-74)
- SOIL
  - EROSION (p55)
  - LEAF FALL (p326)
  - MYCORRHIZA (p277)
- WATER (p92)
  - DROUGHT
  - FLOOD
  - SILTING
- MICROCLIMATE
  - WIND (p75 & 83-89)
  - TEMPERATURE (p77)
  - FROST (p78)
  - SHADE
  - SUNTRAP (pp76-77)

rain but frequent fogs; and preventing the build-up of salt, or reducing existing salinity, in tropical soils.[1]

Multiple Outputs

Every woodland or plantation, existing or planned, has the potential of yielding more than just wood and timber. (See box, Timber and Wood.) The importance of woods for landscape, nature conservation and recreation is widely recognised these days, but from a permaculture perspective we can see an even greater range of useful outputs. These are summarised in the mind map on p303.

This woodland in Somerset is managed for pheasant shooting, with timber, wood and wildlife as by-products. A challenge for permaculture is to make semi-natural woods like this economically viable without relying on blood sports for financial input.

Before looking at these outputs individually it's worth having a look at two general principles which influence how we can get the best value from woodland. The first is integration of the woodland with other land uses and the second is adding value to woodland produce before sale.

Integration

Many of the possible outputs of woodland are indirect ones: they are only realised through the interaction between the woodland and another element in the landscape, such as adjacent farmland or housing. This means that different landscape elements must be carefully integrated in order to get the full value from woodland. This integration is fundamental to the permaculture approach.

It's unfortunate that in Britain, and in some other countries, there's a cultural split between farming and forestry. In part this is due to the history of the countryside. Until the twentieth century most of the land was divided up into estates owned by the aristocracy and gentry. By and large the farms were let to tenant farmers and the woodland was kept in hand and worked by the landowner's agent. So generations of farmers saw woodland as being something quite apart from their interests and experience, and foresters looked on farming in much the same way.

Even now, when a majority of farms are owner-occupied, and many have some woodland on them, the attitudes persist. The following quote from the *Farmers Weekly* is revealing:

"For many farmers the notion of involvement with woods is tantamount to letting the side down," Mr Lloyd Jones told the Forestry and Farming

– Working Together conference. "You only do it because you are no good at what you should be doing. Working woods is rather like keeping horses; good farmers do not do it."[2]

Interestingly, a survey of farmers in northern England and Scotland[3] showed that farmers who actually had woodland on their farm were much more appreciative of woodland in general than those who didn't.

The benefits of woodland which were most highly valued were wildlife and landscape value, with firewood, shelter for animals and farmsteads, and pleasure for the farmer coming next. The disadvantages most commonly cited in the survey were the increased populations of foxes, rabbits and flies. But more farmers said they benefited overall by having trees on the farm than those who said they didn't.

It's encouraging to see that experience can overcome prejudice.

Integrating farmland and woodland has economic advantages as well as ecological ones. Income from farming comes in annually, while that from woodland comes less frequently. Thus a young plantation can represent a farmer's pension fund, needing little expense once established, though some maintenance work is necessary to ensure a high value crop at the end. A standing crop of timber can also act as an insurance against hard times for farmers.

Another advantage of integration is that most woodland work takes place in winter, a relatively slack time on many farms. Some machinery, for example tractors, can be shared by both enterprises.

Adding Value

The financial reward for selling timber or wood as a raw material is pitifully low. Unless the timber is of exceptionally high value, such as veneer-quality logs, the only way to ensure a decent return is to add value to it before sale.

Say a block of standing timber trees is worth £100. It may be worth £200 when felled and stacked by the roadside, £250 delivered to the sawmill, £1000 sawn into planks and maybe twice that when seasoned.

These figures are for relatively low quality timber. It costs the same to fell, transport and process high quality timber, but its intrinsic value is more, so a greater proportion of the price goes to the grower. But the quality of a crop is largely determined by a previous generation of foresters. The best way the owners of small woods can ensure they get a reasonable return from their woods is to do as much of the work as possible themselves.

In many small woods selling timber as a raw material may not be an option anyway. The minimum quantity which a timber merchant will look at is a single lorry load of 20 tonnes, and a small wood may not have this much available at one time.

FASSFERN FOREST FARM[4]

An example of integration in an upland area is the Fassfern estate near Fort William in the Western Highlands of Scotland. Previously it was a hill farm of 2,000ha, carrying 1,200 ewes and 30 cattle. From 1955 to 1986 1,700ha of it, plus an adjoining 700ha were planted with conifers. Most of this is in large commercial blocks on the higher land, which is of little value as grazing. The majority of the lower land has been kept as grazing, but a mosaic of small woods has been planted here specifically to give shelter to the animals and thus increase the value of the grazing land.

In 1988, despite reducing the grazing area from 2,000ha to 300ha, the estate still carried 800 ewes and the lambing percentage had gone up from 50-60% to 110-130%. In other words the total output of lambs had risen by almost half as much again. This was partly due to provision of some winter housing and to fencing of the in-bye land, which allows more intensive grazing. But much of the credit must go to the increase in shelter, and to reducing competition between forest and farm by planting most of the trees on land which is of little value to sheep and cattle. Costs of roads, fencing, drainage and fertiliser have been shared between the two enterprises.

Employment increased from two permanent jobs before afforestation to 18 in 1988. Ten new houses have been built on the estate and there is a small mill turning some of the timber into fence posts. The estate is fortunate in having a manager who is experienced in both forestry and farming.

No doubt there is potential for even more integration and diversity at Fassfern, perhaps including adding more value to timber produce. But what has been done so far serves to illustrate how the sum of two and two is often a good deal more than four.

TIMBER & WOOD

A distinction is often made between timber and wood. Timber is the trunks of trees which have reached planking diameter, while wood includes poles from coppice and pollards, and lop and top from timber trees. In a coppice-with-standards wood the coppice is often referred to as underwood. Thinnings from plantations may or may not be of planking diameter and so can fall into either category.

Ben Law

David Blair has used his own timber to build the work-shop and timber-seasoning polytunnel in his woodland in Argyll. This innovative seasoning structure makes use of solar energy and allows full air circulation. There's also space in it for some food growing and food crop drying.

In any case the buyer of raw timber is usually a much bigger business than the seller, and this puts the producer at a disadvantage. A case in point is the story of all those small poplar plantations you may have seen in parts of lowland Britain. They were planted in odd corners on farms over the 1960s and 70s under contract to a match manufacturer. Then someone invented the disposable cigarette lighter. Before the poplar was ready for harvest the company closed down its British factory and reneged on all its contracts, leaving the farmers with a whole lot of timber for which they had no ready market.

A small business like a farm can't hope to sue a large multi-national company, especially as the plantation would only be a minor part of the farm business. By contrast, doing business face to face with another individual or small firm is empowering, and also helps to develop a thriving local economy.

LOCAL TIMBER FOR LOCAL HOUSES

Neil and Maggie Sutherland farm a traditional croft in Wester Ross. They've planted thousands of trees on their land, carefully matching the species with the varying environments on the croft, and using natural regeneration of the existing native trees where possible. It will be many years before they start harvesting what they have planted, and in the meanwhile they buy in timber, as locally as possible, and mill it and season it at home. Using this timber, they design and build ecological houses in the locality as part of their polycultural living. Their small sawmill and seasoning shed are vital elements in the economy of the croft. (See Colour Section, 34.)

Ways of adding value to coppice are discussed below, under Wood. On the whole these are more time-consuming than processing timber, and may be more suitable for someone who makes their living mainly from the woods than for a farmer with a little woodland on the side. But as it becomes more and more difficult to make a living from small-scale farming, this may no longer be the case. Farmers who now go out to paid work and do their farming in the evenings and at weekends may be better off making rustic furniture or wattle hurdles and staying on the farm full time.

Even if the person who grows the timber is not the same one who adds value to it, local processing leads to local employment, and often helps develop or maintain the kind of community in which mutual help can thrive. On any one farm or smallholding there is unlikely to be enough of the right kind of wood to support continuous production. Traditionally coppice workers and charcoal burners travelled around their district to get the material they needed, and this kind of niche management is still a good way of matching supply and demand. Equally, material can be bought in from surrounding woods and plantations and processed on the farm.

An absolutely crucial aspect of getting value is marketing. This may need as much attention as everything else put together, especially in the early years.

A useful aid to finding local markets is the periodical *Ecolots*, available both on the internet and on paper.[5] It contains advertisements for lots of timber and wood for sale and wanted, mostly in small quantities. It includes standing and felled trees, sawn timber, coppice produce such as hurdles, equipment and services. There has even been an advertisement for pigs for hire to help with natural regeneration. (It also includes advertisments on farming and nature conservation matters.)

Local coppice groups exist largely to help coppice workers with marketing. The Wessex Coppice Group has nationwide contacts and may be able to put you in touch with your local group. (See Appendix D, Organisations.)

Timber[6]

Roundwood has a low sale value, but it can be used as it is for the framework of simple buildings, such as farm sheds and barns. Some innovative architects are now also using it for dwellings.[7] Using timber in the round saves the energy and money cost of milling, and preserves the structural integrity of the tree. It can also be a way of finding a high value use for conifer thinnings. The most suitable species for building purposes are noted in the table, Trees for Timber and Wood, on pp338-340.

For sawlogs, the minimum size is a diameter of around 15-25cm. Trees which are curved or branchy will have little value in conventional markets, especially in small diameters.

Veneer is produced by putting the whole log on a huge lathe, where it's peeled by a knife as long as the log itself. Timber suitable for veneering must be extra straight, free of knots and preferably large.

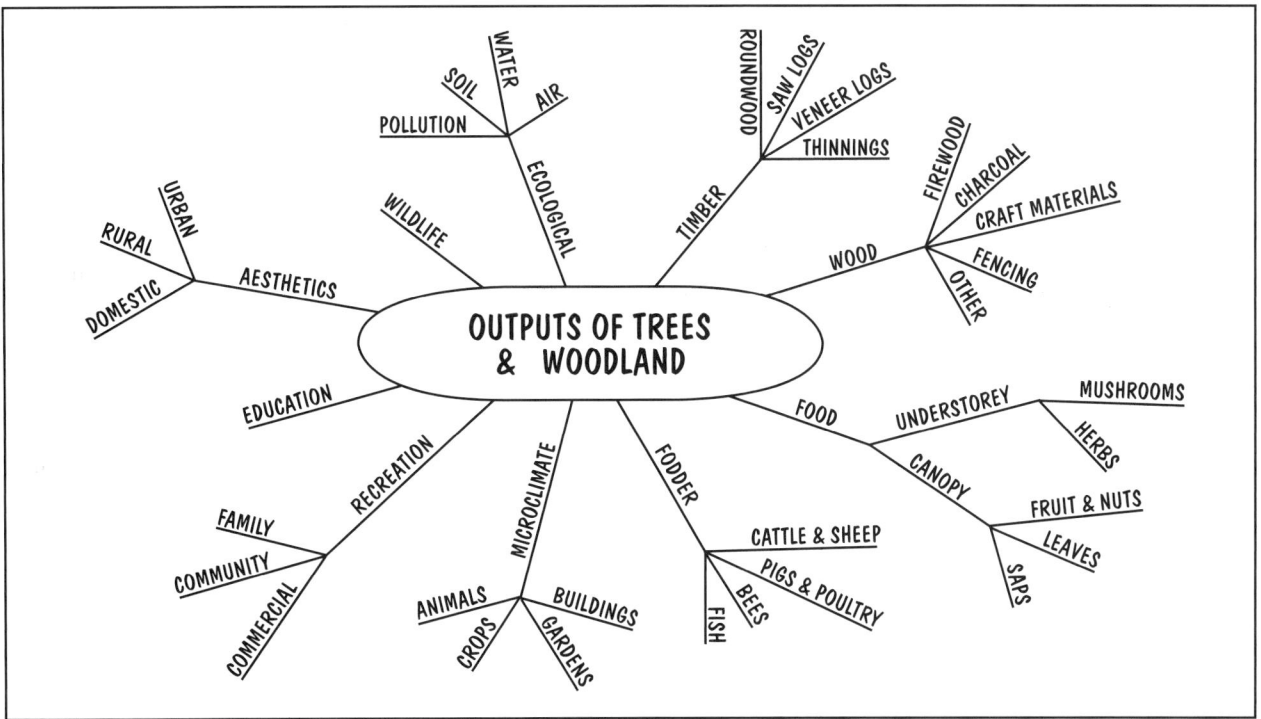

Wild cherry will only make veneer quality if it has been pruned as it grows.

Prices for veneer logs of high quality species, such as oak and walnut, can be ten or twenty times those for sawlogs, but you often can't tell the quality till after felling. A smaller differential is paid for poplar and willow for making vegetable crates, which is done by the same method.

Specialist markets exist for some specific timbers. For example, burr oak is much sought after by wood turners, and some high quality ash thinnings are in demand for sports equipment such as hockey sticks and oars.

Other broadleaved thinnings are of relatively low value, usually going for pulp or firewood. Conifer thinnings may be used for fencing. But unless they are durable species such as larch they will be treated with copper-chrome arsenate before use. (See p306.)

Roundwood frame for a dwelling house under construction at Brithdir Mawr, Pembrokeshire.

Wood

Firewood. In the past the vast majority of coppice wood was used for fuel. Building, fencing and craft materials were always minority uses, except in pure hazel or chestnut coppices. Much the same is true today, especially of mixed coppice, although pulp is another bulk use for coppice wood. Even coppice which is the right species for a particular craft may not be suitable because it has not been cut when it should have been and the poles are too big. This is known as out-of-cycle coppice. In it there may be the occasional pole which is usable by a chair bodger or toy maker, but the majority will only be fit for fuel, and you can often only get at the craft quality material by coppicing the lot.

Firewood is a low value product in cash terms. But if it's seasoned for a summer before selling, and doesn't include trees which don't burn so well in open fires, like willow, a firewood seller can get a good reputation and a loyal clientele.

Alternatively you can use it to heat your own house. To work out how much coppice you need to heat your house you need to know, firstly, what yield you can expect from the coppice and, secondly, how much wood is needed to heat the house. The table overleaf gives typical yield figures for different kinds of coppice.

All the yield figures, except that for SRC, are for material down to 5cm diameter. The smaller twigs can be used by bundling them into faggots, but this is a very laborious job, and one which has not yet been mechanised. On the Quit While You're Ahead principle, it's probably not worth bothering with the twigs.

The yields given are averages, and the actual yield in any one place is greatly affected by climate, microclimate and soil. A traditional mixed coppice in the lowlands can give as much as five times the yield of

TYPICAL YIELD FOR DIFFERENT KINDS OF COPPICE				
Type of Coppice	Yield in tonnes*/ ha/ann	Length of cycle, years	Years to first cut	Notes
Typical farm wood	2.5-4.5	10-30	-	
Purpose-grown, hardwoods	3-5	10-20	10	Full yield by second or third cut
Purpose-grown, willow or poplar	6-7	10-15	7	Ditto. Not suitable for open fires.
Short rotation coppice (SRC)	8-20 (12)**	2-5	3-6	Plantation may last 20 years. For a description of SRC, see p325.
*Air-dry. This is 50% of fresh weight. **Typical yield.				

an oak and rowan wood on an acid soil on a blasted mountainside.

The wood of willow and poplar is much less dense than that of other broadleaved trees. This means that they require more work and more storage space for a given amount of heating value than other broadleaves. They burn well enough in stoves, but neither is suitable for open fires. Poplar tends to produce fewer, thicker poles, more suitable for a log-burning stove, willow more, smaller ones, more suitable for a masonry stove. (See pp152-153.)

Other tree species suitable for firewood production are noted in the table, Trees for Timber and Wood, on pp338-340.

How much wood is needed to heat a house depends on:

- the size and design of the house
- the efficiency of the heating system
- how warm the occupants want to be

Any one of these factors can have a big effect on the amount of wood required, so the two estimates which follow must only be taken as broad indicators.

A Forestry Commission paper published in 1980 gave a figure of seven tonnes of air dry wood to heat a three bedroom house for a year.[8] No information is given about the structure of the house or the heating equipment.

Three tonnes is the figure given to heat "a reasonably draught proofed and insulated small house" for a 200 day heating season with a suitably sized masonry stove. This figure was based on monitoring of the Oxford Eco House by Oxford and Manchester Universities.[9]

Putting together different combinations of woodland type and fuel requirement we get a wide range for the area of land required to heat a house. Around three hectares of relatively poor farm woodland would be needed to heat the nondescript three bedroom house, whereas the small, well insulated house with a masonry stove built into it could manage on one seventh of a

hectare of tip-top short rotation coppice. A seventh of a hectare is a plot of around 50 by 30 metres. This is a difference of nearly thirty times the land area between the least efficient system and the most efficient!

Charcoal. It's possible to make a modest living by cutting coppice and turning it into charcoal, though it is hard work. The advantages of charcoal burning are:

- It can make use of species of trees and shrubs which have little value for other uses.

- It can make profitable use of out-of-cycle coppice, and thus bring it back into cycle.

- It can add value to forestry by-products such as slab wood – the outer slices of a log, discarded when it's planked – though this does make rather low quality charcoal.

- It's a light weight, high value product; when made on site it's easy and economical to extract from a wood with poor access.

- Because conventional ring kilns need constant attention through the hours of the night, it can be an important element in getting planning permission to live in the woods.

- Home grown charcoal replaces ecologically destructive imports (see p134), while coppicing often benefits the ecology of an ancient wood.

Cooking with charcoal is an inherently inefficient way of using wood, as much of the energy in the wood is lost during the charcoal making process. But if people are going to have barbecues anyway, it's much better for the Earth if they use charcoal from local coppice woods than from Indonesian mangrove swamps. There is also a limited market for high quality industrial and artist's charcoal.

A NEW CHARCOAL KILN

A traditional ring kiln must be watched carefully throughout the burn, to gauge the point at which it must be shut down to prevent the charcoal itself from going up in smoke. This can be an advantage if you want to prove to the planning authorities that you need to live in the woods rather than commute there daily, because the crucial moment may come at any time of the day or night. But in practice it can be a pain in the neck.

Another disadvantage of the ring kiln is that the only way to empty it is to get inside it and shovel the charcoal out. This is incredibly dirty work and has been enough to put many people off the whole idea of making a living as a charcoal burner.

The mobile retort kiln, recently developed by Reinhart von Zschock and Matt Dunwell, avoids both these problems. It's designed in such a way that the fire goes out automatically at the point when all the wood has been converted to charcoal. As for unloading, as long as it's sited on a steep bank the finished charcoal can simply be tipped out by rotating the main drum.

Retort kilns are somewhat more expensive than ring kilns, but they greatly reduce the amount of work required for charcoal-making, and the extra cost can soon be recouped in extra output from the same amount of work.

A mobile retort kiln, set up for easy unloading: the chimney is removed, the drum is rolled along the rails, and the charcoal falls out.

Crafts. Making useful things from home-grown wood is one way of reconnecting with the Earth. If you cost the time it takes you to fell a tree and make a chair out of it by hand, compared to buying a manufactured one from a shop, it won't be worth it in purely cash terms. But in a world awash with cash and poor in experiences of truly satisfying work, it might be the most valuable way you could spend your time.

Looking to the future, when the present high input / high output economy may no longer be operating, the habit of self-reliance which grows from this kind of experience may be one of the most valuable assets a person can have.

But craft work can pay even in today's economy. Bill Hogarth, a traditional coppice worker in the Lake District, makes his living by selling some 40 different materials from traditional mixed coppice woods to a network of local craftspeople. But it does take time to set up such a network and learn the exact requirements for each product.[10]

Practising a craft yourself is another way to make coppicing more viable, especially if the area of coppice available is limited so you need to add maximum value to the available wood. At present the various products of hazel are in good demand at reasonable prices. These include wattle hurdles and thatching spars, the wooden staples which are used to keep thatch in place. Minor products from hazel include pea and bean sticks and etherings, the long, flexible binders used in Midland-style hedge laying. At present there's a good demand for little tipis of hazel to support garden flowers.

In 1991 it was calculated that one hectare of well stocked in-cycle hazel coppice in Hampshire, worked on an eight year rotation, could yield a gross income of £2,400 per year in hurdles and sundry products.[11] Nothing like this could be expected from out-of-cycle hazel, nor from mixed coppice, though Ben Law gets a similar yield from his sweet chestnut coppice in Sussex. (See case study, Living in the Woods, p336.)

Making rustic furniture from durable species (see under Fencing below) is another viable way of adding value to coppice wood. It may be worthwhile in out-of-cycle coppice if there is enough reasonably small material there. In the more populated and prosperous parts of the country there is excellent demand for seats, pergolas, arches, trellis and the like.

Other woodland crafts, such as green wood turning, are less profitable, and practitioners usually make their living more from teaching the craft than from selling the goods.

Green woodworking class at Clisset Wood, Herefordshire. Teaching is one way to make a living from the woods.

Fencing. Wooden fencing posts are normally conifer thinnings treated with tanalin, which is copper-chrome arsenate (CCA), to prevent them rotting. The same chemical is used for any outdoor pole constructions, including children's playground furniture. Conventional wisdom has it that it's so tightly bonded to the wood that it never escapes into the environment, but there's plenty of evidence that it does.[12]

A more ecological approach is to use wood which is naturally durable. The table shows the number of years an untreated stake of 5x5cm will last in the ground. A CCA-treated softwood stake may last up to 30 years. (Note that timber of these species will last very much longer when used for building.)

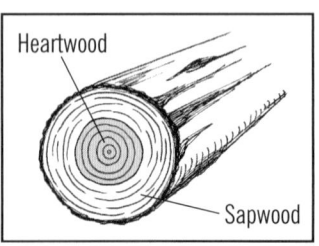

Only the heartwood is durable, so chestnut is more useful at small diameters than oak. (See illustration left.)

Coppicing, with its low renewal cost and high yield compared to planting, is much the best option for stake production. All the broadleaved trees in the table, plus coast redwood, will coppice or sucker. Chestnut is clearly the best choice. It favours sandy soils of low pH and a relatively warm climate. But I have seen it thrive on a soil of pH 7, and it will certainly grow in favoured situations in Scotland.

There can hardly be a farm in the country, other than purely arable ones, that would not benefit from enough sweet chestnut coppice to make it self-sufficient in fencing stakes. It would not only save the use of poisons but keep transport to a minimum and make the farm more self-reliant. These advantages must be balanced against the possibility that the yield will be relatively low if the local climate and soil are not ideal.

On good chestnut land, working on a 15 year rotation, a yield of 1000 fencing stakes per hectare per year can be expected.[13] So a very small patch of chestnut would be enough for most farms. Harvesting can be advanced or delayed by a few years according to the need for stakes, as smaller ones can be used in the round and larger ones cleft to make two or more stakes from each pole.

A hectare or two of chestnut for sale could make a nice sideline for a farmer who could take on a bit of extra work in the winter. At present it's hard to compete on price with mass produced CCA-treated softwoods, but untreated stakes are a premium product and with good marketing may fetch a premium price similar to that paid for organic food. There is also a possibility that CCA may be banned in the future, and then an established chestnut coppice could be very profitable.

There are many other uses for chestnut poles including split rails for post and rail fencing and roundwood for pole barn construction. (See case study, 'Living in the Woods', on p336.)

Other. There is still a small but strong market for oak bark for tanning. Where oak coppice exists it can be profitably worked for this market, with the stripped poles going for rustic furniture. The season for bark peeling is in spring and summer, so it fits in well with

DURABILITY OF UNTREATED STAKES IN THE GROUND		
Durability in Years	Timber	Notes
15-25	Sweet chestnut	Thin sapwood, vigorous and productive coppice
	Oak	Thick sapwood, rather slow growth
	Black locust	Suckers rather than coppices, needs warm climate, can be invasive
	Yew	Slow grower, too valuable for fencing stakes
10-15	Larches	
	Lawson cypress	
	Western red cedar	Not very strong
	Coast redwood	Although a conifer, will coppice
	Turkey oak	Can be invasive in areas of nature conservation value
	Walnut	Too valuable for fencing stakes
5-10	Elm	Suckers regularly grow to fencing size before succumbing to disease
	Most conifers	
< 5	Most broadleaves	

other woodland work, which tends to be in the winter, but not so well with farm work.

An innovative use of woodland, which may have applications here in Britain, is the brushwood composting method pioneered by Jean Pain in Provence. (See pp315-316.) This may be one way of getting good value from out-of-cycle coppice and low value species such as thorn and sallow.

Using coppice wood as a substrate for growing mushrooms is an output which really comes under the next heading.

Food

The concept of forest farming is becoming known in North America, though it is still practically unknown here.[14] It's primarily a matter of growing high value, shade-tolerant crops beneath an established woodland, though fruit and nuts from canopy trees can also be included. The crops need to have a high value because it's virtually impossible to mechanise work inside a woodland.

Understorey. Before embarking on forest farming it's essential to look at what is already growing in the woodland. In an ancient wood much of the biodiversity is in the understorey and ground layer. Replacing a rich semi-natural ecosystem with crops would be an ecological crime. A plantation on land that was previously unwooded is less likely to have any great ecological value in the ground layer.

Understorey crops need to be especially shade tolerant in cloudy British conditions. The most shade-tolerant of crops are mushrooms. Since they don't carry out photosynthesis they need no light at all. Indeed most kinds of mushrooms benefit from the high humidity of woodlands and plantations. Broadly speaking,

Shii-take mushrooms on an oak log.

cultivable mushrooms either live on the humus in the soil or on dead wood. The former include ceps and other boletes and the blewetts, the latter include the oyster mushrooms and shii-take.[15]

At present most attention is on shii-take, as there is a ready market for them and a great deal is known about cultivating them.[16] They are high in protein and B vitamins, including B_{12}, which can be hard to get in a vegetarian or vegan diet. They also help to boost the immune system by encouraging the production of interferon. It's well worth growing a few for home consumption if you don't want to make a business out of it.

Wood-living mushrooms can be grown on fresh sawdust or freshly felled logs. The ideal diameter for the logs is 10-25cm, as they have a higher proportion of sapwood to heartwood than thicker ones. Each kind of mushroom has a preference for certain kinds of wood, but most will grow on more than one kind. The kind of log effects the life cycle of the mushroom. Shii-take are normally grown on oak, where they take a year to start fruiting and continue for around five years. On birch or poplar they will start sooner but only last about three years. Other woods, such as beech or cherry, are intermediate.

Where only coniferous wood is available there are a number of mushrooms which will grow on conifer logs, and humus-feeding ones which will grow beneath coniferous trees. The pine oyster and some of the boletes are examples.[17]

It's hard to identify any edible green plants which are likely to grow well in the interior of a mature British wood or plantation. Even the most shade tolerant, like leafy greens and soft fruit, are unlikely to produce enough to make growing them worthwhile, even under a relatively open canopy. They will grow well enough under the wide-spaced kind of agroforestry described in Chapter 10, but if the only land available for growing fruit and vegetables is woodland, it's really necessary to make a clearing for them. (See pp234-235 and case study, Living in the Woods, p336.)

Medicinal plants may be more suitable for forest farming than food crops. They don't have to produce the same bulk of edible material that we expect of food plants, so the low level of photosynthesis need not be such a great handicap. Some actually prefer shady conditions.[18] In North America one of the leading forest farming crops is ginseng, an oriental medicine of high value which can only be grown in shade.[19] At present there is one farmer growing this crop in Britain on an experimental basis, in an agroforestry combination with poplar trees.[20]

Decorative plants like mosses and ferns are another possibility.

Different undercrops can be grown at different stages in a timber rotation, according to the changing conditions. (See illustration, p317.) In the first few years an intercrop of vegetables can do well and help tree growth at the same time, because trees benefit greatly from having something other than grass growing at their feet. Several years' cropping can be possible before shading from the trees becomes excessive. Since mechanisation is not possible only small areas can be treated like this.

Through the thicket stage there is little space for anything other than the timber crop, with dense shade at ground level. But in the later stages of the rotation, after thinning, there is plenty of space for growing an undercrop between the trunks of the trees, and a varying amount of light, depending on the species of trees and how heavily they are thinned.

Canopy. Food production from the trees themselves can come in one of three forms: fruit and nuts, leaves, and saps.

Combining fruit and nut production with timber production is discussed in Chapter 9. (See pp220-222.)

Tree leaves can be gathered for making leaf curd. (See p260.) Some can also be eaten raw in salads, including beech leaves when they're young in spring, and lime leaves throughout the summer. Picking leaves is only really feasible in coppice woods or on the edge of high forest, where the leaves can be reached from the ground.

Several members of the maple family, including the sugar maple, will grow here in Britain, but in our climate none of them produce sufficiently concentrated sap to make it worthwhile tapping them for sugar. The sugar is there, but the amount of energy needed to evaporate the water would be excessive.

However they are worth tapping for wine, and so is birch. This can be done for domestic consumption or on a small commercial scale.[21]

Fodder

Keeping farm animals in woods is discussed later in this chapter under Working With Existing Woods. (See pp320-321.)

Trees can be an important source of nectar and pollen for bees, but a mixture of trees and herbaceous flowers is needed to give continuity through the year and diversity of diet. (See p246.) Two woody plants which are particularly important sources of bee forage are the spring-flowering or pussy willows and ivy. The former give a high volume of pollen early in the spring when it is needed to build up the colony, and the latter gives food in the autumn when few other plants are flowering.

It has already been noted in Chapter 5 that much of the food for freshwater fish comes from plants on the bankside, and trees being bigger than other plants can provide the lion's share of this. Thus the integration of woodland and aquaculture can be highly productive.

Microclimate

Windbreaks for buildings, gardens, orchards and farms are discussed in detail elsewhere in this book. (See Index.) Other microclimate effects of trees are discussed in Chapter 4. (See mind map on p300.)

As well as being sheltered by a windbreak, farm animals can get shelter inside woods, either as a refuge or as a permanent home. The value of the shelter needs to be balanced against the possible harm done to the woodland by the animals. (See p321.)

Recreation

Often a major reason for owning woodland is the satisfaction and enjoyment which the owner and their family derive from it. This can take different forms for different people. For some it will be walking, bird watching, badger watching or just quietly sitting. For others it's the satisfaction of doing useful and enjoyable physical work in a beautiful environment. For most it will include the experience of getting to know and understand an ecosystem as it changes and develops through the seasons and over the years.

All these benefits can be shared with the local community, either in a community owned wood, or in a private wood where local people are made welcome by the owner. Woodland work is in some ways ideal for community involvement, as it often requires large bursts of energy at infrequent intervals, which can be the subject of communal work days. By contrast garden work requires constant attendance over the growing season, and farm work mostly involves tractors or working with animals, both of which require a fairly high level of skill.

This is not to say that woodland work does not require any skill. Coppice can be ruined by cutting stools at the wrong height, or by cutting too small a coupe. Volunteers should always be guided and supervised by someone who knows what they're doing, and no-one should handle a chain saw without proper training. Woodland owners should be cautious about thinking of volunteers as cheap labour. It can sometimes be quicker and easier to do the job yourself!

The commonest forms of commercial recreation in woods are probably war games and pheasant shooting. But it is possible to get some financial return from more peaceful pursuits. Teaching woodland crafts is one which can amount to a full-time living. At the other extreme, a waymarked nature trail with an honesty box at the end may only bring in a few pounds, but the outlay is correspondingly low.

Education

One of the biggest obstacles to building a sustainable future for our predominantly urban society is the lack of contact most of us have with nature. We depend on living systems for our survival, yet for most of us physical reality is that tiny proportion of the Earth's surface covered by bricks and tarmac. It's hard for people to really care about nature if it's only something they see on the television, yet it's only by forming a deep and meaningful relationship with nature that we will care enough to make the changes in our lifestyle which are essential to our survival.

Planting trees to replace dead elms in a community woodland. This kind of work is one of the most creative ways in which people can reconnect with nature.

Working with trees and woodland is one of the best ways to re-establish this link, both for school children and older people. The size and permanence of trees gives them a physical presence which makes an immediate impression on the senses and emotions. By contrast wild animals are elusive, and the value of herbaceous ecosystems is less obvious to those who are beginners at ecology. Trees offer opportunities for both children and adults to get physically involved, in planting, surveying, and even harvesting and woodland crafts. Yet once trees are well established they can be left alone for long periods without any need for human input.

Aesthetics

It's hard to overestimate the aesthetic value of even small plantings of trees, especially in urban areas and around buildings generally. If ornamental plantings are carefully planned they can also have one or more additional outputs, including microclimate, food, wildlife habitat and more.

In rural areas there's much to be said for using only native trees and shrubs in ornamental plantings. There are ecological reasons for this, (see p368) but there are also aesthetic ones. Exotics, whether broadleaved or conifers, almost always look out of place in the countryside. This can apply as much to trees which are native to another part of the country or to a different environment as to ones from foreign lands. Trees which are truly native to the locality can become a harmonious part of the landscape in a way which is impossible to imitate.

Some suggestions for attractive natives to give visual interest throughout the year are given in Chapter 8. (See p182.)

Wildlife and Ecology

There is almost always more diversity of wild plants and animals in long-established habitats than in newly created ones. Preserving and working sympathetically with existing woodland is thus more beneficial to wild plants and animals than is planting new trees. Indeed planting trees on land which already has a high level of biodiversity will almost certainly be detrimental. (See pp358-359.)

The effects of trees and woodland on soil, water and global climate change are described in the respective chapters. These are effects which should always be borne in mind, both when planning to plant trees and when working with existing woodland. For example, if there's a choice between planting trees on sloping or flat land, choosing the steep land will help to reduce soil erosion. In an existing wood the main water storage 'organ' of the woodland 'organism' is the layer of leaf litter and heavily humified surface soil. This helps to keep water flowing evenly through the landscape throughout the year and thus avoid both drought and floods downstream. This consideration may affect the decision whether to keep pigs or other animals in the wood, as they may disrupt and destroy this layer.

The ability of trees to absorb and neutralise pollution is an output of special importance in urban areas. Woodland with a high proportion of edge and a canopy with an irregular outline can intercept more pollution than a large block of uniform height. Conifers are able to absorb pollutants during the winter, which is an advantage where sulphur dioxide and the oxides of nitrogen are the main problem, as these are worse in the winter. Ozone is more of a problem in the summer, so trees which are most active then are most useful in cleaning up this pollutant. Since water is most often the limiting factor to tree growth, summer-active trees include both drought-tolerant broadleaves and alders and willows planted in permanently wet sites. As road traffic is now the main source of pollution, planting along roadsides is likely to have the greatest effect.[22]

EXISTING WOODLAND

Assessment of Woods

Many permaculturists dream of looking after a piece of woodland. As the purchase price of woodland is much lower than that of farmland it's an achievable dream for many of us. Joining together with some like-minded friends can make it more so.

Whether you're looking at a wood with a prospect of buying it, or making plans for working with one you already have, the kinds of questions you need to ask in assessing it are much the same.

Most woods are quite varied, even small ones. So when making an assessment it's important to examine all parts of the wood, even though some parts of it may be hard to penetrate. Never assume that the rest of it is the same as the bit you've just looked at.

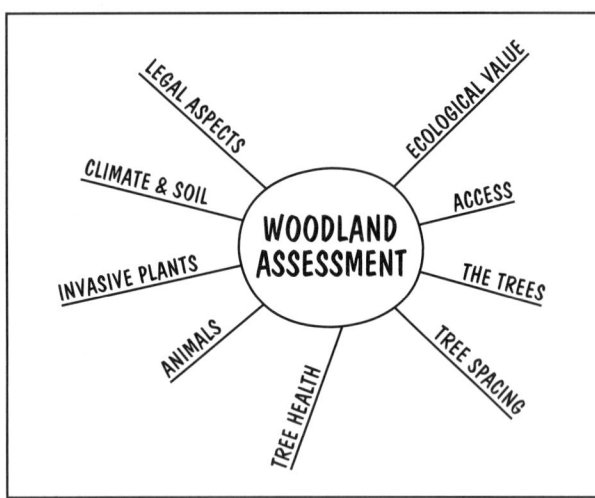

Most woodland work takes place in winter and many woods are poorly drained. Seeing the site in summer only may give an over-favourable impression. Clues like deep ruts and the presence of moisture loving plants such as rushes can help suggest what it's like in winter, but there's really no substitute for seeing the site in all seasons.

Many woods are on steep land and extracting uphill is really a non-starter. A solid track or well drained pasture at the bottom of the wood is ideal. A public road may not be. It may be necessary to get the road closed when you want to extract timber and the administrative cost could be significant when the amount of produce at any one time is small.

The easy extraction route from this wood enhances its value for timber and wood production.

The distance from home may be less important for woodland than for most other types of land. If timber and unprocessed coppice wood are the main products, work tends to be in large and infrequent spells. But any transport consumes time, money and, usually, fossil fuels. If much of the produce is going to be used at home this may be an important factor.

Ecological Value

High ecological value may or may not be a blessing, depending on your aims. If your main aim is to safeguard existing biodiversity a semi-natural ancient woodland would be ideal. But if your plans are more output-oriented you may want to introduce new tree species, or even farm animals or herbaceous plants which are not already present. Any one of these is likely to reduce biodiversity, which above all needs continuity of conditions. A less diverse wood, perhaps a recent plantation, would be a better wood to start with.

Assessment for biodiversity value is dealt with in Chapter 12. (See pp352-356.)

Access

Poor access often limits the value of an otherwise useful wood. There's no point growing wood and timber if it can't be got out, and where the volume of produce is small its value may not be enough to pay for improving the access. The size of the trees and the method of extraction have a bearing on how good the access needs to be. Mobile saws can provide a solution to poor access in many cases. (See p320.)

WOODS, PLANTATIONS & COPPICE

Plantations are, of course, woods which have been planted by human hand, and the term wood or woodland is often reserved for woods which have not been planted. These are either places which have always been wooded or which have regenerated naturally.

In lowland Britain most woods have been coppiced in the past. Coppicing is cutting a tree down and allowing it to regrow, which it does in a multi-stemmed form. When the coppice has regrown for a number of years it can be cut again, yielding a crop of poles, and this cycle can be repeated many times. Most broadleaved trees coppice and most conifers don't.

A wood treated in this way is known as a coppice. Many traditional woods had a coppice-with-standards structure, that is the majority of the trees were coppiced but a few, the standards, were left to grow tall and single-stemmed. (See Colour Section, 36.) Coppice which has not been cut for a long time is known as out-of-cycle coppice.

High forest is a wood or plantation composed entirely of single-stemmed trees. Almost all plantations are high forest, chestnut coppice and short rotation coppice being the commonest exceptions.

Where value is added to coppice material this is best done in the woods, and then it's a great advantage to live close to the wood, or even in it. The same applies if food is one of the outputs of the wood. (See case study, Living in the Woods, p336.)

Also, quiet moments of observation and contemplation in the wood are not only valuable in themselves but will increase your knowledge of the wood and may lead to much more effective decisions. These moments will be more frequent if the wood is a short walk away from home.

The Trees

In the commercial world a higher value is placed on:

- timber than coppice
- large trees than small
- broadleaves than conifers – when mature and of good quality
- straight ones than curved
- clean ones than branchy

But this set of values is really about selling raw materials in bulk, and small woods are rarely financially viable as suppliers of raw materials. (See p301.)

A more permacultural approach is to ask:

- Firstly, how much of the produce can we use at home or add value to ourselves? Firewood may not be very profitable to sell, but it can reduce the need for money to pay for heating, while providing enjoyable and meaningful exercise for the desk-bound. Small-diameter oak logs may have little commercial value, but they're ideal for growing high-value mushrooms on.

- Secondly, is there someone, preferably local, who can use the particular produce of this wood for a specific purpose? Curved trees may be rejected by a timber merchant, but they may be particularly valuable to a green oak frame builder looking for crucks and braces.

The key to making a small woodland viable often lies in building up a local network of mutually beneficial trading connections. The importance of marketing cannot be over-emphasised.

The age of the trees in a woodland affects its value. Both timber and coppice are more useful if the trees are of a range of ages to give a steady supply over the years. Coppice is most likely to be out of cycle. This means the poles will be past the optimum size for craft and other specialist uses and may only be useful for firewood or charcoal until it has been cut and grown again.

Some trees are hard to find a good use for. Pioneers like thorn and sallow may be abundant in woods which have suffered exploitation felling in the past. Such a wood may have a high wildlife value but little cash value in the short or medium term. Charcoal making may be the only financially viable option – indeed sallow can make high quality artist's charcoal. But it's important not to dismiss a wood like this on the basis of a casual inspection. A close look may reveal worthwhile regrowth of valuable species hidden among the scrub.

Non-wood produce, such as wild fruit and nuts or foliage for flower arrangers is worth considering. Wild food is not likely to be there in big enough quantities to make it worth gathering for sale. But it may be a welcome addition to the family diet, and collecting it can be a meaningful part of the relationship between people and woodland. On the other hand decorative foliage, holly for Christmas time and so forth, can make a useful commercial sideline.

Tree Spacing

If timber trees are too widely spaced they will have short trunks and too many branches for quality timber production. If they're too close together they will be tall and spindly and total growth will be reduced. Appropriate spacing in a plantation is achieved by planting close together and progressively thinning as the crop grows.

If thinning has been neglected it may be difficult to get the best from the plantation. The trees will be unstable, and if they're thinned now they're liable to windthrow. If they're left unthinned they will never achieve the size required for high-value timber, especially for broadleaves. It may be necessary to fell the whole crop early for some low-value use, such as pulp, and start again.

In out-of-cycle coppice some of the stools may have been killed by shading from over-big neighbours. If there are too many gaps the potential yield will be reduced, but the gaps can be filled. (See p314.) A spacing of 3m is usually about right, though this depends on the length of the intended rotation – the shorter the rotation the closer together the stools need to be.

Tree Health

There are few diseases which seriously bother trees. Dead or sickly-looking trees are more likely to be suffering from competition due to the neglect of thinning or coppicing than from disease.

Outside of large-scale conifer plantations, the most serious disease of trees in Britain is Dutch elm disease, which is invariably fatal. It's spread by a beetle which infects trees when they get to a certain size, usually around 15cm diameter at breast height. In dense woods there are sometimes much larger elms still alive, while in the surrounding countryside only young suckers survive. This may be because the beetles can't easily find the trees in the wood. When the wood is thinned or part is felled the elms may become infected and die. If they form a significant proportion of the stand this can be a problem. On the whole wych elm is less susceptible than suckering elms and it may escape infection even on an open site.

Ash can suffer from canker, which reveals itself on the surface of the tree as a black blob around 5cm in diameter, which is the fruiting body. The timber of infected trees is stained or rotten. It's not marketable as timber, though it still makes good firewood and I'm told the fruiting bodies make excellent firelighters.

Animals

The population of deer in Britain is probably higher now than it has ever been in historical times. It's relatively easy to protect newly planted trees from being eaten with individual tree shelters. But coppice regrowth, being multi-stemmed, is another matter. In severe cases deer can completely prevent regeneration. There are various ways of reducing deer damage in coppice but all of them are either laborious, expensive or only partially successful. (See pp314-315.)

Deer populations do vary from one locality to another, and they favour certain trees, notably willows and ash. Finding out about the local deer population is an important part of the assessment process, especially for coppice containing palatable trees. The most easily recognisable signs of deer are a browse line, a definite line below which all the foliage has been eaten, and nibbled twigs, especially of young coppice regrowth. (See also illustration on p414.)

Grey squirrels are an introduced species from North America. In their ancestral habitat the young males have the habit of stripping the bark from a tree called the shagbark hickory, so-called because the treatment it gets leaves its bark in tatters. Here they strip the bark of other broadleaved trees, and since these species have not co-evolved with the squirrels they suffer. They are rarely killed, but the leading shoot is often killed or severely damaged and their value for timber is reduced or ruined.

Grey squirrels are found almost everywhere in England and Wales, except parts of the north and the Isle of Wight. Trees are most vulnerable between ten and forty years old. Most damage occurs where there are a relatively small number of young trees near or in a relatively large mature woodland, which provides habitat for the squirrels. Species of broadleaves vary in their susceptibility to grey squirrel damage. (See table, Trees for Timber and Wood, on pp338-340.)

The only way to reduce squirrel damage in established woods is to kill them, and to keep killing them, as populations soon recover. Trapping and poisoning are the only effective methods; shooting and drey poking are ineffective. Poison dispensers are available which cannot be opened by any other species of animal. Even so, many permaculturists will be reluctant to use this method. Trials are currently underway on an oral contraceptive for squirrels, and this may be a practical alternative.[23] Coppice is not often seriously damaged by squirrels and this is an option to consider where their presence makes high forest impractical.

Voles, rabbits and hares can all destroy newly planted trees, but tree guards and fences can prevent this. Rabbits can even be present in large enough numbers to seriously interfere with coppice regrowth.

If farm animals have unrestricted access to a wood they can cause a lot of damage, both by soil compaction and by browsing. Woods or plantations where pigs have been kept or cattle have been fed may have areas of compaction within them, especially near feeders and pig arks. Unrestricted access by grazing animals can eventually destroy the wood altogether by preventing regeneration. (See p321.) So an unfenced wood in an area where there are grazing animals will need fencing. This can be extremely expensive, sometimes costing as much as the purchase price of the wood itself, but grants may be available towards it.

Invasive Plants

The only invasive plant to cause a real problem in managed woodland is *Rhododendron ponticum*, an introduced species from the shores of the Mediterranean and Black Seas. Here in Britain it has no natural restraints on its population, either in the form of pests, diseases, herbivores or competition. Nothing can survive underneath it, as it casts a heavy shade and is allelopathic. Trees which survive above it can't reproduce. Left alone in a woodland where it grows well, it will turn that woodland into 100% rhododendron by the time the oldest trees come to the end of their lives. (See illustration, p369.)

It thrives on any acid soil, from the sands of southeast England to the north-west Highlands of Scotland, but will not grow on alkaline soils. If you see it growing in one part of a wood don't assume that it will spread all over, as there may be more than one soil type in the wood.

It's a difficult plant to kill, and control involves a lot of very hard work, even if a herbicide is used to kill the stumps. It could be possible to eradicate it in a wood which has only a few rhododendron plants and is well isolated from any others which could recolonise it. But where the infestation is heavy eradication would be a huge task, and its mycorrhizal fungus continues to release the allelochemical for several years after the shrubs are gone, so replanting or regeneration with other trees is very difficult. Keeping it under control can make a worthwhile project for a group of people whose aim is to preserve biodiversity and who have plenty of energy to spend.

It burns well, but is impractical for charcoal as the branches are twisty and difficult to pack in the kiln. It can be split and sold as kindling, or used to make funky rustic furniture.

Climate and Soil

Frost pockets, exposure to winds – especially salt winds – thin soil, excessive drainage, poor drainage and extremes of pH will all restrict the range of trees that can be grown. (See table, Trees for Timber and Wood, on pp338-340.) A low level of diversity in a wood may not be a sign of unsympathetic management in the past. It may just be that those are the only trees which will grow there, or at least the only ones which will grow well.

Obviously it's very important to be clear about the limitations imposed by climate and soil before making any plans for the future of the wood.

Legal Aspects

If a wood is a Site of Special Scientific Interest, situated in a National Park or an Area of Outstanding Natural Beauty, or has some other conservation or landscape designation there may be some special restrictions on what can be done with it. Other legal restrictions may be imposed by previous grants paid on the woodland, for example under the now defunct Woodland Dedication scheme.

Working With Existing Woods[24]

A. Coppice

Most coppice woods these days are neglected, or 'out of cycle'. There are a number of ways of working with neglected coppice, each of which has it's particular advantages. In many woods it will be appropriate to treat different areas in different ways. The main options are:

- Leave it alone.
  - It's highly unlikely that any wood will suffer from a few years' neglect while you decide what to do.
  - Non-intervention is one option to consider where biodiversity is the top priority. (See pp356-358.)
  - In upland woods, where growth is often extremely slow, shelter for animals and biodiversity are more viable outputs than wood or timber. But fencing will be necessary to allow for regeneration. (See p321.)

- Restart coppicing.
  - In some woods this may be the best option for biodiversity. (See p356.)
  - Is there a need or a market for the coppice produce, and in what quantities? (See p304 for yields of firewood.)
  - Is regular labour input available? (See box, Work in the Woods.)
  - If the local deer population is high and the tree species are palatable to deer coppicing may be very difficult. (See pp314-315.)
  - If grey squirrels make high forest impractical coppice may be a better option.

- Single to high forest.
  - This means cutting all the stems save one from each coppice stool.
  - It can be viable to convert coppice to high forest where there's a good population of trees of timber species with potentially good form, and coppice is not wanted.
  - It can also be combined with coppicing to turn simple coppice into coppice with standards.
  - It's a low input / low output approach compared to planting.

- Plant at low density.
  - This allows the existing stand to be enhanced with more valuable trees, while preserving most of the existing biodiversity.
  - According to the density of planting it can lead to structures which range from coppice with standards to high forest.

- Plant at high density.
  - This converts the woodland to a plantation. (See Planting New Woods, pp321-328.)
  - It should only be considered where the existing wood has a low ecological value and little productive potential.
  - If locally native species are used it's not an option to be scorned.
  - It may be difficult to kill the coppice and regrowth may swamp the planted trees. (This applies to planting at any density.)

WORK IN THE WOODS

Coppicing can be very labour intensive, especially if value is added to the produce. In Hampshire, where most of Britain's hazel coppice is found, the County Council has calculated that good-quality coppice can support 8 full time jobs per 100 hectares. This is something like three times the average for all kinds of farmland in Britain, and ten times as much as conventional forestry.[25]

The potential for rural job creation is enormous, but at the level of the individual woodland creating all that work may or may not be desirable. Some woodland owners will have bought the woodland in the first place because they want to work it. Others may be able to find a woodland worker who's interested in working it. (See Wessex Coppice Group, Appendix D, Organisations.) But where there is no-one keen to take on the hard physical work of coppicing the high forest option is likely to be the best choice. In many cases this will be the most important factor in the choice between coppice and high forest.

Restarting Coppicing

Trees respond well to recoppicing after a surprisingly long period of neglect. A cycle can be restarted in a wood of mixed native species as much as 80 years after it ceased, or even after 100 years according to one authority.[26] Hazel, on the other hand, often dies out after some 40 years without coppicing.

The first step is to obtain or make an accurate map of the wood and draw up a coppicing plan showing which area is to be coppiced in which year. Coppicing the whole wood in one year would not only be very disruptive to the ecology, but would mean all the work and all the produce was compressed into one year rather than being spread evenly over the cycle.

Each area to be cut at one time is known as a coupe, or by one of a variety of dialect names. The number and size of coupes depends on the length of the cycle and the size of the wood. A reasonable cycle length for firewood in mixed coppice is from ten to fifteen years. For hazel grown for craft purposes the cycle is seven or eight years. (Hazel is too low-yielding to be worth growing for firewood.)

In a very small wood it may be necessary to coppice every other year rather than annually to avoid getting coupes which are too small and shady for good regrowth. As a rule of thumb the diameter of the coupe should not be less than 1.5 times the height of the surrounding trees. The trees on the south side clearly have more influence than those to the east and west.

In a larger wood two medium sized coupes in different parts of the wood may be cut in one year rather than one large one. This is beneficial to wildlife, as it keeps the 'mosaic' size small, which makes it easier for creatures to colonise a coupe which has just reached the stage of growth they need. (See illustration on p317.)

The shape of the coupes can be designed to create beneficial microclimates:

- Wind is minimised by having the long axis at right angles to the prevailing wind, ie a north-west to south-east axis in most places.

- Light is maximised by having it north to south, so each side gets at least half a day's sunshine.

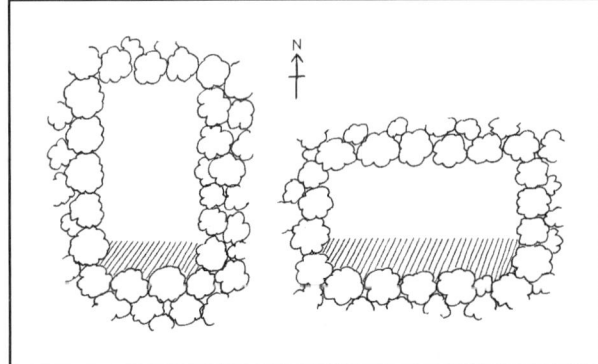

North-south and east-west coupes.

- On a sloping site, having the long axes across the slope will reduce the downward flow of cold air and discourage the formation of frost pockets.

The last also helps to reduce soil erosion, which could be a possibility where ground cover is sparse.

In a neglected coppice some stools will have been suppressed by competition, and some restocking will be necessary, either by layering (see illustration below) or by replanting in the gaps. These can only be done at the time when the wood is coppiced so there is enough light for the young trees to grow.

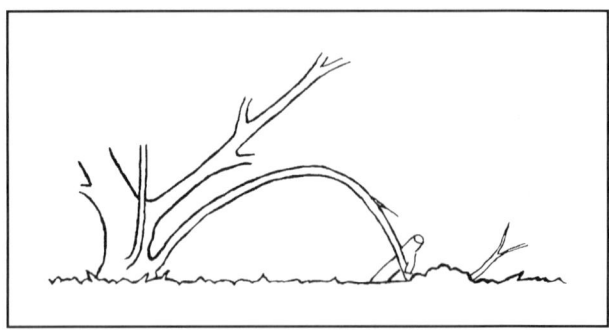

Layering. A young branch is bent over and a part of it buried underground, with the end of the branch re-emerging. A peg helps to keep it in place. The branch will root to form a new plant.

Where the aim is to restore or establish a coppice-with-standards structure, the question arises of how many standards to leave. This somewhat depends on your aims. But if coppicing is the priority the crowns of the standards left after coppicing should cover no more than 5-10% of the ground area. This sounds very little, but the standards will grow fast now they have been released from competition and they will cover a significant proportion of the area long before the coupe is ready for cutting again.

The shorter the rotation the more can be left, but the relative value of the coppice and the standards is critical. Although hazel has a short rotation length anyone wanting to make a living from it in a mixed wood would be tempted to leave no standards at all. (Note that hazel itself does not make a standard tree.)

The standards should be well spaced around the coupe and varied in age. One suggestion for hazel coppice is to leave 50 standards per hectare made up of the following age groups:[27]

20	saplings	0-25 yrs
12	young trees	25-50 yrs
8	semi-mature trees	50-80 yrs
6	mature trees	80-125 yrs
4	ready for felling	110+ yrs

Where deer are a problem there are a number of ways of protecting the regrowth, each with its pros and cons:

- Heaping brash over the freshly cut stools – can cause misshapen growth.

- Fencing the coupe with a dead-hedge made of brash – laborious.

- Fencing with an electric fence – expensive and needs attention.

- Cutting larger coupes – can help when the problem is mild.

- Cull the deer – only works over a large area, not in one wood.

- Raise the coppice to pollards – a drastic measure which greatly increases work and slightly decreases yield; it can't be done with hazel.

- Live in the woodland while coppice regrows, adding value to the coppice produce – may not work if deer are unafraid.

The uses of coppice produce are discussed on pp303-307 above. To summarise, the main uses for the first cut of an out-of-cycle coppice wood are:

- Firewood.
- Charcoal.
- Oak bark.
- Rustic furniture.
- Jean Pain composting. (See box.)

THE JEAN PAIN SYSTEM

Jean Pain lived in Provence and his work was first motivated by the forest fires which plague that area. With the decline of the peasant economy the fires have become a real problem. The shrub layer of the forests used to be kept thin by the grazing of sheep and goats, but now they grow unchecked. When a fire starts, this mass of combustible material is enough to ensure it turns into a devastating blaze which destroys everything. Once the vegetation goes, the soil soon follows.

His idea was to cut brushwood and turn it into compost which could be used to help re-establish trees, while leaving some areas uncut for wildlife habitat. He also used the compost to grow vegetables.

To begin with he made the compost by hand. The raw material was any twigs not more than 8mm thick, always including a mixture of species, some of which must be deciduous. Herbaceous plants were included where available.

The minimum volume for successful composting by this method is 4m³.

The material is soaked in water for three weeks, then built into a heap 1.6m high and 2.2m wide, with the length of the heap varying according to the amount of material. The heap is covered with a 2cm layer of soil, sand or leaf-mould and a rough cover of branches above that.

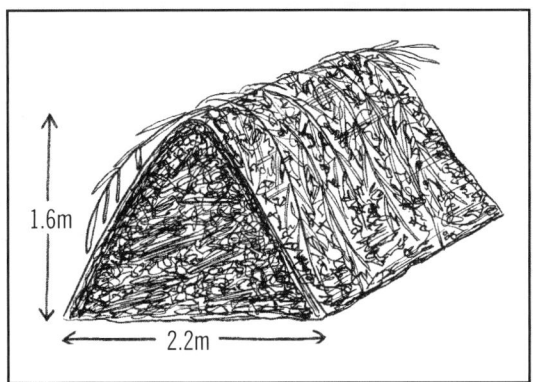

Hand method, cross section of heap.

After three months the compost is ready to use as a surface mulch. Jean Pain and his wife Ida had remarkable yields from vegetables grown with a 7cm mulch of this compost, without any watering after establishment despite the fierce Provencal sunshine. If the heap is turned and allowed to go on composting for a few more months the compost will be sufficiently well rotted for incorporation into the soil. The test to see if it's ready is to take a piece which is still recognisably twig-shaped and squeeze it between finger and thumb. If you can crush it, the compost is well enough rotted to be dug or ploughed in. It can typically take 18 months for the compost to reach this stage.

The next step was to mechanise the process. They started shredding the material so that much larger sticks could be used. For good composting it's important to use a shredder rather than a chipper. A shred has much more edge than a chip. From the shreddings they made a rectangular heap of 50 tonnes of material.

As well as making compost they made use of some of the heat generated by the composting process by running an alkathene pipe full of water through the heap. (See illustration at top of next column). Cold water runs in at the bottom and hot water is pushed out at the top by convection. They used this to heat their house in winter.

The third and final step was to produce biogas from the process. To do this they built a cylindrical heap of 80m³. In the centre of it they placed a drum of 4m³, three-quarters filled

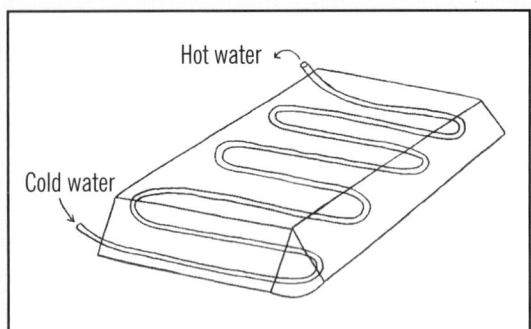

Mechanised method, with hot water production.

with mature compost and topped up with water. As the heap heated up, this mixture produced gas. A pipe led from the drum to a stack of tractor inner tubes which were used to store the gas. The hot water pipe was placed spirally round the drum, and the flow of water could be regulated to keep the temperature in the drum at the ideal level for gas production. By now the hot water was heating a polytunnel for winter vegetable production as well as the house.

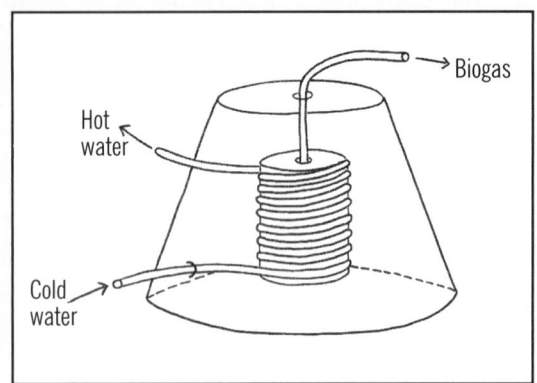

Mechanised method with hot water and biogas production.

In this way they turned what had been a problem – excess brushwood – into not just one solution but a whole series of them.

Note in the diagram at the top of the next column that using the gas to power the machinery that ran the process was not actually done, but it remains as a potential.

As well as costing the process to see that it paid financially, the Pains kept energy accounts. They found that the fuel used to run the machinery was equal to just 12% of the energy output of the system. More importantly, they calculated the total energy input to the system including embodied energy, and this was 26% of the output.

Their work is being carried on by the Comite Jean Pain in Belgium. (See Appendix D,

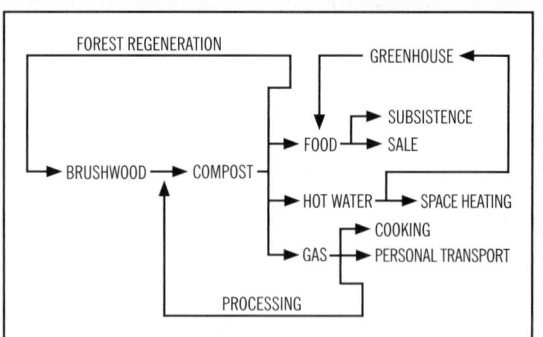

Organisations.) Unfortunately they are concentrating only on the output of compost and not producing hot water or gas from the system. Here in Britain there have been attempts to use 'heat heaps' to provide domestic central heating. They have worked well technically, but find it hard to compete with the low price of fossil fuels. I have recently heard that the method is being used in Portugal with the original aim of reducing forest fires.

Jean Pain composting is the kind of thing you have to get pretty well 100% right if it's going to work. People who want to try it are strongly advised to get the book, *Another Kind of Garden*, which contains full practical details.[28]

As a quick guide, three things you must get right are:

- Use a mixture of material – the remains of a conifer hedge, unmixed, will not work.

- Four cubic metres of uncomposted material is the minimum volume to start with.

- Don't use a chipper – use a shredder or only include material of less than 8mm.

Singling

A possible problem with singling is poor timber quality. Water can collect in the base of a coppice stool and this may have caused rot in the lower part of the singled stem. Also the base of the stem is likely to be curved, and even if it straightens over the years the timber will have uneven stresses in it. Uneven stress is not so important if the timber is to be used in the round.

Many species of tree will eventually single themselves. Oak is a good self-singler and an oak coppice neglected for 100 years can look as though it had always been high forest. Other trees are not so good at it. If chestnut coppice is left too long the stems get over-heavy and start to break off, letting in rot which can kill the tree.

Singled sweet chestnut tree.

Planting at Low Density

This approach has been advocated as a compromise between nature conservation and economic forestry for semi-natural ancient woodland.[29] The planted trees should be those which would grow on that site but which do not regenerate freely of their own accord. In most woods this includes oak, which regenerates poorly within existing woods.

Planting can only be done effectively at the time when the wood is coppiced. At other times there would be too much shade for the newly planted trees to succeed. The coppice regrowth acts as a nurse to the timber crop. Ideally the next coppicing coincides with the first thinning, but often the coppice regrowth will outpace the new planting and an earlier coppicing will be necessary. The final crop is a selection of the best trees, whether they arose from planting, natural seeding or coppice sprouts.

B. High Forest

Almost all plantations in Britain are managed on the clear fell system. A block of land is planted with trees, usually thinned after a number of years, felled when mature, and replanted. It's a relatively profitable system because it makes use of the economies of scale. Large machines can be used to fell and clear large areas, with low labour costs. Management costs are also low because there are very few decisions to be made.

The disadvantages of clearfelling are:

- Soil is exposed to erosion on felling, especially after a dense-shading crop of conifers.

- Wildlife finds it hard to survive the cyclical changes

in habitat occurring over a large area at once. (See illustration below.)

- Mycorrhizas, which are particularly important for trees, may die out over large areas on felling.

- The water table often rises after clearfelling, leaving a waterlogged soil.

- Monoculture of one species is usual.

- Little local employment is generated; the highly mechanised work is often done by contractors from out of the area.

- Landscape continuity is disrupted; plantations which people know and appreciate are suddenly replaced by the moonscape of a clearfell.

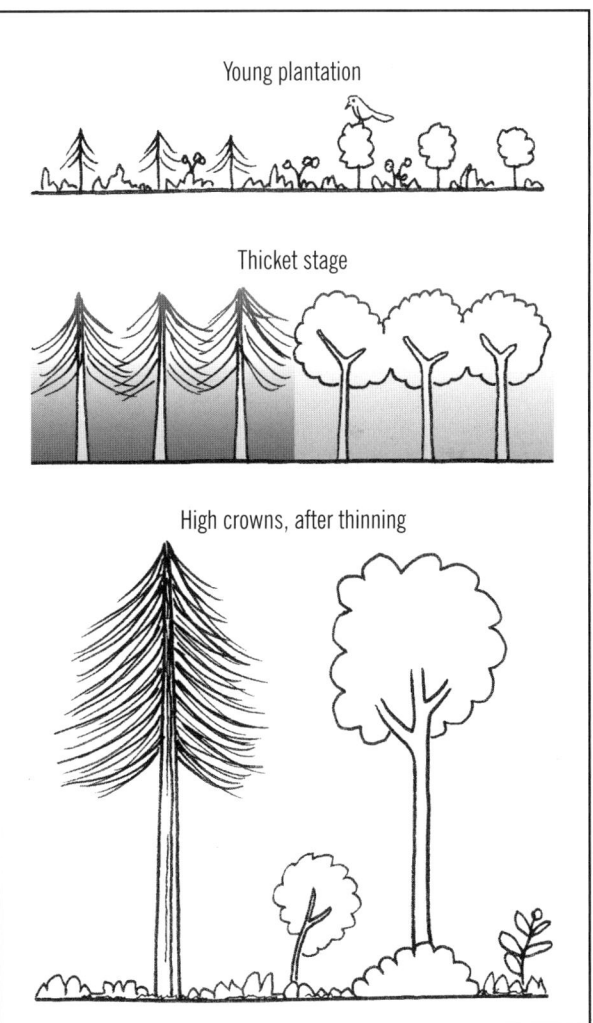

Different species of wild plants and animals can live in a plantation at different stages during the rotation. Where coupes are large the habitat changes over a wide area at the same time, making it difficult for wild plants and animals to find suitable new habitat. Where coupes are small the wild creatures are much closer to suitable new habitat and more likely to colonise it successfully.

Continuous Cover Forestry[30]

There are a number of alternatives to clear felling, which can all be loosely grouped under the heading of continuous cover forestry. The common characteristics of these systems are:

- trees are felled in small groups or individually
- a mixed-age stand is created, ie trees of different ages grow near by each other
- provision is made for replacement, often by natural regeneration but sometimes by planting

A partial felling which doesn't make provision for the replacement of the trees felled is not continuous cover forestry but exploitation felling. It's what happened to many woods and plantations during the two World Wars, leaving them full of trees and shrubs with little economic use. This kind of wood needs remedial treatment before it will start producing again, and the group selection system (see below) has proved a useful tool for this.

In general the advantages of continuous cover forestry are:

- Soil erosion is greatly reduced.

- There is more continuity of habitat for wildlife.

- It's easier for mycorrhizas to pass from old trees to young ones.

- While it doesn't necessarily involve polyculture, the higher level of management required is conducive to more intricate work, including species mixtures.

- The more constant, and higher, labour requirement can lead to steady local employment.

- Cash flow in a small woodland enterprise is more even.

- The more intensive management is conducive to higher timber quality.

- Landscape continuity is greatly enhanced.

The conventional view is that continuous cover systems are usually less profitable than clear felling. Both labour and management costs are higher, as greater skill is needed in felling and the economies of scale are foregone. Also, during the period of conversion from an even-aged stand to the mixed-age stand which is essential for continuous cover, some trees will be harvested before they reach the optimum size. One estimate puts the drop in profit compared to clear felling at 15%.[31]

On the other hand, there is often the opportunity to leave the best individual trees to grow bigger, thus increasing the total value of the crop in the long run.

Also the use of natural regeneration instead of planting can reduce costs and increase the vigour of the trees. Some foresters who practice continuous cover systems reckon they are actually more profitable than clear felling.[32] In Switzerland the consensus is that they are 25% more profitable.[33]

Continuous cover forestry is often practised for the wildlife and landscape values, benefits which accrue to society at large rather than only the woodland owner. The owner can be rewarded financially for these benefits if the wood is certified under the Forest Stewardship Council premium scheme, which rewards more ecologically and socially responsible methods with a premium price. (See Appendix D, Organisations.)

One problem with continuous cover is continuity of management, as the implementation of almost any plan is likely to span the working lives of at least two foresters.

The main physical limitation to the use of continuous cover forestry in Britain is the problem of windthrow. Ours is one of the windiest countries in the world, and many plantations, especially in the uplands, don't get felled by human hand but are blown over by the wind. The trees are then harvested, but the work is much harder than it would be if the trees had been felled, and the trees are less valuable if blown when still young.

Obviously no system of continuous cover can operate in an environment where all the trees could be blown down at some point. It's generally reckoned that there's little scope for continuous cover forestry over 350m above sea level. Even at lower elevations windthrow can be a problem on sites which are especially windy or where root depth is restricted, either by shallow soil or by poor drainage.

Broadly speaking continuous cover systems can be grouped into:

- Shelterwood.
- Individual selection.
- Group selection.

Shelterwood in De Hoge Veluwe National Park, Netherlands. In this case the main motive for using the system is probably visual amenity.

In *shelterwood* systems part of the crop is felled, leaving some of the old trees to provide both seed and shelter for the next crop. This may be done by thinning the old crop very heavily or by felling in strips. The former gives overhead shelter to the young trees, the latter gives side shelter. It's most useful in countries with a very cold winter, such as Norway, and is not much practised in Britain.

In *individual selection* systems trees of all ages grow together in an intimate mixture and individual trees are selected for felling. Only the most shade-tolerant species can be grown, including beech and a number of conifers. It requires the highest input of both management and labour, in terms of both time and skill. In return it gives the most complete soil cover of any system, and is much used on very steep slopes in Switzerland.

Individual selection is not common in Britain, though there is the interesting example of the Hutt-Bradford Plan. This uses a regular grid pattern throughout the forest with each unit being treated in a uniform way. This approach greatly reduces the management time needed.[34]

In *group selection* systems small groups of trees are treated as the unit of felling and regeneration. This probably comes closer than any other forestry system to the structure of the wildwood which existed here in cool temperate Europe in prehistoric times. Regeneration would have occurred after relatively minor windthrow events, rather than after hurricanes or wildfire, both of which can clear large areas at once but do not occur here.[35] (Of native woodland, only the Highland pine forests can burn, and the storm of 1987 was only called a 'hurricane' because we in Britain have never seen a real one.)

Group selection has been used more in Britain than other kinds of continuous cover forestry, and is well suited to small woods in lowland areas. It's suitable both for converting even-aged woods and plantations to continuous cover and for improving unmanaged or scrub woods. It can be used with moderately light-demanding trees as well as shade-bearers. Plantations are suitable for conversion to group selection as long as the trees are not more than 40 years old in the case of conifers and 60-80 years for broadleaves.

Trees are felled in small groups, aiming for a mosaic of groups of diverse ages. The size of groups is typically 0.1-0.5ha, with a minimum diameter of twice the height of the adjacent trees. Light-demanding trees, such as ash, need a relatively large clearing, while natural regeneration of heavier-seeded trees, like beech and oak, need a relatively small clearing to make sure seed reaches the middle. Unfortunately oak is also light-demanding, and if it's included in the mix for a group selection system it may need to be planted.

A possible disadvantage of such small coupes is the relatively high proportion of edge trees with poor form. (See p234.) The poor form of edge trees can be corrected by high pruning. Pruning is rarely considered worthwhile by conventional foresters, but in a small-scale plantation run on a selection system quality rather than quantity is the key to financial success. In addition, pruning is interesting and satisfying work, a task in which the forester develops an intimate relationship with the trees in the wood, and many people would consider it worth doing for reasons other than pure economics.

On the other hand there are also positive edge effects. (See box.)

Group selection lends itself to natural regeneration, though planting allows new species to be introduced if the existing wood has little diversity. Diversity gives flexibility, and it can also help natural regeneration, as

Group selection can be used with coniferous as well as broadleaved trees. Stourhead Estate, Wiltshire. (See also Colour Section, 33.)

THE EDGE EFFECT[36]

The forester JE Garfitt, a pioneer of group selection, observed a marked edge effect in new growth adjacent to older trees. After three or four years, although young trees growing right on the edge of a new group were stunted due to shading, those growing some 10m from the edge were making up to twice as much growth as young trees in the middle of the coupe. There was a progressive decline with increasing distance from the mature trees. The effect is apparently due to a combination of shelter, extra fertility from leaf fall and shared mycorrhiza.

In narrow coupes the edge effect meets in the middle, and the whole area shows enhanced growth. Garfitt recommends narrow shaped coupes aligned north-south to minimise shading. On one site, where a large area was to be felled and replanted, he left narrow parallel strips of the old trees at regular intervals and later felled them to provide access rides. In this way the whole wood benefited from the edge effect.

trees usually regenerate more readily under trees of a different species. But including too many species can make the system hard to operate. A good mix might contain three main timber species, all with similar light requirements, plus a number of secondary species.

Natural regeneration does not happen automatically, and if anything it requires more skill on the part of the forester than does planting. Fellings must be timed to coincide with good mast years or times when plenty of young seedlings are already present. Suckering species – those which produce new trees by vegetative propagation from their roots – are particularly useful. (These are noted in the table, Trees for Timber and Wood on pp338-340.)

Pressure from pest animals is a problem. With small areas of regeneration the number of trees at the vulnerable stage is small in any one locality, so a moderate population of animals will have a major impact. It's generally held that deer, rabbits and grey squirrels must all be controlled for group selection to succeed, though some experienced foresters would disagree.[37] Badgers, on the other hand, can help natural regeneration by burying the seed, as can pigs.

Pigs can be particularly valuable in beech woods, where a thick layer of leaf litter separates the freshly fallen mast from the mineral soil where it can germinate. They will eat as much of the mast as they can, but the amount they miss and bury will be more than enough to provide a new crop of trees. Once the job is done pigs must be kept out of the wood for several years, because they can just as easily root up the seedlings they have sown. Without pigs some kind of mechanical scarification is usually necessary in beech woods.[38]

Extracting Timber

In conventional forestry timber is extracted in whole logs, using heavy wheeled vehicles. This can cause severe compaction, indeed felling in a wet winter can leave a wood looking like the aftermath of the Battle of the Somme. But even mild compaction can take years to heal, and it both seriously impairs future tree growth and makes the soil more vulnerable to erosion. The effect on the biodiversity of the woodland can be devastating. Large vehicles also need large rides and roads, which can be expensive to construct and difficult to fit into intricate group selection systems.

Using cable-crane equipment instead of vehicles can reduce soil erosion during clearfelling by 75%.[39] This does take more time, but the capital cost need not be great if the cable can be supported by pulleys attached to standing trees. This is only an option in selection systems, as there are no nearby standing trees in a clear fell.

Extraction with horses can greatly reduce compaction and erosion. Horse-drawn wheel sledges are now available which make the job as easy as possible for the horse while minimising the log's impact on the soil. Obviously there is a limit to the size of trees which can be extracted by a single horse or a pair. (See Colour Section, 35.)

In many cases the best solution is to take the saw to the tree rather than the tree to the saw. The timber can be planked up in the wood and carried out in loads as small as you like. A wide range of mobile saws are available. Bottom of the range are modified chainsaws, which are cheap, but turn a relatively high proportion of the log into sawdust. More expensive models can be hired by the day.

David and Rob Blair using the portable mill they made from a chainsaw and a ladder.

The Wood Miser, a top-of-the-range mobile sawmill.

Some saws can be taken apart and carried into the wood by hand, and planks can be carried out the same way if necessary. This is particularly worthwhile in woods with a high ecological value and sensitive ground flora. But in any wood it will reduce soil compaction to near zero and do away with the need to construct access rides. A combination of portable saws and horses may be an ideal partnership.

Animals in Woods

The advantages of keeping farm animals in woods are:

• In mast years, pigs can get a significant amount of food from woods.

- They can also help natural regeneration of trees. (See previous page.)

- Grazing animals can get a little food there.

- All animals can get shelter, especially from winter cold, but also from summer heat.

- Animal welfare is better, especially for pigs, which are naturally woodland animals.

The disadvantages are:

- Soil compaction and direct damage to trees can seriously damage the woodland.

- It's more time consuming to look after animals in the woods than in the open.

- Continued grazing destroys the herb and shrub layers and prevents regeneration of trees.

When oak and beech trees have a mast year pigs can, for a few weeks or months, get all or most of their food by free ranging in the woods.[40] But mast is only a seasonal food source, and it's not available every year. Mast years come at intervals of between two and seven years for oak and four to fifteen for beech. The pigs have to be fed some other way the rest of the time.

Pigs are unlikely to do much damage to a wood in the time it takes them to eat up the mast, but only certain woods are suitable for them to live in full time. Firstly, plantations are more suitable than ecologically diverse ancient woods, as the pigs will certainly destroy much of the herb and shrub layers. Secondly, only woods on sandy soils are suitable, otherwise it would be impossible to avoid damage to the trees from soil compaction. In any case the stocking rate should not be too high.

It's more time-consuming keeping pigs in a wood because access is more difficult among the trees. This may make it uneconomic to keep sows and litters in woods, but alright for dry sows, which need less attention.[41] On the other hand the extra cost could be recouped by getting a premium price for high welfare 'forest reared' pigs.[42]

Cattle and sheep usually benefit more from the shelter they can get in woodland than from any food they can find there. Unfortunately the winter, when they need shelter the most, is also the wettest time of the year, when they can cause most compaction. In upland areas the shelter can be very valuable, but it must be used sparingly if the trees are not to be damaged.

Woods can also be used as a food source for grazing animals during serious summer droughts. When the pasture has all dried up, letting them into the wood for a while can be a welcome alternative to feeding them hay. With dairy cows, the possibility of them eating woodland herbs which might taint the milk must be borne in mind.[43]

Giving cattle and sheep unrestricted access to a wood over many years eventually turns that wood into a treeless grassland. Grasses and grassland herbs are uniquely resistant to being repeatedly grazed. Woodland herbs, shrubs and tree seedlings all die if they are nibbled regularly. The herb layer in an ecologically rich woodland can be lost in a few years of grazing, and eventually the woodland itself will disappear as the older trees die and are not replaced. Many upland woods in Britain are unfenced and are being slowly destroyed by unrestricted grazing. Controlled grazing may actually give better regeneration and a more varied ground flora than complete exclusion,[44] but the need for fencing is the same.

Cattle and sheep have free access to this little wood. The ground layer is now reduced to grass, and the mature trees are unable to reproduce. When they reach the end of their lives the wood will be no more.

New Woodlands

When planting a new wood the first thing to get clear is why it is to be planted. Which of the many possible outputs of woodland are expected from this planting, and of these which have the highest priority? A plantation intended primarily for wildlife habitat and beauty will be a very different thing from one intended primarily to supply firewood and building materials – though both woods may well yield all four of these to a greater or lesser extent.

The two main aspects of woodland design are choice of site and choice of species.

Site

On the whole woodland trees are less demanding in terms of soil and climate than farm, orchard and garden crops, and the value of their unprocessed output is less. Thus woodlands have traditionally been sited not so much on land which is good for growing trees as on land which is not much use for anything else. This includes steep slopes, poorly drained soils, infertile sands and harsh upland areas. (The present grant system has somewhat distorted this pattern by paying higher rates of grant for planting on better land, with the specific aim of reducing agricultural surpluses.)

Planting on steep slopes also has the benefit of protecting the soil from erosion. Woodland is a better protection from erosion than even permanent pasture, but only if some kind of continuous cover forestry or coppicing is practised. Clearfelling on a steep slope, especially after a densely-shading crop of conifers which has suppressed all herbaceous ground cover, leaves the soil very vulnerable.

The aspect of a slope can also be relevant. South-

An extreme example of the damage which can be done by sheep grazing on steep land.

These trees may have protected the soil while they were growing, but they have shaded out all the ground cover, and in the year or two after felling the proportion of the soil which has been saved over the rotation can be lost in a few heavy rainstorms.

facing slopes are valuable for crops which benefit from the warmer microclimate and for housing. But trees need moisture more than they need warmth, so north and east slopes, which dry out less in summer, can be better for them.

Woodland usually needs less attention than other forms of land use, so a site far away from the centre of human activity is suggested on zoning grounds. It's interesting to note that ancient woods in fairly flat country are frequently on parish boundaries. However, if some intensive activity, such as forest farming or on-site craft work, is planned for the wood it may be worth considering placing it nearer to hand.

The existing ecological value of the site is an important factor to consider in any land use decision, but it's particularly relevant to planting trees. In many places introducing trees is an unequivocal benefit to the local ecology, but where there is already a rich ecosystem of Sun-loving herbs, such as a herb-rich meadow, planting trees will lead to a reduction in biodiversity. (See pp358-359.)

Edge

In order to get the full range of multiple outputs which are possible from a woodland it needs to be integrated with the rest of the landscape. This can best be done by planting it so that there is plenty of edge between it and other land uses. Large blocks of woodland have very little edge in proportion to their area. Small woods, and especially linear woods, have much more.

The advantages of woods with more edge are:

- Microclimate. A linear wood is in effect a shelterbelt. Small blocks of woodland near buildings can have many beneficial microclimate effects. (See mind map on p83.)

- Food. Growing fruit and nuts in woods is discussed in Chapter 9. (See pp234-235.) But suffice to say that the woodland edge is one of the best places for it.

- Fodder. As we have seen in Chapter 10, grazing animals can be fed tree forage most easily by pollarding trees which grow where they graze. In a woodland this is most easily done on the edge, firstly because trees on the edge produce more leaves than those in the interior (see illustration on p234), and secondly because it avoids the problems of animals entering the wood.

- Fish forage. Leaf fall from waterside trees can provide the bulk of primary food production in fresh water.

- Soil fertility. Leaf fall can contribute plant nutrients and organic matter to adjacent soils. (See under Agroforestry, p277.)

- Soil Erosion. Tree belts are the key to reducing wind erosion. They can also reduce water erosion if they are placed across the slope.

- Access. Extraction of woodland produce is clearly easier if the distance to the middle of the wood is reduced.

- Productivity. There is evidence from work on short rotation coppice (SRC) of a marked edge effect, with stools on the outside rows giving up to 60% greater yield than the mean.[45] The main reason for this edge effect is believed to be the extra light the edge stools receive. It's greater in older, larger stands of SRC, which suggests that it's no less important in other kinds of coppice, which have larger stools and a longer rotation.

The disadvantages of edge are:

- Quality. High-quality sawlogs are produced by growing trees close together, so they grow straight up towards the light and lower branches are suppressed. (See illustration on p234.) The branchy trees which grow on the edge have little value for timber. Coppice wood for craft purposes also needs to be straight, and poles growing on the edge usually have to go for firewood. This problem can often be avoided by planting the edge with coppice for firewood, pollards for animal fodder, or fruit and nut trees.

- Fencing. Woodland must be permanently fenced against farm animals, except in purely arable areas. This is expensive, and the more edge the wood has the more fencing is needed. A hedge which remains impervious to farm animals year after year without any help from fencing is the dream of many permaculturists, but is almost impossible to achieve in practice, especially in the shady conditions of a woodland edge.

One way to minimise the cost of fencing is to establish linear woods by thickening existing hedgerows, and doing so when the hedge needs re-fencing anyway. In this way woods can spread steadily over the countryside over a number of years. This is an example of what's sometimes called 'rolling permaculture'. A second advantage of this approach is that it works with the existing grain of the countryside, developing what is there rather than replacing it.

In many cases the best balance between the advantages and disadvantages of small and linear woods may be had by designing a mixed pattern, with some large to medium blocks of woodland lying in a matrix of linear woods.

Species

> The selection of species to achieve objectives and to match the site and other factors is probably the single most important decision to be made for success in forestry.
>
> Cyril Hart[46]

Although it's not unknown for different authors to flatly contradict each other on the site preferences and qualities of different trees, it's well worth while spending time studying the written information which is available.[47] The table, Trees for Timber and Wood, on pp338-340 gives a pointer to the characteristics of the principal productive species, and can suggest which species are worth investigating in detail for a particular planting plan.

Site Factors

Where soil and climatic conditions are good any trees will grow well. Just because a species will tolerate certain adverse conditions it doesn't mean they won't grow on more favoured sites. Many willows, for example, are quite happy on well drained soils as long as they are reasonably retentive of water.

Where conditions are generally unfavourable for trees the choice is more limited. Most trees will grow almost anywhere after a fashion, but to produce a useful crop trees must grow well. In an intensive situation like an orchard it may be worth investing a lot of energy to improve the soil and microclimate, but in a woodland, where the value of output is much less, this would not be worthwhile. So the trees need to be well tuned to the local conditions.

Although written information and advice from consultants are valuable, it's equally worthwhile to look and see which trees are already growing well in the locality. When doing this a few points should be borne in mind:

- Compare sites with similar microclimates and soils. A neighbouring site with different conditions may tell you less than a more distant one with a similar environment to the proposed planting site. Soil type and microclimate can change significantly over small distances, especially in hilly country.

- Note how well the trees of each species are growing rather than how many there are of them. People have often planted unsuitable species on a large scale by mistake.

- In ancient woods look at the coppice rather than the standards. In the past oaks were almost universally favoured as standards, regardless of whether they were particularly suited to the site. A mixed coppice layer usually reveals the spontaneous local mix, whereas pure hazel or chestnut may have been planted.

Specific areas within a site may have significantly different conditions and require a different species mix from that which is planted overall. In frost pockets, for example, only frost tolerant trees should be planted (see F on the table, Trees for Timber and Wood on pp338-339.), and in areas which stay really wet during the growing season only willows, alders and possibly lodgepole pine will do.

In exposed areas edges may need to be planted with exposure-tolerant trees (X or x on the table). But paradoxically trees in the interior may need to be more windfirm than those on the edge. This is because trees on the edge are exposed to wind and thus develop a firm root system as they grow, while trees on the inside do not. If the plantation is subsequently worked on the group selection system these interior trees may find themselves on an exposed edge, and unprepared for it by their previous life experience. They are then quite likely to be windthrown.

In very exposed areas at high altitude it's difficult to establish most broadleaved trees. The very toughest broadleaves are: birch and rowan among the natives, and Norway maple, grey alder and sycamore among the exotics.[48] In areas where a wider range of native broadleaves could survive but are hard to establish, a crop of a hardy conifer can provide the necessary shelter. (See illustrations in next column.) The choice of species and provenance of the conifer is critical, and the advice of a skilled local forester should be sought.

ORIGIN & PROVENANCE

Exotic trees, especially the commercially grown conifers, often have their origin and provenance quoted. The origin is the place where the strain was originally introduced from, while the provenance is the place where the seed was actually grown. Choosing the ideal origin and provenance for a particular site is a somewhat technical matter.[49]

In coastal areas salt-tolerant trees must be chosen (Sa on the table). How far inland the salt-affected area extends will depend very much on the local topography. Observation of which trees are growing in similar situations nearby, and how well they are growing, is of particular importance here. (For coastal windbreaks, see p89.)

Tolerance of pollution is complex as each species has different reactions to different pollutants. The table on p332 must be taken as a rough guide only. Where air or soil pollution is a possible problem professional advice should be sought.

With such long lived plants as trees the question of global climate change arises. But the answer is not simple. Do we plant heat-loving trees on the basis that

Conifers can be useful to fulfil a function which would be difficult to achieve with broadleaves. Here, on the Isle of Erraid in the Hebrides, a plantation of lodgepole pine has been established on a site lashed by strong salt-laden winds. Native trees can survive here but find it almost impossible to get established on open ground. Even the hardy lodgepole is at its limit: the most seaward trees have failed to survive, and tree height slowly increases as you go inland, each tree gaining a little shelter from its seaward neighbour.

Clearings were left in the conifers, and planted with native broadleaves when the conifers were tall enough to give shelter. When the broadleaves are well enough grown to provide shelter themselves the conifers can be progressively replaced with broadleaves. The broadleaves will not produce an economic crop of timber, but will provide much-needed shelter to a small settlement and increase biodiversity.

the British climate will get warmer, or cold-loving ones on the basis that the North Atlantic Drift will change course? (See p71.) The best solution is to plant a wide range, mixed intimately. Those which subsequently do badly can be removed as thinnings.

Outputs

When considering outputs of timber and wood, some attempt must be made to predict the quantities of produce which will be needed so as to determine the numbers of each tree to plant. This can be difficult, given that some trees have a rotation of 120 years or more, and coppice can last for many centuries. The amount of wood needed to heat today's house, for example, may be quite superfluous in a future of energy efficient houses which only need supplementary heating on a few days in the year.

Nevertheless we probably won't be far wrong if we

SHORT ROTATION COPPICE[53]

Strictly speaking short rotation coppice (SRC) is defined as any coppice rotation of under ten years. In practice the term is used to describe willows or poplars grown on a rotation of around two to five years. This is almost more an arable crop than a form of forestry. Most SRC is grown by farmers, using modified farm machinery, on fields which used to grow arable crops and grass and which may do so again when the coppice reaches the end of its productive life. This may be some 20 years.

This willow windbreak, pictured just three years after planting cuttings, gives an idea of the growth rate which is possible with this remarkable tree.

Apart from their high yield (see p304) willow and poplar have a great advantage in that they are grown from cuttings directly in the field, with no labour-intensive nursery stage. The first harvest is from three to six years after planting, depending on the rotation length.

Weed control at planting time is crucial and herbicides are routinely used, though black plastic mulch or weeder chickens are also possibilities. Once the coppice is established chemical use is low. Mixtures of varieties are always used to reduce the incidence of disease. Manures are not necessary, but SRC is one way to use sewage sludge which may be unsuitable for use on food crops due to contamination with heavy metals.

At present SRC is just moving out of the experimental stage onto farms. The main interest is in energy uses, either burning the chipped wood directly or generating electricity from it. Energy efficiency ratios are good, despite the high level of mechanis-

ation used to handle such a bulky product. There are also bulk markets for pulp, chipboard, mulch etc.

Willows will grow very well on wet land, and there are large areas of wet, rushy pasture in western Britain and Ireland which would yield more value producing energy than producing meat. But on wet ground harvesting is problematic. Although the transpiration of the trees may dry the ground out somewhat, and the roots form a mat which helps to support machinery, soil compaction is likely to be a problem. Also, in some cases wet meadows may be havens of biodiversity which would be lost if coppice was planted.

High rainfall and poor drainage are not necessary for a good yield and SRC can be grown on better-drained land, but here it competes for space with arable crops which have a higher value of output. Steep land is not suitable for the machinery involved.

However there's no reason why SRC should not be grown on a small scale for domestic production, with much of the work done by hand, perhaps assisted by horses or light machinery. Central heating systems can be run on woodchips, but chippers and woodchip-burning boilers are expensive. Masonry stoves, which can burn the sticks whole, need not be expensive and may be the best option. A fraction of a hectare of coppice could heat a small house with a masonry stove. (See p304.)

Willow shoots can also be used for basketry, hurdles, living willow sculpture and bank stabilisation, and any of these may be taken as a supplementary output, or even the main one from a small SRC plot. Woodchips from SRC can be used for mulch, animal bedding and equestrian surfaces, all of which may be worthwhile in small quantities.

Small-scale growing lends itself to integration with other crops, and this is the best way to get the maximum benefit from growing it. This is discussed in Chapter 9. (See p250.)

As it's such a bulky and high-yielding crop SRC is best sited relatively close to the point of use. As a commercial crop it is only feasible if there is a power station or district heating scheme in the locality, or a good market for woodchips. When grown on a smallholding or farm for domestic use the difference in transport distance from one part of the holding to another may not be significant, and its siting may be determined more by the needs of the intercrop or by soil type than by zoning.

assume that all woodland produce will be much more in demand in the future than it is now.

In addition to the outputs discussed earlier in this chapter (see Multiple Outputs, pp300-309), some trees have a beneficial effect on the other trees in the plantation. Broadly speaking these are nurse trees and soil improvers.

A nurse species is a component in a mixture which helps the main crop trees during their early years, by means of:

- giving shelter
- suppressing competing vegetation
- in some cases, providing nutrition

The nurse trees are removed when the plantation needs thinning.

Conifers may be particularly good at the first two of these functions. They are also more usable than broadleaves as thinnings, while broadleaves are more valuable as the final crop. However it must be said that the first thinning of conifers, of fencing stake size, will be treated with copper-chrome arsenate before use, with the possible exception of the more durable species such as larch. (See p306.) Conifers should not be used on semi-natural woodland sites where there is a diverse ground flora.

Alder and other nitrogen fixers can provide nutrition. An example of alder nurse trees with walnut as the main crop is given in Chapter 9. (See p219.) A study in Italy found that interplanting oak with Italian alder increased the height, girth and quality of the oak compared to an oak monoculture. The quality improvement was due to better suppression of lower branches.

All trees cycle organic matter and nutrients to the soil with their leaf fall, but there are important distinctions between species which determine whether they have a positive or negative effect on soil fertility:

- Conifers tend to increase soil acidity, although the effect may be small over the time of a single rotation. Oak and sweet chestnut have a slighter acidifying effect, but this is only likely to be significant on light soils which tend to acidity.

- Birch, which grows well on acid soils, is particularly effective at raising the soil pH, as well as the levels of calcium, phosphorous, nitrogen and earthworms, even to the point of turning a podsol into a brown earth.[50] The deliberate inclusion of some birch in conifer mixtures can prevent podsolisation and lead to higher yields. Work in Finland has shown that including 15% birch in Norway spruce plantations leaves the spruce yield unchanged with the extra 15% of the birch yield as a bonus on top.[51] It has also been observed to increase the regeneration of other species.[52]

- The leaves of broadleaf trees with a high tannin content – such as oak, sweet chestnut and especially beech – decompose slowly on the woodland floor and can leave a rich, deep humus layer for the following crop. Continental foresters sometimes plant a crop of beech with this intention. What a wonderful attitude to forestry – and how sorely lacking on this island!

- Broadleaves with more rapidly decaying leaves – such as ash, limes, willows and poplars – are more suitable for adding fertility to adjacent farmland.

- Nitrogen fixers may be expected to have higher levels of nitrogen in their leaves, as well as transferring nitrogen to neighbouring trees by root turnover and shared mycorrhizas.

Diversity

The level of species diversity possible on any one site may be constrained by climate and soil. A planting plan consisting of a few species which will almost certainly do well is likely to lead to a healthier and more successful plantation than one filled with marginal trees merely for the sake of a high species count.

A distinction can be made between major and secondary species. Major species should be carefully chosen for specific purposes, and with a very clear idea of how they will interact. One may be the final high quality crop and another the nurse crop, but if all the main species are to be grown to maturity together they must have compatible light requirements. If the aim is to develop coppice with standards some species will be more suitable for coppice and others for standards. Major species may make up 60% to 90% of the total trees.

Minor species can be chosen in a more experimental frame of mind. Even in a primarily commercial plantation some of them may be more for wildlife, aesthetic or curiosity value. Some may be for small volume uses, such as lime for woodcarving.

In all but the smallest plantations it's likely that different areas will be planted with different trees. There may be differences of microclimate and soil, and it may also be advantageous to separate functions. For example an area of firewood coppice for home consumption may be distinct from an area of commercial high forest, though each of these can still be a polyculture. Some trees are bad neighbours and need to be grown as a monoculture. Chestnut coppice, for example, is so competitive that other trees can rarely survive in it. Only fast growing trees like birch and alder have a chance, and then only if the coppice is cut on a short rotation.

The question of whether to use native species and varieties is discussed in Chapter 12. (See pp368-371.) As a rule of thumb it's only worth growing an exotic species if it can do something which no native can. For example: sweet chestnut is an exotic and not a good wildlife tree, but it's hard to beat for the production of durable fencing stakes; while exotic conifers can produce an economic crop in severe upland conditions, or help to establish broadleaves in difficult situations. In fact a mixture of broadleaves and conifers can develop just the kind of network of beneficial relationships we look for in permaculture. (See box.)

Exotic varieties of native species may not grow as well as locally native varieties. By one account they can have half the growth rate and twice the mortality of local varieties.[54]

Establishment

It's not always necessary to plant trees in order to establish a new wood. Natural regeneration can be successful under some conditions. Where the existing ground vegetation is sparse trees can get established immediately, but where there's a dense grass sward it may take years or even decades. Where there are no seed parents nearby, as in many urban sites, the new trees will almost all be very light seeded species such as goat willow and birch.

Young trees, whether self-seeded or planted, need protection from browsing animals, which may include deer, rabbits, hares, voles or farm animals. The choice whether to fence the whole plantation or to protect individual trees with tree shelters depends both on the animals in question and the size and shape of the plantation. A fence will be more economical in a large compact plantation, and individual shelters in a small or linear one where the proportion of edge to area is higher.

Equally important is protecting young trees from grass competition. Grass is the worst enemy of young trees. Older trees can co-exist with it quite

happily, but young ones can make little or no growth for years if planted directly into a grass sward. (See illustrations, below.) The exact reasons for this are not fully understood, but allelopathy and competition for water both seem to play a part. Mowing the grass only increases the competition, as the grass is kept young and regrows vigorously after mowing.

A hedge of mixed native species, planted three years before the picture was taken. Before planting the grass was eliminated by a black plastic mulch (recycled silage cover), which was in place over the whole growing season. Note the garden fork for scale.

The same species mix, planted four years before the picture was taken, but without mulch. The same garden fork gives the scale.

Slow growth in the initial years may not be a problem if the trees are being grown purely to increase biodiversity and you're not in a hurry to see the fruits of your labour, though there may be some losses if the first summer after planting is dry. But in a productive woodland a good vigorous start to growth is important.

Mulch kills the plants under it by denying them light, and black plastic is the most effective material. Plastic of other colours is no use at all. For trees planted at normal woodland spacing (in contrast to the closely-spaced hedge plants in the illustration) a spot mulch round each tree is the most practical method. Commercial mulch mats of porous black plastic are designed to allow rain water through but exclude light. (See illustration, p279.) If recycled black plastic sheet is used it should be slightly dished in profile so as to deliver the rain which falls on it to the tree.

A biodegradable mulch of old newspaper or cardboard covered with woodchip or some similar material is an alternative. But it will need replacing, perhaps annually over the three of four years that mulching is needed, which means it's only practical on very small areas of planting. Woodchip alone is not very effective as the grass quickly grows up through it. High-nutrient mulches, such as manure, are unnecessary for woodland trees. They can even be counterproductive, as they soon rot down and the grass and other herbaceous vegetation respond to the extra fertility more than the trees, and then slow down the growth of the trees.

Whatever kind of mulch material is used, the diameter of a spot mulch must be at least 1m and preferably 1.2m.

Bracken, brambles and gorse can all compete successfully with grass, and are better companions for young trees – as long as they don't overtop the trees.

Bracken can swamp the little trees in autumn when it dies down. If trees are planted in bracken it's advisable to put a tall stick in the ground by each one so it can be located when the bracken is tall, then you can go round in autumn and clear away any bracken which is draped over them. (Bracken is allelopathic, but the effect on trees does not seem to be critical.) Bramble and gorse should be cut down to the ground when the trees are planted and cut back again if they grow taller than the trees. This should not need doing more than once, if at all. Given that extra bit of work, these plants are an excellent biological resource, doing away with the need for mulching. Gorse will in addition fix nitrogen, which benefits the trees.

Grants

There are many grants available for tree planting, usually calculated to cover 50% of the costs. But some of them are so generous that they can cover the full cost of establishing a new woodland, including design and labour costs, and still leave the landowner with cash in hand. Some include annual payments for income forgone in the initial years of the plantation. There are also grants for natural regeneration and fencing of woodland to keep out grazing animals.

The grant system is complex and changes frequently. Information on what is available can be had from the Forestry Commission, County Councils and the official nature conservation bodies: the Countryside Council for Wales, Scottish National Heritage and English Nature. There are also consultants who specialise in advice on grants – the subject is that complex.

Not surprisingly, there's a catch in it. The specifications that have to be met in order to qualify for a grant are strict, and they often run counter to what a permaculturist might want to do. To some extent this is because a major aim of the grants is to take land out of food production to reduce agricultural surpluses. Thus food-bearing trees are excluded – except for walnut, which is often grown as a timber tree, and natives like crab and wild service – and premium grants are paid for planting on the best agricultural land. There is also a financial disincentive to plant coppice. Since it brings in earlier returns than a timber crop the payments for income forgone are not paid on it.

Accepting these kind of restrictions can turn a really useful and innovative permaculture design into just another badly sited broadleaved plantation. I would strongly recommend anyone who is thinking of planting woodland to make your design first, before even looking at the grant specifications. Then, if you find you can get a grant for what you want to do, that's excellent news. If you find that accepting the grant money would take the guts out of your design, I would try to think of another way to finance it.

This may involve accepting the lower rates of grant or even managing without any grant at all. Spreading the planting over a number of years can make it more affordable, and may mean you can manage without any paid labour. Involving the local community can work really well. I've done this myself, and got a new wood planted in a single weekend which would have taken me weeks on my own. Everybody likes to plant a tree!

URBAN & COMMUNITY WOODLAND

Most of the information in this chapter so far is equally applicable to urban woods as to rural ones, and to woods where the community at large are involved as to privately owned ones. Yet the situation in towns is different, both in terms of the physical environment and the social context. Here we look at the distinctly urban aspects of trees and woodlands, and at community aspects, which may be relevant in the country as well as in the town.

Urban Trees and Woods

In rural areas we think of forestry in terms of continuous blocks of plantation or woodland, but urban forestry is generally taken to include all urban trees, if only because such a large proportion of trees in towns are in streets, gardens, parks and other dispersed situations.

Outputs
All the outputs of trees and woodland discussed earlier in this chapter (see pp300-309) apply to urban trees as well as rural ones, but the emphasis is different.

The non-physical outputs such as recreation, education and aesthetics are more important, while direct outputs such as wood and timber are less so.

This distinction doesn't apply equally to all urban woodland, or to all parts of an urban wood. People tend to use the edges of urban woods heavily, especially near car parks and housing, but rarely visit the interior. The intensely used part of a wood is best treated as an amenity area, with public safety and aesthetics as the main management aims. The less visited parts can be treated as non-intervention nature reserves or as productive woodland or plantation, much as they would be in rural areas.

Selling off timber as raw material makes even less sense in the city than it does in the country. (See p301.) There are a number of reasons for this:

- As blocks of woodland tend to be smaller the difficulty of making up the minimum 20 tonne load can be greater.

- The costs of urban forestry are higher, and adding value will do more to offset them.

- Cities are where most timber products are used, so transporting it away to a sawmill and back again makes no sense.

- Some city dwellers would like to earn their living in a way which takes them closer to the Earth, without necessarily moving to the country.

- Many others would like to practice woodland crafts in their leisure time as a way of reconnecting with nature.

In most cases the owners of urban woods are the local councils. They may add value to the produce in house, both to provide their own needs for building and fencing materials and so on, and for sale. Or they may sell the timber and wood to self-employed woodworkers and small local businesses. Although the second option is still a sale of raw material it's likely to be more profitable than selling in bulk to a merchant.

A network of purchasers may take some time to build up. But any council with productive woodland should at least consider planking and seasoning their timber rather than selling unseasoned roundwood or standing timber.

As for non-timber produce, the potential for human food is considerable. Anywhere where free-standing trees are grown, in gardens, streets, parks and squares, they can be fruit or nut trees, giving their yield within hand's reach of the consumers. This is discussed in Chapter 9. (See pp236-237.)

Animal fodder from urban woods may be restricted to bee forage, but this output should not be ignored in designing new woods.

Existing Trees

As a general rule preserving old trees is more worthwhile than planting new ones. This is true ecologically, as older trees tend to support more wildlife. (See p355.) It's also true from the point of view of human amenity, as it will be many years before a freshly planted tree will have the same effect on the visual landscape and microclimate that an existing mature tree has. As well as the physical benefits of mature trees there is the important matter of local character and distinctiveness, which trees often contribute a great deal to.

Street trees are under a particular threat these days from the installation of underground services, such as cable television. Trenches which used to be dug by hand are now dug by machines, which make short work of tree roots. Trees rarely die immediately from having a major root cut through. But they are weakened, and may die within a few years. Or they may become susceptible to disorders they would previously have shrugged off, and degenerate to a state where they're considered dangerous and are removed.

There's a voluntary code of practice for this work which states that hand digging should take over from machine whenever the trench reaches a tree root. But it's not kept to, and it's hard to get agreements from the operators because this work is done by a mass of small sub-contractors. This is a situation where direct action on the ground by local residents is sometimes the only real solution. When all else fails the only remedy is to get into the ditch and stand over the threatened roots while someone else phones the local council.

Tree roots can also be damaged by compaction, with similar results. This is especially a problem when an area with existing mature trees in it is redeveloped. Roots growing in previously uncompacted soil can be killed by having building materials stacked over them or by being repeatedly driven over by dumper trucks. An area around each tree of 25% more than the crown diameter should be fenced off and kept free of all materials and vehicles.[56] These fences are often put up,

Added value can be a community activity as much as a commercial one. Volunteers making hurdles at a community woodland near Bristol.

but in the hurly-burly of a building site they tend to get breached and ignored.

Other causes of tree damage during development are: trenching (as above), water table changes, oil leakage, hard surfacing and changes in level. The last can be more a matter of poor design than poor on-site working practice. Exposing roots to the air or covering part of the trunk with soil are both harmful, as the root or trunk is exposed to an environment it is not adapted to cope with. All these things can lead to gradual deterioration and death in a few years.[57]

Acting as a link between local authorities and building contractors is one way in which ordinary people can contribute to preserving the ecology of their locality, either as tree wardens or more informally. (See Community Action, p334.)

Woodland is generally under less threat than individual trees in streets and housing estates. Most urban woodlands are well able to maintain themselves by natural regeneration, perhaps assisted by a little group felling.

PRUNING

Tree surgery is a skilled and potentially dangerous job, and best left to professionals. But many of us often have occasion to prune a few branches, either because they're dead or diseased or to keep the tree to the appropriate size or shape. The ideal way to prune has been established by Dr Alex Shigo, an American arborist, by careful examination of trees which had been pruned in a number of different ways.

- The first step is to take the weight off the branch by means of the first and second cuts, as shown in both the illustrations. Large or awkward branches may need more than two preliminary cuts.

- If there is a branch collar (as in illustration a, right), the third and final cut should be made just outside the collar, as near to it as possible without cutting into it.

- If there's no branch collar (as in illustration b), the final cut should be made just outside the branch bark ridge, at an angle to the main stem which mirrors that of the ridge.

- The final cut is best made in two stages, a small undercut from below, followed by the main cut from above. This avoids the bark below the cut tearing away as the cut is completed.

This kind of cut preserves the special layer of bark cells at the base of a branch which are able to grow and cover the wound. A cut flush with the line of the trunk removes these cells, and a cut further out along the branch leaves a stump which the new bark growth can't cover. In either case pests, disease and rot can enter.

A clean cut reduces the chance of infection getting in before the cut grows over. A sharp bow saw is ideal, with a sharp knife to tidy up any rough fibres. Wound paints are unnecessary if the cut is made correctly, and useless if it isn't.

However carefully it's done, pruning is traumatic for a tree. It's not possible to give general

a) *Natural pruning position on a branch junction with collar.*

b) *Natural pruning position on a branch junction with no collar.*

rules about how much of a tree can be removed without harming it. Every tree is an individual, and it's best to look at it carefully and tune into its needs before taking up the saw. As a rough guide I never remove more than a third of a young tree's canopy in one year.

Even in quite heavily used areas people can be gently guided away from areas of regeneration by suitable path design. The desire to plant trees, often fuelled by the fear that nature can't manage without some help from us, is usually best resisted in a healthy woodland.

Young trees need as much attention as old ones, and very often they don't get it. The budget for planting trees needs to be followed up by a modest allowance for maintenance in the few years after planting, when young trees are especially vulnerable. But all too often it's not. In the voluntary sector too, it's much easier to get enthusiasm for planting trees than for going round the following summer to give them a little care.

Much less maintenance is needed if small transplants are used. (See box, A Subtler Approach, p333.) But larger trees are especially vulnerable to drought. Checking that mulch mats are still in place is well worth while, and, if the first summer is a very dry one, even watering the trees.

One of the commonest, and certainly the most visible, cause of death in the first few years after planting is failing to loosen or remove the plastic ties which hold standard trees to their stake. Personally I always carry a clasp knife and one of it's best uses is to save young municipal trees from strangulation.

Tree Planting

Tree planting in urban areas tends to be both more expensive and less successful than it is in the countryside. The effects of human activity, both in the past and in the present, make the urban environment a difficult place for trees to grow and thrive. Matters which may need extra attention when planting on an urban site are summarised in the mind map below.

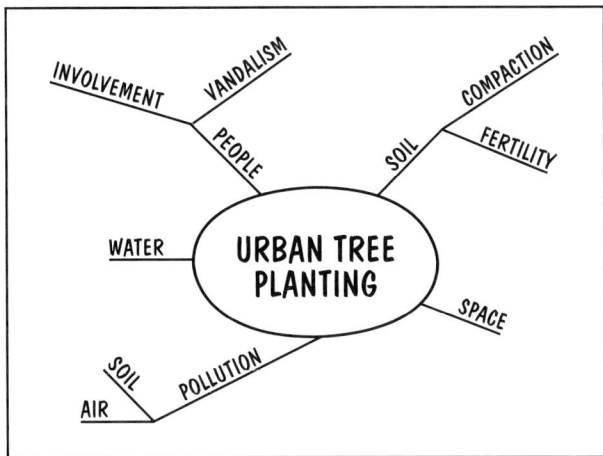

Soil Compaction. Most urban soils are compacted, some severely, and poor drainage is common. It's essential to correct compaction before planting, both to improve drainage and aeration and to allow sufficient root penetration to give the mature tree secure anchorage. In extensive areas where overall treatment is possible, deep ripping with a tracked bulldozer, to at least 50cm and preferably 75cm, is needed. Agricultural machinery is rarely up to the job. If at all possible ripping must be

Planting trees at Filton, Bristol. This project was assisted by BTCV, who can show community groups what to do, lend them tools and even provide a small start-up grant. (See Appendix D, Organisations.)

done in late summer or early autumn when the soil is at its driest, so as to shatter rather than smear the soil. In more restricted spaces, such as roadsides, a large, deep planting hole must be dug for each tree.

Soil Fertility. The soil may be low in nutrients. On sites without topsoil the limiting nutrient is usually nitrogen, and planting nitrogen-fixing trees is much more effective than applying fertiliser. Not all the trees need to be nitrogen fixers, but they must be well mixed with other kinds so that their roots can intermix. Where all nutrients are low sewage sludge can be used. A general account of urban soils is given in Chapter 3. (See p62.)

On soil-less sites overall topsoiling is counter-productive. It gives an advantage to herbaceous plants, including grasses, which can greatly restrict the growth of young trees. (See p327.) Backfilling a planting pit with imported topsoil or compost also has its problems, notably that the tree roots may be reluctant to grow beyond the improved soil of the pit, leading to an un-stable and drought-prone tree. The best solution is usually to use fertiliser containing the nutrients which soil analysis shows to be deficient, applied to each tree. Species choice is particularly important on sites with no real soil. (See table, Trees For Reclaiming Soil-Less Land, overleaf.)

Soil Pollution. This is always a possibility, especially on landfill or ex-industrial sites, and it may be severe enough to affect tree growth. Heavy metals are the most likely pollutants. Testing for pollutants and knowing what to do about them are professional skills, but the history of the site will suggest whether pollution is likely and if so which pollutants should be tested for.

Recent trials by the Forestry Commission suggest that some tree species are more tolerant of heavy metal pollution than others. Eight species were included in the trials and they ranked as follows, in descending order of tolerance:[58]

Birch
Poplar
Sycamore and ash
Willow
Alder
Japanese larch
Oak

city areas, to give maximum impact to the limited opportunities for tree planting in a crowded environment.

Pollarding is an option for large species planted in confined spaces. It has the added advantage of providing a useful output of poles or woodchips. It can be a harmonious way of working with trees, but it must be started when the tree is young, preferably

Air Pollution. This has shown a steady decline in almost all its forms over the last decade, but it can still be a significant problem on some sites. These days road traffic is the main source, but there are still some industries putting out toxic fumes. Some trees are more tolerant of air pollution than others, but different species and varieties are tolerant of different pollutants. The table, Trees for Timber and Wood, on pp338-340 and the table below note which trees are generally tolerant, but these are generalisations and specialised information is needed where specific pollutants are involved.

Space. Lack of space is an important consideration for trees in streets, gardens and other confined sites. In recent years there has been a trend away from the huge limes and London planes towards smaller trees. Many of the latter are highly ornamental, including flowering cherries, crab apples and cherry plums. There is an enormous number of species and cultivars to choose from, with a range of sizes and growth patterns. On the whole the larger, often native kinds are more suitable for tracts of woodland and urban fringe situations, medium sized ones for the suburbs, and small decorative species in inner

The lime is a beautiful tree but also one of the biggest which grows in temperate climates. Not many urban situations have space for one.

TREES FOR RECLAIMING SOIL-LESS LAND			
Tree	N fixer	Air pollution	Notes
Alders, grey & common	✓	t	Fast growing, need moisture
Birches		m	Cold tolerant, fast growing; beautiful, especially in winter
False acacia	✓	t	Dry sandy soils, not cold-hardy; attractive tree
Hawthorn, common		m	Tolerates exposure, including sea winds; good for wildlife; attractive blossom & fruits
Pines, Austrian & Corsican		m	Tolerate sea winds and some exposure; useful timber
Pine, lodgepole		m	Tolerates cold and exposure, including sea winds; useful timber
Sycamore		t	Tolerates exposure, including sea winds; good timber
Willow, goat			Good for wildlife
Willow hybrids			Moist to wet soils; consult a specialist
t = tolerant m = moderately tolerant			

The field maple, with its beautiful autumn colours, can fit in more easily. It grows quickly for the first few years, then slows down, which is very obliging of it from the landscape architect's point of view.

On the other hand, a lime can be kept to the right size by pollarding, which gives a regular yield of wood in addition to its other outputs.

when the trunk is about 15cm diameter at breast height. Starting to pollard a tree when it's already half grown is a shock to its system which will weaken it or even kill it.

Pollarding is much preferable to the practice of letting a tree grow too big in its maiden form and then cutting back the branches in an attempt to maintain a maiden appearance while reducing the overall size. This is not only ugly but gives the tree a weak and unstable structure.

Water. A factor which is specific to trees in streets and other paved areas is water infiltration. Two square metres of permeable surface is needed around the base of each tree. Concrete grids or steel grates can protect the soil from compaction while allowing the area to be trodden on or wheeled over.

People. Direct interactions between trees and people are obviously of paramount importance in urban

forestry. The key subject of involving local people in tree work is discussed in the final section of this chapter. (See overleaf.)

Vandalism is an interaction which can be addressed both by community involvement and by a more enlightened planting plan. (See box below.)

As there's such a wide variety of planting situations in urban areas there must be an equal variety of approaches to the planting plan. For street trees there may be no alternative to planting relatively large trees with expensive individual protection from vandalism and provision for regular watering in the first few years. For small areas of trees in somewhat less intensively used spaces the planting style described in the box below may be the most appropriate. For more extensive areas normal forestry practice can be followed, perhaps making use of any natural regeneration which is present on the site at planting time or which springs up between the planted trees.

A SUBTLER APPROACH

The lone standard tree, sticking up out of a tightly mown lawn like a lollipop, tied to a tall stake which makes it easy to snap the tree in half by bending it over, has become a cliché. It could almost have a sign fixed to it saying "Please vandalise me". A more discreet way to establish trees is to use small forestry-size transplants rather than tall standards, and plant them in a matrix of shrubs which will grow up with them and never let them look like a target. Where the design intention is to end up with a mainly open aspect with a few trees rising above it, rather than a woodland or shrubbery, dwarf shrubs can be used as the nurse crop.

Of course involving people in the planning and planting of trees is the real answer to vandalism, but starting with smaller trees has other benefits. For one thing, trees always suffer a shock when they're transplanted, but the younger the tree the less the shock. Very often a 50cm transplant will catch up with and over-take a 2m standard which has suffered such a planting shock that it's hardly grown at all for a few years after planting. It's also quite easy and economical to mulch the whole area of a mixed planting of shrubs and trees, perhaps with a plastic membrane covered with tree bark, so competition from grass is completely eliminated.

What's more the cost of the trees is much less, and stakes and ties are not needed. Even including the price of the shrubs, the total planting cost can be less. Ongoing maintenance costs are also less because there's no constant grass mowing to be done. (For comparative cost figures see p183.)

New woods can create themselves by natural regeneration on neglected land in towns just as they can in the countryside. These self-sown woods can be seen as the urban equivalent of the ancient semi-natural woodlands of the countryside and they deserve to be cherished wherever they spring up.

Community Action

Tree Planting

It has become a truism that urban forestry can only succeed where the people who live in the area are involved in the process. People can be involved in everything from the decision whether to plant new trees at all, through the design process, to planting and aftercare. When people think of the trees as their own rather than as something that has been imposed on them they will appreciate and enjoy them to the full, be less likely to tolerate vandalism, and hopefully volunteer to help with the work.

In fact many urban forestry projects would never happen at all without the involvement of local people. The work input of a group of volunteers can make all the difference between a project happening and it being shelved. Their help and commitment show how much they care, and reduce the cost to the council. The volunteers can do all the hand work, leaving only the work which requires heavy machinery to be done by contractor.

Getting the people involved, through education, consultation and organising of work days is seen as one of the main tasks of the professional urban forester.[59] But the initiative can also come from the bottom up. We, the people, can initiate tree work rather than waiting to be educated, consulted and invited to participate. One way to do this is to become a local tree warden (see box), but it's not necessary to take on an official role like this. Any of us who care about trees can find out what people in our immediate neighbourhood want and let the council know.

In this community tree planting project in South Gloucestershire contractors cleared the chest-high brambles and volunteers planted the trees.

Communication is just as important in bottom-up tree initiatives as in top-down ones. Council officers are human and can be just as upset as local residents if they're not consulted on something which must concern them. They're also likely to know of any other initiatives going on. Perhaps there's a local tree committee, made up of representatives of voluntary bodies and community groups, which it would be foolish to ignore. Communication should be the very first step.

The next step is to consult people locally. It's always a good idea to start off by talking to people informally before holding a public meeting. It's all too easy to assume that people agree with your aims only to find, at the point where you sit down after making your statement, that most people in the hall do not.

Most tree initiatives are likely to be about planting, either in streets and squares or on open ground. Other projects can include removing rubbish and improving access to woodland, or even coppicing. But planting trees where there were none before is not always popular.

TREE WARDENS

The main job of a tree warden is to be a two-way passer of information on tree matters between the local community and the local council. They may be organised by the borough, county or district council. The aim is usually to have at least one warden in each ward or parish, but there may be more, for example where a large rural parish includes more than one village.

Because they know their local area, both in terms of people and of landscape, tree wardens are more likely to know of suitable planting sites than are council officers. Sometimes they may get to know about inappropriate felling or tree work at an early stage, so things can be discussed in good time rather than at the last minute when the chain saws are already out. They can also keep an eye open for changes of land ownership which may affect the fate of trees.

They can let local landowners know about grants which are available and thus perhaps encourage work which otherwise wouldn't be done. Some wardens take courses in basic tree skills and then pass these on to other local people.

Anyone can be a tree warden. A certain amount of knowledge about trees and some interpersonal skills are advantages, but the main qualification is a keen interest in trees. The first step is to ring up your local council and ask for the Tree Officer, or the equivalent if they don't have one. If the council doesn't have a tree warden scheme it could be worth lobbying them to set one up.

People may be concerned about leaf fall, shading, harbouring midgies or losing the view. They may value an apparently unused open space for walking the dog, an informal playground, a short cut to the shops or just for the feeling of space. Above all, and perhaps on an unconscious, unspoken level, any change to a familiar environment chips away at our feelings of security. In a world of ever faster change the familiarity of a local environment can be a valuable asset.

Where the land in question is at present unused and untended it may have an ecological value which would be spoiled by planting trees. (See Urban Commons, pp349-350.) In fact there may be unobtrusive self-sown trees there which would make planting unnecessary.

Some people specifically object to shrubs, believing that they make ideal cover for muggers and rapists. Beliefs like this are hard to shake, even when they are completely untrue. In fact these crimes happen much more often in the 'concrete jungle' environment than in landscaped areas with shrubs. One way to allay people's fears on this score is to plant only prickly shrubs which no-one could hide themselves in without getting scratched to shreds.

In some places it may turn out that the best solution is not to plant trees. In others people's concerns can be met by a design which leaves crucial parts of the area unplanted. As with all design work, the most important thing is to listen. People who feel they've been really listened to are generally more open to compromise than those who don't.

Once a tree planting scheme has been decided on, many more people may be interested in taking part in the planting than were involved in the planning stage – which is just as well, because it's really the most enjoyable part of the whole process. In my experience people are more willing to plant trees than they are to come to meetings, and the shared experience is one which helps to build a community.

Saving Trees

Something which may receive more unanimous support than planting new trees is saving existing ones which are threatened – though it may elicit less willingness to join in if it gets to the point where physical action is necessary. Physical action can range from large-scale occupations to save whole woods which are threatened by bypasses to small local incidents of a tree here and a clump or row there. The latter don't make the headlines so often but cumulatively they're just as important.

The key to saving threatened trees is to know about the threat as early as possible. Many cases will never get to the point where physical direct action is necessary if other means are tried in time. Often the landowner is thoughtless rather than malicious in planning to remove trees and an amicable agreement is possible. In other situations a Tree Preservation Order can be put on the tree, trees or woodland by the local council, and this can be done quickly if necessary. Hedges now also have some similar protection. The media, especially local media, can be a useful ally: everyone loves a tree.

CASE STUDY

LIVING IN THE WOODS

Living, and making a living, on our own little patch of land is a dream which many of us have. Ben Law has made it a reality in his coppice woodland. In order to do this he has not only had to produce enough income from the woods to support himself, but to prove that he could do so to the local planning authority. This is no mean feat, and Ben's success is as much due to his integration in the local community as to his practical and business skills.

Living on the land, rather than commuting there to work from a house in the local village or town, has given Ben a wealth of opportunities, both to develop the wood permaculturally and to green his own lifestyle.

Prickly Nut Wood is a chestnut coppice of just over three hectares in the Weald of Sussex. In 1992 Ben acquired it by barter from the landowners in exchange for planting trees and hedges on another part of their land. He now lives in the wood and makes his living from it.

Before coming to Prickly Nut he had worked in the woods on and off for a couple of years, but much of the inspiration for his lifestyle came from a spell working in Papua New Guinea as a forest development worker for an overseas aid charity. His job was to help the local people make a living from their woods as an alternative to selling them off to multinational timber companies. The policy of the charity was to encourage international trade of timber products certified as responsibly grown by the Forest Stewardship Council. (See p134.) But Ben saw more future in truly local development, with people building good houses for themselves out of timber and trading non-timber products such as saps, oils and fruits on the local market. When his contract was over he decided to come home and do something similar himself.

The traditional market for chestnut coppice is paling fences. To make a living from palings means at least a 60 hour week of very hard labour, and most of the people still doing it are old men who have been doing it all their lives. It also requires at least 20ha of in-cycle coppice. Ben makes a comfortable living on one seventh of this area, without excessive work, by a policy of adding as much value to the produce as possible and marketing it creatively.

Working with small material like this is not an arduous task, and Ben will turn it into high value produce.

His main product is charcoal, and he also makes rustic furniture, garden pergolas, rose arches, trellises, greenwood buildings, faggots for erosion control on riverbanks, walking sticks and yurt frames. The turnover from this mix of value-added products is in the region of £2,500 per hectare per year. Everything except the yurts is sold locally, mostly within a 6 mile radius. Most of these products require smaller poles than the traditional paling market, so he coppices over a short rotation of four years. This not only makes the work lighter, but means the coppice doesn't compete too strongly with the fruit and nut trees he has planted.

The long-term plan is to move the emphasis from wood products to food. Non-wood products which Ben sells at present include chestnuts and honey. The nuts come from the few standard chestnuts in the

At home in the woods. Ben with one of the yurts he makes frames for.

wood, and Ben can harvest 400kg of top quality nuts from each of the best trees, picking off the ground twice a day over a ten day period. With trees like these he sees no need to plant cultivated varieties. He has five beehives and the woods provide a good sequence of bee forage plants over most of the season: willows, fruit trees, limes and chestnuts, followed by clover in neighbouring organic fields.

When he coppices he leaves a few chestnut standards to provide more nutting trees for the future, and he has planted some walnuts and vigorous cider apple trees as standards in the coppice. But these are outnumbered by the fruit trees he has planted in existing clearings, including a substantial orchard. Overall he has planted more than a hundred fruit trees in the wood. (See p235.)

The orchard is growing well and so far Ben has had no problems with soil-borne diseases and little from predation by birds or squirrels. He reckons that shade is more of a problem for the fruit trees than root competition from the coppice stools. "With fruit trees in the coppice you have to be accurate with felling," he says wryly. "On the orchard edge I do it with wedges so as to be sure of missing the fruit trees. It's good fun."

So far he eats all the fruit he grows, but three local wholefood shops are crying out for local organic fruit and as the orchard matures he will have an increasing quantity to sell. This year he's increasing his beehives from five to ten and plans to sell the honey in the nearest farmers' market. (See p288.) This winter he will fell some large coppice growing in an inaccessible spot, and rather than extract the wood he will inoculate it with shii-take and oyster mushrooms which he will grow for sale.

For his own food he has a forest garden and raised vegetable beds. He has fenced the forest garden with a dead hedge made from brash which keeps out deer and rabbits and makes a trellis for hybrid berries which keep him in delicious fruit through the summer. He finds charcoal fines (pieces of charcoal too small to sell) make a slug-proof path, and the forest garden is criss-crossed with these useful barriers. The coppice immediately to the south of the forest garden is kept strictly to a three year rotation to prevent shading.

He also shoots the occasional wood pigeon or rabbit with an air rifle. Drinking water comes from a spring and he collects rainwater from his roof for watering the gardens. The products of his compost toilet go on the orchard. What little electricity he needs is generated with a solar panel and a couple of small wind generators.

The wood is a Site of Special Scientific Interest on account of the mosses, lichens and ferns which grow there. The mosses and lichens grow on the bark of old trees and are slow to colonise new branches, so he cuts the coppice high, leaving a stump for them to grow on undisturbed. The coppice cycle gives a succession of different habitats, so there is always at least one freshly coppiced area for the nightjars which come to the wood to breed each year. Ben is encouraging more diversity in the coppice by favouring self-sown birch, which he mainly uses for his own firewood.

After seven years of trying, he obtained planning permission to live in the wood, and has built a magnificent timber framed house largely from home-grown materials.

Getting this permission was no mean achievement, as he had to get the support of the inhabitants of the nearby village and the approval of the District Council. Trading locally was one thing which helped him with the villagers. They know him as a useful and respected member of the local community, and the Parish Council supports him wholeheartedly. Interestingly, one of the few objectors was a traditional coppice worker who was convinced Ben couldn't make a living on such a small area. He also had the support of both English Nature and the Sussex Downs Conservation Board.

The District Council see his work and lifestyle as furthering their Local Agenda 21 aims. They have seen that he can make a living from woodland work, and the fact that he makes charcoal has established his need to live on site. (See pp304-305.)

To complement his work at home Ben is a leading member of the local coppice group, which supports people who make all or part of their living from coppicing. The group helps with marketing by putting customers in touch with producers, runs stalls at shows, and encourages owners of coppice woods to rent them to people who are looking for coppice to work. It's currently working with the Forest Stewardship Council to develop a certification scheme for coppice alongside the existing one for timber.

Danny Harling, another member of the coppice group, makes fine greenwood furniture.

Ben has proved that coppice work is not only financially viable but a pleasant and attractive way of life. Much of the work is concentrated in the winter, and every year he gets away for a couple of weeks or more to a secluded cove he knows on the south coast of Devon. He takes plenty of barbecue charcoal with him.

TREES FOR TIMBER & WOOD

See key on page 340. This table can only give approximate information on the characteristics of different trees.
For more information see Further Reading.

Tree	Origin & type	Main uses	Rot-ation	Yield class	Coppice/ sucker	Soil	Micro-climate	N fixer	Pests	Notes
Oaks	N B	S T V Ed	120	2-8 (4)	Co	H	Li		g	On light soils timber may be of poor quality
Ash	N B	T Cr F	70	4-10 (5)	Co	H w	Li		D g	Useful for tool handles etc; Needs fertile soil for timber
Beech	N* B	T V Ed	120	4-10 (6)			Sh x		G	Needs well drained soil; frost-tender; Coppices weakly; *Only native in SE Britain
Sweet chestnut	E B	S T V F Cr Ed	70	4-10 (6) [9-12]	Co	L 1	Li		g	Timber quality may deteriorate if grown to large size; Fast growing as coppice
Wild cherry	N B	T V	70	4-10 (6)	Su	H 3	Li F		R D	Needs fertile soil; Deteriorates if over-mature
Sycamore	E B	T V F	60	4-12 (6)	Co		sh f X Sa		G	Very tolerant of exposure; Called plane in Scotland
Walnuts	E B	T V Ed	40-60			3	Li		g	Need fertile soil and warm, frost-free site
Hazel	N B	Cr Ed			Co	3	sh F x		D	Not a bulk yielder; Only for specialist craft use
Alder, common	N B	Nu	40	6-12	Co	W	F X	N		Can grow very fast near water or in high rainfall areas
Alder, Italian	E B	Nu					F	N		For drier soils
Alder, grey	E B	Nu		6-14	Co Su	w	sh F X	N		For cold sites, with wet or dry soil
Hornbeam	N B	F			Co	w	Sh F		R	For shady, frost-prone sites; rather slow growing
Maple	N B	F			Co	H 4	x		g	Beautiful autumn colour

Species										Notes
Limes	N B	F Ed Cr			Co		sh F x		r D	Bee fodder / Carving wood
Southern beeches	E B	T F	45	10-18	Co				R D	Frost- and cold-sensitive / Timber quality less than beech
Birches, silver and white	N B	Nu F Ed		4-12 (4)	Co	w*	Li F X		g	Very frost- and cold-hardy / Soil improvers / Short-lived, even if coppiced / *White birch
Willows	N B	F Cr		[25]	Co	W	Li		D	Extremely high yield of firewood on fertile soil
Poplars	N B	T V F		6-12 [25]	Co Su*	3 w				Extremely high yield of firewood on fertile soil / *Some species, including aspen
Black locust	E B				Su	L	Li	N		Pioneer on poor dry soils / Needs warm climate / Durable.
Scots pine	N* C	T Nu	60	4-14 (8)		2	Li F X		g	*Only native in Scottish Highlands / Tolerates poor soil.
Larch, European	E CD	S T Nu	50	4-16 (8)			Li f			Needs moderately good soil and microclimate
Larch, hybrid	E CD	T Nu	50	4-16 (8)			Li f			More tolerant of soils and climate than other larches
Douglas fir	E C	S T Nu	50	8-24 (14)			sh		D	Susceptible to windthrow
Lodgepole pine	E C	Nu T	60	4-14 (7)		W	Li F X Sa		D	Tolerates worst climate and soil, including peat
Corsican pine	E C	T	55	6-20 (11)		L	Li F X Sa			Warm, dry climate
Western red cedar	EC	T Nu	55	6-24 (12)		3	Sh f			High rainfall areas, can be used as underplant
Sitka spruce	EC	T	50	6-24+ (12)		w	Li			High rainfall areas, western Britain
Norway spruce	EC	T	60	6-22 (12)			sh f		D	Moderate rainfall areas, eastern Britain

KEY & NOTES FOR TABLE
TREES FOR TIMBER & WOOD

KEY

Tree in bold type		= main broadleaved timber species
Origin & type	N	= native
	E	= exotic
	C	= conifer
	CD	= conifer, deciduous
	B	= broadleaved
Main uses	S	= structural uses in building
	T	= timber
	V	= veneering
	F	= firewood
	Cr	= craft
	Ed	= edible component
	Nu	= nurse tree
Coppice/sucker	Co	= coppices
	Su	= suckers
Soil	L	= best on light soil
	H	= best on heavy soil
	1	= strong preference for low pH
	2	= tolerates/prefers low pH
	3	= tolerates/prefers high pH
	4	= strong preference for high pH
	W	= tolerates wet soil
	w	= tolerates moderately wet soil
Microclimate	Li	= light demanding
	Sh	= shade tolerant
	sh	= moderately shade tolerant
	F	= tolerates spring frosts
	f	= tolerates moderate spring frosts
	X	= tolerates exposed site
	x	= tolerates moderately exposed site
	Sa	= tolerates salty winds
N fixer	N	= nitrogen fixer
Pests	G	= most vulnerable to grey squirrels
	g	= moderately " " " "
	R	= resistant to grey squirrels
	D	= most vulnerable to deer

NOTES

Uses

Firewood: Most species are acceptable for firewood. Those noted in the table have good burning qualities, reasonable growth rate and coppice well. See table, Typical Yield For Different Kinds Of Coppice, p304.

Furniture: Most species listed for timber, plus birches, can be used for furniture making if they're of good quality. Cherry and walnut are the most valuable trees for furniture and veneers.

Turnery: Most broadleaved species can be used for wood turning.

Other markets: Almost all trees can be sold for pulp, chipboard manufacture etc if there is no other use for them.

Rotation length
This is the typical age to which the tree is grown for timber. It will be longer on a poor soil. For coppice cycle lengths see text.

Yield class
This is the typical yield as a timber tree in m^3/ha/ann. The range is given, with the average in brackets.

Figures in square brackets refer to coppice. For sweet chestnut, on a 12-20 year rotation, using material down to 4cm top diameter. For willow and poplar as SRC at a yield of 12t/ha/ann dry weight.

Note that yield class measures volume, not weight. Hardwoods, ie most broadleaved trees, are heavier (and stronger) than softwoods, ie conifers plus willows and poplars. Thus the weight of a hardwood timber crop may equal or even exceed a more bulky softwood crop. In addition, the branches of a broadleaved tree, not included in the yield class measurement, have a value for firewood, charcoal etc.

Microclimate
On the whole, trees which can tolerate shade also cast heavy shade.

OTHER TABLES & LISTS

Bee forage, p246.
Coastal conditions, p89.
Durability for fencing, p306.
Effects on soil, p326.
Firewood yields, p304.
Fodder, p280.
Fruit trees, p228.
Height and growth rate, p87.
Nut trees, pp222-225.
Orchard windbreaks, pp239-240.
Ornamental value, p182.
Trees for difficult urban sites, p332.
Wildlife value, p368.

Chapter 12

BIODIVERSITY

In common with the previous chapter this one takes an overall view of its subject, with the aim of giving general permaculture designers the basic information they need about how to preserve and enhance biodiversity. There is no book as far as I'm aware which covers the whole subject, and those which exist tend to be written from the limited perspective of nature conservation rather than the more holistic one of sustainability. As well as giving an all-round guide to beginners, I hope this chapter will give established conservationists a somewhat new perspective on their subject.

While most of the chapter deals with action we can take on the ground here in Britain, it's important to emphasise that we affect worldwide biodiversity through our consumption patterns. This is discussed in the first part of the chapter, Basics, and though it's short, this is probably the most important part.

BASICS

The aim of permaculture is not to turn the whole surface of the Earth into 'edible ecosystems' designed to meet human needs. It's to make our own productive systems so much more efficient that we need a much smaller area for our own needs, so we can leave a larger area for the other plants and animals with whom we share the planet.

Every permaculture design should have its zone five, that part of the design where the interests of wild species rather than human needs are put first. Even the smallest urban courtyard has space for a bird table or a bat box. But in many cases zone five may not be a separate area but a way of working which permeates the whole design. What matters is that we bear the interests of non-human species in mind whenever we make a decision.

Why?

There are three broad answers to the question, 'Why should we care whether other species survive or not?'

- because we like to look at them
- because it's in our interests from a purely practical point of view
- because it's right

It's perhaps a little trite to characterise the first response as "we like to look at them", but this is the kind of language often used by those who see nature conservation as a privilege of the well-off. Why, for example, should poor farmers in Africa be asked to adopt methods which give less than maximum yield but are beneficial to migratory birds? Is it just so that middle-class Europeans can have the pleasure of bird watching when the migrants move north?

Put like that, the pleasure we get from wild species seems a trivial reason for preserving them and insufficient on its own. But the pleasure, indeed the spiritual experience, we can get from being in beautiful places or from allowing a little of nature's beauty into our everyday environment is something of great value. Every human being has a right to the opportunity to develop a relationship with nature. Indeed developing a personal relationship with nature is an essential component of sustainability. (See p356.)

Young volunteers work on a path in an urban nature reserve in Bristol. Getting involved in nature conservation because we like to look at wildlife can lead to an appreciation of the deeper reasons for preserving biodiversity.

Even if we think that our pleasure in seeing wildlife is a less worthy motive for preserving it than the other two, it is certainly the one which is most attractive to the general public.

The second motive, that of practical self-interest, sees wildlife as useful to us in two distinct ways, as indicators and as resources.

If we create an environment in which other species can flourish it's most likely also one in which we can flourish ourselves. A planet inhabited by nothing but us and a small number of species we consider to be useful is most unlikely to be viable. The complexity of the living systems which make life possible on Earth are way beyond our understanding. We can only see them as wholes, and one way of assessing the health of the whole is to observe how many of its components survive and thrive. (See box, Diversity-Stability vs Species Redundancy, p18.)

Thus it's likely that the African farmers referred to above would create a more sustainable farming system and a more sustainable environment for themselves and their families if they chose the bird-friendly option rather than the maximum yield one.

Some wild species can be used specifically as indicators of ecological health. The amphibians, including frogs, toads and newts, are an example. Since they live both on land and in water they are susceptible to both water and air pollution, all the more so because of their permeable skin which allows them to absorb oxygen in water. Their skin may also make them more susceptible to ultra-violet radiation, which increases as the ozone layer decreases. Their diet of insects makes them vulnerable to insecticide residues, and their watery habitat is easily lost by drainage and over-abstraction from aquifers. Amphibians are declining world-wide. Here in Britain the common frog has declined by 90% over the past 20 years.

Amphibians are also useful to us in the second way, as resources, because of the number of insects they eat. This unofficial pest control is one example of wildlife as a useful resource for humans. Another one lies in the many as yet undiscovered species which could be valuable to us as sources of medicine, food and other products. The rosy periwinkle, found in the rapidly dwindling forests of Madagascar, has yielded a medicine used to treat Hodgkin's disease and leukaemia. We were lucky to find it before the locality where it was found was deforested. How many similar plants have already been lost for ever?

The total number of species on Earth is not known because the vast majority of them have yet to be described by international science. Estimates range from ten to a hundred million. But there is some consensus among scientists that if present trends continue around half of them are likely to be exterminated in the next few decades.[1] Even supposing the Earth's systems can survive such a loss, we are crazy to throw away such an enormous potential resource.

Even counting only the species we know about already, harvesting of natural ecosystems can yield more than destroying them. Tropical rainforests can give fruit, nuts, oils, rubber, fibres and medicines. One study found that the annual value of these products from a hectare of Peruvian rainforest was $422, compared to a one-off value for logging of $1,000, which would destroy the forest ecosystem for ever. Even sustained logging of selected high value trees gave less than a tenth the return of gathering.[2]

The third reason for not destroying wild species is simply because they have as much right to exist and to thrive as we have. This is a fundamental statement of ethics which transcends questions of whether we find wild creatures beautiful or useful. It's part of the ethical bedrock of permaculture, and is discussed in Chapter 1. (See p6.)

Nevertheless it's not necessary to the debate. There is more than enough force in the first two reasons to convince any intelligent person that we need to change our ways.

Ecosystems

All the above assumes that the number of species is the measure of biodiversity. But there is also genetic diversity – the amount of genetic variation between individuals within a species – and ecological diversity

---

EXTINCTIONS IN BRITAIN

Because diversity is much greater in hotter climates, the vast majority of the Earth's biodiversity is in the tropics, and this is where most extinctions are taking place. Here in Britain the total species count is much lower, and there are fewer species which only live in one limited area of the country, so extinctions don't happen so easily here. But our record is not good.

According to a report by WWF UK, 154 species are known to have become extinct here during the twentieth century. Loss of habitat and habitat degradation by contemporary farming methods are given as the main reasons. The latest extinction is the short haired bumble bee. If present trends continue WWF calculate that by 2010 it will be followed by the water vole, the high brown fritillary, pipistrelle bats and skylarks, while the song thrush will disappear from farmland.[3]

Unfortunately the figure for national extinctions tends to minimise the rate of loss. A species which was once common and widespread only needs to hang on in a small population somewhere on the island to keep off the list. Most counties in Britain are experiencing extinctions at ten times the national rate.[4]

– the variety of different ecosystems. Genetic diversity is discussed later in this chapter. (See pp370-371.) Ecological diversity is important for two main reasons.

Firstly, preserving ecosystems is the key to preserving species. Loss of habitat is the main reason for extinctions of species (followed by the introduction of invasive exotic species). The only alternative to preserving self-replicating populations in intact ecosystems is preserving individual animals and plants in zoos, botanical gardens or gene banks. That would be an absurdly high-cost operation and only possible for a minute fraction of the Earth's species and varieties.

From a practical point of view, if we keep examples of all the ecosystem types in an area we can be reasonably sure we've kept all the species too. This doesn't just mean one woodland, one heathland and so on, but all the different kinds of woodland and heath which occur in that area.

Secondly, ecosystems are worth saving in their own right. Natural and semi-natural ecosystems are the best models we have for sustainable systems for our own use. The wisdom they contain, though we may find it hard to understand with our poorly developed knowledge of ecology, is ultimately of more value to us than all the information we have stored in libraries and universities.

How?

There are three broad answers to the question, 'How can we best preserve biodiversity?'

- by modifying our patterns of consumption
- by working with the land as a whole in a sympathetic way
- by means of nature reserves

Consumption

This is far and away the most important of the three. The vast majority of the Earth's species live in the tropical regions, and we in the North have a great deal of influence on how tropical ecosystems are treated by virtue of what we buy.

Few people reading this, I'm sure, would think of buying mahogany furniture. But, in North America at least, much of the low quality beef, such as that used in hamburgers, comes from grassland on the site of former tropical rainforest. Even beef cattle which have been reared here in the North may well have been fed soya grown on recently cleared forest in the south west of Brazil. Large areas of the Amazonian forest have been cleared to provide hydro-electricity and charcoal for steel smelting. Bauxite for aluminium production is quarried from vast areas of tropical subsoils. Regardless of where the actual car or drinks can we buy came from, by buying them we are adding to the demand for the products and making these activities 'economic'.

Virtually everything we consume contains embodied fossil fuel energy, and global climate change is perhaps a bigger threat to the survival of wild species and ecosystems than any other. (See box, Global Climate Change and Biodiversity, p73.) The most effective single thing any of us can do to promote biodiversity, both at home and abroad, is to cut our energy consumption. (See Chapter 4.)

The nature of the problem was illustrated in an example from closer to home, when a quarrying company wanted to divert two rivers in Devon in order to expand their ball clay quarry. Moving the rivers into new channels would turn two vibrant ecosystems into sterile canals.

Ball clay is used to make bathroom ware. Most bathroom ware is sold, not to unwashed people who need a decent level of hygiene, but to people who are bored with their present colour scheme and want a change. The company spokesman said that people could choose not to buy their products if they felt that preserving wildlife was more important. But he didn't say how the average customer entering the showroom would know the facts of the situation.

To a great extent the problem is one of remoteness. We have become so remote from the resources we use and from the consequences of our actions that we can no longer see cause and effect. No doubt there are some people who would still buy the hamburger or new bathroom suite even if they could see at first hand the destruction their purchase was causing, but many would not. If being face to face with the consequences of our actions became the norm there's little doubt that consumption-at-all-costs would soon come to be seen as ethically wrong.

People say education is the answer. Given the global structure of our present economy it's all we can do, but it can never be adequate. It can never compete with advertising, either in terms of the money available for it or in terms of advertising's appeal to the oral side of human nature. The only real remedy is to work towards an economy in which local production for local needs is the norm, and this is one of the main thrusts of permaculture.

The Wider Land

If the kinds of practices described in this book were adopted over the farmland, forest, cities and towns of the world, there would be no crisis of biodiversity and no need for nature reserves.

Simply adopting organic methods is enough to give a dramatic increase in the amount of wildlife on farms, as study after study has shown.[5] Although it is theoretically possible for a conventional farm which is managed to encourage wildlife to have more biodiversity than an organic one which is not, the studies show that in practice this is rare.

This is partly because the kind of farmer who wants to farm organically is usually also the kind who likes wildlife, and the organic standards require farmers to respect wildlife. But there are also structural reasons why an organic farm is likely to support more wildlife:

- The use of rotations to maintain crop health and soil fertility means that there is a greater diversity of habitat on the farm.

- The cultural methods used to reduce disease in cereal crops include growing spring sown cereals, which improves feeding and nesting opportunities for many birds.

- The absence of herbicides means that some wild plants can survive among the crops. These are often endangered species themselves and also provide food and habitat for insects, which in turn also provide food for birds.

- The absence of pesticides allows many more insects, and the birds and other predators which feed on them, to survive.

- The high level of soluble nutrients provided by chemical fertilisers favours a few competitive plants which soon eliminate other species by competition. This is especially important in fertilised grassland. (See box, Eutrophication of Ecosystems, p359.)

Attempts have been made to find a halfway house between conventional and organic farming, the most prominent of which is integrated crop management (ICM). The use of rotations and other cultural methods is combined with some restrictions on the use of chemical fertilisers and pesticides. In theory it should be better for biodiversity than all-out conventional farming, but the only study carried out so far to test this has found no difference between the two.[6]

Using organic methods in home gardens is perhaps as important as it is on farms. Although they cover much less area in total, gardens are potentially high in biodiversity, and the intensity of pesticide use is much greater in gardens than on farms. Farmers are constrained by economics to use no more active ingredient than they need to get the desired effect, whereas a home gardener will hardly notice a few extra pounds a year spent on chemicals applied 'just in case'. In fact I'm sure many get satisfaction from spending the money and applying the chemical regardless of whether it's necessary. For many people gardening is now a consumer activity, where the purchase is the central activity rather than a means to an end.

Permaculture can be expected to be even better for wildlife than organics. Since permaculture is an attempt to make our productive systems as much like natural ecosystems as possible, we may intuitively expect this. But there are also some specific reasons why it should be so:

- The emphasis on diversity is even greater in permaculture than in organics, including simultaneous diversity in the field as well as rotations.

- Agroforestry gives a diversity of structure which is beneficial to many kinds of wildlife. For example, the density of breeding birds in unsprayed orchards can be amongst the highest for any British ecosystem.[7]

- No-till methods preserve high levels of soil organic matter and high populations of soil microbes, earthworms, mycorrhizas and other soil organisms, all of which are vital components of the ecosystem and important to food chains.

- Constant ground cover, such as in the bicrop system, provides habitat for many invertebrates.

- Insects and birds which predate on pests are welcomed. Indeed a low level of pest species is also welcomed in order to maintain a population of predators and thus prevent a serious outbreak of pests. (This point applies to some extent to organics.)

In permaculture a high level of biodiversity is seen as an inherent part of the system, and areas of wild vegetation are valued because they maintain the health of the landscape as a whole. By contrast, in conventional land use systems nature conservation is seen as a constraint on the level of output, a sacrifice which producers make out of the goodness of their heart.

A good example of this contrast in attitudes is continuous cover forestry. (See p318.) According to conventional economics it's less profitable than clear felling, and choosing continuous cover is held to be a mark of public-spirited self-denial in a landowner. But for someone making a living from a small area of woodland continuous cover has silvicultural and economic advantages which outweigh the economies of scale. In a sustainable local economy these advantages would be all the greater.

This hazel nut platt has the ground flora typical of a semi-natural woodland, with a carpet of bluebells in late spring. It's a prime example of how food production and biodiversity go hand in hand in permaculture.

Nature Reserves

Creating nature reserves is an admission that the first two means of preserving biodiversity have failed. The fact that most nature conservation effort is now concentrated on reserves is to some extent an admission of failure.

This is not to disparage nature reserves. Given the circumstances it's vital that we do preserve diverse ecosystems wherever they survive, in the hope that one day they may be able to repopulate the wider landscape. As in every other branch of permaculture, there's a big difference between how we work in the present situation and how we would work in the situation which we are ultimately aiming for.

There's so little biodiversity left in the present landscape that it must be a high priority for any permaculturist to learn both how to identify areas of high diversity and how to work with them sympathetically.

ECOSYSTEMS IN BRITAIN

Rural Ecosystems

There is no wilderness in Britain. In fact, if we define wilderness as land which is unaffected by human activity there is none on the planet.

It used to be assumed that when Europeans arrived in Australia and North America, continents where there had been little or no agriculture before that time, they were seeing wilderness virtually untouched by human hand. But now it's recognised that pre-agricultural people have had significant and sometimes major impacts on ecosystems, especially where fire was regularly used to facilitate hunting.

Equally there is no part of the Earth unaffected by the activities of the present-day industrial economy. DDT has been found in the body fat of penguins and polar bears, and global warming will have its effects everywhere.

The Highlands of Scotland are not a wilderness, as this scene reveals. Where sheep are excluded, as they are here by the surviving fences of a former railway line, the moorland regenerates to woodland.

Nevertheless, in many parts of the world there are areas of vegetation which have been little enough effected by humans to be described as virgin. Wherever such ecosystems survive they are priceless assets, to be preserved at all costs, not least because they are by definition irreplaceable. In many cases they may also act as models for our own productive systems.

Here in Britain we have nothing like that. There is no part of the island that has not been affected by human activity.

The last Ice Age ended around 12,000 years ago. It had wiped most of the island clean of vegetation, with tundra in the south and south-east. Gradually trees recolonised and by about 9,500 years ago they had covered all the land except those parts they were never to reach, the extreme north of Scotland and the highest mountain tops. Some 6,000 years ago the Neolithic culture came to Britain, and with it came agriculture. The conversion of wildwood to arable, grassland, moor and managed woodland began.

The process can be traced in the archaeological record. Pollen grains of trees, preserved in waterlogged deposits, give way to those of arable weeds and grassland herbs, while shells of woodland snails give way to those of open land. The earliest evidence for coppicing comes from wooden causeways constructed over the marshes of the Somerset levels and preserved in the peat. The earliest of these, the Sweet Track, dates to 6,000 years ago, and is made partly of timber and partly of poles which are quite clearly the produce of coppice woodland.[8]

At that time many of the commonest British trees, including beech, hornbeam and ash, were fairly recent immigrants and would not become common over their present range for thousands of years to come.[9] Thus humans have been a major component of British ecosystems for longer than some of the species which we consider to be 'wild'.

By the Bronze Age, 3-4,000 years ago, it is possible that much of lowland England consisted of islands of woodland in a sea of farmland, and this was certainly so by the Roman period. The picture painted by the Domesday Book, compiled on the orders of William the Conqueror some 1,000 years ago, suggests no more than a few scraps of wildwood surviving among the farmland and managed woodland.

Our richest areas of biodiversity are semi-natural ecosystems. These are ecosystems which are the product of a combination of human and natural influences.

A partly cleared, or 'cultural', landscape can have a higher level of biodiversity than one composed of virgin forest. Although the total biomass of wild plants and animals is lower, the total number of species present may be greater because the diversity of structure is greater.[10] This may well have been true here, and humans and wild species have co-existed happily on this island for thousands of years – until the middle of the twentieth century.

A cultural landscape can contain a mosaic of different habitats, including woodland, hedges, fields and moorland, as here, near Brecon in Powys.

Since the end of the Second World War, a combination of powerful new technologies and political support for maximum production of food and timber has led to the destruction of semi-natural ecosystems on an unprecedented scale. In 1984 the then Nature Conservancy Council (NCC) published a study of the loss of semi-natural ecosystems from 1949 to that year.[11] The rates of loss they found varied from 30% to over 90% for different kinds of ecosystem. The actual figures for each kind are noted under the descriptions of ecosystems below.

Since then there has been no such complete survey, but losses continue at a similar rate. Although the NCC study found that 97% of herb-rich lowland meadows had been destroyed, during the mid 1990s they were still being lost, at the rate of 6% per year in Somerset, for example.[12]

Meanwhile the sterility of ordinary farmland which is outside designated nature reserves has increased steadily. This is illustrated by a report from the Joint Nature Conservancy Committee on farmland birds, which recorded drastic falls in the population of common species between 1969 and 1994. Over half of the species studied had fallen by over 60%, and one, the tree sparrow, by 98%.[13]

In writing this I'm not trying to single farmers out for particular blame. It's the norm in our society to disregard the health of the planet in pursuit of one's own profit or pleasure. Farmers are no different from all the rest of us in this. Those of us who fly off on our holidays in an aeroplane are no less destructive than those who plough up old meadows. The only difference is that the consequences of flying are more remote from the action itself and thus less obvious.

The semi-natural ecosystems which remain in Britain can only be seen as an irreducible minimum for the survival of a healthy ecology.

Woodland

Much of the semi-natural woodland in Britain is coppice, mostly out of cycle. The trees were never planted by human hand but are the descendants of ones which colonised the site naturally in the past. Humans have directly influenced the structure of the woodland and this has indirectly influenced the composition of plants and animals which live there, as some species are favoured by the regular cycle of cutting and regrowth and others are discouraged by it. (See p356.)

Other woods have been treated as wood pasture, that is woodland where animals have regularly been allowed to graze. They include deer parks, where the trees were often pollarded, and areas such as the wooded parts of the New Forest. The diversity of trees and herbaceous plants is usually less than in a coppice as only those which are resistant to browsing and grazing survive, but wood pastures are often important for their old trees, which provide a unique habitat for many invertebrates.

The Highland pinewoods are perhaps closer to their original condition than any other woodlands in Britain. But many of them are failing to regenerate at present due to grazing pressure from sheep and red deer.

Ancient woods are defined as those which have been woodland since at least 1600. That date is chosen because it's about as far back as reliable records go. Between 1949 and 1984 50% of the ancient semi-natural woods in Britain were destroyed, mainly by conversion to conifer plantations or farmland. In upland areas many woods are lost by unrestricted grazing preventing regeneration.

Recent woods, those which have arisen by natural regeneration since about 1600, usually have much less diversity. Nevertheless they can be important habitats. Even scrub, young woodland consisting of shrubs and open ground, can be a valuable habitat for birds and insects, though it may be botanically poor.

Grassland

Semi-natural grassland is another ecosystem where people have directly determined the structure of the vegetation and indirectly affected its composition. The grasses and herbs which are present are wild species which are resistant to regular grazing or mowing. Some grasslands can be very diverse, supporting numbers of plant and animal species similar to those found in the most diverse woodlands.

Throughout agricultural times semi-natural grasslands have been a common feature of the landscape, but in the past half-century most of those in lowland areas have been destroyed. Between the 1940s and 1984 97% of herb-rich grasslands on lowland neutral soils were destroyed or damaged, and 80% of chalk or limestone downland. Semi-natural upland pastures, which are more extensive but usually less diverse, had decreased by 30%. Most of these grasslands have been converted to arable or improved pasture.

In some parts of the country, especially the arable east, the only semi-natural grassland remaining is in woodland, along rides and in glades, or on roadsides.

Heath and Moor

These are ecosystems characterised by the presence of heather, though some are dominated by other plants such as bracken or grasses. They are called heath when found in the lowlands and moor in the uplands. They are predominantly on acid soils and the level of diversity is correspondingly low, but they support some species which are not found in other ecosystems.

Lowland heaths declined by 40% between 1950 and 1984, through conversion to arable or improved grassland, afforestation and building, and also by regeneration to woodland through lack of grazing. In fact most lowland heath had already been converted to agriculture in the nineteenth century, so the total loss of heathland is more than this figure suggests.

Heathland and much moorland is semi-natural, maintained by a level of grazing which prevents the regeneration of trees. But some moorland appears to have been formed by a change of climate from drier to wetter, leading to a buildup of blanket bog on which trees do not grow well.

Whether we regard moor as a natural or semi-natural ecosystem, there is no doubt that trees will grow on most of it if the pressure of grazing is reduced. Many people regard moorland as a 'wet desert', low in biodiversity and a shadow of what it could be if woodland was planted or allowed to regenerate. Others point out that although it's widespread in Britain it is less so in other countries and we have a significant proportion of the European total, including some unique ecosystems. So there is some debate about the extent to which moorland is worth conserving.

30% of upland moors and blanket bogs were lost between 1950 and 1980, mostly to conifers and improved grazing.

Water Bodies and Wetlands

Water bodies include rivers, streams, ponds and lakes.

Undisturbed rivers and streams are structurally diverse, with pools, riffles, shoals, islands, both steep and shelving banks, and alternating sunny and shady stretches, providing a diversity of habitats for different plants and animals. But many of them have been transformed into uniformly deep and steep-banked channels, devoid of bankside trees and shrubs, so as to make them into more effective land drains.

Farm ponds have declined enormously over recent decades as they have lost their agricultural functions. Some have been deliberately filled in, others have silted up.

Eutrophication and chemical pollution both lead to a loss of biodiversity, and water bodies are particularly vulnerable to both. (See box, Eutrophication of Ecosystems, p359.) This is partly because nutrients and pollutants tend to be washed down into them from the surrounding landscape, and partly due to the nature of water as a medium: anything entering a body of water rapidly becomes dispersed throughout its volume, and most substances are more soluble in water than in the soil.

Wetlands include acid bogs, alkaline fens, reedbeds and poorly drained fields. Bogs and fens contain unique and diverse assemblies of plants and animals, while wet grassland is particularly valuable for wading birds, which can only feed on ground which is soft enough to penetrate with their long bills. All wetlands have been depleted by draining for agricultural use. Fens and bogs have also been mined for horticultural peat; 50% were lost or damaged between 1949 and 1984 and the destruction continues apace.

Coastal

Marine ecosystems fall outside the scope of this book, but they do deserve mention. Like all shallow water systems coastal waters can be both productive and diverse. As so much human settlement is on the coast there is often a conflict between biodiversity and development.

The sea is the final sink for much of the pollution we create on land. Oil tanker spills are a dramatic reminder of this, but an equal amount of oil ends up in the sea as the result of our day-to-day habit of pouring away anything we no longer have a use for, including used sump oil. In ecological terms there is no such place as 'away'.

Linear Ecosystems

Hedges, roadside verges and railway lines, both used and disused, are the commonest linear ecosystems. They are perhaps less obvious than blocks of woodland and wildflower meadows and we don't usually get so excited about them. But they are very extensive. There are still almost half a million kilometres of hedgerow in Britain, even though we have destroyed 40% of what was here in 1950.[14]

Hedges are nothing but continuous woodland edge, and woodland edges are in some ways the ecologically richest part of the woodland ecosystem. Despite the losses, they are still probably the biggest single wildlife habitat in the country. There's a great deal of difference between an ancient hedge, with several species of trees and shrubs, a bank and a ditch, and a hedge planted in the nineteenth century, consisting of nothing but a double row of hawthorn plants.

There's as much variation in roadside verges, from the margin of a lane which goes back to medieval times to a band of recently laid topsoil sown with a single species of grass. The former may preserve many once common plants which have now been banished from agricultural land. Some local councils are beginning to appreciate the value of verges and put up signs declaring them to be roadside nature reserves. Even fairly recent verges can soon become colonised by wildlife: cowslips and kestrels are both frequently seen by motorways. By one estimate, the total area of roadside verges in Britain is three times as large as that of all the nature reserves put together.

Railway lines, although frequent in the countryside, have more of the characteristics of urban than of rural ecosystems.

Verge, ditch, bank and hedge can each provide habitat for a different range of species. Here the hedge has recently been coppiced, often a good way to deal with an overgrown hedge, and both easier and quicker than laying.

Linear features are often thought of as wildlife corridors, that is passageways down which plants and animals can migrate from one larger area of suitable habitat to another. This is possibly an idea which is sometimes given more credit than is due. Some species are extremely slow to migrate, especially woodland herbs and invertebrates, and progress down a hedge or grassland strip may take centuries rather than years. Other species, especially those of the woodland interior, shun edge habitats and will not make use of hedges.

Nevertheless corridors are very valuable to certain species. Strips of grassland may provide a thoroughfare for animals, including small mammals, and thus make good hunting grounds for predators like barn owls, while overgrown hedges are essential to the ecology of some bat species. Along rivers, a continuous border of semi-natural vegetation is essential for otters, for example.

Urban Ecosystems[15]

Urban ecosystems rarely have the same diversity as the ancient semi-natural ecosystems of the countryside, or contain as many rare and endangered species. They can never compensate for the loss of rich rural ecosystems, but they can have great value in themselves. Firstly, they're near where most people live, so they provide opportunities for a direct experience of nature. Secondly, truly urban ecosystems are unique.

Much urban nature conservation effort has gone into caring for rural ecosystems which have been surrounded by the town but not destroyed –

encapsulated countryside – or in creating copies of rural ecosystems. In his book, *The Ecology of Urban Habitats*, Oliver Gilbert has suggested a radically different approach. (See Appendix A, Further Reading, p443.) He points out that there are semi-natural ecosystems in urban areas which are not found anywhere else on Earth. They are often combinations of native and exotic species which would never exist outside a heavily humanised environment. He suggests that we should value these unique ecosystems more highly, rather than trying to preserve or recreate scraps of countryside in the city.

Some 30-40% of the vegetation in cities is wild, at least in the sense that it has not been deliberately planted by people. This gives rise to semi-natural ecosystems which are different from most rural ones in a fundamental way. Most rural ones are semi-natural in the sense that people have had an influence on pre-existing wild ecosystems over a long period of time. In urban areas it's often the other way round: people have completely destroyed natural conditions and then for one reason or another let go of the land and allowed it to regenerate without deliberate intervention.

Other urban ecosystems, such as parks and gardens, are completely artificial, but they can nevertheless provide great habitat for wild species, especially animals.

The typical characteristics of urban ecosystems are summarised in the mind map at the top of the next page.

Obviously the overall factor which makes urban ecosystems different is the degree of human influence. As we have seen in previous chapters, the soils are often entirely artificial and different from anything which is found in the countryside or in the wild (see p62), and the climate is highly modified, especially in terms of higher temperatures (see p69). Species have been introduced, both deliberately as ornamental plants, and accidentally, for example as weed seeds in imported grain, discarded bird seed, or animals which have hitched a ride on imported produce. Many of these are able to establish themselves in unplanted, semi-natural vegetation.

Given these influences it's not surprising that exotics are a normal and frequent component of semi-natural vegetation in urban areas. Buddleia and Japanese knotweed are well known free-living urban exotics. But less obvious are plants like dandelions and vetches, which may appear to be the familiar native kinds but which in town centres are more often related species from abroad. Even those which are native species are often exotic varieties. Some of the truly native plants may be living far to the north of their natural distribution, making use of the warmer climate cities provide.

Even those animals and plants which are genetically the same as their country relatives may behave differently in town. Foxes, for example, can be found at four times the density in towns than in typical rural habitat, and their life expectancy is much shorter. Both of these factors affect their social and territorial behaviour.[16]

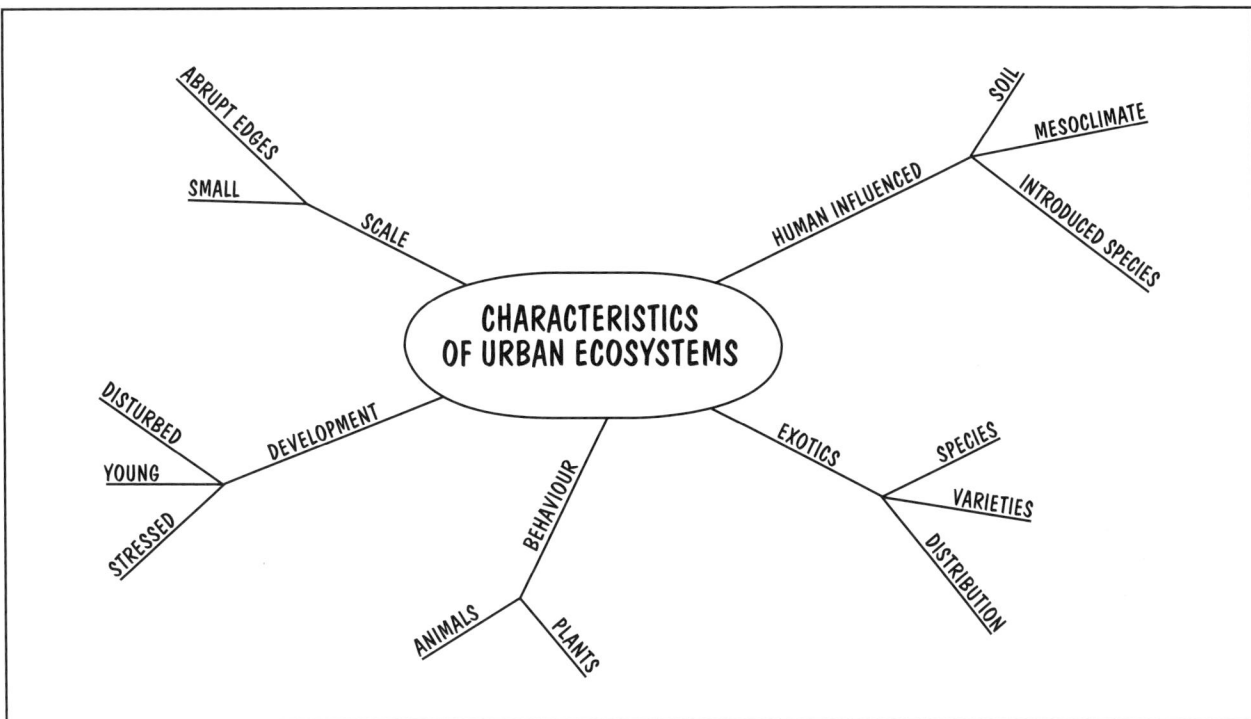

Their diet is also different, with the scavenging of human leftovers being their biggest single source of food.

Native plants in towns usually live in a wider range of habitats than they do in the country. The long-established ecosystems of the countryside are relatively closed. All the space and all the niches have long since been occupied, and the species which occupy them must be very well suited to their situation because they have maintained their place in the system against all comers over time. The high level of disturbance in cities makes for open ecosystems with plenty of space for plants to get established even if they're not perfectly suited to the soil and microclimate. Thus plants normally found in woodland, grassland and wetland can all be found growing side by side on a former demolition site.

Constant disturbance is the norm for urban ecosystems. The cycle of redevelopment is some 50-60 years, which is a very short time in ecological terms. Redevelopment usually means the end for an established ecosystem, but it can also mean an opportunity for a new one to establish, especially if there is a gap in time between demolition and rebuilding. Older ecosystems do exist, for example on railway verges or in scraps of remaining ancient woodland, but they are not characteristic.

This means that species with efficient dispersal mechanisms are at a great advantage. Many urban plants, such as the ubiquitous buddleia, and small animals such as spiders, are wind dispersed. Another characteristically urban means of dispersal is the spread of vegetative material on the wheels of earth moving equipment, especially important for plants which rarely or never set seed, like Japanese knotweed. Other plants arrive on new sites as throwouts from the horticultural trade, or in bark or peat mulches which may contain seeds of rushes, foxgloves or sorrel.

A high level of stress puts a limit on the development of an ecosystem, as the number of species which can survive under stressed conditions is limited. Walls and roofs are extreme examples of stressed ecosystems, but very common in cities. Their rural counterparts are cliffs and scree, which are relatively uncommon. Pollution is a stress which is generally more severe in urban environments.

Urban ecosystems are usually much smaller than rural ones, with frequent abrupt changes. In the countryside the boundaries between one vegetation type and another are more influenced by natural changes in topography and geology, whereas in cities they are determined largely by human activity, and that is kept to a small scale by the high value of urban land.

Urban Commons

This is the name Oliver Gilbert gives to land where buildings have been demolished and levelled and the site is awaiting redevelopment. He has identified a number of distinct ecological stages they pass through, if the site is left undisturbed for long enough.

- Oxford ragwort stage: the pioneer stage, dominated by annuals and short-lived perennials, mostly wind-dispersed, and often characterised by the exotic Oxford ragwort – originally a native of the lava ash of Mount Etna, and thus well adapted to demolition rubble.

- Tall herb stage: after 3-6 years tall perennial herbs, such as rosebay willowherb, Michaelmas daisy and lupin become dominant. Very persistent plants, such as mints, may survive from former domestic gardens on the site.

- Grassland stage: after 8-10 years grasses predominate, with clumps of the more persistent tall herbs remaining. Japanese knotweed may be prominent among these and brambles may start to spread.

- Woodland stage. Few urban commons last long enough to form woodland. Light-seeded trees and shrubs, such as willow, birch and buddleia, may seed themselves in the first year or two. But once the herbaceous vegetation closes only larger-seeded trees, like hawthorn, sycamore, ash and domestic apple can establish. Whether they do depends on the presence of nearby seed parents, or in the case of apple on discarded apple cores.

Urban commons can be valuable habitats for both wildlife and people.

But they are by their very nature short-lived.

The actual vegetation which develops on any one site depends on a number of factors, including the composition of the rubble and which plants happen to be nearby when the chance for colonisation arises. This gives rise to a great deal of variety and there are distinct regional variations, such as the abundance of buddleia in Bristol and the occurrence of alexanders in Norwich, believed to be a survival from monastery gardens.

Although mammals and breeding birds are not numerous on urban commons they are good feeding sites for birds. As an urban common develops there is a succession of invertebrate communities which are often as uniquely urban as the plant assemblies they inhabit.

Woodland

The self-sown woodland that is the end point of urban common succession can account for the majority of the woodland found in some urban areas. A survey of the borough of Sandwell in the West Midlands in the early 1990s showed that the area of woodland had doubled over the previous 12 years, and over half of that new woodland was unplanted. What's more the self-sown woodland was thriving, while the planted woodland was not.[17]

However this may be uncharacteristic of Britain as a whole. An estimate of the origins of urban woodland made in 1978 reckoned that 70% of it was recently planted and only 3% of recent self-sown origin, the remainder being ancient, either semi-natural or replanted.[18] The difference may be in part due to the timing of the two surveys. The Sandwell survey came after the Thatcher years, when much of British industry went to the wall, leaving vacant factory sites to be occupied by urban commons, especially in traditional industrial areas like the West Midlands.

There's rather less distinction between the different kinds of woodland in the city than there is in the country. They are all subject to the same pressures, which may include:

AN ACCIDENTAL FUNGICIDE

Oak trees are relatively intolerant of shade when young. They readily colonise open ground but are less able to regenerate in the shade of existing woodland. In former times they succeeded often enough to maintain their presence in woods, but today they rarely do. There are a number of reasons for this, including increased shade resulting from the decline of coppicing. But a key factor is probably the oak mildew, which was introduced from North America in 1908. Virtually all oaks have it but they can tolerate it unless they are suffering from other stresses, such as shade.

The only woods where oaks now readily regenerate are in urban areas, and this is due to the level of air pollution. The sulphur dioxide from motor exhausts combines with rainfall to make an effective fungicide, effective enough to swing the balance back in favour of the survival of oak seedlings.[19] In some urban woods a whole range of oaks of all ages can be seen, from seedlings and saplings to the tall canopy trees. An example is Queens Wood in Highgate, north London.

- A high level of trampling, which in extreme cases can destroy the herb layer of ancient woods and prevent regeneration in woods of all kinds.

- Exotic trees, which often dominate natural regeneration and may also invade or be planted sporadically in ancient woods.

- Air pollution, which may eliminate sensitive plants, notably most lichens, but can sometimes have a positive effect. (See box, An Accidental Fungicide.)

Linear Ecosystems

Since rapid and effective dispersal is one of the characteristics of the most successful urban plants and animals, wildlife corridors have little relevance for them. Foxes, sparrows, coltsfoot and buddleia can all find their way about the town without the aid of continuous habitat. Nevertheless, there are many species in towns which do need corridors in order to spread or migrate, and given the small size and short lifespan of most urban habitats, corridors can be essential to the survival of these species.

In addition to their role as corridors, roadside verges and railway cuttings and embankments form their own characteristic ecosystems. They can provide habitat for grassland species which were once widespread, and one stretch of chalk grassland in a railway cutting at Winchester was found to be of SSSI quality. The more stressed linear ecosystems, such as the permanent way of railways or the immediate fringes of roads, have a more urban flora, often influenced by the presence of salt. A study of verges around Tyneside found twelve plants which normally grow in saltmarshes, and the strips of white flowers you can often see down the central reservations of motorways in the spring are Danish scurvy grass, a native wildflower which is otherwise only found on the coast.

Disused railway lines are particularly important in urban areas. They often support the nearest thing to

An urban linear habitat. This steep bank is regenerating to woodland. At present it provides habitat for slow worms.

wild vegetation anywhere in the town and are equally important to both wildlife and people.

Water Bodies and Wetlands

Urban streams and rivers have been even more modified than rural ones. Very often streams and smaller rivers have been covered over, which virtually destroys them as wildlife habitat. Even short lengths of culvert are a barrier to the movement of species such as kingfishers, and can cut a viable territory into short lengths which are too small to support a breeding pair. Fortunately, more naturalistic methods of river management are now becoming more popular.

Pollution, including toxic chemicals, eutrophication and untreated or partly treated sewage, is a constant restraint on species diversity. Sudden severe toxic pollution incidents are less damaging than persistent low-level pollution. The underground parts of plants can survive a short-lived incident and vegetation recovers in about three years unless the pollution is repeated. The over-abstraction of water from both rivers and aquifers can reduce the volume of flow and thus increase the concentration of pollutants.

A characteristically urban feature of some riversides, for example in Bristol and Sheffield, is fig trees. The seed came from sewage, and in Sheffield they seem to have become established in the days of heavy industry, when the local air temperature was higher than it is now.

Canals form some of the richest ecosystems in cities. They have existed long enough to have acquired a good diversity of species, and they provide a range of habitats, including deep and shallow water, towpath verges and the undisturbed offside verge. They provide corridors for both aquatic and land species.

There are now more ponds and lakes in towns than there are in the country. Garden ponds have become the saviour of once common amphibians as their rural habitats have disappeared. One survey found 1,000 times more frog breeding sites per square kilometre in the suburbs of Sussex than in the surrounding countryside.[20]

Gardens

Suburban gardens can in some ways be ecologically richer than rural semi-natural ecosystems. They contain a mosaic of mini-habitats and edges, including lawns, vegetable and flower beds, trees, shrubs, hedges, walls, ponds and the houses themselves. This rich mixture particularly favours birds and insects.

The birds are mostly woodland species and, although there may be less diversity of species than in the woods, there may be many more individual birds. For example, a study of blackbirds around Oxford found an average of seven to eight breeding pairs per hectare in the suburbs compared to only one pair per 2.6ha in the woods outside the city. Out of the breeding season, gardens are an important source of food for birds which breed in the surrounding countryside and for migrants from the continent.

Insects are favoured by the sheltered but sunny microclimate and the wide range of food sources, and they in turn are an important food source for the birds. A six-year study of a garden in Leicester recorded such an abundance of insect life that it was calculated that in time the garden would be visited by a third of the entire British insect and spider fauna. The records included eight species never previously recorded in Britain and two which were previously unknown to science.

Parks and Cemeteries

Parks lack the intimate variety of habitats found in gardens. But some of them may contain old native trees, relics of the time when the land was pasture with hedges, perhaps preserved through a period when it was a country house park. The tightly mown grassland has little in the way of wildlife value, but when the frequency of mowing is relaxed diversity increases. This can happen on steep banks, or where bulbs are planted, and is being increasingly done deliberately to create conservation areas.

A more naturalistic approach to public open spaces has been pioneered in the Netherlands and examples can be seen all over Britain. This involves informal plantings of trees, shrubs, grasses and other herbaceous plants, mostly natives, in a way that looks natural and is largely self-maintaining. These plantings are not imitations of rural or natural ecosystems, but are designed in line with the direction that nature might take on that site. (See Colour Section, 39.)

Cemeteries have not usually been so intensively managed and relic communities are more often found there, including ancient woodland, herb-rich grassland (where mowing has not been too intense) and even heathland, for example in some Swansea cemeteries. If they enjoy a certain degree of neglect, cemeteries can develop into unique urban woodlands. They often support a wide range of trees and breeding birds. Woodland ground flora can be deliberately introduced, as it has been in Highgate Cemetery in London, which is a fine example of the genre. (See Colour Section, 42.)

Assessment

Assessment of biodiversity is an essential part of the permaculture design process for any site. Conventional wildlife assessment is often seeking to answer the question, Should this area be preserved as a nature reserve or be destroyed by modern agriculture or development? But the aim of permaculture is to design landscapes where the needs of humans and wild creatures are integrated. So the question is more one of how to work with the whole landscape rather than choosing which bits to trash and which to preserve in aspic.

However, in practice the two approaches may converge. The remaining area of semi-natural ecosystems is now so tiny that we cannot afford to lose any of it, and even permacultural development can reduce biodiversity if it involves a change of land use. For example, a herb-rich grassland will be destroyed if it is planted up with trees. It takes a very long time for a new ecosystem to develop the richness of a long-established one. Given the present scarcity of semi-natural ecosystems which can provide plants and animals to colonise it, it may never happen at all.

Permaculturists are also sometimes faced with questions like, Is it worth campaigning to save this urban common as a nature reserve, or would our energies be better spent on another project?

In fact the best approach to wildlife assessment is to leave your intentions and plans to one side while you actually carry out the assessment and simply allow the place and its inhabitants to speak to you. Decisions about what to do with the land can only be made after you know what's there. If you're looking at a piece of land with the intention of buying it for a specific purpose it's important to make an assessment before you buy it, not after. (See box, Buying Land, p386.)

It's always possible to get advice from people such as the conservation organisations included in Appendix D, Organisations. But such advice needs to be interpreted with some caution. Firstly, although most of these organisations have great expertise at their disposal, you can't guarantee that you'll get perfect advice. For example, I have been advised to plant trees on a wildflower meadow by a man from the county Wildlife Trust. Secondly, different naturalists may have different priorities. A botanist and an ornithologist may have quite different ideas about the same piece of land. So it's a good idea to get advice from more than one source.

But advice is no more than advice. You, as designer, are the one to make the final assessment and the decisions which will arise from it. The biodiversity of the site is only one factor among many which will contribute to the eventual design.

A professional survey can be expensive, though many of the conservation organisations will do one free or for a small donation. Given a keen interest in the subject and some good identification guides you can do the survey yourself. It's a wonderful way to get to know the land intimately, and perhaps the best way to learn about wild plants and animals. You can then get the experts to comment on your findings. Even if you do contract out the whole assessment including the survey, you need to understand the principles on which an assessment is based. These are outlined in the mind map on the next page.

Survey

Finding out what plants and animals are there is the first step. As a general rule it's more useful to record the plants rather than the animals. For one thing the plants stay still, so they're easier to spot. Also, where you have the plants you invariably have the animals. Many animals are totally dependent on one or more specific food plants, while others are pretty exacting in their requirements for food, shelter and nesting sites.

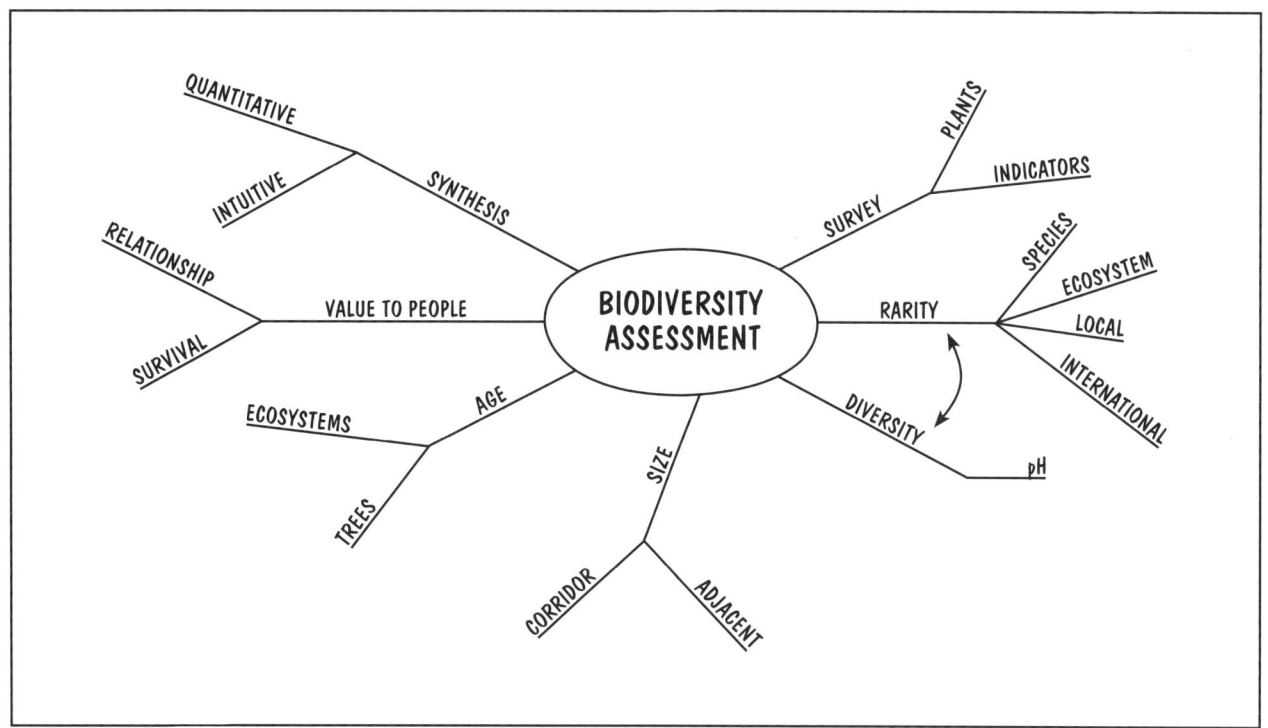

It's important to observe the site through a whole year, although there may be little to record in the winter. Spring and summer inspections should be monthly if not fortnightly. Most of us can only identify plants securely when they are flowering, and some may not flower for more than a few weeks. It's important to visit every part of the site on every occasion, as some plants may be very localised, and differences between different areas will emerge.

It's usually best to wait until you have a full year's observations before seeking advice from experts. But one year is not the maximum. There will almost certainly be new additions to the list in subsequent years and some of these may be really significant. Bee orchids, for example, may remain dormant for years or even decades and then suddenly emerge and bloom in profusion. Observation is something which is never finished.

Taking part in a wildife survey can be fun for children, and open their eyes to the riches around them.

However, if there is some urgency involved, such as a threat to develop the site, or if you're looking at land you're thinking of buying, an expert botanist can be invaluable. They will be able to identify many plants purely by their vegetative parts, and may be able to make a fairly accurate assessment of the site's overall diversity even when there's little visible to a lay person's eye.

An opposite approach to looking at plants only is to look for indicator species, which are often top predators. An example is the spotted owl of the Olympic Peninsula in Washington State, USA. A campaign was waged to save this owl, not just because it was worth saving itself, but because it can only live in a large area of old-growth forest containing a full range of habitats. Where the spotted owl survives there also survive many other plants and animals. On a less grand scale, butterflies have been regarded by some ecologists as indicator species in British woods.

On the whole the plant approach is safer for the non-expert, though any signs or sightings of animals are worth recording too. In old, undisturbed woodland the greatest diversity lies in the inconspicuous invertebrates which may live in the old trees and dead wood. A botanical survey is not a reliable indicator for these creatures, and this really is a job for an expert.[21]

Rarity

It may sometimes be necessary to strike a balance between rarity and diversity. Do we value one place more highly because it's the home of a rare species, or another because there's a greater number of species there although none of them is rare? Of course every case is an individual and must be judged on its merits. Indeed rarity can mean more than one thing.

Firstly there's the distinction between rare species and rare ecosystems. One site may be valued as the home of a single rare plant or animal and another as the only example of that kind of ecosystem in that locality. For example, a field of semi-natural grassland in an arable area may not contain either a great diversity of plants or any particularly rare species, but it may be the sole repository of the local varieties of plants and animals which are now extinct in the wider countryside. Lose them and you lose that much unique genetic variability for ever.

Urban ecosystems can also have rarity value. As we have seen many of them consist of unique assemblies of species found nowhere else in the world.

Then there's the distinction between rarity on a local, national and international scale. There has been some criticism of the amount of resources going into preserving small populations of species which are rare in Britain but not in other countries.

These are often creatures which are at the northern limit of their range here, such as the Dartford warbler and smooth snake which both inhabit the heaths of southern England. These are of course local varieties, and worth consideration as such. But they're not rare species over their range as a whole and don't deserve the share of resources which national pride tends to award them.

Bluebell woods, on the other hand, are a British speciality. Though we may regard them as fairly common they are not so on a world scale and are now officially listed as endangered. (In fact the biggest threat to bluebells is not habitat destruction but global climate change. A warmer climate would favour other, more common plants which could then outcompete them.)

Diversity

The total number of species present is obviously important, but it does need to be seen in the context of the potential diversity on that site. The pH level is particularly relevant in this context. As a general rule alkaline soils can support a greater diversity of plants than acid ones. Thus a chalk downland containing 100 species of plants may not, relatively speaking, be more diverse than an acid meadow in an upland area with a dozen.

This simple idea needs to be used with care. For example, a woodland on a well-drained alkaline soil will often have a herb layer of dog's mercury. This is a very competitive plant and it may succeed in suppressing all other herbaceous plants in the area where it grows. But it's sensitive to trampling and if the wood is coppiced it may be depleted and other woodland herbs which have been lying dormant as seed may reappear. However the trees and shrubs of a wood on an alkaline soil will tend to be more diverse than those of one on an acid soil whatever the herb layer.

Acid ecosystems, such as heath and moor, may be low in diversity but they can support species which are not found in other ecosystems and they deserve consideration on that count.

Size

Other things being equal, larger areas are usually more valuable than smaller areas. There are a number of reasons for this:

- Some predators need a large area of contiguous habitat for a viable territory.

- Some animals need a variety of habitat, eg different kinds of woodland to forage in at different times of year, and this may be difficult to achieve on a small area.

- Both animal and plant species need to be present in large enough populations to give enough genetic variability to prevent inbreeding.

- All populations fluctuate, and a small population can more easily be extinguished by a fluctuation or chance event than a larger one.

- Some woodland species only live in the interior of woods, not on the edge, so large undivided blocks of woodland have a special value.

This doesn't mean that small woods or little ponds are not worthwhile. Smaller areas do support fewer species on average, but not in direct proportion. It typically takes a ten-fold increase in area to double the number of species in isolated ecosystems.[22] However it does mean that any land which is adjacent to another area of similar or higher diversity is all the more valuable. This not only adds to the size of the habitat but provides a source of colonists if the new area is less diverse than the old.

Wildlife corridors which link up semi-natural areas are a second best to large contiguous areas, but where the possibility of doing this exists it can have an important bearing on the assessment. (See pp348 and 351.)

URBAN-RURAL DIFFERENCES

The factors of size and age are more relevant to rural than urban wildlife areas, as urban ecosystems are almost always small and recent. This effects diversity and rarity: we can't expect such a high species count in an urban ecosystem as we may find in the richest rural ones, nor many very rare species.

On the other hand the value to people is much more relevant in an urban context. In cities it may often be appropriate to give as much weight to this factor as to all the others put together.

Age

The longer an area of land has supported the same kind of vegetation the more species it's likely to contain. There are two reasons for this.

Firstly, older ecosystems originated at a time when semi-natural ecosystems were much more common and extensive, so it's likely that there were more plant and animal colonists nearby as it developed.

Secondly, some species are slow to colonise new ground. The ability to colonise varies both between individual species and between different ecosystems.

Aquatic species are among the best colonisers. They have to be because ponds are relatively short-lived in the landscape, gradually silting up as time goes by. Plants, fish and invertebrates may be introduced as seed, vegetative fragments or eggs on the feet of visiting water birds. However there are some aquatic species which are slow to migrate and a new pond is no substitute for an old one.

Woodlands, on the other hand, contain a high proportion of plants and animals which are slow colonisers. Since woodland is the climax ecosystem which all others eventually develop into, under wholly natural conditions it's the most stable and long-lived one. Woodland species survive by persistence rather than by their ability to colonise. While some woodland species, like oaks and lords-and-ladies, are rapid colonisers, others, such as bluebells and dog's mercury can only spread from an ancient woodland to an adjacent new wood at the rate of centimetres per year. Others, such as the native limes and the true oxlip appear never to colonise new woods, even after hundreds of years. Many lichens and wood-living invertebrates are also slow or non-colonisers.

In general, conserving what is already there is much more valuable than creating new wildlife habitats. (See p361.)

Value to People

All the above assessment criteria are concerned with the value of a piece of land to wild plants and animals themselves. But its value as a wildlife site to the local people must be put into the balance too. Obviously the direct benefit to people of the pleasure we get from

---

OLD TREES

Age is as important when assessing individual trees as it is with whole ecosystems, because many of the creatures which live in and on trees only live on old trees. Many of the invertebrates which make up the impressive and oft-quoted tally which are specific to oak (see p368) only start to colonise the oaks when the trees are so old as to be considered senile and worthless by most foresters. In general dead wood supports more life than living wood and there are some species, including rare spiders and beetles, which will only inhabit quite specific parts of the interior of old or dead trees.

Old trees, especially hollow ones with the odd dead branch, standing dead trees and fallen dead wood are all valuable habitats. Yet our passion for tidiness has made them rare, and Britain contains a high proportion of the really old trees which remain in Europe. These trees are a priceless asset, especially on sites where there has been a continuity of old trees since the wildwood, which usually means ancient wood pasture. But in general any old tree will support more biodiversity than a number of younger ones.

Old pollards can be even more valuable for biodiversity than other old trees. Willow pollards need to be cut at fairly regular intervals or the trees will start to collapse under the weight of their branches and die.

One of the greatest values of urban wildlife sites is the opportunity they give us to make a first-hand relationship with nature. These girls are on a Survival Weekend at the Eastwood Farm Woodland, Bristol.

such places is important in itself, but value to people also has indirect benefits to the wild plants and animals themselves.

Firstly, the more a place is valued the more likely it is to survive. With strong economic forces constantly threatening semi-natural areas in both town and country, the strength of active public opinion is often needed to save them from destruction.

Secondly, if an ecosystem is near where many people live it will give far greater opportunities for people to form a first-hand relationship with nature. If we are to start living more sustainably on the Earth we can only do so through a change of heart. If the Earth is nothing more than a theoretical concept we are much less likely to have that change of heart than if we have formed a personal relationship with her. That is unlikely to happen if we don't have regular physical contact with the natural world.

So we may well put a higher value on a much-loved urban ecosystem which is relatively low in diversity than a more diverse rural one. Ultimately it may have more benefit to the survival of wildlife everywhere.

Synthesis

In many cases the process of surveying a site gives you such an intimate knowledge of it that by the end of the survey you have no doubt about how you are going to work with it, or different parts of it. This is often an intuitive or emotional process as much as a rational one.

But sometimes the solution is not so clear. Individual pieces of land may score well under some of the above headings and poorly under others. How do you compare them? There does exist a set of criteria by which the different factors can be quantified and put together to give a comparative index of conservation value. This is really a job for a professional and if you are in doubt as to what to do it's well worth calling in a skilled consultant.

WORKING WITH RURAL ECOSYSTEMS

Historical vs Natural

Broadly speaking there are two approaches to nature conservation in Britain, the historical and the natural.[23]

The historical approach looks at the existing ecosystems in Britain and sees that they have developed with humans as an active part of the system. Coppice woodland, for example, gives ideal habitat for woodland wildflowers, which bloom in profusion in the year or two immediately after coppicing. Many woodland butterflies depend on both these flowers and on the combination of light and shelter found in a coppice woodland. As the coppice regrows it forms thickets which make good nesting habitats for woodland birds such as nightingales, and later in the rotation

it becomes suitable for dormice. Many of these species have become rare as coppicing has declined and some may return when it's resumed.

Proponents of the natural approach note that all the species mentioned in the previous paragraph are attractive ones which people enjoy seeing and hearing in the woods. But most of the really endangered species are invertebrates which need old or dead trees or deep undisturbed shade, and these are often ugly or inconspicuous. They shun edges and thus find little habitat in the fragmented structure of an in-cycle coppice woodland. 65% of the British woodland invertebrates which are classified as threatened are threatened by removal of dead wood and old trees, but only 0.5% are threatened by lack of coppicing.[24]

Therefore the natural approach favours leaving semi-natural woods alone, allowing them to accumulate very old trees and dead wood, and retaining a large undisturbed area with a minimum of edge.

The natural approach has force behind it, but it has a number of limitations:

- It mainly applies to woodland. Moor, heath and grassland would disappear if not grazed by domestic animals, as they would regenerate to woodland.

- Very large areas of non-intervention woodland are needed to ensure that all the stages in the natural cycle of woodland are present, from windfall gaps through regeneration to mature canopy. Many species need one habitat for the larval stage and another for the adult.[25]

- There are few remaining woods in Britain where there has been sufficient continuity of old trees and dead wood back to the days of the wildwood for the species which depend on them to have survived (most are wood pastures, see p346). Indeed many of these species are already extinct in Britain.[26]

- Most of these species have very poor powers of dispersal and colonisation, so the creation of new habitat is of doubtful value to them, even when it is adjacent to existing habitat.

This debate has gone on amongst permaculturists as well as conventional conservationists, though from a somewhat different angle.[27] Permaculture takes a wider view than conventional nature conservation, which concentrates solely on the direct effects of our actions on a particular ecosystem. For example, if a group of volunteers go out into the country to cut scrub which is invading a herb rich grassland, we would consider all the inputs and outputs of the operation, including the pollution caused by driving there by car, the potential uses to which the brushwood could be put and so on, not just the floristic consequences for the piece of land.

Nature conservation for its own sake can seem absurd when looked at in this light. It often involves a lot of car travel to go out scrub bashing, hedge laying or coppicing. The cut wood is sometimes left as habitat for creatures which live on dead wood, but all too often it's burnt, turning what could be a useful output into yet another pollutant. Whether there's a net benefit to the ecology of the planet from these activities is open to question.

It seems all the more absurd when we consider that this work is being done to interfere with a process of nature. It's not a totally natural process since its starting point is a semi-natural ecosystem, but it is the process which Gaia will undertake if we don't intervene. Over a very long time scale there will be no loss of biodiversity, even though in the short term we will lose much that we value. Life has existed on land for hundreds of millions of years, and if some species go extinct they will eventually be replaced.

This permacultural version of the natural approach is often founded more in an intuitive or emotional conviction that leaving nature to her own course is the right approach than in a knowledge of which classes of species are most endangered. It's none the worse for that.

The permacultural version of the historical approach seeks to combine nature conservation with the meeting of our own physical needs. Conventional nature conservationists take it as given that places like coppiced woodlands and herb-rich meadows are uneconomic. But in fact they can be much more efficient in real terms than many of the systems we rely on now, albeit they are not so profitable in the crazy economics of the fossil fuel glut.

Coppicing, for example, compares favourably with plantation forestry in terms of energy input-output ratio, diversity of produce, employment patterns, local economy and soil conservation. Herb-rich grassland may not have the same gross yield as a monoculture of ryegrass treated with chemical fertiliser. But it needs minimal inputs, so its net energy yield is much greater; and the diversity of herbs and grasses gives the animals a diet which positively enhances their health, hence reducing the need for external veterinary inputs.

These ecosystems will be valuable to us in a genuinely sustainable economy. One of the challenges of permaculture is to find ways of making them viable now, not just to provide models for a sustainable future, but also to avoid the ecological absurdities of conservation for its own sake. I would not rule out conventional conservation management where it's the only way to prevent an ecosystem being lost or degraded in the short term. But the key to the long term survival of biodiversity is the realisation that the interests of humans and wild species are really one and the same.

Existing Woodland

In practice there need not be a conflict between the natural and historical approaches. Each is appropriate in different situations. The important thing is to recognise which one is appropriate in any particular situation.

Non-intervention is likely to be the best course in old wood pasture. In woods which have previously been coppiced a completely hands-off policy may be favoured where one or more of the following factors apply.[28]

- There is no immediate need for coppice produce, ie coppicing would be entirely for conservation purposes and the produce would be wasted.

- Poor soil and climate mean that yields would be low.

- The ground flora is inherently poor and would not respond much to coppicing.

- The air is sufficiently clean to allow lichens to grow. Lichens are slow to colonise, and a coppice rotation doesn't give them enough time to get established on trees.

- There are existing old trees with dead wood in them.

- The woodland is large and lies in an area where there is enough adjacent semi-natural woodland to give a viable area, eg middle Speyside, Loch Lomond, much of Deeside, the Chilterns, the lower Wye Valley and parts of the Weald.

- Where continuity of policy into the future can be expected, ie ideally with institutional rather than private owners. The lifetime of one person is not long enough to make non-intervention worthwhile.

Yarner Wood, high up on the edge of Dartmoor, has little diversity in its ground flora but does contain some rare lichens, so it has been left to develop naturally.

Wood anemones and other wild flowers bloom in profusion after coppicing in Lady's Park Wood, which is not far from Yarner but lower down and on a more fertile soil. Coppicing has been restarted here.

Three of these factors, low yields, poor ground flora and clean air, are more likely to be present in the acid-soiled uplands of western Britain than in the lowlands further east.

In small or isolated woods, coppicing, group selection or individual selection forestry are the only ways to get that succession of vegetation structure which can happen naturally in a very large block of woodland. But a small area of non-intervention is almost always worthwhile in a wood worked in one of these ways, ideally sited near the middle of the wood. Any very old trees should certainly be retained. If there are none a few prime trees can be left unharvested to complete their natural lifespan, perhaps those with a poor shape for timber.

The value of leaving a small woodland entirely undisturbed is greatest when: it has been unworked for a long time (more than 40 years); there are old trees within the wood or nearby; or there is a pond or stream in the wood.[29]

It must be said that the majority of broadleaved woodland in Britain is not managed at all, and only a small proportion of the managed woodland is coppiced. So the natural approach is in fact being implemented by default.

Non-Woodland Sites

A hands-off policy on land which is not presently woodland will in almost every case result in the formation of new woodland. In some areas this may be a very slow process, especially where there is a lack of seed parents or grazing pressure from wild animals such as rabbits.

The Trees for Life project, which is bringing back native woodland to the Highlands of Scotland (see Appendix D, Organisations), uses natural regeneration wherever possible, but in areas where there is no native tree for miles around they are compelled to plant. They use seed collected from as close to the planting site as

possible and once the trees are established they are left to do their own thing. Both natural regeneration and planted areas must be fenced to keep out sheep and deer. Although this does involve a great deal of work it's the closest you can get to a non-intervention policy in the circumstances. In a heavily humanised landscape, which the Highlands are, doing absolutely nothing can mean perpetuating the present unnatural situation.

In general leaving land to regenerate naturally may be worthwhile where:

- There is no great biodiversity at present.

- It's adjacent to existing ancient woodland, which will provide plants and animals to colonise the new ground, albeit some of them at a very slow pace.

- The level of nutrients in the soil is not too high. (See box, Eutrophication of Ecosystems.)

- The land is not needed to supply local needs in its present unwooded state, eg as grazing land.

The kinds of unwooded land most likely to have existing diversity are heath and semi-natural grassland. Unfortunately these are just the places which are most often found regenerating to woodland these days, because they have the lowest economic value to present-day agriculture.

Semi-natural grassland is often found on steep slopes. It has survived there simply because the land is too steep to take a tractor, otherwise it probably would have been ploughed and reseeded or at least fertilised. (See Colour Section, 40.) The intensity of grazing is often not enough to prevent regeneration of woodland, and because the grazing is of little financial value people often plant trees on such land. With 97% of some diverse grassland types already destroyed this is a tragic loss to biodiversity.

The brambles on this steep bank represent the first stage in its regeneration to woodland. But is the existing grassland highly diverse? If so allowing the succession to proceed would lead to a loss of diversity.

Permaculturists are not immune to looking on rough grassland as a prime site for natural regeneration. The brambles and thorn bushes which are starting the process often suggest the idea of working with nature by allowing the succession to proceed, and maybe helping it along a bit by planting a few trees. If the grassland is species-poor this can be the best thing to do with it, but if it is diverse it's the worst.

There is a distinction between lowland and upland areas. In the lowlands heathland and unimproved grassland are rare, but in the uplands they are much more extensive, and often rather poor in species even if unimproved. Allowing some upland moor and grassland to regenerate to woodland may be a positive benefit to biodiversity.

In fact the opposite process is much more common in the uplands. Many, if not most, semi-natural woodlands in upland areas are being steadily destroyed by unrestricted grazing, which impoverishes the ground flora and prevents regeneration. (See p321.) Leaving these woods in their present state means losing them. Fencing is needed to exclude grazing animals, or at least to regulate grazing. A carefully regulated level of grazing may actually be beneficial to the woodland ecosystem.[30]

The Short Term

The natural versus historical question is one to consider for the long term future of the land. It need not be rushed and indeed is much better not. But the question of what to do in the meantime, this year or next, remains. The key points are:

- do nothing, or continue as before
- don't eutrophicate
- enjoy a bit of untidiness

Very little harm can be caused by doing absolutely nothing in the short term. How long this holds true for depends to some extent on the kind of ecosystem involved. A grassland which is left unmown and ungrazed will begin to change its character more quickly than a coppice woodland which is left uncut. Both will change with time but the timescales will be different.

Grassland can be left for a couple of years with little harm. Plant diversity will decrease as the more vigorous plants, released from the check of regular defoliation, outcompete the less vigorous. But it will soon return as grazing or mowing is resumed. It is possible to regain diverse grassland after much longer than this. I've even heard of a clifftop grassland which had been bushed over for 30 years and came back with a wide

---

EUTROPHICATION OF ECOSYSTEMS

Eutrophication is the enrichment of an ecosystem with plant nutrients. In extreme cases it can have drastic effects on aquatic systems (see p102), but in general, both in water and on land, it leads to reduced diversity.

The level of soil nutrients which we think of as being normal in a farm or garden is much higher than would be found in most natural soils. Thus the great majority of plants have evolved to live in soils which have a relatively low level of nutrients by our standards, and they don't grow very much bigger or faster in a nutrient-rich soil. A few species have specialised in growing in high-nutrient soils, and when they get the nutrients they like they grow very vigorously, so vigorously that they usually outcompete any other plants growing alongside them.

Nettles are an example. You never find them except where there's an accumulation of phosphate in the soil, usually as the result of some previous human activity, such as an old manure heap. But where they do grow they usually crowd out everything else, and a nettlebed is usually a hundred percent nettles.

Most crop plants, including pasture grasses such as ryegrass, respond well to high nutrient levels. The majority of wild plants, including the grasses

and herbs found in a semi-natural grassland, can't compete if nutrient levels are raised. This means that it's not necessary to plough and reseed a pasture in order to change it from a diverse ecosystem to a very simple one. Merely applying fertiliser year after year will have the same effect.

In addition to direct fertilisation, land in the developed world is subject to general eutrophication, both from nitrate deposition (acid rain) and wind-borne dust containing fertilisers blown from arable land.[31] If this goes on long enough the herbaceous layer of a herb-rich woodland could be reduced to a carpet of nettles.

Where cultivated land is being turned over to wildlife, for example by establishing a wildflower meadow or allowing woodland to regenerate, it's a good idea to choose an area with relatively thin, infertile soil. If that's not possible the next best thing is to grow an exhaustive crop like potatoes or cereals without any manure for a year or two to bring down the level of nutrients. In some cases it can prove difficult to lower the nutrient level by this means, and some conservationists remove the whole layer of topsoil with heavy earth-moving equipment, leaving only the relatively infertile subsoil. The ecological balance sheet on an operation like this is dubious.

diversity of flowers, including orchids, when the scrub was cleared and grazing resumed. But there will always be some loss of species after even a few years, as the viability of seeds and bulbs varies greatly from one species to another.

By contrast woodland will still be a woodland however long it's left, but if regular coppicing or rotational felling of timber have been practised and are discontinued some species can be lost in a relatively few years. For example, violets may be reduced to a few remnant plants on the edges of the wood, in which case the fritillary butterflies which are entirely dependent on them will die out. If coppicing is reintroduced the violets can recolonise the main body of the wood, but if there is no nearby population of fritillaries they may never return.[32]

Continuing with the present management is usually a better option than doing nothing. The present ecosystem will have developed, or at least survived, under the present treatment. Of course you need to be sure that the management you observe in one year is not a recent innovation. For example, species-rich pasture can survive a monoculture of horses for a year or two, but can soon deteriorate under this treatment unless great care is taken. (See p262.) Advice on the care of horse pastures for biodiversity can be had from county Wildlife Trusts.[33]

Planting trees on open ground, tilling or mulching previously uncultivated land, putting in new drains, or resurfacing rides and tracks are the sort of things which should not be done without careful observation and thought. So much harm has been done by thoughtless planting of trees in inappropriate places that I hear there was for a while a band of people known as CATP, pronounced Cat Pee, and standing for Conservationists Against Tree Planting.

The main thing is to keep woodland as woodland and grassland as grassland, but there are finer distinctions than that. For example, a meadow, which is

The front grounds of a small college in Devon. This traditional parkland prospect is valued by the staff of the college, but it's ecologically sterile, and a great deal of energy is expended to maintain it like this.

regularly left to grow tall in spring then mown for hay in summer, will have developed a markedly different flora from a pasture which has been grazed through the spring and early summer. A couple of years of grazing a meadow or mowing a pasture will not have a permanent effect. But a long-term change from one to the other could reduce the existing flora, and the seed of plants which could adapt to the new regime may not be present in the modern landscape.

Boosting soil fertility is another change which should only be made after careful consideration. (See box, Eutrophication of Ecosystems, p359.) In particular care should be taken not to spread manures and fertilisers into hedge bottoms. These are the most diverse habitats remaining on many holdings, and eutrophication can replace a diverse and stable flora with invasive weeds like cleavers, which have little value for wildlife.[34]

Tidiness is one of the biggest enemies of biodiversity in developed countries. It's an old instinct, probably a relic from the time when we were a puny little species at the mercy of natural forces immeasurably greater than ourselves. In that situation any humanising of the environment must have had an element of real survival value and wildness evoked a genuine fear.

For 95% of the time we have existed we have been hunter-gatherers with no more powerful technology than a knapped flint without a handle. Now the boot is on the other foot. Chainsaws, bulldozers and herbicides are wielded by a species whose emotions evolved during the Palaeolithic. In evolutionary terms, the time that has elapsed since we became the most powerful species on the planet is the twinkling of an eye, so we still carry the instincts of the vulnerable underdog. We still have a fear of untidiness.

Short grass punctuated by a few immature trees, and a total absence of weeds seems to be the cultural ideal. But long grass provides habitat for many creatures, including small mammals, which in turn support owls and other birds of prey; brambles provide secure nesting sites for birds and food for insects, birds and mammals; nettles are food for many butterflies; old trees and dead wood are rich and rare habitats; weeds are wild flowers, valuable both in themselves and as sources of food for insects.

I'm not suggesting we should be unneighbourly and release masses of thistledown over nearby fields. But wherever there is no compelling reason to tidy things up I do suggest we learn to revel in the riot of life that expands to fill any space where we have relaxed our controlling hand.

Habitat Creation

Because ecosystems and individual trees become richer in species as they get older, there's a rule of thumb for priorities in nature conservation:

- conserve
- enhance
- create

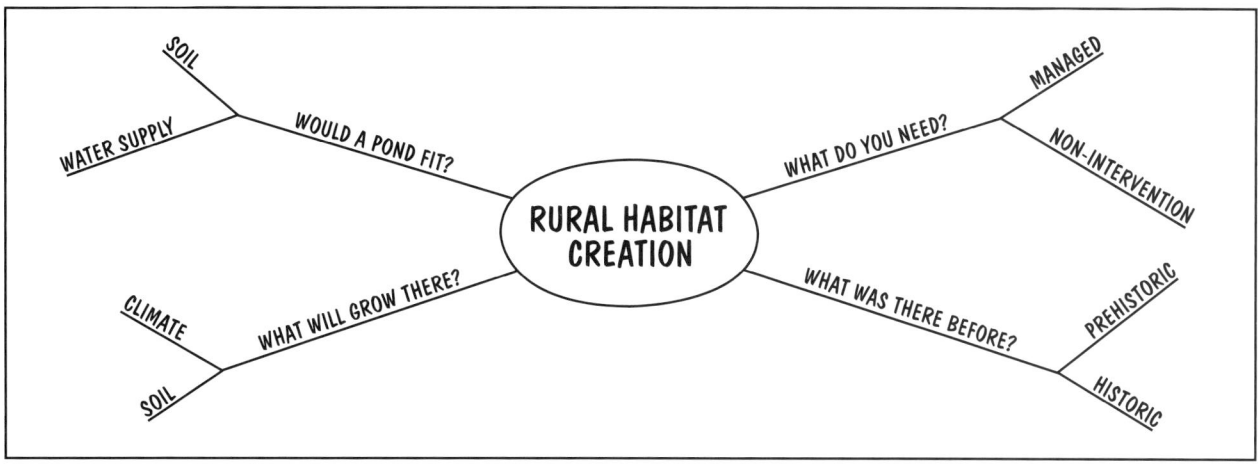

Nothing we can create can ever recompense for an ancient ecosystem destroyed. People have been known to try to justify the destruction of an ancient herb-rich meadow on the grounds that they plan to plant some trees nearby. This is rubbish. Not only is there a net loss of diversity in the short term, but local varieties of plants and animals may become extinct.

Enhancement may consist of enlargement, for example by allowing an ancient wood to colonise nearby land. It could also be a matter of removing something, such as conifers which have been planted in semi-natural woodland, or of discontinuing some practice, such as manuring a herb-rich meadow. It could mean reintroducing some former practice, such as coppicing or grazing.

Many areas of semi-natural vegetation which are only moderately diverse can become more diverse through sympathetic treatment over a period of years. This is especially so if there are other diverse ecosystems in the area, and if the land in question has not been 'improved' for too long or too thoroughly. It's a joy to see wild flowers coming back into a field as it responds to a sympathetic keeper.

Habitat creation should only be considered where there is little or no existing biodiversity. Areas with high soil fertility should be avoided, or have their fertility reduced. (See box, Eutrophication of Ecosystems, p359.) If at all possible local sources of seed or planting material should be used. (See pp370-371.)

Just what to create depends on a number of factors, and these are summarised in the mind map above.

Some kinds of habitat, such as wildflower meadows and coppice woodlands, need regular maintenance work. Others, like non-intervention woodlands and ponds need none or very little. Personally I wouldn't install a habitat which needs work to maintain it unless I needed the produce. If there are grazing animals to feed or a wood stove to fill, a meadow or coppice makes sense, and the needs of the wild creatures become one with ours.

Recreating the kind of vegetation which occupied the site previously is likely to be the most successful approach. This means both the kind of ecosystem, such as heathland, marsh or meadow, and the actual component species. There may be a record of the kind of vegetation that was there in historic times, or it may be possible to deduce it from reading the landscape. The species composition of historic ecosystems can be deduced from visiting similar examples in the locality. Recreating the prehistoric vegetation means in almost all cases establishing woodland, and much can be learnt about the composition of prehistoric woodland in different parts of the country from books.[35]

It's important to assess what will grow well in the specific soil, climate and microclimate of the site. This is a very different matter from assessing which cultivated plants can be grown in a place. We create and maintain the conditions required by cultivated plants, and we resow them or replant them when they come to the end of their lives. Plants which are expected to maintain themselves and reproduce without our help must be as perfectly matched to conditions as possible. If they're not they will sooner or later die out.

There would be no point in planting a wide range of lime-loving trees, shrubs and herbs on an acid soil.

REPLANTING HEDGES

Hedges are one of the most valuable habitats to put back into the landscape. They make a very high contribution to biodiversity in comparison to the small amount of land they take. Indeed they can contribute to the productivity of the land. (See p281.) The course of old hedges may be shown on old maps, and are often visible on the ground, either as a bank in permanent pasture or a soil- or crop-mark in ploughed ground.

The best species to plant are those growing in other hedgerows nearby, and the plants used should definitely be of local provenance. They are best planted in two staggered rows, 60cm apart. Spacing in the row can be 60cm, or 30cm if a small, tight hedge is required.

They would not survive unless the soil was limed both before planting and repeatedly thereafter. That would not be creating an ecosystem; it would be gardening.

One of the most useful habitats to create is a pond. Adding a water body to a landscape with none, or adding still water where there's only a stream, adds another habitat and immediately increases the potential diversity of the area. In addition, as we have seen, many aquatic species are relatively fast colonisers, so a pond can become a diverse ecosystem quite quickly by natural colonisation. (See p355.)

The principles of pond construction are discussed in Chapter 5. Note especially that not every landscape is suitable for a pond. (See p111.)

A seasonal pond can be as valuable for wildlife as one which is there all year round. Amphibians – frogs, toads and newts – can make use of seasonal water to complete their life cycle, but fish, their main predators, cannot survive there. Often a seasonal pond can be made by digging a hole in a marshy area. But think carefully before doing this, as the marsh may have a high biodiversity value already. This could be lost by turning the marsh into a mix of open water and dry land.

WORKING WITH URBAN ECOSYSTEMS

The differences between urban and rural ecosystems have been described earlier in this chapter. But there are other differences between town and country apart from the kinds of ecosystems.

Firstly there is land ownership. In the country some of us may actually own or manage land which has value for biodiversity. In cities and suburbs this is less likely, at least beyond the scale of a domestic garden. However, many areas of spontaneous vegetation in cities are owned by municipal bodies with at least some responsibility to respond to the wishes of the people, or commercial organisations which may value a good public image. The density of population in a city makes it all the more possible to put pressure on these landowners to behave responsibly.

Secondly, the function of urban wildlife sites is somewhat different from that of rural ones. As we have seen, they're usually less valuable from a pure biodiversity point of view but have more value for reconnecting people with nature. Obviously both rural and urban sites can have both functions, but there is a big difference in emphasis.

Thirdly, the historic versus natural debate is not relevant, because urban ecosystems have not been formed by the long-term interaction of wild species and human activities as rural ones have. Practices like coppicing and haymaking may be options worth considering on other grounds, but never on the historical rule of thumb which is usually a safe assumption in rural ecosystems.

Community Action[36]

Community action to preserve biodiversity doesn't just help wild plants and animals; it's often a catalyst which brings people together and plays a part in building true communities. Three ways to get involved are:

- Defending wildlife sites from possible threats of destruction.

- Persuading councils and other landowners to work with the land in a more sympathetic way.

- Becoming actively involved in working on a wildlife site.

A valuable first step to all three is to make a survey of what wildlife habitats there are, and who owns them.

Surveying
In some areas the local Wildlife Trust will have already made a survey. If you're thinking of making one it's as well to ask them before you start or you may find yourself re-inventing the wheel. If no-one has done one yet, the Trust may be a good place to find people who would like to help. A survey conducted under the auspices of a respected body like this can be presented to the local council as a contribution to the overall planning process. To find out about your local Trust contact The Wildlife Trusts. (See Appendix D, Organisations.)

The first stage is mapping the location of the local green spots. The aim at this stage is to find out where they are, not to make descriptive plans of each one. They can be drawn onto an ordinary street map such as the A-Z, or one of the range of Ordnance Survey maps, which are more accurate. (For the range of maps available see pp410-411.) As with all tasks, it's a good idea not to bite off more than you can chew, so the size of the area covered should not be too big. It's better to map a relatively small area thoroughly than to start with a large area and leave gaps in it. It's always possible to expand the area later.

Things to map include:

- Spontaneous vegetation, eg urban commons, railway cuttings, brambles, scrub etc.

- Trees, including tree-lined streets, woodland, parks, well-tree'd gardens, and especially any old trees.

- Wildlife corridors, especially noting small gaps in them which could be filled. (See Linear Eco-systems, p351.)

- Water, including reservoirs, canals, water-filled holes, streams and rivers.

- Special features, eg a fox's earth or a patch of wild flowers.

- Areas with low diversity but plenty of potential, eg close-mown grass.

Winter time can be the best season for this initial survey, because you can see into places which may be hidden by leafy trees and hedges at other times. If you have access to the tops of tall buildings this can be a way of seeing into places which are hidden from the ground.

Patterns may emerge from the initial survey, for example concentrations of green areas which could easily be linked by creating short corridors.

The next stage is to put some more detail on the picture. If many sites have been discovered in the initial survey this could be a big job, so it may be best to concentrate on the few most promising areas rather than to try to survey everything in detail. A large-scale Ordnance Survey map is the best base for this.

Even more important is the species list, which can only be compiled by surveying the sites on a number of occasions through the year. As well as being essential for practical conservation work, knowing which plants and animals live there can be a great help if the site comes under threat. There's a lot more force in saying "Great crested newts live in this pond and it's surrounded by twenty-five species of plants" than "This is a nice pond, please don't wreck it."

An important part of the detailed survey stage is finding out who owns the pieces of land. Landowners have been known to make pre-emptive strikes to destroy what is there before anyone can protest, but most will at least consider modifying their plans to make them a little more wildlife friendly. Sometimes it costs them nothing. For example, when a site is redeveloped streams may be culverted and ponds filled in as a matter of course.[37] If the wildlife benefits of open water are pointed out a stream may be retained as a positive feature.

If threats to local sites do arise, the greatest resource is the local people. It's usually easier to save a site from destruction on the grounds that it's an important amenity which a lot of people use and care about than on the grounds of its ecological importance. Local campaigns with good media coverage can put the wind up councillors who want to be re-elected and have often succeeded.

Parks and other formal green areas are not often under threat, but they have little wildlife value if they're kept in the traditional style. Opinion surveys have shown that most people would prefer a less formal style in municipal parks, with more wildflowers, long grass, undergrowth and songbirds. But the only people who bother to write and complain to the council are the small minority who want things tidier, and these few people have a disproportionate effect on how urban green space is treated. If the rest of us would write to our councils asking for more nature in the parks our cities could start to look very different.[38]

One objection to allowing grass to grow longer is that when it is eventually cut it needs to be removed, whereas gang-mowing of lawns is done frequently

enough that the mowings simply lie on the surface and decompose to nothing in a few days. The cost of removing the long grass can be more than the money saved by mowing less frequently. But these days more and more councils have started composting domestic green waste so as to reduce the volume going to landfill and thus avoid the landfill tax. They sell the finished compost to gardeners. 'Urban hay', as we might call it, could have a cash value as a bulk material to add to this stream. Even if it doesn't show a profit, it could help to cover the costs of its removal.

As well as helping to influence officialdom and other landowners, surveying green space can help to find places which could be taken in hand as urban nature reserves by local voluntary groups.

Hands-On Work

This is almost always done by volunteers. Sometimes a few simple skills are needed and these can readily be gained from organisations such as the British Trust for Conservation Volunteers (BTCV). (See Appendix D, Organisations.) They frequently work in partnership with local community groups, giving advice, lending tools and generally looking after a project till it can continue without outside help.

Very often urban wildlife sites are areas of natural regeneration. The kind of work needed here is usually not so much to 'manage' what is growing there but to change the way people perceive it. Plants will grow and animals will take up residence without any help from us, but they will have a more secure future if people see the site as a nature reserve rather than waste ground.

The three most important tasks are usually to:

- remove rubbish
- make an edge
- put up signs

The effect can be dramatic. I once came across a derelict area on the edge of a new housing estate. Great holes had been gouged out of the clay and had part filled with water, which was acquiring a coating of green scum; willow scrub was invading the space, and the bare ground was rapidly getting covered in weeds. Someone had put a neat fence round it, and a neat painted sign explaining that it was a conservation area. The main feature was a series of ponds, already starting to fill with life, as indicated by the duckweed on the surface; there were the beginnings of a young woodland, and wild flowers were self-seeding abundantly!

In this case the fence and the explanatory sign were rather formal, as befitted a somewhat middle class area. In areas where that kind of formality might be seen as a challenge, as something imposed by 'them' and just asking to be vandalised, a more home-made approach is more appropriate. A hand-painted sign, stating that this wildlife area is being cared for by the local residents is more likely to be respected and have the desired effect.

Alison Malcom

The bridge in Crow Lane Open Space in Bristol used to be regularly vandalised. Then some local school children got together with BTCV and repaired it, and put up posters nearby saying what had been done. It has never been damaged since, and the repair team have become a regular group, with a good reputation locally for tree planting and other environmental improvements.

A fence is not always necessary, but some kind of edging makes it plain that the land is not just abandoned but deliberately left in a wild state because that's what people want. Mowing a metre-wide strip round the boundary is often enough to make this point. If there are stones or bricks on the site they can be used to make a narrow raised bed, like a mini wall with an open centre filled with soil which can be planted up with flowers.

When people think of a place as waste land they usually tip rubbish on it. If possible it's best to get a large enough group of people together to clear all the rubbish at once, because the tipping will probably go on as long as any rubbish is visible on site. But this should not be a reason for small groups to hang back. The best way to attract more volunteers is to start work. People are much more attracted by activity which is already going on than by talk about what is planned.

A creative way to get rid of rubbish is to use it as a support for climbing or scrambling plants. Only inert rubbish, such as pure metal or timber, can be used. Brambles will cover it well, and give food for humans as well as wildlife value. Vigorous varieties of nasturtiums will cover a heap quickly and attractively and will often maintain themselves by self-seeding, though they are slow to start in the spring.

Wildlife Gardening

For many of us, whether we live in town or country, increasing biodiversity in our own gardens is the easiest way to do something practical to help wildlife, both plant and animal. Gardens can never make up for the ecosystems which have been lost in the countryside, but we have already seen how rich in wild species they can be, and the combined effect of millions of gardens could be enormous.

Organic and permaculture gardening are good for wildlife anyway. Maintaining a high level of biodiversity in the garden is the basis of pest and disease control in both approaches. A pond installed primarily to provide habitat for slug-eating frogs will provide habitat for other creatures too. Flowers grown to attract pest predator insects will attract a whole range of other insects. Nettles which host alternative prey for aphid-eating ladybirds also host the larvae of a number of butterflies.

Nevertheless, it's always possible to have some elements in the garden which are there mainly for the benefit of wildlife. This is especially so in larger gardens where there's space left over from the functional things like fruit, vegetables and lawns. In smaller gardens it may be better to think in terms of multiple-function elements with wildlife as one of the functions – though there's always space for a bird table. In fact it's hard to think of anything which has no other use than its wildlife value. Anything which adds diversity is likely to add to the health of the garden and to the aesthetic enjoyment of its human inhabitants.

Some principles of wildlife gardening are given in the mind map on the next page.

Integration. For most of us, our gardens are our main contact with nature, so making it easy for us to see and enjoy the wild plants and animals in it is an important design consideration. The view from inside the house is especially important as the house acts as a hide for watching birds and other animals.

On the other hand, the principle of zoning suggests that the vegetable plot and the lawn should be nearest the house while the wildlife area is best placed furthest away where it will get the least disturbance. In many gardens it's quite possible to place something at the far end and still be able to see it well. In others, especially long narrow ones, there may be a choice to be made here.

A way of integrating many different habitats into a wildlife garden is to model it on a woodland glade. This combines the principles of stacking and edge, and the sunny but sheltered conditions it gives are ideal for a wide range of plants and animals. It's also a way of integrating the different habitats, with a fringe of woodland surrounding grassland and other herbaceous plantings, perhaps with a pond in the centre, where it's visible and in full Sun. In small and medium-sized gardens the 'woodland' is more likely to be a shrubbery, but this is essentially the same kind of habitat.

Many animals need more than one habitat to meet all their needs. For example, the brimstone butterfly, usually the first butterfly to appear in spring, needs buckthorn bushes for food during the caterpillar stage,

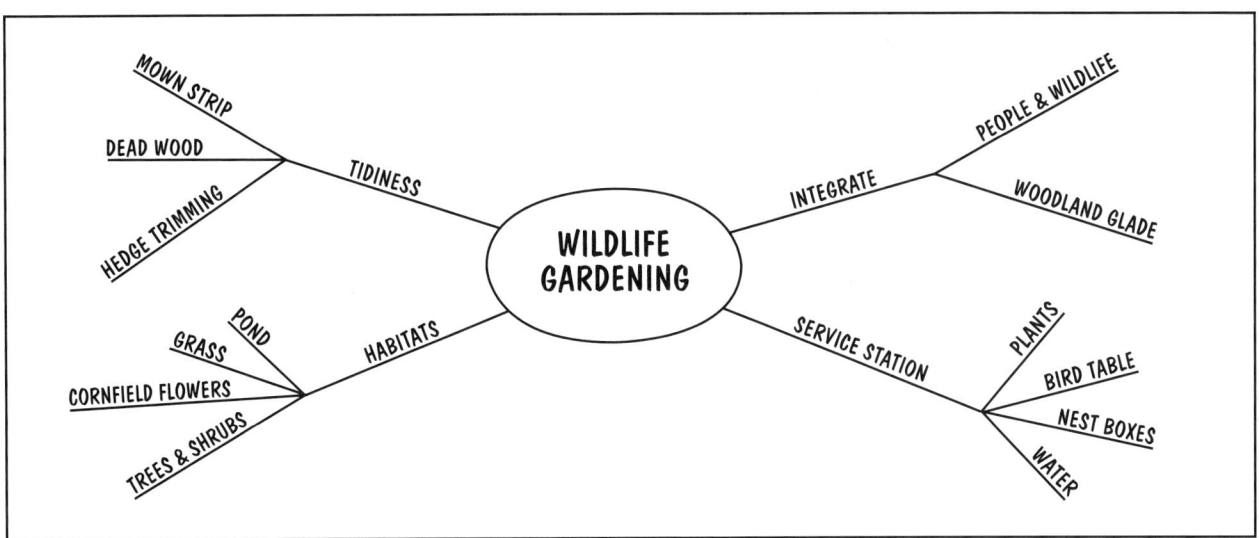

wildflowers of both woodland and grassland for food as an adult, and ivy for hibernation; and they favour a woodland edge habitat. All of these can be integrated into a woodland glade. On the other hand it's not essential to provide everything in one garden, as many creatures will move freely around the locality to find what they want. No-one should be put off wildlife gardening because they only have space for one or two elements.

Service Station. The wildlife service station is a phrase coined by Chris Baines to cover all those things we can do in a garden to provide for passing wildlife. Of course it will be enjoyed by resident creatures too, but the idea is to provide some of the needs of wild animals without necessarily trying to create a complete habitat for them. Some visitors to the service station will be migrants and seasonal visitors, including birds and butterflies. Others will be local residents which find the rest of their needs in the surrounding area.

A major component is plants to provide food, mainly in the form of nectar and pollen for insects and seed and fruits for birds. The main design consideration is to select a range of plants which will provide food at the times it's most needed, or over as long a season as possible. Early and late flowers are important

SOME FAVOURITE SERVICE STATION PLANTS[40]	
Buddleia	Supports a fantastic number of butterflies and bumble bees in high summer.
Cotoneaster	Many species with abundant berries.
Evening primrose	Feeds night-flying moths; beautiful and sweet smelling in the evening.
Grape hyacinth, Honesty, Flowering currant	Three cottage garden classics; early nectar plants; honesty seeds taken by bullfinches.
Guelder rose*	One of the most prolific berry-bearing bushes, and one of the most attractive. (See Colour Section, 25.)
Hawthorn*	Another prolific berry bush; supports a wide range of invertebrates.
Ivy*	The latest nectar plant in the autumn; berries ripe in spring when few others are available; nesting sites for birds; hibernation sites for butterflies.
Japonica (*Chaenomeles*)	Flowers in late winter; fruits taken by thrushes when fallen and bletted.
Michaelmas daisy	A late autumn nectar plant.
Pussy willows*	Catkins provide very early spring pollen for insects, and springtime colour; willows support a wider variety of invertebrates than any other native trees.
Teasel*	Pollen plant; seeds taken by goldfinches; rainwater collects in pockets formed by the base of the leaves, providing a drink for birds.
* Native (though cultivars are commonly sold)	

for insects, to give a life-saving boost in the spring to those which have overwintered as adults, and a good feed before they go to sleep in autumn. (See box on previous page.) Going easy on the dead-heading will leave seed as a food source for birds.

Although the list of plants in the box includes many exotics, there is much to be said for growing native plants in the garden, for two reasons. Firstly, many invertebrates, such as butterflies in their larval or caterpillar stage, can only feed on one species of native plant. Brimstones don't just like buckthorns, they're totally dependent on them, and the small copper is equally dependent on sorrel. Secondly, many cultivated varieties of flowers have 'double' blooms, which, though bright and showy, so crowd the flower with petals that the insects can't get to the nectary.[39]

Bird tables make a significant difference to the survival prospects of some species, especially in hard winters. Many birds which depend on them are winter visitors from distant rural habitats, or indeed from the continent, where winters are much colder on the same latitude.

A wide range of foods, in addition to the standard peanuts and seed, will feed a wide range of birds. Thrush food, crushed grains soaked in water, is a basic alternative for the non-seed-eaters. Not all birds will eat from the table, so some foods should be put on the ground. Security from cats is helped by good all round visibility for the birds, with no close cover. Feeding should stop in March, as the birds will get a much more varied diet containing a wider range of nutrients from natural sources, and they need this for successful breeding. Detailed advice on feeding birds can be had from the Royal Society of the Protection of Birds. (See Appendix D, Organisations).

In freezing weather water is as important to birds as food. They also need to wash, and many other animals, from bees to foxes, need drinking water too. A garden pond can be just as important as a part of the service station as it is as a habitat for resident creatures. When it freezes the ice should be broken daily to allow birds to drink.

Nest boxes are especially important to birds such as blue tits which normally nest in holes in trees. The suburban ecosystem can be ideal for them in all respects except for a lack of trees old enough to have suitable holes. Much the same goes for bat boxes, as the chemicals used to preserve roof timbers have made so many of their traditional roosts uninhabitable for them.

Habitats. A pond is perhaps the most valuable habitat which can be created in gardens. If the water used for the first filling of the pond comes from an old pond with a diverse ecology, the seeds, eggs and spores of many aquatic species will be introduced with it. A pond can be an important food source for bats and birds, which can often be seen hawking for the insects which hover over a pond on summer evenings. Details of pond construction are given in Chapter 5. (See pp117-118.)

Many lawns can soon develop into quite rich ecosystems if mowing is relaxed a bit. Low-growing wildflowers can survive without flowering in a close-mown lawn, and if the lawn is left unmown for a couple of weeks in late May, they will flower. The flowers will in turn attract a range of insects. Another approach is simply to mow a little less often and a little higher than is usual and the low-growing flowers will be able to bloom.

Letting the grass grow too long can lead to the loss of the low-growing flowers by competition from the taller grass, but if it's left to grow all season the taller wildflowers have a chance to grow and flower. They're less likely to be present in a lawn, but they can be introduced. This can only be done with seedlings, preferably in specially prepared plugs. Seed scattered on an established sward will not germinate successfully. Some alternative mowing method will be needed for the long grass at the end of the season, but since it's only once a year doing it by hand is not an excessive amount of work. Scything is a pleasant, rhythmic way to take exercise.[40]

Whichever way a lawn is developed, the mowings must always be removed to prevent the fertility of the soil from building up. If you're starting a wildflower lawn or meadow from scratch on a fertile site it's worth considering removing the topsoil from the lawn area and using it to increase fertility elsewhere in the garden. (See box, Eutrophication of Ecosystems, p359.)

Annual wildflowers used to be common as cornfield weeds, but with modern farming methods they have become some of the rarest of wild plants. A bed of cornfield flowers can give brilliant colour through the summer. It's easy to establish on bare soil in the spring and should self-seed with just a little light raking in the autumn.

The idea of growing native shrubs in the garden has been discussed in Chapter 8. (See pp182-183 and Colour Section, 25-27.) The design theme, from a biodiversity point of view, should be maximum edge. Gardens are edge habitats par excellence and are not big enough to provide habitat for species which shun the edge. Shrubs whose value to wild animals is mainly in their flowers and fruit, such as the wild roses, benefit most from a sunny position. Ones whose main value lies in the leaves, like the buckthorns which feed the brimstone caterpillars, can be in the shadier places.

If there's space for it, at least one tree is worthwhile, if only to give structural diversity to the garden. The birches are a good choice as they are fast-growing, slender, lightly-shading, beautiful at any time of the year, and hosts to a wide variety of insects. Woodland wildflowers can be planted under the shrubs, especially at the edges where they're likely to get enough light to flower. A pile of logs in a shady corner adds the valuable dead wood habitat to the garden.

A shrubbery to benefit wildlife doesn't have to be single-purpose. Although a high proportion of native shrubs and trees will most likely increase the diversity of insects, fruiting trees and shrubs are almost as good. A forest garden with an emphasis on native edible plants can be rich in wildlife. (See box, Some Edible Native Plants.)

SOME EDIBLE NATIVE PLANTS[41]	
Shrubs & Trees	
Bilberry or blaeberry	acid soils only, shade tolerant
Blackberry or bramble	invasive if not pruned
Crab apple	avoid cultivated varieties, fruit a useful source of pectin
Elder	flowers and fruit are both edible, fruit not edible raw
Guelder rose	prefers moister soil, can be used for jam, not edible raw
Hawthorn	haws can be added to hedgerow jam, very young leaves are edible raw
Hazel	all nuts will be taken by grey squirrels if they are present
Raspberry	some wild varieties are very tasty
Rowan	very hardy tree, fruit only edible as a jam
Sweet cherry	same species as the wild cherry, all fruit usually taken by birds
Whitebeam	prefers alkaline soil, fruit edible when bletted
Wild roses	field rose is shrubby, dog rose is a scrambler
Wild service	fruit edible when bletted

Herbaceous Perennials	
Dandelion	minor salad ingredient, needs to be blanched in summer
Deadnettles	white and red, can be used in salads
Musk mallow	mild-tasting salad leaves
Ramsons or wild garlic	very shade tolerant, brilliant white flower, leaves can be eaten cooked or raw
Salad burnet	cucumber-flavoured salad plant
Sea beet	perennial form of leaf beet
Sea kale	stems are blanched and eaten in salads, attractive flowers
Sorrel	sharp, lemony flavour
Stinging nettles	edible when young, excellent butterfly plant
Watercress	can be grown in moist soil
Wild cabbage	more of a kale than a cabbage
Yellow archangel	woodland plant, similar to deadnettles

Herbaceous Annuals & Biennials	
Bittercress, hairy	tiny but delicious salad leaves
Bittercress, wood	a bit bigger, shade tolerant
Chickweed	mild tasting salad plant
Fat hen	salad plant or a tasty spinach when lightly cooked
Jack-by-the-Hedge or garlic mustard	tastes best early in spring, raw or as a pot herb
Shepherd's purse	tasty salad leaves

The wildlife area in my garden. (See case study, pp389-400.)

Tidiness. A wildlife garden doesn't have to be untidy, but an obsession with neatness and straight lines does detract from its value to wildlife. For example, many hedging plants, including the ubiquitous hawthorn, only flower and fruit on second year wood. If a hedge is trimmed every year it has no value as a source of nectar, pollen and berries, though leaf-eating insects and nesting birds will still benefit.

The pile of dead logs in a corner can be as neat as anything, but dead wood left on the tree is an even more valuable habitat, and leaving it there will increase diversity. Questions of safety apart, whether you leave it there is a question of which you value most, biodiversity or tidiness. There's no right or wrong and a different choice will be appropriate in different situations.

The trick of mowing a strip round the edge of an urban common has its place in the garden. If you mow a neat path through the long grassy meadow which your lawn has become, you make the point that the long grass is intentional: it's long because you like wildlife, not because you're lazy. The wild part of my garden is right on the roadside. I keep meaning to put up a sign, "This garden is not neglected, it's a nature reserve." I haven't got round to it yet, but maybe the passers-by understand anyway.

NATIVE OR EXOTIC?

Species

A native species of plant or animal can be defined as one which originally came to the area where it lives by natural means. An exotic is one which was introduced to the area by people. For example, in Britain: hawthorn is native to almost the whole country, Scots pine is only native to the Highlands, beech is only native to parts of southern Britain, and sweet chestnut, which was brought here in Roman times, is an exotic. Only a few exotics are able to maintain themselves as 'wild' plants and animals. These are said to be naturalised.

There are a number of good reasons for growing native plants rather than exotics:

• Natives usually need less help from us in order to grow and reproduce successfully.

• A few exotics become so successful that they outcompete native plants and establish monocultures in place of diverse ecosystems. (See box, Exotics Gone Mad, opposite.)

• Natives are part of the natural biodiversity of the country.

• Native plants support a wider diversity of native animal life than do exotics.

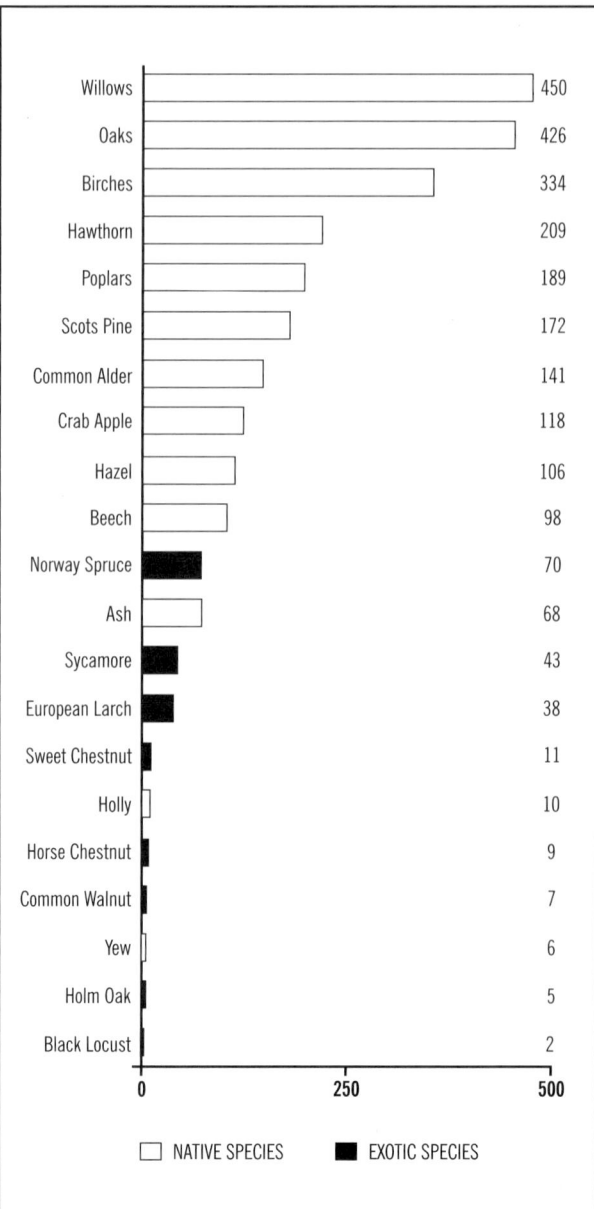

Willows	450
Oaks	426
Birches	334
Hawthorn	209
Poplars	189
Scots Pine	172
Common Alder	141
Crab Apple	118
Hazel	106
Beech	98
Norway Spruce	70
Ash	68
Sycamore	43
European Larch	38
Sweet Chestnut	11
Holly	10
Horse Chestnut	9
Common Walnut	7
Yew	6
Holm Oak	5
Black Locust	2

☐ NATIVE SPECIES   ■ EXOTIC SPECIES

Numbers of herbivorous insects and mites associated with various trees in Britain. Note that nativeness is not the only factor which affects this relationship.[42]

The last point has been dramatically illustrated by a study of the herbivorous insects and mites associated with different trees. (See diagram opposite.) The relationship between nativeness and the number of species supported is striking. It's important to remember that many of these creatures only start to live on the trees when they're well past their fell-by date. (See p355.)

The reason for the difference between natives and exotics is that the native trees have co-evolved with the local fauna while the introduced ones have not. Many invertebrates are specialists, dependent on a single host plant, and even those native trees which only support a few species may be essential to the survival of some specialists. This diversity of herbivorous invertebrates supports a diversity of carnivorous invertebrates, birds and mammals, and these in turn support healthy populations of top predators. Although this study only covered trees it's also reasonable to expect that native shrubs and herbaceous plants will support a wider diversity of insects than will exotics. Many British butterflies are totally dependent on one or a handful of native herbaceous plants for food at the caterpillar stage.

So, in the interests of biodiversity, should we grow only native species? In practice it would be next to impossible to meet all our needs from native plants. However a good rule of thumb is: if we can meet a specific need as well with a native plant as with an exotic, we would do well to plant the native.

Some needs are easier to fulfil with native plants than others. Much depends on whether we're talking about food production, timber, ornamentals, or nature conservation.

EXOTICS GONE MAD

Although most introduced species of plants and animals are incapable of surviving and reproducing in the wild without human help, a few, like pink purslane and muntjac deer, become naturalised and maintain themselves much as though they were natives. But every now and then an exotic plant or animal, released from whatever biological or climatic factors kept its population within bounds at home, becomes extremely successful. This invariably means a loss of local species, either by competition or by predation.

Rhododendron, which has already been described in Chapter 11 (see p312), is an extreme example of a rampant exotic plant. Japanese knotweed is one which is more of a problem in urban areas. It completely takes over wherever it grows and is almost impossible to eradicate.

The grey squirrel (see pp17 and 312) is one animal which is outcompeting its native counterpart, and the recently introduced signal crayfish is another species which replaces its native cousin wherever it becomes established in the wild. Feral mink, escaped from fur farms, are harmful as predators rather than as competitors. They are a major factor in the present catastrophic decline of water voles, along with loss of habitat. In some cases it takes time for a new introduction to become a problem. Rabbits were not able to survive in the wild in Britain for some centuries after their introduction as meat animals. But they eventually became such rampant grazers that they reduced some areas to bare soil, before myxomatosis brought them low.

The introduction of exotics is second only to habitat destruction as a cause of the extinction of wild plants and animals worldwide. Global warming can increase the potential for damage, especially here in the North, where most exotics come from warmer climates and are restricted at present by the cold.

Although a complete moratorium on new exotics is perhaps a shade over-cautious, we should certainly be very careful about growing any plant which is completely new to the area. Plants which spread vegetatively by runners or rhizomes or which are strongly allelopathic are perhaps most likely to become invasive, but in practice it's very hard to predict which will. The fact that an introduced plant will only be grown in gardens is neither here nor there. Rhododendron was only planted in parks and gardens and now it grows where it will.

Nothing can survive under rhododendron, let alone germinate. When the mature trees come to the end of their lives, this woodland will be 100% rhododendron, and virtually devoid of animal life.

Food. How many important food plants can you name which are natives of this country? Wheat, barley, rye and chestnuts come from the Mediterranean or western Asia, and potatoes from the Andes. Cultivated oats and apples are not improved strains of the native wild species but introductions from further south.

Food plants which have been bred from native species, such as cabbage and lettuce, are so far removed from the wild form that they would not survive for a single season without help from us. They can only be considered native in the most theoretical sense.

The list of native food plants given on p367 is not comprehensive, but apart from fungi and water plants it includes most of the important ones. We would be hard put to feed ourselves from them, at least in the foreseeable future. Exotic plants are a normal, necessary part of food growing, though any useful natives are welcome.

Timber and Wood. Here there is much more scope for planting natives. Sometimes the advantages of using exotic trees are almost overwhelming, for example chestnut coppice for durable fencing stakes, or one of the tough exotic conifers for shelter on poor soils in challenging climates. But often there's a choice, for example birch and alder may perform as well as some of the low-value exotic conifers in upland areas, and may maintain themselves by self-seeding to boot. Even where the financial advantage is marginally with the exotics we may choose to put a higher value on ecology and plant natives.

However, the simple division into native and exotic is perhaps over-simple. For example, there is evidence that southern beech (*Nothofagus* spp) supports a richer wildlife than native beech,[43] and the rich ground flora in native woodlands is probably due more to the fact that the trees are deciduous than that they are native.

Sycamore is a vigorous exotic which spreads more successfully than most natives and often replaces them in semi-natural woods. Few invertebrate species live on it and many conservationists regard it as a threat. Others point out that although the diversity of invertebrates it supports is low, the total biomass of the sycamore aphid is very high, and the pH of its bark is ideal for the epiphytes which used to live on elms, now much depleted by disease. They also note that though it is a successful competitor it tends to end up as just one component of mature woods, not the sole dominant.

Nevertheless I would personally only plant sycamore where there was no other tree which could do the required job on the site in question, such as the deciduous component of a coastal windbreak. It would certainly be unwise to plant it in or near semi-natural woods where it doesn't already occur.

To check which tree species are native see the table, Trees for Timber and Wood, on pp338-340.

Ornamentals. There are plenty of native wildflowers, shrubs and trees which are a delight to look at. The reasons why people grow exotic ornamentals or highly bred forms of natives are, firstly, because they increase the range of attractive plants which can be grown, secondly, many of them have a longer flowering season, and thirdly, many have larger, more showy blooms.

Nevertheless, wildflower gardens frequently win awards at the Chelsea Flower Show. There's much to be said for the subtlety and harmony of wild plants, and for the beauty of butterflies and other creatures which need native plants to feed on. An obvious and gaudy display is not to everyone's taste.

Nature conservation. As for places where biodiversity is the top priority, there's a difference between rural and urban situations. Planting exotics would obviously be absurd in a rural nature reserve. In fact the question is usually not whether to plant them but whether it's worth expending a lot of energy removing exotics from semi-natural ecosystems.

In urban ecosystems, as we have seen, a mixture of native and exotic species is the norm. Planting up an urban nature reserve with an assembly of species which might have inhabited the site if it had never become urbanised would seem to be more artificial than using a typical urban mix of natives and exotics. Each city has its own characteristic flora and ideally plants should be chosen which already grow on unmanaged sites in the locality. Exotics like buddleia are quite appropriate in cities and in rural gardens, though they would be out of place in the wider countryside.

The exotic buddleia is excellent for butterflies. It's a welcome addition to the urban flora, and not unduly invasive.

Nevertheless, in general native plants will support a greater diversity of animal life than exotics, and most urban conservationists stick to native species in their planting plans. Perhaps a good middle course for urban areas is to plant only natives, but to accept any self-sown exotics unless they threaten to take over completely.

Varieties[44]

Unfortunately the importance of planting native varieties is not as well known as that of planting native species. But it's equally vital to maintaining biodiversity.

There are few cultivated varieties of native trees and shrubs, but there's a great deal of difference between wild varieties from different parts of the world. At present the majority of 'native' trees in Britain are raised from seed which has been collected in eastern Europe, where labour is cheaper. These trees have evolved in a very different environment from that of Britain and they don't grow well here. What's more they interbreed with native stock, diluting the native gene pool for the future. If this goes on at the present rate the multiplicity of local varieties all over Europe will gradually be replaced by a narrow uniformity.

There are now some nurseries which sell trees of local origin. This is usually stated in the catalogue, though not always, and if the origin is not given it's worth asking. Local should mean not just from this country but from the same region where the trees are to be planted. A hawthorn from Kent is not native in Argyll. The more we demand local origin the more the nurseries will provide it.

An alternative approach is to collect your own seed. It should come from healthy trees and if the aim is a productive woodland they should be of good timber form. Over much of the country there has been severe exploitation felling during the past century, with no provision for replacement, and the trees which are left are often the ones which are useless for timber. So finding good seed parents can require some care.[45]

Avoiding seed which may have cross pollinated with exotic stock can also be difficult. Large blocks of ancient semi-natural woodland are the best bet, though these may be difficult to find in many areas. Small woods are best avoided as the trees may be inbred.

An added advantage of growing from self-collected seed is that it enforces a wait of a couple of years before you can actually plant, time during which you may refine and improve your design greatly.

Most seed of wild flowers and non-commercial grasses also comes from Eastern Europe. There are currently grants paid for large-scale sowings of this seed, for example in the re-creation of chalk downland under the Environmentally Sensitive Areas scheme. Rather than being grant-aided this kind of thing should be illegal. Given the shorter reproductive span of herbaceous plants compared to trees, and that so few native wildflowers and grasses remain, the danger of losing genetic diversity in these species is even greater than it is with trees.

As with trees, it's up to us to specify local, or at least British, seed from suppliers. Collecting local seed is ideal. A good way to create a new meadow or pasture is to take hay from an existing local grassland and spread it over a prepared seedbed in spring or autumn.[46] Seed collecting must always be done with care so as not to harm the ability of the local population to reproduce. The best rules of thumb are: take one in a thousand seeds, and only take from places where there's a thriving population.

PART THREE

DESIGNING

Chapter 13

THE DESIGN PROCESS

So far this book has been about the ingredients of permaculture design. Now we look at how the ingredients may be put together to make a whole. This chapter gives a guide to the process of designing, and Chapter 14 describes some specific methods which can be used at various stages in the process.

These chapters are not meant to be a complete guide to working as a design consultant. That's a skill of its own and if you intend to take up that work you would be well advised to get the appropriate training. These two chapters are intended mainly for people who want to design their own place and perhaps do a little design work for friends and acquaintances on an informal basis.

I've used my own home garden to illustrate the permaculture design process. (See Case Study, pp389-400) I must emphasise that I chose it as an example of the process, not as a model of what a permaculture garden should be like. In fact there's no such thing as a model permaculture garden. The essence of permaculture design is that it recognises the uniqueness of each situation and finds the ideal solution, which will be different in each case.

Design is often thought of as an active process, carried out by a creative person and resulting in a designed object. We could represent this concept diagrammatically:

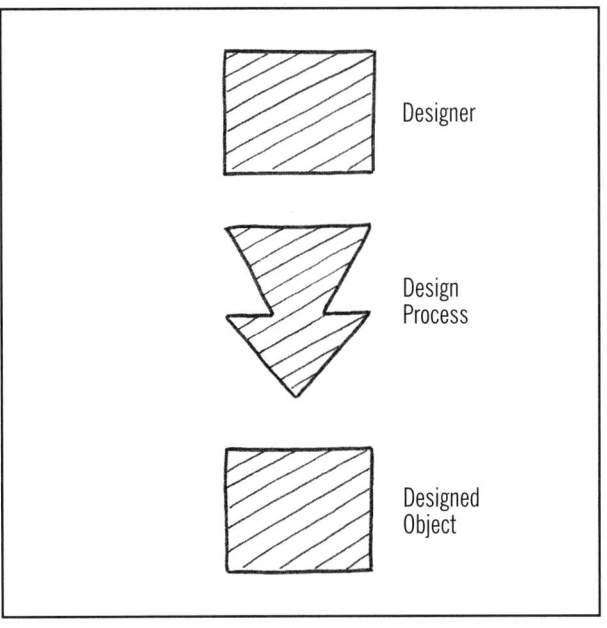

But good design isn't like that. It's much more a matter of listening. In the case of permaculture design there are two main things to listen to, the land and the people. In every design situation we're dealing with a specific place with it's own characteristics, and a specific person or group of people inhabiting it.

Both have things they can offer and things they need. A successful design is a harmonious blend of the characteristics of land and people. We can represent this concept of design diagrammatically too:

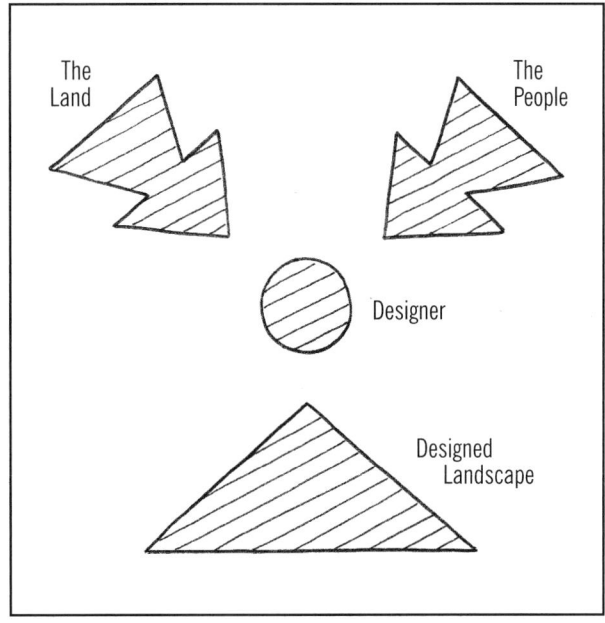

Note that the result is represented by a triangle instead of a square. This indicates that a permaculture design is not a finished object. It's a living system in which people and nature continue to interact. Plants grow and the ecosystem develops, while the people's needs can change with time. A design will

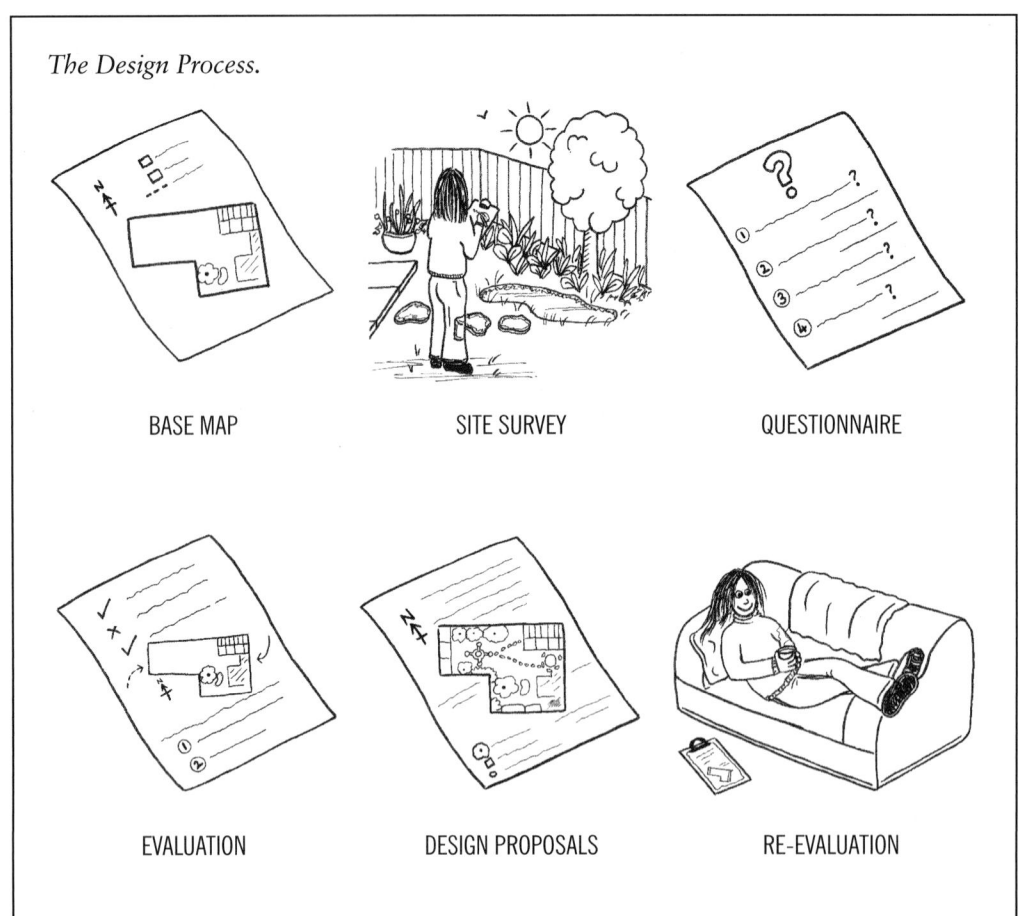

The Design Process.

BASE MAP   SITE SURVEY   QUESTIONNAIRE

EVALUATION   DESIGN PROPOSALS   RE-EVALUATION

never be finished in the sense that it arrives at a final and unchanging state.

We usually present a permaculture design in the form of a map or plan of the design proposals, showing the way it's going to be. This is useful because making a map is by far the most practical way to ensure that everything fits together spatially, and the finished map enables us to visualise the design. But it's important to remember that the design plan is an approximation, at best a snapshot in time which shows how the land will be at one point in its development.

The illustration above shows the main stages of the design process. This basic sequence is the same whatever is being designed, though there will obviously be differences in content. Permaculture design can be used in a wide range of situations, including farms, smallholdings, woodland, parks, community gardens, villages, housing estates, whole settlements and any other kind of land use you can think of. In some of these the buildings are more important than the plants, and in many community projects the focus of the design can be more on the relationships between people than on any physical component.

Some permaculturists would go as far as to say that you can use permaculture principles to design anything whatsoever, from a kitchen to a political system. I'm not quite sure I would go as far as that, but I don't want my choice of example to reinforce the old misconception that permaculture is mainly about gardening.

Before We Start

Whatever the size or complexity of the design project, designing involves both active and receptive modes of working, and the receptive, listening stages of the process come at the beginning. The key to successful design is to stay in the listening mode throughout the early stages and not to shift into the active designing mode too soon.

The first three stages in the chart above – Base Map, Site Survey and Questionnaire – are all essentially listening processes. There is no place for value judgements or decisions during these stages. The Evaluation stage is where we start to make some judgements about what we have learned, and it's only when we get to the Design Proposals stage that we start making decisions about what to do and where to place things.

In practice it's not always easy to stick to this. Ideas start to pop up right from the start. We walk onto a piece of land for the first time and think 'That's a good place for a pond.' Or we start out designing our own place with the idea that the main feature is going to be a forest garden. The place may be ideal for a pond, and a forest garden may be the best solution for our garden, but it's far to early to know that. (See Is a Forest Garden for Me?, p205.)

Ideas like these can easily crystallise into decisions and thus close off other possibilities. The best way to handle them is to regard them as observations, or part of the wish list, saying "I observe that I thought this would be a good place for a pond"; or "I really like the forest garden idea." Handled like this they make valuable contributions to the design, but if they assume the status of decisions the design process has stopped before it's started.

So to some extent the design process is a sequence, where each part needs to be done before the next. But there are times when it's appropriate to depart from the sequence, such as when detailed work at the Design Proposals stage reveals the need for some information which was overlooked in the original Site Survey or Design Questionnaire. The important thing is to resist the temptation to simply impose ourselves on the situation right from the start.

This can be a hard one for us to learn. In our culture we put much more value on the active than the passive. We have sayings like, 'I'd rather be hung for doing something than not doing something.' With this attitude dinned into us we feel we're not contributing anything if we're not putting in something of our own. But listening is in many ways the greatest contribution we can make. Many of the Earth's ills arise from people rushing in and acting before listening to either the land or the people. We need to relearn the value of the receptive mode.

The other important point to make at this stage is that good permaculture design doesn't necessarily have to look unusual or innovative. People often expect to see exotic food plants, forest gardens, alley cropping, circular vegetable beds and so on. These things may be appropriate in particular cases, but they certainly are not the essence of permaculture. The essence of permaculture is that careful observation and thought are put into the initial design. If people say "We're glad we set it up like this. It really works" – then that's permaculture.

BASE MAP

The three listening-mode stages of the design process – making a base map, the site survey and the design questionnaire – can be taken in any order, but it's often a good idea to start with the base map, because mapping a piece of land is one of the best ways of getting to know it.

Making a map takes you to every part of the site and requires you to look at it carefully. The discipline of drawing an accurate map of the whole site often means you see things which, with the best will in the world, would be missed if you were just looking. Because you're concentrating on the mechanics of what you're doing it's also a somewhat meditative, value-free way of observing the site, and this allows the site to speak to you rather than vice versa.

A good base map is essential. As so much of permaculture design is about placing things in the landscape, it's obviously essential to know the extent and nature of the space available. The only effective way to measure this and record it is by means of a map. The base map

Children at Winterbourne Primary School in South Gloucestershire measured the site of their wildlife area and made this map for the management plan. They really enjoyed it.

forms the template for any subsequent maps which may be drawn. Often there is only one other map, that showing the design proposals, but there may be various analysis and concept maps as well, as there are in the example design in this chapter.

Most people find making a simple map both easy and enjoyable. No artistic talent is needed. All you need to do is to work as neatly as you can and follow the steps given in Chapter 14 under Mapping. (See pp408-413.) A few points specially relevant to a base map are:

- Only record what is there. Once when designing for a friend I drew a base map which also had some of her ideas for the site shown on it. It was really confusing because it mixed up what was already there with things which were not. It was useless as a base map and I had to start again.

- Mapping can take forever. However good your map is you can always make it more accurate and add more detail. As none of us has unlimited time we need to know how far to go. The key is to ask the question: Will this extra degree of accuracy or extra piece of detail make the map more useful? Is it relevant to the design?

The most important thing is to get the main dimensions of the site reasonably clear. If you have a garden which is 21.5m long by 3.75m wide at one end and 4m at the other, drawing it as a rectangle of 20x4m is useful. You could get it more accurate, but it may not make much difference to the effectiveness of your design. On the other hand, if you draw it with the same proportions as a piece of A4 paper it will be completely useless.

It's easy to get bogged down measuring and mapping one intricate part of an otherwise simple site. Very often the exact details of a small area are not going to make much difference to the design and they can be drawn in freehand. The relevance of a

A BASE MAP CHECK LIST

Note:
Not every item will be relevant in every design.

Boundaries

External & Internal	Hedges
	Walls
	Fences

Buildings

	Houses
	Other buildings
	Greenhouses

Access

External & Internal	Roads
	Paths
	Gates
Small-scale designs	Doors
	Windows

Plants

| *Broadscale* | Vegetation |
| *Small scale* | Individual plants |

Water

Broadscale	Ponds
	Streams
	Springs
	Mains
	Marshes
	Drains
Small scale	Taps
	Downpipes
	Water butts
	Roof catchments
	Grey water pipes
	Garden ponds
	Drains

Earth Resources

	Sand
	Gravel
	Potting clay

Landform

| | Degree of slope |
| | Aspect |

Microclimate etc

	Wind
	Shade
	Frost pockets
	Views, off site

detail can depend on your intentions. For example, the exact height of a retaining wall or small flight of steps may be significant if you're planning to work with water but irrelevant if you're not.

Decide what's relevant and quit while you're ahead.

- Too much clutter reduces the usefulness of a map. You need to be able to look at it and get an immediate visual image of what the place is like, and too much detail obscures this image.

  If your map is getting too cluttered you could consider drawing more than one map. Subsidiary maps can concentrate on specific themes, such as microclimate, or on small areas which are more detailed and need to be drawn at a larger scale. For example, if you're mapping a farm it may be worth drawing a separate map of the farmyard at a larger scale than the whole farm.

  Alternatively you could consider making one or more transparent overlays which fit over an outline base map, each one covering a different theme. These become useful design tools in themselves. (See The McHarg Exclusion Method, p422.)

- The items on the checklist under Landform and Microclimate etc are usually best treated on subsidiary maps or overlays.

- Features which are outside the design area are not normally included, but they can be if they have a big influence on the site, for example neighbouring trees or buildings which cast significant shade onto the site. If you don't draw them in you may have to keep reminding yourself they're there as you read the map and this reduces its value as a visual tool.

- In some designs it may be appropriate to include a profile in addition to the base map. This is a cross-section of the land showing how steep the slopes are along the line of the cross-section. This is most likely to be useful where water is involved in the design, or on steep sites where some of the land is too steep for bare-soil cultivation and other parts are not.

- The map is not the land. No matter how good a map is it's always a simplification of the actual landscape. It's never possible to design from a map alone. It's not a substitute for being there.

SITE SURVEY

Like the base map, the site survey deals only with what is already there at the beginning of the design process. It covers every physical aspect of the site which may be relevant to the design, from the soil type to the state of the local bus service. The information comes both from observing the land and from asking people who know.

The distinction between the Site Survey and the Design Questionnaire is that the former is about the land while the latter is about the people who inhabit it. There may be some items which could arguably go in either, and it really doesn't matter as long as they're covered at some point.

A great deal can be learned from neighbours, former inhabitants, old maps and old photographs. Sometimes they can tell you things you couldn't find out any other way, such as what plants were growing on the land previously, so you can avoid the mistake of following like with like – something we failed to do in our design. (See p400.)

However it can be a mistake to put total reliance on people's memories. Oliver Rackham, the great authority on woodland history in Britain, writes that: "More than once I have been told by someone familiar with a wood that 'nothing had been done to it in his time'; yet the tree-rings have shown indisputably that it must have been felled well within his recollection!"[1]

If at all possible the land should be observed through all the seasons so that all the microclimate effects can reveal themselves. Taking a full year over the site survey also gives time for quiet contemplation, which tends to lead to the best designs. We call this the Twelve Month Rule, and it's discussed in more detail in Chapter 4. (See p81.)

Very often the site survey will bring up observations you can't fully explain. For example, there may be trees of a certain species which are not doing well on the site, but you can't say for certain whether this is due to an inappropriate soil or microclimate, disease or some other cause. This means that some research is necessary. It's important not to guess the cause if you're not sure you know.

Quantities are important. Not every observation about the land can be quantified, but those which can are often much more useful once a figure has been

put on them. The rate of flow of a water source is an example. Working out quantities can take quite a bit of time and it may be worth waiting to see whether you will actually need that information before you work it out. For example, at this stage you may be interested in collecting rainwater from the roof but not be sure whether you want to aim for self-sufficiency, so it could be premature to calculate the annual yield of the roof.

When designing your own place it's rarely necessary to write down everything about your site, but it is a good idea to jot down anything which you might easily forget, and certainly any quantities which you measure.

The mind map below summarises the categories of things to look for, and the check list overleaf can help to ensure that nothing relevant has been missed out.

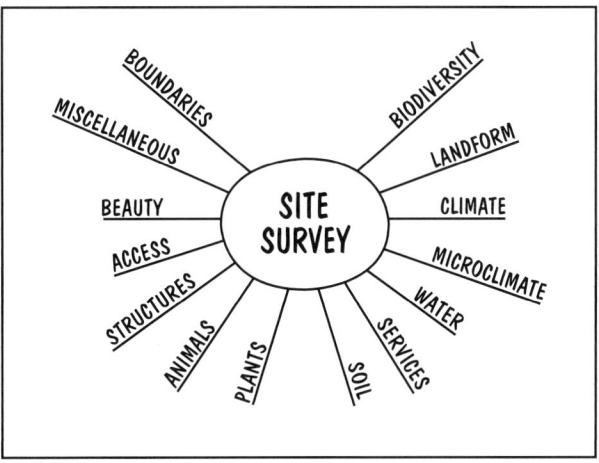

Biodiversity. Here we are looking at wild plants and animals which might be affected if the existing land use was changed or modified. It's never safe to assume that a piece of land is devoid of wildlife value and that any changes we make will necessarily be an improvement for biodiversity. Even the most unpromising areas may hold surprises, and the less diversity there is in the locality the more valuable is whatever remains. Assessment for biodiversity is discussed in Chapter 12. (See pp352-356.)

Landform. The height above sea level has a dominant influence on what can be grown on any site. (See p69 and p226.) In hilly areas there may be a significant difference between the highest and lowest points within the site, and these should be noted.

The soil erosion potential of the land will be affected by the steepness and length of any slopes and the area of catchment. (See pp52-55.) The aspect will affect what crops can be grown there. (For fruit see p238.) Both aspect and steepness will affect the lighting of buildings and the passive solar heating potential. (See p150.)

The landform can give clues about microclimate, especially potential frost pockets. This is second best to observing them first hand, but when it's not possible to observe the site through all the seasons deducing microclimates from the landform is the only option.

THE VALUE OF OBSERVATION

Mike Fisher has an organic market garden in Hampshire. He told me that if there's one thing he would have done differently with the benefit of hindsight it would have been to observe the place for a full twelve months before setting up, and at all times of the day.

He has a serious rabbit and deer problem. If he'd been there at six in the morning a few times he would have seen them, but in the year before he bought the land he only paid a few visits, and those around midday. He reckons that in his first year on the land, before he put in a full rabbit and deer fence, the amount of produce he lost was worth at least as much as the cost of the fence.

Water collection and use is also affected by landform. For example, a garden which is downhill from the house can make use of roof water without pumping. On a broad scale it's always worthwhile to keep an eye open for possible water storage sites. (See pp112-115.)

Climate. Detailed climatic information can be very costly and it's often of doubtful value. The nearest weather station, even if it's quite close, may have significantly different conditions to the design site, especially in hilly or coastal areas. The information you can get from the maps on p70 is good enough for most purposes, and the Met Office web site has a useful series of maps.[2] Observing what's already growing in the district, and how well it's growing, is often more use than a comprehensive set of figures. Where solar and wind energy systems are contemplated, professional advice should be sought.

Microclimate. This is usually much more important than the regional climate and on many sites it's the most important part of the site survey. Both the microclimate of the site as a whole and differences between different parts of it are significant. The microclimate factors are discussed in detail in Chapter 4. (See pp74-83.)

Water. The check list gives a range of water sources which may be available on both broadscale and domestic sites. It's worth noting which of these are actually in use at the time of the survey and which are potential sources.

At this stage in the design process it may not be known just where on the site water will be needed nor in what quantities. So the Needs section of the water survey, as shown on the check list, may be tentative for the time being.

Services. In addition to water mains, the location of electricity and telephone cables and gas mains need to be known. Underground services are important if any earth moving is contemplated and overhead wires may affect tree planting.

Soil. The soil can vary in different parts of even a small site, so every part of it should be checked, even where there are no obvious clues such as changes in slope.

Indicator plants can give clues to the soil type, but reading them is a fine art and no substitute for digging a few inspection pits. A pH test is almost always worthwhile, and sometimes a test for phosphorous and potassium status too. Soil texture tests are easy and quick once you've got the hang of the finger method. A useful habit to develop is that of constantly taking the texture of the soil as you walk over any land you're getting to know. As well as giving useful information about the soil it's a very tactile way of making contact with the land you're walking on.

In urban areas, or anywhere where there may have been buildings in the past, it's important to check for old foundations, tarmac, concrete yards and the like

under the surface. Sometimes an apparently healthy grass sward can be growing on a few centimetres of soil over solid masonry. Systematically piercing the soil with a garden fork or iron spike is the quickest way to check this.

Soil assessment is covered in detail in Chapter 3. (See pp58-66.)

Plants. The base map should show what plants, or on larger sites what vegetation types, are present. But it's also worth noting how well the plants are growing, both to aid the Evaluation stage later on and because the condition of plants can indicate much about the soil and microclimate. Examples are wind-flagged trees and yellow patches in crops due to poor drainage.

Animals. The presence or absence of slugs, rabbits, deer etc can be a key factor in determining what can be grown on the site.

Structures. This includes buildings, walls, fences and any miscellaneous structures. Surveying buildings is a specialist skill, but some indications of what to look for can be had from the Renovation section in Chapter 7. (See pp162-165.)

The relationships between buildings and the land need to be noted. In a garden design these include: rainwater and grey water potential, the most frequently used doors, the windows which are most often looked out of, and the aspects of walls, including walls which are not part of a building.

Access. This includes openings (gates, doors etc) and ways (tracks, paths etc), both those giving access to the site and those giving internal access. These should be clearly shown on the base map, but a number of factors will affect how useful each one is:

* Siting – which places are linked by each opening or way.

* The kind of traffic it can take, eg wheelbarrows or feet only.

* Steepness.

* Whether a path or track channels storm water and thus causes erosion.

* Which seasons of the year it can be used in – important in woods and on farms.

* Whether it needs repair.

A desire line is the line people actually want to take between two points, as opposed to the paths provided for them, which may have been placed inappropriately. They are often seen on school and college campuses as paths of bare earth leading across lawns. Where they

A SITE SURVEY CHECK LIST

Note: Not every item will be relevant in every design. For explanation see text.

Biodiversity

	Ecosystems
	Plant species
	Animal species
	Old trees

Landform

Altitude		
Slope	Erodibility	
	Aspect	
Microclimates	Frost	
	Exposure to wind	
	Suntraps	
Water	Gravity feed	
	Storage sites	

Climate

Data	Rainfall
	Temperature
	Length of growing season
	First and last frosts
	Wind rose
	Sunshine
Local observations	What's growing
	How well

Microclimate

Whole site & sub-areas	Wind
	Light
	Temperature
	Frost
	Moisture
	Combined effects

Water

Broadscale: <u>Sources</u>	Mains	
	Springs	
	Streams	
	Ponds	
	Pond sites	
<u>Needs</u>	Irrigation	
	Animals	
	Aquaculture	
	Domestic	
	Food processing etc	
Domestic: <u>Sources</u>	Taps	
	Roofs & downpipes	
	Grey Water	
<u>Needs</u>	Drinking	
	Other indoor	
	Vegetables	
	Other plants	

Sewage system

Land drains

Services

	Underground
	Overhead wires

Soil

	Differences
	Indicator plants
	Profiles
	Texture
	pH & nutrient tests

Plants

	See Base Map
	Condition
	Indicators

Animals

	Pests

Structures

Buildings	Condition
	Energy potential
	Relationships to land
Walls etc	Aspect
	Condition

Access

Paths, roads, tracks, gates, doors, bridges etc	Siting
	Kind of traffic
	Steepness
	Erosion potential
	Seasonality
	Condition
Desire lines	

Beauty

Views	From site
	Within site
	Of site from outside

Miscellaneous

Resources	Minerals
	Energy potential
Hazards	Midgies
	Fire risk
Special areas	Archaeological
	Sacred
Off-site	Transport
	Other services
	Markets

Boundaries

	Interactions

can't be seen they can often be inferred, though they are quite likely to change in response to decisions made later in the design process.

Beauty. Three kinds of views may be important: those of the surrounding landscape from the site, those within the site, and the appearance of the site itself from outside. Both the good and the bad need to be noted.

Miscellaneous. If minerals such as potting clay or building sand are present it's worth noting and mapping their position, so you can avoid planting trees or building on them if you want to keep them as potential resources.

The aspect and angle of roofs has a bearing on solar energy potential. (See p163.) Wind and water energy potential are only likely to be relevant in remote rural locations.

In some north-western areas, especially the Scottish Highlands, midgies can be a problem. They tend to be restricted to sheltered areas, and it's valuable to know just what the boundaries of these areas are. Fire is very rarely a hazard in British vegetation, except for heathland and conifer plantations in dry areas. If the site abuts on either of these the fire risk must be considered in the design.[3]

Some design sites may contain archaeological features or sacred sites, and these need to be respected.

External features, from bus services to markets for produce, may be very important in some designs and not figure at all in others.

Boundaries. The interactions between the design site and neighbouring land, or water, can be as important as the internal interactions. Walking the boundary and looking to see what's over it is an important part of the site survey. In fact it may be the best way to start. (See Listening to the Landscape, pp416-417.)

DESIGN QUESTIONNAIRE

If you're designing someone else's place for them you clearly need to sit down with them at some point and ask them a whole series of questions about what they want for their site and the personal resources they can bring to it. When you're designing your own place the need to deliberately state what you want is less obvious, but it's just as important.

Going through a set of questions systematically is the best way to make sure that you've covered everything which is relevant to the design and given it sufficient thought. It can also give a chance for the quieter and less assertive members of the family or group to have their say – something which often doesn't happen otherwise. It may bring up some differences, and reconciling these can often be one of the main functions of the designer.

When designing for other people it's a good idea to prepare a tailor-made questionnaire in advance.

When working with your own family this may be unnecessarily formal and it may be more appropriate to have a general discussion followed by a look through the check list to make sure nothing has been left out.

Many different kinds of questions are appropriate, some rational and down-to-earth, others more nebulous and feeling-based. Among the former some can be answered with a quantity and others can't. Anything which can be quantified should be. This will be valuable information later on in the design process.

Basics

A good place to start is with the question: 'What's your overall vision for the site or project?' Some people are quite clear about this while others have never even thought about their place in that way. Sometimes the answer only emerges as you consider the more detailed questions later on. Either way it's an important thing to be aware of from the start, especially if different people in a group have different visions.

Whether a place is rented or owned clearly has a big impact on what can be done. Land may also have legal designations which place restrictions on what can be done or which make specific grants available, for example being in an Area of Outstanding Natural Beauty or an Environmentally Sensitive Area.

Wants

If you've already been living or working on your design site for some time the outputs you already get

---

ONLY LISTEN

When I first started to design for other people I found it hard to keep to the receptive role of listening and not to start making recommendations immediately. I felt I had to justify my presence by telling my clients something. After all, how could anyone value my input if I just sat there and listened? I could not have been more wrong.

On one occasion I completely alienated a client on our first meeting by telling her about gardening techniques, when she knew much more about gardening than I did, and I was not asked to continue with that design. Another time, a couple of years later, I had a sceptical client who was only seeing me because his wife had persuaded him to have me walk round their farm and make a few suggestions. By now I had learned a little more about designing, and felt more confident in myself. I walked round with him for an hour and a half and just listened, occasionally asking a question for clarification. By the end of that time his scepticism had gone and he was open to the suggestions I had.

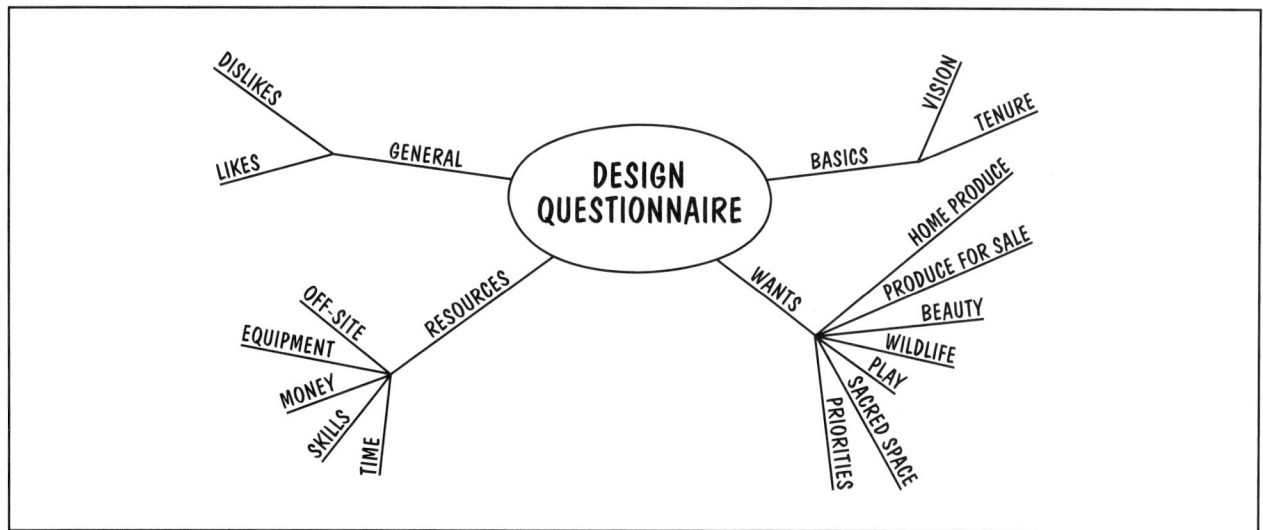

from it are part of the baseline from which the design starts. These outputs may include both physical produce and non-physical things like beauty. You need to note which ones you want to keep and which you don't mind losing.

Moving on to the required outputs, there's a big difference between produce for subsistence, that is home consumption, and produce for sale. Not only does produce for sale have to show a reasonable profit, but there's usually a quantum leap between the quantities needed for subsistence and those needed for sale. A small surplus from a domestic garden can often be traded but it's usually not enough to make a significant difference to the family income.

Produce For Home Use. As well as food, this can include things like firewood, household water and feed for animals.

There's a difference between aiming for self-sufficiency in any of these products and just producing part of what you need. If the aim is self-sufficiency, or something near it, quantities are important. You're going to need to know whether self-sufficiency is possible – Is the roof big enough to supply all our water? Is the garden big enough to grow all our vegetables? – and then allocate the right amount of space for each land use, enough but not too much. These calculations come later in the design process, but at this stage we need to know how much produce is required.

If you're only going to grow a proportion of the fruit and vegetables you eat the quantity is not critical. You will probably eat everything you grow and whatever the quantity it will be welcome. The main exception to this is fruit trees. A single tree of a non-keeping kind of fruit can easily produce more than the family can eat before it goes bad. So you need to have an idea of how much fruit the family eats in a week in order to select the appropriate size of tree. This is discussed in Chapter 9. (See p232.)

The best way to calculate the consumption of fruit, vegetables and eggs is to actually measure how much the family eats over a period of weeks and multiply up to get the annual requirements. Whether you intend to store food, by freezing, bottling or any other method, will have an influence on the quantities of different produce which can be used. Some typical figures for household firewood consumption are given in Chapter 11 (see p304), and for water in Chapter 5 (see p97).

When designing for people who have little experience of growing their own food it can be helpful to have a list of all the possible fruits and vegetables to run through. If you simply ask them what they want they may not have a clear idea of which fruit and vegetables they can grow.

Produce For Sale. If all or part of an income is required from the land or project, the first thing to do is to put a figure on it. This could be expressed as, say, "a living for a family of three", or an actual sum of money.

Ideally the kind of produce to be grown or otherwise produced for sale should be something which flows from the design process, determined by a blend of the nature of the land and the skills of the people involved, rather than something given from the beginning. But many of us have an idea of what we want to do, often a long-cherished ambition. If so it's as well to note it down at this stage, but with the proviso that it may not prove to be the best choice when all things have been considered. (See box, Buying Land, p386.)

Along with physical produce we can consider any other income-generating activities, such as an off-site job.

Other Outputs. In some designs things like beauty, play space and wildlife are more important than the material produce of the land.

Two questions about beauty are: 'What do you find beautiful?' and 'Where do you want to see it from?' You may want a beautiful view from the place where you spend most of your time or you may like the idea of having to deliberately go somewhere to see it. The view your place gives to the outside world may also be important.

Space for play and recreation can be quantified more easily than other non-produce outputs. The number of children and their ages, and the number of adults who may relax on the lawn at one time are relevant quantities. Most lawns are much bigger than needed for recreation. (See p181.)

Priorities. Before leaving the Wants section it's necessary to put an order of priorities on the things which are wanted. It's not always possible to include everything which has been asked for in a design, and without a clear sense of priorities it's possible to lose something really important in favour of something which was only mentioned as an afterthought. This is especially important if you're designing for someone other than yourself and your family.

Asking about priorities may highlight differences between family members. This is not the time to resolve any such differences. They should be simply recorded at this stage and hopefully resolved at the Design Proposals stage.

Resources
Having looked at the outputs we would like to get we now look at the inputs we can offer in order to get them. These are our own personal resources, as distinct from the resources offered by the land.

Time. When you're in the process of designing your own place you're probably at the peak of your enthusiasm. It's easy to overestimate the amount of time you will actually devote to maintaining it once the design has been implemented and working on the land has to compete with all the other claims of a busy modern life. (For figures of time needed, see p177.)

I'm thinking here primarily of a domestic design, where the work will be mainly gardening, but much the same applies to a community project run by volunteers, and it's not so different in a design where you're aiming to earn your living. The time available for productive work is always less than the total time available for work. The hours spent on the phone, filling in forms, doing accounts and nipping into town to get a spare part all have to be taken out before a single plant is planted or seed is sown. Add to that the fact that work always takes longer than the time allowed for it and you can see the need for caution in estimating how much work you can take on.

As well as the amount of time, the pattern of that time can be important. For example, if you want to keep chickens you need to be there every day of the year twice a day to let them out and shut them up, or be able to arrange for someone to be there for you. This will be important if you are often away from home.

Along with the time you have for work you may need to consider your general level of health and in some cases a specific disability. It may also be appropriate to design in a level of flexibility which allows you to gradually wind things down as you get older.

The time needed for implementing the design in the first place, as opposed to maintaining it once it's up and running, needs to be considered separately. It's often appropriate to implement the design over a number of years rather than all at once. (See p388.)

Skills. Clearly you need a much higher level of skill if you want to make your living at something than if you're simply aiming to meet some of your own needs. I wouldn't want to discourage anyone from getting stuck into a bit of home gardening, but if you want to make a living as a market gardener I'd strongly advise you to apprentice yourself to a practising market gardener for a year or two.

Specific skills can often be bought in, especially if they are needed once-off to set up a new project. One skill which is often sadly lacking, in both commercial and charitable projects, is marketing and publicity. The old saying, "If you invent a better mousetrap the world will beat a path to your door," is sadly mistaken. Marketing often needs more time and attention than making the mousetraps themselves.

No-one ever succeeded by being afraid of making a few mistakes, but when it comes to keeping animals there's an ethical dimension to this. Taking on the responsibility for the welfare of a sentient being is a serious matter. Specific skills are needed, and no amount of good will can make up for the lack of them.

Money. As with time so with money, both the quantity and the pattern are important. Some people may have a lump sum to spend on developing their place, while others may be able to spare a regular amount from their income, and some grants may have to be spent within a particular period of time. These considerations will effect the nature of the design, especially the implementation plan.

In a commercial venture you need to consider what you're going to live on during the period before any significant returns start coming in. This leads on to cash flow forecasts and business plans, which are beyond the scope of this book.

The equipment you already have is another thing particularly relevant to commercial projects, as it can make a significant difference to start-up costs.

Off-site. These may overlap with the off-site resources included in the Site Survey, but the focus here is on human resources which can supplement your own skills or work capacity and give support. They may include helpful or knowledgeable neighbours, local gardening clubs or national organisations to which you already belong.

General
Most of the above questions are of the rational kind, looking for specific answers. Some suggestions for more open questions, which can lead anywhere and may engage your feelings about the site, are given in the Check List.

A DESIGN QUESTIONNAIRE CHECK LIST
Note: Not every item will be relevant in every design. For explanation see text.

Basics		Resources	
	Vision	*Time*	For implementation
	Tenure etc		For maintenance:
			Amount
Wants			Pattern
Present outputs	What to keep		Health
	What to get rid of	*Skills*	Domestic level
	Order of priority		Professional level
Produce for	Edible produce:	*Money*	Amount
home use	Kinds of fruit,		Pattern
	vegetables, poultry etc;		Equipment
	Quantities required of	*Any other resources*	
	each;	*Off-site*	Local
	Consider storage &		National
	preserving		
	Other, eg water, firewood,		
	electricity, animal feed		

General
What do you particularly like about the site?
What do you particularly dislike about it?
Which aspects do you most want to keep as they are?
Is there anything you particularly want to get rid of?
What would you most like to see change?
For people already working or living on the site:
What things about it work best?
What things work least well?
Are there any things which might work better if
moved within the site?

Produce for sale	Income required
	Existing plans, if any
Other outputs	Beauty
	Preferences
	Viewpoints:
	Windows
	Lawn/patio
	From outside
	Play
	Numbers
	Ages
	Interests
	Wildlife
	Meditation space
	Education
Any other wants	
Priorities	Order of priority
	Any disagreements

These questions sometimes bring up points which the other questions failed to elicit. In the case of our example design the important subject of privacy only came up in response to one of the questions in this section.

EVALUATION

With the last few questions in the Design Questionnaire some value judgements have come in, in contrast to the pure gathering of facts we have concentrated on up to now. But in the Evaluation phase proper we look systematically at the information collected so far and put some values on what we have found.

One way of doing this, particularly on larger-scale projects, is the SWOC design method – an acronym for Strengths, Weaknesses, Opportunities and Constraints. This is described in Chapter 14. (See p420.) Analysis using the Key Planning Tools, as described in Chapter 2, may also be appropriate at this stage. (See pp420-421.)

Using a definite method of this kind can be helpful but it's not essential. The important thing is to consider what we already have before starting to bring in ideas for change. The two broad categories of things to be evaluated are the site itself and the people's wants.

The Site
It is possible to be too ruthless about removing unwanted plants. Some gardens are full of plants which are, to say the least, inappropriate: conifers, laurels, sycamores and the like. The first impulse of a staunch permaculturist is often to get rid of the lot of them. But this can turn into a bit of a scorched earth policy,

which leaves the garden bare and unsheltered with little visual interest or habitat for wildlife. In many cases a better approach is to remove the unwanted plants bit by bit, if possible planting replacements before the old plants are taken out.

With structures, like sheds, paths, tracks and fences, it's always worth remembering that time, energy and money have been invested in them. Sometimes a creative use can be found for a structure which at first seems inappropriate for the new design. It's a matter of striking a balance between achieving the best possible design and conserving embodied energy.

The People's Wants

There can often be discrepancies between what the people want and the resources available. There may not be enough land, or it may be unsuitable, or the human resources of time, skill and money may be insufficient.

Some of the more obvious discrepancies will have been weeded out during the design questionnaire. Rather than sticking too religiously to the rule that the questionnaire stage is for listening only, it's only common sense to avoid spending time discussing something which is clearly a non-starter. For example, if someone suggests growing blueberries on an alkaline soil it's just as well if a more knowledgeable member of the design group points out that this is not going to work.

Other things may need more careful consideration, especially where quantities are involved, and the evaluation stage is the time to make a few back-of-an-envelope calculations. These may be to do with land resources, such as whether a garden is big enough to provide all the family's fruit and vegetables, or human resources, such as whether two hours' work a week would be enough to produce that food. This is not the time to make detailed calculations, but rather to get an idea of what is likely to be feasible and what is not.

DESIGN PROPOSALS

This part of the design process can be divided into three parts:

* aims
* concepts
* details

The distinction between these three is best explained by means of an example. Let's take that of the design for a new woodland.

Aims: self-sufficiency in firewood
        soil protection on slope
        wildlife habitat
        beauty

Concept: coppice wood, using only native species
        approximate layout of main features, eg rides, glades

Details: exact boundaries of planted area
        glades, rides and ponds, with structural details
        species list
        species layout

BUYING LAND

The evaluation stage of the design process is sometimes the point at which people realise they've bought the wrong piece of land. It may be too small for what they want to do, or too large, in which case the purchase price may have left them short of money to develop it. The climate may be wrong, the soil may be unsuitable, or the existing level of biodiversity may be so rich that to change the vegetation would be to lose much of value.

I was once involved as a consultant in a design process on a farm which had been bought to develop as a demonstration permaculture site. It was a grassland farm, producing mainly sheep, and the vision was to make it much more diverse, with tree crops and a greatly increased proportion of woodland cover.

Unfortunately the climate was unsuitable for top fruit and nuts, being at a moderately high altitude in a wet part of the country. Also the grassland turned out to be exceptionally rich in wild flowers. This meant that any new plantings of woodland would have eaten into what is now one of the rarest and most endangered semi-natural ecosystems in Britain.

The best option for that particular land would be to keep it all as grassland so as to maintain its priceless biodiversity. This would be a perfectly sound permaculture design. Preserving biodiversity is an important part of permaculture, and there are thousands of farms with low diversity where agroforestry can be developed. But it would not be a design which visibly demonstrates permaculture.

It was a dilemma, one which we never really resolved, and one which could have been avoided if a suitable farm had been bought in the first place.

People who are in a position to buy land, whether privately or as part of a community project, are in a strong position. They can actually choose the land to suit their aims. But it's a position which is all too easy to throw away. The design process needs to start before looking for land to buy, not after it has already been bought.

plant spacing, and numbers of plants required
size of transplants to be used
tree protection and mulching
expected yield of firewood
supplier of plants and materials
costing
implementation plan
maintenance plan

This working order is a useful discipline. It can often be tempting to get stuck right into the juicy details. Many people have their enthusiasms, perhaps for certain plants or certain technologies such as wind generation. Even if we don't have favourites of this kind, it's easy to drift off into discussion of details before taking a look at what the major options are. At worst this can lead to the ideal solution being overlooked altogether, and at best to time being wasted on details which later turn out to be irrelevant.

When a group of people are designing together it's often appropriate to work on aims and concepts together, then split down into sub-groups or individuals to work on detail, and reconvene to make final decisions on what has been worked out.

As with the design process as a whole, it's sometimes appropriate to go back to an earlier stage. For example, in the woodland design it may emerge at the Details stage that the yield of firewood will not be enough for self-sufficiency. One response could be to go back to the concept and question the natives-only policy. The inclusion of some exotics such as sweet chestnut or hybrid poplars might make all the difference. Another response could be to go back further and look at the energy performance of the house. Reducing the need for firewood may be the best solution.

This brings up the permaculture principle of designing in wholes. (See pp35-37.) It's important to be aware that what we are designing is rarely the whole system, and things we propose may have implications beyond the boundary we're looking at, either locally or globally.

Aims

Although the aims of the design will have been discussed quite a lot at earlier stages in the process, they will not so far have been stated definitively. It's always worth doing so at this stage. We need to know exactly what we're trying to achieve before we look at the means of achieving it, and we're often not as clear about our objectives as we think we are.

Concepts

There are two kinds of design concept: themes and layouts. The woodland example above shows one of each. But often there are a number of optional themes and layouts which can be considered. Themes tend to be more fundamental than layouts and should usually be considered first.

For example, in a garden design there may be a choice of themes between a formal approach, with separate orchard, fruit cage and vegetable plot, on the one hand, and a more naturalistic forest garden which combines all three on the other. Clearly there's no point in working on layouts till this choice has been made.

The best way to choose between different layout concepts is to make a number of photocopies of the base map and sketch the different options on them. This is what we did in our garden design. (See pp395-396.)

It's important to use an accurate base map of known scale and sketch things in with a reasonable degree of accuracy. Some things need to be drawn more accurately than others. For example, our greenhouse looked too big to fit in the space where the redcurrants were, in the south-east corner of the garden. But drawing the greenhouse accurately to scale revealed that it would indeed fit. We wouldn't have known that that was an option if we hadn't been working to scale.

Other features, for example the lawn and fruit areas in our example, can be sketched in more freely, but they still need to be of a realistic size. If you sketch in pencil rather than pen it's easy to check the size of what you've drawn with a ruler and adjust it if it's way out of proportion.

Developing a number of alternative concepts is a good way to work when designing for someone else, or when one member of a family or group is doing most of the design work. The concepts are then presented to the clients or the rest of the group for discussion. Sometimes one option will be selected more or less as it is, and sometimes the presentation can start up a creative discussion which leads to new ideas or a 'pick and mix' from the existing concepts. People who are unused to landscape design may find it difficult to contribute to the process until they have something concrete to react to.

In general, presenting people with alternative concepts draws them more actively into the design process. The participation of the people who will use the site or project which is being designed is important to the success of a design. While a consultant designer may have more technical knowledge, the clients usually know the site better and certainly have a better feel for what will work for them. They will also feel ownership for a design they have participated in, and a have greater commitment to seeing it succeed.

Details

This part of the process is mostly a matter of putting flesh on the skeleton of the chosen design concept, but there are also a number of specific things which may need to be considered, including:

- Yield estimates.
- Costings.
- Suppliers.
- Implementation plan.
- Maintenance plan.

In general, the larger and more complex the design the more relevant these become.

- Yield estimates are important either where the intention is to make some income from the design or where self-sufficiency in one or more products is wanted. These estimates should be compared with the requirements given at the Design Questionnaire stage. (See p383.)

  Typical yield figures for vegetables are given in Chapter 8 (see p179), for fruit in the Fruit Table (see p230), for nuts in Chapter 9 (see p216), for firewood in Chapter 11 (see p304), and for rainwater in Chapter 5 (see pp97-98).

- Much the same applies to costings: the bigger and more commercial the design, the more important it is to estimate the cost of implementing it. In our example design we didn't do a formal costing.

- When designing for other people, a list of suppliers of the plants and materials recommended is among the more useful things you can give them. (See Appendix E.) It's always worthwhile to go for the highest quality you can afford, and trust more to personal recommendation than advertisements.

- An implementation plan is almost always worthwhile. There are two reasons for making one.

  Firstly, some things need to be done in a certain order or at certain times. For example, a windbreak needs to be well established before planting an orchard on a windy site; and bare-rooted trees can only be planted in winter. Combining these two factors may show how the plan needs to be implemented through time.

  Secondly, you may not have the time, energy or money to implement the design all at once and the job may need to be spread over a number of months or years.

  This can be particularly important if you're designing for other people, especially if they have not participated much in the design themselves. At first sight even a relatively simple garden design can look like a mammoth task, and all too often designs are left unimplemented simply because they look daunting. Breaking it down into bite-sized chunks can help people to see that it is in fact quite achievable.

  It can also be a help to give priorities, perhaps even to say "If you only implement one part of this design, make it this part." Without this guidance it may be hard for the client to recognise which are the really important bits and which are optional extras.

  Where people are making a living from the land which is being designed, the need for income in the short term needs to be balanced with implementing the design, and this will often dictate a step-by-step implementation plan. Cash flow forecasts and workload patterns both need to be considered.

- An actual maintenance plan may only be appropriate on larger designs. But maintenance requirement is always something to bear in mind while designing, whatever the size and nature of the design job. Keeping work to a minimum is one of the main aims of permaculture.

  Spreading the workload throughout the year is especially important on farms, smallholdings and other design projects where people will be working full time. Woodland work, much of which is done in winter, can mesh well with farming and gardening. Resources can be spread in the same way. For example a building which is used for food processing in summer and autumn can be used for wood working in the winter.

RE-EVALUATION

The process of designing never really stops. As you implement the design and start living with it you can learn from experience. You may see improvements which can be made, discover mistakes in the original plan, and react to changing circumstances. This can be an extra reason for implementing a design over a number of years. Experience in the early years may suggest improvements which can be made before the later parts of the design are implemented.

If you're designing for other people, re-evaluation means revisiting the project in later years to see how it turned out. This can take a bit of nerve: your design may not be successful, or not be implemented at all, and both situations can be embarrassing. But it's worth risking this, because seeing what has worked and what hasn't is the best way to improve our design skills.

CASE STUDY

DESIGN PROCESS

for No 37 Leg of Mutton Road

Home of Patrick and Cathy Whitefield

Site Survey

See Base Map overleaf.

The site is a rented bungalow and garden on the edge of a small country town in Somerset. It's situated on the western crest of a steep hill, 65m above sea level, with flat land stretching from the foot of the hill to the sea. There's a belt of trees on the other side of the road, but a gap in the trees directly opposite the house gives a fine view over the countryside to Exmoor, some 30 miles away, and allows the prevailing west wind to blow strongly over the front garden, to the west of the house.

This wind is the dominating microclimate factor in the front garden. The back garden is sheltered from all directions at ground level, but the two pear trees can suffer from wind at blossom and fruiting time. It seems to be too windy a site for top fruit, really. The great advantage of the hill-crest position is that early and late frosts are never a problem and winter temperatures are relatively mild.

The Microclimate Map overleaf shows the incidence of direct shade in the garden. Most of the shady areas receive good indirect light, especially the front lawn, which is fully exposed to the afternoon Sun. The northern part of the back garden, immediately to the south of the low south-facing wall, is warmer than the rest of it. This was shown by a crop of over-wintering broad beans planted in different parts of the back garden. (See Colour Section, 11 and 12.) Whether the same is true of the eastern strip of the front garden, immediately to the west of the house wall, is not known, but given the windiness there it may not be.

The local climate is moderately maritime by British standards, with average temperatures of 17°C in July and 5.5°C in January, and a rainfall of 750mm.

The soil over the whole garden is a fine sandy loam, with a pH of 7. At present it's low in organic matter. The garden has been partially terraced and all parts of it are flat enough to cultivate except for parts of the two strips nearest the road. (See Key Planning Tools Analysis Map on p393.)

That part of the garden is also the least accessible, being at the bottom of a long, steep flight of steps. The gate here gives access to the road, which is the route to the town centre by foot or bicycle. The gate at the opposite end of the garden, the east, gives access to the car park, bus stop, post box and local shop, and is the one most often used. The back door, on the east side of the house, is more often used than the front door on the south side.

The existing plants are shown on the Base Map. The two blackberry plants are pruned right down to stumps. Also present but not shown on the map are some decrepit gooseberry bushes in the bare soil area and a productive grape vine in the greenhouse. Ash trees are regenerating freely in the western part of the front garden.

The most significant animal species present are slugs and snails. These may limit the range of crops grown and/or the methods of growing.

There's no particular biodiversity value on the site at present. The two pear trees are some thirty years old, which is as old as any trees in the immediate vicinity.

However the garden is part of a mosaic of habitats on the rural/suburban edge, and provides food and other resources for mobile animals such as birds and flying insects. 24 bird species have been observed from the house over a 12 month period. The garden is not accessible to less mobile animals, such as reptiles and small mammals, because it's separated from other habitats by inhospitable terrain such as roads and close-mown lawns.

There's a high population of cats in the neighbourhood, at least three of which regularly visit the garden. This means that there's no point in providing nesting sites for birds. It also means that it will not be possible to establish frogs for slug control.

Garden water sources are shown on the Base Map: an outside tap and two downpipes, each draining one half of the house roof. Water supply, sewage, gas and electricity are all on the mains. The water supply is not metered. There is gas central heating and no fireplace or chimney. A possible resource is a scattering of stones in the herbaceous border in the front garden, evidently put there in an unsuccessful attempt to make it look like a rockery.

The windows from which the garden can be seen from the house are shown on the Base Map. Both west-facing windows also have the long view over the countryside.

There is an hourly bus service to the town centre from a bus stop 300m from the house on a fairly level walk. Long-distance transport is less well provided, with the nearest station some 15 miles away with no regular bus link. The other main off-site resource is our field, 5 miles away. This is mainly herb-rich meadow, kept as a nature reserve, and there is space there for a small orchard.

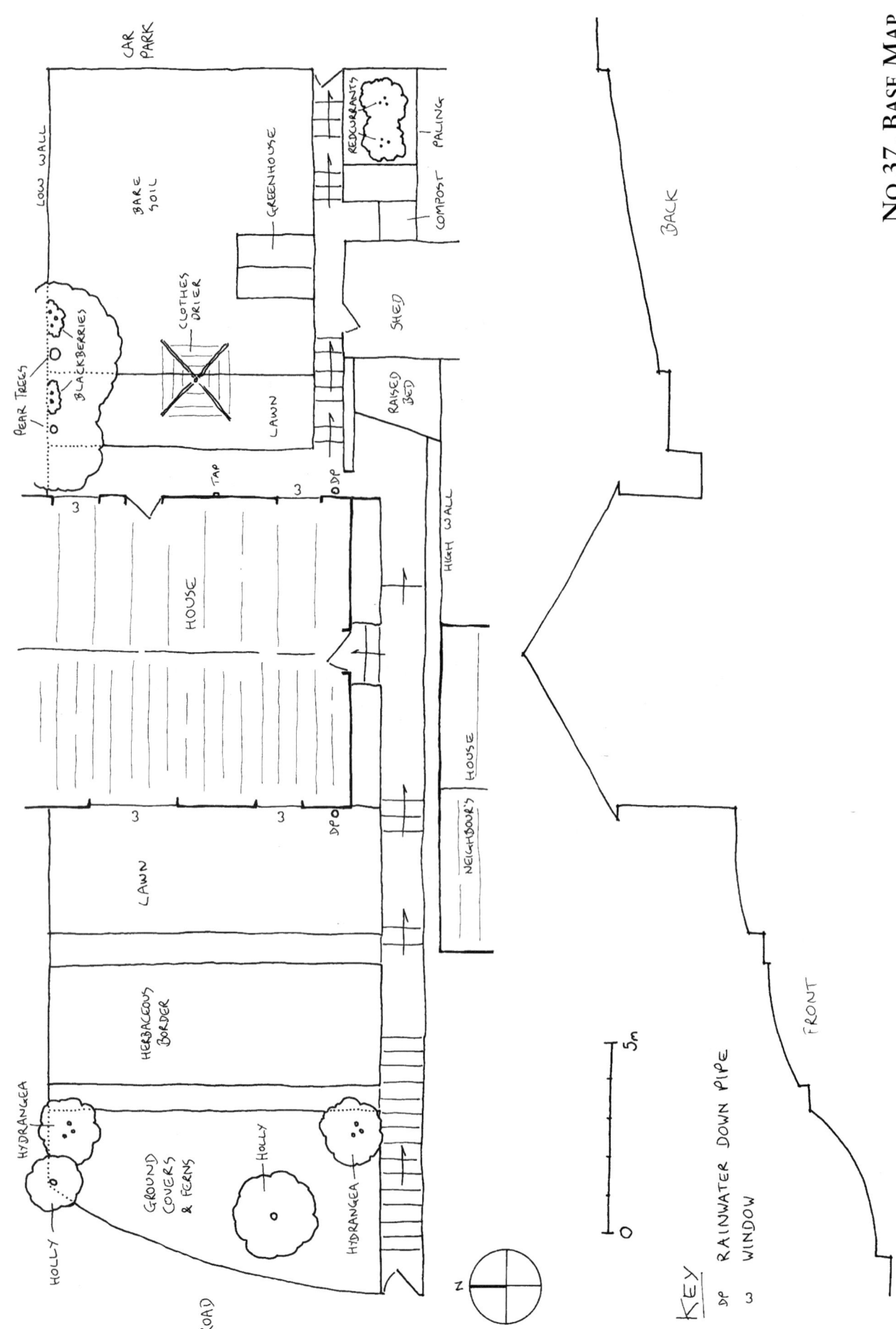

No 37, BASE MAP

CAR PARK

LOW WALL

BARE SOIL

GREENHOUSE

REDCURRANT?

COMPOST

PALING

PEAR TREES

BLACKBERRIES

CLOTHES DRIER

LAWN

SHED

RAISED BED

TAP

3

DP

3

HIGH WALL

HOUSE

NEIGHBOUR'S HOUSE

3

3

DP

LAWN

HERBACEOUS BORDER

HYDRANGEA

HOLLY

GROUND COVERS & FERNS

HOLLY

HYDRANGEA

ROAD

N

BACK

FRONT

5m

0

KEY

DP   RAINWATER DOWN PIPE

3   WINDOW

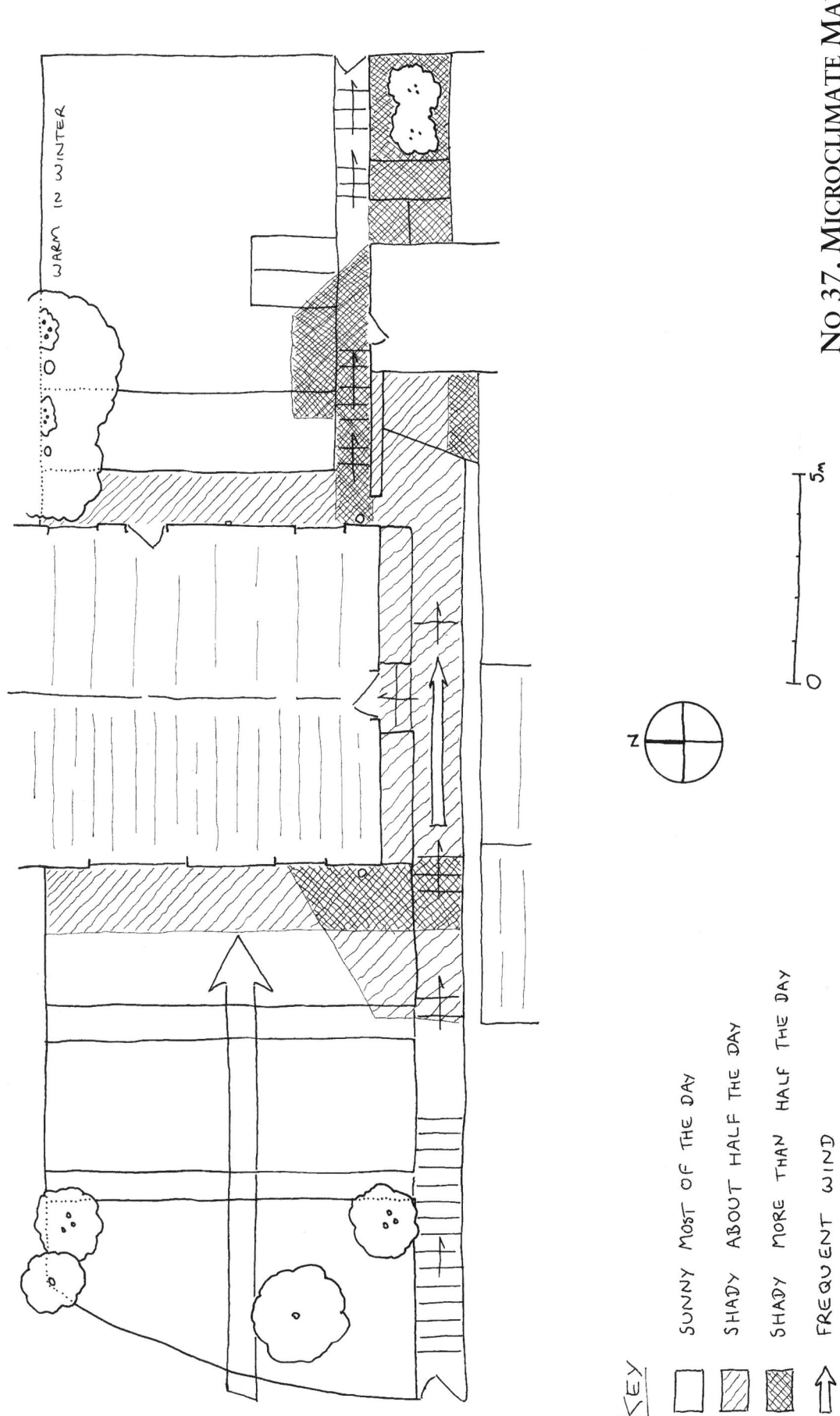

No 37, Microclimate Map

WARM IN WINTER

5m

0

N

Key

SUNNY MOST OF THE DAY

SHADY ABOUT HALF THE DAY

SHADY MORE THAN HALF THE DAY

FREQUENT WIND

Design Questionnaire

Two of us live at No 37, myself and my wife Cathy. At the time of the design we had lived here for a year. We had not made any changes to the garden in that time, except to plant out our collection of perennial vegetables in the part of the back garden marked Bare Soil on the Base Map, which had been the previous inhabitants' fruit and vegetable area.

Basics
Our vision for the garden
First and foremost we want our garden to be part of a sustainable lifestyle. We would like whatever we do at home, in house or garden, to make a positive contribution to that end. Beyond that, each of us has a slightly different vision for the garden itself:
Cathy – to be both beautiful and productive.
Patrick – a garden which produces food with a minimum maintenance requirement.

Tenure
Rented shorthold. We're not sure how long we'll be here for.

Wants
Produce for subsistence
We aim to grow a proportion of our own vegetables. We can easily buy organic vegetables, so there's no need to grow a complete range ourselves. The most important ones to have from the garden are salads and greens, as these don't keep well and we have one or the other with virtually every meal. We would like them to be available all year round.

We would like to grow a range of soft fruit. Raspberries are our favourite, and currants and gooseberries our least favourite. We would consider growing top fruit in the garden.

We would like to collect water from the house roof, mainly for garden use.

Beauty
This is a higher priority for Cathy than for Patrick.
Cathy – I would like plenty of flowers in the garden. I particularly like the potager idea, with flowers and attractive varieties of edibles grown together. I would like to be able to see colour from the west and north-eastern windows (see Base Map), and from a lawn or other sitting area in the garden. The area to the south of the sitting area is important as I like to sit facing the Sun and to have a colourful view in front of me.
Patrick – I particularly value the long view to the west from the house. I would like a pleasant view from the window of my work room, the south-east window. I find as much beauty in the shape of different leaves as I do in flowers.

Other
We like to eat outside whenever possible and ideally would like a patio, but failing that we will eat on the lawn. The lawn should be big enough to take a group of eight to ten adults comfortably. This is a compromise between Cathy, who likes the feeling of space you get from a relatively large lawn, and Patrick, who would like the smallest possible area of lawn.

We would like to provide wildlife habitat and 'services', eg pond, bird table.

Cathy would ideally like an outdoor meditation space but she feels she couldn't get what she wants in this garden, which is small and surrounded by other people.

Patrick enjoys experimenting with perennial vegetables.

Priorities
Cathy – lawn, flowers, edibles, in that order.
Patrick – edibles, interesting plants, low maintenance requirement, in no particular order.

Resources
Time
Neither of us has very good health and our work sometimes takes us away from home for a couple of weeks at a time, often at crucial times in the gardening year, so low maintenance is a high priority. Patrick does have spells when his health is good enough for moderately hard work and he can use these times for implementation work.

Skills
Patrick is a fairly competent gardener and capable of carrying out simple construction projects. Cathy has an eye for aesthetic design.

Money
We can't put an exact figure on what we are prepared to spend on the garden. But as this is a rented house and we don't know how long we will be here, the upper limit will be hundreds rather than thousands.

General
Cathy – The thing I would most like to keep is the ornamental aspect of the front garden. I would like to lose the rectangularity of the garden, especially at the back, and to move the greenhouse so that it's not right over the sitting area.

The thing I particularly dislike is the lack of privacy. Being adjacent to the car park is worst, but the back garden is also overlooked from both sides. We get on well with the neighbours, but I'd rather be able to use the garden in privacy. The front garden is in full view of people walking up and down the road, and it's noisy from the traffic.

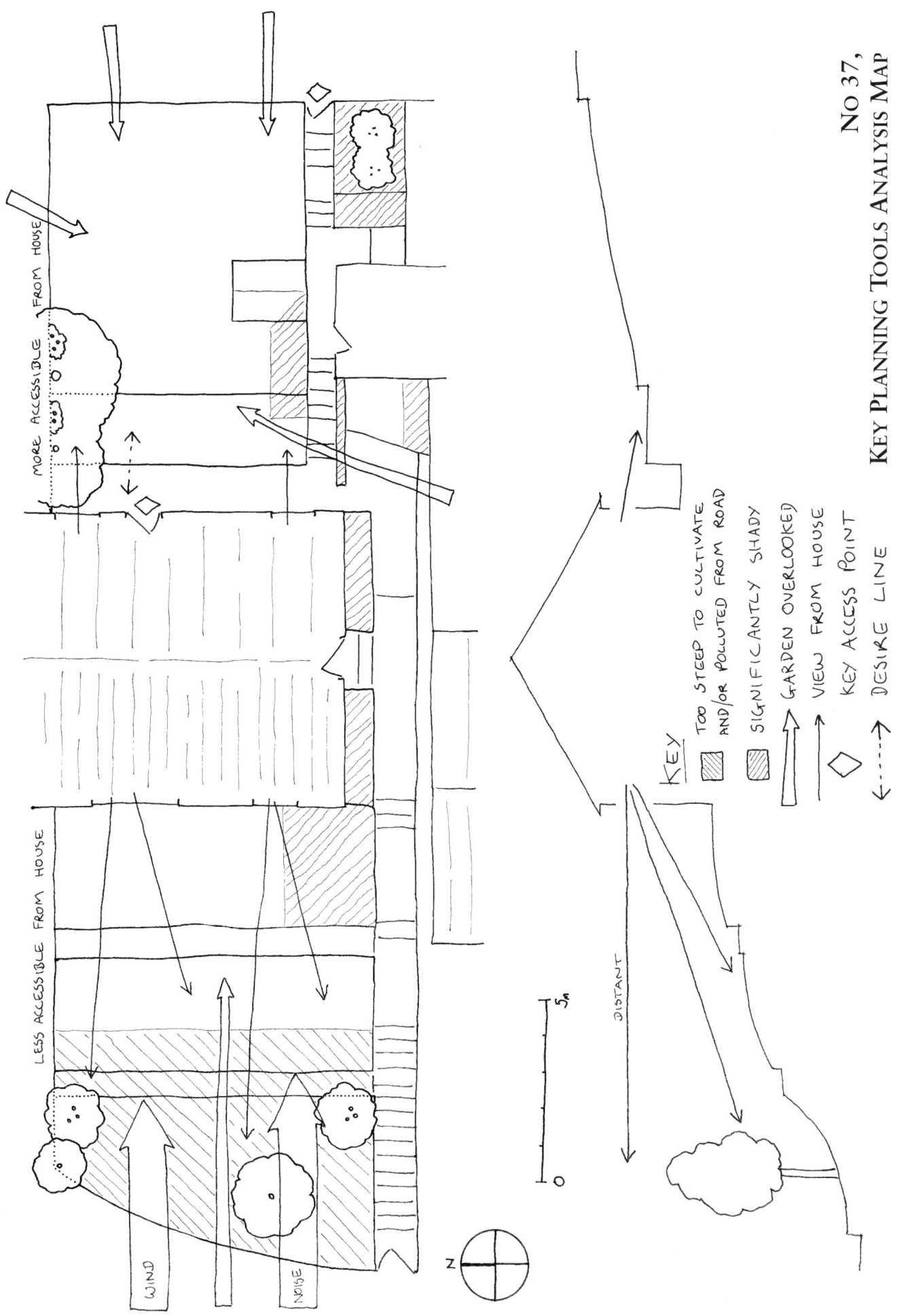

No 37,
Key Planning Tools Analysis Map

Patrick – I love the view from here, and being on the edge of town and country. We can get to the town facilities without using the car and can walk to the nearest fields in two minutes. The garden itself has a good combination of mature plantings, which we can keep, and open space, which gives us scope for action. The greenhouse has already proved its worth for growing winter salads.

I dislike the rectangularity, the concrete paths and the excessive amount of lawn.

Evaluation

The site, existing plants and structures:
All the existing plants are wanted, except the gooseberry bushes and possibly the redcurrants. The amount of lawn we require is less than the total which is there now. The long, narrow shape of the lawns is impractical for social gatherings.

The concrete paths and steps are unsightly, but replacing them is out of the question in a rented place. The retaining walls emphasise the rectangularity of the garden, a feature of it which neither of us like, but at least some of them are needed to keep the soil in place. The shed is a solid, rather ugly structure, useful and not removable.

Our wants
These don't seem to be out of keeping with the resources available. The need for low maintenance combines well with Patrick's interest in perennial vegetables.

We looked at the garden from the viewpoint of the Key Planning Tools: zone, network, sector and elevation. What we found is summarised in the Key Planning Tools Analysis Map and the following notes.

Zone
Although the front and back gardens are equally visible from the house, the back is much more accessible. We use the back door more than the front door and the gate at the back of the garden more than that in the front, so we naturally visit the back garden much more often than the front. We rarely spend time in the front garden, but often potter in the back, even though the metre-high retaining wall makes direct access to the back garden from the back door awkward.

Network
In theory there is a desire line from the back door to the back gate, but the existing path and steps, though indirect, give rapid and easy access. There is a strong desire line for direct access from the back door to the back garden.

Sector
The two priorities are improving privacy and reducing wind in the front garden.

Elevation
Water can be collected from the downpipe in the front garden, the one in the back garden, or both. From either of these points it can be used by gravity in the front garden, but not in the back. A tank beside the front downpipe would be too far from the back garden for carrying water by hand, but the raised bed by the back downpipe offers a site for a tank which would be close enough. The bed may need to be lowered somewhat to accommodate a tank, given the elevation of the gutter.

Aims of the Design

- To grow some of our own vegetables and fruit.

- A low maintenance requirement, allowing for absence at critical gardening times of the year.

- A lawn for sitting out on, with improved privacy.

- Ornamentals, to look good both from the lawn and from inside the house.

- To maintain the view of the countryside from the front of the house.

- If possible, to relieve the rectangular appearance of the garden.

- To grow unusual perennial edibles.

- To provide some wildlife habitat.

This is not an order of priorities, as the two of us have different priorities. It's more a collection of aims which must be blended together.

Design Concepts

Themes
It should by now be obvious that perennial vegetables will play a large part in our garden, both because of my interest in them and our wish for a low maintenance requirement. However, we will include a small area of annual vegetables to give some extra variety to our output. Perennial flowers also fit the bill better than annuals. Given the windy site, and the fact that we have other land where we can plant fruit trees (in addition to the existing pears) the fruit plantings will be soft fruit only.

The potager concept is one that attracts Cathy. But, although we don't intend to have a very clear distinction between edible and ornamental areas, we won't go for the rather formal layout and strict planting plan which is usually associated with that style. I am the main gardener and I'm much more at home with a higgledy-piggledy layout, determined more by intuition and chance, such as self-seeding, than by a formal plan. It's what I call a minimalist garden.

Layouts

The back garden has so many advantages over the front (see Evaluation, above) that it will certainly be the main area both for food growing and for outdoor recreation.

The front garden more or less designs itself:

- The lowest terrace will be developed as a windbreak and wildlife area.

- The middle terrace will remain an ornamental border.

- Most of the lawn will be converted to an annual vegetable plot.

The back garden presents more choices. The chief elements to be placed in it are: lawn, soft fruit, perennial vegetables, flowers and greenhouse. We have come up with three main options and these are shown in the concept maps, Options A, B and C. In all three cases we will screen the garden to give all-round privacy.

Note: The dashed line indicates the present retaining wall, which we want to remove in order to reduce the rectangular appearance of the garden.

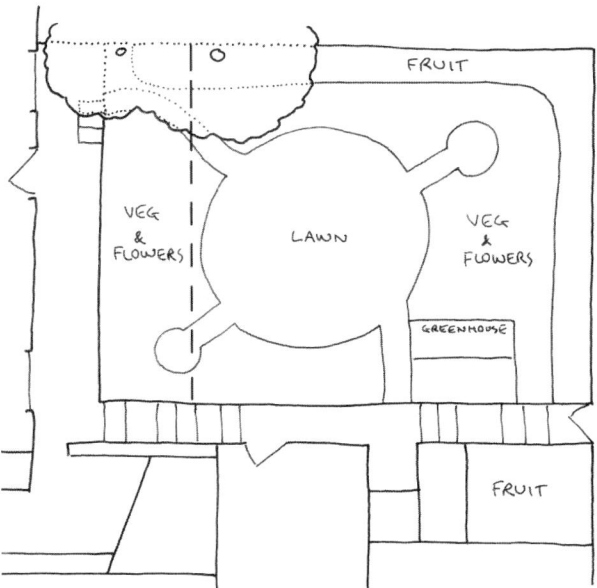

Option B. *This requires more work: the whole of the lawn must be relaid; a low retaining wall is needed on both uphill and downhill sides of the new lawn; considerable earthmoving is needed to turn the existing retaining wall into a slope; and the greenhouse must be moved. The idea of a circular lawn surrounded by a mix of ornamental and productive plants is attractive, but both the shed and the greenhouse are visually intrusive. The greenhouse could be positioned as in Option C, and the shed could be softened with a creeper.*

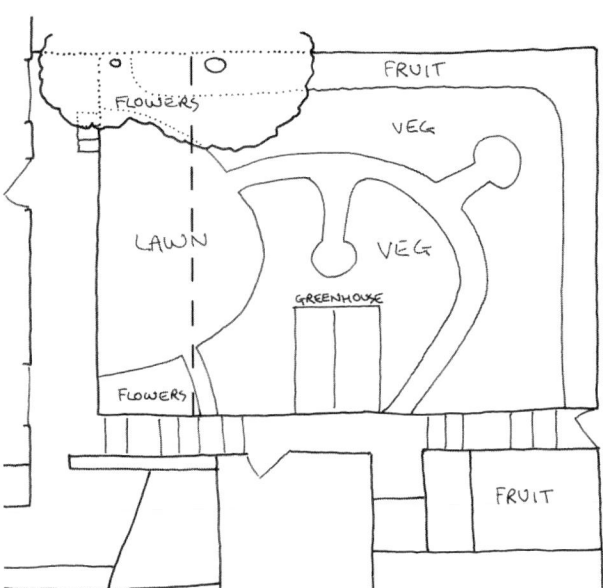

Option A. *One advantage of this option is that by simply changing the line of the retaining wall we both change the appearance of the garden and make the lawn a more usable shape, with a minimum of earth moving and relaying of turves. The disadvantage is that the lawn loses the Sun earlier in the evening than it would under the other options - at five or six o'clock in the summertime. The blank wall of the house is not pleasant to look at close to, but this can be planted up with a creeper. The greenhouse could be positioned as in Option B or C, but this would mean that the vine is left in the open rather than under glass and the grapes would not ripen. Moving the greenhouse would also be a big job.*

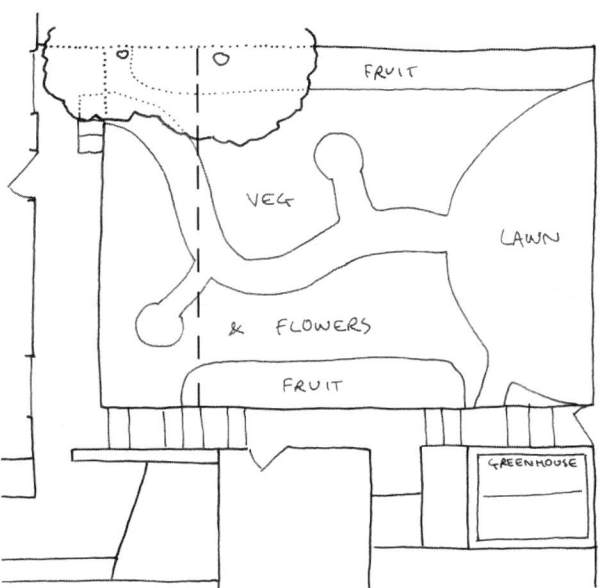

Option C. *The amount of work required would be similar to that in Option B. An advantage is that the soft fruit to the south could partially screen the unattractive shed, while tolerating the shade it casts. But the greenhouse would be in one of the shadiest parts of the garden, which would be bad for the winter salads we grow in it. It could be moved to the north side of the garden, but would then*

be visually intrusive from the lawn. The lawn will keep the Sun till seven or eight o'clock on summer evenings, but it will be right by the car park. Even though we plan to screen the garden from the car park with a hedge or fence, this would not be a barrier to sound and there would be less feeling of privacy.

We chose Option A. From the point of view of growing fruit and vegetables there's little to choose between the three, as long as the greenhouse is not placed in the south-east corner, as in Option C. From the recreational point of view there was not enough advantage in either B or C to justify the extra labour – an important point with our low physical energy and low budget.

Design Details

See map, Design Proposals, opposite.

Back Garden Reconstruction
The existing turf is just the right amount for the new lawn. The new, curved retaining wall is to be built of the natural stones which are presently lying around in the herbaceous border in the front garden. The concrete blocks from the existing wall are to be used to construct two things: a much-needed flight of steps up from the back door to the lawn, and a planter against the back wall of the house. We will plant a *Clematis montana* here and train it over the back wall to make the view from the lawn more pleasant.

The paths in the main fruit and vegetable area will be allowed to grass down naturally. They will be supplemented with stepping stones so that every part of the growing area can be reached without stepping on the soil.

Fruit
The blackberries, redcurrants and the grape vine in the greenhouse are to be retained. Two varieties of summer fruiting raspberries are to be planted along the eastern boundary. We decided against autumn-fruiting varieties because our work takes us away from home each year at the height of their season.

We also chose a veitchberry, a hybrid berry which is said to fill the gap between the harvest time of raspberries and blackberries. This will go on the north side of the back garden. Along with the blackberries, it will be trained on a trellis, made of locally grown larch and hazel, on that part of the boundary to the east of the pear trees.

On the partially-shaded south facing wall of the house we will train four hardy kiwis (*Actinidia kolomikta*). This is an experiment, as this species of kiwi has rarely been grown for fruit in this country. It should tolerate the semi-shade conditions. As the plants are grown from seed we won't know if we have the appropriate mix of females and males for pollination and fruit production.

Our other growing site is our field (not shown on any of the design maps). Here we will plant five apple trees, three pears, two walnuts and four hazels. The fruit varieties are chosen to give the longest possible eating season, to pollinate each other, for disease resistance and for suitability to the soil and climate. (See box.) There is less information about nut varieties and our selection is more of a random choice from what is available.

Boundaries
The fruit trellis on the northern boundary is intended to yield privacy as well as fruit. On the eastern boundary, by the car park, we will fix two-metre-high hazel wattle hurdles to the existing chain link fence. We will fill the gap between the house and the shed with a willow 'fedge' – a cross between a hedge and a fence, made of living willow. These three structures should give all-round privacy in the back garden.

Perennial Vegetables etc
We already have a collection of perennial vegetables and I am constantly getting hold of new kinds and trying them out. Our species list is similar to the list given in Chapter 8. (See p200.) It includes a wide range of salad plants. Most of the ground area in the back garden is already planted to perennial and self-seeding vegetables and these will remain.

OUR APPLE TREES

As an example of choosing varieties, this is our selection of apples:

Worcester Pearmain. Eating season Sept-Oct; the most delicious early apple; a close neighbour has a very productive one, grown organically.

Charles Ross. Oct-Nov; keeps well for an early variety and tastes quite good; grown locally, scab resistant and does well on a limy soil, which is what we have.

Egremont Russet. Oct-Dec; a rare and delicate flavour; very disease resistant and recommended for organic growing.

Kidd's Orange Red. Nov-Jan; the most delicious apple on Earth, and beautiful to look at; grown commercially in the locality, but not organically, resistant to scab but susceptible to canker – a bit of a risk.

Pixie. Dec-Mar; the only one we haven't tasted, but we wanted a good late keeper and don't know one personally; resistant to scab.

All the apples are in pollination groups 2, 3 & 4.

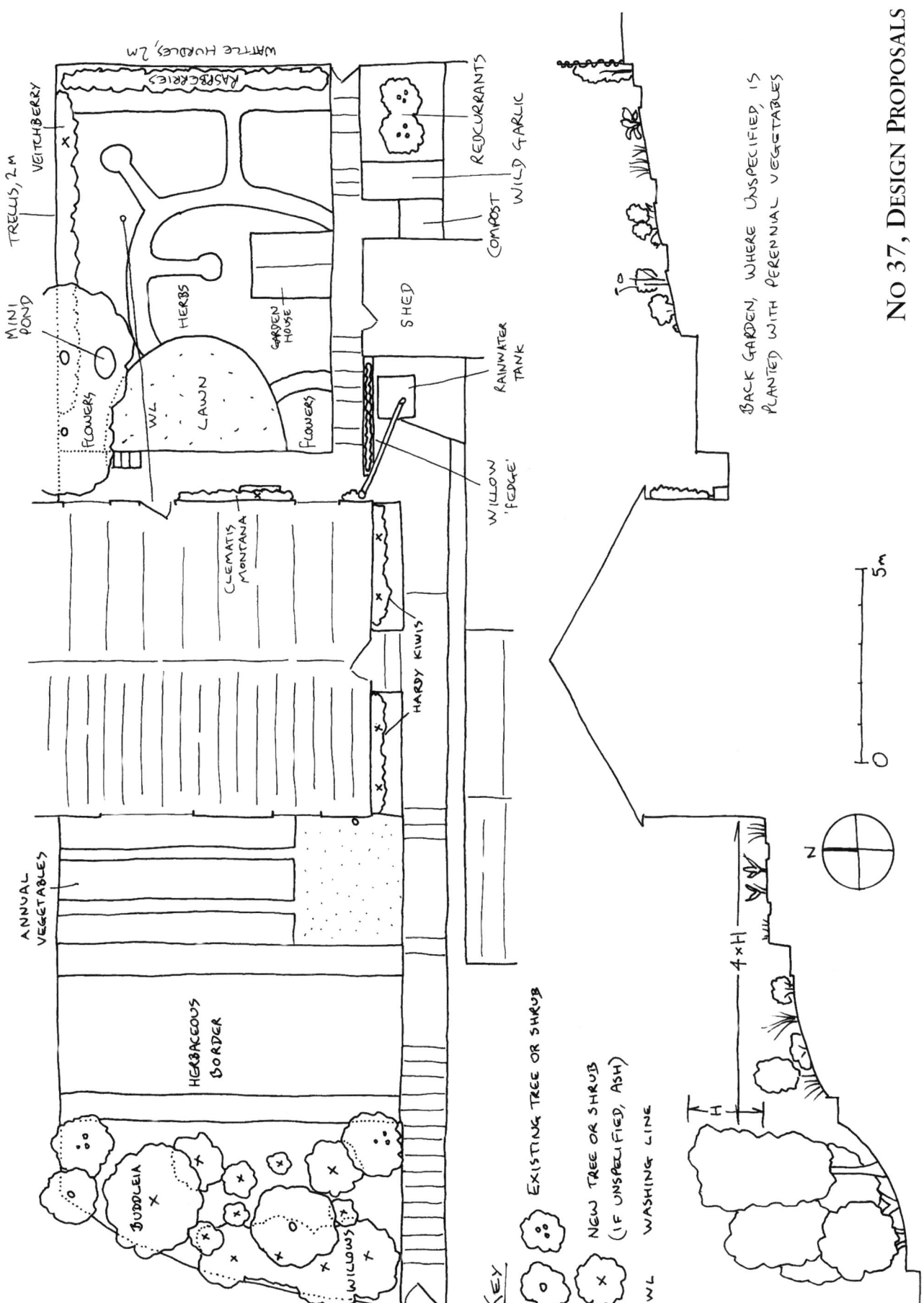

No 37, Design Proposals

Back garden, where unspecified, is planted with perennial vegetables

Key
- Existing tree or shrub
- New tree or shrub (if unspecified, Ash)
- WL   Washing line

Wattle hurdles, 2m
Raspberries
Veitchberry
Trellis, 2m
Mini pond
Flowers
WL
Lawn
Herbs
Garden house
Clematis Montana
Flowers
Shed
Hardy kiwis
Willow 'fence'
Rainwater tank
Redcurrants
Compost
Wild garlic
Annual vegetables
Herbaceous border
Buddleia
Willows
5m
N
4 × H
H

The greenhouse will continue to produce self-seeded winter salads. The shadiest part of the garden will be planted to wild garlic. (See Design Proposals Map.) The warm area to the north of the plot will be planted to a similar mix of vegetables as the rest of the garden and will hopefully give a somewhat earlier crop in the spring. We will also try the climbing perennial tuber crop, American groundnut (*Apios americana*) here, as it needs a somewhat warmer climate than we can usually provide in this country.

The flowers, all perennials, are placed to give maximum visual impact both from the house and from the lawn. The aromatic herbs add beauty and scent to the lawn environment and are relatively accessible from the house for easy picking. (See Design Proposals Map.)

A bird feeding station and a mini pond, of 70cm diameter, will be placed in view of the north-east window of the house. This will give us a view of the birds feeding and bathing. As the pond is so tiny, cleaning the leaves which will fall into it from the pear trees will be no problem, and the branches are too high to shade it.

Front Garden
Some of the ash trees which self-seed so prolifically in the front garden will be left to grow up into a windbreak in the section of the garden closest to the road. These will be pollarded every year to keep them from obstructing the view or casting too much shade. This should give excellent shelter to the whole of the front garden, as none of it is further away from the windbreak than four times its effective height. (See Design Proposals Map.) Although the windbreak is short from end to end, it lies directly opposite the gap in the belt of trees on the far side of the road, so eddies round the ends of the windbreak should not be a problem. (See p88.)

Ornamental willows, coppiced annually, and a buddleia will be added to the windbreak both to give it 'bottom' and to add to the wildlife value – both are excellent wildlife plants. (See illustration, p368.)

The herbaceous border will be added to as necessary, preferably with native plants such as meadow cranesbill.

The selection of annual vegetables is based on three criteria: what we like to eat, what's easy to grow, and the need for balance in the rotation. It is: Year one – leeks and garlic; Year two – carrots; Year three – broad beans; Year four – squash. This will not supply us with all our requirements even for these vegetables, and all other kinds will be bought in.

The small patch of lawn remaining at the front, though shady for much of the day, catches the Sun on a summer evening when all the back garden is in shade. Once the windbreak screens it from the road it may be an acceptable place to sit on a sunny evening.

IMPLEMENTATION				
	Back Garden	Front Garden	Field	House
Already installed	Perennial vegetables			
Year 1 *Winter*	Remodel lawn Plant fruit and 'fedge'			Radiator reflectors and extra thermostat
Spring	Erect fence and trellis	Implement ¼ of vegetable plot		
Year 2 *Winter*	Install rainwater tank	Plant willows and buddleia.	Plant 6 nut trees	('Green' electricity becomes available to domestic customers)
Spring		Implement ¼ of vegetable plot		
Year 3 *Winter*			Plant 5 apple trees	
Spring		Implement ¼ of vegetable plot		
Year 4 *Winter*			Plant 3 pear trees	
Spring		Implement ¼ of vegetable plot		
Ongoing	Plant flowers	Plant flowers		Buy house plants

This may look like an excessively cautious plan, but it's realistic in terms of the state of our health, even with a little help from friends now and then. With an easy programme like this we will have time and energy to do the work really well. If a tree is going to bear fruit for the next hundred years I reckon it's worth planting it as well as we possibly can, and we can only do that if we don't feel we have to hurry.

The back garden, in the first summer after the initial design was implemented.

Soil Care

Most of the plants in the garden are perennials and the annuals will be grown on a no-dig system. As the soil is currently rather low in organic matter, generous mulches will be applied, and whenever a new plant is established, either fruit or herbaceous perennial, the opportunity will be taken to dig in plenty of organic matter. The composting system is described in Chapter 3. (See p46.)

Rainwater Collection

A 1,000l tank will be placed on the raised bed and fed from the downpipe at the back of the house by a pipe running over the path, well above head level. It will be necessary to lower the bed by 30cm, but this will still leave plenty of space to fill a can beneath the tap. The front garden can be watered from it with a hose.

The House

As it's rented there's a limit on what we can do to improve the ecological performance of the house. Steps we will take include: ensure the landlord services the boiler regularly; fit an individual thermostat on the one radiator which hasn't got one; put reflectors behind the radiators (see p130); change to a renewables-only electricity supplier; introduce a wide range of house plants; use only ecological paints etc; practice water economy; recycle used materials. In general our level of material consumption is low, and this is probably as important in reducing our ecological impact as any of the specific things listed above.

Costing

We don't feel the need to cost our proposals as we are quite sure we can afford the plants and materials needed. The possible exception is the hazel wattle fencing on the eastern boundary, which will cost much more than the equivalent larch-lap fencing. The choice is between supporting a local craftsperson working a coppice woodland, or buying a chemically treated product from a distant conifer plantation and saving ourselves some money. (In the event, I'm glad to say our finances ran to the hazel hurdles.)

Re-evaluation

See map, Changes, overleaf.

On the whole the design has been successful. The amount of fruit and vegetables we have harvested and the amount of work the garden has required are both just what we envisaged. There is some colour in the garden throughout the year, and it's a much more pleasant place to spend time in than it was. The bird feeding station, mini pond and wildlife plantings have attracted more birds into the garden, and the herbs and flowers are often visited by butterflies and wild bees. But our attempt to establish frogs in the little pond was thwarted by the local cats, as we had feared.

The back garden, in the third summer, after the re-evaluation and subsequent changes.

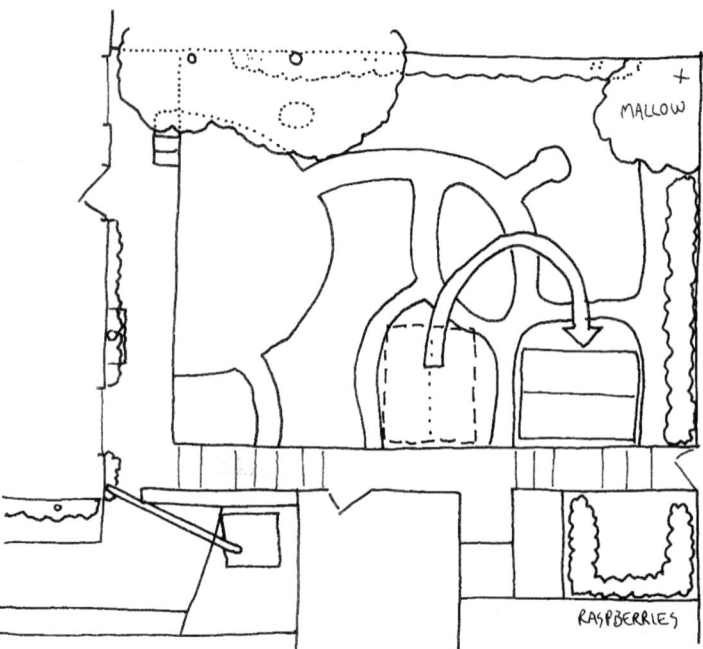

No 37, Changes

The windbreak in the front garden has developed as planned, and the three screening structures in the back garden have been effective, except that the trellis with berries has taken three years to become effective as a visual screen. Growing berries like this for both fruit and privacy is an experiment, and we still don't know how successful it will be in the long run. (See p233.)

Two of the hardy kiwis have died and the remaining two have grown slowly and have yet to flower after three years' growth. They may dislike the rather windy microclimate between the two houses. The American groundnut we planted in the warmest part of the back garden has maintained itself but grown weakly. I've decided not to dig it up and investigate tuber production till it shows more vigorous growth. (I've seen it growing really well in polytunnels locally.)

In the first year after planting, the raspberry canes at the northern end of the row started to die. We then learned from a neighbour that the previous inhabitant had had his raspberries along the northern boundary. The place where they were dying was where the present and previous plantings overlapped. It was a case of replant disease. (See p227.) We could have avoided this failure by involving our neighbour at the site survey stage of the design process.

We replaced the dead canes with a bush mallow, which provides a backdrop of colour to the garden for a good four months of the year. It has grown bigger than the space allowed for it and has killed off more of the raspberries by competition, though the veitchberry has held its own. So we decided to take out the redcurrants, which we never much liked, and replace them with more raspberries, choosing two varieties different from the ones we already have.

The garden is too sluggy for carrots and we substituted Swiss chard for them in our annual rotation. We already grow it as a self-seeder in the back garden, but it's one of our favourite vegetables so we are not over-producing it. Including a few salad vegetables with the annual maincrops in the front garden has not worked, because they don't get picked. It's worth going round the front for a whole dish of greens, but when I'm picking a salad in the back garden I don't often make it round to the front for a single ingredient.

Another change we have made is to move the greenhouse away from the lawn. It was a looming rectangular presence and the lawn is a more pleasant place to be now it's further away. I couldn't have done this on my own, but we had some help from a couple of friends. The plan is to keep the old grape vine where it is, with one branch trained inside the greenhouse so that we can continue to get a dessert grape crop, and another branch trained over the shed to hide its ugliness somewhat.

Chapter 14

DESIGN SKILLS & METHODS

This chapter is a supplement to Chapter 13, The Design Process. The various techniques described here can be used at different stages in the design process. Some will be appropriate to one design, others to another. You can take your pick.

However I believe the sections on listening, to both land and people, are relevant to everyone, from the most skilled designer to the person who has no interest in designing anything.

SURVEYING SKILLS

Surveying simply means measuring the land. This is a necessary part of making a base map, either when making one from scratch or when updating or modifying an existing map. Surveying the land accurately enough for permaculture purposes is very easy, using just a few simple techniques and some inexpensive equipment. None of the techniques described in this chapter takes more than a few minutes to learn and most of them require little or no purchased equipment.

The three kinds of measurements which can be made are distances, angles and elevation. Distances and angles need measuring for every map. Elevation is only needed in some cases, mainly where the design involves working with water.

The first thing to do is to decide just what needs to be measured. It's all too easy to rush in and spend time on work which later turns out to be unnecessary. Then make sure there's enough time to work in an unhurried way, especially if you're using the techniques for the first time.

The measurements are recorded on a sketch map in the field and this is later converted to a scale map on the desk top. A sketch map is one which is not drawn to scale, but in this case it has all the measurements written in on it.

Distance

For accurate measurement of relatively small spaces, such as gardens, a tape measure is best. But for larger-scale designs pacing is accurate enough for permaculture design. It's also much easier and quicker, especially when one person is working on their own. I sometimes even measure a garden by pacing if a rough idea of the dimensions is good enough for my purposes.

Pacing

Each person's normal, relaxed pace is remarkably constant in length. Once you know how long yours is you can measure distances simply by walking along and counting your steps. In order to find out how long your own pace is:

- Measure out 100 metres on the ground. This should be in as straight a line as possible. If it's on an incline that's OK, though steep hills are best avoided.

- Walk along this measured distance in a normal, relaxed way, counting your paces as you go. It's easier to count every other pace and double the result than to count every pace as you go. If you do this be sure to start counting on the second pace, not the first.

- When you reach the end, remember how many paces you took, then walk back along the measured way, counting again.

- Compare the two figures you get. Most people find they get a difference of not more than two or three paces. If you get more than that do it again one or two times till you get a better agreement. The first time you do it it can be difficult to relax and not modify your normal pace in any way.

Once you're reasonably confident you know the length of your pace compared to 100 metres you can start to use it as a unit of measurement. As an example, take my pace, which is 120 to the 100m. When making a sketch map in the field I write everything down in paces. When I get to drawing up the scale map from my sketch map I simply convert my paces to metres with a calculator. The formula is:

Distance measured on the ground x 100 ÷ 120

which is the same as: Distance x $\frac{5}{6}$

which is the same as: Distance x 0.83

Either of the latter two can be very quickly keyed in on a pocket calculator, and the total time taken both to measure in the field and convert on the desktop is much less than it would take with a tape measure and two people.

Pacing can be used on any relatively smooth surface, including roads, paths, grassland and young arable crops. In taller grass and arable crops it becomes less accurate and it's useless in bracken or in woodland. The best way to measure a wood is from the outside and along the rides, but if there are measurements which must be made among the trees you have to use a tape.

On steep hills it's a good idea to pace once up and once down and take an average. Contrary to expectation, we often take longer paces going uphill than down.

Recording Distance

Whether you're measuring with a tape or by pacing, the distances should always be noted on the sketch map cumulatively, as shown in the illustration below.

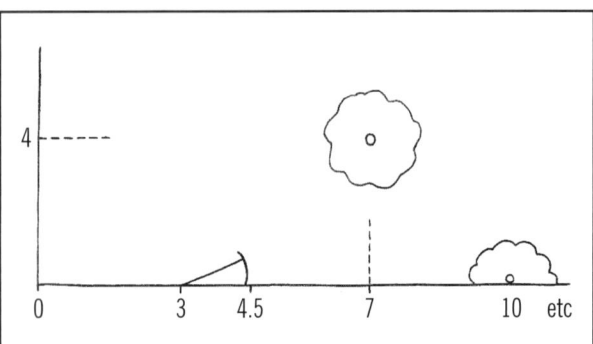

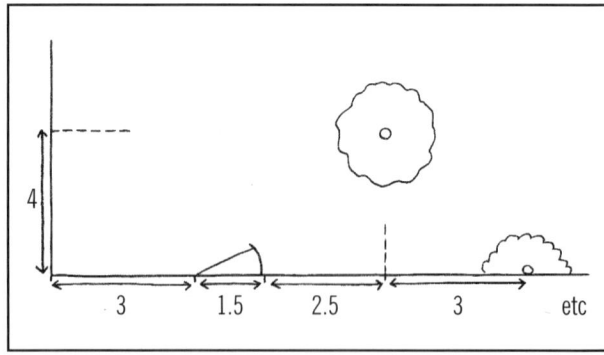

This cumulative method is much more sure than measuring a series of short distances individually, as in the illustration above. There's absolutely no doubt what each number on the sketch map means, which there easily can be when a whole lot of short distances are noted down. It's also more accurate: any errors are one-off and are not carried forward to make all subsequent measurements inaccurate by the same amount.

Angles

Two easy ways to measure angles are triangulation and a chain survey, and these are described below.

In a perfectly rectangular landscape there's no need to measure angles because every one's a right angle. Many gardens are like this, an example being our own garden which was used as an example in the last chapter.

TREE CANOPIES

Perhaps the commonest error in mapping, at least on a garden scale, is drawing in the canopies of trees too small. This applies both to base maps and design maps.

When surveying existing trees, people usually measure the position of the trunks accurately and draw them in carefully on the map. But all too often the canopy is added, almost as an afterthought, without any attempt to measure it. Very often a group of trees whose canopies meet and form one continuous canopy are drawn as though they were isolated from each other, with canopies only two or three times as wide as the trunk. This is an important mistake. Amongst other things it gives a completely untrue picture of how much shade the trees will cast.

The same thing is often done with proposed trees on the design map, and it makes the map more or less useless, either as a design tool or as a means of presenting the design. Drawing trees at a realistic diameter is the best way of finding out how many can be fitted into the available space.

Fruit trees must always be shown at the diameter

they will grow to when mature. Drawing them smaller than this may lead to too many of them being planted, then some will have to be taken out when they are part-grown, which is a very wasteful thing to do with a fruit tree.

Ornamental trees and shrubs are sometimes drawn in at the size they will be after a certain number of years, often ten years. If the map shows them filling all the available space at this age it implies that some thinning will have to take place at that point.

Planting trees and other perennials too close together is one of the commonest mistakes in garden design. It can be hard to imagine that the little plant which arrives from the nursery will one day have a canopy of three or even ten metres. Accurate mapping is the best way to avoid this mistake.

In broadscale designs it's unlikely that every tree will be drawn in. A large orchard may simply be represented by symbolic rows of trees. If there is any doubt as to whether these trees are a symbol or an accurate representation of individual trees, there needs to be a marginal note stating which they are.

In a basically rectangular garden, any features which are not rectangular – including irregular stretches of boundary, internal features or rectangular features which are not aligned with the boundaries – can be mapped by reference to the sides of the garden, as shown in the illustration on the previous page.

It can be worth checking that what appears to be a rectangle really is one. Gardens and other enclosures which look rectangular to the eye often turn out not to be when they're measured. This is most easily checked by triangulation. In landscapes which are clearly irregular either triangulation or a chain survey can be used.

Whichever method is used, it's also necessary to align the map with north. This can often be done by reference to an existing map of the area. If not an ordinary magnetic compass, used with care, is accurate enough. Remember never to use a compass near a large iron or steel object, such as a car or steel gate, as these will influence the needle and give you the wrong bearing.

Triangulation

If there are two points on the ground whose positions you know, you can always find out the position of a third point by measuring the triangle which connects those three points.

For example, in the illustration below, we have measured the house, so we know the positions of the two house corners, A and B. We can now find the position of the tree, C, by measuring the distances A to C and B to C. A triangle which has its base on the line AB and sides AC and BC of the lengths we have measured can only have point C in one place – the position of the tree. When it comes to making the scale map, we can draw in the line AB and find point C by using a pair of compasses.

This is triangulation. As well as being used to locate the position of an isolated object it can be used to determine the shape of an irregular area of land. The method is just the same except that the unknown point

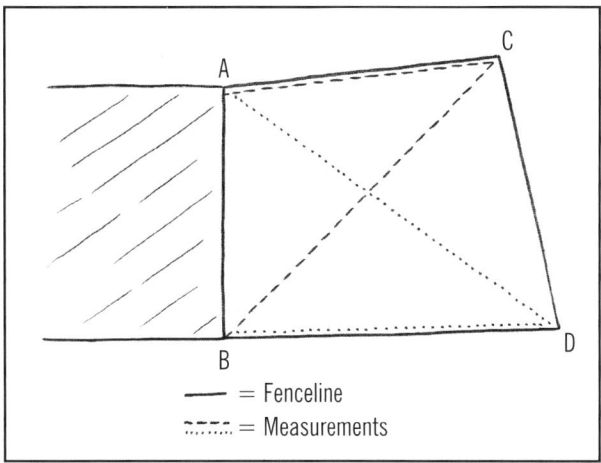

—— = Fenceline
᠁᠁ = Measurements

is a corner of the boundary rather than an isolated feature. In the illustration above, first the triangle ABC is plotted, to find the position of the corner C, then the triangle ABD, to find the position of the corner D.

When choosing the baseline – that is the line between two known points which is needed to start triangulation – it's best to use the longest one available from which all or most of the area can be seen. On many sites it's necessary to use more than one base line. There's no reason why a second baseline should not be a line which itself has just been found by triangulation.

An alternative way of using triangulation is to measure angles rather than distances. If you know the baseline, AB, and the angles ABC and BAC, you can find the position of C just as surely as by measuring the sides. The disadvantage of measuring angles is that you need a sighting compass, which at the time of writing can range in cost from £30 to £65. A sighting compass is a special kind of magnetic compass which you can sight on a distant object at the same time as reading the bearing. The bearing is the angle between the line of sight and north. An ordinary magnetic compass is not accurate enough.

The advantage of triangulation with a compass is that it's quick and easy to use over longer distances. Only the original baseline needs to be measured. The original Ordnance Survey of the whole of Britain was done by compass triangulation.

Chain Survey

This takes its name from the days before tape measures had been invented and surveyors used chains with distinctive links to mark the feet and inches. It can be done with two tape measures or by pacing. It may take a little longer than working with a compass, but it's an easy way of mapping irregular areas accurately without any special equipment.

First the base line is chosen. This is the line along the longest dimension of the area to be surveyed. It's important that it is the longest dimension, otherwise there will be parts of the area which cannot be surveyed from it.

When pacing is used to measure the distances, the base line is not marked out on the ground but two

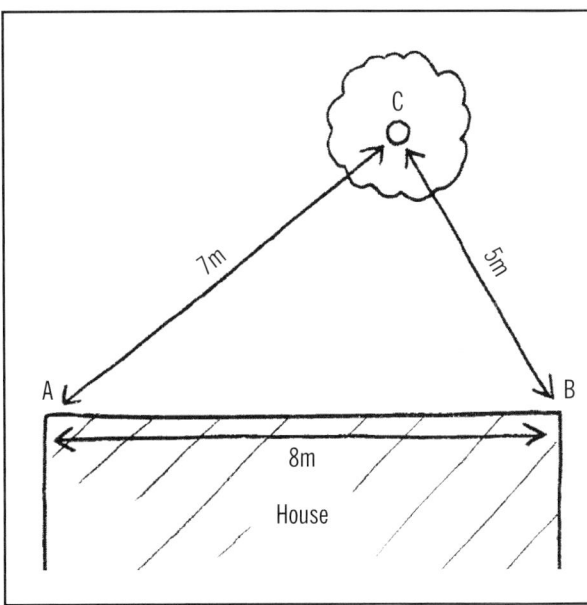

sighting poles are set up at one end of it, exactly on the line. These are straight sticks some 2m tall which can easily be seen from a distance. Painting them white can be helpful. Two people are needed to do this.

Person A places the first sighting pole at one end of the line. Person B, standing at the other end of the line, tells the A where to put the second sighting pole so they are perfectly in line from where B is standing. Now anyone standing within the survey area can place themselves exactly along the line by looking at the pair of sticks: if the two sticks are lined up with each other you are standing on the base line. (See illustration below.)

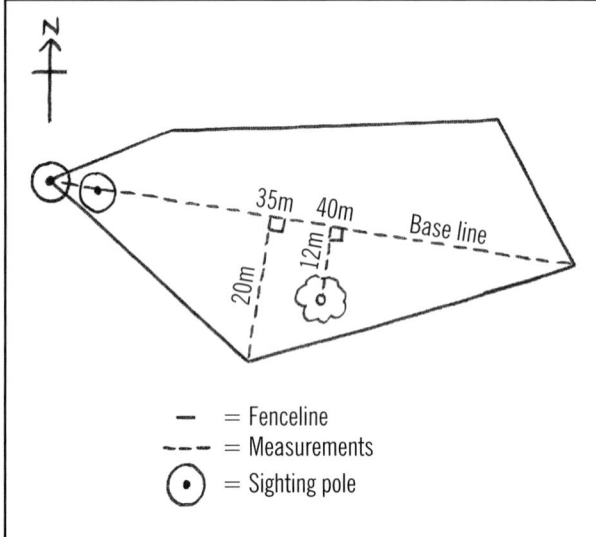

Chain survey.

If tape measures are used and one of them is long enough to cover the whole length of the base line, this can be securely fixed along the base line and no sighting poles are then needed.

Whether pacing or tape measures are used, everything is mapped in relation to the base line, including the boundaries and any features within the survey area. This is done as follows:

- Choose one end of the line as the zero point and mark this down on the sketch map.

- Walk along the central line till you're opposite a point you want to record, either a corner in the boundary or an internal feature.

- Stand facing the point, with your shoulders aligned with the base line; look at the pair of sticks to check you are aligned, and look straight ahead to check you are exactly opposite the point you want to record (The human body is a remarkably good instrument for measuring a right angle.).

- Measure the distance between the point where you're standing and zero.

- Measure the distance from the point where you're standing to the point you want to record.

As with all surveying work, the important thing is to get the main dimensions right. Detail can be added freehand. In an area with a complex boundary it's a good idea to start the survey by deciding which points are to be measured and which to be drawn in freehand.

It's often possible to identify part of the boundary which is marked on an existing map of the area, and this can give you the direction of north. If a compass is used to determine north it's best sighted along the base line.

A chain survey can be remarkably accurate. I once took part in an exercise involving some 25 people, none of whom had used the method before, all measuring with their paces. The site was a field, bounded on one side by a meandering river and on the other by an irregular hedge. When the resulting map was superimposed on the existing Ordnance Survey map it was almost identical.

Elevation

Elevation measurements may be needed where water works are contemplated or where soil erosion may be a problem.

The water applications include:

- swales
- ditches
- ponds
- other water storage or delivery systems

Swales need to be laid out exactly on the contour, while ditches need to have a constant fall on them. Ditches can include land drains, diversion drains (see pp54, 99 and 114) and irrigation ditches.

The detailed design of a pond is usually carried out by the contractor who installs it, but a general designer needs to be able to make elevation measurements before that stage is reached, often in order to check whether a particular site is suitable. One factor which can help determine a pond site is whether water from a point source, such as a spring, can be led to it by gravity. Another is the landform of the pond site itself, which will determine how much earthmoving will be necessary, how much water can be stored, and where the shoreline will come. (See pp113, 116 and 117.)

Other water systems may include ram pumps and rainwater or grey water systems in gardens. For example, it may be necessary to find out how much of a garden can be reached by rainwater from the roof without the need for pumping, or if a pump is needed, how high and how far it will have to pump the water.

On a site with erosion potential the applications include:

- Finding which areas are flat enough for tillage and which should be put down to perennial vegetation.

- Marking out contours for cultivation strips or beds across the slope.

- Designing and laying out the position of terraces.

MEASURING SLOPES

There are three ways to measure a slope: proportion, percentage and degrees.

- Proportion is a direct comparison of the horizontal difference with the vertical difference. Thus if a slope is said to be one in four, or written down as 1:4, it means that for every four metres you travel horizontally you rise, or fall, by one metre vertically.

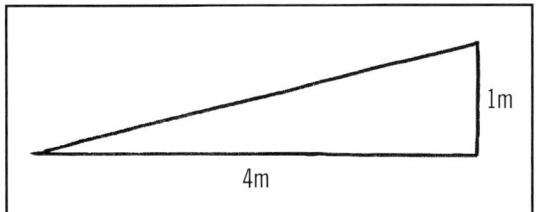

- Percentage slope is based on the concept that a one in one slope is a 100% slope. Our 1:4 slope turns out to be 25% when measured as a percentage.

- Degrees of slope are based on the concept that a vertical fall is 90°, as measured by a protractor. By this measure 1:4 comes out at 14°.

Proportion is much the simplest of the three to work with, as it can easily be calculated from measurements taken in the field or from a contour map. Conversions between proportion and percentage are straightforward, but neither of them can be easily converted to degrees.

Two simple home-made pieces of equipment can be used for measuring elevation: the A-frame and the 'bunyip' water level.

A-frame
This is a simple, cheap and reasonably accurate tool which can be made in a few minutes with three sticks, a piece of string and a small weight, as shown in the illustration in the next column. It can be operated by one person alone, but can only be used do one thing, to mark out a contour line on the ground.

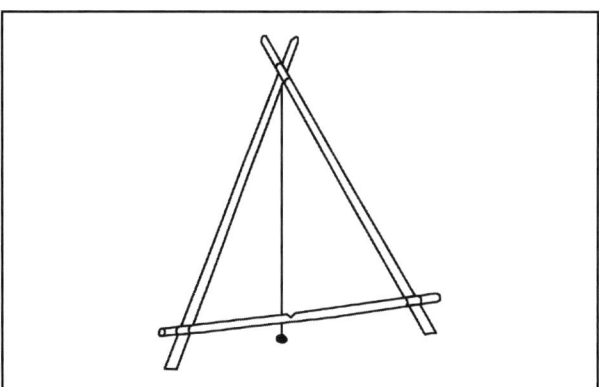

A-frame.

It works on the principle that, when the string hangs exactly over the mark on the horizontal stick, the places where the feet of the two diagonal sticks stand are the same height above sea level. The position of the mark is critical. It's determined as follows:

- Hold the A-frame upright, its feet touching the ground, on any relatively level piece of ground.

- Mark where the string intersects the horizontal stick with a temporary mark.

- Reverse the position of the A-frame so each foot stands exactly where the other was before.

- Make a second temporary mark where the string intersects the horizontal stick.

- Make a permanent mark exactly half-way between the two temporary marks, and erase the temporary marks.

The A-frame can now be 'walked' across the land to mark out a line exactly on the contour. The method is as follows:

- Place one foot of the A-frame at the point where the line is to start.

- Keeping that foot in place, move the other until the string matches up with the mark on the horizontal stick when both feet of the A-frame are firmly on the ground.

- If the string swings wildly back and forth, lean the A-frame back till the string rubs against the horizontal stick and the friction will slow it down.

- Stick a marker peg in the ground where the first foot is.

- Swing the A-frame round through 180°, keeping the second foot in place on the ground.

- Repeat as before.

The result is a line of pegs stretching across the land on the contour.

A variation on the A-frame can be made with a spirit level, tied to a horizontal piece of wood with a short vertical leg at either end, as in the illustration. This must of course be constructed accurately, so a line between the bottoms of the two legs is exactly parallel to the spirit level.

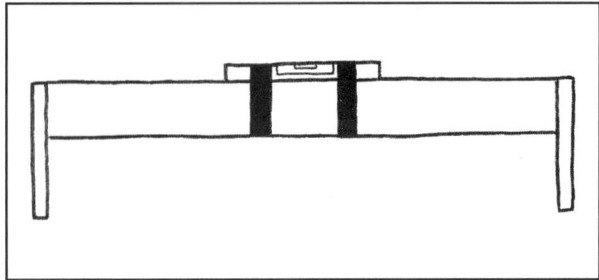

Spirit level used as an A-frame.

Bunyip

The original Bunyip was a character from Australian Aboriginal mythology, a monster who lived in a billabong and made life difficult for other creatures. I don't know why such a name was given to a piece of equipment which makes life so easy for people who want to measure elevation. Two people are needed to operate it, but in addition to marking out contours it can also be used to mark out a line with a predetermined fall on it and to measure the steepness of the land in profile.

To make one you need: two pieces of wood around 5x2cm and 130cm long, a length of transparent hose of 1-2cm diameter (half-inch) and just over 12m long, and four jubilee clips the right size to fix the hose to the sticks.

Both sticks are marked out as follows:

- the bottom 20cm is left unmarked
- the next 100cm is marked out in centimetres, with 0 at the bottom and 100 at the top
- the top 10cm is left unmarked

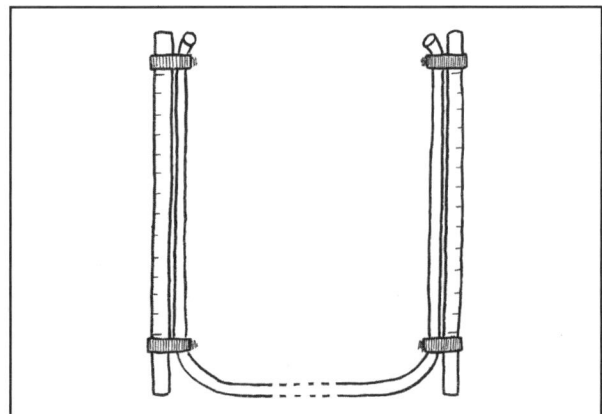

A bunyip water level.

The hose is fixed to the sticks with the jubilee clips as shown in the illustration at the bottom of the previous column.

The hose is then filled with water until the level at either end is approximately opposite the 50cm mark on both sticks. Care must be taken that there are no air pockets in the hose. This can best be done by raising the two ends high up so that there is a continuous fall from the ends to the lowest point, which is the middle of the hose. It may take a little time for the air pockets to rise to the top and disappear.

Marking a contour. A bunyip works on the principle that water always finds its own level. Thus, when the water level at both ends of the hose is opposite the same number on the two sticks, the bottoms of the sticks are the same height above sea level. It needs two people to operate it. Each one holds one of the sticks. To mark a contour the bunyip is walked across the land in much the same way as an A-frame:

- One operator stands still, holding one of the sticks vertically at the point where the line starts.

- The second operator walks forward and places the other stick vertically on the ground.

- Each looks at the water level in the hose on their stick.

- If there is a difference in the numbers opposite the water level, the second operator moves their stick up or down the slope until the two are the same.

- When both numbers are the same a marker peg is stuck in the ground at the base of the first stick and the first operator carries their stick beyond the second, who keeps their stick in the same position.

- The process is repeated.

When moving the bunyip from place to place the operators put their thumbs over the ends of the hose to prevent water escaping. When reading the level they must take them off, otherwise the level will not move. The horizontal distance between one measurement and the next is not critical. On a uniform slope they can be relatively far apart, and on a more intricate slope closer together so as to follow small irregularities.

Marking a Line with a Fall. The method is almost the same as for marking a contour, but the length of free hose between the zero mark on both sticks must be known, and measurements are only taken when this part of the hose is fully extended in a straight line. Then the operators simply have to move their sticks so that they show a certain difference in level rather than the same level. This can best be illustrated by an example:

- Length of hose between the base of the two sticks: 10m = 1,000cm

- Required fall: 1:500

- Divide the length of the hose in cm by the second figure in the fall ratio: 1,000 ÷ 500 = 2

- When there's a difference of 2cm between the level indicated at either end of the bunyip there is a fall of 1:500 between the points at which the two sticks stand.

Working like this, two people can mark out several

kilometres of ditch or drain in a day with only a few minutes training.

Measuring a Profile. A profile is a cross-section of the land, showing the slope in section. This is useful in two situations: firstly, when looking at the profile of a potential pond site, to see how much water the pond would hold and where the shoreline would come to for a given height of dam wall; secondly, to see which parts of a site are sufficiently flat for annual cultivation.

In the second case, suppose that, after considering the erodibility of the soil and the kind of cultivation system planned, it's decided that all areas with a slope of less than 1:8 (around 5°) are suitable for growing annual crops. On a farm scale the best way to identify the suitable areas in the first place is to calculate the slopes from a contour map, but once located they should be checked with a bunyip. (See box, below.) On a garden

---

MEASURING SLOPES FROM A CONTOUR MAP

The 1:10,000 Ordnance Survey maps are at a large enough scale to be useful for calculating slope. To illustrate the method let's assume we're looking for areas of land with a slope of more than 1:8.

The vertical distance between each contour line on this series of maps is 10m, so we're looking for land on which the contours are not less than 80m apart horizontally. Since the scale of the map is 1:10,000, and a millimetre is a thousandth of a metre, each millimetre on the map represents 10m on the ground. Therefore 8mm on the map represents 80m on the ground.

The easiest way to find where the contour lines on the map are this distance apart is with a little circular piece of card with a small arm projecting

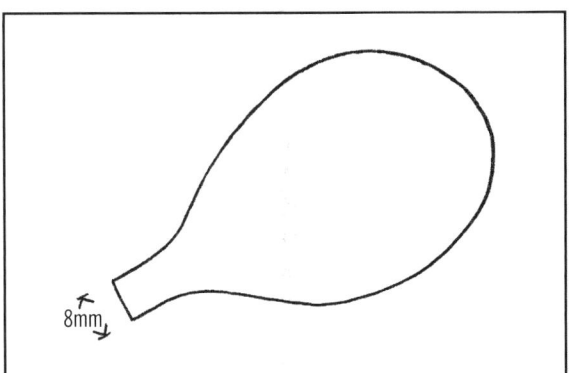

Contour measuring card.

from it. The end of the arm is cut square, precisely 8mm across. Holding the card by the circular part, with the 8mm edge on the surface of the map, use it to measure the horizontal distance between the

contour lines. The 8mm edge must be as nearly as possible at right angles to the contour lines.

Wherever the contour lines are exactly this distance apart the slope is 1:8. Where they are further apart it's less steep than that and where they're closer it's steeper. Make a mark at every point where the slope is 1:8 and these will enclose the area which is steeper than that.

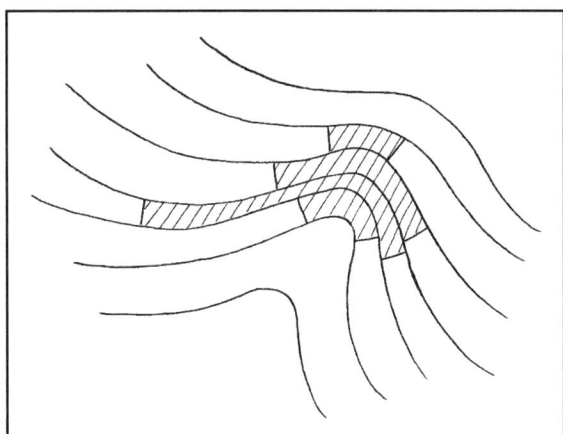

Approximate area with a slope of more than 1:8 marked on a contour map after using the contour measuring card.

This is not a very accurate method. The boundaries of the area will only be approximately right and changes of slope which amount to less than 10m will be missed by the contour lines. It will be necessary to check things on the ground, perhaps with a bunyip. But this is a useful and quick way of getting an idea of the extent and distribution of steep land.

scale no contour map will be sufficiently detailed and a profile from top to bottom of the garden can be made quickly and easily with a bunyip.

The method is as follows:

• The first operator holds one stick of the bunyip vertically at the bottom of the slope.

• The second operator moves the other stick upslope until the difference between the water level against the scales on the two sticks is 50cm.

• The horizontal distance between the two sticks is measured and recorded.

• The second operator keeps their stick in the same position and the first moves theirs uphill beyond the second.

• The process is repeated.

Strictly speaking the horizontal difference should be measured exactly on the level. But on all but the steepest slopes there won't be a significant difference between the horizontal distance and the distance measured along the ground.

The profile can be recorded straight onto squared graph paper in the field or onto a sketch profile which is later copied onto graph paper on the desktop. In either case the method is the same, as shown in the illustration below. The profile will show which sections of the land along the chosen line have a slope of 1:8 or less. If more detail is required a smaller vertical difference than 50cm may be chosen.

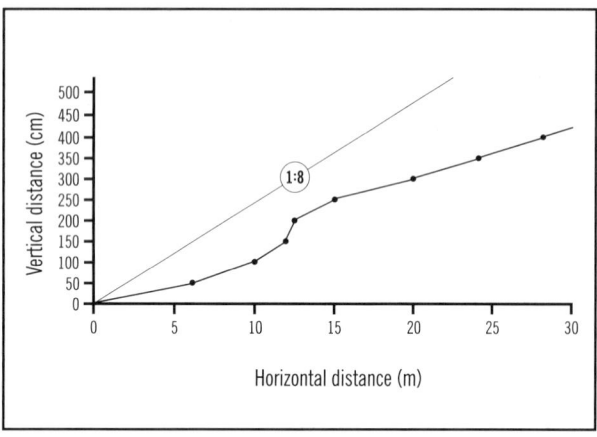

A bunyip can be used to make a contour map. The starting point for each contour line can be measured by the profiling method and the lines marked out with the contouring method. Once that has been done the position of the contours must be plotted onto a two-dimensional map. Even on a relatively small area this is an enormous job and rarely worth the effort. A few well-placed profiles can give a much better return on effort invested.

MAPPING

Basics

Drawing a map to the standard required for good permaculture design is really quite easy. In order to be useful a map doesn't have to look as though it comes from a professional drawing office. It just needs to be tidy, reasonably accurate, and give a clear visual message. In fact a professional appearance can be counterproductive, giving the design proposals an air of untouchable authority rather than a welcoming feel which invites people to work with it and adapt it as necessary.

I've already discussed the degree of accuracy that's appropriate for a base map in Chapter 13. (See pp377-378.) Accuracy for its own sake is not a virtue. You need to ask what the purpose of the map is and then work to the accuracy needed for that purpose. For example, if you want to use your map to work out how many trees will fit into the space you've allocated to a new orchard, the map needs to be accurate to the diameter of one tree when fully grown, however big that may be.

Clarity in a map comes partly from careful working and partly from understanding how a map works.

A map communicates to us through the visual part of the brain, which perceives things in a different way from the verbal part. The verbal mind builds up an idea of what is being communicated bit by bit as separate packages of information come in to it. The visual mind primarily sees wholes, and wholes are seen in one moment.

One look at a map gives you an overall idea of what the place it depicts is like. If the map is so full of detail that the overall picture is obscured it ceases to work like a map. It has to be read like a page of text, and a page of text where all the 'words' are spread about the place irregularly rather than arranged in neat rows. This is not very useful.

A map is not a literal picture but a symbolic one. The value of using conventional symbols is that they have an instant meaning to everyone who knows maps. Using unconventional symbols means the map has to be read like text, understood bit by bit. The best source for conventional symbols is the key of an Ordnance Survey map. (See pp410-411.)

An important part of this symbolic language is the convention that north is always at the top of the map. Keeping to this convention means that the moment we look at a map we know where the Sun rises and sets, which are north-facing slopes and walls and which are the south-facing, which corners will be shady in the morning, which in the afternoon and which will be sunny or shady for most of the day.

It can be tempting to make the site fit the paper. For example, if a house is aligned north-east to south-west it can be easier to draw it with the walls of the house parallel to the sides of the paper, with the symbol indicating north pointing diagonally to the right.

But this immediately reduces the usefulness of the map. At first glance we get an impression which is misleading, possibly seriously misleading for permaculture purposes. (See illustrations below.)

Getting basics like this right is necessary if the map is to be useful in permaculture design. A professional-looking finish is not.

There are three things which every map must have:

- north point
- scale
- title and date

The direction of north is normally indicated by an arrow pointing north, a cross showing the four

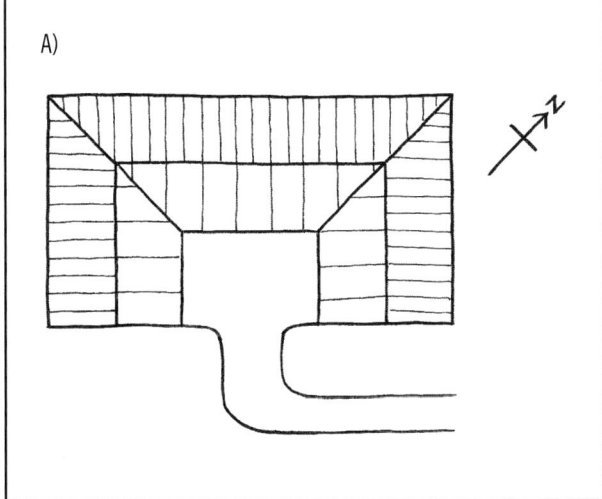

A)

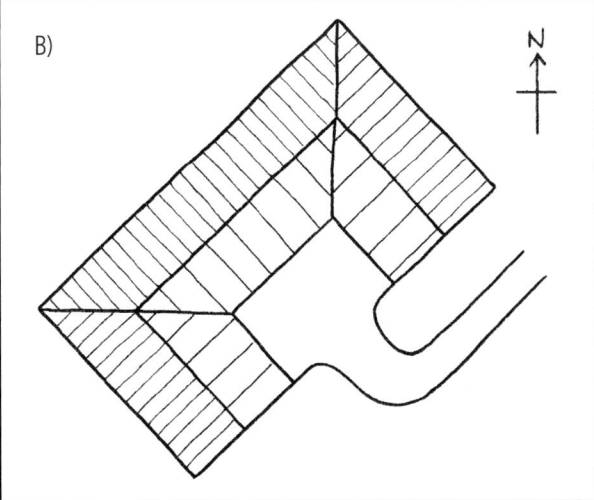

B)

The immediate impression given by A) is that the two internal corners of the house receive about the same amount of light during the day. B) shows the true situation at a glance: the difference in the microclimate of the two corners will be significant. To use A) effectively for permaculture design you must constantly remind yourself that it's misleading. This greatly reduces its usefulness as a visual message.

THREE KINDS OF NORTH

There are three kinds of north: true north, magnetic north and grid north.

- True north is the direction of the North Pole from any place on Earth.

- Magnetic north is the direction a compass needle will point to. It varies from year to year and at present is about 5° west of true north.

- Grid north is the alignment of the grid of lines which are the basis of the Ordnance Survey and are drawn in blue on their maps. Because the Earth's surface is curved and the maps are flat, grid north only coincides with true north at the Greenwich meridian, but the difference is only a degree or two at most.

For permaculture purposes any one of the three is accurate enough.

---

directions or a combination of both. As well as acting as confirmation that the convention has been kept to, it provides the map maker with an opportunity for a decorative feature.

The scale may be represented by a statement, such as 'Scale, 1:200'. This means that one centimetre on the map represents 200cm, or two metres, on the land. It may also be represented by a bar, drawn to represent actual distances as they appear on the map.

Certain scales are regarded as standard scales and in theory all maps should be drawn to one of them. They are: 1:10, 1:2, 1:5 and multiples of all these, such as 1:200 or 1:5,000. All other scales are known as bastard scales. The advantage of using standard scales is familiarity: when you see a map in a scale

---

SCALES

As a quick guide to some of the commoner scales, here are the distances in centimetres on the map which represent one metre on the ground.

At a scale of:	1m is shown on the map as:
1:100	1cm
1:150	0.67
1:200	0.5
1:250	0.4
1:300	0.33
1:500	0.2

you're used to you have a visual idea of the size of the features shown on it. But it's not nearly as important as putting north at the top, and the best scale to use is usually the largest one which will comfortably fit on the size of paper you're using. (See box, Scales, at bottom of previous page.)

The terms 'large-scale' and 'small-scale' refer to the way the map is drawn, not to the size of the area depicted. Thus a large-scale map depicts a relatively small area and shows it large, while a small-scale map depicts a larger area and shows it small. Thus a 1:50,000 map is smaller scale than a 1:1,000.

Putting a title and the date on a map is more than a matter of presentation. It means the map can be useful to people other than the map-maker and those who know the territory so well that they immediately recognise it. Maps often survive long after they have been made and if they're dated they can become useful historical documents – possibly of interest to perma-culture designers of the future.

In addition to these three essentials, most maps also have a key, showing the meaning of the symbols used on the map. It is possible to imagine a map so simple that it doesn't need a key, but it's relatively unusual.

A map can be made either from scratch or by updating and correcting an existing one, in Britain usually one of the Ordnance Survey range, described in the next section. Adapting an existing map is usually appropriate when the design site is relatively large. When the area is small there may not be a map available at a large enough scale to show sufficient detail, at least not without spending a disproportionate amount of money on an Ordnance Survey Superplan map. Surveying and mapping a typical domestic garden can be done in an afternoon.

Ordnance Survey Maps

The Ordnance Survey makes a range of maps covering the whole of Britain. They include both sheet maps, published on paper, and computerised maps, which can be printed out when they are purchased. The sheet maps are at relatively small scales, and the computerised ones can be printed at a range of scales including very large.

Sheet Maps. The Pathfinder series of 1:25,000 maps, in green covers, used to cover the whole country but it's being progressively replaced by the Explorer, in orange covers, which covers a larger area per map. At the time of writing Pathfinders have replaced Explorers over the southern half of England and all of Wales except the north coastal strip. North of that are Pathfinders. The process should be complete by 2004. In certain special tourist areas, from South Devon to the Cairngorms, the 1:25,000 map is published in the Outdoor Leisure series, in yellow covers, instead of either the Pathfinder or Explorer.

These maps can be used for designing large areas of land such as big farms or extensive woodland, but are better enlarged for all but the biggest areas. Note that all Ordnance Survey maps are copyright and repro-duction is prohibited without permission.

The largest scale maps published in sheet form with contours are those in the 1:10,000 series. These are more expensive per sheet than the 1:25,000 maps, but more accurate and are at a more useful scale for designing. They are gradually being replaced by the Landplan computerised maps and are only available in the areas not already covered by those.

Sheet maps in the Siteplan series are available in a choice of three larger scales, but they don't have

ORDNANCE SURVEY MAPS					
Name(s)	Scale(s)	Sheet or Computerised	Contours	Full Colour	Available from
Pathfinder Explorer Outdoor Leisure	1:25,000	Sheet	Yes	Yes	Many bookshops and stationers
1:10,000 Series	1:10,000	Sheet	Yes	Yes	Superplan Agents
Siteplan	1:2,500 1:1,250 or 1:500	Sheet	No	No	By post from Superplan Agents
Landplan	1:10,000 or 1:5,000	Computerised	Optional	Yes	Superplan Agents
Superplan	1:200 to 1:10,000	Computerised	No	No	Superplan Agents

contours. They are somewhat cheaper per sheet than the 1:10,000 series, and additional copies can be had at the time of purchase for a nominal extra fee. They can be bought over the phone and delivered by post.

Computerised. These maps are available at a limited number of shops, called Superplan Agents by the Ordnance Survey. They are tailor-made to the customer's requirements. You tell the shop exactly what area you want covered and at what scale, and they print a one-off map to your specifications. You can look at it on screen before it's printed off. In fact if you use Computer Aided Design you can buy the data on disk and not even receive a physical map at all.

The standard computerised map is the Superplan, which can be reproduced at any scale from 1:10,000 to 1:200. It's in monochrome and doesn't show contours. The Landplan is a full colour map, with optional contours, and can be printed at either 1:10,000 or 1:5,000.

As a general rule it's best to work at the largest convenient scale. The largest possible scale may give you a very unwieldy map, as Superplan maps can go up to a size of 82cm x 3m! For most permaculture purposes contours are useful. But these are not available at the largest scales, and it may be preferable to go for the appropriate scale rather than for contours.

The different kinds of Ordnance Survey maps are summarised in the table. For a full catalogue and a list of approved stockists and Superplan Agents contact their head office in Southampton. (See Appendix D, Organisations.)

Geological and soil maps may be useful in some permaculture work, especially if you're working on a large area which would take a long time to examine on the ground. Most parts of the UK are covered by geological maps at 1:50,000 and a few areas at 1:25,000 or 1:10,000. Soil map coverage is patchy and less than half the country is covered by maps at either 1:50,000 or 1:25,000.[1]

Drawing Skills

You don't need any talent for drawing in order to make a good, clear map. You just need to take care and concentrate on what you're doing. Give yourself plenty of time: don't try to draw to a deadline. Make sure you're quite comfortable, with a table and chair of the right height. Create the right atmosphere. I like total silence; some people may find unobtrusive music provides the right background. Allow yourself to become totally absorbed in what you're doing. Observe carefully, and don't put pencil to paper till you're quite sure of the size and shape of what you're drawing. Be forgiving of your mistakes, and draw faintly the first time round so rubbing out is easy.

The highest quality drawings can be done with a combination of graph paper and tracing paper. The method is as follows:

- Fix a sheet of A1 graph paper to a drawing board with masking tape.

- Draw the map on this in pencil, making corrections and alterations as necessary, leaving a clear, traceable outline map.

- Tape a sheet of tracing paper over this and trace the fair copy in pen.

- Have this photocopied. (Some architects' offices have plan copiers which can handle large sizes of paper and will make copies for the general public.)

- The photocopy is the final map, and colour can be added to it if wanted.

The advantages of this method are that any pencil corrections are completely invisible on the final copy, and you can run off several copies in case you make any mistakes in the colouring.

Personally I don't go to such lengths. I work straight onto an A2 pad of cartridge paper, drawing faintly in pencil at first, then in bold pencil, and inking in when I feel I've got it right. I don't often use colour. This may give a slightly less precise and clean presentation than the method outlined above, but perhaps a warmer and more personal one.

Some people may like to use computer aided design. I haven't tried this myself, but I'm told the available software is only really suitable for ornamental gardening.[2] However this may be a good way to present a design to people who are really into computers and would feel more at home with a design on a screen than one on paper.

Tools
Scale Ruler. This is a ruler marked out not in centimetres and millimetres but in metres as they appear at the scale of the map. Thus a 1:100 scale ruler has metres marked on it at one-centimetre intervals. Some scale rulers have four edges, the same as a normal flat ruler, with a different scale on each. But usually they are like a three-pointed star in section and have six edges, giving a choice of six scales. Different models have a different selection of scales on them and you can choose one with the scales which you are most likely to use.

Using a scale ruler can save a great deal of time in making a scale map from a sketch map. Without one, every measurement must be calculated separately. Having said that, if you're converting from paces to metres you have to convert each measurement anyway. If you work out a combined factor which takes you straight from paces to the scale of the map, you can dispense with a scale ruler without giving yourself any extra work.

A scale ruler can also be used for making a larger scale map from an existing smaller scale one. This may be useful if a small, detailed part of the site needs to be shown at a larger scale.

Pens. Ball-point pens are not good enough for mapping, though I have sometimes used an ordinary fountain pen to good effect.

The very best drawing pens are Rotring isographs. These are rather expensive, and you need a range of pens of different thicknesses to make the most of the high quality finish they can give. If they're not in constant use they must be dismantled after use and the parts carefully washed, which can be laborious. The next best thing are disposable drawing pens, such as the Pilot range. These are cheap and maintenance-free and do a very good job. Far be it from me to suggest anyone buys disposable goods! But they last a long time and are very small compared with the amount of rubbish even a permaculturist produces.

Copies and Enlargements
Where there is an existing map of the design site which can form the basis of the base map, it's often at a small scale and may be a general map of the locality, much of which is not needed for the design. Such a map needs to be copied and enlarged, with irrelevant areas left out. Copying is also useful to reproduce the completed base map so you have a number of copies on which you can sketch out design concepts.

Copies and enlargements can both be done on a photocopier, though it's sometimes hard to get an enlargement to exactly the scale you want, and irrelevant features will still be there. They can also be done quite easily by hand. This not only gives the designer more control over the output but can save a trip to the nearest photocopier.

Copying. Take the existing map, fix a blank sheet of paper of the same size over the top of it with blu-tack at the corners, and fix it to the inside of a window. You can then easily trace the map onto the blank sheet. This has to be done during daylight, of course: at night you would have to go outdoors and do it on the outside of the window!

Enlarging. The method is as follows:

*   First copy the existing map to the centre of a larger piece of paper; draw it faintly so it can be rubbed out later.

*   Mark a point anywhere inside the boundary of the site on the map.

*   Place a ruler so that the zero is at this point and the edge of the ruler runs through an important point on the map, such as a corner on the boundary.

*   Read off the distance between the two points.

*   Multiply this distance by the proportion of the enlargement, eg if you want to double its size multiply by two.

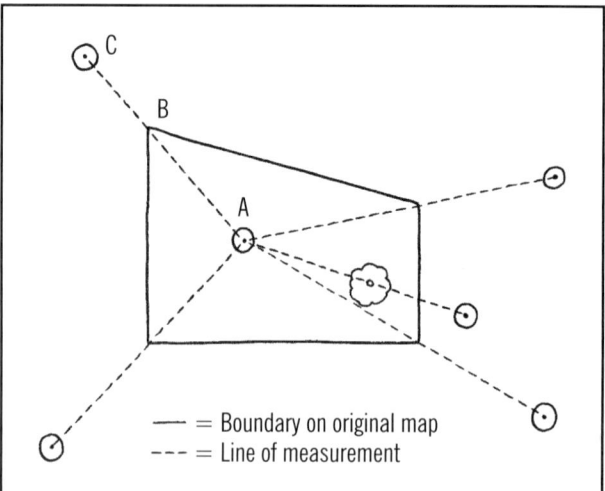

Enlarging. A-B = xmm, therefore A-C = 2xmm. The four corners can now be joined up to give the boundary at twice the original scale.

*   Without moving the ruler, make a mark at that distance from zero.

*   Repeat for other important points on the map.

*   Join up the marks.

This is much easier with a simple map than a complex one. With a complex map it's best to copy in stages, joining up the marks which indicate the boundary before starting to make marks for the internal features. It may be best to only enlarge the main features of the map by this method and draw in the minor features, perhaps using a scale ruler, once you have an outline map.

Drawing Style
There are many different ways of drawing trees, buildings and other features on a map and each person has their own preferences. If you get a chance to look at a few large-scale maps you can get an idea of the possible range of styles and practice a few before settling on what feels best for you.

The following are a few suggestions for simple ways of drawing things which frequently occur in a permaculture design.

*   *Trees.* First draw the position of the trunk, then in faint pencil draw the outer edge of the canopy as a continuous circle, accurately to scale. Then draw in the outline of the canopy in bolder pencil, giving it a conventional shape.

    Existing trees and trees proposed in the design can be differentiated by using different symbols for the trunk. Different outlines can be used for broadleaf and coniferous trees.

    Shrubberies and continuous woodland can be drawn in as a block.

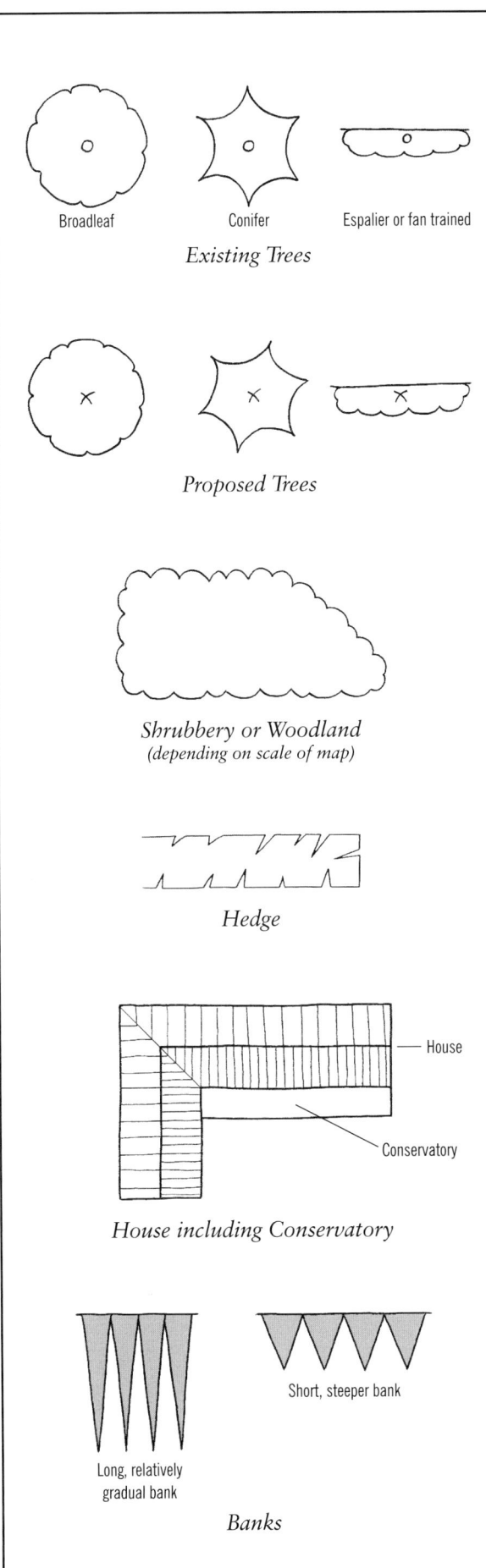

Broadleaf   Conifer   Espalier or fan trained

Existing Trees

Proposed Trees

Shrubbery or Woodland
(depending on scale of map)

Hedge

House

Conservatory

House including Conservatory

Short, steeper bank

Long, relatively
gradual bank

Banks

- *Hedges.* Draw the outline accurately in faint pencil, then more boldly giving it a conventional shape. It can be useful to use a different shape for hedges from that used for trees and shrubberies. On a small-scale map a single line is best for a hedge.

- *Buildings.* A simple way to draw a building is to first draw the outline, then a line for the roof ridge, and then add thinner parallel lines representing the pitch of the roof, close together on one side and further apart on the other, as in the illustration. Some roofs are made up of a complex of different pitches. Where this complexity is relevant to the design – as it might be if the volume of water catchment from different parts of the roof needs to be calculated separately – the complexity of the roof must be reproduced faithfully. If it's not relevant the roof can be simplified.

  To distinguish a greenhouse from other buildings the parallel lines denoting pitch can be left out.

- *Slopes.* Contour lines, where they are known, can be represented by broken lines or lines of a different colour, to distinguish them from other lines on the map. Otherwise slope can be assessed visually and represented by arrows, pointing downhill. A longer arrow represents a longer slope and a thicker arrow a steeper slope.

  In most designs it's not necessary to represent slope. Where it is a significant factor the base map can easily become cluttered with arrows, and a separate slope map, or a transparent overlay with only the slope lines on it, is often worthwhile.

  Steep banks can be represented by the conventional triangular symbols used on Ordnance Survey maps for cuttings and embankments. The narrow point of the triangle points downhill, and once again the length and steepness of the bank can be suggested by the length and thickness of the triangles.

- *Lawns, paths, ponds etc.* These need not always be represented by symbols. Lawns and paths are usually self-evident and a pond can be labelled as such. You may like to experiment with effects such as stippling to distinguish lawns and paths. Patios can be labelled or suggested by a few lines indicating paving slabs.

- *Lettering.* Carefully drawn capital letters usually look good. An alternative is to use stencils, which can be had in various sizes from stationers. But they are slow to use and don't really look any better.

Colour can be effective, but when used inexpertly it can make a map look worse. It's a good idea to make a copy to experiment on rather than risk spoiling the only one. It's most effective when applied lightly with coloured pencils.

LISTENING SKILLS

A basic question that can be asked in two ways is:
'What can I get from this land, or person?'
or
'What does this person, or land, have to give if
I co-operate with them?'

Bill Mollison[3]

We can only cooperate with a person or a place if first
we listen to them. I use the word listen here in it's
broadest sense, to include all the ways we can learn
about places and people, not just those which involve
our ears. To listen seems to me altogether more intimate
and patient than to observe, which carries a cool and
detached connotation. Ask someone to listen to the
landscape and they're more likely to go quiet and still;
ask them to observe it and they're more likely to stride
about with a clipboard. Both of these modes are
necessary, but we're so inclined to miss out the former
that I like to emphasise it.

The Land

Very often our first question about any piece of land is
'What can I do with it?' It's easy to get so absorbed with
answering this question that we never get to ask the
question which should come before it: 'What are the
intrinsic characteristics of this land?'

One important way in which we can answer
this question is by understanding the landscape in
a rational, information-based way. For this I use the
metaphor 'reading the landscape'. This is a basic skill
of permaculture and I deal with it first in this section.
Later I describe a more complete way of getting to
know the land, using all our faculties.

Reading the Landscape

Learning to understand how the landscape works can
be an endless source of fascination and delight. It
certainly is to me. The main thing you need in order to
read the landscape is simply a keen interest in it. Once
you start to look about you with an enquiring mind you
will soon begin to notice things which would otherwise
pass you by, and then start to ask why a particular
landscape is the way it is and not some other way.

The answer to the question 'Why?' is not always
simple. There are many influences at play and it's the
interaction of those influences which is responsible for
any feature we may observe, whether it's a wood, a
stream or a pattern of fields.

These influences can be divided into three main kinds:

- The climatic – including both climate and microclimate.

- The edaphic – strictly speaking the influence of
the soil, but when we're looking at whole landscapes

rather than individual plants and animals we can
include geology.

- The biotic – the influence of living things, both on
each other and on the landscape as a whole.

All these factors interact with each other. Thus, while
climate, microclimate and soil largely determine which
plants and animals can live in a particular place, the
plants and animals help to form and modify the climate,
microclimate and soil. For example, trees only grow
where there is adequate moisture, but they rehumidify
the atmosphere and thus cause more rain to fall;
coniferous trees tend to acidify the soil, while some
broadleaves tend to make it more alkaline; earthworms
can only live in soils of a certain pH and drainage
status, but they have a huge influence on the fertility of
a soil where they are present. The plants and animals
also affect each other directly. For example, compare
the ground flora under deciduous woodland and a
coniferous plantation.

Since human beings are living things, human
influences are usually included in the biotic. In most
parts of the world, and certainly here in Europe, we're
almost always dealing with landscapes which have been
heavily modified by human activity. Often the influence
of people seems to be greater than that of all other
organisms put together, so it can be useful to put our
influence on the landscape in a separate category. We
can even think of the landscape as the product of just
two groups of influences, the natural and the human,
rather than the three given above.

The interplay of natural and human influences
over thousands of years has produced the landscape
we live in now. The way people have worked with
the land has varied from place to place according to
the local climate, geology, soils and landform, and this
has resulted in the variety of landscapes we can see
around us. In recent times, as our technology has

*Signs of deer: a roe buck has been rubbing the velvet off
his antlers on this coppice stool. The presence of deer is
an important biotic factor for a permaculture designer
to note.*

become more powerful, we have increasingly ignored local conditions and imposed ourselves on the land in a more uniform way. This requires more energy and produces more undesirable consequences than does working with the landscape. Reading and understanding the landscape is the first step towards working with it in a sensitive and sustainable way.

It can get quite complex, but for the purposes of practical permaculture design it's not always necessary to understand every nuance of the landscape. Most of the important processes and relationships are described in various parts of this book. Many of them have to do with the soil and microclimates, and these are described in Chapters 3 and 4 respectively. Looking out for soil indicator plants and the effects of microclimate on the vegetation are good places to start.

Other themes to look out for in the landscape are to be found in: Chapter 2, The Principles of Permaculture, especially the part on Diversity; Chapter 11, Woodland; and Chapter 12, Biodiversity. Of particular interest to

---

THINGS, PROCESSES & RELATIONSHIPS

Features in the landscape can be seen in two ways, either as things, or as processes and relationships.

Take for example a lawn. If we regard it as a thing, we can measure it, describe it, count the species of grasses and herbs in it and so forth. So far so good; but it doesn't get us very far.

If on the other hand we look at it as a process we can see that the grass is growing. If it's not mown, the grass and any other plants which are present will continue to grow and eventually flower and set seed. In winter it will stop growing and the top growth will die down. If it's only mown once a year for a few years the species composition will change. Rosette-forming herbs which can stand repeated mowing, like common daisy, will die out and be replaced by taller ones, such as ox-eye daisy. If it's left unmown altogether woody plants will become established and it will eventually develop into woodland.

But of course it is mown. The lawn as we know it represents a relationship between the plants which are present and the people who mow it. The actual nature of the lawn will vary according to the nature of the relationship. A lawn which is mown 30 times a year is different from one which is mown half a dozen times, and one which is fertilised, irrigated and treated with weedkiller will be different again.

This is a simple example and I have only mentioned some of the processes and relationships which go on even in a lawn. Everything we can see in the landscape, both rural and urban, can be seen as a complex of processes and relationships. This is true whether humans are involved or not, though here in Britain we almost always are.

Even the most humanised of landscape features, buildings and roads, are in a constant process of erosion and colonisation by plants and animals, a process which is only arrested because of our constant relationship with them.

DON'T JUMP TO CONCLUSIONS

Recently I was doing some design work with a client who had just bought some land. It was November and we were walking across a field of maize stubble. Along one edge of the field water stood on the surface, though the rest of the field appeared dry enough. The obvious conclusion was that this part of the field is badly drained and collects water, and is perhaps a good place for a pond. However, more careful observation revealed that the area where water was standing was covered with tyre tracks and the stubble there was completely flattened, in contrast to the rest of the field where there were few tracks and the stubble still stood. It was clear that during harvest the trailers, heavy with maize silage, had repeatedly run over this strip of land on their way out of the field to the clamp, and it was suffering from compaction in the surface layers.

Not a permanently wet part of the field, just an area of compaction.

This surface compaction can easily be corrected by cultivation and then the wet area will be no wetter than any other part of the field. It's no more suitable for a pond than any other part of the field, and no less suitable for crops.

This little story illustrates how easy it is to jump to conclusions and how important it is to carefully examine what's going on in the landscape.

urban permaculturists is the book *The Ecology of Urban Habitats* by Oliver Gilbert. (See Appendix A, Further Reading, Chapter 12.)

If ecology is the language of landscape reading, the alphabet is the individual plants and animals which inhabit the landscape. It's the presence or absence of different species, and the way they are growing, which tell us so much about what's going on. Learning to identify enough species to get started is not difficult. There's a relatively small number of important indicator species, and it's not even necessary to know all of these right off if you have a good field guide, that is an identification book.

There's a wide range of guides available. They vary greatly in quality but to some extent the choice is a matter of personal preference. I would advise prospective landscape readers to have a good look at as many as possible before deciding which one or ones feel right for them. A wildflower guide and a tree guide will cover most of the relevant species. An alternative is a general guide, which usually includes the most common plants, animals and fungi. My personal preferences are given in Appendix A, Further Reading.

The best way to start is to become familiar with the common plants and animals of your own area. Some of these will be found in the list of Soil Indicator Plants in Chapter 3. (See p59.) But not all of the indicator plants will occur in any one area and there's no point learning to identify plants which you're not likely to encounter. Learning is easy once you become interested, and it's never too late to start. At the age of thirty I only knew some half dozen each of native trees, wild flowers and birds. Now I can read the countryside like a book, but I never had to work at it, and I'm still learning.

'Listening to the Landscape'

This is a comprehensive way of listening to what the land has to say to us, which I learnt from Christopher Day, an ecological architect. He developed it along with Margaret Colquhoun, a biologist and landscape architect, and I am grateful to them both for inventing a really useful tool. I have somewhat adapted it for use in permaculture design and what follows is my own version of their method.

It's a process of tuning into the landscape in four distinct ways, or modes: the intuitive, the objective, the imaginative and the subjective. It's best done alone, or if in a group, in silence. The aim is to listen to the land, not to each other, and talking gets in the way of listening. When working in a group it's always good to get together immediately after a listening session and share impressions.

It's important to remember that in each mode we are asking what the land says to us, not what we can do with it.

Intuitive. The best time to listen in the intuitive mode is when we visit a piece of land for the first time. First impressions are unrepeatable, and they can have a clarity which later gets overlaid by other knowledge. How often have you had the experience of meeting someone for the first time, forming an instant first impression of them, then dismissing it, saying "No, I must get to know this person well before jumping to conclusions," only to find in the end that your first impression was spot on? On the other hand, sometimes first impressions do turn out to be mistaken. Intuition is not always right, but it's not always wrong either, and we need to respect it and put it on an equal standing with rational thought.

The best way to start is to walk round the boundary. Our normal reaction on meeting a piece of land for the first time is to stand in the middle and look about us. This gives one perspective of the land, but it's only one among many. Walking round the boundary gives us many others, and it's usually the quickest way to get to know a place. The boundary is also somewhere where many things happen. It's an edge. Looking over the boundary gives an idea of how this land relates to its immediate neighbours and to the locality in general.

This is not a time to observe specific facts about the land. That will come later. Having a relaxed mind and an absence of expectations is the best way to let the intuition flow.

Objective. In this mode we observe facts about the land. It's never possible to be completely objective, but the aim here is to simply look at what is there without putting any judgement on it. The best way to do this is to observe the site on a number of different occasions and choose a different theme to work with on each. Themes can include: plants, animals, microclimates, water features, soils, structures and so on. Exactly which themes are appropriate will vary from site to site.

A person working alone can cover the whole range of themes on a number of occasions. This is preferable to observing a range of different things one after the other on the same day. That can lead to hurrying if one theme takes longer than was allowed for it, or to looking at everything at once, which throws away the clarity which comes from concentrating on one thing at a time. A group of people can split the themes up among them and work simultaneously. In either case it's best if at all possible to observe the site on a number of occasions through the year. The plants and animals which are in evidence will be different at different times, and of course the microclimates will vary through the seasons.

Imaginative. Not everything about a place can be observed or experienced directly. This particularly applies to things which have happened in the past or may happen in the future. For these we need to use our imagination.

The first thing to imagine is what the place was like in prehistoric times, before people started to interact with it. This is not entirely an exercise of the imagination. Most of us have some idea of the prehistory of the landscape, if only that almost all of this country was originally woodland. But however well informed we are we can never work out exactly what a particular place was like. This is an opportunity to still our minds and allow the original spirit of the place to speak to us.

Another useful thing to imagine is how the place would change if people stopped interacting with it now. What stages of succession would the vegetation go through, what would be the fate of existing structures and water features?

If you have to make all your observations over a short period of time you will also need to imagine what the place is like in the other seasons of the year. This is a poor second to actually being there in all four seasons, but sometimes it's unavoidable.

Subjective. This is a good way to end a session which has been done mainly in one of the other modes. Allow yourself to be drawn to whichever part of the site you feel you want to be in. This is not necessarily the most pleasant part – sometimes I find myself drawn to quite uncomfortable, windy places – and it will not necessarily be the same place each time. It's not a matter of thinking of the kind of place you normally like and then finding one, but of letting your body take you wherever it will.

When you get there, express yourself in some non-rational way. Sing a song, write a poem, draw a picture, make a sculpture out of what is there or do a dance. Or if you prefer just sit there.

People who are designing their own place, whether individuals, families or community groups, can use the method of listening to the landscape over a period of months to build up a really intimate knowledge of the land they are working with. The amount you can learn about the place by intentional listening of this kind, compared to what you can learn simply as a by-product of going about your daily business, cannot be over-stated.

It's more difficult for a visiting designer to use the method to the full. Nonetheless I would recommend anyone who is thinking of working as a permaculture designer to use it whenever possible. It's not necessary to use it in full on every design in order to get a benefit from it, and there's no reason why it shouldn't be adapted to meet the needs and preferences of different individuals or groups.

I have found that using it has sharpened and developed my powers of observation. Now I look and listen much more fully even when I'm not consciously using the method. Practising it as an exercise can be worthwhile. (See box, An Exercise in Listening to the Landscape.)

AN EXERCISE IN
LISTENING TO THE LANDSCAPE

Listening to the Landscape can be done as a short exercise, not necessarily as a part of an actual design but in order to gain the experience of observing in this quiet and receptive way. I include this exercise in the permaculture design courses I teach.

I ask people to observe in each of the four modes in turn for ten minutes each. In the intuitive mode it's not necessary to get right round the boundary; this is a learning experience and the important thing is to experience the mode. In the objective mode I ask people to observe the plants, making a list of all they see. Any plant which they can't identify can be given a brief description, such as 'yellow flower'. In the imaginative mode I ask them to imagine the distant past, and the future without people, but not the other seasons. The subjective mode is just as described above.

I let them know when each ten minutes is up by ringing a bell or blowing a whistle. At the end we get together, either in small groups or all together, to share our impressions.

I recommend this exercise to any group of people, or to individuals, who want to practice permaculture design, or simply to deepen their relationship with the land. It can be done in any reasonably quiet place, from a suburban garden to fields of two or three hectares. Larger spaces can be used if more time is allowed. In small spaces the number of people needs to be restricted so that each person can have enough space to work as an individual.

Students on a permaculture design course listening to the landscape.

The People

Good listening is central not just to permaculture design but to most other things in life as well, from personal relationships to professional activities. Yet we tend to be very bad at it. It's a skill like any other and can be learnt. In fact once we become aware of the way we're listening – or not listening as the case may be – change can happen fast and we can often transform the way we listen to people with little effort. The first step is to have some understanding of what makes for good listening. The main points are summarised in the mind map.

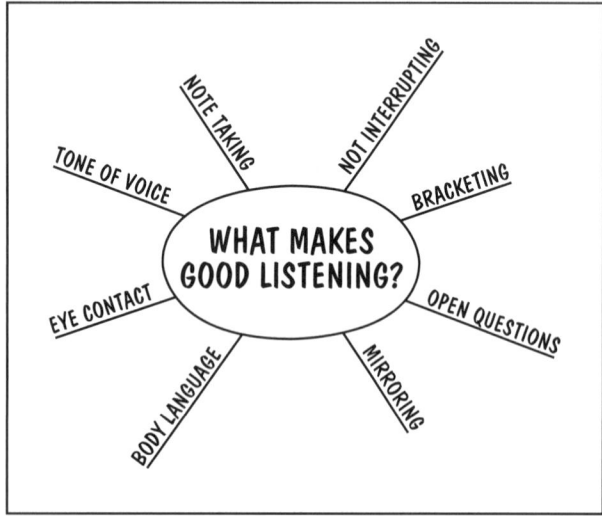

There is a difference between active listening, as for example when a permaculture designer is listening to people talking about what they want in their design, and ordinary day-to-day listening, as in a conversation. The points in the mind map refer primarily to active listening. With a little modification they apply to day-to-day listening too, and this is discussed below.

- Not interrupting is obviously fundamental to good listening. Interrupting is not always a matter of butting in while another person is actually speaking. Changing the subject can be an interruption too, especially in an active listening situation. In other words we interrupt as much when we break into someone's train of thought as we do when we break into their flow of words.

  The three most common kinds of interruption are: attempting to finish another person's sentence for them, changing the subject to our own concerns, and giving unasked-for advice. All three of these break into the person's train of thought, and once that's done they may never return to it and we may never get to hear what they were going to say.

  Not listening is similar to interrupting – the effect is much the same, though the speaker is unaware that it's happening. We stop listening as soon as we start to think about what we're going to say in response to what the speaker is saying, or when we feel we

know what they're going to say next. We're still there, still silent and apparently listening, but we might as well not be there.

Most of us are culturally conditioned to believe that giving someone advice is more valuable than just listening to them. In fact nothing can be further from the truth. Just listening to someone is very often the most valuable thing we can do for them – and it's certainly the appropriate thing in the design questionnaire stage of a permaculture design. Giving unasked advice can be a form of interruption.

- Bracketing means putting one's own concerns on one side for the time being. It's a concept used by counsellors. It amounts to saying: "I accept that it's not my turn to talk at the moment. That will come at some other time in the future. Just now my function is to actively listen." It can help prevent feelings of frustration and resentment. It's also an acknowledgement that we all need listening to at some point, even those of us who are professional listeners.

- It can sometimes be appropriate to speak when we're actively listening, for example if the person who's speaking dries up. Sometimes silence can be worthwhile, but probably not during a permaculture design questionnaire session, and usually a question will get the person going again. Open questions are usually more useful in this situation than closed ones. A closed question can be answered by Yes or No, and an open one cannot. For example, "Do you like pears?" is a closed question, "What kind of fruit do you like?" is an open one. The answers to an open question can sometimes lead off into unexpected directions.

- Mirroring means reflecting back to the speaker what they've just said, but in your own words. This may be necessary when you haven't understood what they said. To remain silent when we don't understand is not good listening; in fact it's not listening at all. It may be appropriate to say "Hang on a minute. When you said that did you mean... ?" At other times mirroring can be useful to sum up what the person has said. Hearing a summary of what they've been saying can help a person clarify their own ideas to themselves.

- Words are only one of the ways we communicate with each other. We tend to underestimate the amount we communicate through body language because by and large we're not conscious that we're doing it. But an active listener, who is keeping silent most of the time, is giving out more information by this means than by any other. Three ways we communicate with our bodies are: the degree of openness, the distance from the other person, and the angle our body makes with theirs.

Body language is closed when we cover the front of our bodies. Folded arms, crossed legs, a clipboard held across the heart or hands touching the face are all ways of closing down communication, while a relatively uncovered front indicates openness to what is being said. The distance between us and the other person gives a similar message: the closer we are to them the more interest, warmth and intimacy we express. The same goes for the angle at which we stand or sit: the more nearly facing them we are the more open and warm the body language.

- Eye contact is a particularly powerful form of body language. It's hard to believe that someone who never looks at you is really listening, while too much eye contact can be intimidating.

  In all of these there is a balance to be struck. Hunching down in your chair, legs crossed, with a hand over your face and looking in the opposite direction definitely gives the impression you're not interested. But sitting right in front of someone, legs apart, arms akimbo, staring unblinking into their eyes, is more likely to put them off than reassure them that you're really listening.

  There's no prescription. Much depends on the character of the person you're listening to, your relationship with them and the situation in which you're listening to them. The meaning of body language can be different from one social group to another, too. In some sub-cultures eye contact means sincere interest; in others it's a challenge to physical violence. The important point is to be aware of body language and remember that we're communicating with it, whether we or the speaker are conscious of it or not. The aim, at least in a permaculture design situation, is to help the speaker feel relaxed and confident that they're being heard, but the exact means will vary according to circumstances.

- Tone of voice can convey much the same messages as body language. The same sentence spoken in different tones of voice can mean two or more completely different things. Once again there's no prescription and simply being aware that it's important can improve the quality of communication.

- The way we take notes of what a person is saying can also affect the quality of our listening. Too much writing while a person is talking can leave them feeling out on a limb, as though there's no-one there. They may be right, too, because it's pretty well impossible to simultaneously write down what they just said and listen to what they're saying now. It can sometimes be appropriate to ask them to pause a moment while you write something down.

  It's equally disconcerting for them if you never write anything down during a design questionnaire session – and of course you do need a record of what they've said. You may have a super-power memory, but it's more likely that you're not as interested in what they have to say as you might be. When listening to a number of people in turn and taking notes it's good to take at least some notes of what every person has to say. No-one likes to feel ignored.

  A useful way of combining good listening with note-taking is the mind map technique. The essence of mind mapping is to write down only key words, not whole phrases or even sentences. These key words act as a memory prod when looked at later on and remind you of what was said. This won't work if you make a mind map one day and don't look at it again till you pick it up a month later. You would be unlikely to remember what many of the key words stand for. If there's going to be a gap in time between asking the questions and working on the design you need to read through your mind map between times to fix the associations in your memory. Ideally this should be done within a day or two of making the mind map.[4]

Ordinary everyday listening is different from full active listening only in degree. Sometimes it's appropriate to bracket our own stuff and give another person a bit of really good attention. This doesn't have to be done in any formal way. It can be appropriate to slip into listening mode just when we notice the person we're with needs attention and the situation is right. Also, practising active listening can improve the way we listen in our normal day-to-day interactions. Having experienced active listening we become more aware of the way we listen in ordinary conversation and more able to choose to do it really well.

DESIGN METHODS

Options and Decisions

This is an overall approach to design which is more relevant to designers who are designing for other people than those who are designing their own place. In essence it's about giving your clients some advice but not actually telling them what to do.

The classic model of landscape design is essentially that presented in Chapter 13, culminating with a set of proposals as shown in the Design Proposals Map. The options and decisions model follows the same course up to the Evaluation stage, but instead of a definite set of proposals the clients are presented with a series of options they can take. The decisions are then up to them.

The options may include overall concepts, and/or more detailed recommendations for different parts of the site or for different aspects of the design, such as food production, wildlife and so on. They must be accompanied by appraisals of the possible outcome of taking each option. Often the most valuable function of the designer is to identify those options which certainly won't work rather than those which may.

A particularly useful part of the presentation is a resource list, giving contacts for organisations and individuals who could be useful in designing or implementing one option or another. These can include specialist advisers, people with relevant experience, suppliers of plants and equipment, and buyers of produce. Lists of plant and animal species which may be useful in the design, with notes on the needs and outputs of each, are also usually included.

In some ways options and decisions is the ideal way for a permaculture consultant to work. It enables them to give some expert guidance while leaving the actual decisions to the people who know the place intimately and who will have to live and work with the consequences of them. But it's not an easy way of designing.

The reason for this is its complexity. In any design each element must be looked at in relation to every other one. When some decisions are left open and there are several options for each one, the total number of possible combinations is very high. The designer must make every attempt to check these out and identify those which will combine well and those which won't. In practice this means that it's not possible to go into as much detail in an options and decisions design as in a design where a single solution is presented on a map.

There's also more of a need for continuing contact between designer and clients, as there's much more likelihood of unforeseen questions coming up as the design develops.

Design Tools

These are specific ways of working which may be useful at one time or another during the design process.

Analysis Tools

The three methods described here – SWOC, the key planning tools and input-output analysis – are ways of giving structure to the questions which come up in the process of permaculture design. Perhaps the biggest advantage of using them is that they're systematic, so they can help you be sure you've covered everything. When working in an unstructured way it's easy to miss something, and that something may be important.

The three have much in common and you may come up with much the same information and conclusions by using any one of them. But they do differ and one may suit a particular design or designer more than another. SWOC comes from the world of business and may be of most use in larger-scale designs where a living is to be made from the project. The key planning tools specifically relate to the land and may be of less use in designs which are not land-based. Input-output analysis is most relevant in fairly complex designs where there are many different productive elements which need to

be linked with each other or with off-site elements in the locality. (See for example the case study, A Permaculture Farm, p294.)

Having said that, all of them can be useful in any kind of design. I would suggest that new designers try them all and find out what works best for them.

SWOC. This is an acronym, standing for:

> Strengths
> Weaknesses
> Opportunities
> Constraints

Some readers may be familiar with this as SWOT, with 'threats' rather than 'constraints' as the last word. The change of word is not vital, but it avoids a certain sense of paranoia.

SWOC can be used to evaluate a site, a design idea or a whole design. Thus it can be used both at the evaluation stage and when working on design proposals.

Strengths and weaknesses are about inherent characteristics. For example, looking at a potential market garden site we might find that a strength is the inherent fertility of the soil and a weakness is its lack of shelter from strong winds. Opportunities and constraints are more about relationships. Taking the same example, an opportunity might be that there is an unfilled local demand for organic vegetables and a constraint that the owners lack working capital to develop it.

It may be hard to decide whether a certain characteristic is inherent or a matter of a relationship, but it doesn't really matter. The aim is not to classify things, but to make sure no points have been missed. Similarly, some things may be neither particularly positive or negative. For example the site may be larger than needed for the garden. This could be seen as an advantage, as it gives room for later expansion, or a disadvantage, if it means spending some of the purchase price on something which isn't wanted. In that case it should go down on both sides of the equation.

During the design proposals stage SWOC can be used to look at an area or an aspect of the site which is hard to know what to do with. A thorough analysis may come up with an idea that would not have emerged from a less structured approach. Towards the end of the process it can be used to assess the whole design. It may reveal important points which have been missed.

Key Planning Tools. These four concepts – zone, network, sector and elevation – incorporate some of the fundamental ideas of permaculture and are described in detail in Chapter 2. (See pp27-30.) Any permaculture designer will bear them in mind all the time, and they will be part of the rationale of any design, but they can also be used in a structured way as a specific method of design.

The key planning tools can be used to:

- Evaluate the site as a whole at the evaluation stage, leading to the generation of design proposals. (See the example on p394.)

- Decide on the location of specific features which are wanted in the design (see the example on p395.); if there are a number of possible sites for a particular element a KPT analysis can be done on each to see which is most suitable.

- Analyse the design proposals themselves to check that no important factors have been missed.

Input-Output Analysis. This design tool makes use of the permaculture principle of linking, that is combining elements so that the output of one becomes the input to another. (See p32.) It can be used both to analyse an existing situation and to assess a design proposal or a whole design.

In both cases the method is much the same. First a list is made of the elements – plants, animals and structures – present either on the existing site or in the design proposal. For each of these a list of inputs and outputs, either actual or potential, is made.

INPUT-OUTPUT ANALYSIS	
Inputs can include:	Outputs can include:

space
soil conditions
microclimate
feed
shelter
materials
energy
work
skills
capital
information
markets
etc
	all kinds of physical produce
	beauty
	recreation
	indirect outputs:
	    microclimate effects
	    soil improvement
	    predator habitat
	    etc
	undesirable outputs:
	    pollution
	    shade
	    etc

Any important characteristic of the element which isn't really an input or an output can be listed in a third column, headed characteristics.

Poultry are notable for the number of connections which can be made between them and other elements. (See pp248-250.) Most other elements will yield shorter lists than poultry, and there is little to be gained from making the list as long as possible by adding fanciful or trivial inputs and outputs. Corn dollies as an output of wheat, for example, is not worth listing unless you happen to be a craftsperson who might realistically make this a significant output.

Stating the obvious is not much help either. For example, air, sunshine and water are needed by all plants and including things like these only clutters the list. But inputs like "needs watering in summer" or "tolerates part shade" are useful. Figures must be given for any input or output which can be quantified. For example, if the element is a fruit tree, under 'space' the size – or range of sizes – of the mature tree should be given, and under produce the expected yield and the number of years before full yield is expected.

These lists may bring to light a number of possibilities:

- Linking some of the existing inputs and outputs within the system which are presently not linked; a simple example comes from the design in Chapter 13, where the unused stones in the front garden were used to build an attractive wall in the back, and the blocks from the unsightly wall at the back were used to build steps and a planter. (See p396.)

- Moving things around so as to facilitate these links; for example, placing a chicken house so that the chickens can be easily let into an orchard for winter pest control.

- Making links outside the system with local partners; for example, it may be possible to share equipment with a neighbour rather than buying it, or a local box scheme may be able to market a small surplus of eggs or mushrooms along with their vegetables.

- Introducing new elements which can make use of unused outputs or provide inputs which are presently imported into the system; a classic example is the rainwater harvesting system which both solved a drainage problem and provided summer irrigation water on Cindy Engel's smallholding. (See p189.)

Input-output analysis can also be used to evaluate a completed design, to see how close the design comes to the aim of self-reliance and minimum ecological footprint. Here the method can be simpler. Just looking at what is imported from outside and at any unused outputs will give an idea of how successful the design has been from the point of view of the linking principle.

It's not always appropriate to make every single link between input and output which can be made. Some needs may be best met by an off-site input and some outputs may be best left unused, especially when they are small and the net result will not repay the effort put into making the link. This design tool is intended to highlight missed opportunities, not to impose a dogma on the designer.

Mapping Tools
Card Cutouts. Imagine you're designing a large forest garden or a mixed orchard. You've already decided how many trees you want to plant, and made a list of the species, varieties and rootstocks. The next task is to decide which tree goes where.

Tim Harland with the design of his garden showing the base map drawn on card and a tracing paper overlay with scaled paper cut-outs of trees at mature size. These can easily be moved around until the ideal design layout has been found.

Each one needs to be placed in relation to the various microclimates provided by the site and relative to the other trees around it. You don't want a vigorous cooking apple in a prime site, nor a peach in a shady corner. You don't want a relatively tall tree shading a smaller one, while a tough damson may be well placed if it protects a tender gage from the prevailing wind.

The number of different arrangements which are possible is enormous. It increases geometrically with the number of trees. In order to work out what goes where you could run off dozens of copies of the base map and pencil in a series of possible alternatives on them. But it would be tedious, especially if you took care to draw the diameter of each tree accurately, which is necessary if the plans are to be realistic.

The alternative is to use a single base map and discs of card representing the trees. One card is needed for each tree. They must be cut out accurately to the size the tree will be when mature, at the scale of the base map. It's best to use a different colour of card for each species, with the variety of the individual tree written on each disc. The cards can be moved about the base map until the ideal combination is found.

The McHarg Exclusion Method.[5] This is a method of placing a feature in the landscape by deciding not where it should go, but where it shouldn't.

It was developed by Ian McHarg, a Scotsman who was professor of landscape design at an American university. One day he was approached by some local residents who were concerned about plans to build a new road right past where they lived. The ideal solution would probably have been not to have built the road at all,

but that was not an option. The best he could do was to suggest a route which would do the least harm all round, and to do this he set about identifying those routes which were unsuitable for one reason or another.

He made a base map of the land through which the road was to pass, and a series of transparent overlays which fitted over it. On each of the overlays he mapped out areas through which the road should not pass for one specific reason. The reasons included: too close to housing, high wildlife value, woodland, marshy areas, steep land which would be too expensive to grade, and so on. When all the overlays were superimposed on the base map at the same time the least unsuitable route for the road became apparent.

This method has more application for large-scale planning issues than for the design of individual properties. It can be used to identify suitable sites for housing. When potentially damaging development proposals are put forward it can be more productive to suggest a less damaging alternative than to oppose the development full stop. If, for example, an important wildlife site is threatened it may be possible to show that the site is unsuitable for other reasons. The McHarg Exclusion Method may be very effective in putting a message across to planning committees, as it's both clear in its presentation and professional in its approach.

Planning for Real®

This is a comprehensive design system developed by The Neighbourhood Initiatives Foundation (NIF) for involving people in the redesign of their own neighbourhoods.

It's not the only model for participatory design, but it is the most accessible. The big problem with participatory design methods is that they're not easy to use successfully without proper training. Planning for Real has the advantage that there's an organisation devoted to teaching it and organising it on the ground. Anyone who is interested in using it should contact the Neighbourhood Initiatives Foundation and not try to

A planning For Real® session in full swing.

work on the basis of the brief description given here. (See Appendix D, Organisations.)

The aim of Planning for Real is to make the design process equally accessible to everybody. It does this by getting away from speech as the main means of communication. Consultation meetings which are based on speech get dominated by the few people who feel comfortable about speaking up in a public meeting. Everyone else keeps silent or doesn't bother to attend, though their input may be equally valid. For most people the speech-based approach is disempowering, and action which comes as a result of it often feels more like something imposed by 'them' than something generated by 'us'.

The first step is to find a few moving spirits in the community. These are not necessarily the leading lights, but just ordinary people who feel motivated to do something to improve their neighbourhood. They make up the core group which organises the process.

A big three-dimensional model of the neighbourhood, including the good bits and the bad bits, is made by local schoolchildren. It's made in sections so it's easy to move around, and is displayed in places where people gather, such as schools, markets and sports centres. On a side table are a series of problem cards, each one with a specific problem written on it, which anyone viewing the model can place on it. There are also blanks which people can write on themselves. This stage of the process generates interest and people begin to come up with ideas for improving the neighbourhood.

Meanwhile some of the moving spirits begin a house-to-house survey of local skills. Each resident is invited to tick off various skills they have, from painting and decorating to sports and music, rating themselves as either keen beginners or experienced.

Then comes the big day. The model is set up in a public gathering space where everyone feels equally at home, often the local school. Around the model are suggestion cut-outs which people can add to the model where they would like an improvement to be made. These include things such as: meeting places, changes in traffic management, play areas, greenery, housing improvements and so on. Again, blanks are available.

People can disagree with a suggestion which has been made simply by turning a card over: on the back of each one is written the word Disagree. This can be done discreetly or even anonymously, so as to avoid confrontations.

Next the suggestions which most people favour are sorted out according to: when they could be achieved; what kind of support is needed; and who could help to do each job. A chart is drawn up, as shown below. From this an action plan is negotiated between all those concerned.

This will include officials from the local council. Many of the suggestions will need their involvement, both from a legal and a practical point of view. They attend the meeting, but take no part unless a resident asks them for information. 'Experts on tap, not on top' is the phrase used to describe their role.

Over the past ten years hundreds of Planning for Real schemes have been developed in the UK and it has been adapted for use in other countries, from South Africa and Jamaica to Germany and Russia.

Although it has so far been used mainly in urban neighbourhood design, NIF reckon that with little change it could be used for anything from designing a living room to trying to reconcile conflicting interests in the use of the countryside. An example of a project which did not involve a neighbourhood comes from the Centre for Alternative Technology in Wales. The design of their new Autonomous Environmental Information Centre building started with a Planning for Real process involving the staff of the centre.

Wherever and however it's used, the aim of Planning for Real is to enable things to happen in communities by turning 'us and them' into 'us plus them'.

NOTES

Some references are given simply by author and date. These refer to books listed in Further Reading, Appendix A. Those followed by † are listed in the General section, those without † are listed in the By Chapter section under the relevant chapter. Note that photocopies of articles from journals are available from the public library service.

Chapter 1
WHAT IS PERMACULTURE?

1　See table on p22, source: Whittaker, Robert H. (1975) *Communities and Ecosystems*, 2nd edn, Macmillan.

2　Mollison and Holmgren (1978)†.

3　See Mollison (1988)†, pp40-41.

4　Mollison and Slay (1991)†.

5　Norberg-Hodge, Helena (1991) *Ancient Futures – Learning from Ladakh*, Rider.

6　Rahman, Atiq, Nick Robins and Annie Roncerel, eds. (1993) *Consumption versus Population: Which is the Climate Bomb?* Climate Network Europe, 44 rue de Taciturne, 1040 Brussels, Belgium.

7　Whitelegg, John (1997) *Critical Mass*, Pluto Press, p129.

8　Wackernagel, Mathis and William Rees (1996) *Our Ecological Footprint, reducing human impact on the Earth*, New Society Publishers, Canada, p149.

9　von Weizsäcker, Ernst, Amory B Lovins and L Hunter Lovins (1997) *Factor Four, doubling wealth, halving resources use*, Earthscan.

10　For a full account see: Wackernagel & Rees, op cit; and Chambers, Nicky, Craig Simmons and Mathis Wackernagel (2000) *Sharing Nature's Interest, ecological footprints as an indicator of sustainability*, Earthscan. Both books are accessible and authoritative.

11　Giradet, Herbert (1999) *Creating Sustainable Cities*, Green Books.

12　Wackernagel and Rees, op cit, pp128-129.

13　Harper, Peter (1996) 'Getting Through to Sweden' in *Clean Slate* No 20, pp10-11; and 'Tigers and Yoghurt Pots; the "Sweden" debate continues' in *Clean Slate* No 21, pp10-12.

14　McLaren, Duncan, Simon Bullock and Nusrat Yousuf (1998) *Tomorrow's World, Britain's share in a sustainable future*, FoE/Earthscan.

15　I am indebted to Peter Harper for most of the figures quoted in this section, as well as the original idea.

16　Wackernagel and Rees, op cit.

17　See Chapter 2, note 23.

Chapter 2
THE PRINCIPLES OF PERMACULTURE

1　Bazzaz, FA and DD Ackerly (1992) 'Reproductive Allocation and Reproductive Effort in Plants', in Fenner, M, ed. *Seeds: the Ecology of Regeneration*, CAB International; and: Jackson, Laura L and Chester L Dewald (1994) 'Predicting Evolutionary Consequences of Greater Reproductive Effort in *Tripsacum dactyloides*, a Perennial Grass,' in *Ecology* 75 (3), the Ecological Society of America.

2　Recent advances in population genetics have shown that the concept of race, at least with regard to humans, is scientifically meaningless, and even the concept of species is now regarded as more of a useful category than a physical reality. Nonetheless both concepts are real enough in the day to day world, as here in the discussion of biodiversity.

3　For a detailed account of the niche concept see Miller (2002).

4　One factor in the replacement of the red squirrel by the grey may be a disease which the greys carry and tolerate but which is fatal to the reds, but this is unlikely to be the only factor involved.

5　These three woodpeckers make up what ecologists call a guild: a group of organisms which occupy very similar niches but avoid competition by virtue of each species using a slightly different part of the resource base. It is unfortunate that Bill Mollison chose to use the same word, guild, to signify a quite different concept: a group of organisms occupying completely different niches in the same habitat and having mutually beneficial interactions.

6　Andy Daw, personal communication.

7　Tilman, David, David Wedin and Johannes Knops (1996) 'Productivity and Sustainability Influenced by Biodiversity in Grassland Ecosystems', in *Nature* Vol 379, 2nd Feb 1996.

8　Tilman, David and John A Downing (1994) 'Biodiversity and Stability in Grasslands' in *Nature* Vol 367, 27th Jan 1994.

9　Altieri, Miguel (1995) *Agroecology, the science of sustainable agriculture*, Intermediate Technology Publications.

10　Kourik, Robert (2004) *Designing and Maintaining Your Edible Landscape Naturally*, Permanent Publications, pp237-244.

11　For a full account of this subject, including solutions, see Velvé, Renée (1992) *Saving the Seed*, Earthscan.

12　'Crop Genetic Resources' in *Biodiversity for Food and Agriculture*, FAO, 1998.

13　For some examples of farmer/scientist co-operation see: Almekinders, Conny and Walter de Boef (2000) *Encouraging Diversity, the conservation and development of plant genetic resources*, Intermediate Technology.

14　For an enthusiastic view of the potential of teleworking in rural repopulation see: Body, Richard (1992) 'The 21st-Century Peasant' in Conford, Philip, ed. (1992) *A Future for the Land, organic practice from a global perspective*, Green Books.

15 Source, Whittaker (1975).

16 Strictly speaking the term used to describe annual growth patterns is phenology, while succession refers to year-on-year changes in the composition of an ecosystem.

17 Source: Whittaker (1975).

18 For a discussion of the role of home gardening in food production see: Holmgren, David (1991) *Gardening as Agriculture, a Permaculture Perspective*, Holmgren Design Services, Australia.

19 Calculated by David Pickles, Newark and Sherwood District Council.

20 See Holmgren (1996)[†], p31.

21 Ibid, p34.

22 Lampkin, Nicolas (1994) *Organic Farming*, Farming Press, p585.

23 Jackson, Wes, Wendell Berry and Bruce Coleman (1983) *Meeting the Expectations of the Land*, North Point Press, USA; Lampkin, Nicolas (1994) *Organic Farming*, Farming Press. The sources for these figures date back to the 1960s and 1970s. As far as I'm aware no more recent studies have been made, but it's safe to assume that the general picture remains the same. Any increases in mechanical efficiency will have been swallowed up by increases in mechanisation, and especially in long-distance transport.

24 'World Census on Agriculture,' FAO Census Bulletins, Rome 1980, quoted in Anderson, Luke (1999) *Genetic Engineering, Food and our Environment, a brief guide*, Green Books, pp61-62.

25 Best, Robin H, and JT Ward (1956) *The Garden Controversy, a critical analysis of the evidence and arguments relating to the production of food from gardens and farmland*, Dept. of Agricultural Economics, Wye College, Kent.

26 Lovelock, James (1991) *Gaia, the practical science of planetary medicine*, Gaia Books. For a brief summary see: Allen, Paul (1998) 'Gaia Theory: Gaia Practice' in *Clean Slate* No 27.

27 Miller (1992) *Living in the Environment*, Wadsworth.

Chapter 3
THE SOIL

1 Mollison (1988).[†]

2 Lampkin (1994) p82.

3 Ibid.

4 Ibid.

5 Bracken can be carcinogenic in early summer and when spores are produced. The latter is in September, but only after a very hot summer, around one year in five. Avoid cutting it at these times. Also beware of tics, which can transmit Lyme disease.

6 Lampkin (1994).

7 For more information on these flatworms contact the HDRA information service. See Appendix D, Organisations.

8 For detailed information on nitrogen fixers see *Agroforestry News* Vol 3, No 3.

9 Lists of plants belonging to the various groups are given in ibid.

10 See in particular *The Organic Gardening Catalogue*, Tel: 01932 253 666; and Suffolk Herbs, Tel: 01376 572 456.

11 For detailed information on dynamic accumulators see *Agroforestry News* Vol 3, No 3.

12 For more information on planting dynamic accumulators for grazing animals, including a chart showing the minerals most effectively accumulated by each, see: Younie, David and Aslam P Umrani (1997) 'Forage Herbs and Dietary Minerals' in *New Farmer and Grower* No 56 pp14-15.

13 For a detailed discussion of this question in a farming context see Lampkin (1994) pp36-48.

14 See for example: Brady, Nyle C (1990) *The Nature and Properties of Soils*, Tenth Edition, Macmillan, USA; and *HDRA Newsletter*, Nos 89, 93, 97 & 119.

15 For design details see: Coleman, Eliot (1995) *The New Organic Grower*, Chelsea Green.

16 Ibid.

17 A detailed prescription for using the Soil Reconditioning Unit in pasture is given in Mollison (1988)[†], pp218-221.

18 Note that Bill Mollison refers to the Soil Reconditioning Unit as a chisel plough. This may be the Australian terminology, but in Britain the term chisel plough usually refers to an implement which causes maximum soil disturbance at the surface but does not penetrate deeply.

19 Much of the information in this section is from: Hodges, RD and C Arden-Clarke (1986) *Soil Erosion in Britain*, Soil Association; and Lampkin (1994) pp14-17.

20 BBC Radio 4, Farming Today, 9th Aug 1994.

21 SSLRC and Silsoe College (1994) Submission to the Royal Commission on Environmental Pollution: *Environmental problems associated with soil in Britain*, Soil Survey and Land Research Centre and Silsoe College, Bedfordshire, quoted in: Baldock et al (1996) *Growing Greener, sustainable agriculture in the UK*, CPRE &WWF-UK.

22 For a more complete account of the ethylene cycle see Widdowson, RW (1987) *Towards Holistic Agriculture, a scientific approach*, Pergamon Press, pp21-23.

23 Miller, G Tyler (1992) *Living in the Environment*, Wadsworth. p318.

24 Morgan, RPC (1979) *Soil Erosion*, Longman, p25.

25 van der Werff, Peter and Jan Kouwenhoven (1996) 'De Ekoploeg' in *Ekoland* No 7/8, July/Aug 1996, Netherlands.

26 For an account of the biology of mycorrhizas see Widdowson, op cit, pp16-17.

27 Zarb, John (1999) 'Another Look at Soil Fungi' in *Organic Farming* No 63, Autumn 1999.

28 Lampkin (1994) p40.

29 Martin, NW and J Keable, in Stonehouse, B, ed. (1981) *Biological Husbandry*, Butterworths, p138.

30 For a comprehensive list of garden/farmland weeds as indicators see Lampkin (1990) pp165-166. For a North American list see Kourik, Robert (2004) *Designing and Maintaining Your Edible Landscape Naturally*, pp36-38.

31 The system of soil classification and much of the information in this section, though not the names I have used for the soils, is from Gilbert (1991) Chapter 4.

32 After Gilbert (1991).

33 Ibid, pp51-52.

Chapter 4
CLIMATE & MICROCLIMATE

1 The reasons for this are complex and not fully understood. But for a brief explanation see: Whittaker, Robert H (1975) *Communities & Ecosystems*, 2nd edn, Macmillan pp96-98; and for a discussion of the various hypotheses: Leakey, Richard and Roger Lewin (1996) *The Sixth Extinction, biodiversity and its survival*, Phoenix.

2 EO Wilson, quoted in ibid, p104.

3 See: www.metoffice.gov.uk/climate/uk/averages/index

4 These figures are taken from Gilbert (1991).

5 The main sources for this section are: Miller, G Tyler (2002) *Living in the Environment*, twelfth edn, Brooks/Cole (USA); and Rahman, Atiq, Nick Robins and Annie Roncerel, eds. (1993) *Consumption versus Population: Which is the Climate Bomb?* Climate Network Europe, 44 rue de Taciturne, 1040 Brussels, Belgium.

6 IPCC (2000) *Climate Change 2001: The Scientific Basis*, Intergovernmental Panel on Climate Change. See: www.ippc.org.ch/press/pr.htm

7 Lester Brown of the Worldwatch Institute, quoted in: AtKisson, Alan (1999) *Believing Cassandra*, Chelsea Green.

8 Allen, Paul (2001) 'Gaia Theory and Climate Change', in *Clean Slate* No 39.

9 Watson, Robert T et al (2000) *Land Use, Land Use Change and Forestry*, Intergovernmental Panel on Climate Change. See: www.grida.no/climate/ipcc/land_use/index.htm

10 Pearce, Fred (1998) 'Down to Earth' in *New Scientist* Vol 160, Issue 2161; see also: Mollison, Bill (1995) 'Carbon Dioxide and the Soil' in *Magazine* No 10; and Samways, Greg (1995) 'Further Thoughts on CO_2 and Soil,' in ibid.

11 Whitelegg, John (1997) *Critical Mass*, Pluto Press, p129.

12 For a general discussion of this subject see: Markham, Adam, Nigel Dudley and Sue Stolon (1993) *Some Like it Hot*, WWF International.

13 Van Gleder, Berry, and Phil O'Keefe (1995) *The New Forester*, Intermediate Technology Publications.

14 Borer, Pat and Cindy Harris (1998) *The Whole House Book, ecological building design & materials*, CAT, p84.

15 Miller, G Tyler (1992) *Living in the Environment*, Wadsworth.

16 Rutter, Phil (1986) 'The Hope of Woody Agriculture: fixing carbon, building soils, growing food,' in *The Permaculture Activist* No 40. See also: www.badgersett.com

17 Borer and Harris, op cit, p84.

18 For details of working with light and shade in forest gardens see Whitefield, Patrick (1996) *How to Make a Forest Garden*, Permanent Publications.

19 Adapted from Mollison (1988)[†].

20 Research done in California, mentioned in: Larkcom, Joy (1995) *Vegetables for Small Gardens*, Hamlyn.

21 Kourik, Robert, (2004) *Designing and Maintaining Your Edible Landscape Naturally*, Permanent Publications.

22 These figures are derived from: Hodges, RD and C Arden-Clarke (1986) *Soil Erosion in Britain*, Soil Association; Caborn (1965); Kourik op cit; Johnston, Jacklyn, and Newton, John, *Building Green*, London Ecology Unit, 1993; Barton, Hugh et al, *Sustainable Settlements*, University of the West of England, 1995.

23 Kourik op cit.

24 Caborn (1965).

25 Emma Planterose, personal communication.

26 For more ideas see: Leach, Eric (1991) *Coastal Gardening with Trees and Shrubs*.

Chapter 5
WATER

1 Miller, G Tyler (1989) *Resource Conservation and Management*, Wadsworth (USA).

2 Ibid.

3 Miller, G Tyler (2002) *Living in the Environment*, twelfth edition, Brooks/Cole (USA)

4 McLaren, Duncan et al (1998) *Tomorrow's World*, FoE/Earthscan.

5 Stauffer (1996) p52.

6 Ibid. There is considerable variation in the figures quoted in different sources.

7 Source: Barton, Hugh, Geoff Davis and Richard Guise (1995) *Sustainable Settlements, a guide for planners, designers and developers*, University of the West of England / Local Government Management Board.

8 Miller, G Tyler (2002) *Living in the Environment*, twelfth edition, Brooks/Cole (USA)

9 This is adapted from: Harper, Peter (1998) 'Water Efficiency in the Home', in *Clean Slate* No 28.

10 For detailed information on the water requirements of different vegetables see: Bleasdale, JKA et al (1991) *The Complete Know and Grow Vegetables*, Oxford University Press.

11 For an appraisal of the possibilities for grey water use in Britain see Harper and Halestrap (1999). If having read that you think you would like to go ahead, read Ludwig (1998).

12 For a thorough discussion of rainwater quality see Gould and Nissen-Petersen (1999). For professional advice see the *CAT Resource Guide*. (See p444.)

13 Taken from Stauffer (1996).

14 Baldock, David and Kevin Bishop (1996) *Growing Greener, sustainable agriculture in the UK*, CPRE/WWF.

15 The keyline system is described in detail in: Yeomans, PA (1981) *Water for Every Farm*, Second Back Row Press, Australia, and more briefly in Mollison (1988)[†], pp155-164 & 218-221.

16 Mollison (1988)[†].

17 Information on hydraulic rams, including suppliers, is available from the Centre for Alternative Technology. (See Appendix D, Organisations.)

18 Harper, Peter and Dave Thorpe (1994) *Fertile Waste*, CAT.

19 Aitken, MN (1996) 'Sustainable Use of Sewage Sludge on Agricultural Land' in Taylor, Andrew G et al, eds, *Soils, Sustainability and the Natural Heritage*, Scottish Natural Heritage, HMSO, paragraph 9.9.

20 Ibid, paragraph 9.1.

21 For the method see Harper and Halestrap (1999).

22 A National Rivers Authority (NRA) report dated 23rd Sept 1993 on the three-stage reedbed system at Marsh Country Hotel, Leominster, Herefordshire, gives an output of <2mg/l biochemical oxygen demand (BOD) and 2.0mg/l suspended solids (SS). This compares with the minimum standards required by the NRA of 50mg/l BOD and 60mg/l SS.

23 See Harper and Halestrap (1999).

24 From ibid.

25 Mollison (1988)[†].

26 Greer, RB (1979) 'A Tree Planting Trial at Loch Garry (Tayside Region) Aimed at Habitat Improvement for Fish', in *Scottish Forestry* Vol 33, No 1; and Whittaker, Robert H (1975) *Communities and Ecosystems*, Macmillan, p241.

27 See Hutchinson, (in preparation).

28 Jimmie Hepburn, personal communication.

29 Information from: Sinclair, David (1993) 'When Basic Biology Pays' in *Country Life*, 11th Feb 1993. Roy Watkin can be contacted at Reedbed Technology Ltd on 0161 865 0155.

30 Claassen, PW (1919) 'A Possible New Source of Food Supply' in *Scientific Monthly* Vol 185, Part 9, pp179-185.

31 Vegetable and craft uses are taken from: Fern, Ken (1997) *Plants For A Future*, Permanent Publications, Chapter 6.

32 This list is based on ibid, Chapter 6, where more detail and a wider range of species may be found.

33 Mollison (1988)[†], pp155-166.

Chapter 6
ENERGY & MATERIALS

1 For a more detailed account of energy fundamentals and our cultural approach to energy see: Rifkin, Jeremy (1985) *Entropy, a New World View*, Paladin.

2 Also known as the first and second laws of thermodynamics.

3 These figures are broad averages, and should not necessarily be taken to apply to any specific animal. Some species and individuals are more efficient, others less so. See also note 3 to Chapter 10.

4 The exceptions are geothermal energy, which makes use of the hot rocks beneath the Earth's surface, and tidal energy, which is driven by both Sun and Moon.

5 Agnew, PW (1994) *The Efficient Alternative*, Tarragon Press; and Harper, Peter (1993) 'Standard of Living Versus Quality of Life, we are designers too,' in *Clean Slate* No 10.

6 Lovins, Amory B, 'Technology Is the Answer (But What Was the Question?)' in Miller, G Tyler (1992) *Living in the Environment*, seventh edition, Wadsworth (USA).

7 Rifkin, op cit.

8 Miller (2002).

9 Agnew, op cit.

10 Figures from the National Energy Foundation. See Appendix D, Organisations.

11 Miller, G Tyler (1992) *Living in the Environment*, seventh edition, Wadsworth (USA).

12 Source: OECD (1993) *Cars and Climate Change*, Organisation for Economic Co-operation and Development, Paris, quoted in: Whitelegg, John (1997) *Critical Mass*, Pluto Press.

13 See Miller (2002) for a wide-ranging account of the different forms of pollution and what we can do about them.

14 Agnew, op cit.

15 Survey by the Department of Trade and Industry, quoted in Borer, Pat and Cindy Harris (1998) *The Whole House Book*, CAT.

16 Miller, G Tyler (1992) *Living in the Environment*, seventh edition, Wadsworth (USA), p446.

17 Agnew, op cit.

18 Source, Dept of Trade and Industry, *Digest of UK Energy Statistics 2000*, HMSO.

19 Source: Littler, John and Randall Thomas (1984) *Design with Energy*, CUP.

20 Estevan, Antonio and Alfonso Sanz, 'Hacia La Reconversión Ecologica del Transporte en España', Centro de Investigacion para la Paz, Madrid, quoted in Douthwaite (1996)[†].

21 McLaren, Duncan, Simon Bullock and Nusrat Yousuf (1998) *Tomorrow's World, Britain's share in a sustainable future*, Friends of the Earth/Earthscan, pp 114-115.

22 Whitelegg, op cit.

23 Seifried, D (1990) *Gute Argumente: Verkehr, Beck'sche Reihe, Beck, Munich*, quoted in Whitelegg, op cit.

24 Bowers, Chris (1993) *Getting the Prices Right*, short version, European Federation for Transport and Environment, Rue de la Victoire 26, 1060 Bruxelles, Belgium.

25 Pearce, D (1995) *Blueprint 5: The True Cost of Road Transport*, Earthscan.

26 Also: Edinburgh City Car Club, 0131 555 4010; London sMiles, 0171 435 9064; European Car Sharing, +49 421 71045, www.carsharing.org; Munich StattAuto, www.eltis.org/data/68e.htm.

27 Penner, Joyce E et al (1999) *Aviation and the Global Atmosphere*, IPCC; and Miller (2002), p457.

28 The total tax break to the aviation industry in Britain has been calculated at £6.8 billion per year. This is equivalent to a subsidy of £182.45 for every man, woman and child in Britain. Source: Dr Caroline Lucas MEP, quoted in Ward, Phil (2002) 'Terminal Decline', in *Going Green*, No 42. (*Going Green* is the magazine of the Environmental Transport Association. See Appendix D, Organisations.)

29 *The Food Miles Report: the dangers of long distance food transport*, quoted in: Garnett, Tara (1996) *Growing Food in Cities, a report to highlight and promote the benefits of urban agriculture in the UK*, SAFE Alliance.

30 Whitelegg, op cit, p40.

31 Source Centre for Alternative Technology.

32 Much information in this section is supplied by the Centre for Alternative Technology.

33 For a brief guide to buying and using fridges and washing machines from an ecological perspective see Anon (1997) 'Cool & Clean' in *Clean Slate* No 24.

34 Allen, Michael (1994) 'Ecosystems for Industry' in *New Scientist*, 5th Feb 1994.

35 McLaren et al, op cit.

36 See for example: Pearce, Fred (1997) 'Burn Me' in *New Scientist*, 22nd Nov 1997.

37 For a detailed proposal for using annual fibre crops see: Desai, P, ed (1996) *Local Paper – waste paper and non-wood fibres for a sustainable paper cycle in the south east*, Bioregional Development Group.

Chapter 7
BUILDINGS

1 Stamp, LD (1948) *The Land of Britain, Its Use and Misuse*, Longmans, quoted in Best, Robin H, and JT Ward (1956) *The Garden Controversy, a critical analysis of the evidence and arguments relating to the production of food from gardens and farmland*, Dept. of Agricultural Economics, Wye College, Kent, p13.

2 Jeavons, John (1995) *How to Grow More Vegetables*, Ten Speed Press.

3 Density can also be measured in people per hectare, but I have used dwellings per hectare throughout to make comparisons easy.

4 Holmgren, David (1993) 'Energy and Permaculture' in the *Permaculture Edge* Vol 3, No 3; and idem (1991) *Gardening as Agriculture*, pamphlet, Holmgren Design Services, Hepburn, Victoria, Australia.

5 Sherlock, Harley (1991) *Cities are Good for Us*, Paladin; for a concise summary of the argument see McLaren, Duncan et Al (1998) *Tomorrow's World*, Foe/Earthscan, pp128-129.

6 Barton et al (1995).

7 Condensed from Preston, John (1999) 'Living in the City, the minimal ecological footprint,' in *Permaculture Magazine* No 20.

8 Barton et al (1995).

9 Ibid.

10 Ibid, p80.

11 National Society of Allotment and Leisure Gardeners, www.ncare.co.uk/nsalg/2035.htm

12 Andrews, Sophie (2001) *The Allotment Handbook, a guide to promoting & protecting your site*, Eco-Logic Books.

13 Mollison and Holmgren (1978)[†]; Wrench, Tony, (1994), 'Permaculture Land', in *Permaculture Magazine* No 5; Fairlie (1996).

14 For an enthusiastic view of the potential of teleworking in rural repopulation see Body, Richard (1992) 'The 21st-Century Peasant' in Conford, Philip, ed (1992) *A Future for the Land, organic practice from a global perspective*, Green Books.

15 Barton et al (1995).

16 For a detailed account of applying ecological principles to settlement design see ibid.

17 Ibid, p189.

18 Ibid

19 Borer, Pat, et al (1992). *Environmental Building*, pamphlet, CAT.

20 Desai, Pooran (1993) 'Taking the Lead, low energy housing in Schiedam', in *Permaculture Magazine* No 3.

21 Miller, G Tyler (2002) *Living in the Environment*, Brooks Cole.

22 Borer and Harris (1998) p77

23 Ibid, p35.

24 For a checklist comparing the pros and cons of heavy (masonry) and lightweight (timber) houses see Barton et al (1995) p225.

25 Borer and Harris (1998).

26 von Zschock, Reinhart (1997) 'Ceramic Stoves' Pt 1 & Pt 2, in *Permaculture Magazine* Nos 13 & 14. For detailed plans of masonry stoves see Aalders, Willem (1991) *Gemetselde Kachels, fornuizen en open haarden*, Zuid Boekprodukties, Netherlands.

27 Johnston and Newton (1993), referring to Ulrich, RS (1984) 'View Through a Window May Influence Recovery from Surgery', in *Science* Vol 224, April 27th, pp420-421.

28 Higher figures which have been published turn out to apply to very old houses or to be miscalculations.

29 Clarke, David (1997) 'Shelter Trees for Energy Conservation' in Palmer, Harriet et al *Trees for Shelter*, Technical Paper No 21, Macaulay Land Use Research Institute.

30 Ibid.

31 Caborn, JM (1965) *Shelterbelts and Windbreaks*, Faber & Faber.

32 Barton et al (1995).

33 Johnson and Newton (1993).

34 Tony Andersen, personal communication.

35 For a full account of green roofs see Johnson and Newton (1993). Much of the information in this section is taken from that source.

36 See Guerra (2000).

37 The books recommended for further information in this section are by no means the only ones which may prove useful. I have given those which seem the best to me or which are the most highly recommended.

38 Sobon, Jack and Roger Schroeder (1984) *Timber Frame Construction*, Storey Books.

39 Broome, Jon and Brian Richardson (1996) *The Self Build Book*, Green Earth Books; Borer, Pat and Cindy Harris (1994) *Out of the Woods, ecological designs for timber frame self-build*; Clarke, James (1995) 'House of the Future, a Design Outline', in *Permaculture Magazine* No 8.

40 Seddon, Leigh (1985) *Practical Pole Building Construction, with plans for barns, cabins and outbuildings*, Williamson.

41 Lacinski, Paul and Michelle Burgeron (2000) *Serious Straw Bale*, Chelsea Green (USA), detailed information; Jones, Barbara (2002) *Building with Straw Bales, a practical guide for the UK and Ireland*, Green Books, much briefer; also see Appendix D, Organisations.

42 Carpenter, Peter, ed (1994) *Sod It, an introduction to earth sheltered development in England and Wales*, British Earth Sheltering Association.

43 Johnson and Newton (1993).

44 Dear, Neville (1997) 'A Re-Tyrement Home, New Barn's Earthship' in *Permaculture Magazine* No 15.

45 Bee, Becky (1997) *The Cob Builders Handbook*, Groundwork, USA.

46 Figures taken from Borer, Pat et al op cit.

47 Stanton, WI (1971) 'Mendip Quarries: their past, present and future,' in WG Hall ed, *Man and the Mendips*, The Mendip Society.

48 Anink, David, Vhiel Boonstra and John Mak (1996) *Handbook of Sustainable Building*, James and James; and Wooley, Tom and Sam Kimmins (2000) *The Green Building Handbook*, Vols 1 & 2, SPON. New publications on this subject are coming out all the time; for up to date information contact the Association for Environment Conscious Building. See Appendix D, Organisations.

49 See: Johns, Howard (1999) 'Bathing in the Sun's Rays' in *Permaculture Magazine* No 19, for some examples ranging from the home made to the professionally installed.

Chapter 8
GARDENS

1 Garnett, Tara (1996) *Growing Food in Cities, a report to highlight and promote the benefits of urban agriculture in the UK*, SAFE Alliance.

2 Christensen, Karen (1995) *The Green Home*, Piatkus.

3 See Larckom, Joy (1997) *Creative Vegetable Gardening*, Mitchell Beazley.

4 For detailed information on edible flowers see Guerra (2000), p145.

5 Lists of ornamentals for different microclimates and soil conditions is given in: Greenoak, Francesca (1998) *Natural Style for Gardens*, Mitchell Beazley; for soils only see Hamilton (1991), pp108-109.

6 For more details see: Wolverton, BC (1997) *How to Grow Fresh Air, 50 house plants that purify your house or office*, Penguin (also published as *Eco-Friendly House Plants* (1996)).

7 Taken from ibid.

8 For details of sprouting technique see: Bruce, Elaine (1995) 'The Indoor Garden', in *Permaculture Magazine* No 9.

9 For details of the cut-and-come-again technique see: Larkcom (2002), pp 156-158.

10 Johnston, Jacklyn, and John Newton, *Building Green*, The London Ecology Unit, 1993.

11 Steele, Judy (1999) 'Grow Your Own', in *Growing Organically*, HDRA Newsletter, No 156.

12 See: Larckom, Joy (2001) *The Organic Salad Garden*, Frances Lincoln for a wide range of annual salads, including edible flowers; and Whitefield (1996) for perennials and self-seeders.

13 For information on how to achieve continuity see: Salter, PJ (1991) 'Planning continuity of supply' in Bleasdale, JKA et al, *The Complete Know and Grow Vegetables*, Oxford Paperbacks.

14 See: Larkcom (1992) *The Vegetable Garden Displayed*, Royal Horticultural Society; and Cunningham, Sally (1999) 'Storing the harvest' in *The Organic Way* (HDRA newsletter), Issue 157.

15 Yield figures are mainly adapted from: Jeavons, John (1995) *How to Grow More Vegetables*, Ten Speed Press, USA; and Hessayon, DG (1997) *The New Vegetable and Herb Expert*, Transworld Publishers.

16 Mollison (1988),[†] pp 434-435.

17 Harper, Peter (1997) 'A Friendly Critique of Permaculture,' in *Permaculture Magazine* No 14.

18 For details of native trees and shrubs see: de la Bédoyère, Charlotte (2001) *Plant a Natural Woodland, a handbook of native trees and shrubs*, Search Press. See also Whitefield (1996) pp112-114.

19 Source, Rotterdam City Council, 1980, quoted in Barton, Hugh et al (1995) *Sustainable Settlements*, University of the West of England.

20 Pretty, Jules (1998) *The Living Land*, Earthscan.

21 Based on Brown, Maggi (2000) 'A Shady Character' in *The Organic Way* (HDRA newsletter) No 159, with some additions.

22 Whitefield (1996), pp73-74.

23 Larckom, Joy (1984) *The Salad Garden*, Frances Lincoln

24 For a detailed discussion of windbreaks and greenhouses, including the shade cast at various times of day and year by windbreaks of various orientations see: Caborn, JM (1965) *Shelterbelts and Windbreaks*, Faber & Faber, pp40-44.

25 Ibid.

26 For information on *Eleagnus* spp, see Fern (1997).

27 Mollison and Slay (1991)[†].

28 For more detailed information on the relationship between trees and other garden plants see Whitefield (1996).

29 Russian State Committee on Standards and Cornell Univeristy, New York, quoted in: United Nations Development Programme (1996) *Urban Agriculture, food, jobs and sustainable cities*, UNDP.

30 See: Gilbert, OL (1991), *The Ecology of Urban Habitats*, Chapman & Hall, p118.

31 HDRA have done a long series of Members Experiments comparing 'standard', no-dig, and double-dug raised bed methods of gardening, reported in various issues of their newsletter. See also Kourik, Robert (2004) *Designing and Maintaining Your Edible Landscape Naturally*, Permanent Publications.

32 For double digging technique see: Larkcom (2002), p67.

33 For a concise introduction to raised beds see ibid; for more information see: Pears, Pauline (1992) *Beds*, HDRA/Search Press.

34 For example JLH Chase, reported in Kourik, op cit, pp109-110.

35 For various accounts of practical experience with mulch see: Whitefield, Patrick (1992) *Practical Mulching*, self-published, distributed by Permanent Publications.

36 Harper, Peter (1998) 'Recycled Paper Mill Sludge, a promising raw material', in *Clean Slate* No 29.

37 Source: *Organic Gardening* (USA) Jan-Feb 1999, quoted in 'Gleanings', in *Organic Farming* No 62, Summer 1999.

38 See Kourik, op cit, p96.

39 A minimalist vegetable garden is hardly different from the vegetable layer of a forest garden, and the most comprehensive account of the subject is in Whitefield (1996). See also my series of articles, 'The Minimalist Garden', *Permaculture Magazine* Nos 10, 11 & 12; and Kourik, op cit, pp 95-98.

40 Details of these plants and many other perennials are on the Plants For A Future database, which can be accessed at www.pfaf.org or by contacting Plants For A Future. See Appendix D, Organisations.

41 For a concise account of beneficial insects in the garden see: Pears, Pauline (1999) 'Know Your Friends' in *Growing Organically*, HDRA Newsletter 156.

42 What follows is taken largely from his article, 'Garden Polycultures' *The Permaculture Activist* No 27, reprinted in *Permaculture Magazine* No 4.

43 Condensed from: Guerra, Michael (1994) 'Growing Potatoes in Tyres', in *Permaculture Magazine* No 5; and idem (1997) 'Spuds in Tyres Update', in *Permaculture Magazine* No 13.

44 See Hart (1991).[†]

45 See: Whitefield, Patrick (1998) 'Ground Cover Plants for Forest Gardens', in *Permaculture Magazine* No 17.

46 Kourik, op cit.

47 (1995) Ten Speed Press, USA.

48 For complete instructions for making this kind of hot box see Guerra, Michael (1994) 'The Hot-Box' in *Permaculture Magazine* No 7. For further ideas on the same principle Michael recommends: Pierce, John H (1992) *Home Solar Growing*, Key Porter Books.

Chapter 9
FRUIT, NUTS & POULTRY

1 Much of the information on these three nuts, including the yield figures, is from Crawford (1995a, 1995b & 1996).

2 Rutter, Phil (1986) 'The Hope of Woody Agriculture: fixing carbon, building soils, growing food,' in *The Permaculture Activist* No 40. See also: www.badgersett.com

3 For a comprehensive account see: Rice, EL (1984) *Allelopathy*, Academic Press.

4 Kourik, R (2004) *Designing and Maintaining Your Edible Landscape Naturally*, Permanent Publications.

5 Dawson, JO and JW van Sambeek (1993) 'Interplanting Woody Nurse Crops Promotes Differential Growth of Black Walnut Saplings' in *Northern Nut Growers Association 1993 Annual Report*, 84, p38-46.

6 For a more comprehensive list see: Crawford, Martin (1995) 'Fertility in Agroforestry and Forest Gardens: Nitrogen', in *Agroforestry News* Vol 3, No 3.

7 Rutter, op cit.

8 Jobling, J and ML Pearce (1977) *Free Growth of Oak*, Forestry Commission Forest Record 113, HMSO.

9 See Whitefield (1996) p100 for a discussion of squirrels and hazels.

10 Available from Nutwood Nurseries. See Appendix E, Suppliers of Permaculture Requisites.

11 Especially Nutwood Nurseries and Clive Simms. See Appendix E, Suppliers of Permaculture Requisites.

12 For a detailed look at oaks as a food crop, including a comprehensive species list, see: Crawford, Martin (1997) 'Edible Acorns from Oaks' in *Agroforestry News* Vol 5, No 7.

13 For more on eating acorns, including recipes see: Tillstone, Gavin (1990) 'Oak Wise', and Smith, Owen (1990) 'Acorns, a Source of Human Food?' both in *Permaculture News* No 21. I have personally tried the first method of leaching tannins given here and can vouch for it, but not the others.

14 de Foresta, H, and G Michon (1996) 'Tree improvement research for agroforestry: a note of caution', in *Agroforestry Forum* Vol 7, No 4.

15 Fern (1997).

16 For a discussion of pine species for growing in Britain see: Crawford, Martin (1994) 'Nut Pines' in *Agroforestry News* Vol 3, No 1, pp28-34; and Fern (1997) pp34-36.

17 By Horticulture Research International, East Malling, and HDRA. See Appendix D, Organisations.

18 Hall, Peter (1997) 'Apples & Pears, a step at a time', in *New Farmer & Grower* No 54.

19 Baker (1991).

20 For details of the climate and microclimate requirements of different fruits see Whitefield (1996), pp73-74.

21 Crawford, Martin: (2001) *Directory of Apple Cultivars*; (1996) *Directory of Pear Cultivars*; (1996) *Plums, production, culture and cultivar directory*, Agroforestry Research Trust.

22 For more detail on replant disease see: 'Replant Disease' in Whitefield (1996), p26.

23 Source: *The Maine Organic Farmer & Gardener* Vol 27, No 4, Dec 2000 - Feb 2001, reported in *Organic Farming* No 69, spring 2001.

24 For a fuller discussion of dwarf and vigorous trees, and details of the rootstocks available for the common fruit trees, see Whitefield (1996).

25 For a full discussion of pollination see Whitefield (1996). All good nursery catalogues state which varieties are self-infertile and their blossoming times.

26 This was done by Phil Sumption of Radford Mill Farm, Timsbury, Somerset.

27 For a detailed account of honey fungus, including lists of resistant species and rootstocks and notably susceptible species, see: *Agroforestry News* Vol 7, No 4, pp17-21.

28 Some information on saving old orchards and establishing community orchards, with case studies, is to be found in: Keech, Dan et al (2000) *The Common Ground Book of Orchards, conservation, culture and community*, Common Ground.

29 For full details on selecting fruit for different microclimates, including walls of different aspect, see Whitefield (1996) pp17-19 and 73-75.

30 Phil Corbett, personal communication.

31 For more detail see Caborn JM (1965) *Shelterbelts and Windbreaks*, from which much of this information on orchard windbreaks comes.

32 Haward, Robert (2000) 'Habitat Management for Crop Production' in *Organic Farming* No 66. This article contains a case study of an integrated planting scheme to increase biodiversity on an organic nursery.

33 For a discussion and selected range of plants see: Whitefield, Patrick (1997) 'Ground Cover Plants for Forest Gardens' in *Permaculture Magazine* No 17. For a comprehensive range of plants see: Crawford, Martin (1997) *Ground Cover Plants*, Agroforestry Research Trust.

34 For an illustrated account of the most important insects involved see: *Agroforestry News* Vol 7, No 4, pp29-35.

35 For a more complete list, including some annuals, see: Pears, Pauline and Sue Stickland (1995) *Organic Gardening*, Mitchell Beazley, p25.

36 Haward op cit.

37 Crowder, Bob (1994) 'An Organic Monoculture, Can it Work?' in *Permaculture Magazine* No 5; and for illustrations, Lampkin (1994) *Organic Farming*, Farming Press, plates 7, 16 & 17.

38 See for example Phillips (1998).

39 Matheson, Andrew (1996) *Bumble Bees for Pleasure and Profit*, IBRA.

40 From Hughes, Melanie (2000) 'A Helping Hand for the "Humble Bee"' in *Stewardship News* (the Countryside Stewardship newsletter) MAFF.

41 Azeez, Gundula (2000) *The Biodiversity Benefits of Organic Farming*, Soil Association.

42 Most of the information in this section is taken from Hooper (1997).

43 Francis, Charles A, Cornelia Butler Flora and Larry D King, eds (1990) *Sustainable Agriculture in Temperate Zones*, John Wiley.

44 Brownlow, Mark JC (1992) 'Acorns and Swine: historical lessons for modern agroforestry', in *Quarterly Journal of Forestry* Vol 86, Part 3, referring to: Spencer, S (1921) *The Pig; Breeding, Rearing and Marketing*, C Arthur Pearson Ltd.

45 For more details contact Dereck Jackson, the Pure Chicken Co, 01626 774744.

Chapter 10
FARMS & FOOD LINKS

1 For a concise introduction to ICM see: Jordan, VWL, JA Hutcheon & DM Glen (1993) *Studies in Technology Transfer of Integrated Farming Systems, Considerations and Principles for Development*, Institute of Arable Crops Research, Long Ashton Research Station, Dept of Agricultural Sciences, University of Bristol, Long Ashton, Bristol BS18 9AF. 0117 939 2181.

2 See for example: IGER (1996) *Conversion to Organic Milk Production*, Institute of Grassland and Environmental Research, Aberystwyth, quoted in Pretty, Jules (1998) *The Living Land, agriculture, food and community regeneration in rural Europe*, Earthscan, p96.

3 This is not the same as the Food Conversion Ratio used by farmers as a measure of efficiency. The FCR compares the weight of dry food with the gross live weight of the animal. It doesn't allow for the fact that most of the weight of the animal is water, nor for the

50-60% of the animal's weight composed of inedible bones, skin etc. Nor does it allow for the food fed to the ewes, sows and beef cows who gave birth to the meat animals. When all's taken into account a FCR of 3:1 is very much the same as a ratio of edible dry matter in to edible dry matter out of 10:1.

4 Figures from Find Your Feet. See Appendix D, Organisations.

5 Percentages calculated from figures in: MAFF (1997), *Agriculture in the UK 1996*, Stationary Office.

6 Mainly by Elm Farm Research Centre. For a practical farming example see Lampkin (1994), pp420-424.

7 *Elm Farm Research Centre Bulletin No 37* (June 1998) quoted in *Agroforestry News*, Vol 6, No 4.

8 Francis, Charles A, Cornelia Butler Flora and Larry D King, eds (1990) *Sustainable Agriculture in Temperate Zones*, John Wiley.

9 Mollison (1988)[†]. Neither source states whether the pigs are breeding sows or growing pigs.

10 The information in this section comes mainly from: Wolfe, MS and MR Finckh (1996) 'Diversity of Host Resistance Within the Crop: Effects on Host, Pathogen and Disease', in Hartleb, H, R Heitfuss and HH Hoppe, *Plant Resistance to Fungal Diseases*, Gustav Fischer Verlag, Jena; Finckh, MR and MS Wolfe (1997) 'The Use of Biodiversity to Restrict Plant Diseases and Consequences for Farmers and Society', in *Ecology of Agriculture*, ed Jackson, LE, Chapman & Hall; Lampkin (1994), pp 232-233; and Martin Wolfe, personal communication. See also Wolfe, Martin (1998) 'Disease? Keep Taking the Mixture' in *New Farmer & Grower* No 57.

11 Elings, Anne, et al, 'Use of Genetic Diversity in Crop Improvement' in Almekinders, Conny & Walter de Boef (2000) *Encouraging Diversity, the conservation and development of plant genetic resources*, Intermediate Technology Publications.

12 Much of the information in this section comes from Lampkin (1994) and: Altieri, Miguel A (1995) *Agroecology, the science of sustainable agriculture*, Intermediate Technology Publications. More detail can be found in both sources

13 Caborn, JM (1965) *Shelterbelts and Windbreaks*, Faber & Faber.

14 Lampkin (1994).

15 Martin, MPLD and RW Snaydon (1982), 'Root and Shoot Interactions Between Barley and field Beans when Intercropped' in *Journal of Applied Ecology* No 19, pp263-272.

16 Altieri, op cit, p 209.

17 Salter, PJ, Jayne M Akehurst and GEL Morris (1985) 'An Agronomic and Economic Study of Intercropping Brussels Sprouts and Summer Cabbage', in *Experimental Agriculture* Vol 21, pp153-167.

18 Detailed information on this kind of planting, with a case study of an integrated planting scheme for biodiversity on an organic farm, is to be found in: Haward, Robert (2000) 'Habitat Management for Crop Production' in *Organic Farming* No 66.

19 Lampkin (1994) pp60-63.

20 Köpe, Ulrich and Guido Haas (1996) 'Farming, Fossil Fuels and CO_2' in *New Farmer & Grower* No 50. For a detailed discussion, including energy use in transport, see: Cormack, Bill and Phil Metcalfe (2000) 'The Efficiency Test' in *Organic Farming*, No 67.

21 A number of articles have been published on bicropping over the past few years. Most of what they say is contained in: Clements, Bob and Guy Donaldson (1997) 'Clover and Cereals – Low Input Bi-Cropping' in *Farming & Conservation* Vol 3, No 4. Dr Clements can be contacted at the Institute of Grassland & Environmental Research, North Wyke, Okehampton, Devon EX20 2SB.

22 Ferguson, CM, BIP Barratt and PA Jones (1988) 'Control of the grey field slug (*Deroceras reticulatum* (Muller)) by stock management prior to direct-drilled pasture establishment', in *Journal of Agricultural Science*, Cambridge, 111, 443-449.

23 Jones, Lewis and RO Clements (1993) 'Development of a low input system for growing wheat (*Triticum vulgare*) in a permanent understorey of white clover (*Trifolium repens*)', in *Annals of Applied Biology* No 123, pp109-119.

24 Bob Clements, BBC Farming Today, 11th Nov 1995.

25 Figures from Clements and Donaldson, op cit. This percentage profitability resulted from spreading the establishment costs of the clover over five years. When these costs were spread over three years profitability dropped to 60%.

26 Finckh, MR and MS Wolfe (1997) 'Population Biology of Airborne Pathogens and Consequences for Disease Control' in *Ecology of Agriculture*, ed Jackson, LE, Chapman Hall.

27 Fukuoka, Masanobu (1978) *The One-Straw Revolution*, Rodale.

28 For a description of the Bon Fils method see: Moodie, Mark (1991) *The Harmonious Wheatsmith*, self-published booklet.

29 For a general account of this system, plus a critique of present-day agriculture, and a concise introduction to the science of ecology as it applies to agriculture see: Soule, Judith D and Jon K Piper (1992) *Farming in Nature's Image*, Island Press.

30 Much of what follows is taken from: Piper, Jon K (1993) 'A Grain Agriculture Fashioned in Nature's Image: The Work of the Land Institute', in *Great Plains Research*, 3 (Aug 1993). Most of the papers cited in this section are available from the Land Institute. See Appendix D, Organisations.

31 The Land Institute prefer to use the term Natural Systems Farming for the kind of grain culture they are developing. But this is a rather unspecific name, which could equally be applied to other systems which mimic natural vegetation, such as agroforestry practised in areas where forest is the natural vegetation.

32 Van Tassel, David (1997) 'Plant Breeders Develop Perennial Crops', in *The Land Report* No 58. See also: Jones, Stephen et al (2000) 'A Wheat to Hold Landscape Together: Breeding in Perennialism from Wild Grass', in *The Land Report* No 67.

33 Piper, Jon K (1993) 'Neighbourhood Effects on Growth, Seed Yield, and Weed Biomass of Three Perennial Grains in Polyculture' in *Journal of Sustainable Agriculture* Vol 4(2). See also: Soule and Piper, op cit.

34 Piper, Jon K, Mary K Handley and Peter A Kulakow (1996) 'Incidence and Severity of Viral Disease Symptoms on Eastern Gamagrass Within Monoculture and Polycultures' in *Agriculture, Ecosystems & Environment* No 59.

35 Much of the information in this section comes from: Finch, S and E Brunel eds (1994) 'Integrated Control in Field Vegetable Crops', *IOBC/WRPS Bulletin* Vol 17(8).

36 See: Waterman, Andy (1997) 'Life in Clover' in *Permaculture Magazine* No 14.

37 *American Journal of Alternative Agriculture* Vol 15, No 2 (2000), quoted in *Organic Farming* No 67

38 Much of the information in this section comes from: *Agroforestry Forum*, the newsletter of the UK Agroforestry Research Forum, editor, Dr Fergus Sinclair, School of Agricultural and Forest Sciences, University of Wales, Bangor, Gwynedd LL57 2UW. 01248 382 459. It is a technical journal, containing detailed information on experimental work carried out both in the UK and overseas. It has now ceased publication but the back copies contain a rich store of experimental results.

39 Gordon and Newman (1997), pp198-199.

40 Haycock, Nicholas, Farming Today, BBC Radio 4, 25th April 1994.

41 Hislop and Claridge (2000), pp20-21.

42 Caldwell, Martyn (1996) 'Hydraulic Lift in Agroforestry systems', in *Agroforestry Forum*, Vol 7, No 3.

43 Hislop and Claridge (2000), pp109.

44 Crawford, Martin (1998) *Agroforestry Options for Landowners*, Agroforestry Research Trust.

45 Nature Conservancy Council (1983) 'The Food and Feeding Behaviour of Cattle and Ponies in the New Forest', quoted in: Mabey, Richard (1998) *Flora Britannica*, the concise edition, Chatto & Windus.

46 See for example: Incoll, LD et al (1997) 'Temperate Silvoarable Forestry with Poplar' in *Agroforestry Forum*, Vol 8, No 3, p14.

47 Caborn, JM (1957) *Shelterbelts and Microclimate*, Forestry Commission Bulletin 29, HMSO.

48 Andersen, PChr. (1943) *Laeplantnings-Bogen*, 5th edn. Danish Heath Society, Viborg, quoted in: Palmer, H et al, eds (1997) *Trees for Shelter*, Forestry Commission Technical Paper 21, HMSO.

49 After: Caborn, JM (1965) *Shelterbelts and Windbreaks*, Faber & Faber.

50 Palmer et al, op cit.

51 Douthwaite (1996),† pp 277 & 283; and Pretty, op cit, p154. Each gives somewhat different figures.

52 Douthwaite (1996),† p283.

53 We could debate whether present organic methods are indefinitely sustainable or not. But they're certainly the most sustainable practical alternative available now, and a great improvement on conventional practice.

54 Whitelegg, John (1997) *Critical Mass*, Pluto Press, p43.

55 Garnett, Tara (1996) *Growing Food in Cities, a report to highlight and promote the benefits of urban agriculture in the UK*, National Food Alliance / SAFE Alliance, p22.

56 Tutt, Pat et al (1988) *Bath Farmers' Market, a case study*, Eco-Logic Books.

57 Furusawa, Koyu (1992) 'Co-operative Alternatives in Japan', in Conford, Phillip, ed, *A Future of the Land, organic practice from a global perspective*, Green Books; and: Festing, Harriet (1992) 'The Worldwide Producer/Consumer Movement', in *New Farmer & Grower*, Summer 1992.

58 Groh, Trauger and Steven McFadden (1997) *Farms of Tomorrow Revisited, community supported farms, farm supported communities*, Biodynamic Farming and Gardening Association (USA).

59 Taken mainly from: Groh, Christina (1991) 'Community Supported Agriculture, a new approach to marketing', in *New Farmer and Grower*, Spring 1991.

Chapter 11
WOODLAND

1 See Mollison (1988),† especially Chapter 6.

2 Davies, Robert, 'Farmers Wary of Pitfalls Lurking in Woodlands', in *Farmers Weekly*, 7th November 1997.

3 Sidwell, C, M Phil. thesis, quoted in Blyth et al (1991).

4 McPhillimy, Donald (1989) *Conservation in forests, case studies of good practice in Scotland*, Countryside Commission for Scotland.

5 www.ecolots.co.uk. For paper copy apply to: Beacon Forestry, 2a Rutland Square, Edinburgh EH1 2AS, Tel: 0131 228 4176, Email: mb@beacon forestry.a-i-s.co.uk.

6 For more detail on markets for timber and wood see Hart (1991).

7 See for example: Borer, Pat and Cindy Harris (1998) *The Whole House Book*, CAT, p254.

8 Crowther, RE and Patch, D (1980) *How Much Wood for the Stove?* Forestry Commission Research Information Note.

9 von Zschock, Reinhart (1997) 'Ceramic Stoves, Part 2, Fuel and Fuel Consumption', in *Permaculture Magazine* No 14.

10 Gardner, Michael, (undated) *The Micro-economics of Coppice Management (in the Furness area of*

Cumbria), the New Woodmanship Trust, PO Box 12, Carnforth, Lancs LA5 9PD. This booklet gives notes on 37 different coppice products with market value where known, labour time and likely profit margin.

11 Howe, Jonathan (1991) *Hazel Coppice, past, present and future*, the Hampshire experience, County Planning Department, Hampshire County Council.

12 See Borer and Harris, op cit, p91.

13 Calculated from information in: Rollinson, TJD and J Evans (1987) *The Yield of Sweet Chestnut Coppice*, Forestry Commission Bulletin No 64, HMSO.

14 See: Crawford, Martin (1989) *Agroforestry Options for Landowners*, Agroforestry Research Trust; and: Gordon, Andrew M, and Steven M Newman (1997) *Temperate Agroforestry Systems*, CAB International.

15 For a complete list of edible native fungi see: Mabey, Richard (1989) *Food for Free*, Collins.

16 Books on mushroom cultivation, including shii-take, are available from www.fungiperfecti.com. A title from this supplier recommended to me by a British shii-take grower is: Kozak, ME and J Krawczyk, *Growing Shiitake Mushrooms in a Continental Climate*.

17 For details of a full range of mushrooms which can be grown see: Steineck, Hellmut (1984) *Mushrooms in the Garden*, Mad River Press, USA.

18 For a list of possible plants see Crawford, Martin (1998) *Agroforestry Options for Landowners*, Agroforestry Research Trust.

19 For details see: Crawford, Martin (1997) 'Forest Farming of Ginseng' in *Agroforestry News*, Vol 5, No 3.

20 Simon Tunnard of Biggleswade, Bedfordshire, 01767 313 291/316 232.

21 See: Whitefield, Patrick (1996) *How to Make a Forest Garden*, Permanent Publications.

22 For detailed information see: Broadmeadow, Mark SJ and Peter H Freer-Smith (1996) *Urban Woodland and the Benefits for Local Air Quality*, Forestry Commission Research Division, HMSO.

23 'Squirrel Management' in *Forest Research Annual Report and Accounts 1999-2000*, Forestry Commission, reported in Agroforestry News, Vol 9, No 3.

24 An overview from a biodiversity point of view is given in: Peterken, George (1996) *Natural Woodland, ecology and conservation in northern temperate regions*, Cambridge University Press, sections 16.7 & 16.8.

25 Pretty, Jules (1998) *The Living Land*, Earthscan.

26 Hart, Cyril (1995) *Alternative Silvicultural Systems to Clear Cutting in Britain: a Review*, Forestry Commission.

27 Howe, op cit.

28 Pain, Ida and Jean (1977) *Another Kind of Garden*, self-published.

29 Peterken, op cit, p474.

30 For detailed information on continuous cover forestry, including British examples, see: Hart, Cyril (1995) *Alternative Silvicultural Systems to Clear Cutting in Britain: a Review*, Forestry Commission; and Garfitt, JE (1995) *Natural Management of Woods, Continuous Cover Forestry*, John Wiley. There is also an organisation, the Continuous Cover Forestry Group, which promotes continuous cover and organises field visits and training events; contact David Pengelly, 01963 364 067, or John Niles, 01234 881 443.

31 Hart, op cit.

32 David Pengelly, Stourhead (Western) Estate, Wiltshire, personal communication.

33 Ditto, from a study tour of Switzerland.

34 See Hart (1991) for a detailed description of this system.

35 Peterken op cit.

36 Garfitt op cit pp85-87 and Hart (1991) p258.

37 Tickell, Oliver (1999) '40 Years of Continuous Service' in *Forestry & British Timber*, Sept 1999 – an inspiring article about Talis Kalnars, who has practised continuous cover forestry in Wales since 1959.

38 For a discussion of the art of natural regeneration see, for example: Hart (1991) pp275-285.

39 Grieve, IC and JA Hipkin (1996) 'Soil Erosion and Sustainability', in Taylor, Andrew G et al, eds, *Soils, Sustainability and the Natural Heritage*, Scottish Natural Heritage.

40 See for example Garfitt op cit.

41 de Selincourt, Kate (1997) 'Pigs in the Forest', in *Agroforestry News*, Vol 5 No 3.

42 For those interested in keeping pigs in woods, Holme Lacey College in Herefordshire run swine-herding courses.

43 Detailed recommendations for grazing farm animals in woods of all kinds are given in: Mayle, Brenda (1999) *Domestic Stock Grazing to Enhance Woodland Biodiversity*, Forestry Commission. This paper is summarised in *Agroforestry News* Vol 8, No 1, pp 2-4.

44 Peterken, op cit, p422; and Mayle, op cit.

45 Stott, KG et al (1983) 'Productivity of Coppice Willow in Biomass Trials in the UK', in *The Proceedings of the Second European Communities Conference on Energy from Biomass, Berlin 1982*, Applied Science Publishing. See also: Cannell, MGR (1982) 'Short Rotation Coppice', in Malcom, DC, J Evans and PN Edwards, *Broadleaves in Britain*, Institute of Chartered Foresters.

46 In Hart (1991).

47 For example Blyth et al (1991) and especially Hart (1991); Beckett (1979) gives site requirements for native trees and maps of Britain showing where each species is found.

48 Greer, RB (1979) 'A Tree Planing Trial at Loch Garry (Tayside Region) Aimed at Habitat Improvement for Fish', in *Scottish Forestry* Vol 33, No 1.

49 For guidance on this see Hart (1991).
50 Miles, J (1977) 'The Influence of Trees on Soil Properties' in *Annual Report of the Institute of Terrestrial Ecology*, 1977.
51 Mentioned by Tallis Kalnars in Tickell, op cit.
52 Ibid.
53 For detailed information on this subject see Macpherson, George (1995) *Home-Grown Energy from Short-Rotation Coppice*, Farming Press. Information is also available from the Department of Trade and Industry; contact: Renewable Energy Enquiries Bureau, ETSU, Harwell, Oxfordshire OX11 0RA. Tel 01235 432 450.
54 Dr Richard Worrell of Edinburgh University, quoted in Lean, Geoffrey, 'Foreign Invaders Threaten UK Trees' in *The Independent on Sunday*, 15th Nov 1998.
55 Tickell, op cit.
56 Hibberd BJ ed (1989) *Urban Forestry Practice*, Forestry Commission, p108. British Standard 5837 gives guidance on fencing to protect trees during building work.
57 For details of this syndrome see: Gilbert, OL (1991) *The Ecology of Urban Habitats*, Chapman & Hall, pp64-65.
58 'Woodland Restoration of Contaminated Land' in *Forest Research Annual Report and Accounts 1999-2000*, Forestry Commission, reported in *Agroforestry News* Vol 9, No 3.
59 See for example Hibberd, op cit.

Chapter 12
BIODIVERSITY

1 Leakey, Richard and Roger Lewin (1995) *The Sixth Extinction, biodiversity and its survival*, Phoenix.
2 Wilson, EO (1992) *The Diversity of Life*, Allen Lane, pp303-304.
3 www.wwf-uk.org, 14th Dec 1998.
4 Marren, Peter (2001) 'Where Have All the Colours Gone?' in *Plantlife*, Spring 2001.
5 Azeez, Gundula (2000) *The Biodiversity Benefits of Organic Farming*, Soil Association; and Lampkin, Nicolas (1994) *Organic Farming*, Farming Press, pp574-579; also, for further discussion: Everett, Mike (1999) 'Strictly for the Birds' in *Organic Farming* No 64.
6 Ogilvy, S (2000) *Link Integrated Farming Systems (a field scale comparison of arable rotations)* Vol 1, Experimental Work, Home Grown Cereal Authority Project No 173, ADAS.
7 Simms, Eric (1971) *Woodland Birds*, Collins, pp83 & 219. For measurements of bird and insect responses to developing agroforestry systems see: *Agroforestry Forum* (1996) Vol 7, No 3; and ibid (1998) Vol 9 No 1. (See Chapter 10, note 38.)
8 See Rackham (1986), p73.
9 Rackham (2003) p98.
10 Peterken (1996) pp 276-278.
11 Anon (1984) *Nature Conservation in Great Britain*, Nature Conservancy Council.
12 Smart, Jane (1999) 'Exectutive Director's Report', in *Plantlife Annual Review 1998-1999*, Plantlife.
13 Campbell, LH and AS Cooke (1997) *Review of the Indirect Effects of Pesticides on Birds*, Joint Nature Conservancy Committee.
14 Pretty, Jules (1998) *The Living Land*, Earthscan.
15 Much of the information in this section is taken from Gilbert (1991).
16 See Gilbert (1991), pp253-255.
17 Chris Baines, lecture, Lanscape Architecture Students' Spring School, 1993.
18 Peterken, George (1981) *Woodland Conservation and Management*, Chapman & Hall.
19 Rackham (2003), p297.
20 See Gilbert (1991), p287.
21 Thompson, Richard (1999) *A Review of Somerset's Mature Woodland Invertebrates and their Habitats, with Particular Reference to Woodland-interiors*, Somerset Wildlife Trust, p4.
22 Macarthur, RH & EO Wilson (1967) *The Theory of Island Biogeography*, Princeton University Press.
23 See: Hambler, Clive and Speight, Martin, R (1995) 'Biodiversity conservation in Britain: Science Replacing Tradition', in *British Wildlife* No 6, pp137-147; and Fuller, Robert J, and Martin S Warren (1995) 'Management for Biodiversity in British Woodland – Striking a Balance,' In *British Wildlife* No 7, pp26-37. The first is a challenge to the historical orthodoxy and the second a reply to it. See also: Letters in *British Wildlife* No 6, pp337-338 & 405, & No 7, p66, being responses to Hambler and Speight.
24 British Red Data Books, quoted in Hambler and Speight, op cit.
25 See Peterken (1996) especially pp286-289 and p414.
26 Ibid pp268-269.
27 For presentations of the natural and historical points of view respectively see: Evans, Martin (1991) 'Letting Go of the Countryside', in *Permaculture News* No 22; and Thomas, Bryn (1992) 'Wildlife Habitats and Permaculture', in *Permaculture Magazine* No 2.
28 Based on Peterken (1996).
29 Thompson, op cit, gives a useful discussion of the pros and cons of reintroducing coppicing, probably relevant all over lowland Britain, not just in Somerset.
30 Mayle, Brenda (1999) *Domestic Stock Grazing to Enhance Woodland Biodiversity*, Forestry Commission.
31 Peterken (1996) p474.
32 Tait, Joyce et al (1988) *Practical Conservation, site assessment and management planning*, Open University, p50.

33 Ask for the leaflet: *Horses, Grasslands & Nature Conservation*, published by English Nature and the British Horse Society.

34 A good general summary of hedge management for wildlife is: Thorne, Ben and Sarah Butler (1997) 'Hedge management for Wildlife', in *New Farmer & Grower* No 53.

35 Such as Rackham (2003) and Peterken (1996).

36 Many of the ideas in this section come from Baines (1986).

37 See Gilbert (1991) p58.

38 Baines (1986).

39 For planting suggestions see: Hamilton, Jill Duchess of, and Franklyn Perring (2000) *Scottish Plants for Scottish Gardens*, Mercat Press; and Hamilton, Jill Duchess of, Penny Hart and John Simmons (2000) *English Plants for Your Garden*, Frances Lincoln; or www.nhm.ac.uk/fff For comprehensive lists of wildlife gardening plants for every situation see Baines (2000).

40 Detailed advice on converting a lawn to a wildflower meadow, including various mowing regimes, are given in Baines (2000).

41 Many more edible natives are described in Mabey (1989), though he doesn't always distinguish between native and naturalised plants.

42 Kennedy CEJ and TRE Southwood (1984) 'The Number of Species of Insects Associated with British Trees: a Re-analysis,' in *Journal of Animal Ecology* No 53, pp455-478.

43 Wigston, DL (1980). 'The insect fauna of nothofagus,' in *Institute of Terrestrial Ecology Annual Report for 1980*, pp50-53, Natural Environment Research Council.

44 See: Herbert, Roger et al (1999) *Using Local Stock for Planting Native Trees and Shrubs*, Forestry Commission practice note, available at: www.forestry .gov.uk/publications/index.html; and: Akeroyd, John (1994) *Seeds of Destruction? non-native wildflower seed and British floral biodiversity*, Plantlife booklet; guidance for people and organisations involved in planting native species in the countryside is available from Flora Locale. See Appendix D, Organisations, for both Plantlife and Flora Locale.

45 For details of how to propagate native trees see: Beckett, Keneth and Gillian (1979) *Planting Native Trees and Shrubs*, Jarrold.

46 For details of this technique see Andrews and Rebane (1994).

Chapter 13
THE DESIGN PROCESS

1 Rackham, Oliver (1997) *The History of the Countryside*, Phoenix, p22.

2 www.metoffice.gov.uk/climate/uk/averages

3 See Mollison (1988);[†] and Holmgren, David (1993) *The Flywire House*, a case study in design against bushfire, Holmgren Design Services.

Chapter 14
DESIGN SKILLS & METHODS

1 For catalogues of the available maps apply to: Sales Desk, British Geological Survey, Kingsley Dunham Centre, Keyworth, Nottingham NG12 5GG, 0115 936 3100; and: Publications Officer, Soil Survey and Land Research Centre, Cranfield University, Silsoe, Bedford MK45 4DT, 01525 863 242.

2 Michael Guerra, personal communication.

3 Mollison (1988).[†]

4 For full details of the theory and practice of mind mapping see: Buzan, Tony and Barry Buzan (2000) *The Mind Map Book*, BBC Books.

5 For a full account of this method see: McHarg, Ian (1969) *Designing with Nature*, Doubleday.

Appendix A

FURTHER READING

The following titles are the best available at the time of writing, but new books come out all the time. For an up-to-date selection contact the suppliers at the end of this Appendix, especially the *Green Shopping Catalogue*. Titles marked * are available from the *Green Shopping Catalogue*. For books which are out of print, the public library service are committed to finding any book which has ever been published in Britain.

GENERAL

Books

Anon (2001) *Go MAD, 365 daily ways to save the planet*, The Ecologist. A very accessible guide to greening our lifestyles.

Douthwaite, Richard (1996) *Short Circuit, strengthening local economies for security in an unstable world*, Lilliput. This book approaches sustainability from the point of view of economics. It's comprehensive, authoritative and very readable.

Goldring, Andrew, ed (2000) *Permaculture Teachers' Guide**, WWF/Permaculture Association (Britain)/ Permanent Publications. A how-to book for permaculture teachers, with contributions from a wide range of British teachers.

Hart, Robert (1991) *Forest Gardening**, Green Books. The personal testament of the originator of forest gardening in the temperate world.

Holmgren, David (1996) *Hepburn Permaculture Gardens**, Holmgren Design Services. A detailed case study of the Holmgren family's permaculture holding over eleven years of development.

Mollison, Bill (1988) *Permaculture, a Designer's Manual**, Tagari. The standard work, long and encyclopaedic. Stronger on permaculture practice for tropical and desert climates than for temperate regions.

Mollison, Bill & David Holmgren (1978) *Permaculture One, a perennial agriculture for human settlements*, Tagari. The book that started it all. Although now superseded from a technical point of view, an important text for anyone who wants an in-depth understanding of permaculture and where it comes from.

Mollison, Bill and Reny Mia Slay (1991) *Introduction to Permaculture**, Tagari. A concise and accessible account of Bill Mollison's work.

Periodicals

Agroforestry News, quarterly, published by the Agroforestry Research Trust. (See Appendix D, Organisations.) 'News' is a misnomer. This is an encyclopaedia, published in parts, of detailed information on useful trees and shrubs and agroforestry design, much of it not readily available elsewhere. Perhaps more for serious students and researchers than for general readers.

*The Permaculture Activist**, quarterly, USA, distributed in the UK by Permanent Publications. In-depth articles on permaculture in North America and other parts of the world.

*Permaculture Magazine – solutions for sustainable living** is the leading sustainable lifestyle magazine. Published quarterly in full colour by Permanent Publications, it offers the best ideas, advice and inspiration from people who are working towards a more sustainable world.

Videos

Global Gardener, gardening the world back to life, with Bill Mollison, 1991, ABC. Four half-hour episodes: Tropical, Dryland, Temperate and Urban.

In Grave Danger of Falling Food, the permaculture concept, with Bill Mollison, 1992, ABC. An inspiring tour de force, introducing permaculture both in Australia and internationally. 50 minutes.

*The Synergistic Garden, a step by step guide to the methods of synergistic agriculture**, with Emilia Hazelip, 1998. An introduction to this highly permacultural method of gardening. 30 minutes.

By Chapter

Chapter 1
WHAT IS PERMACULTURE?

See General section.

Chapter 2
THE PRINCIPLES OF PERMACULTURE

Fern, Ken (1997) *Plants For A Future, edible & useful plants for a healthier world*, Permanent Publications. A comprehensive account of useful perennial plants which may be grown in Britain, many of them little known.

Marten, Gerald G (2001) *Human Ecology, basic concepts for sustainable development*, Earthscan, 2001. A concise account of the interactions between human and natural systems worldwide.

Miller, G Tyler (2002) *Living in the Environment*, twelfth edition, Brooks/Cole (USA). A comprehensive and accessible account of environmental science, much of which is relevant to permaculture.

Whittaker, Robert H (1975) *Communities & Ecosystems*, 2nd edition, Macmillan. An introduction to the science of ecology, giving the basic science on which much of permaculture is based.

Chapter 3
THE SOIL

Gilbert, OL (1991) *The Ecology of Urban Habitats*, Chapman & Hall. Chapter 4 gives a detailed account of urban soils.

Lampkin, Nicolas (1994) *Organic Farming*, Farming Press. Chapters 2, 3 and 4 give a detailed overview of soil and crop nutrition from an organic perspective. Assumes some basic knowledge.

Pears, Pauline and Sue Stickland (1995) *Organic Gardening*, Mitchell Beazley. Especially good on organic matter, composting, including worm composting, and green manuring.

Readman, Jo (1991) *Soil Care & Management*, HDRA/ Search Press. A concise, accessible, illustrated guide to soil for gardeners.

Roulac, John (1998) *Backyard Composting*, Green Earth Books. Handbook for hot and cold composting, though not worm composting.

Smillie, Joe and Grace Gershuny (1999) *The Soul of Soil, a soil-building guide for master gardeners and farmers*, Chelsea Green, USA. An all-round guide of medium length, more practical than the title might suggest.

Chapter 4
CLIMATE & MICROCLIMATE

Burke, Steven (1998) *Windbreaks*, Inkata Press. A concise guide for farmers. Although Australian, there is much here that is relevant in temperate countries.

Caborn, JM (1965) *Shelterbelts & Windbreaks*, Faber & Faber. A comprehensive and detailed account of the subject for both domestic and broadscale applications, not bettered since it was published.

Gilbert, OL (1991) *The Ecology of Urban Habitats*, Chapman & Hall. Chapter 3 gives a detailed account of urban climates in Britain.

Palmer, Harriet et al, eds (1997) *Trees for Shelter*, Forestry Commission Technical Paper 21. Information for shelterbelt design in both urban and rural situations.

Chapter 5
WATER

Gould, John and Erik Nissen-Petersen (1999) *Rainwater Catchment Systems for Domestic Supply*, Intermediate Technology. Comprehensive information for collecting rainwater off your roof.

Grant, Nick, Mark Moodie and Chris Weedon (1996) *Sewage Solutions, answering the call of nature*, CAT Publications. An overview of ecological approaches to sewage, giving an assessment of different systems. Includes information on using less water in the home.

Harper, Peter and Louise Halestrap (1999) *Lifting the Lid, an ecological approach to toilet systems*, CAT Publications. The state-of-the-art compost toilet book, with case studies of different systems. Also covers low flush toilets, septic tanks and grey water use.

Horváth, László, Gizella Tamás and Chris Seagrave (1992) *Carp & Pond Fish Culture*, Fishing News Books. Detailed cultural information on common carp and a number of other species (although it includes some intensive practices which would not fit into permaculture). Complements Swift, below.

Hutchinson, Laurence (2005) *Ecological Aquaculture, a sustainable solution*, Permanent Publications. Detailed prescriptions for sustainable trout farming in all but the most extreme climates.

Ludwig, Art (1998) *Create an Oasis with Grey Water*, Oasis Design. A realistic account of grey water options by the master of the art.

Stauffer, Julie (1996) *Safe to Drink? the quality of your water*, CAT Publications. An overview of the water system, including information on alternatives to the mains.

Swift, Donald R (1993) *Aquaculture Training Manual*, Fishing News Books. Outlines basic aquaculture techniques.

Chapter 6
ENERGY AND MATERIALS

Miller, G Tyler (2002) *Living in the Environment*, twelfth edition, Brooks/Cole (USA). Gives an accessible account of all aspects of energy from an ecological perspective.

Noonan, Brendan and Steve Cousins (1999) *The Car Club Kit*, Smart Moves, available from Community Car Share Network. (See Appendix D, Organisations.) All about car clubs, including how to set one up.

Salomon, Thierry and Stephane Bedel (2003) *The Energy Saving House*, CAT Publications. A guide to saving energy and saving money in the home. Gives the detailed information needed to put many of the ideas in this chapter into practice.

Semlyen, Anna (2000) *Cutting Your Car Use*, Green Books. A short but comprehensive guide to greening your transport choices.

Smith, Peter F (2004) *Eco-Refurbishment, a guide to saving & producing energy in the home*, Architectural Press. Contains information on energy saving and home generation of electricity.

Chapter 7
BUILDINGS

Barton, Hugh, ed (2000) *Sustainable Communities, the potential for eco-neighbourhoods*, Earthscan. Looks at what can be done to make new or existing settlements more sustainable. Contains a chapter on food production by permaculturist Rob Hopkins.

Barton, Hugh, Geoff Davis and Richard Guise (1995) *Sustainable Settlements, a guide for planners, designers and developers*, University of the West of England & the Local Government Management Board. A practical workbook, it includes principles and covers everything from overall settlement design to the individual dwelling.

Borer, Pat and Cindy Harris (1998) *The Whole House Book, ecological building design & materials*, CAT Publications. A comprehensive and accessible account of the subject, for both builders and householders who want to understand the issues involved.

Fairlie, Simon (1996) *Low Impact Development, planning and people in a sustainable countryside*, Jon Carpenter. A handbook for rural resettlement, both visionary and practical. A basic primer for anyone who wants to get permission to live on their own land in the country.

Guerra, Michael, (2000) *The Edible Container Garden*, Gaia. Includes information on roof gardens and other food-growing opportunities on buildings.

Harland, Edward (1997) *Eco-Renovation, the ecological home improvement guide*, Green Books. A guide for householders.

Johnston, Jacklyn and John Newton (1993) *Building Green*, London Ecology Unit. An introduction to biotecture.

Smith, Peter F (2004) *Eco-Refurbishment, a guide to saving & producing energy in the home*, Architectural Press. Less comprehensive than Harland, but complements it.

Chapter 8
GARDENS

Biggs, Tony (1999) *Growing Vegetables*, Royal Horticultural Society/Mitchell Beazley. A clear, step-by-step guide. Not organic.

Fern, Ken (1997) *Plants for a Future, edible & useful plants for a healthier world*, Permanent Publications. Describes a wealth of perennials, most of which can be grown in the garden.

Guerra, Michael (2000) *The Edible Container Garden, fresh food from tiny spaces*, Gaia. All the information you need to grow food in very small gardens, patios, balconies and on rooftops. Lavishly illustrated.

Hamilton, Geoff (1991) *Organic Gardening, pocket encyclopaedia*, Dorling Kindersly. Covers the whole range of the subject, including: basics, fruit, vegetables, herbs and ornamentals; therefore brief on each subject, but packs in most of the essential information.

Hemenway, Toby (2001) *Gaia's Garden, a guide to home-scale permaculture*, Chelsea Green. A North American guide to 'original permaculture' in the garden; includes a clear explanation of garden ecology.

Larkcom, Joy (2002) *Grow Your Own Vegetables*, Frances Lincoln. The best guide to growing vegetables organically. Compact and inexpensive but utterly comprehensive.

Pears, Pauline and Sue Stickland (1999) *Organic Gardening*, Royal Horticultural Society/Mitchell Beazley. Not a general gardening guide, but concentrates on specifically organic techniques. Good on soil care, composting, pests, diseases and weeds.

Readman, Jo (1993) *Muck & Magic*, Search Press. For children – a lively, entertaining book to help them get started in gardening.

Steineck, Hellmut (1984) *Mushrooms in the Garden*, Mad River Press, USA. A guide to growing a wide range of species.

Whitefield, Patrick (1996) *How to Make a Forest Garden*, Permanent Publications. Detailed information on forest gardening design and maintenance, including a comprehensive account of perennial and self-seeding vegetables.

Chapter 9
FRUIT, NUTS & POULTRY

Baker, Harry (1992) *The Fruit Garden Displayed*, Royal Horticultural Society. Covers a slightly narrower range of fruits and nuts than Baker (1999) but in more detail. The book most professionals use.

Baker, Harry (1999) *Growing Fruit*, Royal Horticultural Society & Mitchell Beazley. The best beginners' guide to growing a wide range of fruit and nuts. Not organic.

Bevan, Josie and Knight, Stella (2001) *Organic Apple Production, Pest & Disease Management*, HDRA. A booklet with full information on variety choice, pests and diseases and other cultural information for British growers.

Crawford, Martin (1995a) *Chestnuts, Production and Culture*; (1996) *Hazelnuts, Production and Culture*; (1995b) *Walnuts, Production and Culture*, Agroforestry Research Trust. Three booklets with comprehensive information on the main nut crops for temperate regions.

Feltwell, Ray (1980) *Small-Scale Poultry Keeping, a guide to free-range poultry production*, Faber & Faber. For domestic or small-scale commercial production.

Fern, Ken (1997) *Plants for a Future, edible & useful plants for a healthier world*, Permanent Publications. Describes a wealth of perennials, including many unusual fruits and nuts.

Hooper, Ted (1997) *Guide to Bees & Honey*, Marston House. An in-depth guide to beekeeping.

Lind, K et al (2003) *Organic Fruit Growing*, CABI Publishing. Comprehensive and authoritative, though the climatic and variety information is more applicable to continental Europe than to Britain.

Phillips, Michael (1998) *The Apple Grower, a guide for the organic orchardist*, Chelsea Green, USA. The first full-length book on organic fruit culture, inspiring and informative. Covers all aspects of organic apple growing, including a good chapter on marketing.

Whitefield, Patrick (1996) *How to Make a Forest Garden*, Permanent Publications. Contains design information for the domestic fruit grower, whether you're growing a forest garden or simply fruit on its own, including a guide to choosing species and varieties.

Chapter 10
FARMS & FOOD LINKS

Blake, Francis (1994) *Organic Farming & Growing*, Crowood. A brief guide to: the principles of organic farming, conversion to organic methods, fruit and vegetable growing, and animal farming. Cereals are not covered.

Coleman, Eliot (1995) *The New Organic Grower*, Chelsea Green. The best guide to small-scale organic market gardening, also useful for large-scale domestic gardeners. The author's deep personal experience in New England can be transposed to other climatic regions, including Britain, with little modification.

de Selincourt, Kate (1997) *Local Harvest, delicious ways to save the planet*, Lawrence & Wishart. An account of the current food system and the part which direct food links can have in reducing our ecological impact and reviving the countryside.

Egerton, Liam (1998) *Local Food for Local People*, Soil Association. A concise guide to the various kinds of food links, with brief case studies and a contacts list.

Gordon, Andrew M and Steven M Newman (1997) *Temperate Agroforestry Systems*, CAB International. Describes the results so far achieved with agroforestry in temperate parts of the world, including Britain, Continental Europe and North America.

Henderson, Elizabeth, with Robyn Van En (1999) *Sharing the Harvest, a guide to community supported agriculture*, Chelsea Green. Essential reading for anyone planning to start subscription farming or CSA, whether farmers or consumers. Draws on the experience of many North American CSAs, but very relevant to Britain.

Hislop, Max and Jenny Claridge (2000) *Agroforestry in the UK*, Forestry Commission Bulletin 122. Gives the results of experimental work in the UK to date, with recommendations for combining trees with grazing animals, pigs, poultry, and arable, with financial forecasts.

Lampkin, Nicolas (1994) *Organic Farming*, Farming Press. The standard work, covers every aspect of the subject, except fruit growing, in considerable detail.

Pilley, Greg (2001) *A Share in the Harvest, a feasibility study for community supported agriculture*, Soil Association. A short report on the potential for subscription farming in England, with case studies of existing schemes.

Pullen, Mandy (2004) *Valuable Vegetables, growing for pleasure & profit*, Eco-logic. A guide to small-scale organic vegetable box scheme production in Britain.

Sattler, Friedrich and Eckard von Wistinghausen (1992) *Bio-Dynamic Farming Practice*, Bio-Dynamic Agricultural Association. An accessible guide, useful to anyone interested in sustainable farming, whether they want to go fully biodynamic or not.

Chapter 11
WOODLANDS

Beckett, Kenneth and Gillian (1979) *Planting Native Trees and Shrubs*, Jarrold. Detailed information on species choice from an ecological perspective.❋

Blyth, John, Julian Evans, William ES Mutch and Caroline Sidwell (1991) *Farm Woodland Management*, Farming Press. A handbook for farmers.

Broad, Ken (1998) *Caring for Small Woods*, Earthscan. A concise guide to all aspects of caring for existing woods and planting new ones.

Agate, Elizabeth (2002) *Woodlands, a practical handbook*, BTCV. Detailed information on how to carry out all kinds of woodland work, mainly with hand tools.❋

Evans, J (1984) *Silviculture of Broadleaved Woodland*, HMSO. Detailed information on working with existing woods and planting new ones, covers high forest, coppice, neglected woods and species choice.❋

Hart, Cyril (1991) *Practical Forestry for the agent and surveyor*, Alan Sutton. Detailed information on all aspects of the subject. The reference book to consult when all others fail.❋

Hodge, Simon J (1995) *Creating and Managing Woodlands Around Towns*, Forestry Commission. Covers every aspect of urban woodland, except for individual trees in streets, parks etc.❋

Law, Ben (2001) *The Woodland Way, a permaculture approach to sustainable woodland management*, Permanent Publications. About working with woods and the woodland way of life, written from a permaculture perspective by a practising coppicer. (See case study, Living in the Woods, p336.)

Tabor, Ray (2000) *The Encyclopaedia of Green Woodworking*, Eco-Logic Books. Information on coppicing and on making a wide range of products from the wood. (Not actually in encyclopaedia format.)

❋ Best books for information on species choice.

Chapter 12
BIODIVERSITY

Andrews, John and Michael Rebane (1994) *Farming and Wildlife, a practical management handbook*, RSPB. A comprehensive guide to working with all kinds of rural ecosystems, including woodland, for all kinds of wildlife, not just birds.

Baines, Chris (2000) *How to Make a Wildlife Garden*, Frances Lincoln. The classic guide to wildlife gardening, recently reissued.

Baines, Chris (1986) *The Wild Side of Town*, BBC Books. A guide to action for preserving and enhancing biodiversity in cities and towns.

Gilbert, Oliver (1991) *The Ecology of Urban Habitats*, Chapman & Hall. A fascinating and comprehensive account of urban ecosystems, with an innovative approach to urban nature conservation.

Mabey, Richard (1989) *Food for Free*, Collins. The best guide to wild edible plants.

Peterken, George (1996) *Natural Woodland, ecology & conservation in northern temperate regions*, Cambridge University Press. A comprehensive guide to non-intervention woodland.

Rackham, Oliver (2003) *Ancient Woodland*, Edward Arnold. Basic text for the historical approach to woodland conservation.

Rackham, Oliver (1997) *The History of the Countryside*, Phoenix. The authoritative account of the history of rural ecosystems. Very readable.

Chapter 13
THE DESIGN PROCESS

Holmgren, David (1996) *Hepburn Permaculture Gardens*, Holmgren Design Services. A very detailed account of David Holmgren's design for his family smallholding.

Whitefield, Patrick (1996) *How to Make a Forest Garden*, Permanent Publications. Contains a detailed example of a forest garden design.

Chapter 14
DESIGN SKILLS & METHODS

Alexander, Rosemary (1994) *A Handbook for Garden Designers*, Ward Lock. An accessible guide for beginners, well illustrated and especially good on mapping.

The Neighbourhood Initiatives Foundation publish a range of books and other materials on Planning for Real. See Appendix D, Organisations.

My personal favourite field guides:

Fitter, Richard, Alastair Fitter and Marjorie Blamey (1997) *The Wild Flowers of Britain & Northern Europe*, Harper Collins.

Fitter, Richard, Alastair Fitter and Norman Arlott (1981) *Collins Complete Guide to British Wildlife*, Harper Collins.

Vedel, Helge and Johan Lange (1960) *Trees & Bushes in Wood and Hedgerow*, Methuen.

SUPPLIERS OF BOOKS

The Green Shopping Catalogue
Permanent Publications
The Sustainability Centre
East Meon
Hampshire
GU32 1HR
UK
0845 458 4150 (local rate UK only) or 01730 823 311
info@green-shopping.co.uk
www.green-shopping.co.uk

As well as *Permaculture Magazine* and books on permaculture and related subjects, Permanent Publications publishes the *Green Shopping Catalogue*. Containing over 500 titles, videos and other products relevant to sustainable living, it covers a broad range of subjects including: permaculture, organic gardening, biodynamics, natural history, foraging, mushroom cultivation, food and drink, farming and smallholding, agroforestry, ecological architecture and building, appropriate technology, crafts, human-scale economy and development, community and groups, useful resources, natural medicine, earth medicine, videos, body products and tools.

As well as being published in paper form an extended version including a greater number of titles and products is available at the web site:
www.green-shopping.co.uk

CAT Publications
Centre for Alternative Technology
Machynlleth
Powys SY20 9AZ
01654 702 400
www.cat.org.uk
info@cat.org.uk

CAT publishes books, booklets and 'tipsheets' (single page) on renewable energy, ecological building, water supply, sewage treatment and organic growing. These are of high quality and include some titles for children and teachers.

Their Resource Guides are a unique and valuable service. Covering the same range of subjects, they are derived from the centre's constantly updated database. Titles range from the specific, eg Woodstoves Resource Guide, to the general, eg Water Treatment and Supply. Each one gives: sources of information, organisations, consultants, research bodies, training courses, suppliers of equipment, and other contacts as appropriate. These guides are the first port of call if you want to implement some of the more technical ideas described in this book and need specialist help.

Others
Many of the organisations listed in Appendix D publish and/or supply books on their subject. (See pp455-459.)

Appendix B

GLOSSARY

Actinorhizal: group of nitrogen-fixing plants including alders and *Eleagnus*. *See Legume*.

Active solar heating: system involving moving parts. *See Passive solar heating*.

Aerobic: (micro-) organism which uses oxygen for respiration; condition where a medium such as soil or compost heap contains oxygen-rich air; biological process involving aerobic micro-organisms. *See Anaerobic*.

Agroecosystem: the ecosystem of a cultivated area. *See Ecosystem*.

Agroforestry: growing tree crops and herbaceous crops on the same piece of land at the same time.

Allelopathic plant: one which releases chemicals which have an effect on the growth of other plants, usually detrimental.

Alley cropping: agroforestry planting in which rows of trees alternate with strips of field crops.

Amphibian: animal which spends part of its life cycle in water and part on land, including frogs, toads and newts.

Anaerobic: (micro-) organism which does not use oxygen for respiration; condition where a medium such as soil or a compost heap is deficient in oxygen-rich air; biological process involving anaerobic micro-organisms. *See Aerobic*.

Ancient wood: (strictly) woodland which has been woodland continuously for at least 400 years; (generally) woodland which has been in situ for long enough to develop a diverse ecology. *See Semi-natural*.

Annual plant: one which completes its life cycle within one year. *See Biennial and Perennial*.

Aquaculture: the growing of human food, both plant and animal, in water.

Aquifer: layer of porous rock or sand which is saturated with water, often used as a source of human water supply.

Aspect (of a slope or wall): direction which it faces in relation to north.

Bicrop: two crop species grown on the same area of land at the same time. *See Monocrop*.

Biennial plant: one which completes it's life cycle in two years, making only vegetative growth in the first year and flowering and setting seed in the second. *See Annual and Perennial*.

Biodynamic: system of farming and gardening developed by Rudolf Steiner, integrating organic and spiritual principles.

Biological resource: animal or plant, living or dead, which performs a function which would otherwise be performed by mechanical or chemical means.

Biomass: weight of organic material, plant and animal, living and non-living, within a system.

Bletting: softening process, akin to rotting, which must take place before certain fruits are edible, eg wild service and medlar.

Broadscale: any landscape on a larger scale than a home garden, eg farm or woodland.

Brown earth: a fertile soil type which develops under deciduous woodland.

Capping: process in which the soil surface loses structure and forms a thin impermeable layer.

Carnivore/carnivorous: meat-eating (animal).

Chinampas: productive system in which narrow strips of land alternate with narrow strips of water.

Climax: relatively stable ecological community which is the final stage of succession on a particular site, in most temperate regions some kind of woodland. *See Succession.*

Combined heat and power: thermal power station in which the output of heat, as well as that of electricity, is used.

Contour line: a line connecting points which are the same height above sea level.

Conventional: (in food and farming) non-organic.

Coppice: tree which has been felled and allowed to regrow in a multi-stemmed form; woodland in which the trees have been coppiced.

Coppice with standards: woodland consisting mainly of coppice with some standard, ie single-stemmed, trees.

Coupe: area of woodland felled or coppiced at one time.

Cultivar: cultivated variety of plant.

Dioceous: having male and female flowers on different plants.

Direct drilling: sowing seeds into agricultural land with no or very little cultivation beforehand, using a specially designed seed drill.

District heating: central heating system which heats a number of houses and other buildings from a single heat source.

Double digging: loosening the soil to two spades' depth.

Downwind: the direction towards which a wind is blowing. *See Upwind.*

Dynamic accumulator: plant which has a particular ability to take up certain nutrients from the soil.

Ecological footprint: a way of expressing the ecological impact of a human population, lifestyle or activity in terms of the amount of land required to provide the resources consumed and absorb the pollution caused.

Ecosystem: community of different species interacting with one another and with the chemical and physical factors, eg soil and climate, making up its non-living environment.

Edge effect: the tendency for productivity and diversity in ecosystems, natural and cultivated, to be greater at the edge of the ecosystem than in the interior.

Elevation: (in permaculture) analysis of the land according to altitude, slope and land form. *See Key planning tools.*

Embodied energy: the energy which has been used in the production of something, eg a building, machine or foodstuff. *See Energy in use.*

Energy in use: the energy which a structure, machine or system requires for its operation. *See Embodied energy.*

Entropy: the loss of usable energy which results when energy is transformed from one form to another.

Epiphyte: plant which grows on another plant, eg ferns and mosses on trees.

Ethylene cycle: the natural cycle in the soil in which aerobic conditions alternate with anaerobic conditions.

Eutrophication: enrichment of an ecosystem with mineral nutrients; (commonly) the excessive eutrophication of water bodies due to human action, causing rapid growth of algae followed by oxygen depletion and death of most organisms in the water body.

Exotic (plant or animal): species or variety which is not native to the country where it grows or lives.

Exploitation felling: removal of the best trees from a woodland with no provision for their replacement, either by natural regeneration or planting.

External cost: cost of production which is not paid for by the consumer of a product, eg the pollution costs of present-day transport and farming.

Fall: (in surveying) the amount by which a line descends over a given horizontal distance.

Food chain: series of organisms each of which consumes the one preceding it, eg plant, herbivore, carnivore, top carnivore.

Forest garden: garden in which fruit trees, soft fruit and perennial vegetables are grown together in a three-layered structure.

Frost pocket: area of land more prone to frosts than other nearby places.

Fungicide: chemical substance used to control fungus diseases.

Gaia: the Earth, seen as a self-regulating biological system similar to a single living organism.

Green manure: short term crop grown not to produce food but to add organic matter to the soil, hold nutrients against leaching, protect the soil surface and, if a legume, to fix nitrogen.

Grey water: water which has already been used in the home, other than that used to flush toilets.

Habitat: place or type of place where an organism or a population of organisms live. *See Niche.*

Heavy soil: clay or silty soil.

Herbaceous: (plant) composed entirely of non-woody material, ie all plants other than trees and shrubs

Herbicide: chemical substance used to kill weeds.

Herbivore/herbivorous: (animal) which eats plants.

High forest: a wood or plantation composed entirely of single-stemmed trees, as opposed to coppice.

Holistic: the way of thinking which attempts to understand systems as wholes rather than as merely the sum of their parts. *See Reductionist.*

Host-specific insect: one which feeds only on one kind of plant.

Humus: well-decomposed organic matter in the soil.

Hybrid: plant or animal bred from parents of two different varieties or breeds, or, rarely, of two closely related species.

In-bye land: in upland areas, relatively fertile pasture near the homestead as opposed to less fertile moorland.

Invertebrate: animal without a backbone; includes insects, molluscs etc.

Key Planning Tools: (in permaculture) a design system which integrates the concepts of Zone, Network, Sector and Elevation.

Key point: the point on sloping land just below where the convex upper slope changes to the concave lower slope.

Lambing percentage: the number of lambs reared in one year per 100 ewes in a flock.

Land equivalent ratio (LER): a measure of the yield level resulting from growing crops in polyculture, expressed as the area of monoculture, in hectares, required to produce the same yield as one hectare of polyculture; a LER of more than one is said to be a positive LER and indicates that the polyculture outyields the monoculture.

Leaching: process by which mineral nutrients are carried downwards through the soil by excess water and lost from the ecosystem.

Legume: member of a family of nitrogen-fixing plants, including beans and clovers. *See Actinorhizal.*

LER: see Land equivalent ratio.

Ley: temporary grassland, in organic systems a mixture of grasses and clovers.

Light soil: sandy soil.

Linking: (in permaculture) designing a system so that an output of one element, eg crop, structure or activity, becomes an input to another.

Loam: soil with a well-balanced mixture of sand, silt and clay.

Mast: seed of woodland trees.

Mast year: year in which woodland trees produce abundant seed.

Meadow: grassland used mainly for producing hay.

Mesoclimate: the distinctive climate of a region or moderate-sized area. *See Microclimate.*

Microclimate: the distinctive climate of a small area.

Monocrop: crop consisting of one species only. *See Monoculture and Bicrop.*

Monoculture: the practice of growing only one crop species on an area of land, especially growing the same species on the same land year after year; a crop grown in this way. *See Polyculture.*

Mycorrhiza: combined structure made up of a plant root and a fungus.

Network: (in permaculture) the arrangement of elements in a system relative to each other where there is more than one centre of human activity; analysis of land on this basis. *See Zone and Key Planning Tools.*

Niche: the function of an organism in an ecosystem in relation to other species; the sum total of its relationships with other species and with the soil and climate. *See Habitat.*

Niche management: system of land management in which a person leases the right to carry out a particular enterprise over an area of land rather than renting the land itself.

Nitrogen fixing: process by which bacteria convert nitrogen in the air into mineral forms in the soil, mostly in association with certain plants, including the legume family; these plants are known as nitrogen fixers.

Omnivore/omnivorous: (animal) which eats both plant and animal foods.

Organism: living thing, whether plant, animal or micro-organism.

Original permaculture: (in this book) the kind of permaculture which directly imitates natural ecosystems, as set out in the book *Permaculture One*, in contrast to 'design permaculture' where the emphasis is on designing sustainable systems which may or may not resemble natural ecosystems.

Out-of-cycle coppice: coppice which has not been cut for a long time and thus the stems have grown too big for craft uses.

Oxidation: chemical process in which oxygen is combined with another substance; respiration and burning are processes of oxidation in which oxygen is combined with organic matter to produce carbon dioxide and water, releasing the energy stored in the organic matter.

Passive solar heating: system with no moving parts, usually a building which receives extra heating from the Sun simply by virtue of its orientation, placement of windows etc. *See Active solar heating.*

Pasture: grassland used mainly for grazing.

Pathogen: disease-causing organism.

Perennial plant: one which lives for more than two years. *See Annual and Biennial.*

Pesticide: (in this book) chemical substance used to kill weeds, crop pests or disease organisms. (Strictly speaking these are herbicides, pesticides and fungicides respectively, but the single word is used here for convenience.)

Photosynthesis: process in which green plants use the energy of the Sun to convert carbon dioxide and water into organic food, thus turning solar energy into storable chemical energy; oxygen is released into the atmosphere. *See Respiration.*

Phytoplankton: plant plankton. *See Plankton.*

Plankton: very small plants and animals which float in aquatic ecosystems.

Plant population: (in farming) the density of crop plants in a field, expressed as so many plants per hectare.

Podsol: an acid, leached soil of low fertility.

Pollard: tree which has been cut at around 2m above ground level and regrows with multiple branches arising from this point.

Polyculture: the practice of growing a number of different crop species on the same land at the same time; an assembly of plants grown in this way. *See Monoculture.*

Potable: drinkable.

Potager: vegetable garden, also including flowers, herbs and fruit, laid out so as to be beautiful as well as productive.

Precipitation: rain, hail, sleet and snow.

Reductionist: the way of thinking which assumes that we can understand complex systems by understanding the parts of which they are composed. *See Holistic.*

Respiration: process in which living organisms combine oxygen with organic food to release energy for their use; carbon dioxide and water are released into the atmosphere. *See Photosynthesis.*

Rhizome: horizontal root-like structure from which roots and stems can sprout.

Rhizosphere: the part of the soil solution which is in intimate contact with plant roots, usually extending some three millimetres from the roots.

Ride: trackway in a woodland or plantation.

Rotation: (in gardening and farming) a series of different crops grown on the same piece of land in successive years; (in forestry) the length of time between planting and harvest, or in the case of coppice the time between one felling and the next on the same coupe.

Sector: (in permaculture) area of land which is affected by a particular influence other than human attention, eg shade or wind; analysis of land on the basis of these influences. *See Key planning tools.*

Semi-natural ecosystem: one which has developed under a blend of natural and human influences; in rural areas, usually an originally natural ecosystem which has been affected by human action, eg coppicing or grazing; in urban areas, usually spontaneous colonisation of land previously much influenced by human actions, eg a demolition site.

Short rotation coppice: coppice grown on a rotation of less than ten years; commonly taken to mean willow or poplar coppice grown on a rotation of two to five years, usually as an energy source.

Slabwood: the outer, curved portion of a sawlog, discarded after it has been milled.

Soft fruit: fruits of shrubs, including both bush fruits, eg gooseberries, and cane fruits, eg raspberries. *See Top fruit.*

Soil structure: the way in which individual mineral particles in a soil are bound together to form aggregates. *See Soil texture.*

Soil texture: the relative size of the individual mineral particles in a soil, including sand, silt and clay. *See Soil structure.*

Species: group of organisms which resemble each other and can interbreed with one another to produce fertile offspring, eg pig, potato. *See Variety.*

Spring crop: (in farming) crop sown in the spring of the year in which it is harvested. *See Winter crop.*

Stacking: (in permaculture) growing two or more plants of different heights on the same area of land at the same time; making use of vertical space, especially for growing plants.

Stocking rate: (in farming) the number of animals per hectare kept on the land.

Succession: the natural process whereby the assembly of organisms in an ecosystem changes over time, typically with small annual plants giving way to large perennial plants; a similar process of change in a designed permaculture system.

Suckering: process by which certain trees and shrubs spontaneously produce new plants from their roots.

Sustainable: able to be continued indefinitely in time.

Swale: broad, shallow furrow, running exactly along the contour, designed to intercept moving surface water and allow it to infiltrate into the ground.

Symbiotic relationship: one in which two kinds of organisms live together in an intimate association to the benefit of one or both of them; in common speech it is often restricted to relationships in which both organisms benefit.

Tilling: cultivating the soil, in the sense of physically moving and disturbing it, especially by methods which turn the soil upside down, such as digging and ploughing.

Timber (as opposed to wood): tree trunks of planking diameter.

Top fruit: tree fruits. *See Soft fruit.*

Topography: landform; the shape of hills, valleys and other features.

Transpiration: the flow of water through plants, from the roots to the leaves and out into the atmosphere as water vapour.

Upwind: the direction from which a wind is blowing. *See Downwind.*

Variety: group of organisms which are of the same species but form a recognisable group which is distinct from other members of the same species, eg Tamworth pig, King Edward potato. *See Species.*

Volunteer: (in farming) plant of crop species which has grown from seed accidentally shed from a previous crop rather than being intentionally sown.

Whole-crop silage: cereal crop cut for silage when the grain has formed but is still immature.

Wildwood: the wholly natural woodland which existed before human intervention.

Wind rose: diagram showing average frequency and strength of winds from different directions for a particular place.

Windthrow: the uprooting of trees by strong winds.

Winter crop: (in farming) crop sown in the autumn previous to the year in which it is harvested. *See Spring crop.*

Wood (as opposed to timber): coppice poles and other woody material of less than planking diameter.

Yurt: circular Mongolian-style tent with walls based on a lattice of small poles.

Zone: (in permaculture) area of land which receives a certain level of human attention: zone 0, the house, and zone I, the home garden, receive most, while zone V, wilderness, receives the least; analysis of land on the basis of the amount of attention it needs. *See Zoning and Key planning tools.*

Zoning: (in permaculture) arranging elements in a design so that those which need the most human attention are placed closest to the centre of human activity and those which need least are furthest from it. *See Zone.*

Zooplankton: animal plankton. *See Plankton.*

Appendix C

SCIENTIFIC NAMES
of Plants and Invertebrate Animals
Mentioned in the Text

Alder, common	*Alnus glutinosa*
grey	*Alnus incana*
Italian	*Alnus cordata*
Alexanders	*Smyrnium olustratum*
Almond	*Prunus dulcis*
Angelica	*Angelica archangelica*
Apple sawfly	*Hoplocampa testudinea*
Apple scab	*Venturia inaequalis*
Apple, domestic	*Malus domestica*
crab	*Malus sylvestris*
Apricot	*Prunus armeniaca*
Areca palm	*Chrysalidocarpus lutescens*
Arrowhead	*Sagittaria sagittifolia*
Artichoke, globe	*Cynara scolymus*
Jerusalem	*Helianthus tuberosus*
Ash	*Fraxinus excelsior*
Asparagus	*Asparagus officinalis*
Aubergine	*Solanum melongena*
Autumn olive	*Eleagnus umbellata*
Bamboo palm	*Chamaedorea seifrizii*
Bay	*Laurus nobilis*
Bean, broad & field	*Vicia faba*
French	*Phaseolus vulgaris*
runner	*Phaseolus coccineus*
Beech, common	*Fagus sylvatica*
southern	*Nothfagus* spp
Beetroot	*Beta vulgaris* ssp *vulgaris*
Bilberry	*Vaccinium myrtillus*
Bindweed, garden	*Calystegia sepium*
Birch, silver	*Betula pendula*
white	*Betula pubescens*
Birdsfoot trefoil	*Lotus corniculatus*
Bittercress, hairy	*Cardamine hirsuta*
wood	*Cardamine flexuosa*
Black horehound	*Ballota nigra*
Black locust	*Robinia pseudoacacia*
Blackberry	*Rubus fruticosus* agg
Blackcurrant	*Ribes nigrum*
Blackthorn	*Prunus spinosa*
Bladder senna	*Colutea arborescens*
Blaeberry	*Vaccinium myrtillus*
Blewits	*Lepista* spp
Bluebell	*Hyacinthoides non-scriptus*
Blueberries	*Vaccinium* spp

Bog myrtle	*Myrica gale*
Bolete mushrooms	*Boletus* spp
Boston fern	*Nephrolepis exaltata Bostoniensis*
Bracken	*Pteridium aquilinum*
Bramble	*Rubus fruticosus* agg
Brassy beetle	*Phratora vitellinae*
Brimstone butterfly	*Gonepteryx rhamni*
Broccoli,	*Brassica oleracea, Italica* Group
Nine-Star	*Brassica oleracea botrytis aparagoides*
Broom	*Cytisus scoparius*
Brussels sprouts	*Brassica oleracea, Gemmifera* Group
Buckthorn	*Rhamnus catharticus*
Buckwheat	*Fagopyrum esculentum*
Buddleia	*Buddleia davidii*
Bullrush	*Scirpus lacustris*
Bumble bee (general name for wild bees)	
short haired	*Bombus subterraneous*
Burdock	*Arctium minus* agg & *A. lappa*
Buttercup, creeping	*Ranunculus repens*
Butternut	*Juglans cinerea*
Cabbage	*Brassica oleracea Capitata* Group
Calabrese, see broccoli	
Cane midge	*Resseliella theobaldi*
Caraway	*Carum carvi*
Carrot	*Daucus carota*
Catsear	*Hypochaeris radicata*
Cauliflower	*Brassica oleracea Botrytis* Group
Cedar, western red	*Thuja plicata*
Ceps	*Boletus edulis*
Chamomile,	*Anthemis nobilis*
corn	*Anthemis arvensis*
Cherry plum or myrobolan	*Prunus cerasifera*
Cherry, bird	*Prunus padus*
sour or Morello	*Prunus cerasus*

sweet	*Prunus avium*	Fennel	*Foeniculum vulgare*
wild	*Prunus avium*	Fenugreek	*Trigonella foenum-graecum*
Chervil	*Anthriscus cerefolium*	Ficus alii	*Ficus macleilandii Alii*
Chestnut, horse	*Aesculus hippocastanum*	Fig	*Ficus carica*
sweet	*Castanea sativa*	Filbert, see hazel	
Chickweed	*Stellaria media*	Fir, Douglas	*Pseudotsuga menziesii*
Chicory	*Cichorium intybus*	Fleabane	*Pulicaria dysenterica*
Chives	*Allium schoenoprasum*	Floating sweetgrass	*Glyceria fluitans*
Cleavers	*Galium arapine*	Flowering rush	*Butomus umbellatus*
Clover, alsike	*Trifolium hybridum*	Foxglove	*Digitalis purpurea*
red	*Trifolium pratense*	Fringed water lily	*Nymphoides peltata*
subterranean	*Trifolium subterraneum*	Fritillary butterflies	various spp of family
white	*Trifolium repens*		*Nymphalidae*
Clover rot	*Sclerotinia trifolii*	Garlic	*Allium sativum*
Cocksfoot	*Dactylis glomerata*	Garlic cress	*Peltaria alliacea*
Codlin moth	*Cydia pomonella*	Golden rod	*Solidago virgaurea*
Comfrey, common	*Symphytum officinale*	Golden saxifrage	*Chrysosplenum* spp
Russian	*Symphytum x uplandicum*	Good King Henry	*Chenopodium bonus-*
Cornflower	*Centaurea cyanus*		*henricus*
perennial	*Centaurea montana*	Gooseberry	*Ribes uva-crispa*
Couch grass	*Elymus repens*	Gorse	*Ulex europaeus*
Courgette	*Cucurbita pepo*	Grape hyacinth	*Muscari armeniacum*
Cow parsley	*Anthriscus sylvestris*	Grape vine	*Vitis* spp
Cress	*Lepidium sativum*	Guelder rose	*Viburnum opulus*
Cucumber	*Cucumis sativus*	Hawthorn	*Crategus monogyna*
Currant, flowering	*Ribes sanguineum*	Hazel, filbert	*Corylus maxima*
red	*Ribes rubrum & R. spicatum*	native & cobnut	*Corylus avellana*
white	*Ribes rubrum & R. spicatum*	Heartnut	*Juglans ailantifolia*
Cypress, Lawson	*Chamaecyparis lawsonia*		*cordiformis*
Leyland	*Cupressocyparis x leylandii*	Heather, bell	*Erica cinerea*
Daisy, common	*Bellis perennis*	common	*Caluna vulgaris*
Michaelmas	*Aster novae-anglia,*	cross-leaved heath	*Erica tetralix*
	A. novi-belgii, etc	Hedge woundwort	*Stachys sylvatica*
ox-eye	*Leucanthemum vulgare*	Hogweed	*Heracleum sphondylium*
Shasta	*Chrysanthemum x superbum*	Holly	*Ilex aquifolium*
Damson	*Prunus domestica ssp insitita*	Honesty	*Lunaria annua*
Dandelion	*Taraxacum officinale*	Honey fungus	*Amillaria mellea*
Day lily	*Hemerocallis* spp	Honey locust	*Gleditsia triacanthos*
Deadnettle, red	*Lamium purpureum*	Honeysuckle	*Lonicera periclymenum*
white	*Lamium album*	Hop	*Humulus lupulus*
Dill	*Antheum graveolens*	Hornbeam	*Carpinus betulus*
Dogs mercury	*Mercurialis perennis*	Horseradish	*Cochlearia armoracia*
Dogwood	*Cornus sanguinea*	Hybrid Berries	*Rubus* hybrids
Dracaena Janet Craig	*Dracaena deremensis Janet*	Iceplant	*Mesembryanthemum*
	Craig		*crystallinum*
Dwarf date palm	*Phoenix roebelenii*	Iris, yellow flag	*Iris pseudacorus*
Egg plant, see Aubergine		Ivy, common	*Hedera helix*
Elder, American	*Sambucus canadensis*	Boston	*Parthenocissus tricuspidata*
common	*Sambucus nigra*	Jack by the hedge	*Alliaria petiolata*
red	*Sambucus racemosa*	Japanese knotweed	*Reynoutria japonica*
Eleagnus	*Eleagnus* spp	Japanese wineberry	*Rubus phoenicolasius*
Elephant grass	*Miscanthus sinensis*	Japonica	*Chaenomeles* spp
	& M. sacchariflorus	Juneberry	*Amelanchier canadensis*
Elm	*Ulmus* spp	Kale, annual	*Brassica oleracea Acephala*
wych	*Ulmus glabra*		Group
Endive	*Cichorium endivia*	perennial	*Brassica oleracea acephala*
Evening primrose	*Oenothera erythrosepala*		*Daubenton*
False acacia	*Robinia pseudoacacia*	Kiwi	*Actinidia chinensis*
Fat hen	*Chenopodium album*		or *A. delicosa*

Kiwi, hardy	*Actinidia arguta*	Pine, Austrian	*Pinus nigra* var *austiaca*
	& *A. kolomikta*	Corsican	*Pinus nigra* var *maritima*
Knapweed, black	*Centaurea nigra* agg	lodgepole	*Pinus contorta*
Kohl rabi	*Brassica oleracea Gongylodes*	maritime	*Pinus pinaster*
	Group	Monterey	*Pinus radiata* or *P. insignis*
Lady palm	*Rhapis excelsa*	Scots	*Pinus sylvestris*
Ladybirds	Family *Coccinellidae*	Plantain, great	*Plantago major*
Lamb's lettuce	*Valerianella locusta*	ribwort	*Plantago lanceolata*
Land cress	*Barbarea verna*	Plum sawfly	*Hoplocampa flava*
Larch, European	*Larix decidua*	Plum yews	*Cephalotaxus* spp
hybrid	*Larix x eurolepis*		& *Torrya* spp
Leek	*Allium porrum*	Plum, beach	*Prunus maritima*
Lettuce	*Latuca sativa*	domestic	*Prunus domestica*
Limes	*Tilia* spp	Poplars	*Populus* spp
Ling,		Potato	*Solanum tubersosum*
see Heather, common		Pumpkin	*Cucurbita maxima*
Loganberry	*Rubus idaeus loganii*	Purslane, pink	*Montia sibirica*
Lords-and-ladies	*Arum maculatum*	summer	*Portulaca oleracea*
Lovage	*Levisticum officinale*	winter	*Claytonia perfoliata*
Lucerne or alfalfa	*Medicago sativa*	Quince	*Cydonia oblonga*
Mallow, common	*Malva sylvestris*	Radishes	*Raphanus sativus*
musk	*Malva moschata*	Ramsons	*Allium ursinum*
tree	*Lavatera* sp	Rape, oilseed	*Brassica campestris*
Maple, field	*Acer campestre*		var *oleifera*
Marigold, pot	*Calendula officinalis*	salad	*Brassica napus*
Marjoram,	*Origanum majorana*		var *napus*
wild	*Origanum vulgare*	Raspberry beetle	*Byturus tomentosus*
Medlar	*Mespilus germanica*	Raspberry	*Rubus idaeus*
Midgie	*Culicoides punctatum*	Reedmace	*Typha latifolia*
	& others		& *T. angustifolia*
Mint	*Mentha* spp	Rhubarb	*Rheum x cultorum*
Mitsuba	*Cryptotaenia japonica*	Richard the Lionheart	*Reichardia picroides*
Monkey puzzle	*Araucaria araucana*	Rocket, annual	*Eruca sativa*
Mulberry	*Morus nigra*	Turkish	*Bunias orientalis*
	& other *Morus* spp	Rose, dog	*Rosa canina* agg
Mustard	*Sinapis alba*	field	*Rosa arvensis*
Nasturtium	*Tropaeolum majus*	ramanas	*Rosa rugosa*
Nettle, stinging	*Urtica dioica*	Rosemary	*Rosmarinus officinalis*
Oak, holm	*Quercus ilex*	Rowan	*Sorbus aucuparia*
pedunculate	*Quercus robur*	Rubber plant	*Ficus robusta*
sessile	*Quercus petraea*	Ruby chard	*Beta vulgaris* ssp *cicla*
Onion, common	*Allium cepa*	Russian olive	*Eleagnus angustifolia*
everlasting	*Allium perutile*	Russian vine	*Polygonum bauldschianicum*
tree	*Allium cepa proliferum*	Rye	*Secale cereale*
Welsh	*Allium fistulosum*	Ryegrass	*Lolium perene*
Oregon grape	*Mahonia aquifolium*	Sage	*Salvia officinalis*
Oxlip	*Primula elatior*	Salad burnet	*Sanguisorba minor*
Oyster mushroom, pine	*Pleurotus colombinus*	Sallow, see Willow, pussy	
Oyster mushrooms	*Pleurotus* spp	Salmonberry	*Rubus spectabilis*
Pansy	*Viola* spp	Salt bush	*Atriplex halamis*
Parsley, inc. sheep's	*Petroselenium crispum*	Scabious, field	*Knautia arvensis*
Parsnip	*Pastinaca sativa*	small	*Scabiosa columbaria*
Passionfruit	*Passiflora edulis*	Scurvy grass, Danish	*Cochlearia dancia*
Pea	*Pisum sativum*	Sea beet	*Beta vulgaris* ssp *maritima*
Peace lily	*Spathiphyllum* spp	Sea buckthorn	*Hippophae rhamnoides*
Peach	*Prunus persica*	Sea holly	*Eryngium maritimum*
Pear midge	*Contarinia pyrivora*	Sea kale	*Crambe maritima*
Pear scab	*Venturia pirina*	Service, wild	*Sorbus torminalis*
Pear, Asian	*Pyrus serotina*	Shepherd's purse	*Capsella bursa-pastoris* agg

Shii-take mushroom	*Lentinus edodes*
Siberian pea shrub	*Caragana brevispina*
	& C. arborescens
Skirret	*Sium sisarum*
Slugworm	*Caliroa cerasi*
Small copper butterfly	*Lycaena phlaeas*
Sorrel, common	*Rumex acetosa*
Spider plant	*Chlorophytum capense*
	or *C. elatum*
Spinach, common	*Spinacea oleracea*
perpetual	*Beta vulgaris* ssp *cicla*
Spindle	*Euonymus europaeus*
Spruce, Norway	*Picea abies*
Sitka	*Picea sitchensis*
Squash	*Cucurbita* spp
Stonecrop	*Sedum* spp
Strawberry, domestic	*Fragaria* spp
alpine	*Fragaria vesca alpina*
Sunflower	*Helianthus annuus*
Swede	*Brassica napus*
Sweet cicely	*Myrrhis odorata*
Sweet corn	*Zea mays*
Swiss chard	*Beta vulgaris* ssp *cicla*
Sycamore	*Acer pseudoplanatus*
Tamarisks	*Tamarix* spp
Tansy	*Tanacetum vulgare*
Teasel	*Dipsacus fullonum*
Thistle, creeping	*Cirsium arvense*
Thyme	*Thymus* spp
Tomato	*Lycopersicon esculentum*
Triticale	*Triticosecale*
Turnip	*Brassica campestris*
Vetches	*Vicia* spp (not *V. faba*)
Violets	*Viola* spp
Vipers bugloss	*Echium vulgare*
Virginia creeper	*Parthenocissus quinquefolia*
Walnut, black	*Juglans nigra*
common	*Juglans regia*
Water lily, yellow	*Nuphar lutea*
Watercress	*Nasturtium officinale*
Wayfaring tree	*Viburnum lantana*
Whitebeam	*Sorbus aria*
Wild garlic, see Ramsons	
Willow	*Salix* spp
purple	*Salix purpurea*
pussy	*Salix caprea* agg
Willowherb, rosebay	*Epilobium angustifolium*
Winter moth	*Operophthera brumata et al*
Wood anemone	*Anemone nemorosa*
Yarrow	*Ahillea millefolium*
Yellow archangel	*Lamiastrum galeobdolon*
Yew	*Taxus baccata*
Yorkshire fog	*Holcus lanatus*

ORGANISATIONS

Agroforestry Research Trust
46 Hunters Moon, Dartington, Totnes, Devon TQ9 6JT
01803 840 776
mail@agroforestry.co.uk
www.agroforestry.co.uk
Researches into temperate tree and shrub crops and agroforestry systems. Informative publications and open days.

Alternative Technology Association
Supporters' organisation for CAT (see below). Informative magazine and annual conference.

Aquavision
Claycombe Cottage, Burleigh, Stroud, Gloucestershire GL5 2PS
01453 887 500
info@aquavisiononline.com
www.aquavisiononline.com
Consultancy and project management in sustainable and organic aquaculture, also water gardening and backyard fish farming.

Association for Environment Conscious Building
Nant-y-Garreg, Saron, Llandysul, Dyfed SA44 5EJ
01559 370 908
admin@aecb.net
www.aecb.net
Information service on ecologically sound building materials and methods.

Bio-Dynamic Farming & Gardening Association
199 Rudolf Steiner House, 31 Park Road,
London NW1 6XT

Bioregional Development Group
BedZED Centre, Helios Road, Wallington,
Surrey SM6 7BZ
020 8404 4880
info@bioregional.com
www.bioregional.com
Promotes and develops sustainable enterprises, eg hemp fibre and locally produced charcoal.

British Beekeepers Association
c/o Mrs Betty Showler, Riverside, Newport Street,
Hay-on-Wye, Herefordshire HR3 5BG
Will put you in touch with local beekeepers who can help you to start beekeeping; send SAE for contact details of your local association.

British Trust for Conservation Volunteers (BTCV)
36 St Mary's Street, Wallingford, Oxfordshire OX10 0EU
01302 572 244
information@btcv.org.uk
www.btcv.org
Offers help and advice for people in local communities wishing to improve their environment, practical nature conservation opportunities for volunteers, training and a range of handbooks.

Brogdale Horticultural Trust
Brogdale Road, Faversham, Kent ME13 8XZ
01795 535 286
info@brogdale.org.uk
www.brogdale.org.uk
Very extensive collection of fruit and nut varieties, open to the public, with guided tours.

Centre for Alternative Technology (CAT)
Machynlleth, Powys SY20 9AZ
01645 702 400
info@cat.org.uk
www.cat.org.uk
Demonstration site, publications and telephone/internet information service covering: energy, building, water supply and sewage, and organic gardening. Publish resource guides on these subjects. (See p444.)

Chapter 7
The Potato Store, Flaxdrayton Farm, South Petherton,
Somerset TA13
01460 249 204
chapter7@tlio.demon.co.uk
www.oneworld.org/tlio/chapter4/
Gives planning advice for low impact homes and self-builders. Works within the planning system to encourage laws which allow for sustainable development.

Comite Jean Pain
Avenue Princesse Elizabeth 18, B1030 Brussels, Belgium
+ 32 22 41 08 20
Promotes and teaches the Jean Pain method.

Common Ground
Gold Hill House, 21 High Street, Shaftesbury,
Dorset SP7 8JE
info@commonground.org.uk
www.commonground.org.uk
Empowers people to celebrate what is distinctive and
special about their own locality. Gives advice and
support on community orchards.

Community Car Share Network
The Studio, 32 The Calls, Leeds LS2 7EW
0113 234 9299
office@carshareclubs.org.uk
www.carshareclubs.org.uk
Facilitates the development of car sharing clubs in the
UK, in both urban and rural areas.

Community Composting Network
67 Alexandra Road, Sheffield S2 3EE
0114 258 0483
ccn@gn.apc.org
www.othas.org.uk/ccn
Gives advice and support to composting groups and
helps new ones to get started.

Domestic Fowl Trust
Honeybourne, Nr Evesham, Worcestershire WR11 5QG
01386 833 083
dft@honeybourne.demon.co.uk
www.mywebpage.net/domestic-fowl-trust
Conserves rare poultry breeds and gives advice.

Ecological Design Association
The British School, Slad Road, Stroud,
Gloucestershire GL5 1QW
01453 765 575
ecological@designassociation.freeserve.co.uk
www.edaweb.org
Concentrates mainly on architecture and building.

Ecology Building Society
18 Station Road, Cross Hills, Nr Keighley,
West Yorkshire BD20 7EH
01535 635 933
info@ecology.co.uk
www.ecology.co.uk
Lends money on properties and projects which are
ecologically sound, often ones which other lenders
won't cover. A good ethical place to invest any spare
cash.

Elm Farm Research Centre
Hamstead Marshall, Newbury, Berkshire RG10 0HR
01488 658 298
elmfarm@efrc.com
www.efrc.com
Researches into organic farming; offers advisory and
soil analysis services.

Environmental Transport Association
10 Church Street, Weybridge KT13 8RS
01932 828 882
eta@eta.co.uk
www.eta.co.uk
A road rescue service like the AA which campaigns for
more ecologically sustainable transport policies.

Federation of City Farms & Community Gardens
The Green House, Hereford Street, Bedminster,
Bristol BS3 4NA
0117 923 1800
admin@farmgarden.org.uk
www.farmgarden.org.uk
Promotes city farms and community gardens and offers
a range of services to its members.

Find Your Feet
37-39 Gt Guildford Street, London SE1 0ES
020 7401 8794
fyf@fyf.org.uk
www.fyf.org.uk
Promotes the development of leaf curd.

Flora Locale
36 Kingfisher Court, Hambridge Road, Newbury,
Berkshire RG14 5SJ
floralocale@naturebureau.co.uk
www.floralocale.org
Promotes the conservation of the genetic diversity
of native plants; gives advice to businesses and other
bodies.

Forest Stewardship Council
Unit D, Old Station Building, Llanidloes,
Powys SY18 5EB
01686 413 916
amy@fsc-uk.demon.co.uk
www.fsc-uk.demon.uk
Runs certification scheme for timber, similar to that for
organic food.

Friends of the Earth
26-28 Underwood Street, London N1 7JQ
020 7490 1555
info@foe.co.uk
www.foe.co.uk

Genetic Engineering Network (GEN)
PO Box 9659, London N4 4YJ
020 7837 9229
geneticsforum@gn.apc.org
www.geneticsforum.org.uk
Provides information on genetic engineering to the public.

Global Action Plan (GAP)
9 Fulwood Place, London WC1V 6HG
020 7405 5633
all@gapuk.demon.co.uk
www.globalactionplan.org.uk
Helps people take practical environmental action in their homes, at work and in the community.

Green Party
1a Waterlow Road, Archway, London N19 5NJ
020 7272 4474
office@greenparty.org.uk
www.greenparty.org.uk

Greenwood Trust
Station Road, Coalbrookdale, Telford TF8 7RD
01952 432 769
thegreenwoodtrust@thegreenwoodtrust.demon.co.uk
Teaches traditional woodland management and green woodworking skills.

Henry Doubleday Research Association (HDRA)
Ryton-on-Dunsmore, Coventry CV8 3LG
024 7630 3517
enquiry@hdra.org.uk
www.hdra.org.uk
The largest and most active organic gardening organisation in Europe, with demonstration centre and members' information service. Its Heritage Seed Library works to preserve the genetic diversity of garden vegetables.

Intermediate Technology Development Group (ITDG)
Schumacher Centre, Bourton Hall, Bourton-on-Dunsmore, Rugby, Warwickshire CV23 9QZ
01788 661 100
itdg@itdg.org.uk
www.oneworld.org/itdg
Mainly involved with overseas projects, but could help with sourcing intermediate equipment which is hard to obtain in northern countries; also publishes books.

Kentish Cobnuts Association
Clakkers House, Crouch, Sevenoaks, Kent TN15 8PY
01732 780 038
Hazelnut growers association, open to non-growers.

The Land Institute
2440 E. Water Well Road, Salina, KS 67401, USA
+1 785 823 5376
theland@landinstitute.org
www.landinstitute.org
Researches natural systems agriculture (domestic prairie).

Learning Through Landscapes
Third Floor, Southside Offices, The Law Courts, Winchester, Hampshire SO23 9DL
01962 846 258
schoolgrounds-uk@ltl.org.uk
www.ltl.org.uk
Helps school improve their grounds for the benefit of children, and develops ways of using school grounds to teach subjects in the National Curriculum.

Middle Wood Trust
Roeburndale West, Wray, Lancaster LA2 8QR
01524 221 880
Research and education centre for permaculture.

Met Office
London Road, Bracknell, Berkshire RG12 2SZ
0845 300 0300
enquiries@meto.gov.uk
www.metoffice.gov.uk
Provide climatic data useful for planning rainwater collection, renewable energy systems and crop choice.

National Association of Farmers' Markets
South Vaults, Green Park Station, Green Park Road, Bath BA1 1JB
01225 787 914
nafm@farmersmarkets.net
www.farmersmarkets.net
Promotes and supports farmers' markets and helps set up new ones.

National Society for Allotment Gardeners
Hunters Road, Corby, Northamptonshire NN17 1JE
01536 266 576

NEF Renewables
The National Energy Foundation, Davy Avenue, Knowlhill, Milton Keynes MK5 8NG
0800 138 0889
renewables@greenenergy.org.uk
www.greenenergy.org.uk
Gives independent advice on green energy for the home, garden, farm or community building.

Neighbourhood Initiatives Foundation
The Poplars, Lightmoor, Telford TF4 3QN
01952 590 777
nif@cableinet.co.uk
www.nif.co.uk
Promotes and organises Planning for Real events.

Ordnance Survey
Romsey Road, Southampton SO16 4GU
0845 605 0505
www.ordsvy.gov.uk
custinfo@ordsvy.gov.uk
Supplier of maps.

Patrick Whitefield Associates
37 Leg of Mutton Road, Glastonbury,
Somerset BA6 8HH
01458 832 317
ka.tonga@virgin.net
www.patrickwhitefield.co.uk
Permaculture education provider.

Permaculture Association (Britain)
BCM Permaculture Association, London WC1N 3XX
07041 390 170
office@permaculture.org.uk
www.permaculture.org.uk
Promotes permaculture; can supply local and international contact lists, details of courses and permaculture sites which may be visited; puts on annual convergence or get-together of permaculturists.

Plantlife
14 Rollestone Street, Salisbury, Wiltshire SP1 1DX
01722 342 730
enquiries@plantlife.org.uk
www.plantlife.org.uk
Conservation organisation for wild plants in Britain.

Plants For A Future
The Field, Higher Penpol, St Veep, Nr. Lostwithiel,
Cornwall PL22 0NG
01208 873 554
ken.fern@lineone.net
www.pfaf.org
Assesses the suitability of plants for permaculture in Britain, especially trees and other perennials.

Radical Routes
Cornerstone Housing Co-operative,
16 Sholebroke Avenue, Chapeltown, Leeds LS7 2HB
0870 733 2538
cornerstone@gn.apc.org
www.radicalroutes.org.uk
Helps set up housing co-operatives.

Reforesting Scotland
62-66 Newhaven Road, Edinburgh EH6 5QB
0131 554 4321
reforscot@gn.apc.org
www.gn.apc.org/reforestingscotland
Working to restore woodland and the woodland economy all over Scotland.

Royal Society for the Protection of Birds (RSPB)
The Lodge, Sandy, Bedfordshire SG19 2DL
01767 680 551
bird@rspb.demon.co.uk
www.rspb.org.uk

Small Woods Association
The Cabins, Malehurst Estate, Minsterley,
Shropshire SY5 0EQ
01743 792 644
enquiries@smallwoods.org.uk
www.smallwoods.org.uk
Aims to conserve small woods by education, encouraging productive use, and involving local communities; membership open to woodland owners and interested individuals.

Soil Association
Bristol House, 90 Victoria Street, Bristol BS1 6DF
0117 929 0661
info@soilassociation.org
www.soilassociation.org
Promotes organic farming and certifies organic producers. Produces the *Organic Directory Yearbook*. Promotes producer-consumer links; keeps lists of producers looking for consumers and vice versa.

Straw Bale Building Association
PO Box 17, Todmorden, Lancashire OL14 8FD
01706 818 126

Sustain
94 White Lion Street, London N1
020 7837 1228
sustain@sustain.org
www.sustainweb.org
Umbrella organisation for food and farming issues, including growing food in cities.

Sustrans
PO Box 21, Bristol BS99 2HA
0117 929 0888
info@sustrans.org.uk
www.sustrans.org.uk
Promotes walking and cycling; involved in Safe Routes to School and in creating national network of cycle paths.

Transport 2000
The Impact Centre, 12-18 Hoxton Street,
London N1 6NG
020 7613 0743
join@transport2000.demon.co.uk
Campaigning for greener transport policies.

Trees for Life
The Park, Findhorn Bay, Forres, Moray IV36 3TZ
01309 691 292
trees@findhorn.ogr
www.treesforlife.org.uk
Reforesting the Scottish Highlands; tree-planting opportunities for volunteers.

Triodos Bank
Brunel House, 11 The Promenade, Bristol BS8 3NN
0117 973 9339
mail@triodos.co.uk
www.triodos.co.uk
Lends money to ethical and ecological projects; can help with innovative financing ideas.

Walnut Club
c/o Karen Russell, Horticulture Research International, East Malling, West Malling, Kent ME19 6BJ
01732 843 833
karen.russell@hri.ac.uk
Walnut growers organisation, giving information and advice.

Walter Segal Self-Build Trust
15 High Street, Belford, Northumberland NE70 7NG
01668 213 544
info@segalselfbuild
www.segalselfbuild.co.uk
Co-ordinates and supports self-build groups all over Britain.

Waste Watch
Europa House, Ground Floor, 13-17 Ironmonger Row, London EC1V 3QG
020 7253 6266
info@wastewatch.org.uk
www.wastewatch.org.uk
Provides information on waste reduction, reuse and recycling.

Wessex Coppice Group
Vale Farm, Smugglers Lane, Monkwood, Hampshire SO24 0HD
01962 772 030
linda.glynn@coppice.org.uk
www.coppice.org.uk
Supports coppice workers, especially with marketing, training and contacts with woodland owners. Keeps a nationwide database of coppice workers.

Willing Workers on Organic Farms (WWOOF)
PO Box 2675, Lewes, Sussex BN7 1RB
01273 476 286
fran@wwoof.org
www.wwoof.org
Links up organic farms and gardens with volunteer workers, both nationally and internationally.

WI Country Markets Ltd
Reada Court, Vachel Road, Reading RG1 1NY
0118 939 4646
info@wimarkets.co.uk
www.wimarkets.co.uk

The Wildlife Trusts
The Kiln, Waterside, Mather Road, Newark NG24 1WT
01636 677 711
info@wildlife-trusts.cix.co.uk
www.wildlifetrusts.org.uk
Umbrella organisation for county and urban wildlife trusts in Britain. Contact them to find out about your local trust.

Women's Environmental Network (WEN)
PO Box 30626, London E1 1TZ
020 7481 9004
wenuk@wen.org.uk
www.wen.org.uk
Provides information, training and campaigns on ecological issues which effect women's daily lives, eg the link between health and environment.

SUPPLIERS OF PERMACULTURE REQUISITES

Most of the plants, seeds and equipment needed for permaculture are widely available.
More specialised requirements should be available from one or more of the following.

Agroforestry Research Trust
46 Hunters Moon, Dartington, Totnes, Devon TQ9 6JT
01803 840 776
mail@agroforestry.co.uk
www.agroforestry.co.uk
Seeds and plants of trees, shrubs and herbaceous perennials for agroforestry, including many unobtainable elsewhere. Profits go to research into agroforestry.

Biologic Designs
Archenhills, Stanford Bishop, Bringsty,
Herefordshire WR6 1TZ
01886 884 721
Wetland Ecosystem Treatment (WET) systems.

BTCV Enterprises Ltd
Conservation Centre, Balby Road, Doncaster DN4 0RH
01302 572 200
www.btcv.org
Native trees, shrubs and wildflowers of British provenance. (If you have a local supplier of plants of local provenance, that's even better.)

Butterworths' Organic Nurseries
Garden Cottage, Auchinleck Estate, Cumnock,
Ayrshire KA18 2LR
01290 551 088
butties@webage.co.uk
www.webage.co.uk/apples/
Organically grown apple trees.

The Ceramic Stove Company
4 Earl Street, Oxford OX2 0JA
01865 245 077
info@ceramicstove.com
www.ceramicstove.com
Suppliers of masonry stoves.

Chiltern Seeds
Bortree Stile, Ulverston, Cumbria LA12 7PB
01229 581 137
chilternseeds@compuserve.com
www.edirectory.co.uk/chilternseeds
A wide range of unusual plant seeds.

Cool Temperate
39 Maple Avenue, Beeston, Nottingham NG9 1PU
0115 917 0416
Fruit trees and shrubs, nitrogen-fixers etc. Profits go to research into own-root fruit trees.

Domestic Fowl Trust
Honeybourne, Nr Evesham, Worcestershire WR11 5QG
01386 833 083
dft@honeybourne.demon.co.uk
www.mywebpage.net/domestic-fowl-trust
Poultry breeding stock, equipment and books.

Edulis
Flowers Piece, Ashampstead, Reading,
Berkshire RG8 8SG
01635 578 113
edulis2000@hotmail.com
Edible ornamental specialists.

Future Foods
Luckleigh Cottage, Hockworthy, Wellington,
Somerset TA21 0NN
01398 361 347
enquiries@futurefoods.com
www.futurefoods.com
Seeds, tubers and spawn of unusual edibles, including annual and perennial vegetables, trees, shrubs, fungi and ferments.

George Latham
2 Wyck Rissington, Cheltenham,
Gloucestershire GL54 2PN
01451 822 098
Walnut specialist, can supply 50-70 varieties.

The Green Shop
Bisley, Stroud, Gloucestershire GL6 7BX
01452 770 629
mailorder@greenshop.co.uk
www.greenshop.co.uk
A wide range of ecological products, including rainwater collection systems. "If we don't have it... we'll tell you where to go."

Heritage Seed Library
HDRA, Ryton-on-Dunsmore, Coventry CV8 3LG
024 763 03517
enquiry@hdra.org.uk
www.hdra.org.uk
Join this seed club and you'll get seeds of old vegetable
varieties now unobtainable commercially, and support
the valuable work of preserving genetic diversity at the
same time.

Nutwood Nurseries
2 Millbrook Cottages, Lowertown, Helston, Cornwall
TR13 0BZ
01326 564 731
Nut tree specialists, with the best range in Britain.

Solartwin
01244 403 404
www.solartwin.com
Simple solar hot water systems, DIY or professionally
installed.

Appendix F

METRIC-IMPERIAL CONVERSIONS

Length

1cm = 0.4in
1m = 1.1yd
1km = 0.6 miles

Area

$1m^2 = 1.2^2$ yd
1ha = 2.5 acres

Volume

1l = 1.8 pints (0.9 quart US)
$1m^3$ = 220gall (265gall US)

Weight

1kg = 2.2lb
1tonne = 1ton (1.1ton US)

Crop Yield

1t/ha = 0.4t/ac

Temperature
(*Farenheit-Celsius*)

$°F = (°Cx1.8)+32$

SOME BASIC EQUIVALENTS

If you're not used to using the metric system the easiest way to understand it is to become familiar with a few basic equivalents. They don't have to be totally accurate, but they give you enough idea to make sense of a text which uses metric measurements. It certainly beats getting out a pocket calculator every time you come across a quantity! Here are some you may find useful.

5cm is 2 inches
30cm is 1 foot
1 metre is 1 yard
100m is 300 feet

1 square metre is 1 square yard
1 hectare is 2½ acres

1kg is 2 pounds
2kg is 5 pounds
1 tonne is 1 ton

0°C is freezing
10°C is 50°F
20°C is 70°F

1 litre is 1 American quart, or 2 British pints
1000l is 250 American gallons, or 200 British gallons
this is a cubic metre

INDEX

Abrahams, Jay 104
Actinorhizal plants 48-49
Agroecology 19
Agroforestry 274, 275-284
    advantages of 276-278
    alley cropping 280-281
    definition 275
    disadvantages of 278
    grassland with trees 278
    hedgerows 281
    principles of 275-278
    protecting trees 279
    shelterbelts 282-284
    timber 280-281
Air travel 129
Allelopathy 19, 218, 273
Allotments 143, 144, 179-180,
    212-213, 286-287
Almonds 225
Alternative Technology
    135-137
Altieri, Miguel 19
Animals (for individual species
see separate entries) 258-264
    choice of 261-263
    foggage system 263-264
    grazing 50-51, 263-264
    rotation of pasture 261-262
    ruminants 259
    welfare 258
Aquaculture 25-26, 106-118
    advantages of 106-107
    domestic scale 110-111
    edible plants 110
    fish 107-108
    plants, broadscale 108-109
Aspect 80
    for fruit 238

Beckett, Gillian & Kenneth
    236
Beech
    as nutting tree 225
Bees 244-246
    forage plants for 246
Beetles 47, 266, 269
Beneficial relationships 3-4,
    107-108, 248-250,
    294-297
Biodiversity (see also
ecosystems) 16-18
    assessment 352-356
    consumption & 343
    eutrophication 359
    exotic species 368-370

exotic varieties 370-371
extinctions 342
habitat creation, rural
    360-362
historical vs natural
    approach to 356-359
importance of 341-343
integrated crop
    management & 344
native species 368-370
native varieties 370-371
natural regeneration
    358-359
nature reserves 345
old trees 355
organic farming & 343-344
organic gardening &
    343-344
permaculture & 344
planting hedges 361
short-term care 359
tidiness 360, 363, 364, 368
trees for 368
urban 362-368
urban survey 362-363
wildlife corridor 348, 351,
    354
wildlife gardening
    364-368
Biological resources 32
Biotecture 153-156
Black locust (see False acacia)
Black walnut 225
Blair, David 302
Bon Fils, Marc 271
Box scheme 289-291
British Trust for Conservation
    Volunteers 363, 364
Broadfork 51
Building materials 159-162
    embodied energy 35, 160,
        161
    pollution 160, 162
Buildings
    condensation 164
    costs 147-148
    energy use 147-153
    insulation 149-150,
        164-165
    obstruction angle 150
    passive solar 123, 148-152,
        163-164, 166-167
    renovation 147, 162-167
    wood heating 152-153,
        304

Building styles 157-159
    cob 158-159
    earth sheltered 158
    earthship 158
    green framing 157
    pole building 158
    Segal system 157-158
    straw bale 158
Bunyip 100, 113, 116, 117,
    406-408
Buschberghof Farm 293
Buying land 386

Carbon cycle 72
Carbon dioxide 71, 72,
    101-102
Carbon:nitrogen ratio 44
Carp 107-108
Cattle 17, 263-264
Centre for Alternative
    Technology 63, 105, 135
    444
Cereals (see Grain)
Charcoal 295-297, 336-337
Chestnuts 223, 336-337
Chicken-greenhouse 26
Chickens 246-247, 248-253
    chicken tractor 248-249
    forage plants 252
    forage system 251-253
    in short rotation coppice
        250
    on farms 250
    zoning 249-250
Chinampas 115-116
Clare, Richard 212-213
Clements, Bob 268
Climate
    altitude effects on 69
    British averages 70
    coastal 69
    growing season 69-70
    regional 68-70
    types of 67-68
    urban 69
Climate change 27, 71-74
    action against 73
    air travel 129
    carbon dioxide 71, 101-102
    causes 71
    consequences 71-73
    biodiversity & 73, 369
    forests & 74
    methane 71, 101-102
    peat 72

plantation forestry 72
sewage treatment 101
tree crops & 74
Cohousing 146-147
Colquhoun, Margaret 414
Comfrey 50
Community
    agriculture 292-293
    designing 422-423
    gardens 183-184, 212-213
    orchards 236
    planning 420-421
    wildlife 362-364
    woodland 334-335
Community supported
    agriculture 292-293
Companion plants 242-244
Competition 17
Compost 43-46
    choosing a system 45-46
    example system 46
    how to make 45-46
    Jean Pain system 255,
        307, 315-316
    reasons for making 43-45
    urine 44-46, 105-106
Compost heap placing 30
Conservatory 3-4, 151-152,
    163, 174
Consumption 7-9, 129-130,
    132-134, 343
Corbett, Phil 254
Cotton 93
Crowder, Bob 243
Cultivation (see also No-till)
    deep tined 51, 52
    double digging 56
    ploughing 56

Day, Christopher 416
Deep ecology 6
Deer 312, 314-315, 414
de Selincourt, Kate 288
Design examples 296-297,
    387-398
Designing (Case Study
references in brackets)
    base map 377 (390)
    design aims 387 (395)
    design concepts 387
        (394-396)
    design details 387-388
        (396-399)
    design proposals 386-388
        (394-399)

design questionnaire
382-385 (392-394)
evaluation 385-386
(393-394)
implementation plan (398)
in wholes 36
listening 382
microclimate map (391)
observation 379
re-evaluation 388
(399-400)
site survey 378-382
(389-401)
Design skills & methods
card cutouts 421-422
computer aided design 411
drawing 411-413
input-output analysis 421
key planning tools 27-30,
385, 393-394, 420-421
listening 382, 414-419
listening to the landscape
416-417
mapping 408-413
McHarg exclusion
method 422
options & decisions
419-420
Ordnance Survey maps
410-411
Planning for Real 422-423
reading the landscape
414-417
slope measurement
404-408
surveying 399-406
SWOC 385, 420
Diversity (see also Polyculture)
16-22, 30-31
benefits of 17-18
climate & 67
cultural 22
disadvantages of 20, 22
ecological 21-22
fertility & 17-18
food links & 290-291
genetic 20-21, 370-371
in agroforestry 277
scale of production &
30-31
species 18-20
stability & 18, 19
Drainage
garden 189-190
Ducks 111, 250
Dunwell, Jan 294
Dunwell, Matt 288, 294-297,
305
Dynamic accumulators 49-50

Earthcare 6
Earthworms 47, 57, 269
as soil indicators 60
Ecological footprint 8
Ecological impact 7-10
Ecolots 302
Ecosystems
coastal 347
gardens 351-352
grassland 346
heath & moor 347
linear, rural 347-348
linear, urban 351
natural 3

parks & cemeteries
351-352
rural 345-348
semi-natural 345-346
urban 348-352
urban commons 349-350
water & wetland 347, 351
wilderness 345
woodland, rural 346,
356-358
woodland, urban 350-351
Edge 24-26, 115
disadvantages of 26
water/land 106
woodland 319, 322-323
Elevation planning 28-30
gardens 190-191
Employment, polycultural 22
Energy (see also Heating &
Transport)
biofuels 123, 124, 315-316
buildings 147-153
combined heat & power
123, 125
conservation 121-122
domestic 130-132
efficiency 122-123
electricity generation 119
embodied 33-34, 35, 120
embodied in food 10, 31,
129-130
hydrogen 125
in use 33
laws of 119
nuclear 124
on farms 57, 130, 268,
269
passive solar 123, 148-152,
163-164, 166-167
pollution 123-124
quality 124
scale 125
solar 124-125
solar vs fossil 121
storage 124-125
use of 126-132
windpower 124-125
Engel, Cindy 189
Entropy 119-121
Ermen, Hugh 254
Ethics 5
Ethylene cycle 55-56
Eutrophication 359
of water 102
Evans, Ianto 202
Exotic species 368-370
Exotic varieties 370-371

Fair shares 7-9
False acacia 33
Farmers Weekly 300
Farming
animals (see also Animals)
258-264
arable & vegetable crops
264-275
bicropping of grain
268-270
bicropping of vegetables
273-275
Bon Fils system 270-272
conventional 267
crop mixtures (see
Polyculture)

crop rotation 259-261,
267-268, 275
disease control 264-265
domestic prairie 272-273
foggage (see Animals)
Fukuoka system 271
grain 264, 267-273
integrated crop
management 257-258,
344
minimum tillage 268
non-crop diversity 266
organic 257-258, 267-268
pest control 265-266
Farm shop 287-288
Fassfern Forest Farm 301
Firefighting 115
Fisher, Mike 379
Fish farming 106-108
Food chain 119-120
Food co-op 286-287
Food links 27, 285-293
box scheme 289-291
case study 294-295
community supported
agriculture 292-293
farm shop 287-288
food co-op 286-287
Japan 290
local directory 288
markets 288-289
subscription farming
291-292
Food miles 11, 285-286
Food production
British 258-261
world 37
Forest farming 307-308
Forest garden 204-206
definition 3
Forest Stewardship Council
134, 336
Frogs 110-111
Frost 68, 78-79
aspect & 80
in gardens 191
observation 81-83
pocket 78-79, 191
Fruit (see also Orchard)
225-244
choice of 226-230, 396
commercial growing 225,
227, 246-247, 295
community orchards 236
companion plants
242-244
dual-purpose trees
220-222, 254-255
in gardens 232-233, 235
in market gardens
233-234
in schools 237
in towns 235-237
in woodland 234-235,
336-337
microclimate 237-240
on farmland 233
own-root trees 254-255
pest control 242-244,
246-247, 248
pollination 232, 241,
243-246, 281
polyculture 240-244
replant disease 227

rootstocks 231, 233
soft fruit 233, 237
table of species 228-230
tree forms 231
wall-trained 23, 26, 233
windbreaks 238-240
Fukuoka, Masanobu 271
Fungi 49, 56-57, 172,
295-297, 307

Gaia theory 35-36
Gandhi, Mahatma iii
Gardening
animals 202-203
balconies 173-174
beds 52, 192
bicropping of vegetables
273-275
chickens 248-249
children & 210-211
choosing what to grow
177-180
community gardens
183-184, 212-213
companion plants
242-244
container growing
173-174, 175-176
crop rotation 180, 275
cut and come again
173-174
drainage 189-190
edible ornamental 171
forest garden 204-206
fruit 232-233
golden rules 206-208
indoor 172
large gardens 181
lawn 181, 366
meadow 181-182
microclimate 184-190
minimalist 199-201
mulching 194-197
no-dig 191-194
patio 175
perennial vegetables 15,
68, 174, 197-201
pick and pluck 173-174
pollution 190
polyculture 201-203
potato tower 204
public gardens 183
roof gardening 173-174
school gardens 184
shrubbery 182-183
small gardens 176-177
soil fertility 43
stacking 203-204
steep 190-191
table of vegetables 179
terracing 190-191
vertical space 203-204
wildlife 364-368
window boxes 173-174
Garfitt, JE 319
Geese 247
Gene banks 21
Genetic engineering 20-21
Genetic diversity 20-21,
370-371
Gilbert, Oliver 348, 349, 416
Ginseng 307
Global warming (see Climate
change)

Goats 262
Grain 264, 267-273
    bicropping 268-270
    Bon Fils 270-272
    conventional 267
    domestic prairie 272-273
    Fukuoka 271
    organic 267-268
    perennial 15
Grants 328
Greenhouse 3-4
Greenhouse effect (see Climate change)
Green roofs 155-157
Groh, Christina 293
Growing season 69-70
Guerra, Julia & Michael 209-211
Guilds 425

Harper, Peter, 101, 135-137
Hazel
    as nutting tree 222
Heating
    active solar 123
    combined heat & power 123,125, 143-144
    convected 149
    domestic 130, 131, 165
    domestic water 130
    fuels compared 123
    Jean Pain system 315-316
    masonry stove 152-153
    passive solar 123, 148-152, 163-164, 166-167
    radiant 149
    solar water 135-137
    wood 152-153, 304
Hedges 55, 266, 281
    fruit in 233
    planting 361
Henderson, Elizabeth 292
Henry Doubleday Research Association 21
Hogarth, Bill 305
Hollins, Arthur 263
Holmgren, David 4, 28, 30, 150
Homeworking 127, 147
Horses 262, 360
Houseplants 172
Housing
    cohousing 146-147
    density 142-144
    location 141
    planning policy 144-145
    sectoring 145-146
    zoning 145
How To Grow More Vegetables 208
Hydraulic ram 100-101

Industrial ecology 133
Input-output 32, 419
Integrated crop management 257-258
    biodiversity & 344
Introduction to Permaculture 5

Jackson, Wes 272
Jacobs, Carol 287
Jeavons, John 208
Joshua, 'Josh' 295

Kalnars, Tallis 327
Key planning tools 27-30, 385, 393-394, 420-421
Kramer, Peter 294

Land equivalent ratio 19-20, 266, 273
Landslips 55
Law, Ben 336-337
Leaf curd 260
Legumes 48-49
Ley 43, 52, 259, 261, 263, 267-269, 271, 273, 275
Lifestyle changes 9-10, 343
Light & shade 76, 77
    aspect & 80
    in gardens 184-186
    observation 81-83
    windbreaks & 83
Linking 32, 248-250, 419
    in aquaculture 107-108
    in poultry 248-250
Listening 414-419
Lister, Jim 287
Local food directory 288
Local distinctiveness 22, 147, 159
Local uniqueness ix, 13, 22
Loss, Carl-August 293
Lovelock, James 36
Lovins, Amory 122

Manure 44
Maps (see Designing)
Market garden 193
    soil fertility 43
Marketing 285-286
Markets 288-289
Maryon, Darrell 212-213
McHarg, Ian 420
Meat eating 119-120, 258-261
Mechanisation 30-31
Methane 71, 101-102
    digestor 103
Microclimate 28, 74-83
    albedo 76-77
    aspect 80
    aspect for fruit 238
    benefits of 75
    combined effects 81
    courtyard 176
    definition 74
    fruit 237-240
    frost 78-79, 191
    gardens 184-190
    heat storage 77
    in agroforestry 277
    light 76, 80
    moisture 79-81
    observation 81-83
    patio 175
    plant indicators 81-82
    shade 77
    strip cropping & 265
    suntrap 76-77, 79
    temperature 76-77, 188-189
    temperature of water 106
    trees & temperature 77
    wall 79, 80
    wind 75-76
Mixed cropping (see Polyculture)
Mollison, Bill 4, 5, 39-40, 106, 115

Monkey puzzle 224
Monocarpic plants 51
Mulching 44, 46, 52, 54
    garden 194-197
    trees 327-328
Multiple causes 36, 414
Multiple outputs 32-33, 316
Mushrooms (see Fungi)
Mycorrhiza 49, 56-57, 277, 319

Native species 368-370
Native varieties 370-371
Nature conservation (see Biodiversity)
Naturewise 237
Neighbourhood Initiatives Foundation 422-423
Nettles 50
Networking 28
New Zealand flatworms 47
Niche 16-17, 23
    aquatic 107
    definition 16
    in agroforestry 276
Niche management 235, 261
Nitrogen 48-49, 62, 105, 267
    fixers 48-49, 219, 242, 251-252, 266, 268
    leaching 267, 268, 269
No-till 14, 50-58
    climate change & 72
    gardens 191-194
    organic matter & 43
    reasons for 52
    reasons for tilling 50-52
Nuts (for individual species see separate entries) 216-225
    climatic range 217 (see also individual species)
    designing a nuttery 217-222
    dual-purpose trees 220-222
    food value 216
    microclimate (see individual species)
    nitrogen fixers & 219
    pests (see individual species)
    pollination (see individual species)
    polycultures 219
    soil (see individual species)
    squirrels 222-224
    yields 216 (see also individual species)

Oak
    as nutting tree 223
Observation 376-377, 378-382, 414-417
    microclimates 81-83
Orchard (see also Fruit) 233
    animals in 244-248
    bees 244-246
    chickens 246-247
    establishment 24
    geese 247
    ground cover plants 242
    layout 219-220
    nitrogen fixers 219, 242
    pigs 247-248

wild birds 248
windbreak 244
Ordnance Survey maps 408-409
Organic food 10-11, 128, 258, 285-286
Organic farming 257-258
    biodiversity & 343-344
    soil erosion & 54-55

Pain, Jean 315-316
Peoplecare 6-7
Perennial plants 14-16
    grains 272-273
    vegetables 68, 174, 197-201
    yield of 15, 272-273
Permaculture
    biodiversity & 344
    can it feed the world? 37
    definition 3, 5
    design 4
    green movement & 10
    history 4-5
    original 4, 191
    vision of the future 32, 215, 257
Permaculture, a Designers' Manual 5, 115
Permaculture One 4
Pest control, 281
    in gardens & orchards 242-244, 246-247, 274
    on farms 265-266, 269, 273, 274
Phosphorous 56-57, 105
Pigs 109, 262-263
    in orchards 247-248
Pine nuts 225
Planning for Real 422-423
Planning permission 304, 336-337
Plantation forestry 72, 134, 317-321
Plant health 269
    bad neighbours 84
    compost & 47
    ethylene cycle & 55-56
    mycorrhiza & 56
Planting Native Trees & Shrubs 236
Plant nutrients (see also Nitrogen & Phosphorous) 40-41, 48-50
    dynamic accumulators 49-50
    in human excreta 105
Plants
    experimental 67-68
    hardiness 68
Plants For A Future 1, 18, 67
Plum yews 224
Pollution
    acid rain 124
    aircraft 129
    building materials 160, 162
    energy use & 123-124
    in gardens 190
    nuclear waste 124
    of rainwater 98
    trees & 331-332
    water 113

Polyculture (see also Diversity)
16-20, 96
farm 264-266
fish 106, 107
fruit 240-244
garden 201-203
grain 272
market garden 266
nuts 219
species mixtures 265
variety mixtures 264-265
woodland 326-327
Ponds
chinampas 115-116
construction, broadscale
115-116
construction, domestic
117-118
depth 25, 116
dew 81
edge 25-26, 115, 117
fish 107-108
island 115
keyline 99
liner 116
pollution 113
puddling 117
siting, general 111
siting, broadscale
112-115, 415
siting, domestic 117
spillway 116
suitable soil 113
trees & 115, 117
water supply 112-113,
117,118
wildlife 366
Population 7-8
Poultry 248-253
choosing breeds 251
Preston, Jessica & John
Prickly Nut Wood 326-327
Pullen, Mandy 294-295

Rabbits 301, 312, 320, 327,
358, 369, 379
Rackham, Oliver 379
Ragmans Lane Farm 294-297
Rainfall
British average 70, 98
Rainwater harvesting 93,
97-98
Recycling 132-134
Reedbeds 97, 103-105
Reedmace 109
Relative location 4, 26-27
Rhizosphere 47
Rhododendron 312, 369
Rotovator 51
Rural lifestyle 141, 145
Rural repopulation 144-145
Rutter, Phil 217

Scale of production 30-32,
257
School gardens 184, 237
Sectoring 28
example 389-394
gardens 184-190
housing 145-146
Self-sufficiency
land needed 142
water 97-98
wood 304

Sewage 101-106
aerobic & anaerobic
101-102
compost toilet 105-105
mains 102-103
methane digestor 103
reedbed 97, 103-105
septic tank 103
WET system 104
Sheep 17, 261, 269
Sheffield Organic Food
Initiative 212-213
Shigo, Alex 330
Short rotation coppice 250, 325
chickens in 250
Slay, Reny 5
Slugs 110-111
Soil 14-16, 30, 39-66
acidity 41, 47
aeration 40, 55
assessment 58-61
brown earth 61
capping 40, 56
chemical analysis 64-66
clay 41-42
colour 60-61
compaction 51, 52, 191-193
composition of 39
covered 14, 55
depth 40
drainage 40
erosion 52-55
ethylene cycle 55-56
fertility 39-50
fertility on farms 259-261
gley 61
life 46-47, 56, 61
loam 42
manure 43
micro-organisms 56
organic matter 42-46
pan 52
permeability 40
plant indicators 58-60
podsol 61
profile 60-61
sandy 41-42
sewage sludge 102-103
silty 41-42
structure 40, 41, 56, 60
temperature 40
texture 41-42, 64, 65
tilth 52
types of 61-62
urban 62-64
water economy 95-96
water-holding 40
waterlogging 40
Soil erosion 52-55
causes & solutions 53
off-farm costs 54
in Britain 53-54
in forestry 54
in USA 53, 54
trees & 55
wind 53
Small scale 30-32
disadvantages of 32
diversity & 30-31
yield & 30-32
Smith, Richard 263
Species
definition 16
native & exotic 368-370

Spiritual energy 120
Squirrels 222-224, 312, 369
Stacking 22-24
in gardens 203-204
in water 107
Strip cropping 25, 265-266
Subscription farming 291-292
Subsoiler 51
Succession 24, 241
Sunshine
British average 70
Sutherland, Maggie & Neil 302
Swales 100

Temperature 76-77
British average 70
garden 188-189
water 106
trees & 77
The Ecology of Urban Habitats
348, 416
Timber
adding value 301-302
building material 161-162
consumption 8, 133-134
definition 301
sources of 134
uses 302-303
Thermodynamics, laws of
(see Energy)
Thorpe, Dave 101
Toilets 94-95
compost 105-105, 167
Torgut, Alpai 237
Transport 126-130
air 129
car sharing 127-129, 147
cycling 128, 129
external costs 128
homeworking 127
housing density & 143
reducing need 127
safe routes 127
school run 127
shopping 127
Trees (see also Fruit)
cash value 311
ecological functions
299-300
effect on soil 326
farming & 278, 300-301
fodder 280
growth rates 87
health 311-312
heights 87
nurse trees 326
old 355
origin 324
pollution 331-332
ponds & 115
protecting 279, 327
provenance 324
pruning 330
soil erosion & 55
table of timber & wood
species 338-340
temperature & 77
tree wardens 334
urban 329-331
wildlife value 368
Trophic pyramid 120
Trout 106-107
Turf roofs 155-157
Twelve month rule 81

Urban food growing 23, 26-27,
28, 142-144
case studies 209-213
productivity of 32, 142

Vale, Brenda & Robert 98
Variety, plant
definition 16
loss of 20
mixtures 20
native & exotic 370-371
Vegetables (see also Gardening)
bicropping 273-275
choice of 177-180
mixed 23, 24, 201-203
perennial 15, 174, 197-201
polycultures 264-266
self-seeding 199-201
Veganism 258-261
Vegetarianism 258-261
von Zschock, Reinhart 305

Walnut 222
Ward, Mark & Paul 275
Warren, Andy 105
Water (see also Aquaculture &
Ponds)
consumption 93-98
cycle 91-93
domestic consumption
93-94, 97
drainage 92-93
embodied 93
global resource 91-92
grey 96-97
hydraulic ram 100-101
in the garden 95-96, 97
in the house 94-95, 96
in the landscape 92-93
irrigation 96, 99-100
keyline system 99-100
on farms 99-101
purification (see Sewage)
rainwater harvesting 93,
97-98
rainwater quality 98
reusing 96-97
swales 100
trees & 92
using less 94-96
Watkins, Roy 108
Weeds 50-51
as indicators 58-59
on farms 270, 273, 274
in gardens 193, 197, 274
geese 247
Wetlands 92
WET systems 104
Wildlife (see Biodiversity)
Wildlife gardening 364-368
Wilson, Edward O 67
Wind 75-76
gardens 186, 389-398
observation 81-83
Windbreaks 83-89
coastal 88-89
definitions 83
design principles 84-88
disadvantages 83
effect on yield 282
for buildings 153-154
for nuts 217-218
for orchards 238-240, 244
in gardens 186-188

on farms 282
on hill farms 284
outputs 83
shelterbelts 282-285
siting 85-87
urban 153-154
Windspeed
British average 70
Wolfe, Martin 264, 274-275,
268
Wood
adding value 301-302
definition 301
durability 306
fencing materials 306-307
firewood 303-304
uses 303-307
Woodland (see also Woodland,
new)
access 310-311
aesthetics 309

animals, domestic 320-321
animals, wild 312, 320
assessment 309-313
charcoal 304-305, 336-337
chestnut coppice 306,
336-337
clear felling 317
community 334-335
continuous cover
318-320
coppice 310, 313-317,
336-337
crafts 305
definitions 310
ecological functions
299-300
ecology 309, 310
education 308-309
extraction 320
fencing materials 306-307
firewood 303-304

firewood yields 304
fodder 308
food production 307-308,
336-337
group selection 319-320
hazel coppice 305
high forest 317-321
individual selection 319
invasive plants 312
legal 313
marketing 301-302
mobile saws 320
oak bark 306-307
outputs 300-309
recreation 308
shelterwood 319
short rotation coppice
250, 325
timber 302-303
tree spacing 311
urban 328-334

Woodland, new 321-328
choice of site 321-322
community 334-335
design example 386-387
diversity 326-327
edge 322-323
establishment 327-328
grants 328
planting 327-328
species choice 323-327
urban 331-334
table of trees 338-340
World food production 37

Yeomans, PA 99

Zoning 27-28
chickens 249-250
gardens 170-171
housing 145
woodland 310-311